Quantitative
Organic
Microanalysis

Quantitative Organic Microanalysis

Al Steyermark
Microchemical Department
Hoffmann-La Roche Inc.
Nutley, New Jersey

2ND EDITION
revised and enlarged

1961

ACADEMIC PRESS
New York and London

COPYRIGHT © 1961, BY ACADEMIC PRESS INC.

ALL RIGHTS RESERVED

NO PART OF THIS BOOK MAY BE REPRODUCED IN ANY FORM
BY PHOTOSTAT, MICROFILM, OR ANY OTHER MEANS,
WITHOUT WRITTEN PERMISSION FROM THE PUBLISHERS.

ACADEMIC PRESS INC.
111 FIFTH AVENUE
NEW YORK 3, N. Y.

United Kingdom Edition
Published by
ACADEMIC PRESS INC. (LONDON) LTD.
17 OLD QUEEN STREET, LONDON S.W. 1

Library of Congress Catalog Card Number 61-12278

PRINTED IN THE UNITED STATES OF AMERICA

*This book is dedicated
to
the memory of my father,
the late*
M. BENJAMIN STEYERMARK

Preface to Second Edition

Since its introduction in the early part of this century, microanalysis has become a much used tool, particularly as an aid to research where the amounts of material to be analyzed are both scarce and costly. Many articles on the subject have been published with the purpose of making it possible for others than those trained by the "old masters" to satisfactorily perform the work. These articles have been reviewed both for individual determinations over long periods of time and for all phases over short periods. For the benefit of the experienced analyst who wishes to use variations of old procedures or to put into use newly published methods, a large number of articles, representing a high percentage of all of those published through 1959, and some in 1960, are referred to in tabular form at the end of each chapter. The above-mentioned reviews have been most helpful in collecting this data.

To be successful the beginner, as well as the experienced microanalyst, must have correctly proportioned apparatus and, in addition, the former must have specific directions for performing the work—methods known to be successful in the hands of those with comparatively little experience. Several groups in the United States have worked toward this goal, namely, the following:

(a) Committee on Microchemical Apparatus of the Division of the Analytical Chemistry of the American Chemical Society.

(b) Subcommittee No. 29 on Microchemical Apparatus under Committee E-1 on Methods of Testing of the American Society for Testing Materials.

(c) The Association of Official Agricultural Chemists.

(d) Commission on Microchemical Techniques of the Section of Analytical Chemistry of the International Union of Pure and Applied Chemistry.

Extensive work on the standardization of apparatus has also been in progress for a number of years in Great Britain.

The first edition of this book* included the various pieces for which the American Chemical Society group had recommended specifications and methods adopted by the Association of Official Agricultural Chemists up to that time. Since the writing of the first edition, additional material has been published by the above two organizations and the American Society for Testing Materials has come into the picture reviewing and changing, if necessary, the recommendations of the American Chemical Society group. The International Union of Pure and Applied Chemistry group also has been active since the days of the first edition. There is close contact between these groups—in fact certain individuals are associated with three of the groups and the author with all of them. The second edition is up to date on all of the latest recommendations

---

* Published in 1951 by The Blakiston Company.

of the American (or American participating) groups, since the author firmly believes that, through the use of these, best results can be obtained by both the beginner and the experienced analyst.

In the first edition, *only* those procedures were described in detail which had been used extensively by the author and his assistants in the microchemical laboratories of their company during the preceding twelve years. These laboratories act as a service department for the company's large research division. During the past ten years, a number of new procedures have been developed for various determinations; however, in this second edition, the author presents, for the most part, basically the same procedures contained in the first edition since these, after nearly twenty-two years of use by him and his assistants, are still considered to be the best guide for the beginner as well as time-tested reliable methods for the experienced analyst. Where more than one method for a determination has met these qualifications, a choice is included and the beginner will do well to become acquainted with each inasmuch as experience has shown the importance of having available referee methods for proving the reliability of analytical data, particularly in connection with the extensive research programs in progress in many organizations.

Considerable material has been added to that which was presented in the first edition. This includes discussions of the following: test samples, blank tests, description of a second type of efficient vibration-absorbing balance table, enlargement of the section on microchemical balances, new Kjeldahl procedures to determine nitrogen in compounds in which nitrogen is connected to nitrogen or to oxygen, oxygen flask combustions, determination of fluorine, and microhydrogenation.

The procedure for the determination of oxygen has been modified to a gravimetric one, which eliminates the errors due to interfering elements. One of the alkoxyl procedures has been changed to coincide with that developed through the joint efforts of the Committee on Microchemical Apparatus of the Division of Analytical Chemistry of the American Chemical Society and the Association of Official Agricultural Chemists.

Long experience has shown that the procedures give best results when the pieces of apparatus are in constant use. In some cases, after a determination has been giving reliable results for months and is then suddenly not used for a period of time, trouble will be experienced on trying to resume the work. The encountered difficulties are not always easily corrected and at times even appear to correct themselves. Naturally, the more experience the operator has had, the fewer will be these instances.

The order in which the various chapters are presented is the same order in which the work has been taught to new members of the author's department.

During the past years, chemists holding bachelor's degrees, but without previous knowledge of microchemistry, have been employed and given a training period, the length of which depends upon the ability of the individual.* On the average, after a few weeks, the newcomer is able to accept research samples for the determination of moisture, metals, nitrogen, and, in some cases, halogens and sulfur. After about a month's experience with the above, the novice is ready to begin work on the carbon-hydrogen determination, which is considered to be the most important of all. After a few more weeks he is able to accept research samples and perform the carbon-hydrogen analyses with the desired accuracy. Within the following months the new analyst is taught all of the other determinations in this book, as the author is of the opinion that each member of the staff should be able to do all of the various analyses. Of course, the above schedule is not a practical one for teaching a college course, but it must be remembered that the author uses it to train microanalysts who must, within as short a time as possible, perform the work in an industrial laboratory with the apparent responsibilities involved. Naturally, this text is equally adaptable to either the one-semester or one-year courses offered in colleges to advanced students.

In closing, the author wishes to emphasize that few microanalysts carry out the determinations in exactly the same manner, each having his own modification. Despite this, the beginner must have reliable methods as a guide, particularly those known to give good results in the hands of many analysts. This book presents such procedures, and its use, as a college text, as well as a guide for industry, is suggested.

In general, the apparatus required is commercially available from the various scientific apparatus dealers. Suggestion by the author of a certain source of supply should *not* be interpreted as an endorsement of the company in question. The same applies to the various instruments used. The author is merely attempting to be helpful by informing the readers of some sources of supply and, in certain cases, the type of instruments used.

AL STEYERMARK

March 1961

---

* The late Rev. F. W. Power of Fordham University once stated that "microchemistry is an art as well as a science." This statement is proved by the fact that many well-trained chemists with considerable experience have been failures at this work, while others, seemingly less equipped, have been very successful. The author has found that persons who have developed the use of their hands through piano playing, typing, art, mechanics, etc., are best suited for this work.

Foreword to First Edition

Microchemistry during the past thirty years in the United States has been developing at an ever-increasing pace. Only a few years ago every microanalyst or teacher of microchemistry either had been a student of or had been influenced by such leaders as Emich or Pregl in this new field of analytical chemistry. This is no longer true. Today we find microchemical laboratories or microchemical methods employed in some manner in almost every chemical industry. It is not strange then that many of the chemists using microchemical methods have had but little and often no training in microchemistry. These people must rely upon a good text for an understanding of microchemical techniques and procedures. This book, because of its simplicity of style and its attention to minute manipulative details, will prove to be of immeasurable help to the beginning microanalyst. The procedures and the directions not only are easy to follow but also are described in such a way as to inspire confidence. The reader soon learns that the procedures are not the results of extensive library work but have been proved by constant use in the author's own microchemical laboratory. The author has spared no effort to aid the beginner. Typical is the chapter on the microbalance which in addition to describing balance construction gives the practical details of cleaning and maintaining the balance and complete instructions for making a micro weighing. This, as in other procedures throughout the book, is explained through the use of carefully chosen examples.

Dr. Steyermark has in no way limited this book to the beginner. Instead, it will be of great value to the trained microanalyst for the procedures are those of the American school of microchemistry which, while based upon the techniques of Pregl, have been modified in typical American fashion.

The methods which are included in the chapters on elemental and group analyses recognize and use mechanized equipment, the apparatus recommended by the Committee for the Standardization of Microchemical Apparatus and methods which have been adopted by the Association of Official Agricultural Chemists. Of particular importance to the research microchemists is the extensive and up-to-date bibliography.

Dr. Steyermark has been well prepared for this task for in addition to supervising one of the largest microchemical laboratories in America he has served as chairman of the American Chemical Society's Committee for the Standardization of Microchemical Apparatus and as an Associate Referee of the Association of Official Agricultural Chemists for the standardization of microchemical methods.

C. O. WILLITS

Philadelphia
August 1951

Acknowledgments

The illustrations in this book, aside from those taken from the author's personal files, are from a variety of sources. Reproduction of the following illustrations and tables is authorized through the courtesy of the persons and publications named herewith:

Academic Press Inc., New York, *"Semimicro Quantitative Organic Analysis,"* by E. P. Clark (Figures 205 in part, 206, and Table 39 in part).

Wm. Ainsworth & Sons, Inc., Denver, Colorado (Figures 13, 14, 15).

Dr. Herbert K. Alber (Figure 57).

American Instrument Company, Silver Springs, Maryland (Figures 80, 107, 163).

American Society for Testing Materials, Philadelphia, Pennsylvania (Figures 23-45, 47-49, 65, 92, 93, 97-102, 105, 106, 109 in part, 110, 112, 113 in part, 114, 121, 153, 165, and Tables 11 and 12).

Analytical Chemistry, American Chemical Society, Washington 6, D. C. (Figures 2-10, 23-43, 45, 47, 48, 58-61, 65 in part, 68, 74-79, 92 in part, 93 in part, 98, 99 in part, 100 in part, 102 in part, 105, 106, 110, 112, 115B, D, E, F, G, H, J, K, L, 116-118, 120, 121, 127-129, 132-134, 137, 143-145, 147, 150, 151, 152 in part, 154, 156, 158, 160, 162, 164, 166, 167, 172 in part, 172C, 175, 177, 179, 182, 195-197, 217, Table 21).

Analytica Chimica Acta, Elsevier Publishing Co., Amsterdam (Figures 207-209).

Christian Becker, Clifton, New Jersey (Figure 16).

C. A. Brinkmann & Co., Inc., Great Neck, Long Island, N. Y. (Figure 125).

P. Bunge, Hamburg, and Pfaltz & Bauer, Inc., New York, New York (Figures 11, 12, 19).

Coleman Instruments, Inc., Maywood, Illinois (Figure 104).

Fisher Scientific Company, New York-Pittsburgh (Figures 126, 215, 216).

J. & G. Instrument Corporation, Milltown, N. J., U.S.A. Agents for L. Oertling Ltd., London (Figure 20).

Journal of the Association of Official Agricultural Chemists, Association of Official Agricultural Chemists, Washington, D. C. (Figure 157).

Journal of Biological Chemistry, Williams & Wilkins Co., Baltimore, Maryland (Figures 186-194, Tables 30, 31).

Mettler Instrument Corp., Hightstown, New Jersey (Figures 17, 18, 21).

Microchemical Journal, published under the auspices of The Metropolitan Microchemical Society by Interscience Publishers, Inc., the copyright owner, New York (Figures 1, 15, 16, 18, 103, 123).

Mikrochimica Acta (*Mikrochemie vereinigt mit Mikrochimica Acta*), Springer-Verlag, Vienna (Figures 57, 67).

Pure and Applied Chemistry, International Union of Pure and Applied Chemistry, Butterworths Scientific Publications, London (Table 3 in part).

E. H. Sargent & Co., Chicago 30, Illinois (Figures 124, 184).

Scientific Glass Apparatus Company, Bloomfield, New Jersey (Figures 52, 63).

Arthur H. Thomas Co., Philadelphia, Pennsylvania (Figures 22, 46, 50, 51, 53, 55, 69-72, 81-84, 87, 88, 90, 96, 108, 111, 135, 136, 138, 139, 146, 148, 149, 152 in part, 159, 170, 171, 173, 174, 176, 178, 180, 181, 185, 201, 202, 203, 210-213, 218, and Table 17B).

John Wiley & Sons, New York, New York, *"Micromethods of Quantitative Organic Analysis."* Second Edition, by J. B. Niederl and V. Niederl, 1942. (Tables 6 and 43).

The author also gratefully wishes to acknowledge assistance from a number of departments of his Company. These include the Engineering Department for preparing various figures, General Office for its typing assistance, the Librarians, and the Photo Laboratory (particularly C. Auricchio and S. White). Without such assistance the task of writing this book would have presented many more difficulties.

Contents

Preface to Second Edition vii

Foreword to First Edition xi

Acknowledgments xiii

1. Introduction 1
   Recommendations, Collaboration, and Standardization. Blank Tests and Corrections. Test Substances. Critical Determinations. Laboratory Report Sheets. Laboratory Set-up. References.

2. The Microchemical Balance 30
   Essential Parts. Mounting the Balance. Assembling and Cleaning the Balance. Additional Information. References.

3. Weighing on the Microchemical Balance 50
   Determination of Zero Reading and Deflection. Determination of Sensitivity and Precision. Calibration of Weights. Apparatus. Weighing of Samples. Additional Information. References.

4. Preparation of Samples for Analysis 83
   Homogeneity of Samples. Apparatus. Determination of Moisture. Additional Information. References.

5. Standard Solutions (Microtitration Techniques) 102
   Reagents. Apparatus. Procedures. Additional Information. References.

6. Microdetermination of Metals by the Ashing Technique 133
   Reagents. Apparatus. Procedure. Additional Information. References.

7. Microdetermination of Nitrogen by the Dumas Method 151
   Reagents. Apparatus. Procedure. Calculation of Results. Additional Information. References.

8. Microdetermination of Nitrogen by the Kjeldahl Method 188
   Reagents. Apparatus. Regular Procedure. Procedure for Determination of Nitrogen in Azo and Nitro Compounds, Oximes, Isoxazoles, Hydrazines, and Hydrazones. Procedure for Determination of Nitrogen in Nitrates. Additional Information. References.

9. Microdetermination of Carbon and Hydrogen 221
   Reagents. Apparatus. Assembling the Combustion Train. Procedure. Additional Information. References.

10. Microdetermination of Sulfur ... 276

CARIUS COMBUSTION. *Volumetric Carius Method.* Reagents. Apparatus. Procedure. *Gravimetric Carius Method.* Alternate Gravimetric Carius Procedure. SCHÖNIGER COMBUSTION. *Volumetric Schöniger Method.* Reagents. Apparatus. Procedure. PREGL CATALYTIC COMBUSTION. Reagents. Apparatus. Assembly and Gravimetric Procedure. Acidimetric (Direct Neutralization) Procedure. Volumetric Procedure. Additional Information. References.

11. Microdetermination of Halogens ... 316

DETERMINATION OF BROMINE, CHLORINE, AND IODINE. *Carius Method.* Reagents. Apparatus. Procedure. *Pregl Catalytic Combustion Method.* Reagents. Apparatus. Assembly and Procedure. *Simultaneous Determination of Chlorine and Bromine.* DETERMINATION OF FLUORINE. Reagents. Apparatus. Procedure. Additional Information. References.

12. Microdetermination of Phosphorus ... 354

Kjeldahl Gravimetric Method. Reagents. Apparatus. Procedure. *Carius Gravimetric Method.* Reagents. Apparatus. Procedure. *Simultaneous Determination of Barium and Phosphorus.* References.

13. Microdetermination of Arsenic ... 367

Carius Volumetric Method. Reagents. Apparatus. Procedure. *Carius Gravimetric Method.* Reagents. Apparatus. Procedure. References.

14. Microdetermination of Oxygen ... 377

Gravimetric Method. Reagents. Apparatus. Assembling the Apparatus. Procedure. *Volumetric (Iodometric) Method.* Reagents. Apparatus. Assembling the Apparatus. Procedure. References.

15. Microdetermination of Neutralization Equivalent, Ionic Hydrogen, or Carboxyl Groups ... 410

Reagents. Apparatus. Procedure. References.

16. Microdetermination of Alkoxyl Groups (Methoxyl and Ethoxyl) ... 422

Volumetric (Iodometric) Method. Reagents. Apparatus. Procedure. *Gravimetric Method.* Reagents. Apparatus. Procedure. References.

17. Microdetermination of Acyl Groups (Acetyl and Formyl) ... 444

Reagents. Apparatus. Procedure. References.

CONTENTS xvii

18. **Microdeterminations Carried Out on the Van Slyke Manometric Apparatus** 454

    MANOMETRIC CARBON DETERMINATION. Reagents. Apparatus. Procedure. MANOMETRIC DETERMINATION OF PRIMARY AMINO NITROGEN IN THE ALIPHATIC α-AMINO ACIDS. Reagents. Apparatus. Procedure. OTHER DETERMINATIONS CARRIED OUT ON THE MANOMETRIC APPARATUS. *Ampul and Vial Testing with Manometric Apparatus. Manometric Micro-Kjeldahl Determination of Nitrogen. Blood and Urine Analysis. Gas Solubility Studies. Indirect Methods Based on Manometric Combustion of Organic Precipitates.* Determination of Phosphorus. Determination of Sulfur. Determination of Magnesium. References.

19. **Microdetermination of Unsaturation (Double Bonds)—Hydrogen Number** 495

    Reagents. Apparatus. Procedure. References.

20. **Microdetermination of Other Groups** 508

    DETERMINATION OF ALKIMIDE GROUPS (N-METHYL AND N-ETHYL). Reagents. Apparatus. Procedure. DETERMINATION OF HYDROXYL GROUPS. *Micromethod.* Reagents. Apparatus. Procedure. *Semimicromethod.* MICRODETERMINATION OF ACTIVE HYDROGEN. MISCELLANEOUS MICRODETERMINATIONS. References.

21. **Microdetermination of Molecular Weight** 528

    THE RAST METHOD FOR THE DETERMINATION OF MOLECULAR WEIGHT. Reagents. Apparatus. Procedure. THE SIGNER (ISOTHERMAL DISTILLATION) METHOD FOR THE DETERMINATION OF MOLECULAR WEIGHT. Reagents. Apparatus. Procedure. Additional Information. References.

22. **Microdetermination of Some Physical Constants** 548

    MICRODETERMINATION OF MELTING POINT. Apparatus. Procedure. MICRODETERMINATION OF BOILING POINT. Apparatus. Procedure. MICRODETERMINATION OF SPECIFIC GRAVITY. *Using the Gravitometer.* Reagents. Apparatus. Procedure. *Using Micro Weighing Pipettes, Density-Type (Pycnometers).* MICRODETERMINATION OF OTHER PHYSICAL CONSTANTS. References.

23. **Calculations** 565

    Calculation of Factors. Calculation of Percentages from the Empirical Formula and Vice Versa. References.

AUTHOR INDEX 603

SUBJECT INDEX 639

CHAPTER I

Introduction

Nearly a half a century ago, quantitative organic microanalysis which deals with samples ranging in size from one to 10 mg. (1 mg. = 10^{-3} gram) was developed as a means of both convenience and necessity by Fritz Pregl, a physician, who was the medical examiner of the city of Graz, Austria. While working on bile acids, he obtained crystalline degradation products in quantities too small to be analyzed by the existing methods. Two alternatives were his, either to work for several years repeating his experiments on a much larger scale or to adapt the existing methods to the milligram quantities he had on hand. He chose the latter. He became so interested in this new field that he never returned to the work on the bile acids, but devoted his entire time to the development and teaching of microanalysis. Advanced students of chemistry from all parts of the world went to Graz to learn the microtechniques. For his contribution to science, Pregl received the Nobel Prize in 1923.[14] Articles relating to Pregl, as well as the roles played by Friedrich Emich and others in the development of this field have been published and to these the reader is referred.[28,65,96,159] Early events of importance in the development of microchemistry are shown in Table 1, while Table 2 shows the approximate size of the samples used for various determinations.

TABLE 1
EARLY WORK OF IMPORTANCE IN THE DEVELOPMENT OF MICROCHEMISTRY

Date	Author	Subject matter
1867	T. G. Wormley[214]	"Microchemistry of Poisons"—book
1903	W. Nernst, and E. H. Riesenfeld[117,118]	Determination of atomic weights
1904	G. Barger[17]	Determination of molecular weights of organic compounds
1909	F. Emich[50-52]	Determination of halogens and sulfur by the Carius method
1911	F. Emich[50,51]	Determination of nitrogen by the Kjeldahl method
1911	J. Donau[47]	Recommended use of the Kuhlmann microchemical balance. Previous to this, the Nernst microbalance was used.
1911	F. Pregl[136]	Quantitative organic microanalysis was developed through the use of the Kuhlmann microchemical balance

1. Introduction

TABLE 2
SAMPLE SIZE FOR VARIOUS DETERMINATIONS[24,25,44,61,62,119,120,136,144-147,174]

Determination	Sample size (mg.)
Carbon (manometric)	1– 5
Carbon-hydrogen	4– 6
Nitrogen (Dumas or Kjeldahl)	1– 6
Halogens	5– 8
Sulfur	2– 8
Oxygen	3–10
Phosphorus	4– 6
Metals	4– 6
Groups	5– 8

Modifications of the original methods have since been named semimicroanalysis[43] (milligram procedures using sample sizes upwards of 10 mg.) and ultramicroanalysis[77] (samples sizes down to 10^{-6} gram). The present trend,[37,38,105,177] however, is to substitute these terms with more descriptive ones such as, (a) gram methods for the so-called macro- or classical; (b) centigram for the present semimicro-; (c) milligram for micro-; (d) microgram for ultramicro-; (e) nanogram for submicro- (samples down to 10^{-9} gram) and; (f) picogram for subultramicro- (samples down to 10^{-12} gram). Likewise, the terms gamma (γ), and lambda (λ) are to be substituted with microgram (μg.) and microliter (μl.) respectively as follows:

$$1 \text{ microgram} = 1 \text{ }\mu\text{g.} = 0.001 \text{ mg.} = 10^{-6} \text{ gram}$$
$$1 \text{ microliter} = 1 \text{ }\mu\text{l.} = 0.001 \text{ ml.}$$

Recommendations, Collaboration, and Standardization

With the acceptance of microchemical methods by industry, it became necessary for others than those trained by Pregl, or his students, to perform this type of work and to eventually train assistants and/or students. Due to the very nature of the work, the beginner must have correctly proportioned apparatus and specific directions for performing the work, particularly methods known to be successful in the hands of those with comparatively little experience. At least as far as apparatus is concerned, the above is of equal importance to the experienced microanalyst. Recognizing these needs, several groups have been actively engaged in work along these lines.

The initial sources of supply of equipment were from Europe and the various pieces received were not always interchangeable. The American manufacturers had no experience along these lines and no specifications to act as a guide. In 1937, the Division of Microchemistry (later known as the Division of Analytical and Micro Chemistry and now the Division of Analytical Chemistry) of the American Chemical Society, appointed the Committee for the Standardization of Microchemical Apparatus, composed of experienced microchemists whose purpose "was to investigate the wide variations in dimensions of various common pieces of apparatus which are employed in organic microanalysis and to suggest to the Division recommended specifications for such apparatus in so far as it seems practical to do so."[150] This action came at a very opportune time inasmuch as European supplies were cut off soon afterwards due to World War II. The committee concentrated on the apparatus for the determinations most widely performed and in 1941 was able to recommend specifications for the pieces used in the carbon-hydrogen and Dumas nitrogen set-ups.[148,150] Two years later, recommendations were published for those pieces used in connection with the determinations of halogens and sulfur by the catalytic combustion method.[149] This committee then ceased to function. After the end of World War II, attention again was called to the need for continued work along these lines, and in 1947, the Division of Analytical and Micro Chemistry (now the Division of Analytical Chemistry) of the American Chemical Society appointed a new Committee for the Standardization of Microchemical Apparatus (now the Committee on Microchemical Apparatus). The new committee declared its intentions to be the following[36,182]:

1. Where needed, to revise the recommendations of the former committee.

2. To recommend specifications for other items of quantitative micro-, semimicro-, and ultramicroapparatus and finally of the qualitative field.

3. All recommendations were to be made with the understanding that the specifications represented the best thought at that time and that revisions would be made when necessary. Also that primary consideration would be given to glass apparatus.

The new committee has been very active[163,175-184] in successfully carrying out its original intentions as shown in the following respective chapters. All recommendations have been made only after considerable experimental work by members of the committee and cooperating chemists. (Alber[5] in an address to the First International Microchemical Congress at Graz, Austria, in 1950, discussed the general subject of standardization in the United States in progress up to that time.) The British Standards Institution's accomplishments[35] have been along the same lines and have been of great value to the British manufacturers. The Division of Analytical Chemistry of the American Chemical Society also has functioning the Committee on Balances and Weights which makes recommendations along these lines.[103,140]

1. Introduction

The American Chemical Society does not sponsor standardization to the extent that the recommended specifications of the Committee on Microchemical Apparatus could become "standards" instead of recommendations. Therefore, this committee now also functions as the nucleus of Subcommittee No. 29 on Microchemical Apparatus of Committee E-1 on Methods of Testing of the American Society for Testing Materials. This subcommittee reviews the recommended specifications a year or more *after* these have been *published* in *Analytical Chemistry*. In this interval, if changes in dimensions or design have been found necessary, which is rarely the case, they are made. These, together with the recommendations not needing revision, are incorporated into a report to the American Society for Testing Materials and these specifications are adopted as tentative specifications of this Society.[13] Where such specifications exist, they are included in the following respective chapters.

The Association of Official Agricultural Chemists has been sponsoring a program of collaborative studies[2,41,42,124-129,165-170,170a,185,186,188,205-207] on microchemical methods for a number of years in efforts to develop methods with which good results can be obtained by relatively inexperienced microanalysts who carefully adhere to the directions. For the first year of each study, samples are submitted to a number of collaborating chemists who analyze the substances by whatever methods they normally employ and report *all* values obtained. The chemists also provide detailed descriptions of their procedures. Statistical comparisons are then made between the values obtained by the various methods to determine which one is the most accurate and precise. Similarly, comparisons are made to determine whether or not variations in procedure produce significant differences in results. For the second year's study, all the collaborators are requested to analyze the same samples by a specific procedure based on that method which gave the best results the first year. If the returns justify such action, the method is adopted as "First Action," and then eventually as "Official." In the event that the method does not prove satisfactory, additional studies are conducted. Those in charge of these collaborative studies are also members of the Committee on Microchemical Apparatus and of Subcommittee No. 29 so that each group has the advantage of the experiences of the others, and the programs are so arranged. The work accomplished is covered in the following respective chapters.

The Commission on Microchemical Techniques of the Section of Analytical Chemistry of the International Union of Pure and Applied Chemistry is now actively engaged in recommending test substances for determining the applicability of apparatus and methods for the various determinations when dealing with different types of structures.[45] This is covered in detail in the section of this chapter dealing with test substances and is mentioned here only for the sake of complete coverage of the general subject under discussion.

A great deal of time, effort, and expense is involved in these programs on the part of all parties involved—those conducting the work and those collaborating, both chemists and manufacturers of laboratory apparatus. In spite of this, the benefit to the beginner, the experienced analyst, and the manufacturers is so great that these programs will be continued indefinitely and the reader is advised to follow the literature dealing with these.

The author is a strong advocate of *all* collaborative work, as shown by his affiliation for years with each of the above-mentioned programs and many of the methods and pieces of apparatus recommended in the following chapters are based on such collaborative efforts.

Blank Tests and Corrections

Entire determinations carried out in the *absence* of the sample are called blank tests.* The same term applies if a pure sample is used which does *not* contain the element(s) or group to be determined; for example, a Dumas determination done on benzoic acid, which more nearly duplicates the conditions of the determination than if no sample is used. Blank tests show the amount of error resulting from contamination of the reagents used or from faulty apparatus. The volume of standard solution consumed or amount of end product (gas or precipitate) obtained during a blank test is applied as a correction for actual determinations.

Although blank tests have been recommended in connection with the carbon-hydrogen determination, the author does not agree with this practice since a definite hydrogen value will always be obtained due to dehydration of the lead dioxide portion of the combustion tube filling, as explained in the chapter in question. A *test sample* provides the necessary information as to whether or not the carbon-hydrogen train gives satisfactory results. Except for the few determinations, such as the manometric carbon, that specifically† call for blank tests, the author prefers to use these only as a means of determining the cause of poor results and then eliminate this cause, if possible, before proceeding with samples. For example, if high nitrogen results are obtained on a *test sample* by the Kjeldahl method, *blank tests* on the various *individual* reagents will show which one(s) is at fault. Blank values should be applied *only* if a number of blank tests show that a *constant* error is present and all attempts to correct the condition have failed. In the opinion of the author, this is an emergency measure rather than a practice that should be encouraged.

* The number of references to the use is too great to list here so that only a few are cited.[44,61,62,119,120,136,144-147,174]

† The reader is referred to the particular chapters in question.

1. Introduction

For a few determinations, standard corrections are applied regardless of the results of blank tests.[136] Two per cent of the volume of gas collected during a micro-Dumas determination is automatically deducted as being essentially due to the volume of potassium hydroxide adhering to the walls of the nitrometer and to the alteration in vapor tension. For the gravimetric procedures for the determination of alkoxyl and alkimide groups, an empirical correction of 0.06 mg. is added to the weight of silver iodide *for each milliliter* of silver nitrate used. Empirically, Pregl established the factor used in the determination of phosphorus which differs from that used in the determination on the macroscale.

*Test Substances**

Compounds in a state of high purity are used to determine the applicability and reliability of methods and pieces of apparatus to the types of compounds to be analyzed. Such materials are known as test substances or test compounds and should be used at *frequent* intervals. In the selection of a test substance, consideration should be given to the *composition, structure,* and *percentages* of the various elements or/and groups present, and, if possible, a compound should be selected which most closely approximates on all points those of the samples which have been submitted for analysis. This, of course, assumes that the analyst has been supplied with all data concerning these samples, including structures. *Any compromise to this practice places the analyst at a disadvantage* inasmuch as methods are often varied according to structural types. Obviously, when many compounds representing all types must be analyzed routinely, for practical purposes, a variety of types of test substances can be used *only* when questions of doubt arise regarding the validity of certain results. Otherwise, for general routine checking, the author recommends the use of test substances which contain relatively high percentages of the element(s) or group(s) in question.

The Commission on Microchemical Techniques of the Section of Analytical Chemistry of the International Union of Pure and Applied Chemistry is in the process of recommending test substances for the various determinations. The list for use with the carbon-hydrogen determination has been completed.[45] Work is in process on similar lists for other microchemical determinations and the reader is advised to watch for *all* of these which are expected to be published during the next several years. In the meantime, the author recommends the use of the compounds shown in Table 3, most of which are expected to be included in the above lists, and most have been taken from the list for the carbon-hydrogen determination.[45] These substances are stable, solid at room

---
* Also referred to as test compounds or test samples.

temperature, nonhygroscopic, and nonvolatile. The last three properties have been included as requirements so that differences in techniques of the analysts in handling samples with other properties does not become a factor. All of these compounds are either commercially available in a sufficiently pure state to be used for test purposes, based on the accuracy of present-day methods, or may be purified or prepared from commercially available material by conventional laboratory means to meet these standards.

Critical Determinations

A discussion on the subject of critical determinations is in order at this time. On too many occasions, meaningless analyses are requested by the research groups and the microanalyst should be able to recognize this condition when confronted with it. Habitually, the research chemist requests carbon, hydrogen, nitrogen, etc., without considering the values of such. In general, since we are dealing with organic compounds, the determination of carbon and hydrogen is always in order but is of no value if one is trying to differentiate between two compounds having these values so close together that they fall within the limits of allowable error. For example, suppose two such compounds have theoretical values of the following:

(a) Compound I
$C = 66.23\%$
$H = 4.76\%$

Compound II
$C = 65.73\%$
$H = 4.36\%$

If a carbon-hydrogen determination gave

$C = 66.03\%, \qquad H = 4.56\%$

it would be impossible from this to say which of the two compounds, I or II, had been analyzed since the allowable error on both carbon and hydrogen are $\pm 0.3\%$. In such a case, other determinations, such as nitrogen, halogen, methoxyl, acetyl, etc., often give more information. In many cases, either carbon *or* hydrogen, but not both, is critical for differentiating as shown by the following:

(b) Compound III
$C = 79.86\%$
$H = 4.23\%$

Compound IV
$C = 73.21\%$
$H = 4.60\%$

(c) Compound V
$C = 75.00\%$
$H = 7.86\%$

Compound VI
$C = 75.42\%$
$H = 3.62\%$

In (b) the carbon value would be critical while in (c) it would be the hydrogen value.

1. Introduction

TABLE 3

LIST OF TEST SUBSTANCES SHOWING PERCENTAGES OF ELEMENTS AND GROUPS PRESENT

Class and compound	Empirical formula	Molecular weight	C (%)	H (%)	O (%)	N (%)	Hal (%)	S (%)	Element (Me) (%)	Group (%)
(1) *CH*										
Anthracene[45]	$C_{14}H_{10}$	178.234	94.34	5.66	—	—	—	—	—	—
Naphthalene*[45]	$C_{10}H_8$	128.174	93.71	6.29	—	—	—	—	—	—
(2) *O (C,H)*										
Anisic acid	$C_8H_8O_3$	152.152	63.15	5.30	31.55	—	—	—	—	20.40% CH_3O 29.59% COOH
Anthraquinone	$C_{14}H_8O_2$	208.218	80.76	3.87	15.37	—	—	—	—	—
Benzoic acid[45]	$C_7H_6O_2$	122.125	68.85	4.95	26.20	—	—	—	—	36.85% COOH
Cholesterol[45]	$C_{27}H_{46}O$	386.665	83.87	11.99	4.14	—	—	—	—	—
d-Glucose (dextrose)[48]	$C_6H_{12}O_6$	180.162	40.00	6.71	53.29	—	—	—	—	—
Phthalic acid anhydride[45]	$C_8H_4O_3$	148.120	64.87	2.72	32.41	—	—	—	—	—
Stearic acid[45]	$C_{18}H_{36}O_2$	284.486	76.00	12.76	11.25	—	—	—	—	15.82% COOH
Vanillin	$C_8H_8O_3$	152.152	63.15	5.30	31.55	—	—	—	—	20.40% CH_3O
(3) *N (C,H, . . .)*										
Acetanilide[45]	C_8H_9ON	135.168	71.09	6.71	11.84	10.36	—	—	—	—
Azobenzene	$C_{12}H_{10}N_2$	182.228	79.09	5.53	—	15.37	—	—	—	—
2,4-Dinitrophenylhydrazine[45]	$C_6H_6O_4N_4$	198.146	36.37	3.05	32.30	28.28	—	—	—	—
Diphenylamine[45]	$C_{12}H_{11}N$	169.228	85.17	6.55	—	8.28	—	—	—	—
Hexamethylenetetramine[45]	$C_6H_{12}N_4$	140.194	51.40	8.63	—	39.97	—	—	—	—
Picric acid[45]	$C_6H_3O_7N_3$	229.114	31.45	1.32	48.88	18.34	—	—	—	—
Histidine hydrochloride monohydrate	$C_6H_{12}O_3N_3Cl$	209.643	34.38	5.77	22.90	20.05	16.91% Cl	—	—	6.68% N from NH_2
Benzocaine (ethyl p-aminobenzoate)	$C_9H_{11}O_2N$	165.195	65.44	6.71	19.37	8.48	—	—	—	27.28% C_2H_5O
Methyl p-aminobenzoate	$C_8H_9O_2N$	151.168	63.56	6.00	21.17	9.27	—	—	—	20.53% CH_3O
(4a) *F (C, . . .)*										
p-Fluorobenzoic acid[45]	$C_7H_5O_2F$	140.117	60.00	3.60	22.84	—	13.36% F	—	—	—
Perfluorodicyclohexylethane[45]	$C_{14}F_{26}$	662.154	25.39	—	—	—	74.61% F	—	—	—
Trifluoroacetanilide[45]	$C_8H_6ONF_3$	189.144	50.80	3.20	8.46	7.41	30.14% F	—	—	—
m-Trifluoromethyl benzoic acid	$C_8H_5O_2F_3$	190.128	50.54	2.65	16.83	—	29.98% F	—	—	—
(4b) *Cl (C, . . .)*										
Chloroacetamide[45]	C_2H_4ONCl	93.519	25.69	4.31	17.11	14.98	37.91% Cl	—	—	—
p-Chloroacetanilide	C_8H_8ONCl	169.617	56.65	4.75	9.43	8.26	20.90% Cl	—	—	—

Test Substances

TABLE 3 (*Continued*)

Class and compound	Empirical formula	Molecular weight	C (%)	H (%)	O (%)	N (%)	Hal (%)	S (%)	Element (Me) (%)	Group (%)
1-Chloro-2,4-dinitrobenzene[45]	$C_6H_3O_4N_2Cl$	202.563	35.58	1.49	31.60	13.83	17.50% Cl	—	—	—
Hexachlorobenzene[45]	C_6Cl_6	284.808	25.30	—	—	—	74.70% Cl	—	—	—
Hexachlorocyclohexane[45]	$C_6H_6Cl_6$	290.856	24.78	2.08	—	—	73.14% Cl	—	—	—
Tetrachloro-*p*-benzoquinone (chloranil, tetrachloroquinone)[45]	$C_6O_2Cl_4$	245.894	29.31	—	13.01	—	57.68% Cl	—	—	—
(4c) *Br* (*C,H,O, . . .*)										
p-Bromoacetanilide[45]	C_8H_8ONBr	214.076	44.88	3.77	7.47	6.54	37.33% Br	—	—	—
2,4,6-Tribromophenol[45]	$C_6H_3OBr_3$	330.838	21.78	0.91	4.84	—	72.47% Br	—	—	—
(4d) *I* (*C,H,O, . . .*)										
Erythrosin (iodeosin, 2,4,5,7-tetraiodofluorescein)[45]	$C_{20}H_8O_5I_4$	835.924	28.74	0.96	9.57	—	60.73% I	—	—	—
o-Iodobenzoic acid[45]	$C_7H_5O_2I$	248.027	33.90	2.03	12.90	—	51.17% I	—	—	—
(5) *S* (*C,H, . . .*)										
S-Benzylthiuronium chloride[45]	$C_8H_{11}N_2ClS$	202.715	47.40	5.47	—	13.82	17.49% Cl	15.82	—	—
L-Methionine	$C_5H_{11}O_2NS$	149.217	40.25	7.43	21.45	9.39	—	21.49	—	9.39% N from NH_2 30.17% COOH
Sulfanilamide[45]	$C_6H_8O_2N_2S$	172.212	41.85	4.68	18.58	16.27	—	18.62	—	—
Sulfanilic acid (anhydrous)[45]	$C_6H_7O_3NS$	173.196	41.61	4.07	27.71	8.09	—	18.51	—	—
Sulfonal [2,2′-bis(ethylsulfonyl)propane][45]	$C_7H_{16}O_4S_2$	228.337	36.82	7.06	28.03	—	—	28.09	—	—
Thiourea[45]	CH_4N_2S	76.125	15.78	5.30	—	36.80	—	42.12	—	—
(6) *P, As* (*C,H,O, . . .*)										
o-Arsanilic acid (*o*-Aminophenylarsonic acid)[45]	$C_6H_8O_3NAs$	217.048	33.20	3.72	22.11	6.45	—	—	34.51% As	—
5-Chloro-4-hydroxy-3-methoxybenzyl isothiourea phosphate[45,157]	$C_9H_{14}O_6N_2ClSP$	344.725	31.36	4.09	27.85	8.13	10.29% Cl	9.30	8.99% P	—
(7) *Me* (*C, . . .*)										
Calcium oxalate[45]	C_2O_4Ca	128.102	18.75	—	49.96	—	—	—	31.29% Ca	—
Potassium acid phthalate[45]	$C_8H_5O_4K$	204.228	47.05	2.47	31.34	—	—	—	19.15% K	—
Sodium oxalate[45]	$C_2O_4Na_2$	134.004	17.93	—	47.76	—	—	—	34.31% Na	—
Mercuric acetate	$C_4H_6O_4Hg$	318.702	15.07	1.90	20.08	—	—	—	62.95% Hg	—

* **May** be used for carbon-hydrogen if used IMMEDIATELY after weighing.
Notes: The atomic weights proposed by the Commission on Atomic Weights of the IUPAC, 1957[76] have been used for the calculation of molecular weights and percentages. The percentages are given to the second decimal. Where the third decimal is less than 0.005, it has been disregarded; where it is greater than 0.005, the second decimal has been increased and underlined, i.e., 0.375 = 0.38.

1. Introduction

In general, low theoretical values are not critical ones unless in one substance the element or group is present while in the other it is absent. Even so, the determination of such elements or groups gives no indication as to the purity. For example, if a compound theoretically has 4% of nitrogen, it could be contaminated with 5% of some substances containing no nitrogen or with considerable amounts of one having a relatively high amount of nitrogen and still give an acceptable analysis (within ±0.2% of the theory).

Laboratory Report Sheets

Good laboratory report sheets are essential in a well-organized laboratory and they are quite important to the analyst, particularly the information contained on them. The sample shown on the next page is that used in the author's laboratory. It is filled out in duplicate so that both research chemist and analyst have a permanent record. It is designed to assist in the bookkeeping of both the research chemist and the analyst. Whenever possible *all* of the information requested on the sheet should be filled in by the research chemist requesting the analysis. A knowledge of both the physical and chemical properties is helpful to the analyst in handling the sample as discussed in the following chapters. The theoretical percentages of the elements or groups to be determined are valuable inasmuch as a comparison between these and the found values often serve as a guide as to the performance of the apparatus being used. In other words, a pure research compound serves the same purpose as does a test sample for checking.

Laboratory Set-up

PLAN

Plans for microchemical laboratories have been published previously by Niederl and Niederl,[119,120] Alber and Harand,[7] Kirner,[78] Boos,[34] and the author.[164,174] Any one of a number may prove satisfactory but in the author's opinion certain features, included in his laboratories, are a necessity and are discussed below. The plan of the author's laboratories which have proven to be most satisfactory is shown in Fig. 1 and serves as an example. The unit consists of an entrance hallway which contains filing space, two offices, a separate filing room, balance room, and three laboratories. At present, these are occupied by nine chemists, a secretary, and a cleaner.

VOLTAGE REGULATION

Modern laboratories should be equipped with electric furnaces for carrying out the combustion procedures described in the following chapters and for most of these procedures, the operating temperatures are most critical, particularly in regard to constancy. Some pieces have thermostatic control while others do not; therefore, it is most desirable to have the electricity supplied at a constant voltage to accomplish the above. Either constant voltage transformers or me-

---

F 21*
HOFFMANN-LA ROCHE INC.
NUTLEY, N. J.

Microchemical No. _____

MICROCHEMICAL REPORT

Date _____ Submitted by _____

Author's Exp. No. _____ Code _____ Sample No. _____

Analysis Requested _____

Formula _____

Theory _____

Structural Formula

m.p. _____ after crystallization from _____

m.p. _____ after _____ times recrystallized from _____

Dry at _____ °C, in vacuo _____ hours

b.p. _____ °C at _____ mm

Properties (hygroscopic, etc.) _____

Determination of Carbon and Hydrogen
  Weight of sample _____
  Weight of H_2O _____
  Weight of CO_2 _____
  % Hydrogen _____
  % Carbon _____

Determination of Nitrogen
  Weight of sample _____
  ml. N_2 or N/100 acid _____
  Temperature and pressure _____
  % of Nitrogen _____

Determination of _____
  Weight of sample _____
  Weight of (or ml. of N/100) _____
  % _____

Determination of _____
  Weight of sample _____
  Weight of _____
  % _____

Determination of _____
  Weight of sample _____
  Weight of _____
  % _____

Analyst's Remarks _____

Date Completed _____

1. Introduction

chanical voltage regulators may be used. In the author's laboratories, the line voltage to all table tops is kept constant to within the limits of ±1% by means of a 40 KVA regulator of the latter type.

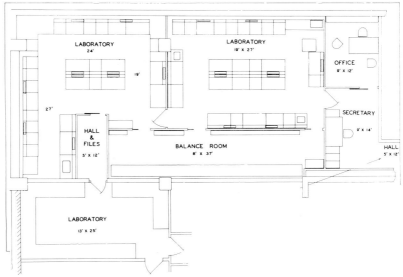

FIG. 1. Plan of author's laboratory.

AIR-CONDITIONING

The laboratories (and offices) are air-conditioned to maintain constant temperature and humidity 24 hours per day, 7 days per week, throughout the year, because, by so doing, the balances are not subjected to radical changes.[56,119,120, 144-147,164,174] The temperature is kept *constant* at approximately 75° F. (23.9° C.) and the relative humidity held in the 40 to 50% range,[70] preferably in the low forties. The exact temperature at which the laboratories are kept is a matter of personal preference, but it is very important that there is no variation either in the rooms or between the different rooms. The relative humidity, however, is closely controlled in the above-mentioned range to reduce weighing errors. When the percentage is below forty, glassware acquires electrostatic charges while being handled which are not readily lost or dissipated and make weighing either difficult or impossible. The object being weighed *appears* to rapidly lose weight. If the relative humidity is above 50%, certain results, particularly carbon and hydrogen have a tendency to be high. The outlets of the air-conditioning system are quite large and of the diffusing type so that there is no apparent motion of air. This is of particular importance in the balance room, since balances are seriously affected by air currents.

WEATHER-PROOFING

As an aid to the air-conditioning system and to prevent drafts in winter, all windows in these laboratories are the storm type. The outside frames contain Thermopane glass (double panes with a dehydrated air space between, manufactured by Libbey-Owens-Ford Glass Co., 570 Lexington Avenue, New York, N.Y.). Inside, and approximately one foot away, are plate glass sections extending the entire height of the outside panes. Along the entire base between these two is a vane type of radiator which is used (in winter) solely for the purpose of eliminating drafts from the windows. Experience has shown that these "extras" are necessary when laboratory tables are situated under or near outside windows.

BALANCE ROOM

In the author's opinion, a balance room is a "must." Balances should *not* be placed out in the laboratory. They are the analyst's basic tool and should be so treated and placed accordingly, protected from laboratory conditions. Since a large percentage of the analyst's time is spent at the balance, sufficient space should be provided to prevent crowding. No amount of space should be considered "too much." The author's present balance room is thirty-seven feet long and eight feet wide and contains balance table space for nine balances or approximately four feet per balance which the author considers to be the minimum requirement. The balance room has only inside walls so that no weather-proofing is necessary. However, where outside walls and windows do exist in the balance room, the former should be *thoroughly* insulated and the latter should be of the heated storm type.[164] Figure 2 shows an electrically heated storm window with thermostatic control, which proved satisfactory for twelve years in a balance room previously occupied by the author. In addition, these balance room windows should be fitted with Venetian blinds and drapes as further protection from the sun's rays. Radiators, if present, should *not* be in the vicinity of the balances which are *very* sensitive to temperature changes and for the same reason, fluorescent lighting should be used for illumination. A temperature change of 1–2° C., *no matter what the cause,* produces a shift of the balance reading of about 10 µg.[119,120,144–147,174]

BALANCE TABLES

The type of tables upon which the balances are mounted is most important. It was emphasized above that excellent balances will not perform properly in a drafty location subjected to temperature changes. Neither will they give reliable service unless protected from shocks and vibration which affect their sensitivity and precision (see Chapter 3). The cost of a good balance alone should suggest its being properly mounted to increase its lifetime. The location

1. Introduction

of the balance room, particularly with reference to machinery *in* the building as well as disturbances *outside* the building, such as railroads, trucks, etc., governs the needs for the particular type of balance table which should be used. Two types are described in detail in this book, both of which the author has used over periods of years. However, in his opinion, the table described in the next section, "Inertia Block-isolator Type Table" is far superior, *in any location,* to any other which has appeared in the literature. The other, described under "Rigid, Combination Type Table," although satisfactory under certain

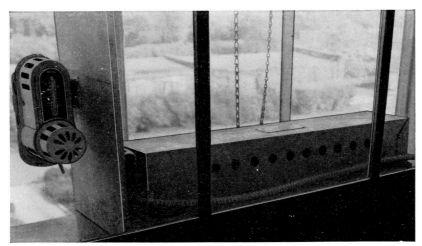

FIG. 2. Electrically heated storm window for balance room.

conditions, is a *poor,* cheaper substitute. Also, it should be borne in mind that future new equipment may create new vibration problems with which the rigid type cannot cope. This was the experience of the author.

Inertia Block-Isolator Type Table[187]

This type of table consists of three parts, namely, isolators, inertia block, and protective woodwork. The isolators and inertia block form a unit, the size of the former is *entirely dependent* on the weight of the latter. Each location presents its own vibration problem which should be solved independently, but the same principles are applicable and the selection of the proper combination of springs and inertia block (mass) should make it possible to construct a suitable table. A particularly bad vibration problem existed in the author's laboratory due to the existence of all types of machinery including six 40-inch basket-type centrifuges on the floor beneath the balance room. The principles involved in the construction of the table in the author's laboratory, shown in Figs. 3–8, will serve as an example. Provision is made for nine balances and for safety, the total weight of the table(s) was limited to 9000 pounds, including isolators, inertia blocks, and woodwork. Of these, the isolators and woodwork approximates 200 pounds per balance leaving 800 pounds for each block. (Note: It is advantageous to use the maximum permissible weight for the blocks, but lighter blocks together with lighter isolators would be as efficient as far as vibration absorption is concerned, as shown in the following discussion.)

Vibration isolation depends upon the ratio of the frequency of the disturbing vibration, F_D, and the natural frequency of the isolation mountings, F_N, the efficiency being expressed by the formula,

$$\% \text{ efficiency} = 100 \left(1 - \frac{1}{(F_D/F_N)^2 - 1} \right)$$

where

$$F_N = 188 \sqrt{\frac{1}{D}} \text{ cycles per minute}$$

and D, the deflection of the loaded isolation mounting, is given by

$$D = \frac{\text{load}}{K}$$

K, the spring constant, is the number of pounds required to make the mounting deflect one inch. If a force is applied to the loaded mounting to give a slight additional displacement and is then released, as during normal operation of the balance, the mounting will oscillate at a definite frequency (the natural frequency, F_N) which depends upon the static deflection of the mounting shown in the equation above. From the formula for the efficiency, it can be seen that a mounting having a natural frequency that will isolate the lowest expected disturbing frequency with a particular efficiency will isolate higher frequency disturbances with a higher efficiency.

In these laboratories, the closest source of disturbance is the six centrifuges on the floor beneath the balance room, which operate at 900 r.p.m. If unbalanced, the centrifuges would generate 900 cycles per minute disturbances; therefore, it was decided to protect the balances against these. Isolators were selected suitable for a 200-pound load per isolator, with approximately 1.2 inches deflection and a natural frequency of 172 cycles per minute.[91] Four of these, uniformly loaded under an 800-pound block, would have, according to the formula above, a theoretical isolation efficiency of 96% for disturbances of 900 cycles per minute. Actually, in practice, an efficiency of 93% was achieved.[187] When *properly adjusted,* the tables return to their original position after being displaced. *This is an absolute necessity;* otherwise, the precision of the balance (Chapter 3) would be greatly affected. (In spite of careful handling, a balance, and the table, are displaced *every time* it is put

1. Introduction

into operation.) The springs of the isolators should be *completely undamped* or the efficiency of vibration isolation is *decreased* and, what is more important, the table will *not* return to its original position after being deflected or displaced.

Isolators. The isolators are commercially available. Korfund Type[91] LK/D-52 Vibro-Isolators were recommended by the manufacturer after consideration of the various pieces of equipment causing vibrations in the building. A set of four isolators is required for each reinforced concrete inertia block, one at each of the four corners. Figures 3 and 4 show the isolators in place.

Reinforced concrete inertia blocks. One of these, weighing approximately 800 pounds, is used for each balance. Figure 4 shows the blocks, and Figs. 5 and 6 give the details of construction. Dimensions of the plateau (12 by 20 inches)

FIG. 3. Close-up of isolator.

FIG. 4. Inertia blocks mounted on isolators.

Laboratory Set-up

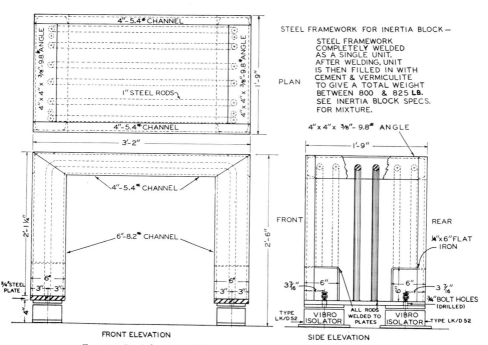

FIG. 5. Steel framework for inertia block—details of construction.

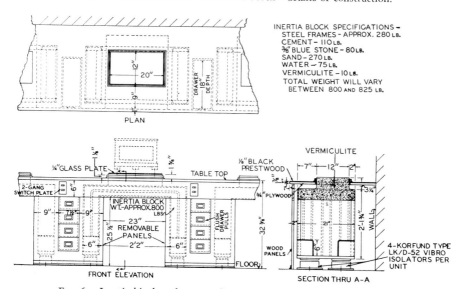

FIG. 6. Inertia block and surrounding woodwork—details of construction.

(Fig. 6) may be varied to accommodate the balance in question. Those shown are suitable for the Ainsworth, Becker, and Bunge balances. A plateau measuring 15 by 16 inches is suitable for a Mettler microchemical balance. On top of the concrete plateau is placed an equal size piece of ¼ inch plate glass, held in place by resting it on thin strips of plastic tape.

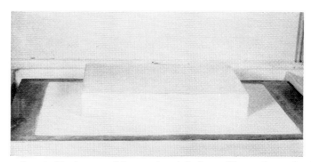

FIG. 7. Inertia block encased in woodwork before table top is put in place.

FIG. 8. Section of finished balance table.

Woodwork. The woodwork construction of the table is designed to protect the inertia block from being touched by the operator. The details of construction are given in Fig. 6. A clearance of one inch exists between the inertia block and the woodwork at all points. This is shown in Fig. 7, a photograph taken during construction before the table top was added. The table top has an opening to allow the small concrete plateau to protrude through it, but not touch it. (Upon this plateau rest a glass plate ¼ inch thick and, finally, the balance.) A small molding is attached to the table top to prevent the plateau from coming in contact with notebooks and other objects. Removable panels are used in the

FIG. 9. Rigid type of balance table showing balance mounting.

knee space, so that, if necessary, adjustment of the isolators can always be made with the minimum of difficulty. Figure 8 shows a section of the finished balance table which accommodates nine balances.

Lighting. Since illumination is necessary in the immediate vicinity of the balances, fluorescent fixtures are used.[164,174,187] A number of satisfactory fixtures are commercially available. For some balances, such as the Becker and Bunge types, two fixtures are used, one in front and the other directly over the beam. Figure 8 shows the manner of illuminating the author's balance table, while Figs. 6 and 8 show the arrangement of the instantaneously acting switches which control the lighting.

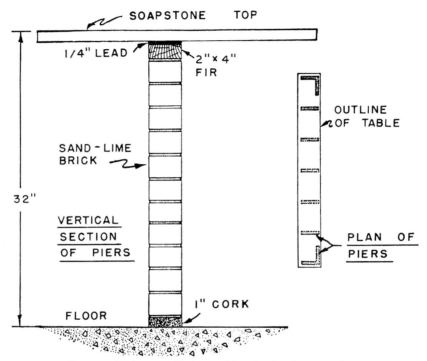

FIG. 10. Rigid type of balance table—details of construction.

Rigid, Combination Type Table[164,174]

This table is constructed of cork, brick and mortar, wood, lead, and stone, in the order named and rests on a concrete floor. The table must *not* have contact with the walls. The use of several different materials produces a table capable of damping out a variety of vibrations, *but in no way can it approach the inertia block-isolator type.* Figures 9 and 10 are a photograph and details of construction, respectively, of the rigid combination table formerly used by the author for a period of twelve years, upon which were placed six balances. Experience has shown that *individual* tables, four feet long, would be more satisfactory from the standpoint of one operator disturbing another. The cork sections on the floor should be protected in some way from contact with water and washing compounds when the floor is washed. Figure 9 shows the placement of the fluorescent lighting fixture and the instantaneously acting switch used to control lighting.

TABLE 4

ADDITIONAL INFORMATION ON REFERENCES* RELATED TO CHAPTER 1

In order to present the reader with a broader view of the subject, the material in this chapter is supplemented with the data in Table 4 which lists references to published work along the lines indicated. No attempt has been made to *completely* cover the field nor to give *detailed* information. The purpose is to furnish additional information, mostly in the form of clues that the reader may follow up by referring to the original literature. This same plan is also used in the following chapters.

Books
Association of Official Agricultural Chemists, 16
Belcher and Godbert, 24, 25
Boëtius, 33
Clark, E. P., 43
Clark, S. J., 44
Emich, 50
Emich and Schneider, 52
Fritz and Hammond, 55
Furman, 56
Grant, 61, 62
Kirk, 77
Milton and Waters, 109, 110
Mitchell, Kolthoff, Proskauer, and Weissberger, 111–113
Natelson, 116
Niederl and Niederl, 119, 120
Niederl and Sozzi, 121
Pregl, 136
Proceedings of the International Symposium on Microchemistry 1958, 138
Rodden, C. J., 139
Roth, 144–147

Nomenclature
Benedetti-Pichler, 32
Cheronis, 37–40
Malissa and Benedetti-Pichler, 105
Steyermark, 174

General
Abrahamczik, 1
Adibek-Melikyan and Sarkisyan, 3, 4
Alber, 5
Alimarin and Petrikova, 9, 10
Ambrosino, 12
Belcher, 21

General (Cont.)
Benedetti-Pichler, 30
El-Badry and Wilson, 49
Gillis, 58, 59
Gysel, 63
Haagen-Smit, 65
Haslam, 68
Hecht, 71
Ingram, 73
Kahane, 76
Kirsten, 79–85
Kohn, 87, 88
Kono, 89, 90
Korshun and Chumachenko, 92
Kuck, 93
Lévy, 95
Lieb, 97
Ma, 98–101
Ma and Sweeny, 102
Milner and Edwards, 108
Otter, 132
Parks and Lykken, 133
Rogers, 141
Schöniger, 151, 153–155
Schumacher and Streiff, 156
Sobotka, 158
Stephen, 161, 162
Steyermark, 164, 171, 173
Stock and Fill, 189
Vagnina, 192
Večeřa and Šnobl, 193
Wagner, 194
Wenger, 196
West, 197, 198
Wilson, 208–212
Zacherl, 215

* The numbers which appear after each entry in this table refer to the literature citations in the reference list at the end of the chapter.

1. Introduction

TABLE 4 (*Continued*)

Historical
Analytical Chemistry Editorial, 1950, 14
Barger, 17
Benedetti-Pichler, 28
Donau, 47
Emich and Donau, 51
Haagen-Smit, 65
Lieb, 96
Nernst, 117
Nernst and Riesenfeld, 118
Niederl and Niederl, 119
Pregl, 136
Rosenfeld, 142
Soltys, 159
Steyermark, 171
Wilson, 212
Wormley, 214

Reviews
Alicino, 8
Alvarez Querol, 11
Batalin, 18
Beaucourt, 19
Becker, 20
Belcher, 21
Belcher, Bevington, Stephen, and West, 22
Belcher, Gibbons, and Sykes, 23
Belcher, Nutten, West, and Whiffen, 26
Belcher, Sheridan, Stephen, and West, 27
Benedetti-Pichler, 29
Dory and Messmer, 48
Fort, 53, 54
Gorbach, 60
Hallett, 66
Haslam and Squirrell, 69
Ingram and Waters, 74
Kirsten, 83, 84
Lamo and Doadrio, 94
Ma, 98–101
Milner and Edwards, 108
Mitchell, Kolthoff, Proskauer, and Weissberger, 111–113
Moelants, 115
Ogg, 122, 123

Reviews (*Cont.*)
Ogg and Willits, 130
Olleman, 131
Parks and Lykken, 133, 134
Pella, 135
Prévost and Souchay, 137
Roth, 143
Schöniger, 153, 154
Stagg, 160
Steyermark, 171–173
Steyermark and McGee, 188a
Terent'ev and Terent'eva, 190
Thomson, 191
Waldmann, 195
Willits, 200, 201
Willits and Ogg, 202, 203
Wilson, 208–210
Witekowa, 213

Test samples (*Test substances*)
Benedetti-Pichler, 31
Committee of the Microchemistry Group of the British Chemical Society, 46
Moelants, 114
Smith, 157
Williams, 199
Willits and Ogg, 204

Balance tables
Alber and Harand, 6, 7
Ani, 15
Boëtius, 33
Clark, S. J., 44
Emich, 50
Gage and Sullivan, 57
Gysel and Strebel, 64
Hallett, 67
Howard, 72
Kirner, 78
Kissa, 86
Malissa, 104
McIlwraith, 106
Mettler Instrument Corporation, 107
Pregl, 136
Roth, 144–147
Schöniger, 152

REFERENCES

1. Abrahamczik, E., *Mikrochim Acta*, p. 756 (1959).
2. Acree, F., Jr., *J. Assoc. Offic. Agr. Chemists*, **24**, 648 (1941).
3. Adibek-Melikyan, A. I., and Sarkisyan, R. S., *Zhur. Anal. Khim.*, **12**, 265 (1957); *Chem. Abstr.*, **52**, 170 (1958).
4. Adibek-Melikyan, A. I., and Sarkisyan, R. S., *J. Anal. Chem. U.S.S.R. (English Translation)*, **12**, 267 (1957); *Chem. Abstr.* **52**, 7944 (1958).
5. Alber, H. K., *Intern. Microchem. Congr., 1st Congr.* Graz, Austria, 1950; *Mikrochemie ver. Mikrochim. Acta*, **36/37**, 75 (1951).
6. Alber, H. K., and Harand, J., *Ind. Eng. Chem., Anal. Ed.*, **10**, 403 (1938).
7. Alber, H. K., and Harand, J., *J. Franklin Inst.*, **224**, 729 (1937).
8. Alicino, J. F., *Microchem. J.*, **2**, 83 (1958).
9. Alimarin, I. P., and Petrikova, M. N., *Zhur. Anal. Khim.*, **7**, 341 (1952).
10. Alimarin, I. P., and Petrikova, M. N., *Khim. Nauka i Prom.*, **4**, 223 (1959); *Chem. Abstr.*, **53**, 15854 (1959).
11. Alvarez Querol, M. C., *Inform. quim. anal. (Madrid)*, **5**, 209 (1951).
12. Ambrosino, C., *Analyst*, **78**, 258 (1953).
13. American Society for Testing Materials, *ASTM Designations*, **E 124-56T** (1956), **E 124-57T** (1957), **E 147-59T** (1959), **E 148-59T** (1959).
14. *Anal. Chem.* (Editorial), **22**, 375 (1950).
15. Ani, M., *Machine Design*, p. 134 (November 1951).
16. Association of Official Agricultural Chemists, "Official Methods of Analysis", 8th ed., Washington, D. C., 1955.
17. Barger, G., *J. Chem. Soc.*, **85**, 309 (1904).
18. Batalin, A. Kh., *Zhur. Anal. Khim.*, **4**, 308 (1949).
19. Beaucourt, J. H., *Metallurgia*, **38**, 353 (1948).
20. Becker, W. W., *Anal. Chem.*, **22**, 185 (1950).
21. Belcher, R., *Sci. Progr.*, **47**, 250 (1959); *Chem. Abstr.* **53**, 12951 (1959).
22. Belcher, R., Bevington, J. C., Stephen, W. I., and West, T. S., *Ann. Repts. on Progr. Chem. (Chem. Soc. London)* **52**, 353 (1955).
23. Belcher, R., Gibbons, D., and Sykes, A., *Mikrochemie ver. Mikrochim. Acta*, **40**, 76 (1953).
24. Belcher, R., and Godbert, A. L., "Semi-micro Quantitative Organic Analysis," Longmans, Green, London, New York, and Toronto, 1945.
25. Belcher, R., and Godbert, A. L., "Semi-micro Quantitative Organic Analysis," 2nd ed., Longmans, Green, London, 1954.
26. Belcher, R., Nutten, A. J., West, T. S., and Whiffen, D. H., *Ann. Repts. on Progr. Chem. (Chem. Soc. London)*, **51**, 347 (1954).
27. Belcher, R., Sheridan, J., Stephen, W. I. and West, T. S., *Ann. Repts. on Progr. Chem. (Chem. Soc. London)*, **53**, 354 (1956).
28. Benedetti-Pichler, A. A., *Mikrochemie ver. Mikrochim. Acta*, **35**, 131 (1950).
29. Benedetti-Pichler, A. A., *Angew. Chem.*, **63**, 158 (1951).
30. Benedetti-Pichler, A. A., *Mikrochemie ver. Mikrochim. Acta*, **36/37**, 38 (1951).
31. Benedetti-Pichler, A. A., *Mikrochemie ver. Mikrochim. Acta*, **39**, 319 (1952).
32. Benedetti-Pichler, A. A., "Introduction to the Microtechniques of Inorganic Analyses," Wiley, New York, Chapman & Hall, London, 1942.
33. Boëtius, M., "Über die Fehlerquellen bei der mikroanalytischen Bestimmung des Kohlenstoffes und Wasserstoffes nach der Methode von Fritz Pregl," Verlag Chemie, Berlin, 1931.

1. Introduction

34. Boos, R. N., *in* "Laboratory Design," (H. S. Coleman, ed.), p. 188, Reinhold, New York, 1951.
35. British Standards Institution, *Brit. Standards,* CW (LBC) 8189: **1428**, Pt. G2; CW (LBC) 6476: **1428**, Pt. K1; CW (LBC) 6812: **1428**, Pt. A5; 1428, Pt. I1 (1953), and Pt. A1 (1950); 797 (1954) (also listed as **1428**, Pt. D4): **1428**, Pt. D5 (1955); Pt. D2 (1950); Pt. A2 and Pt. A3 (1952); Pt. A4 (1953); Pt. B1, and Pt. B2 (1953), Pt. C1, and Pt. C2 (1954); Pt. D1 (1952), Pt. D3 (1950), Pt. D6 (1955), Part E1 (1953), Pt. E2 (1954), Pt. E3 (1953), Pt. G1 (1954), Pt. H1 (1952), Pt. J1 (1954), Pt. A5, Pt. F1, and Pt. G2 (1957), and Pt. A1, and Pt. K1 (1958).
36. *Chem. Eng. News,* **26**, 883 (1948).
37. Cheronis, N. D., *Microchem. J.,* **1**, 167 (1957).
38. Cheronis, N. D., *Microchem. J.,* **3**, 477 (1959).
39. Cheronis, N. D., "Micro and Semi-micro Methods," Interscience, New York, 1954.
40. Cheronis, N. D., *Mikrochim. Acta,* p. 202 (1956).
41. Clark, E. P., *J. Assoc. Offic. Agr. Chemists,* **15**, 136 (1932).
42. Clark, E. P., *J. Assoc. Offic. Agr. Chemists,* **22**, 622 (1939).
43. Clark, E. P., "Semimicro Quantitative Organic Analysis," Academic Press, New York, 1943.
44. Clark, S. J., "Quantitative Methods of Organic Microanalysis," Butterworths, London, 1956.
45. Commission on Microchemical Techniques, Section of Analytical Chemistry, International Union of Pure and Applied Chemistry, *Pure & Appl. Chem.,* **1**, 143 (1960).
46. Committee of the Microchemistry Group of the British Chemical Society, *Analyst,* **78**, 258 (1953).
47. Donau, J., *Monatsh.,* **32**, 35 (1911).
48. Dory, I., and Messmer, A., *Magyar Kém. Lapja,* **6**, 66 (1951).
49. El-Badry, H. M., and Wilson, C. L., *Mikrochemie ver. Mikrochim Acta,* **40**, 141 (1953).
50. Emich, F., "Lehrbuch der Mikrochemie," 2nd ed., Bergmann, Munich, 1926.
51. Emich, F., and Donau, J., *Monatsh.,* **30**, 753 (1909).
52. Emich, F., and Schneider, F., "Microchemical Laboratory Manual," Wiley, New York, 1934.
53. Fort, R., *Chim. anal.,* **39**, 319 (1957); *Chem. Abstr.,* **52**, 972 (1958).
54. Fort, R., *Chim. anal.,* **39**, 366 (1958); *Chem. Abstr.,* **52**, 4397 (1958).
55. Fritz, J. S., and Hammond, G. S., "Quantitative Organic Analysis," Wiley, New York, Chapman & Hall, London, 1957.
56. Furman, N. H., ed., "Scott's Standard Methods of Chemical Analysis," 5th ed., Vol. II, Van Nostrand, New York, 1939.
57. Gage, D. G., and Sullivan, P., *Anal. Chem.,* **28**, 922 (1956).
58. Gillis, J., *Mededel. Vlaam. Chem. Ver.,* **12**, 15 (1950).
59. Gillis, J., *Experientia,* **8**, 865 (1952).
60. Gorbach, G., *Fette u. Seifen,* **53**, 3 (1951).
61. Grant, J., "Quantitative Organic Microanalysis, Based on the Methods of Fritz Pregl," 4th ed., Blakiston, Philadelphia, Pennsylvania, 1946.
62. Grant, J., "Quantitative Organic Microanalysis," 5th ed., Blakiston, Philadelphia, Pennsylvania, 1951.
63. Gysel, H., *Mikrochim. Acta,* p. 1456 (1956).

64. Gysel, H., and Strebel, W., *Mikrochim. Acta,* p. 782 (1954).
65. Haagen-Smit, A. J., *J. Chem. Educ.,* **28**, 496 (1951).
66. Hallett, L. T., *Ind. Eng. Chem., Anal. Ed.,* **14**, 956 (1942).
67. Hallett, L. T., *Metropol. Microchem. Soc. 1st Symposium,* New York, March 1946.
68. Haslam, J., "Proceedings of the International Symposium on Microchemistry 1958," p. 486, Pergamon Press, Oxford, London, New York, and Paris, 1960.
69. Haslam, J., and Squirrell, D. C. M., *Ann. Repts. on Progr. Chem. (Chem. Soc. London),* **60**, 425 (1959).
70. Hayman, D. F., *Ind. Eng. Chem., Anal. Ed.,* **8**, 342 (1936).
71. Hecht, F., *Mikrochemie ver. Mikrochim. Acta,* **36/37**, 1181 (1951).
72. Howard, H. C., *Ind. Eng. Chem.,* **13**, 231 (1921).
73. Ingram, G., *Chemist and Druggist,* **155**, 615 (1951).
74. Ingram, G., and Waters, W. A., *Ann. Repts. on Progr. (Chem. Soc. London),* **46**, 280 (1949) (Pub. 1950).
75. International Union of Pure and Applied Chemistry, Commission on Atomic Weights, "Table of Atomic Weights, 1957." Comptes Rendus, 19th Conference in Paris.
76. Kahane, E., *Bull. soc. chim. France,* **17**, 1 (1950).
77. Kirk, P. L., "Quantitative Ultramicroanalysis," Wiley, New York, Chapman & Hall, London, 1950.
78. Kirner, W. R., *Ind. Eng. Chem., Anal. Ed.,* **9**, 300 (1937).
79. Kirsten, W. J., *Anal. Chem.,* **25**, 74 (1953).
80. Kirsten, W. J., *Anal. Chem.,* **27**, 23A (1955).
81. Kirsten, W. J., *Anal. Chim. Acta,* **5**, 458 (1951).
82. Kirsten, W. J., *Anal. Chim. Acta,* **5**, 489 (1951).
83. Kirsten, W. J., *Chim. anal.,* **40**, 253 (1958).
84. Kirsten, W. J., *Microchem. J.,* **2**, 179 (1958).
85. Kirsten, W. J., *Mikrochim. Acta,* p. 836 (1956).
86. Kissa, E., *Microchem. J.,* **4**, 89 (1960).
87. Kohn, M., *Anal. Chim. Acta,* **5**, 337 (1951).
88. Kohn, M., *Anal. Chim. Acta,* **9**, 249 (1953).
89. Kono, T., *Nippon Nôgei-kagaku Kaishi,* **31**, 622 (1957); *Chem. Abstr.,* **52**, 12662 (1958).
90. Kono, T., *Mikrochim. Acta,* p. 461 (1958).
91. Korfund Co., Inc., Westbury, Long Island, New York, Chart SA-2028-0, and Drawing SC-32-2.
92. Korshun, M. O., and Chumachenko, M. N., *Doklady Akad. Nauk S.S.S.R.,* **99**, 769 (1954).
93. Kuck, J. A., *Anal. Chem.,* **30**, 1552 (1958).
94. Lamo, M. A. de, and Doadrio, A., *Anales real soc. españ. fís. y quím. (Madrid)* **44B**, 723 (1948).
95. Lévy, R., *Bull. soc. chim. France,* p. 672 (1952).
96. Lieb, H., *Mikrochemie ver. Mikrochim. Acta,* **35**, 123 (1950).
97. Lieb, H., "Proceedings of the International Symposium on Microchemistry 1958," p. 73, Pergamon Press, Oxford, London, New York, and Paris, 1960.
98. Ma, T. S., *Anal. Chem.,* **30**, 760 (1958).
99. Ma, T. S., *Anal. Chem.,* **30**, 1557 (1958).
100. Ma, T. S., *Microchem. J.,* **2**, 91 (1958).
101. Ma, T. S., *Microchem. J.,* **3**, 415 (1959); **4**, 373 (1960).

1. Introduction

102. Ma, T. S., and Sweeny, R. F., *Mikrochim. Acta,* p. 191 (1956).
103. Macurdy, L. B., Alber, H. K., Benedetti-Pichler, A. A., Carmichael, H., Corwin, A. H., Fowler, R. M., Huffman, E. W. D., Kirk, P. L., and Lashof, T. W., *Anal. Chem.,* **26**, 1190 (1954).
104. Malissa, H., *Österr. Chemiker-Ztg.,* No. 13/14 (1952).
105. Malissa, H., and Benedetti-Pichler, A. A., "Anorganische qualitative Mikroanalyse," Springer, Wien, 1958.
106. McIlwraith, C. G., *Anal. Chem.,* **23**, 688 (1951).
107. Mettler Instrument Corp., Hightstown, N. J., *Mettler News,* pp. 13–18 (December 1959).
108. Milner, G. W. C., and Edwards, J. W., *Met. Revs.,* **2**, 109 (1957); *Chem. Abstr.,* **52**, 9848 (1958).
109. Milton, R. F., and Waters, W. A., "Methods of Quantitative Microanalysis," Longmans, Green, New York, and Arnold, London, 1949.
110. Milton, R. F., and Waters, W. A., "Methods of Quantitative Microanalysis," 2nd ed., Arnold, London, 1955.
111. Mitchell, J., Jr., Kolthoff, I. M., Proskauer, E. S., and Weissberger, A., "Organic Analysis," Vol. I, Interscience, New York, 1953.
112. Mitchell, J., Jr., Kolthoff, I. M., Proskauer, E. S., and Weissberger, A., "Organic Analysis," Vol. II, Interscience, New York, 1954.
113. Mitchell, J., Jr., Kolthoff, I. M., Proskauer, E. S., and Weissberger, A., "Organic Analysis," Vol. III, Interscience, New York, 1956.
114. Moelants, L. J., *Ind. chem. belge,* **21**, 207 (1956).
115. Moelants, L., *Mededel. Vlaam. Chem. Ver.,* **12**, 120 (1950).
116. Natelson, S., "Microtechniques of Clinical Chemistry for the Routine Laboratory," Thomas, Springfield, Illinois, 1957.
117. Nernst, W., *Göttinger Nachr.,* p. 75 (1903).
118. Nernst, W., and Riesenfeld, E. H., *Ber.,* **36**, 2086 (1903).
119. Niederl, J. B., and Niederl, V., "Micromethods of Quantitative Organic Elementary Analysis," Wiley, New York, 1938.
120. Niederl, J. B., and Niederl, V., "Micromethods of Quantitative Organic Analysis," 2nd ed., Wiley, New York, 1942.
121. Niederl, J. B., and Sozzi, J. A., "Microanálisis, Elemental Orgánico," Calle Arcos, Buenos Aires, 1958.
122. Ogg, C. L., *Anal. Chem.,* **26**, 116 (1954).
123. Ogg, C. L., *Anal. Chem.,* **28**, 766 (1956).
124. Ogg, C. L., *J. Assoc. Offic. Agr. Chemists,* **35**, 305 (1952).
125. Ogg, C. L., *J. Assoc. Offic. Agr. Chemists,* **36**, 335, 344 (1953).
126. Ogg, C. L., *J. Assoc. Offic. Agr. Chemists,* **38**, 365, 377 (1955).
127. Ogg, C. L., *J. Assoc. Offic. Agr. Chemists,* **41**, 294 (1958).
128. Ogg, C. L., *J. Assoc. Offic. Agr. Chemists,* **42**, 319 (1959).
129. Ogg, C. L., *J. Assoc. Offic. Agr. Chemists,* **43**, 682, 689, 693 (1960).
130. Ogg, C. L., and Willits, C. O., *Anal. Chem.,* **23**, 47 (1951).
131. Olleman, E. D., *Anal. Chem.,* **24**, 1425 (1952).
132. Otter, I. K. H., *Nature,* **182**, 393 (1958).
133. Parks, T. D., and Lykken, L., *Anal. Chem.,* **22**, 1444 (1950).
134. Parks, T. D., and Lykken, L., *Petrol. Refiner,* **29**, No. 8, 85 (1950).
135. Pella, E., *Mikrochim. Acta,* p. 687 (1958).

136. Pregl, F., "Quantitative Organic Microanalysis" (E. Fyleman, trans., 2nd German ed.), Churchill, London, 1924.
137. Prévost, C., and Souchay, P., *Chim. anal.*, **37**, 3 (1955).
138. "Proceedings of the International Symposium on Microchemistry 1958," Pergamon Press, Oxford, London, New York, and Paris, 1960.
139. Rodden, C. J., Editor-in-chief, "Analytical Chemistry of the Manhattan Project," National Nuclear Energy Series, Manhattan Project Technical Section, Division VIII, Vol. I, McGraw-Hill, New York, 1950.
140. Rodden, C. J., Kuck, J. A., Benedetti-Pichler, A. A., Corwin, A., and Huffman, E. W. D., *Ind. Eng. Chem., Anal. Ed.*, **15**, 415 (1943).
141. Rogers, L. B., *J. Chem. Educ.*, **29**, 612 (1952).
142. Rosenfeld, B., *Microchem. J.*, **3**, 135 (1959).
143. Roth, H., *Angew. Chem.*, **53**, 441 (1940).
144. Roth, H., "Die quantitative organische Mikroanalyse von Fritz Pregl," 4th ed., Springer, Berlin, 1935.
145. Roth, H., "F. Pregl quantitative organische Mikronanalyse," 5th ed., Springer, Wien, 1947.
146. Roth, H., "Pregl-Roth quantitative organische Mikroanalyse," 7th ed., Springer, Wien, 1958.
147. Roth, H., "Quantitative Organic Microanalysis of Fritz Pregl," 3rd ed. (E. B. Daw, trans., 4th German ed.), Blakiston, Philadelphia, Pennsylvania, 1937.
148. Royer, G. L., Alber, H. K., Hallett, L. T., and Kuck, J. A., *Ind. Eng. Chem., Anal. Ed.*, **15**, 476 (1943).
149. Royer, G. L., Alber, H. K., Hallett, L. T., and Kuck, J. A., *Ind. Eng. Chem., Anal. Ed.*, **15**, 230 (1943).
150. Royer, G. L., Alber, H. K., Hallett, L. T., Spikes, W. F., and Kuck, J. A., *Ind. Eng. Chem., Anal. Ed.*, **13**, 574 (1941).
151. Schöniger, W., *Angew. Chem.*, **67**, 261 (1955).
152. Schöniger, W., *Mikrochim. Acta*, p. 382 (1959).
153. Schöniger, W., *Mikrochim. Acta*, p. 670 (1959).
154. Schöniger, W., *Mikrochim. Acta*, p. 1456 (1956).
155. Schöniger, W., *Mitt. Gebiete Lebensm. u. Hyg.*, **43**, 105 (1952).
156. Schumacher, E., and Streiff, H. J., *Helv. Chim. Acta*, **41**, 1771 (1958).
157. Smith, W. H., *Anal. Chem.*, **30**, 149 (1958).
158. Sobotka, M., *Mikrochemie ver. Mikrochim. Acta*, **36/37**, 407 (1951).
159. Soltys, A., *Mikrochemie*, **20**, 107 (1936).
160. Stagg, H. E., *Ann. Repts. on Progr. Chem. (Chem. Soc. London)*, **47**, 388 (1950).
161. Stephen, W. I., *Mfg. Chemist*, **25**, 300 (1954).
162. Stephen, W. I., *Mfg. Chemist*, **25**, 339 (1954).
163. Steyermark, Al, *Anal. Chem.*, **22**, 1228 (1950).
164. Steyermark, Al, *Ind. Eng. Chem., Anal. Ed.*, **17**, 523 (1945).
165. Steyermark, Al, *J. Assoc. Offic. Agr. Chemists*, **38**, 367 (1955).
166. Steyermark, Al, *J. Assoc. Offic. Agr. Chemists*, **39**, 401 (1956).
167. Steyermark, Al, *J. Assoc. Offic. Agr. Chemists*, **40**, 381 (1957).
168. Steyermark, Al, *J. Assoc. Offic. Agr. Chemists*, **41**, 297, 299 (1958).
169. Steyermark, Al, *J. Assoc. Offic. Agr. Chemists*, **42**, 319 (1959).
170. Steyermark, Al, *J. Assoc. Offic. Agr. Chemists*, **43**, 683 (1960).
170a. Steyermark, Al, *J. Assoc. Offic. Agr. Chemists*, **44**, in press (1961).
171. Steyermark, Al, *Microchem. J.*, **2**, 21 (1958).

172. Steyermark, Al, *Microchem. J.,* **3**, 399 (1959).
173. Steyermark, Al, "Proceedings of the International Symposium on Microchemistry 1958," p. 537, Pergamon Press, Oxford, London, New York, and Paris, 1960.
174. Steyermark, Al, "Quantitative Organic Microanalysis," Blakiston, Philadelphia, Pennsylvania, 1951.
175. Steyermark, Al, Alber, H. K., Aluise, V. A., Huffman, E. W. D., Jolley, E. L., Kuck, J. A., Moran, J. J., and Ogg, C. L., *Anal. Chem.,* **28**, 112 (1956).
176. Steyermark, Al, Alber, H. K., Aluise, V. A., Huffman, E. W. D., Jolley, E. L., Kuck, J. A., Moran, J. J., and Ogg, C. L., *Anal. Chem.,* **28**, 1993 (1956).
177. Steyermark, Al, Alber, H. K., Aluise, V. A., Huffman, E. W. D., Jolley, E. L., Kuck, J. A., Moran, J. J., and Ogg, C. L., *Anal. Chem.,* **30**, 1702 (1958).
178. Steyermark, Al, Alber, H. K., Aluise, V. A., Huffman, E. W. D., Jolley, E. L., Kuck, J. A., Moran, J. J., Ogg, C. L., and Pietri, C. E., *Anal. Chem.,* **32**, 1045 (1960).
179. Steyermark, Al, Alber, H. K., Aluise, V. A., Huffman, E. W. D., Jolley, E. L., Kuck, J. A., Moran, J. J., and Willits, C. O., *Anal. Chem.,* **23**, 1689 (1951).
180. Steyermark, Al, Alber, H. K., Aluise, V. A., Huffman, E. W. D., Jolley, E. L., Kuck, J. A., Moran, J. J., Ogg, C. L., and Willits, C. O., *Anal. Chem.,* **26**, 1186 (1954).
181. Steyermark, Al, Alber, H. K., Aluise, V. A., Huffman, E. W. D., Kuck, J. A., Moran, J. J., and Willits, C. O., *Anal. Chem.,* **21**, 1283 (1949).
182. Steyermark, Al, Alber, H. K., Aluise, V. A., Huffman, E. W. D., Kuck, J. A., Moran, J. J., and Willits, C. O., *Anal. Chem.,* **21**, 1555 (1949).
183. Steyermark, Al, Alber, H. K., Aluise, V. A., Huffman, E. W. D., Kuck, J. A., Moran, J. J., and Willits, C. O., *Anal. Chem.,* **23**, 523 (1951).
184. Steyermark, Al, Alber, H. K., Aluise, V. A., Huffman, E. W. D., Kuck, J. A., Moran, J. J., and Willits, C. O., *Anal. Chem.,* **23**, 537 (1951).
185. Steyermark, Al, and Faulkner, M. B., *J. Assoc. Offic. Agr. Chemists,* **35**, 291 (1952).
186. Steyermark, Al, and Garner, M. W., *J. Assoc. Offic. Agr. Chemists,* **36**, 319 (1953).
187. Steyermark, Al, Ingalls, E. D., and Wilkenfeldt, J. W., *Anal. Chem.,* **28**, 517 (1956).
188. Steyermark, Al, and Loeschauer, E. E., *J. Assoc. Offic. Agr. Chemists,* **37**, 433 (1954).
188a. Steyermark, Al, and McGee, B. E., *Microchem. J.,* **4**, 353 (1960).
189. Stock, J. T., and Fill, M. A., *Mikrochim. Acta,* p. 89 (1953).
190. Terent'ev, A. P., and Terent'eva, Ev. A., *Khim. Nauka i Prom.,* **4**, 242 (1959); *Chem. Abstr.,* **53**, 15870 (1959).
191. Thomson, M. L., *Metallurgia,* **39**, 46 (1948).
192. Vagnina, L. L., *Mikrochim. Acta,* p. 221 (1956).
193. Večeřa, M., and Šnobl, D., *Chem. listy,* **51**, 1482 (1957).
194. Wagner, H., *Chimica,* **13**, 213 (1959).
195. Waldmann, H., *Mikrochemie ver. Mikrochim. Acta,* **36/37**, 973 (1951).
196. Wenger, P. E., *Mikrochemie ver. Mikrochim. Acta,* **36/37**, 94 (1951).
197. West, T. S., *Chem. Age (London),* **80**, 193 (1958).
198. West, T. S., *Research (London),* **7**, 60 (1954).
199. Williams, M., *Ind. Chem. Mfr.,* **32**, 492 (1956).
200. Willits, C. O., *Anal. Chem.,* **21**, 132 (1949).
201. Willits, C. O., *Anal. Chem.,* **23**, 1565 (1951).
202. Willits, C. O., and Ogg, C. L., *Anal. Chem.,* **22**, 268 (1950).

203. Willits, C. O., and Ogg, C. L., *Anal. Chem.*, **24**, 70 (1952).
204. Willits, C. O., and Ogg, C. L., *Ind. Eng. Chem., Anal. Ed.*, **18**, 334 (1946).
205. Willits, C. O., and Ogg, C. L., *J. Assoc. Offic. Agr. Chemists*, **32**, 561 (1949).
206. Willits, C. O., and Ogg, C. L., *J. Assoc. Offic. Agr. Chemists*, **33**, 179 (1950).
207. Willits, C. O., and Ogg, C. L., *J. Assoc. Offic. Agr. Chemists*, **34**, 608 (1951).
208. Wilson, C. L., *Ann. Repts. on Progr. Chem.* (*Chem. Soc. London*), **48**, 328 (1951).
209. Wilson, C. L., *Ann. Repts. on Progr. Chem.* (*Chem. Soc. London*), **49**, 318 (1952).
210. Wilson, C. L., *Ann. Repts. on Progr. Chem.* (*Chem. Soc. London*) **50**, 355 (1953).
211. Wilson, C. L., *Mikrochim. Acta*, p. 58 (1953).
212. Wilson, C. L., *Mikrochim. Acta*, p. 91 (1956).
213. Witekowa, S., *Wiadomości Chemi.*, **4**, 174 (1950).
214. Wormley, T. G., "Microchemistry of Poisons," Lippincott, Philadelphia, Pennsylvania, 1867.
215. Zacherl, M. K., *Mitt. chem. Forschungs-insts. Ind. Österr.*, **4**, 27 (1950).

CHAPTER 2

The Microchemical Balance

The microchemical balance* is generally accepted as the instrument used for making the weighings in connection with quantitative organic microanalysis† and the first one was built by Kuhlmann[51] for Pregl[73] as listed in Table 1. Since then, a number of companies[2,9,16,53,56,59,62,69,71,84-86] have produced these instruments. They are similar, but smaller, in construction to good analytical balances and often possess the same type of extras: weight carriers, optical projection, damping devices for reducing the weighing time, etc. The essential differences between the various types of balances are in the capacities and

TABLE 5
Types of Balances, Their Capacities, and Claimed Sensitivities

Type	Capacity	Claimed sensitivity
Analytical	200 gram	0.1 and 0.05 mg. (some 0.025 mg.)
Assay	1 gram	0.02 mg.
Semimicrochemical	100 gram	0.01 and 0.005 mg.
Microchemical	20 and 30 gram	0.001 mg.
Microbalance (Microgram balance)	Usually a few mg. (some up to several tenth gram)	Some down to 0.001 and 0.005 µg.

the sensitivities (and precisions or reproducible sensitivities) as shown in Table 5. (See Chapter 3 for Determination of Sensitivity and Precision. Also "Additional Information" for this chapter.)

Theoretically,[12] if the precision (or reproducible sensitivity) of a balance is

* Please see references 1, 5, 10, 11, 12, 20, 21, 29, 30, 34, 35, 37, 45, 49, 65, 66, 68, 73, 74, 76, 78-81, 87, 88, 93.

† This is the analysts' basic tool and the caliber of work performed is dependent upon its condition. For this reason, the author strongly advocates that each analyst be provided with his (or her) own balance. With proper care, a balance will give excellent results for twenty years or more before reconditioning is necessary. In addition, the arrangement of one balance per person is a necessity for those doing carbon-hydrogen determinations. Any other arrangement places the analyst at a distinct disadvantage from the standpoints of both quantity and quality.

0.001 mg., it is possible to work with samples which weigh as little as 0.3 mg., but if the precision is only 0.020 mg., the sample must weigh at least 6.0 mg. The practical weights, however, are much greater.[67] Table 6 shows the relationships which exist between these variables[18, 66] and from it can be seen that the microchemical balance is suitable for samples in the size range used in the procedures in the following chapters.

TABLE 6
SIZE OF SAMPLE REQUIRED WHEN USING BALANCES WITH VARIOUS PRECISIONS

Precision of balance (mg.)	Weight of sample required	
	Theoretical minimum (mg.)	Practical average (mg.)
0.001	0.3–0.4	3–5
0.002	0.6–0.8	4–6
0.005	1.5–2.0	5–8
0.010	3.0–4.0	6–10
0.020	6.0–8.0	8–12

Microchemical balances have a reputed sensitivity of 0.001 mg., but the precision of the best instruments is of the order of 0.003 mg.[76] at no load. At the maximum load of 20 grams (30 in some cases) the sensitivity and precision of a good balance drops off about 5%.[5*] A good semimicrochemical balance[2,9,56,85] can be used if the sample size is increased slightly, but the author approves of such procedures for emergency measures only. Although some of the microbalances[23,25,26,44,52,63,64,70,82] (microgram balances) are more sensitive than microchemical balances the former are *not* suitable because of the lower capacities.

It is taken for granted that the beginner in microanalysis is familiar with the details of construction of, as well as has had experience with, analytical balances. Therefore, microchemical balances will be discussed rather generally here, with emphasis only on certain details.

Essential Parts

The essential parts of the balances are the beam, stirrups, and pans or hangers. A particular balance might have particular features regarding the parts, but the following general information will acquaint the reader with basic information which will cover all cases. Additional information regarding several different makes and types of balances follows this basic material.

* Two pan type.

2. The Microchemical Balance   32

BEAM

The *beams* of modern balances are usually prepared from special alloys (aluminum), bronze, or platinum plated metal, approximately four or five inches in length. This is in direct contrast to the tiny beam of the original Kuhlmann[51] balance which was 70 mm. in length. Attached to the beam is a *central knife edge* usually made of agate, sapphire, or other extremely hard material such as boron-carbide. In addition, there are *end knife edges* for each of the pans or hangers (two for the Ainsworth,[2] Becker,[9] Bunge,[16] etc., types and one for the constant load types represented by the Mettler[56]). Attached to

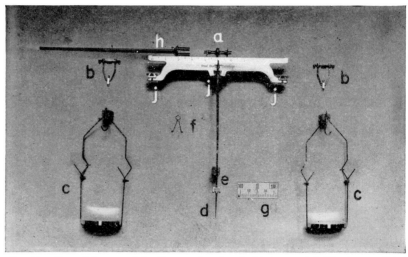

FIG. 11. Essential parts of the Bunge microchemical balance. (a) The beam. (b) The stirrups (two). (c) The pans (two). (d) The pointer. (e) The scale or reticle. (f) The rider. (g) The image of the reticle projected on a screen. (h) Zero point adjuster, not attached, used only when balance is first assembled. (j) Knife edges (three).

the beam, on the two pan type, is a *pointer* [Fig. 11(d)] which swings in front of a fixed scale, usually for the purpose of making rough readings. A separate small scale, or *reticle* [Fig. 11(e)], is sometimes attached to the pointer, above the point. As the pointer swings, this reticle is viewed either through a telescope or microscope (see Fig. 9) or the *image of the reticle is projected on a screen* for magnification [see Figs. 8, 11(g), 12, 15, 16, 17, and 20]. Toward the top of the pointer rod or on the beam, is a small weight (either in the form of a threaded knurled knob or sliding weight with a locking bolt) used for adjusting the sensitivity. Raising the weight raises the center of gravity and increases the sensitivity, while lowering the weight lowers the center of gravity and decreases the sensitivity. This adjustment should be made, only by an expert, if the sensitivity does not fall within the range of 95 to 105 (see Chapter 3). (If the center of gravity is raised too much, an unstable pendulum system results and the balance becomes erratic.) The *knife edges* are either held in place by a set of tension bolts (Fig. 11) or are swaged into place (Figs. 13 and 14). When tension bolts are used, they *never* should be tampered with as the setting of these are most vital to the sensitivity and precision of the balance. If any of the knife edges become chipped, even though only detected microscopically, the edges must be reground by the manufacturer. The central knife edge, in operation, rests upon the flat surface or

FIG. 12. Bunge microchemical balance, model 25 MPN.

plane at the top of the column of the balance, while the end knife edges are in contact with the planes of the stirrups as explained below. All of the planes are prepared of the same material as the knife edges.

STIRRUPS

As mentioned in the above paragraph, during operation the end knife edges are in contact with the stirrups, or suspensions [Figs. 11(b) and 14], and that the flat surfaces or planes are prepared of the same material as the knife edges. The metal portions are prepared from various metals, often gold or rhodium plated. When the flat surfaces or planes on the stirrups or column become worn so that microscopic grooves are present, the sensitivity and precision are greatly affected, just as when the knife edges become chipped, and the balance must be returned to the manufacturer for reconditioning.

2. The Microchemical Balance

PANS

The pans or hangers [Fig. 11(c)] are suspended from the stirrups. These are equipped with hooks for the purpose of supporting some of the objects to be weighed.

RIDER AND RIDER CARRIER

Most balances are equipped with notched beams [Figs. 11(a), 13, and 14], and, therefore, require *riders*. A number of shapes are used,[2,9,49,74] each with the purpose of obtaining perfect rider placement which is absolutely necessary or the balance will have poor precision.[2,9,49,65,66,74] Riders are generally made of aluminum, bronze, or in some instances, of quartz. Their shapes are usually either wishbone [Fig. 11(f)], U-, or rod-shaped. The *rider carrier*, of course,

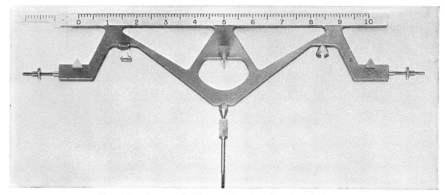

FIG. 13. Beam of Ainsworth microchemical balance.

must be designed to fit the rider. Some of these, as shown in Figs. 15 and 16 are equipped with a magnifier as an aid in manipulating. Figure 14 shows the carrier on an Ainsworth balance equipped with a rod-shaped rider.

CASES AND RELEASING MECHANISMS

Balance cases are prepared from mahogany or metal, the present trend being definitely towards metal. Some have only front doors (Fig. 15), some have only side doors (Figs. 12 and 17), while some have both (Figs. 16 and 20). The beam- and stirrup-, and pan-releasing mechanisms on some balances operate simultaneously from a single control (Figs. 12, 17, and 19), while others use separate mechanisms similar to those found on the ordinary analytical balance (Fig. 15).

A number of excellent balances are commercially available, but space here does not permit their detailed description. In addition, the author prefers to adhere to the policy adopted throughout this book of going into detail *only* in regard to apparatus and procedures with which he is familiar. Therefore, descriptive material is included for Ainsworth, Becker, Bunge, and Mettler balances. Instruments of the first three named have been used by the author

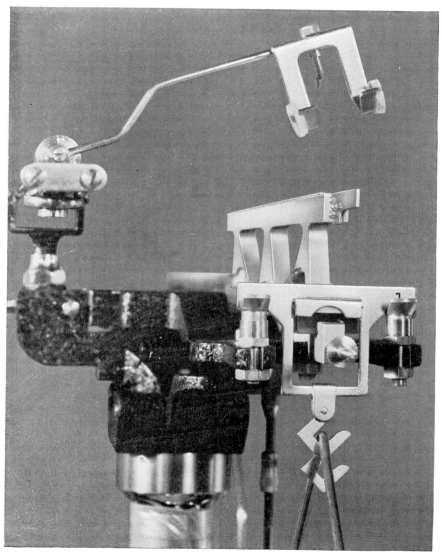

Fig. 14. Enlarged view to show details of construction of stirrup and rider carrier of Ainsworth microchemical balance.

2. The Microchemical Balance   36

FIG. 15. Ainsworth microchemical balance. Model shown is FHM having keyboard operated weight carrier employing 0.5 mg. rider giving beam 0 to 1 mg. range. Model FH is similar, but without keyboard operated weight carrier and employing 5 mg. rider giving beam 0 to 10 mg. range. (Note: The author suggests discarding the equilibrium or zero adjusting riders furnished with this balance and making the adjustment by means of the knurled screw bolts on the beam.)

FIG. 16. Becker microchemical balance, model EM 1.

for periods ranging from fifteen to twenty-two years, although not necessarily the models shown in Figs. 12, 15, and 16, which obviously are the current ones. The description of these individual balances should *not* be interpreted by the reader to mean their endorsement over all other makes. Likewise, the omission of descriptions of other makes of balances should *not* be interpreted

FIG. 17. Mettler microchemical balance, model M-5.

as an indication of their inferiority. Before selecting a balance, the analyst should take into consideration his (or her) particular needs regarding weight carriers, damping,* price, etc.

Figures 12, 15, 16, and 17 show the Bunge (model 25 MPN), Ainsworth (model FHM), Becker (model EM-1), and Mettler (model M-5) microchemical balances, respectively. Table 7 shows the capacities, sensitivities, lengths and compositions of beams, number of rider notches, weights of riders, materials from which knife edges and planes are made, types of optical systems,

---

* Some manufacturers contend that with damping a certain amount of precision is sacrificed for the sake of speed.[2,9]

2. The Microchemical Balance

TABLE 7
DESCRIPTIVE INFORMATION REGARDING BALANCES SHOWN IN FIGURES

Balance	Figure	Capacity (grams)	Sensitivity	Length of beam	Composition of beam	Number of rider notches	Rider weight
Ainsworth, FHM	15	20	0.001 mg. (1 μg.)	5 inches	Aluminum alloy	201 (including 0)	0.5 mg.
Ainsworth, FH	15	20	0.001 mg. (1 μg.)	5 inches	Aluminum alloy	201 (including 0)	5 mg.
Becker, EM-1	16	20	0.001 mg. (1 μg.)	4 inches	Bronze alloy	101 (including 0)	5 mg.
Bunge, 25 MPN	12	30	0.001 mg. (1 μg.)	130 mm. (5⅛ inches)	Special metal alloy	101 (including 0)	5 mg.
Mettler, M-5	17	20	Readability 0.001 mg. (1 μg.)	—	Aluminum alloy	None	Riderless

Essential Parts

TABLE 7 (*Continued*)

Balance	Composition of knife edges and planes	Method of holding knife edges	Weight carrier	Optical system	Damping	Case
Ainsworth, FHM	Synthetic sapphire	Swaged	Keyboard up to 221 mg.	Projection inside case, direct reading in µg.	Undamped	Aluminum, front door
Ainsworth, FH	Synthetic sapphire	Swaged	None	Projection inside case, estimation* of µg.	Undamped	Aluminum, front door
Becker, EM-1	Agate	Swaged	None	Projection inside case, estimation* of µg.	Undamped	Mahogany, 1 front door 2 side doors
Bunge, 25 MPN	Agate	Adjusting screws	Dials, 10–1210 mg.	Projection outside case, at top, estimation* of µg.	Undamped	Mahogany, hexagonal form with 2 side doors
Mettler, M-5	Synthetic sapphire	Adjusting screws	Dials, complete built-in weights	Optical scale, range 20 mg., read at front of case, 1/5 micrometer graduation = 1 µg.*	Air	Aluminum, 2 side sliding doors

* Estimation of micrograms (µg.) on these models is, in the opinion of the author, as accurate as direct reading of micrograms on most models.

2. The Microchemical Balance

damping, weight carriers, cases, etc., of the above-mentioned balances. The Ainsworth, Becker, and Bunge balances are of the two pan type while the Mettler is of the one pan, constant load variety. With the former, weights are *added* to the right side to counterbalance the load on the left. With this type, the sensitivity decreases with the load, usually about 5%[5] at full load (20 or 30 grams). With the Mettler balance, the weights, which are on the *same* side

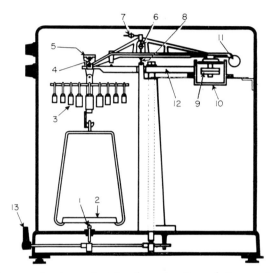

FIG. 18. Diagram showing principle of construction of the constant load Mettler microchemical balance (damped). (1) Pan brake. (2) Pan. (3) Sets of weights. (4) Sapphire knife edge. (5) Stirrup for hangers and weights. (6) Movable weight for adjustment of sensitivity. (7) Movable weight for adjustment of zero point. (8) Main knife edge (sapphire). (9) Counterweight. (10) Air damper. (11) Engraved optical scale. (12) Lifting device. (13) Arrest lever.

of the main* knife edge as the pan, are *removed* to compensate for the load added (substitution). Since there is a constant load, the sensitivity remains constant. Figure 18 is a diagram of the Mettler instrument.

In closing the discussion on balances, it should be remembered that many factors affect their performance and the analyst should consult the manufacturer immediately in case of dissatisfaction.

* The constant load balances have two knife edges instead of the customary three.

Mounting the Balance

The table upon which the balance should be placed was described in Chapter 1. When the inertia block-isolator type is used, balance feet made of metal, glass, or resin are placed on top of the glass plate with small pieces of thin plastic tape between the bases of the feet and the plate to prevent slipping. (Any rubber or other compressible material attached to the balance feet should be removed as this detracts from the system.) The balance is placed on the feet and made level by means of the adjusting thumb screws* until the bubble(s) in the spirit level(s), or the plumb at the end of the plumb line, is centered. The balance is allowed to remain undisturbed for some time before using—a day or two if brought in from the outside in cold weather.

If the rigid combination type of table is used, four large regular rubber stoppers (at least two inches in diameter) are placed on the table, and on top of these is placed a piece of $1/4$ inch plate glass which is slightly larger than the base of the balance. On the glass are placed balance feet made of metal, glass, or resin from which any soft spongy rubber or easily compressible material has been removed. The rest of the mounting is performed as above. Depending upon the type of vibrations encountered, other mountings, such as lead or *tightly* compressed paper, *might* prove more effective than the above. However, nothing should be used which can be depressed without immediately returning to its original position or the precision of the balance will be affected. Regardless of the mounting, the same extent of vibration isolation *cannot* be obtained with the rigid type table that is possible with the inertia block-isolator type and this fact must be accepted.

Assembling and Cleaning the Balance

Balance manufacturers always furnish detailed instructions[3] for assembling (and cleaning) their balances, but a brief description of the operations in connection with the simpler types, such as that shown in Fig. 16 is in order here to serve as an example and will serve as a guide when dealing with the more complicated models. Since a microchemical balance is a very delicate instrument, its assembly should be done with extreme care and, if possible, by someone with previous experience. Because the cleaning of a balance includes both dismantling and reassembling, its description will cover both the setting up of a new instrument and the servicing of an old one. After a balance has been assembled and is operating satisfactorily, it is wise to accept the advice given

---

* Two or three screws, if the balance case has three legs, and four screws, if the balance has four legs.

2. The Microchemical Balance

by the saying, "let sleeping dogs lie." If the atmosphere in the balance room is kept comparatively dust-free, cleaning need be done only once or twice per year, unless dust particles can actually be seen on the beam and stirrups. On the other hand, the pans should be frequently removed and dusted with a clean camel's-hair brush. This can be done by either of two methods. First, it should be made certain that the arresting mechanism for the beam is locked. The pans or hangers are then removed with the aid of a clean, dry chamois skin,* dusted and replaced. Or, if preferred, the pans need not be removed, but can be held by means of the chamois skin or a pair of bone-tipped forceps to keep them from swinging, while being brushed. *No part of the balance should ever be touched with unprotected fingers.* Likewise, the base plate of the balance should be kept free from dust by cleaning almost daily with a camel's-hair brush as this will help keep the working parts dust-free. Occasionally, the base plate should be wiped off with a clean cloth dampened with alcohol. The disassembling, cleaning and reassembling of the balance should be done in the following manner: The rider is first removed from the beam and allowed to remain on the carrier which is pulled to the extreme right. If the rider shows dust particles it should be gently cleaned with a small camel's-hair brush. *After making certain that the beam and stirrup arrestments are in the locked position,* the pans or hangers are removed and dusted with a camel's-hair brush and, if necessary, a chamois skin. They should then be placed on clean chamois skins on a table. The stirrups are next removed using a piece of chamois skin. (Caution: Again make certain that the beam arrestment is in the locked position). They are gently dusted with the camel's-hair brush and set beside the pans or hangers. Finally, the beam is removed using chamois skin and cleaned by brushing the beam, notches, and all three knife edges. It is then set beside the other parts, all of which are continually protected by resting on chamois skin. The flat surface or plane on the column on which the central knife edge of the beam rests during operation is then gently brushed. Occasionally, if the balance has become very dusty it may be necessary to wipe all the knife edges on the beam and all the flat surfaces or planes on the stirrups and column with chamois.† After each knife edge and flat surface is cleaned, it should be examined with a magnifying glass to make certain that no brush hair or lint particle still adheres. Then all beam and stirrup arrestment supports are cleaned with the camel's-hair brush, or if necessary with the chamois skin.

The balance is now ready for reassembling and all parts are replaced in the reversed order from which they were removed. Chamois skin is used at all

---

* Chamois skins are best cleaned by repeated washing with acetone or by washing with soap, water, and dilute ammonia, rinsed with water and air dried.

† In extreme cases, acetone may be used on the parts.

times to protect the parts from the fingers. First the beam is replaced. (Caution: Arresting mechanism locked.) Next, the stirrups are replaced, making sure that they are put in their respective places. (The left-hand one is marked with a single dot, the right-hand one with two dots.) The pans or hangers are then hung onto their respective stirrups. They are marked in the same manner as the latter (one dot for the left, two dots for the right). Finally, the rider is replaced on the beam. After cleaning, the balance should be allowed to rest overnight before being used. It should then be tested for sensitivity as described in Chapter 3 (Weighing). If the sensitivity is lower than before cleaning, or if the balance does not swing freely, it should be dismantled and recleaned as it probably contains brush hairs or lint which interfere with the knife edges.

ADDITIONAL INFORMATION FOR CHAPTER 2

Additional Bunge microchemical balances, of the same capacity and sensitivity, are available in mahogany cases with air damping built into the base plates and reading devices which consist of a telescope without objective. The

FIG. 19. Bunge microchemical balance, model 25DKO.

2. The Microchemical Balance

Fig. 20. Oertling microchemical balance, model 147.

Fig. 21. Mettler microbalance, model M-5/SA (lower section).

ocular is fitted with a microscale. Either of two types of scales are available, namely a regular type with readings to both sides of the zero line and the other of entirely different design described by Zimmermann.[94] To avoid any possible errors, the scale bears no markings on the left side, with the exception of a single graduation mark "E. 0.1 mg." which is used in the sensitivity test only. Two metal case models are available in hexagonal cases, one with the same conventional beam (model 25DKL) used on all other Bunge models and the other designed for use as a riderless balance, with an additional fractional weight loader for weights from 1–10 mg. Both metal case models have fractional weight loaders (10–990 mg.) and a special collimation reading device for viewing either of the above types of scales. Figure 19 shows a photograph of the new, riderless balance, model 25DKO.

Another riderless,* air-damped balance with weight carrier, is available from Oertling, their model 147. It has a capacity of 20 grams and a sensitivity of 0.002 mg. per division, direct reading, using optical projection. The reticle is divided 0–1000 in 500 divisions, the range being 1.0 mg. The case is mahogany with a separate glass-enclosed beam compartment. The beam, which is 5 inches in length is made of nickel chromium alloy, and the knife edges, which are held in place by tension screws, are agate (Fig. 20).

A slightly modified Mettler instrument, known as model M-5/SA, was designed for handling small samples, with means of loading the sample outside the weighing chamber. This model is used as a microbalance (Fig. 21).

The Cahn electrobalance[17,61] is a portable microbalance designed for weighing small samples. Current through a coil applies a torque to the beam and this current is so adjusted to make the torque equal and opposite to that caused by the sample. Since the beam is always returned to the same position, the torque is proportional to the current.

A new ultramicro balance is now available from Mettler, their model UM 7. It has a microscope reading system and a claimed reproducibility of $\pm$ 0.1 µg.

A new automatic recording vacuum semimicrochemical balance is available from Ainsworth. It has a capacity of 100 grams, sensitivity of 0.01 mg., and a range of automatic weight operation of 400 mg. It will weigh samples in air or inert gases, at atmospheric or reduced pressures, at room or higher temperatures, on the balance pan or suspended below the balance in a furnace for thermogravimetry or differential thermal weighing. It has remote control, built-in weights. The recorder charts weight, or weight and temperature against time.

* A rider type is also available.

TABLE 8

ADDITIONAL INFORMATION ON REFERENCES* RELATED TO CHAPTER 2

Following the plan used in Chapter 1, this table lists additional references to which the author wishes to call the attention of the reader. (See statement at top of Table 4 of Chapter 1, regarding completeness of this material.)

Books

Belcher and Godbert, 10, 11
Clark, E. P., 18
Clark, S. J., 20
Emich, 25, 26
Furman, 29
Grant, 34, 35
Niederl and Niederl, 65, 66
Niederl and Sozzi, 68
Pettersson, 70
Pregl, 73
Roth, 78–81
Steyermark, 88

Microchemical balances

Brown, 15
Corwin, 21
Furter, 30
Hodsman, 38–40
Hull, 41
Kuck and Loewenstein, 49
Macurdy, Alber, Benedetti-Pichler, Carmichael, Corwin, Fowler, Huffman, Kirk, and Lashof, 54
Martin, 55
Mitsui, 57, 58
Pfundt, 72
Ramberg, 74
Tuttle and Brown, 90
Wilson, 92

Microbalances

Alber, 4
Ambrosino, 6
Asbury, Belcher, and West, 7, 8
Bradley, 13
Cahn, 17
Czanderna and Honig, 22

Microbalances (Cont.)

Donau, 23
Edwards and Baldwin, 24
Emich, 25, 26
Gorbach, 31, 32
Graham, 33
Hales and Turner, 36
Ingram, 42, 43
Kirk, Craig, Gullberg, and Boyer, 44
Komárek, 46
Korenman and Fertel'meister, 47, 48
Kuck, Altieri and Towne, 50
Müller, G., 60
Müller, R. H., 61
Pettersson, 70
Richards, 75
Rodder, 77
Sarakhov, 83
Wiesenberger, 91

Electronic balances

Cahn, 17
Clark, 19
Feuer, 27, 28

Assay balance

Bromund and Benedetti-Pichler, 14

Semimicrochemical balances

Ainsworth, 2
Becker, 9
Clark, E. P., 18
Seederer-Kohlbusch, 85
Stock and Fill, 89

Analytical balance

Niederl, Niederl, Nagel and Benedetti-Pichler, 67

* The numbers which appear after each entry in this table refer to the literature citations in the reference list at the end of the chapter.

REFERENCES

1. Ainsworth, A. W., *Ind. Eng. Chem., Anal. Ed.,* **11**, 572 (1939).
2. Ainsworth, Wm., & Sons, Inc., Denver, Colorado.
3. Ainsworth, Wm., & Sons, Inc., "Installation, Care and Use of Ainsworth Balances," Denver, Colorado, 1943.
4. Alber, H. K., *Ind. Eng. Chem., Anal. Ed.,* **13**, 656 (1941).
5. Alber, H. K., Personal communication (1940).
6. Ambrosino, C., *Chim. e ind. (Milan),* **33**, 775 (1951).
7. Asbury, H., Belcher, R., and West, T. S., *Mikrochim. Acta,* p. 598 (1956).
8. Asbury, H., Belcher, R. and West, T. S., *Mikrochim. Acta,* p. 1075 (1956).
9. Becker, Christian, Clifton, New Jersey.
10. Belcher, R., and Godbert, A. L., "Semi-micro Quantitative Organic Analysis," Longmans, Green, London and New York, 1945.
11. Belcher, R., and Godbert, A. L., "Semi-micro Quantitative Organic Analysis," 2nd ed., Longmans, Green, London, 1954.
12. Benedetti-Pichler, A. A., and Paulson, R. A., *Mikrochemie ver. Mikrochim. Acta,* **27**, 339 (1939).
13. Bradley, R. S., *J. Sci. Instr.,* **30**, 84 (1953).
14. Bromund, W. H., and Benedetti-Pichler, A. A., *Mikrochemie ver. Mikrochim. Acta,* **38**, 505 (1951).
15. Brown, L. E., *Anal. Chem.,* **23**, 388 (1951).
16. Bunge, Paul, Hamburg, Germany.
17. Cahn Instrument Co., Downy, California.
18. Clark, E. P., "Semimicro Quantitative Organic Analysis," Academic Press, New York, 1943.
19. Clark, J. W., *Rev. Sci. Instr.,* **18**, 915 (1947).
20. Clark, S. J., "Quantitative Methods of Organic Microanalysis," Butterworths, London, 1956.
21. Corwin, A. H., Personal communication; meeting of Metropol. Microchem. Soc., New York City, May 23, 1940; *Ind. Eng. Chem., Anal. Ed.,* **16**, 258 (1944).
22. Czanderna, A. W., and Honig, J. M., *Anal. Chem.,* **29**, 1206 (1957).
23. Donau, J., *Mikrochemie,* **9**, 1 (1931); **13**, 155 (1933).
24. Edwards, F. C., and Baldwin, R. R., *Anal. Chem.,* **23**, 357 (1951).
25. Emich, F., *in "Handbuch der biologischen Arbeitsmethoden"* (E. Abderhalden, ed.), Abt. I, Tl. 3, p. 183, Urban u. Schwarzenberg, Berlin, 1921.
26. Emich, F., *Ber.,* **43**, 10 (1910); also "Lehrbuch der Mikrochemie," pp. 71–80, Bergmann, Munich, 1926.
27. Feuer, I., *Anal. Chem.,* **20**, 1231 (1948).
28. Feuer, I., *Mikrochemie ver. Mikrochim. Acta,* **35**, 419 (1950).
29. Furman, N. H., ed., "Scott's Standard Methods of Chemical Analysis," 5th ed., Van Nostrand, New York, Vol. II, 1939.
30. Furter, M. F., *Mikrochemie,* **18**, 1 (1935).
31. Gorbach, G., *Mikrochemie,* **20**, 254 (1936).
32. Gorbach, G., *Mikrochim. Acta,* p. 352 (1954).
33. Graham, I., *Mikrochim. Acta,* p. 746 (1954).
34. Grant, J., "Quantitative Organic Microanalysis, Based on the Methods of Fritz Pregl," 4th ed., Blakiston, Philadelphia, Pennsylvania, 1946.
35. Grant, J. "Quantitative Organic Microanalysis," 5th ed., Blakiston, Philadelphia, Pennsylvania, 1951.

36. Hales, J. L., and Turner, A. R., *Lab. Practice*, **5**, 245 (1956).
37. Hallett, L. T., *Ind. Eng. Chem., Anal. Ed.*, **14**, 956 (1942).
38. Hodsman, G. F., *Mikrochemie ver. Mikrochim. Acta*, **36/37**, 133 (1951).
39. Hodsman, G. F., *Roy. Inst. Chem. (London), Lectures, Monographs, Repts.*, **4**, 5 (1950).
40. Hodsman, G. F., *Mikrochim. Acta*, p. 591 (1956).
41. Hull, D. E., *Anal. Chem.*, **29**, 1202 (1957).
42. Ingram, G., *Metallurgia*, **39**, 224 (1949).
43. Ingram, G., *Metallurgia*, **40**, 231, 283 (1949).
44. Kirk, P. L., Craig, R., Gullberg, J. E., and Boyer, R. Q., *Ind. Eng. Chem., Anal. Ed.*, **19**, 427 (1947).
45. Kirner, W. R., *Ind. Eng. Chem., Anal. Ed.*, **9**, 300 (1937).
46. Komárek, K., *Chemie (Prague)*, **4**, 6 (1948).
47. Korenman, I. M., and Fertel'meister, Ya. N., *Zavodskaya Lab.*, **15**, 785 (1949).
48. Korenman, I. M., Fertel'meister, Ya. N., and Rostokin, A. P., *Zavodskaya Lab.*, **16**, 800 (1950).
49. Kuck, J. A., and Loewenstein, E., *J. Chem. Educ.*, **17**, 171 (1940).
50. Kuck, J. A., Altieri, P. L., and Towne, A. K., *Mikrochim. Acta*, p. 254 (1953).
51. Kuhlmann, Wm. H. F., Hamburg, Germany.
52. Lenz, W., *Apotheker Ztg.*, No. 21/23 (1912).
53. Longue, C., Paris, France.
54. Macurdy, L. B., Alber, H. K., Benedetti-Pichler, A. A., Carmichael, H., Corwin, A. H., Fowler, R. M., Huffman, E. W. D., Kirk, P. L., and Lashof, T. W., *Anal. Chem.*, **26**, 1190 (1954).
55. Martin, F., *Chim. anal.*, **30**, 4, 37 (1948).
56. Mettler Instrument Co., Hightstown, New Jersey, and Zurich, Switzerland.
57. Mitsui, T., *Kyoto Daigaku Shokuryô Kagaku Kenkyujo Hôkoku*, **No. 18**, 8 (1956).
58. Mitsui, T., *Kyoto Daigaku Shokuryô Kagaku Kenkyujo Hôkoku* **No. 18**, 18 (1956).
59. Moriya, K., Tokyo, Japan.
60. Müller, G., *Mikrochemie ver. Mikrochim. Acta*, **36/37**, 143 (1951).
61. Müller, R. H., *Anal. Chem.*, **29**, 49A (1957).
62. Nemetz, J., Vienna, Austria.
63. Nernst, W., *Göttinger Nachr.*, p. 75 (1903).
64. Nernst, W., and Riesenfeld, E. H., *Ber.*, **36**, 2086 (1903).
65. Niederl, J. B., and Niederl, V., "Micromethods of Quantitative Organic Elementary Analysis," Wiley, New York, 1938.
66. Niederl, J. B., and Niederl, V., "Micromethods of Quantitative Organic Analysis," 2nd ed., Wiley, New York, 1942.
67. Niederl, J. B., Niederl, V., Nagel, R. H., and Benedetti-Pichler, A. A., *Ind. Eng. Chem., Anal. Ed.*, **11**, 412 (1939).
68. Niederl, J. B., and Sozzi, J. A., "Microanálisis, Elemental Orgánico," Calle Arcos, Buenos Aires, 1958.
69. Oertling, L., (Ltd.), London.
70. Pettersson, H., "New Microbalance and Its Use," Thesis, Göteborg, Stockholm, 1914.
71. Pfaltz & Bauer, Inc., New York.
72. Pfundt, O., *Mikrochim. Acta*, p. 539 (1954).
73. Pregl, F., "Quantitative Organic Microanalysis" (E. Fyleman, trans., 2nd German ed.), Churchill, London, 1924.

74. Ramberg, L., *Svensk Kem. Tidskr.,* **41**, 106 (1929); **44**, 188 (1932); *Arkiv. Kemi, Mineral. Geol.,* **11-A**, 7 (1933).
75. Richards, F. M., *Rev. Sci. Instr.,* **24**, 1029 (1953).
76. Rodden, C. J., Kuck, J. A., Benedetti-Pichler, A. A., Corwin, A. H., and Huffman, E. W. D., *Ind. Eng. Chem., Anal. Ed.,* **15**, 415 (1943).
77. Rodder, J., *Chem. Eng. News,* **37** (43), 9 (1959).
78. Roth, H., "Die quantitative organische Mikroanalyse von Fritz Pregl," 4th ed., Springer, Berlin, 1935.
79. Roth, H., "F. Pregl quantitative organische Mikroanalyse," 5th ed. Springer, Wien, 1947.
80. Roth, H., "Pregl-Roth quantitative organische Mikroanalyse", 7th ed. Springer, Wien, 1958.
81. Roth, H., "Quantitative Organic Microanalysis of Fritz Pregl," 3rd ed. (E. B. Daw, trans., 4th German ed.), Blakiston, Philadelphia, Pennsylvania, 1937.
82. Salvioni, E., "Misura di masse comprese fra g. 10^{-1} e.g. 10^{-6}," Messini, 1901.
83. Sarakhov, A. I., *Doklady Akad. Nauk S.S.S.R.,* **86**, 989 (1952).
84. Sartorius-Werke, A. G., Göttingen, Germany.
85. Seederer-Kohlbusch, Inc., Englewood, New Jersey.
86. Starke-Kammerer-A.G., Vienna, Austria.
87. Steyermark, Al, *Ind. Eng. Chem., Anal. Ed.,* **17**, 523 (1945).
88. Steyermark, Al, "Quantitative Organic Microanalysis," Blakiston, Philadelphia, Pennsylvania, 1951.
89. Stock, J. T., and Fill, M. A., *Metallurgia,* **40**, 232 (1949).
90. Tuttle, C., and Brown, F. M., *Ind. Eng. Chem., Anal. Ed.,* **16**, 645 (1944).
91. Wiesenberger, E., *Mikrochemie,* **10**, 10 (1932).
92. Wilson, C. L., *Metallurgia,* **31**, 101 (1944).
93. Willits, C. O., *Anal. Chem.,* **21**, 132 (1949).
94. Zimmermann, W., *Mikrochemie ver. Mikrochim. Acta,* **31**, 149 (1944).

CHAPTER 3

Weighing on the Microchemical Balance

Almost all microchemical procedures begin at the balance. Therefore, it is of utmost importance that the beginner in this field first master the technique of weighing. It must be stressed at this point that any bad weighing habits that might have been acquired through the misuse of analytical balances must be stopped immediately. Otherwise, the beginner will not be successful and, what is more important, the balance will be ruined within a short time. The delicate nature of this instrument was pointed out in Chapters 1 and 2 and this point must be constantly in the mind of the operator, regardless of his experience. In fact, the more experienced the chemist is in this field, the greater will be his respect for his balance, as he realizes his great dependency on it and the manner in which it is functioning. All movements by the operator should be slow and deliberate as this will prevent accidents which shorten the life of the balance.

It should be needless to state that carelessness has no place around this instrument. Dust particles, if allowed to accumulate on the balance pans, weighing vessels, weights, or tares, often will represent weighing errors or contamination representing large percentages, since the weights and amounts of material dealt with are relatively small.

Four methods [7,8,59,62] are used for weighing objects on the ordinary analytical balance, namely, the so-called (a) ordinary, (b) deflection, (c) transposition, and (d) substitution. However, only one of these, the deflection method, is used when working with a microchemical balance.[7,8,11,18,22,24,25,41,42,44-46, 48,51-54,62] Consequently, this method alone will be described.

Determination of Zero Reading* and Deflection†

The determination of the zero reading of the balance should be the first operation performed by the beginner. If an unloaded balance (rider in the 0.0-mg. notch and both pans empty) is allowed to swing freely, it would eventually come to rest and the position on the scale, at which the pointer would remain, is called the rest point of the unloaded balance or the *zero point*. However, the balance is allowed to swing only until equilibrium is reached.

---

* Please see references 7, 8, 22, 44, 45, 59, 62.
† Please see references 7, 8, 11, 18, 22, 24, 25, 41, 42, 44–46, 48, 51–54, 62.

Determination of Zero Reading and Deflection

The deflection of the balance, under these conditions, is called the deflection of the unloaded balance or the *zero reading*. The zero reading is *two times* the zero point. Similarly, a loaded balance swinging in equilibrium would eventually come to rest and the position on the scale, at which the pointer would remain, is called the rest point of the loaded balance or merely the *rest point*.[7,8,22,24, 25,41,42,44,45,48,51–54,59,62] The deflection of the balance under these conditions is called the deflection of the loaded balance or merely the *deflection*—(deflection = 2 × rest point). The following method of obtaining the zero reading and the deflection deals with the free-swinging (undamped) types of balances, but also applies (with obvious variations) to all varieties. [On balances with damping devices (chapter 2), the pointer actually comes to rest. Obviously, on these balances the zero point and rest point are obtained. However, the scale has been altered so that the zero reading and the deflection are read off directly.]

METHOD

Before a balance is used each day, the doors of the case should be left partly open for 15–20 minutes. This permits air to circulate, equalizing the atmospheric conditions inside and outside the case. Likewise, during the day when the balance is not in use, the case doors should be left partly open so that the balance can then be used without delay. However, the balance is never allowed to swing when the doors are open.

With the rider in the zero notch (0.0 mg.) and the doors of the case closed, the beam- and pan-arresting mechanisms are released (in the order named, if separate). The amplitude of swing should be small, but great enough so that the balance is actually swinging; the pointer should swing an arc equivalent to about 3 to 8 scale divisions, that is, 30 to 80 deflection units.* If on releasing the arresting mechanisms the pointer does not swing a large enough arc, lock the beam-arresting mechanism just as the pointer is swinging towards the center and when it is as near as possible to it. This produces the minimum jar. Where the pan arrests act separately from that of the beam, finally lock the pans in place. Then set the balance swinging again as above—first releasing the beam arrest and then that of the pans.† The first few swings to each side should be ignored so as to be certain that equilibrium is closer to being obtained. The

---

* Readings on the models similar to the Ainsworth FH, Becker EM-1, and Bunge 25 MPN balances (Chapter 2) are always recorded as whole numbers instead of as fractions. Consequently, if the deflection is 3.5 divisions to the left (each division = 0.01 mg. = 10 deflection units) the reading is recorded as —35 (35 µg. or deflection units)—see diagrams under cases 1 to 6 below. However, on the Ainsworth FHM, direct readings are in µg.

† Caution: The balance should never be set to swinging by the practice of releasing the arresting mechanisms and gently fanning one pan with the hand.

3. Weighing on the Microchemical Balance

scale is then observed by whatever means is provided on the balance—lens, telescope, microscope or projection device. The readings are recorded (separating those to the left and to the right) until the decrement, or decrease, on each side is equal as shown in the following examples. All values to the left of the zero are negative (—) and all values to the right of the zero are positive (+). The zero reading is simply the algebraic sum of the average of the last three readings to the left and the average of the last two readings to the right.

Rider at 0.0 mg.		Rider at 0.0 mg.	
swing to left deflection units or µg.	swing to right deflection units or µg.	swing to left deflection units or µg.	swing to right deflection units or µg.
−28		−27	
−25	+30	−25	+16
−22	+28	−23	+13
−20	+26	−22	+11
−18	+24	−21	+10
		−20	+9
−20 Average of last three	+25 Average of last two	−21 Average of last three	+10 Average of last two
∴ Zero reading = +5 deflection units or µg.		∴ Zero reading = −11 deflection units or µg.	

Rider at 0.0 mg.		Rider at 0.0 mg.	
swing to left deflection units or µg.	swing to right deflection units or µg.	swing to left deflection units or µg.	swing to right deflection units or µg.
−40		+4	
−37	−11	+7	+40
−35	−13	+8	+38
−33	−15	+9	+37
−35 Average of last three	−14 Average of last two	+8 Average of last three	+38 Average of last two
∴ Zero reading = −49 deflection units or µg.		∴ Zero reading = +46 deflection units or µg.	

Determination of Zero Reading and Deflection

The zero reading should be taken at least six times in immediate succession to determine the reproducibility of the balance (see Determination of Sensitivity and Precision below). During the course of each day, it changes about 5–10 deflection units due to temperature changes[22,28,44,45,50,54,61,62] and various other causes. (See Chapter 1.) Consequently, correction has to be made for these changes if an object is weighed several times during the course of a determination and there is a time lapse of hours or days between weighings. If, however, all of the weighings necessary can be accomplished within a matter of minutes, such as when weighing a sample, the zero reading is disregarded and, in fact, is assumed to be zero.

The deflection of the loaded balance (objects and weights on the pans and rider usually in some notch other than 0.0 mg.) is done in the same manner as that used for the zero reading.

For the weighings where the zero reading may be assumed to be zero, a deflection to the right (+) of the zero signifies that not enough weight is on the rider and, therefore, the deflection is added to the weight of the rider; likewise, if the deflection is on the left (—) of the zero, too much weight is on the rider and the amount must be subtracted. In other words, the deflection is added algebraically to the rider. For example, if the rider is in the 6.2 mg. notch and the deflection is —21, the rider reading is 6.179 mg. Similarly, if the rider is in the 1.3-mg. notch and the deflection is +2, the rider reading is 1.302 mg.

The beginner usually has little trouble with the deflections until he is confronted with the problem of correcting for the zero readings when an object must be reweighed several times at later intervals. The author believes that no confusion will result if the beginner records in his notebook the weights *corrected for the zero reading* each time a weighing is made rather than to deal with the *change* in zero reading. To make this correction the zero reading is subtracted algebraically from the deflection, observing all signs and the result is added algebraically to the rider quantity. If the following diagrams of the scale and examples are kept in mind, no confusion will exist regarding how the zero reading is to be used in the correction. All possible combinations are represented. Let us suppose that the zero reading (unloaded balance, rider at 0.0 mg.) is +20 and the deflection for a weighing (loaded balance) is +42:

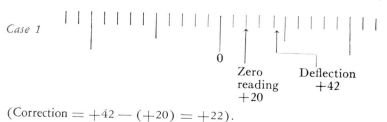

(Correction = +42 — (+20) = +22).

3. Weighing on the Microchemical Balance 54

This means that not enough weight is on the rider and 0.022 mg. must be added to the rider weight. If the deflection is +20 and the zero reading is +42,

Case 2

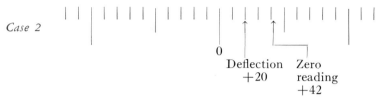

(Correction = +20 — (+42) = —22),

the opposite would apply, 0.022 mg. would be subtracted from the rider weight. Using similar reasoning, the following diagrams are self-explanatory.

Case 3

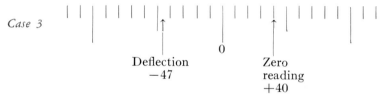

(Correction = —47 — (+40) = —87)   ∴ 0.087 mg. to be subtracted from rider weight.

Case 4

(Correction = +40 — (—47) = +87)   ∴ 0.087 mg. to be added to rider weight.

Case 5

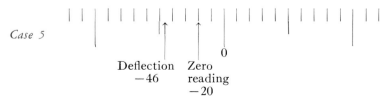

(Correction = —46 — (—20) = —26)   ∴ 0.026 mg. to be subtracted from rider weight.

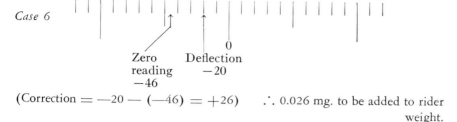

Case 6

Zero reading −46

Deflection −20

(Correction = −20 − (−46) = +26)   ∴ 0.026 mg. to be added to rider weight.

In the examples shown, it will be noticed that no deflection of 50, or more, is recorded. If the deflection is 50, or more, the rider should be moved over one notch so that one of less than 50, of opposite sign, will be obtained.

Determination of Sensitivity and Precision*

The sensitivity and precision of the balance should be determined frequently, since it is an indication of the manner in which the instrument is performing. The zero reading is first determined as described above. The rider is then moved to the 0.1-mg. notch and the deflection obtained. The displacement caused by the load of 0.1 mg. is called the sensitivity. The value is obtained by subtracting, algebraically, the deflection (rider at 0.1 mg.) from the zero reading. Table 9 shows three different examples which include the various possibilities with a balance having a sensitivity of 98.

TABLE 9
EXAMPLES OF THE DETERMINATION OF SENSITIVITY

	(No weights on pans)		
	Zero reading (rider at 0.0 mg.)	Deflection (rider at 0.1 mg.)	Sensitivity (No. of deflection units equal to 0.1 mg.)
Example 1	+10	−88	+10 − (−88) = 98
Example 2	0	−98	0 − (−98) = 98
Example 3	−5	−103	−5 − (−103) = 98

In each of these examples, the addition of 0.1 mg. caused a displacement of 98 units or 1 unit = 0.00102 mg. The sensitivities of good balances are rarely 100 (1 deflection unit = 0.001 mg. or 100 units = 0.1 mg.), although the manufacturers' specifications often claim this. By raising or lowering the center of gravity of the beam, by means of the weight on the pointer (see Chapter 2), it is possible to obtain this figure with a good instrument. However, this is not necessary as a balance with a sensitivity between 95

* Please see references 7, 8, 22, 44, 45, 50, 62.

3. Weighing on the Microchemical Balance

and 105 gives excellent results without applying a correction[44,45] (too small to be significant). When it is not within these limits, correction should be made as shown by the following examples. If the sensitivity is only 75, 0.1 mg. is equal to 75 units or each unit equals $^{100}/_{75}$ or 0.00133 mg. instead of 0.001 mg. Similarly, if the sensitivity were 125, each deflection unit would represent $^{100}/_{125}$ or 0.0008 mg. Obviously, the deflection units must be multiplied by these factors to yield the correct values.

The procedure of taking the sensitivity should be repeated at least six times in rapid succession to determine the precision or reproducibility.[44,45,50,62] This will include a zero reading, followed by the deflection when loaded with 0.1 mg., followed by a zero reading, and so on. The values obtained for the six determinations will usually vary and the standard deviations are calculated. For example, let us suppose the following set of readings in Table 10.

TABLE 10
DETERMINATION OF PRECISION OF BALANCE

Zero reading	Deflection	Sensitivity	Deviation of zero readings from mean	Deviation of deflections from mean
+20	—78	98	3	3
+15	—81	96	2	0
+17	—82	99	0	1
+21	—76	97	4	5
+15	—82	97	2	1
+16	—85	101	1	4
Mean +17	Mean —81	Mean 98	$s = 2.6$	$s = 3.2$

(In this Table,

$$s = \text{standard deviation}^{60,72}$$

$$= \sqrt{\frac{\Sigma (x - \bar{x})^2}{n - 1}}$$

where n is the number of readings, x is equal to the individual readings, and $\bar{x}$ is the mean of the individual readings.)

It can be seen from Table 10 that this balance cannot be expected to reproduce either the zero readings or the deflections closer than approximately 3 deflection units. Since the balance has a sensitivity of 98, each unit is equal to 0.00102 mg. (see above). Therefore, the precision is approximately 3×0.00102 mg. $= 0.00306$ mg. or simply, 0.003 mg. (3 µg.). This figure coincides with that established[20,28,50] years ago, as being acceptable.* (If the

---

* In spite of their being used constantly for periods ranging from six to twenty-two years, all of the balances in the author's laboratories have precisions under 2.7 µg. (from 1.1 to 2.6 µg.).

reproducibility of the balance is poor, it might be due to poor rider placement which can be determined by taking a series of zero readings without moving the rider between determinations. Also for the zero reading a number of successive determinations with the rider on the left-hand pan instead of on the beam will prove to be informative. If the reproducibility is poor when the rider is in the zero notch on the beam, but is good when the rider is on the left-hand pan, the zero notch on the beam allows the rider to move.)

After obtaining the sensitivity and precision with a load of 0.1 mg., the procedure should be repeated with a 10-gram weight in the center of each pan.* The zero reading cannot be taken when weights are on the pans, so, instead two deflections are determined. Since the two 10-gram weights will vary somewhat, the rider is placed in the notch giving the closest conditions to equilibrium and the deflection determined. The rider is then moved over 0.1 mg. on the beam so that a deflection of opposite sign will be obtained and the procedure repeated. This will give the number of divisions equivalent to 0.1 mg. with a 10-gram load in each pan. To determine the reproducibility of the balance under these conditions, the above should be repeated at least six times moving the rider back and forth as was done above. A balance will rarely have the same sensitivity and precision or reproducibility under a load of 10 grams as it does under one of 0.1 mg. (except the constant load type) but the decrease, with a good instrument, will not be over 5%.[1]†

Calibration of Weights

Only a few weights out of the ordinary set receive constant use in this field of work as will be explained later in this chapter. However, those few which are used must be accurately calibrated. For this calibration, the rider of the balance is used as the standard. (If necessary, this could be checked against a Bureau of Standards 10-mg. weight.) During the process of making a weighing, the weights, W, are always placed on the right-hand pan. Consequently, they should be similarly placed during calibration and, therefore, the substitution method[7,8,17,44,45,59,62] is recommended. A second set of weights, T, is used and these are always placed on the left-hand pan. It is only necessary to calibrate two 10-mg., one 20-mg., and one 50-mg. weights.

The tare 10-mg. weight, T_{10}, is placed on the left-hand pan and the rider is placed in the 10-mg. notch. The deflection‡ is determined, checked several

---

* All weights, tares, and objects to be weighed are always centered.

† Errors due to inequalities in the length of the beam arms may be determined in a manner similar to that used for analytical balances.[59]

‡ The zero reading of the balance need not be determined since the weights are merely being compared to the rider.

3. Weighing on the Microchemical Balance

times, and the average taken. (Note: For the calibrations, all values should be checked several times, using the average.) The rider is then placed in the zero notch (0.0 mg.) and the 10-mg. weight, to be calibrated, W^1_{10}, is placed on the right-hand pan and the deflection again determined. The difference between these is the calibration to be applied to the weight. Suppose, for example, the deflection with the rider in the 10-mg. notch is $+8$, and when the weight, W^1_{10}, is on the right-hand pan (rider in the zero notch) the deflection is $+10$. This would mean that the weight, W^1_{10}, is 0.002 mg. less than the rider weight. Consequently, the exact weight of W^1_{10} is 9.998 mg.

The weight W^1_{10} is then replaced by the weight W^2_{10} (on the right-hand pan), the rider kept in the zero notch, the deflection obtained, and the correction applied. Again suppose for example that the deflection is -4. This would mean that the weight W^2_{10} is 0.012 mg. heavier than the rider, or 10.012 mg.

Next, the 10-mg. tare weight, T_{10} is replaced on the left-hand pan by the 20-mg. weight, T_{20}. On the right-hand pan is allowed to remain the 10-mg. weight, W^2_{10}, and the rider is placed in the 10-mg. notch. The deflection is again taken. For example, suppose it is $+5$. The weight W^2_{10} is then replaced by the 20-mg. weight, W_{20}, and the rider placed in the zero notch. The deflection is again determined. Let us suppose it is then $+15$. This would mean that the weight, W_{20}, is 0.010 mg. lighter than the rider plus the weight, W^2_{10} or,

$$\begin{aligned} W_{20} &= \text{Weight of rider} + \text{weight of } W^2_{10} - 0.010 \text{ mg.} \\ &= 10.000 + 10.012 - 0.010 \\ &= 20.002 \text{ mg.} \end{aligned}$$

Using the rider and calibrated weights W^1_{10}, W^2_{10}, and W_{20}, the 50-mg. weight, W_{50}, is then calibrated. If desired, the 100-mg. weight, W_{100}, may be checked against the rider $+ W^1_{10} + W^2_{10} + W_{20} + W_{50}$.

Apparatus

COMBUSTION BOATS*

(See Table 11)[6,62-64]

Size A

This boat (Fig. 23), referred to in the literature as the Hayman boat,[31,62,63,65] weighs approximately 0.45 gram. Because of its small size, it is used for samples weighing from 1–5 mg. For drying procedures, it should be used with the weighing bottle, pig type, metal, size A (Fig. 31).

---

* For some procedures, such as those described in Chapters 7, 8, and 15, porcelain boats of comparable size are recommended (Fig. 22); usually the dimensions are 17 mm long, 6 mm wide, and 4 mm high.

Fig. 22. Porcelain combustion boat.

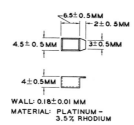

Fig. 23. Platinum combustion boat, size A—details of construction.

TABLE 11
COMBUSTION BOATS[6]

Size	Figure	Use	Approximate weight (grams)
A	23	For samples of 1 to 5 mg.[a]	0.45
B	24	For samples of 5 to 25 mg.[a]	0.7
C	25	For samples over 25 mg.[a,b]	1.5

[a] For drying procedures, the size A boat is used with the weighing bottle, Fig. 31. Sizes B and C can be used either with the metal weighing bottles (Figs. 32 and 33, respectively) or with the glass weighing bottles, pig-type (Figs. 29 and 30).

[b] The size C boat is especially suitable for bulky materials and explosive substances, and for holding glass capillaries containing liquid samples.

Size B[6,62-64]

This boat (Fig. 24) weighs approximately 0.7 gram. It is the most commonly used size for handling samples in the range of 5 to 25 mg.

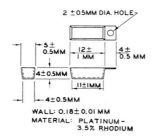

Fig. 24. Platinum combustion boat, size B—details of construction.

Size C[6,63]

This boat (Fig. 25) weighs approximately 1.5 grams and is used for semimicro purposes with samples that weigh up to 50 mg. and for micro purposes with bulky material or explosive substances. Capillaries containing liquid samples are placed in this boat, and the combination is introduced into a combustion tube.

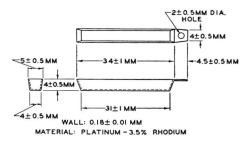

FIG. 25. Platinum combustion boat, size C—details of construction.

WEIGHING CUP[6,63]

This item (Fig. 26) is designed as a sample container to fit into the weighing bottle, outside cap (Fig. 28). This combination is useful for weighing hygroscopic materials.

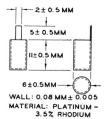

FIG. 26. Weighing cup—details of construction.

WEIGHING BOTTLE[6,63]

This type of weighing bottle was originally recommended by Roth.[25,52,63] It may be inserted directly into such an item as a Carius combustion tube. It should be made of borosilicate glass, so that it will withstand elevated temperatures and may be used repeatedly (Fig. 27). It may be used for either non-hygroscopic or hygroscopic samples, liquid or solid (see Table 12).

WEIGHING BOTTLE, OUTSIDE CAP[6,63]

The microweighing bottle originally was described by Hayman[30] (Fig. 28). It should be made from soda-lime glass in order to reduce accumulation of electrostatic charges.[6,62,63] The weighing cup (Fig. 26) should fit inside for use as a liner, if so desired. The bottle has been designed with an outside cap, which permits the use of a lubricant with less danger of contaminating the sample than if an inside stopper were used. For the sake of simplicity, the capillary is straight (see Table 12).

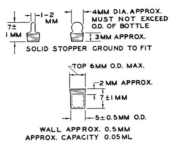

FIG. 27. Weighing bottle—details of construction.

TABLE 12
WEIGHING BOTTLES[6]

Figure	Approximate capacity (ml.)	Approximate weight (gram)	Material
27[a]	0.05	0.6	Borosilicate glass
28	0.75	2.5	Soda-lime glass
29[b]	3	5.0	Soda-lime glass
30[b,c]	3	5.7	Soda-lime glass
31[d]	0.5	1.5	Aluminum alloy
32[d]	1	4.5	Aluminum alloy
33[d]	2	6	Aluminum alloy

[a] This bottle is intended for introducing samples into containers of small diameters.

[b] The cap of the bottle shown in Fig. 28 may be used instead of the caps with rod handle shown in Figs 29, and 30.

[c] This bottle (Fig. 30) is used for extremely hygroscopic materials; only the small cap is removed for vacuum drying, and it is replaced as soon as the vacuum is released. The bottle containing the combustion boat with sample may be attached to a combustion tube by means of a rubber adapter, and the boat pushed into position in the tube by passing a wire through the small joint.

[d] These bottles are designed for use in conjunction with the combustion boats (Figs. 23 to 25), as follows:

Bottle	Use with boat
Fig. 31	Fig. 23
Fig. 32	Fig. 24
Fig. 33	Fig. 25

The combination is useful for weighing and drying extremely hygroscopic material. Sample, boat, and bottle are weighed, and the combination, without removing the cap, is placed in the drying apparatus (Fig. 65). During drying, vapors escape through the capillary. After drying, the combination is reweighed.

3. Weighing on the Microchemical Balance

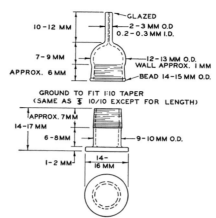

FIG. 28. Weighing bottle, outside cap—details of construction.

WEIGHING BOTTLE, PIG-TYPE, WITH OUTSIDE CAP[6,62,63]

This is the so-called Friedrich-type[21] weighing bottle (Fig. 29). It should be made preferably from soda-lime glass in order to reduce accumulation of electrostatic charges. The outside cap for the weighing bottle (Fig. 28) may be used instead of the cap with the rod handle. The capacity of this bottle is approximately 3 ml. (see Table 12).

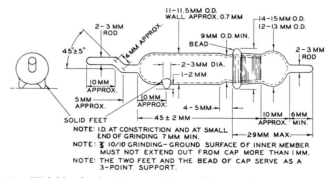

FIG. 29. Weighing bottle, pig-type, with outside cap—details of construction.

WEIGHING BOTTLE WITH TWO CAPS[6,63,71]

This bottle (Fig. 30) should also be made preferably from soda-lime glass. The approximate capacity of the main body is 3 ml. This weighing bottle is used for extremely hygroscopic materials[70,71]; only the small cap is removed for vacuum drying, and it is replaced as soon as the vacuum is released. For the determination of hydrogen or oxygen in extremely hygroscopic materials, the weighing bottle containing the combustion boat with sample is attached to a combustion tube by means of a rubber adapter, and the boat pushed into position in the combustion tube by passing a wire through the small joint. The outside cap for the weighing bottle (Fig. 28) may be used instead of the cap with the rod handle (see Table 12).

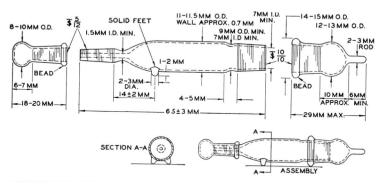

Fig. 30. Weighing bottle with two caps—details of construction.

WEIGHING BOTTLE, PIG-TYPE, METAL[6,62,63]

Size A

This weighing bottle (Fig. 31), referred to as the Hayman[30,31,62,63,65] type, is designed for use in conjunction with the boat, size A (Fig. 23), as shown in the assembly drawing. The bottle should be made of an aluminum alloy, and the cap should be lapped to fit the body joint. The combination is useful for weighing and drying extremely hygroscopic material. Sample, boat, and bottle are weighed, and the combination, without removing the cap, is placed in the modified Abderhalden drying apparatus (Fig. 65). During the drying

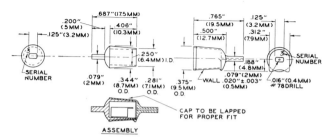

Fig. 31. Weighing bottle, pig-type, metal, size A—details of construction.

operation, water leaving the sample passes through the capillary. The combination is finally removed from the dryer and again weighed. The approximate capacity is 0.5 ml.; the approximate weight is 1.5 grams (see Table 12).

Size B[6,63]

This weighing bottle (Fig. 32) should be made from an aluminum alloy, and the cap should be lapped to fit the body joint. It is designed for use with the combustion boat, size B (Fig. 24), as shown in the assembly drawing. The manner in which it is used is described under weighing bottle, pig-type, metal, size A. The approximate capacity of the bottle is 1 ml.; the approximate weight is 4.5 grams (see Table 12).

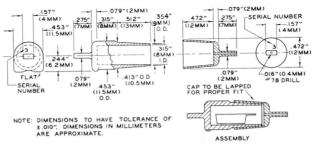

FIG. 32. Weighing bottle, pig-type, metal, size B—details of construction.

Size C[6,63]

This weighing bottle (Fig. 33) should be made of an aluminum alloy, and the cap should be lapped to fit the body joint. It is designed for use with the combustion boat, size C (Fig. 25), as shown in the assembly drawing. The manner in which it is used is described under weighing bottle, pig-type, metal, size A. The approximate capacity is 2 ml.; the approximate weight is 6 grams (see Table 12).

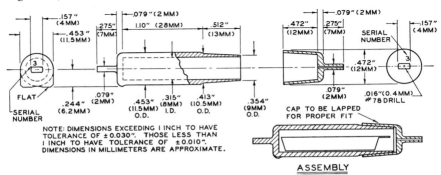

FIG. 33. Weighing bottle, pig-type, metal, size C—details of construction.

TARE FLASKS[6,62,63]

Three types of tare flasks are recommended, one with, and two without a ground-in stopper (Figs. 34, 35, and 36). They should be made preferably of soda-lime glass. On all three flasks the serial numbers should be etched in

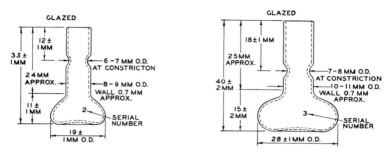

Fig. 34. (*Left*) Tare flask, without stopper, small—details of construction.
Fig. 35. (*Right*) Tare flask, without stopper, large—details of construction.

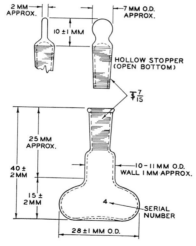

Fig. 36. Tare flask, with stopper—details of construction.

order to avoid rough surfaces. They are used as counterweights for such items as absorption tubes, crucibles, filter tubes, etc. Their weights are adjusted by adding small lead shot or glass beads until these closely approximate those of the objects which are to be counterbalanced. The tare flasks are marked and stored in the balance case.

SPATULA, METAL[6,63]

Type A

This type of spatula[4,6,62,63] (Fig. 37), in addition to being useful as a general spatula, can be used as a preparative tool, the bottom end for crushing crystals and the bent blade for scraping containers. It is to be made preferably of stainless steel.

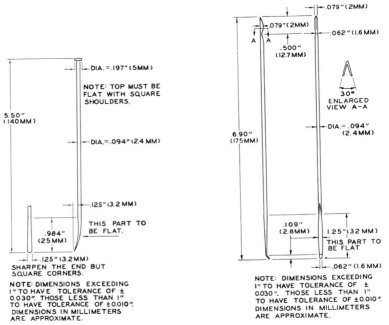

FIG. 37. (*Left*) Spatula, metal, type A—details of construction.
FIG. 38. (*Right*) Spatula, metal, type B—details of construction.

Type B

This spatula[4,6,62,63] (Fig. 38) has a flat, bent portion at one end, and a V-shaped scoop at the other. It is made preferably of stainless steel and is particularly useful in the weighing of samples.

Type C

The spatula[6,62,63] (Fig. 39) is suitable for the larger samples commonly encountered in semimicro and preparative work. It has a U-shaped scoop at one end and a V-shaped scoop at the other, and is made preferably of stainless steel. The spatula can be used for adding lead shot or beads to the tare flasks

Apparatus

(Figs. 34, 35, and 36) and for measuring and introducing the solid reagent into the combustion tube of the apparatus for the manometric determination of carbon[62,66] (see Chapter 18).

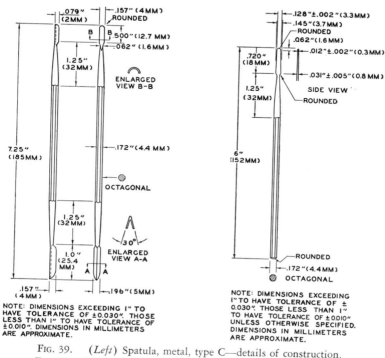

FIG. 39. (*Left*) Spatula, metal, type C—details of construction.
FIG. 40. (*Right*) Spatula, metal, type D—details of construction.

Type D[6,63]

This spatula (Fig. 40), which is actually a dental spatula, has been found useful by the author and other experienced analysts. It is made preferably of stainless steel.

STEEL FORCEPS (NICKEL PLATED) WITH PLATINUM—5% RUTHENIUM TIPS[6,63]

The tips are sturdy and are made from the platinum–5% ruthenium alloy. A pin has been included. This serves as a stop to prevent the forceps from being depressed to such an extent that the tips can open, allowing the held object to drop. When the forceps are pressed together, the tips make contact for a distance of 0.25 to 0.5 inch. Figure 41 shows the forceps with platinum–5% ruthenium tips. The forceps may be made without rare metal tips, using any other metal such as nickel-plated or stainless steel, depending upon their in-

3. Weighing on the Microchemical Balance

tended use. The construction and over-all dimensions should be identical to those shown in Fig. 41.

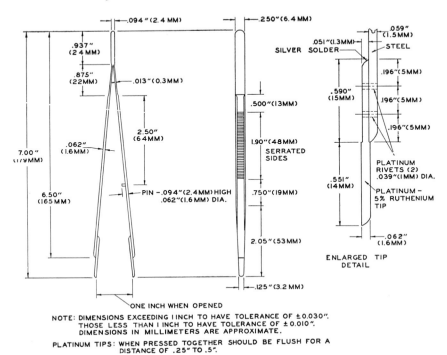

FIG. 41. Steel forceps (nickel plated) with platinum–5% ruthenium tips—details of construction.

STEEL FORCEPS (NICKEL PLATED) WITH CONICAL TAPERED HOLDERS[6,63]

Until the introduction of the forceps[2,62,63] of the type shown in Fig. 42, absorption tubes, filter tubes, etc., were handled with the fingers covered with chamois cots, with metal forks,[44,45,48,51-54] or with forceps prepared by soldering sections of metal tubing to the tips of dissecting forceps.[62] The forceps (Fig. 42), which have been commercially available for a number of years, are of superior construction and provide a better means of handling objects. Because they are made in one piece from spring steel, they are also more durable.

DESICCATORS[6]

Either of the desiccators shown in Figs. 43, 44, and 46 may be used for cooling, storage and safe transportation of microweighing equipment. They are used in combination with the metal cooling block[6,62,64] which is made of

Apparatus

a metal or alloy with a high heat conductivity and whose surface is highly resistant to abrasion and corrosion (Fig. 45). One desiccator (Figs. 43 and 46) consists of an aluminum body and a borosilicate glass cover,[6,62,64] and the other desiccator (Fig. 44) consists of a complete glass desiccator with cover[6] and contains an aluminum insert for holding a metal cooling block.[64] In general,

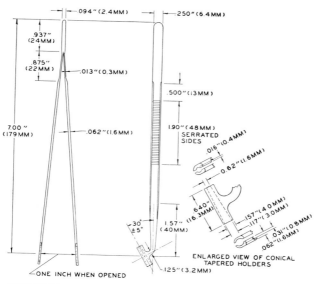

FIG. 42. Steel forceps (nickel plated) with conical tapered holders—details of construction.

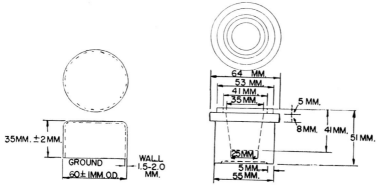

FIG. 43. Metal crucible container with glass cover (desiccator)—details of construction (*left,* glass part, *right,* metal part).

the author prefers the use of the former, except in such cases where the use of desiccants is required.

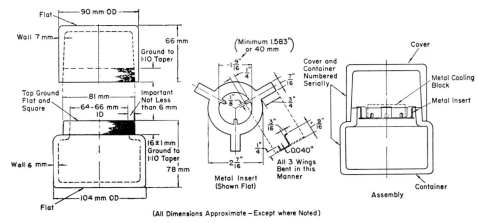

FIG. 44. Glass desiccator with metal insert.

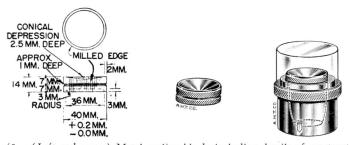

FIG. 45. (*Left and center*) Metal cooling block, including details of construction.
FIG. 46. (*Right*) Metal crucible container with glass cover and metal cooling block.

WEIGHING (OR CHARGING) TUBES[6,62,64]

Three types of weighing or charging tubes (Figs. 47, 48, and 49) are used for weighing samples by difference. The two types with ground caps are used for hygroscopic samples. These may be inserted into the drying apparatus for additional drying purposes.

RACK[7,8,44,45,51–54,62]

A rigid wire rack of the type shown in Fig. 50 is used to support such items as absorption tubes (Chapter 9) and filter tubes (Chapters 11 and 16) while waiting for them to come to equilibrium before weighing.

FORK[7,8,44,45,48,51-54,62]

A metal fork (Fig. 51) prepared from twisted wire is used for transferring absorption tubes and filter sticks from the rack to the balance. It catches the end of the object whose weight holds it in place.

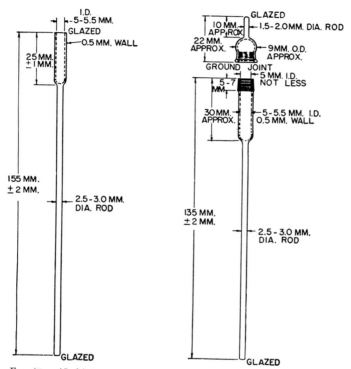

FIG. 47. (*Left*) Weighing (charging) tube—details of construction.
FIG. 48. (*Right*) Weighing (charging) tube with cap—details of construction.

KNURLED IRON WIRE[24,25,44,45,48,51-54,62]

Knurled iron wires (Fig. 52) about 1 mm. O.D. and 120–150 mm. in length are used as swab sticks with cotton tips for cleaning out the ends of various objects previous to weighing.

LEAD SHOT[24,25,44,45,48,51-54,62]

Small lead shot of various sizes are used in the tare flasks (Figs. 34, 35, and 36) to increase the weight of the latter to within a few milligrams of that desired. Glass beads may be used instead.

3. Weighing on the Microchemical Balance

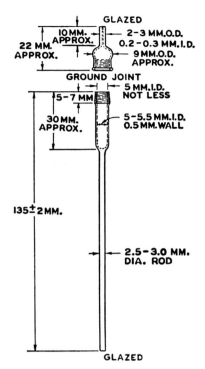

FIG 49. Weighing (charging) tube with vented cap—details of construction.

FIG. 50. (*Left*) Wire rack.
FIG. 51. (*Right*) Fork, showing method of use.

FIG. 52. Knurled iron wire.

Weighing of Samples

NON-HYGROSCOPIC SAMPLES WEIGHED IN BOATS

A large portion of the samples dealt with are weighed in small combustion boats, either platinum[6,62-64] (Figs. 23–25) or porcelain[44,45,62] (Fig. 22). If the samples are not hygroscopic, these may be placed directly on the left-hand pan. If, however, the sample is hygroscopic the weighing is accomplished as described below. The boats are first cleaned by immersion in boiling dilute nitric acid, washed with distilled water, and then heated to redness in the flame of a burner. The boat is best held by means of a platinum wire[44,45,62] hook (sealed into a glass rod, Fig. 53) which fits into the small hole in the former. After

FIG. 53. Platinum hook (sealed into glass rod).

ignition, the boat is allowed to cool on a metal block in a small desiccator[6,41,42,44,45,62,64] (Figs. 43–46). Platinum boats may be weighed after about 10 minutes, but porcelain boats should not be weighed for at least 30 minutes. Tare counter weights may be prepared for each boat from aluminum, bronze, or glass rod by carefully filing or grinding away material until the weight is within a milligram of that of the boat.[22,24,25,41,42,44,45,48,51-54,62] The tare and boat then should be kept together. The exact weight of the boat need not be known, since the weight of a sample is obtained by the *gain* as described below.

The boat is placed on the left-hand pan of the balance by means of a platinum-tipped forceps (Fig. 41), and the tare counter weight placed on the right-hand pan using ordinary bone-tipped weight forceps. The balance is brought as close to equilibrium as possible by moving the rider to the nearest notch in a manner similar to that employed with an analytical balance. Each time the beam is set down on its knife edges, it should not be allowed to actually swing. The release should be only enough so that the direction of swing of the pointer can be noted. After the rider has been placed in the notch necessary for equilibrium, the balance is allowed to swing and the deflection noted as described above. If it is found to be ±50, or greater, the rider should be moved one notch so that a value of less than ±50 is obtained. The weight as shown on the *rider plus or minus the deflection** is recorded

---

* It is not necessary to take the zero reading of the balance since it will not change during the course of the few minutes required to weigh the sample—it is assumed to be zero in these weighings.

as the weight of the boat, as for example, 3.068 mg. The boat is then removed from the left pan and placed on the metal block of the microdesiccator.* The sample is then introduced into the boat with the aid of a microspatula[6,41, 42,44,45,62,63] (Figs. 37–40). (Care must be exercised so that the outside of the boat does not have any attached particles of sample. If so, these are removed by either tapping the boat on the metal block or by holding it with the forceps and carefully brushing the outer sides and bottom with a small camel's-hair brush, making certain that no hairs become detached from the brush and adhere to the boat.) The boat containing the sample is placed on the left-hand pan and the rider again moved to obtain equilibrium as above. Again, the combined rider and deflection are recorded, as for example, 9.873 mg. Subtracting the weight of the empty boat from this, i.e., 3.068, gives 6.805 mg. as the weight of the sample. Besides the commonly used size B platinum combustion boat (Fig. 24), two other sizes may be used as shown in Table 11.

HYGROSCOPIC SAMPLES WEIGHED IN BOATS

When a hygroscopic sample is weighed in a boat, it must be protected from the moisture in the atmosphere. The boat is placed in the weighing bottle, pig-type, with outside cap (Fig. 29; also see Table 12).[6] (This weighing bottle or pig should be made preferably from soda-lime glass in order to reduce accumulation of electrostatic charges after being wiped clean with a chamois skin. Practically all glass articles that are to be weighed are prepared from this glass in preference to borosilicate glass, because the latter retain such charges†).

The pig should be wiped with a chamois skin, grounded against a light fixture or pipe, and then let stand for about 15 minutes before being placed on the balance. (During this time the charge will be lost.‡) The pig must be handled by means of a chamois skin from this point, but should not be stroked lest it again becomes charged. It is then placed on the balance along with a boat. The combination is weighed, removed from the balance, sample added to the boat as described previously in this section, the boat inserted into the stoppered pig and the combination reweighed and then preserved in this

---

* Boats should always be removed from the balance before attempting to add samples. Samples should *never* be added to a boat on the balance pan.

† All glassware develops a charge (when wiped with the chamois skin), detectable on the microchemical balance. As stated in Chapter 1, when the relative humidity is kept at 40–50%,[29,54,62] the charges are rapidly lost. Grounding[45,61,62] exposure to U.V. light,[45,49,61,62] contact with a high-frequency discharge[45,61,62,67] or with radioactive material[29,45,61,62] also accomplishes the same aim.

‡ If the *electrostatic charge* is not lost, on the balance the object decreases in weight so rapidly that weighing is impossible. The rate of decrease is often as much as 0.3 mg. in several minutes.

condition until analyzed. If the sample is very hygroscopic it will be necessary to weigh the boat and pig, correcting for the zero reading of the balance. An approximate amount of sample is then added to the boat, and the open pig, boat, and sample are dried several hours in a drier (see Chapter 4, Preparation of Sample). The combination is then handled with a chamois skin and weighed, taking into account the zero reading of the balance. This same technique is used with samples which readily absorb carbon dioxide from the air, as for example, some amines.

Very hygroscopic samples are best handled by means of either the combination of platinum boats and the various weighing bottles shown in Figs. 30–33, or the Friedrich-type (Fig. 29) but with a cap having a capillary opening such as that of the weighing bottle shown in Fig. 28. The boat and container are first accurately weighed. The approximate amount of very hygroscopic sample is added and the unit, with the stopper attached, is placed in a drier. Moisture leaves the sample by way of the capillary opening. After drying, the unit is again weighed. The capillary opening, being small, does not allow an appreciable amount of moisture to re-enter the system during the weighing time. (A series of experiments[63] established the following: (a) Water vapor diffuses through a dry ground joint and (b) There is little difference in the amount of diffusion whether the cap or stopper is closed or contains a capillary, straight or with bulbs. Benedetti-Pichler[9,63] and Bromund[14,63] recommended lubrication and the Committee on Microchemical Apparatus of the Division of Analytical Chemistry, American Chemical Society, established that the only way to prevent absorption of moisture is by means of a lubricated closed cap. However, the author does not recommend the use of lubricants in any drying procedure due to possible contamination.)

WEIGHING OR CHARGING TUBES[6,7,8,44,45,51–54,62,64]

Three types of so-called weighing (charging) tubes are used for weighing samples. For non-hygroscopic samples the open end variety (Fig. 47) may be used but for hygroscopic samples the glass-stoppered types (Figs. 48 and 49) are necessary. The charging tubes are generally used, for samples that are to be placed eventually in Carius combustion tubes (see Chapters 10 and 11) or Kjeldahl digestion flasks (see Chapters 9 and 12). With charging tubes, samples are weighed by difference. The charging tube is wiped with the chamois skin, grounded against a light fixture or pipe, and set aside on a wire rack (Fig. 50) for about 15 minutes.* It is then held with a chamois skin or special forceps (Fig. 42), the approximate weight of sample added

---

* If the electrostatic charge is not lost, on the balance the object decreases in weight so rapidly that weighing is impossible. The rate of decrease is often as much as 0.3 mg. in several minutes.

3. Weighing on the Microchemical Balance

by means of a microspatula,* and then placed horizontally on the hooks on the left-hand hanger of the balance. (As a tare counterweight, either aluminum, glass or brass rod may be used—see paragraph above regarding weighing with boats.) The charging tube plus sample is accurately weighed (zero reading of balance need not be determined—it is assumed to be zero in this case). By means of the special forceps or chamois skin, the charging tube is removed from the balance, the sample dumped out into the desired container (making certain that no particles adhere to the edges) and the empty charging tube reweighed.

LIQUID SAMPLES

(a) High-boiling liquids with a low vapor pressure should be weighed in boats in the same manner as non-hygroscopic solids. (b) Low-boiling liquids may be handled either by (1) or (2).

(1) *Capillaries.* Low-boiling liquids must be weighed in glass capillary tubes[22, 24,25,44,45,48,51-54,62] of the type shown in Fig. 54. These are prepared by drawing out a piece of molten tubing so that the inside diameter is about 1 mm. They are cut in lengths of 2.5 to 5 cm. depending upon the purpose for which they will be used (5 cm. for Carius combustions, 2.5 cm. for all others—see following chapters). One end is sealed by heating in a flame. To the tubes which are to be used in connection with the carbon-hydrogen and Dumas nitrogen determinations, about 2–3 mm. of powdered potassium chlorate†[22,28,45,62] should be added by the method used for filling melting-point capillaries. After the chlorate is in the bottom (sealed end), the powder is melted by warming over a small flame and then immediately allowed to resolidify. This prevents loss on handling and subsequent weighing errors. Tubes both with and without added chlorate may be further constricted at the open end to a finer capillary if so desired. The additional constriction is desirable when handling very low-boiling liquids, but introduces the danger of breaking the fragile tip while handling.

The tube is wiped clean with a chamois skin and allowed to set some minutes before placing on the balance pan. It is then placed on the left-hand pan with the aid of the forceps and accurately weighed (assume zero reading to be zero), using the fractional weights as tares. The tube is removed from the balance with the aid of the forceps (and chamois skin) and held for a few seconds about an inch above a heated hot plate or burner of a combustion furnace to expand the air inside. The open end is immediately placed below the surface of the sample and held there until the liquid ceases to rise in the

---

* Caution: Remove any adhering particles on outside as above described.

† The chlorate decomposes on heating, yielding oxygen. This furnishes an atmosphere of oxygen in the capillary and also sweeps out the volatilized sample.

capillary. The tube is then placed, closed end down, in a centrifuge and whirled for a few seconds to force the liquid to the bottom. The tube is gently wiped and returned to the balance for an approximate weighing. If too little sample is present, the tube may be gently warmed again, making certain not to vaporize the liquid already in the capillary, and the above process repeated as many times as necessary. If, however, too much sample is obtained the

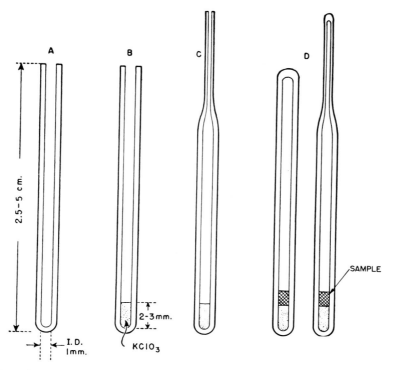

FIG. 54. Glass capillaries. (A) Plain capillary tube, unconstricted. (B) Tube containing KClO$_3$, unconstricted. (C) Constricted tube containing KClO$_3$. (D) Sealed off tubes, both unconstricted and constricted, containing KClO$_3$ and sample.

capillary may be partly emptied by shaking while inverted or by warming to the extent that some sample volatilizes. After the desired approximate amount is in the tube, it is sealed by holding the open end in a flame, first making certain that no liquid is near this end. The tube is again wiped as above and accurately weighed. Before placing such samples in the necessary apparatus for the determination, the tip is scratched with a file and broken off—then both portions are introduced.

(2) *Capsules*.[65] Low-boiling liquids may also be weighed in methylcellulose or gelatin capsules, depending, of course, whether or not these capsules contain the element(s) or group to be determined. They do, however, present the problem of destroying considerably more organic material than ordinarily encountered. They are best weighed when placed in the type of small aluminum cup shown in Fig. 55. The cup and *two caps* are weighed. All parts should be handled by means of forceps, preferably those shown in Fig. 42. To the one

FIG. 55. (*Left*) Aluminum cup for holding methylcellulose (or gelatin) capsules. Enlarged (approximately 2 × actual dimensions) to show details.

FIG. 56. (*Right*) Methylcellulose (or gelatin) capsules showing method of fitting together for effective closure. Enlarged (approximately 2 × dimensions) to show details—preferably capsule size No. 4 or No. 5.

cap, supported in the aluminum cup is placed the required amount of sample and then the second cap, *closed end down,* is forced into the first cap (Fig. 56). This forms a tight seal, preventing escape of the sample. The combination is reweighed to obtain the weight of sample.

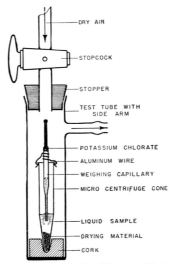

FIG. 57. System for filling capillaries.

ADDITIONAL INFORMATION FOR CHAPTER 3

Alber[3,28,62] described the system shown in Fig. 57 for filling capillary tubes with liquids, particularly hygroscopic ones. The system is momentarily evacuated causing a reduced pressure in the capillary. On breaking the vacuum, the liquid is forced up into the tube. A similar system is described by Furman.[22]

TABLE 13

ADDITIONAL INFORMATION ON REFERENCES* RELATED TO CHAPTER 3

Following the system used in the preceding chapters, this table lists additional references the author wishes to call to the attention of the reader. (See statement regarding such information at top of Table 4 of Chapter 1.)

Books
Belcher and Godbert, 7, 8
Benedetti-Pichler, 11
Clark, E. P., 17
Clark, S. J., 18
Furman, 22
Grant, 24, 25
Milton and Waters, 41, 42
Niederl and Niederl, 44, 45
Niederl and Sozzi, 46
Pregl, 48
Roth, 51–54
Steyermark, 62

Performance
(errors, sensitivity, effect of temperature fluctuations, vibrations, etc.)

Ambrosino, 5
Bradley, 12
Bromund and Benedetti-Pichler, 15
Corner, 19
Gysel, 26
Hirayama, Ieki, and Miyahara, 32
Hodsman, 33
Kuck, Altieri and Towne, 36
Lashof and Macurdy, 37

Performance (Cont.)
Loscalzo and Benedetti-Pichler, 38
Macurdy et al., 40
Mitsui, 43
Sarakhov, 55
Singer, 57
Smith, 59
Waber and Sturdy, 68

Calibration of weights
Benedetti-Pichler, 10

Weighing vessels
British Standards Institution, 13
Cahn and Cadman, 16
Goldberger and Pöhm, 23
Jones, 35
Ma and Eder, 39
Schuele and McNabb, 56
Skinner, 58

Weighing techniques
Benedetti-Pichler, 11
Gysel, 26, 27
Hull, 34
van Nieuwenburg, 47
Williams and Park, 69

* The numbers which appear after each entry in this table refer to the literature citations in the reference list at the end of the chapter.

REFERENCES

1. Alber, H. K., Personal communications (1940).
2. Alber, H. K., *Mikrochemie,* **18**, 92 (1935).
3. Alber, H. K., *Mikrochemie ver. Mikrochim. Acta,* **25**, 47, 168 (1938).
4. Alber, H. K., *Mikrochemie ver. Mikrochim. Acta,* **29**, 294 (1941).
5. Ambrosino, C., *Chimica e industria (Milan),* **33**, 775 (1951).
6. American Society for Testing Materials, ASTM Designations, **E124-57T** (1957).
7. Belcher, R., and Godbert, A. L., "Semi-Micro Quantitative Organic Analysis," Longmans, Green, London, New York, and Toronto, 1945.
8. Belcher, R., and Godbert, A. L., "Semi-Micro Quantitative Organic Analysis," 2nd ed., Longmans, Green, London, New York, and Toronto, 1954.
9. Benedetti-Pichler, A. A., Personal communication (1953).
10. Benedetti-Pichler, A. A., *Mikrochemie ver. Mikrochim. Acta,* **34**, 241 (1949).
11. Benedetti-Pichler, A. A., "Handbuch der Mikrochemischen Methoden. Waagen und Wägung," Springer, Vienna, 1959.
12. Bradley, R. S., *J. Sci. Instr.,* **30**, 84 (1953).
13. British Standards Institution, *Brit. Standards* **1428**, Pt. H1 (1952); Pt. E1, and Pt. I1 (1953), Pt. G1 (1954).
14. Bromund, W. H., Personal communication (1953).
15. Bromund, W. H., and Benedetti-Pichler, A. A., *Mikrochemie ver. Mikrochim. Acta,* **38**, 505 (1951).
16. Cahn, L., and Cadman, W. J., *Anal. Chem.,* **30**, 1580 (1958).
17. Clark, E. P., "Semimicro Quantitative Organic Analysis," Academic Press, New York, 1943.
18. Clark, S. J., "Quantitative Methods of Organic Microanalysis," Butterworths, London, 1956.
19. Corner, M., "Proceedings of the International Symposium on Microchemistry 1958," p. 64, Pergamon Press, Oxford, London, New York, and Paris, 1960.
20. Corner, M., and Hunter, H., *Analyst,* **66**, 149 (1941).
21. Friedrich, A., and Lacourt, A., "La Pratique de la Microanalyse Organique Quantitative," 2nd ed., p. 335, Dunod, Paris, 1939.
22. Furman, N. H., ed., "Scott's Standard Methods of Chemical Analysis," 5th ed., Vol. II, Van Nostrand, New York, 1939.
23. Goldberger, H., and Pöhm, M., *Mikrochemie ver. Mikrochim. Acta,* **39**, 73 (1952).
24. Grant, J., "Quantitative Organic Microanalysis, Based on the Methods of Fritz Pregl," 4th ed., Blakiston, Philadelphia, Pennsylvania, 1946.
25. Grant, J., "Quantitative Organic Microanalysis," 5th ed., Blakiston, Philadelphia, Pennsylvania, 1951.
26. Gysel, H., *Mikrochim. Acta,* p. 267 (1953).
27. Gysel, H., *Mikrochim. Acta,* p. 577 (1956).
28. Hallett, L. T., *Ind. Eng. Chem., Anal. Ed.,* **14**, 956 (1942).
29. Hayman, D. F., *Ind. Eng. Chem., Anal. Ed.,* **8**, 342 (1936).
30. Hayman, D. F., *Ind. Eng. Chem., Anal. Ed.,* **10**, 55 (1938).
31. Hayman, D. F., Personal communication.
32. Hirayama, H., Ieki, T., and Miyahara, K., *Yakugaku Zasshi,* **73**, 103 (1953).
33. Hodsman, G. F., "Proceedings of the International Symposium on Microchemistry 1958," p. 59, Pergamon Press, Oxford, London, New York, and Paris, 1960.
34. Hull, D. E., *Vortex* (California Sect. of Am. Chem. Soc.), p. 342 (September 1948) (abstract).

References

35. Jones, A. R., *Rev. Sci. Instr.*, **24**, 230 (1953).
36. Kuck, J. A., Altieri, P. L., and Towne, A. K., *Mikrochim. Acta*, p. 254 (1953).
37. Lashof, T. W., and Macurdy, L. B., *Anal. Chem.*, **26**, 707 (1954).
38. Loscalzo, A. G., and Benedetti-Pichler, A. A., *Mikrochemie ver. Mikrochim. Acta*, **40**, 232 (1953).
39. Ma, T. S., and Eder, K., *J. Chinese Chem. Soc.*, **15**, 112 (1947).
40. Macurdy, L. B., Alber, H. K., Benedetti-Pichler, A. A., Carmichael, H., Corwin, A. H., Fowler, R. M., Huffman, E. W. D., Kirk, P. L., and Lashof, T. W., *Anal. Chem.*, **26**, 1190 (1954).
41. Milton, R. F., and Waters, W. A., "Methods of Quantitative Microanalysis," Longmans, Green, New York, and Arnold, London, 1949.
42. Milton, R. F., and Waters, W. A., "Methods of Quantitative Microanalysis," 2nd ed., Arnold, London, 1955.
43. Mitsui, T., *Kyoto Daigaku Shokuryô Kagaku Kenkyujo Hokoku*, **No. 18**, 1 (1956).
44. Niederl, J. B., and Niederl, V., "Micromethods of Quantitative Organic Elementary Analysis," Wiley, New York, 1938.
45. Niederl, J. B., and Niederl, V., "Micromethods of Quantitative Organic Analysis," 2nd ed., Wiley, New York, 1942.
46. Niederl, J. B., and Sozzi, J. A., "Microanálisis, Elemental Orgánico," Calle Arcos, Buenos Aires, 1958.
47. Nieuwenburg, C. J., van, *Anal. Chim. Acta*, **20**, 127 (1959).
48. Pregl, F., "Quantitative Organic Microanalysis" (E. Fyleman, trans., 2nd German ed.), Churchill, London, 1924.
49. Rodden, C. J., *Ind. Eng. Chem Anal. Ed.*, **12**, 693 (1940).
50. Rodden, C. J., Kuck, J. A., Benedetti-Pichler, A. A., Corwin, A. H., and Huffman, E. W. D., *Ind. Eng. Chem. Anal. Ed.*, **15**, 415 (1943).
51. Roth, H., "Die quantitative organische Mikroanalyse von Fritz Pregl," 4th ed., Springer, Berlin, 1935.
52. Roth, H., "F. Pregl quantitative organische Mikroanalyse," 5th ed., Springer, Wien, 1947.
53. Roth, H., "Pregl-Roth quantitative organische Mikroanalyse," 7th ed., Springer, Wien, 1958.
54. Roth, H., "Quantitative Organic Microanalysis of Fritz Pregl," 3rd ed. (E. B. Daw, trans., 4th German ed.), Blakiston, Philadelphia, Pennsylvania, 1937.
55. Sarakhov, A. I., *Doklady Akad. Nauk S.S.S.R.*, **86**, 989 (1952).
56. Schuele, W. J., and McNabb, W. M., *Chemist Analyst*, **46**, 101 (1957).
57. Singer, E., *J. phys. radium*, **12**, 554 (1951).
58. Skinner, C. G., *Anal. Chem.*, **28**, 924 (1956).
59. Smith, G. McPhail, "A Course of Instruction in Quantitative Chemical Analysis for Beginning Students," rev. ed., p. 1, Macmillan, New York, 1922.
60. Snedecor, G. W., "Statistical Methods," 4th ed., Iowa State College Press, Ames, Iowa, 1946.
61. Steyermark, Al, *Ind. Eng. Chem. Anal. Ed.*, **17**, 523 (1945).
62. Steyermark, Al, "Quantitative Organic Microanalysis," Blakiston, Philadelphia, Pennsylvania, 1951.
63. Steyermark, Al, Alber, H. K., Aluise, V. A., Huffman, E. W. D., Jolley, E. L., Kuck, J. A., Moran, J. J., Ogg, C. L., and Willits, C. O., *Anal. Chem.*, **26**, 1186 (1954).

64. Steyermark, Al, Alber, H. K., Aluise, V. A., Huffman, E. W. D., Kuck, J. A., Moran, J. J., and Willits, C. O., *Anal. Chem.,* **21**, 1555 (1949).
65. Thomas, Arthur H., Company, Philadelphia, Pennsylvania.
66. Van Slyke, D. D., Folch, J., and Plazin, J., *J. Biol. Chem.,* **136**, 509 (1940).
67. Van Straten, F. W., and Ehret, W. F., *Ind. Eng. Chem., Anal. Ed.,* **9**, 443 (1937).
68. Waber, J. T., and Sturdy, G. E., *Anal. Chem.,* **26**, 1177 (1954).
69. Williams, A. F., and Park, T. O., *Analyst,* **85**, 126 (1960).
70. Willits, C. O., *Anal. Chem.,* **23**, 1058 (1951).
71. Willits, C. O., and Ogg, C. L., Personal communication (1953).
72. Youden, W. J., "Statistical Methods for Chemists," Wiley, New York, 1951.

CHAPTER 4

Preparation of Samples for Analysis

It is assumed that a sample submitted for microanalysis is pure. If it is a liquid, it should have been subjected to fractionation and be a constant boiling fraction. If it is a solid, it should have been crystallized, preferably from more than one type of solvent, until additional crystallization did not raise the melting point. The purification of a sample really has no place in a book on microanalysis. Therefore, very little about the methods should be given here. However, during the existence of every large microchemical laboratory, a few purification problems are encountered. The research chemist with but a few milligrams of precious material often consults the microchemist for suggestions for purification. Such emergencies are handled almost always by merely adapting macromethods to the microsize samples using filter tubes (Chapter 11), filter sticks and crucibles (Chapter 10), centrifuge tubes, etc. For example, crystallization may be made in centrifuge tubes and the crystals collected by centrifuging[20,65,101,107,128] instead of by filtration. The microcentrifuge tubes[118] shown in Figs. 58, 59, 60, and 61 are well adapted to micropurification work—collecting precipitates and separation of liquid fractions. The tubes with the cylindrical bottoms will be found to be particularly useful when working with small quantities of precipitates. The graduated tube shown in Fig. 61, as well as the microvolumetric flasks[118] (Fig. 74) and pipettes[118,119,120] (Figs. 75, 76, and 77), all described in Chapter 5, also will be found to be useful.

Two methods of purification which have proved useful in the author's laboratory are presented below.

DRY CRYSTALLIZATION[115,116]

One method of crystallization, while little known, is extremely useful for handling a few milligrams of material. It is the Hooker[115,116] method of dry crystallization which is as follows: The material is placed in a small Erlenmeyer flask (about 25–50 ml. capacity), one or two drops of solvent are added, and solution is effected by heating the contents with a microburner* while the flask is held in an almost horizontal position. The flask is then placed, in this position, on a suberite ring and the contents allowed to cool and evaporate (Fig.

---

* See Fig. 181, Chapter 16.

62). As evaporation occurs, the resulting rim of crystals is washed down, from time to time, into the mother liquor. When the contents go to dryness, the purified crystals are found in the center of a circle of resinous material. The crystals are loosened and the flask turned so as to move the crystals to a clean

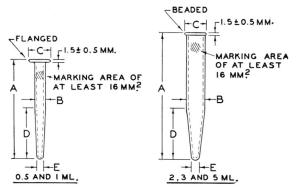

FIG. 58. Microcentrifuge tube with conical bottom, plain—details of construction.

Nominal capacity (ml.)	A Over-all height (mm.)	B Outside diameter cylindrical portion (mm.)	C Outside diameter top finish (mm.)	D Length of taper (mm.)	E Outside diameter at bottom (mm.)
0.5	58 ± 2	6.0 ± 0.25	13 ± 1.0	30 ± 2	3.5 ± 0.5
1	61 ± 2	8.25 ± 0.25	13 ± 1.0	30 ± 2	3.5 ± 0.5
2	66 ± 2	10.75 ± 0.25*	13.5 ± 1.0	30 ± 2	4.0 ± 0.5
3	74 ± 2	10.75 ± 0.25*	13.5 ± 1.0	30 ± 2	4.0 ± 0.5
5	101 ± 2	13.0 ± 0.50	16.25 ± 0.75	40 ± 2	4.0 ± 0.5

Wall thickness of all sizes to be approximately 1 mm.

* The outside diameter of 2 and 3 ml. must not exceed 11.0 mm.; otherwise tubes will not fit into centrifuge shields.
Bottoms to be rounded.

place in the flask. The resinous rim is removed by a cotton swab moistened with a solvent and the crystals are again dissolved and the process repeated a number of times.

SUBLIMATION

The microsublimator of the type[52,63,115] shown in Fig. 63 is useful for purifying samples which sublime either at atmospheric pressure or *in vacuo*. The crude material is placed in the lower container which is heated and can be evacuated. The sublimate collects on the upper water-cooled jacket and can be removed on dismantling.

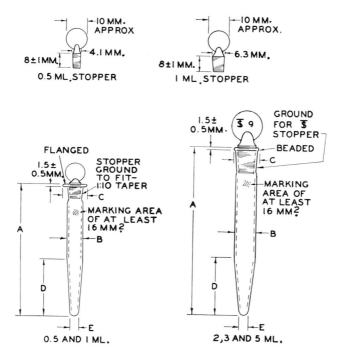

FIG. 59. Microcentrifuge tube with conical bottom, stoppered—details of construction.

Nominal capacity (ml.)	A Over-all height (mm.)	B Outside diameter cylindrical portion (mm.)	C Outside diameter top finish (mm.)	D Length of taper (mm.)	E Outside diameter at bottom (mm.)
0.5	66 ± 2	6.0 ± 0.25	13 ± 1.0	30 ± 2	3.5 ± 0.5
1	69 ± 2	8.25 ± 0.25	13 ± 1.0	30 ± 2	3.5 ± 0.5
2	80 ± 2	10.75 ± 0.25*	13.5 ± 1.0	30 ± 2	4.0 ± 0.5
3	88 ± 2	10.75 ± 0.25*	13.5 ± 1.0	30 ± 2	4.0 ± 0.5
5	115 ± 2	13.0 ± 0.50	16.25 ± 0.75	40 ± 2	4.0 ± 0.5

Wall thickness of all sizes to be approximately 1 mm.

* The outside diameter of 2 and 3 ml. must not exceed 11.0 mm.; otherwise tubes will not fit in centrifuge shields.
Bottoms to be rounded.

4. Preparation of Samples for Analysis

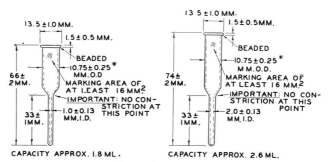

FIG. 60. Microcentrifuge tube with cylindrical bottom, plain.

Wall thickness of both sizes to be approximately 1 mm.

* The outside diameter of both sizes must not exceed 11.0 mm.; otherwise tubes will not fit into centrifuge shields.

Bottoms to be rounded.

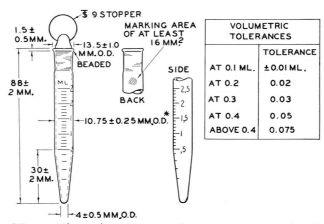

FIG. 61. Microcentrifuge tube with conical bottom, stoppered, graduated, 2.5 ml. in 0.1 ml.—details of construction.

Wall thickness to be approximately 1 mm.

* Outside diameter must not exceed 11.0 mm.; otherwise, tube will not fit into centrifuge shields.

Bottom to be rounded.

FIG. 62. Hooker dry crystallization.

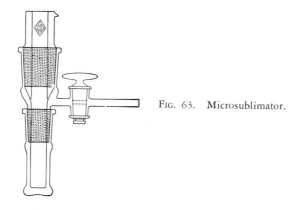

Fig. 63. Microsublimator.

Homogeneity of Samples

Obviously, a *pure* sample is homogeneous as to composition. However, the question of purity often arises when the analyst fails to obtain values in agreement with the theoretical ones for the compound *believed* to have been submitted for analysis. Since the "analyst is always wrong" in the opinion of the chemist submitting the sample, the analyst must be able to defend his results. The analyzing of test substances is the analysts' means of defense. However, the question often arises as to why several analyses on the submitted sample do not check each other. Obviously if the sample is not homogeneous, each analysis will differ unless the components have the same percentage compositions. The lack of homogeneity of solid samples can be demonstrated often with the aid of a microscope, particularly if a Kofler micro hot stage is used for the determination of the melting point as explained in connection with the operation of this instrument (see Chapter 22). The nonhomogeneity of liquid samples cannot be as easily demonstrated, unless, of course, portions change color or precipitates are present. Although a liquid shows no change, *each* time the sample bottle is opened lower-boiling components evaporate at a faster rate than the high boiling ones, surface oxidation or absorption of carbon dioxide may occur, etc. Remembering the above can be helpful to the analyst on such occasions.

Apparatus

SAMPLE CONTAINERS

Figure 64 shows the types of amber glass vials* in which samples for analysis are submitted to the author's laboratories. The ground glass stoppered vial is used for liquids. It is approximately 45 mm. in length and 13 mm. in outside diameter, and is fitted with a solid ⚭ 8 stopper. The polyethylene stoppered vial

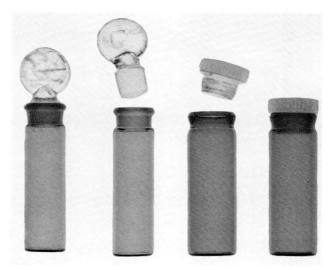

FIG. 64. Sample containers: ground glass stoppered type for liquids; plastic cap type for solids.

is used for solids. It is approximately 45 mm. in length and 14.5 mm. in outside diameter. Both types have been in use for the past twelve years and have proved to be quite satisfactory.

MODIFIED ABDERHALDEN DRYING APPARATUS[5]

The drying apparatus shown in Fig. 65 is a modification of the original Abderhalden pistol drier,[26,27,44,49,50,80-82,88,89,102-105,115,121] a battery of which is shown in Fig. 66. The original model obtained the name of "pistol drier" because of its shape. It permitted drying only *in vacuo,* while the modified apparatus shown in Fig. 65 permits the drying of samples by the passage of dry air at reduced pressure, in accordance with the newer methods of drying,[121, 126,127,] as well as eliminates structural disadvantages of the early models.

* Available from Kimble Glass Company, Subsidiary of Owens-Illinois Glass Company, Vineland, New Jersey.

Apparatus

The drying apparatus consists of the following parts whose functions are described:

Boiling Flask

Into this is placed whatever liquid (acetone, alcohol, water, acetic acid, toluene, xylene, *p*-cymene, etc.) is required for the vapors, produced by boiling, to give the desired drying temperature.

Drying Chamber (Part A, Fig. 65)

This is composed of the sample compartment, which connects on the left to the desiccator bulb (Part B, Fig. 65), and the vapor compartment, which acts as a hot envelope for the sample compartment. The lower joint connects to

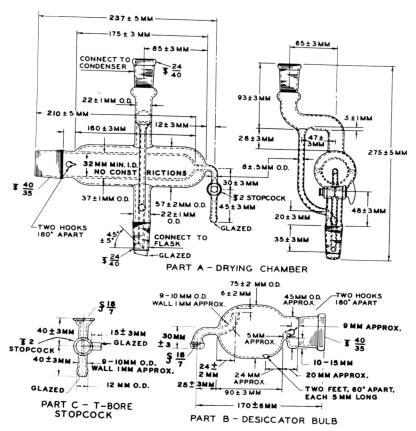

FIG. 65. Modified Abderhalden drying apparatus—details of construction.

the boiling flask and the upper joint connects to the reflux condenser. The reflux returns into the flask, but the upward indentation in the vapor tube (end view) prevents cooling of the drying chamber by cold condensate. At the right end of the chamber is a stopcock through which *dried* air (passed over a desiccant) at reduced pressure may be passed over the sample, if desired. Otherwise, this stopcock is kept closed for conventional vacuum drying.

Desiccator Bulb (Part B, Fig. 65)

Into this is placed the desiccant, such as phosphorus pentoxide. The desiccator bulb connects to the drying chamber. The bulb is so designed that it may be detached and rested on the table without additional support. The shape of the tube attached to the ball joint is intended to prevent the desiccant from being carried into the sample when the vacuum is broken. The T-bore (three-way) stopcock (Part C, Fig. 65) connects the system to the source of vacuum and allows the vacuum to be broken at the source without breaking it in the flask or vice versa. A cap for the ball joint and a stopper for the ⚵ 40/50 joint may be used to protect the desiccant when the desiccator bulb is temporarily disconnected from the drying chamber and stands alone.

During the drying process, the vapor from the boiling flask passes up

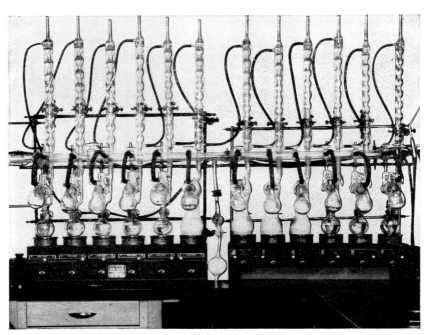

FIG. 66. Battery of Abderhalden drying apparatus.

through the drying chamber and then into the water-cooled condenser where it condenses and returns to the boiling flask. A number of these should be set up, with a variety of solvents, having different boiling points, so that a number of drying temperatures are always available. Acetone, alcohol, water, acetic acid, toluene, xylene, *p*-cymene, etc., are recommended for providing drying ranges from about 55 to 175° C. and over. Although phosphorus pentoxide is generally used as the desiccant, others[1,7] may be used. For the removal of certain nonpolar solvents, such as benzene, from samples, paraffin chips have been used. Figure 66 is a photograph of the setup in the author's laboratory. It consists of a battery of twleve electrically heated driers, connected to a manifold, which in turn is connected to vacuum pumps (backing and diffusion) and gauges. With this battery, it is possible to dry at the abovementioned range of temperatures and at pressures as low as 1×10^{-5} mm. of Hg. The use of three-way stopcocks on the manifold allows the vacuum to be broken on one flask with no disturbance to the others or to the rest of the vacuum system.

VACUUM OVENS

Small electrically heated vacuum ovens can also be used for drying samples. The temperature is thermostatically controlled to that desired.

ORDINARY LABORATORY OVENS

Ordinary laboratory drying ovens usually found in chemical laboratories may be used for drying filter tubes, crucibles, etc., but usually not samples.

Determination of Moisture

To be perfectly safe, all solid compounds should be dried to constant weight before being analyzed. However, in a large laboratory handling thousands of samples, this becomes almost an impossibility and actual moisture determinations are done on but a comparative few.

Ordinarily, samples are merely dried for several hours *in vacuo* (0.1–0.5 mm. of Hg*) over phosphorus pentoxide (or some other desiccant),[1,7] at some temperature well below their melting points previous to their being analyzed. (With some materials, more efficient drying is accomplished by allowing *dry* air to be sucked, at reduced pressure, through the *partly* open stopcock on the drying chamber. The passage of the dry air over the sample aids the removal of the water vapor.[121,126,127]) This will suffice to remove moisture and solvents in the majority of cases. The exact time of drying and the temperature re-

---

* High vacuums are rarely needed and increase the possibility of loss by sublimation.

quired to do the job is anybody's guess. The safest rule to follow is to dry for a number of hours in a good vacuum and at a relatively low temperature, since the lower the temperature the less possibility of decomposition. In spite of the fact that a sample melts rather high (as, for example, 225° C.), when heated rapidly in a melting point determination, heating at 80° C. for many hours *might* cause decomposition, yielding a tar. Fortunately, such cases are the exception, but the possibility must be considered when dealing with valuable materials.

The following method is used for drying to constant weight or for determining the moisture present. If available, 50–75 mg. of material are weighed in a pig (Fig. 29) (taking into account the zero reading) and dried at some temperature well below its melting point *in vacuo* over phosphorus pentoxide. After several hours, the sample should be cooled and weighed. (The sample should be observed for signs of decomposition and the cold portion of the drying chamber should be examined for sublimate. In either case, the determination is valueless and an attempt should be made to dry at a lower temperature.) It should then be returned to the drier and again heated for several hours and then reweighed. The process should be repeated until no further loss is noted. (Some samples decompose at such a slow rate that a loss of 0.1% will be noted after each successive drying of 6–8 hours which will go on for days unless the experiment is stopped.)

Calculation:

$$\frac{\text{Loss in wt.}}{\text{Wt. of sample}} \times 100 = \% \text{ loss* on drying}$$

SPECIAL CASES

When the water or solvents are merely absorbed by the material the matter of driving them off is much easier than if they are bound up as water or solvent of crystallization.[115,117,125] In these latter cases the substance often decomposes in the process of becoming anhydrous and the sample must be analyzed with the substance of crystallization attached. The idea of an organic compound crystallizing with water or solvent of crystallization such as ethanol, methanol, acetic acid, acetone, or benzene is a hard one for the organic chemist to "swallow" particularly when these are present in such amounts as $\frac{1}{2}$, 1, $1\frac{1}{2}$, or 2 molecules in the compound. In the author's laboratory, many such cases have been encountered.[115,117] If the sample contains water of crystallization, and will not yield it on heating, proof will have to rest on the elementary analyses unless a Karl Fischer determination[40,69,84,111] is successful. If, however, ethanol,

---

* Loss may be due to water or solvents, either mechanically held or as water or solvent of crystallization.

methanol, acetic acid or acetone of crystallization is present, a determination of ethoxyl, methoxyl, acetyl, or acetone, respectively, will furnish proof of the existence.[115,117] Determination of ethoxyl and methoxyl are carried out as described in Chapter 16, the determination of acetyl as described in Chapter 17, while acetone of crystallization may be determined by the colorimetric method of Czonka.[32] The determination of oxygen, as described in Chapter 14, often furnishes conclusive data.

Combinations of solvents are also sometimes found attached to organic molecules. Twenty years ago in the author's laboratory a substance was found to have water, methanol, and acetone of crystallization.[115,117] Fortunately, it yielded the solvents, on heating, which were collected and analyzed qualitatively as follows. The presence of methanol was proved by (a) conversion to formaldehyde,[4] (b) Vorisek test,[4] and (c) conversion to m-nitromethyl benzoate which was identified by melting point (78° C.) and mixed melting point with a known sample.[56,61] The presence of acetone was proved by (a) iodoform reaction,[61] (b) sodium nitroprusside test,[57,79] and (c) conversion to the p-nitrophenyl hydrazone which was identified by melting point (148°–149° C.) and mixed melting point with a known sample.[57,61] Neither methanol nor acetone could have resulted from decomposition of the compound. Analyses were then done on the solid material. The methanol was obtained quantitatively by a methoxyl determination and the acetone quantitatively by the method of Czonka referred to above.

If a sample containing alcohol of crystallization is placed over water in a closed system, such as under a bell jar or in a desiccator, there often occurs a replacement of the alcohol in the molecule by water.[115,125] In many such cases, the water of crystallization is then more easily removed than is the alcohol.

Besides the presence of solvents of crystallization, the analyst might encounter other special cases which alter the analyses. Amine hydrochlorides and phosphates have been analyzed which contained an additional molecule of hydrochloric and phosphoric acids, respectively, than was required for the formula (3 mol HCl although only 2 nitrogens were present and 2 mol H_3PO_4 although only 1 nitrogen was present, respectively). The analyses of carbon, hydrogen, nitrogen, and halogen or phosphorus substantiated these theories.[117]

ADDITIONAL INFORMATION FOR CHAPTER 4

When it is necessary to concentrate small amounts of solutions by evaporation of the solvent in a dust-free atmosphere and with the prevention of creeping, the apparatus[93] shown in Fig. 67 will prove helpful. It consists of a boiling flask, A, into which is placed whatever liquid is necessary to produce the desired evaporation temperature. Heating is accomplished by means of an electric immersion heater. On the way to the reflex condenser, the vapors surround the evaporation vessel, E, which is secured in a heating well in the body, B. The suction tube, D, fits into a rubber sleeve and can be raised or lowered. Air enters the drying chamber through the cotton filters at C. The

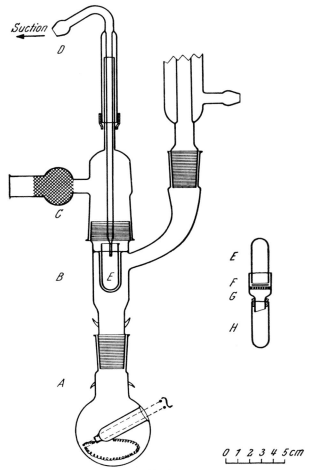

FIG. 67. Perold microevaporator.

tip of the suction tube, D, is kept just above the surface of the liquid in E at all times. Air is drawn down along the inner wall of E, which prevents creeping.

The determination of water has been done by means of the Karl Fischer reagent.[40,69,84,111] This method involves the direct titration of the wet material with a solution of iodine, sulfur dioxide, and pyridine in methanol. Levy, Murtaugh, and Rosenblatt[69] adapted the method to the microscale (Fig. 68). They used a small titration vessel made from a 16-mm. diameter test tube into which are sealed platinum electrodes. These electrodes are connected with an

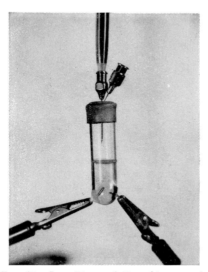

FIG. 68.   Levy-Murtaugh-Rosenblatt vessel.

electrometric apparatus by means of clamps. The vessel is closed with a tightly fitting serum bottle sleeved rubber stopper. Through the stopper are placed two hypodermic needles, one attached to a microburette for titration and the other open to the air and acting as a valve to equalize the pressure during titration.

For additional material in regard to apparatus, procedures, etc., related to the subject of this chapter, the reader is referred to Table 14, page 96.

4. Preparation of Samples for Analysis

TABLE 14
ADDITIONAL INFORMATION ON REFERENCES* RELATED TO CHAPTER 4

Following the plan used with the previous chapters, this table lists additional references the author wishes to call to the attention of the reader. (See statement regarding such information at top of Table 4 of Chapter 1.)

Books
Belcher and Godbert, 10, 11
Benedetti-Pichler and Spikes, 13
Bernhauer, 15
Cheronis and Entrikin, 23, 24
Clark, E. P., 26
Clark, S. J., 27
Emich and Schneider, 36
Furman, 44
Grant, 49, 50
Huntress and Mulliken, 57
Kamm, 61
Lieb and Schöniger, 70, 71
Milton and Waters, 81, 82
Niederl and Niederl, 88, 89
Niederl and Sozzi, 90
Pregl, 96
Roth, 102–105
Schneider, 107
Steyermark, 115

Beakers, pipettes, centrifuge accessories
British Standards Institution, 18
Lieb and Schöniger, 70, 71

Distillation
Alber, 3
Babcock, 6
Belcher and Godbert, 10, 11
Bering, 14
Bernhauer, 15
Cheronis and Entrikin, 23, 24
Conolly and Oldham, 29
Dubbs, 33
Dubowski and Shupe, 34
Erdos, 37
Gould, 48
Grant, 49, 50
Haendler, 51
Hallett, 52
Irlin and Bruns, 58

Distillation (Cont.)
Lappin, 67
Lesesne and Lochte, 68
Lieb and Schöniger, 70, 71
Morton and Mahoney, 85
Oldham, 91
Podbielniak, 95
Rose and Rose, 99
Schneider, 107
Shrader and Ritzer, 108

Crystallization
Benedetti-Pichler and Spikes, 13
Blount, 17
Cannon, 21
Cheronis and Entrikin, 23, 24
Craig, 31
Emich and Schneider, 36
Hallett, 52
Lieb and Schöniger, 70, 71
Schneider, 107
Shead, 109

Sublimation
Clarke and Hermance, 28
Erdos, 37
Fischer, 41
Flaschenträger, Abdel-Wahhab, and Habib-Labib, 43
Grant, 49, 50
Hallett, 52
Hickman and Sanford, 54
Hippenmeyer, 55
Klein and Werner, 63
Lieb and Schöniger, 70, 71
Marberg, 75
McDonald, 77
Meyer, 78
Morton, Mahoney, and Richardson, 86
Soltys, 113
Werner, 123

* The numbers which appear after each entry in this table refer to the literature citations in the reference list at the end of the chapter.

TABLE 14 (*Continued*)

Filtration
British Standards Institution, 18
Dubbs, 33
Emanuel, 35
Feldman and Ellenburg, 38
Lieb and Schöniger, 70, 71

Evaporation
Kurtz, 66
Lieb and Schöniger, 70, 71

Extraction
Cheronis and Entrikin, 23, 24
Craig, 30
Dubbs, 33
Kirk and Danielson, 62
Lieb and Schöniger, 70, 71
Neustadt, 87

Fractionation, separation, purification
Batt and Alber, 9
Bickford, 16
Cheronis and Entrikin, 23, 24
Ferguson, 39
Gettler, Umberger, and Goldbaum, 45
Lieb and Schöniger, 70, 71
Malissa, 74
Perold, 93
Rosenthaler, 100
Soltys, 114

Ovens, desiccators, drying devices
Alber, 1, 2
Barraclough, 7
British Standards Institution, 18
Clark, 26
Furman, 44
Grant, 49, 50
Knight, 64
Maurmeyer and Ma, 76
Milner and Sherman, 80

Ovens, etc. (Cont.)
Milton and Waters, 81, 82
Niederl and Sozzi, 90
Pavelka, 92
Pregl, 96
Roth, 102–105
Smith, 112
Ten Eyck Schenck and Ma, 122

Drying and moisture in general
Bunzell, 19
Central Scientific Company, 22
Heckly, 53
Sandell, 106
Willits, 125, 126

Determination of moisture by neutron scattering
Johnson, 60

Karl Fischer
Bastin, Siegel, and Bullock, 8
Ciusa and Moroni, 25
Fischer, Karl, 40
Johansson, 59
Levy, Murtaugh, and Rosenblatt, 69
Mitchell, 83
Mitchell and Smith, 84
Peters and Jungnickel, 94
Ricciuti and Willits, 97
Roberts and Levin, 98
Smith, Bryant, and Mitchell, 111
Wiberley, 124

Cinnamyl chloride, naphthyl phosphorus oxychloride, succinyl chloride, calcium carbide, etc.
Belcher, Thompson and West, 12
Gorbach and Jurinka, 46, 47
Fischer, R., 42
Lindner, 72, 73
Sirotenko, 110

REFERENCES

1. Alber, H. K., *Mikrochemie ver. Mikrochim. Acta*, **25**, 47 (1938).
2. Alber, H. K., *Mikrochemie ver. Mikrochim. Acta*, **25**, 167 (1938).
3. Alber, H. K., *Z. anal. Chem.*, **90**, 100 (1932).
4. "Allen's Commercial Organic Analysis" (S. S. Sadtler and E. C. Lathrop, eds.), 5th ed., Vol. I, p. 95, Blakiston, Philadelphia, Pennsylvania, 1923.
5. American Society for Testing Materials, *ASTM Designations*, **E124-56T** (1956), **E124-57T** (1957).
6. Babcock, M. J., *Anal. Chem.*, **21**, 632 (1949).
7. Barraclough, K. C., *Metallurgia*, **31**, 269 (1945).
8. Bastin, E. L., Siegel, H., and Bullock, A. B., *Anal. Chem.*, **31**, 467 (1959).
9. Batt, W. G., and Alber, H. K., *Ind. Eng. Chem., Anal. Ed.*, **13**, 127 (1941).
10. Belcher, R., and Godbert, A. L., "Semi-Micro Quantitative Organic Analysis," Longmans, Green, London, New York and Toronto, 1945.
11. Belcher, R., and Godbert, A. L., "Semi-Micro Quantitative Organic Analysis," 2nd ed., Longmans, Green, London, 1954.
12. Belcher, R., Thompson, J. H., and West, T. S., *Anal. Chim. Acta*, **19**, 148 (1958).
13. Benedetti-Pichler, A. A., and Spikes, W. F., "Introduction to the Microtechnique of Inorganic Qualitative Analysis," Microchemical Service, Douglaston, New York, 1935.
14. Bering, P., *Svensk. Kem. Tidskr.*, **61**, 10 (1949).
15. Bernhauer, K., "Einführung in die organisch-chemische Laboratoriumstechnik," p. 59, Springer, Berlin, 1934.
16. Bickford, C. F., *J. Am. Pharm. Assoc., Sci. Ed.*, **38**, 356 (1949).
17. Blount, B. K., *Mikrochemie*, **19**, 162 (1935).
18. British Standards Institution, *Brit. Standards*, **1428**, Pt. D2 (1950), Pt. D4 (1954) (or **797** (1954)), Pt. D5 (1955), Pt. E2 (1954), Pt. E3 (1953), Pt. F1, and Pt. G2 (1957).
19. Bunzell, H. H., *Cereal Chem.*, **12**, 54 (1935).
20. Burton, F., *Metallurgia*, **32**, 285 (1945).
21. Cannon, J. H., *J. Assoc. Offic. Agr. Chemists*, **38**, 844 (1955).
22. Central Scientific Company, "Summary of Methods, Suggested List of Equipment for the Determination of Moisture," Chicago, Illinois, 1957.
23. Cheronis, N. D., and Entrikin, J. B., "Semimicro Qualitative Organic Analysis," Crowell, New York, 1947.
24. Cheronis, N. H., and Entrikin, J. B., "Semimicro Qualitative Organic Analysis," Interscience, New York, and London, 1957.
25. Ciusa, W., and Moroni, E., *Mikrochemie ver. Mikrochim. Acta*, **36/37**, 273 (1951).
26. Clark, E. P., "Semimicro Quantitative Organic Analysis," Academic Press, New York, 1943.
27. Clark, S. J., "Quantitative Methods of Organic Microanalysis," Butterworths, London, 1956.
28. Clarke, B. L., and Hermance, H. W., *Ind. Eng. Chem., Anal. Ed.* **11**, 50 (1939).
29. Conolly, J. M., and Oldham, G., *Analyst*, **76**, 52 (1951).
30. Craig, L. C., *Anal. Chem.*, **26**, 110 (1954).
31. Craig, L. C., *Ind. Eng. Chem., Anal. Ed.*, **12**, 773 (1940).
32. Czonka, F. A., *J. Biol. Chem.*, **27**, 209 (1916).
33. Dubbs, C. A., *Anal. Chem.*, **21**, 1273 (1949).
34. Dubowski, K. M., and Shupe, L. M., *Am. J. Clin. Pathol.*, **22**, 709 (1952).

35. Emanuel, C. F., *Chemist Analyst,* **45**, 52 (1956).
36. Emich, F., and Schneider, F., "Microchemical Laboratory Manual, pp. 12, and 29, Wiley, New York, 1934.
37. Erdos, José, *Mikrochemie ver. Mikrochim. Acta,* **36/37**, 417 (1951).
38. Feldman, C., and Ellenburg, J. Y., *Anal. Chem.,* **29**, 1557 (1957).
39. Ferguson, B., *Ind. Eng. Chem., Anal. Ed.,* **14**, 493 (1942).
40. Fischer, Karl, *Angew. Chem.,* **48**, 394 (1935).
41. Fischer, R., *Mikrochemie,* **15**, 247 (1934).
42. Fischer, R., *Mikrochemie ver. Mikrochim. Acta,* **31**, 296 (1944).
43. Flaschenträger, B., Abdel-Wahhab, S. M., and Habib-Labib, G. *Mikrochim. Acta,* p. 390 (1957).
44. Furman, N. H., ed., "Scott's Standard Methods of Chemical Analysis," 5th ed., vol. II, Van Nostrand, New York, 1939.
45. Gettler, A. O., Umberger, C. J., and Goldbaum, L., *Anal. Chem.,* **22**, 600 (1950).
46. Gorbach, G., and Jurinka, A., *Fette u. Seifen,* **51**, 129 (1944).
47. Gorbach, G., and Jurinka, A., *Mikrochemie ver. Mikrochim. Acta,* **34**, 174 (1949).
48. Gould, C. W., Jr., U.S. Patent 2,459,375 (1949).
49. Grant, J., "Quantitative Organic Microanalysis, Based on the Methods of Fritz Pregl" 4th ed., Blakiston, Philadelphia, Pennsylvania, 1946.
50. Grant, J., "Quantitative Organic Microanalysis," 5th ed., Blakiston, Philadelphia, Pennsylvania, 1951.
51. Haendler, H. M., *Anal. Chem.,* **20**, 596 (1948).
52. Hallett, L. T., *Ind. Eng. Chem., Anal. Ed.,* **14**, 956 (1942).
53. Heckly, R. J., *Science,* **122**, 760 (1955).
54. Hickman, K. C. D., and Sanford, C. R., *J. Phys. Chem.,* **34**, 637 (1930).
55. Hippenmeyer, F., *Mikrochemie ver. Mikrochim. Acta,* **39**, 409 (1952).
56. Hodgman, C. D., and Lange, N. A., "Handbook of Chemistry and Physics," 9th ed., p. 246, Chemical Rubber, Cleveland, Ohio, 1922.
57. Huntress, E. H., and Mulliken, S. P., "Identification of Pure Organic Compounds," Wiley, New York, 1941.
58. Irlin, A. L., and Bruns, B. P., *Zhur. Anal. Khim.,* **5**, 44 (1950).
59. Johansson, A., *Acta Chem. Scand.,* **3**, 1058 (1949).
60. Johnson, W. B., Personal communication. 1959.
61. Kamm, O., "Qualitative Organic Analysis," p. 53, Wiley, New York, 1923.
62. Kirk, P. L., and Danielson, M., *Anal. Chem.,* **20**, 1122 (1948).
63. Klein, G., and Werner, O., *Z. physiol. Chem.,* **143**, 141 (1935).
64. Knight, C. A., *Chemist Analyst,* **44**, 108 (1955).
65. Kolthoff, I. M., and Amdur, E., *Ind. Eng. Chem., Anal. Ed.,* **12**, 178 (1940).
66. Kurtz, L. T., *Ind. Eng. Chem., Anal. Ed.,* **14**, 191 (1942).
67. Lappin, G. R., *J. Chem. Educ.* **25**, 657 (1948).
68. Lesesne, S. D., and Lochte, H. L., *Ind. Eng. Chem., Anal. Ed.,* **10**, 450 (1938).
69. Levy, G. B., Murtaugh, J. J., and Rosenblatt, M., *Ind. Eng. Chem., Anal. Ed.,* **17**, 193 (1945).
70. Lieb, H., and Schöniger, W., "Arbeiten mit kleinen Substanzmengen," *in* "Houben-Weyl's Methoden der organischen Chemie," Band I, Tl. 2, Allgemeine Laboratoriumspraxis, Band II, Tl. 2, Müller, E., Thieme, Stuttgart, 1959.
71. Lieb, H., and Schöniger, W., "Präparative Mikromethoden in der organischen Chemie," *in* "Handbuch der mikrochemischen Methoden," (F. Hecht, and M. K. Zacherl, eds.), Springer, Wien, 1954.

72. Lindner, J., *Mikrochemie ver. Mikrochim. Acta*, **32**, 155 (1944).
73. Lindner, J., *Mikrochemie ver. Mikrochim. Acta*, **32**, 133 (1944).
74. Malissa, H., *Mikrochemie ver. Mikrochim. Acta*, **34**, 393 (1949).
75. Marberg, C. M., *J. Am. Chem. Soc.*, **60**, 1509 (1938).
76. Maurmeyer, R. K., and Ma, T. S., *Mikrochim. Acta*, p. 563 (1957).
77. McDonald, E., *J. Franklin Inst.*, **221**, 132 (1936).
78. Meyer, F., *Süddeut. Apotheker-Ztg.*, **88**, 337 (1948).
79. Meyer, H., "Nachweis und Bestimmung organischer Verbindunger," p. 61, Springer, Berlin, 1933.
80. Milner, R. T., and Sherman, M. S., *Ind. Eng. Chem., Anal. Ed.*, **8**, 427 (1936).
81. Milton, R. F, and Waters, W A., "Methods of Quantitative Microanalysis," Longmans, Green, New York, and Arnold, London, 1949.
82. Milton, R. F., and Waters, W. A., "Methods of Quantitative Microanalysis," 2nd ed., Arnold, London, 1955.
83. Mitchell, J., Jr., *Ind. Eng. Chem., Anal. Ed.*, **12**, 390 (1940).
84. Mitchell, J., Jr., and Smith, D. M., "Aquametry, Applications of the Karl Fischer Reagent to Quantitative Analyses, Including Water," Interscience, New York, 1948.
85. Morton, A. A., and Mahoney, J. F., *Ind. Eng. Chem., Anal. Ed.*, **13**, 494 (1941).
86. Morton, A. A., Mahoney, J. F., and Richardson, G., *Ind. Eng. Chem., Anal. Ed.*, **11**, 460 (1939).
87. Neustadt, M. H., *Ind. Eng. Chem., Anal. Ed.*, **14**, 431 (1942).
88. Niederl, J. B., and Niederl, V., "Micromethods of Quantitative Organic Elementary Analysis," Wiley, New York, 1938.
89. Niederl, J. B., and Niederl, V., "Micromethods of Quantitative Organic Analysis," 2nd ed., Wiley, New York, 1942.
90. Niederl, J. B., and Sozzi, J. A., "Micronanálisis Elemental Orgánico," Calle Arcos, Buenos Aires, 1958.
91. Oldham, G., *Analyst*, **77**, 542 (1952).
92. Pavelka, F., *Mikrochemie ver. Mikrochim. Acta*, **32**, 141 (1944).
93. Perold, G. W., *Mikrochim. Acta*, p. 251 (1959).
94. Peters, E. D., and Jungnickel, J. L., *Anal. Chem.*, **27**, 450 (1955).
95. Podbielniak, W. J., *Ind. Eng. Chem., Anal. Ed.*, **13**, 639 (1941).
96. Pregl, F., "Quantitative Organic Microanalysis," (E. Fyleman, trans. 2nd German ed.), Churchill, London, 1924.
97. Ricciuti, C., and Willits, C. O., *J. Assoc. Offic. Agr. Chemists*, **33**, 469 (1950).
98. Roberts, F. M., and Levin, H., *Anal. Chem.*, **21**, 1553 (1949).
99. Rose, A., and Rose, E., *Anal. Chem.*, **26**, 101 (1954).
100. Rosenthaler, L., *Mikrochemie ver. Mikrochim. Acta*, **35**, 164 (1950).
101. Roswell, C. A., *Ind. Eng. Chem., Anal. Ed.*, **12**, 350 (1940).
102. Roth, H., "Die quantitative organische Mikroanalyse von Fritz Pregl," 4th ed., Springer, Berlin, 1935.
103. Roth, H., "F. Pregl quantitative organische Mikroanalyse," 5th ed., Springer, Wien, 1947.
104. Roth, H., "Pregl-Roth quantitative organische Mikroanalyse," 7th ed., Springer, Wien, 1958.
105. Roth, H., "Quantitative Organic Microanalysis of Fritz Pregl," 3rd ed. (E. B. Daw, trans. 4th German ed.), Blakiston, Philadelphia, Pennsylvania, 1937.
106. Sandell, E. B., *Mikrochemie ver. Mikrochim. Acta*, **38**, 487 (1951).

107. Schneider, F., "Qualitative Microanalysis," Wiley, New York, 1946.
108. Shrader, S. A., and Ritzer, J. E., *Ind. Eng. Chem., Anal. Ed.,* **11**, 54 (1939).
109. Shead, A. C., *Ind. Eng. Chem., Anal. Ed.,* **9**, 496 (1937).
110. Sirotenko, A. A., *Mikrochim. Acta,* p. 917 (1955).
111. Smith, D. M., Bryant, W. M. D., and Mitchell, J., Jr., *J. Am. Chem. Soc.,* **61**, 2407 (1939).
112. Smith, G. F., *Anal. Chim. Acta,* **17**, 192 (1957).
113. Soltys, A., *Mikrochemie, Emich Festschrift,* p. 275 (1930).
114. Soltys, A., *Mikrochemie, Molisch Festschrift,* p. 393 (1936).
115. Steyermark, Al, "Quantitative Organic Microanalysis," Blakiston, Philadelphia, Pennsylvania, 1951.
116. Steyermark, Al, *Record of Chem. Progr. (Kresge-Hooker Sci. Lib.)* **1**, 46 (1946).
117. Steyermark, Al, Unpublished results.
118. Steyermark, Al, Alber, H. K., Aluise, V. A., Huffman, E. W. D., Jolley, E. L., Kuck, J. A., Moran, J. J., and Ogg, C. L., *Anal. Chem.,* **28**, 1993 (1956).
119. Steyermark, Al, Alber, H. K., Aluise, V. A., Huffman, E. W. D., Jolley, E. L., Kuck, J. A., Moran, J. J., and Ogg, C. L., *Anal. Chem.,* **30**, 1702 (1958).
120. Steyermark, Al, Alber, H. K., Aluise, V. A., Huffman, E. W. D., Jolley, E. L., Kuck, J. A., Moran, J. J., Ogg, C. L., and Pietri, C. E., *Anal. Chem.,* **32**, 1045, (1960).
121. Steyermark, Al, Alber, H. K., Aluise, V. A., Huffman, E. W. D., Jolley, E. L., Kuck, J. A., Moran, J. J., Ogg, C. L., and Willits, C. O., *Anal. Chem.,* **26**, 1186 (1954).
122. Ten Eyck Schenck, R., and Ma, T. S., *Mikrochemie ver. Mikrochim. Acta,* **40**, 236 (1953).
123. Werner, O., *Mikrochemie,* **1**, 35 (1923).
124. Wiberley, J. S., *Anal. Chem.,* **23**, 656 (1951).
125. Willits, C. O., *Anal. Chem.,* **21**, 132 (1949).
126. Willits, C. O., *Anal. Chem.,* **23**, 1058 (1951).
127. Willits, C. O., and Ogg, C. L., United States Department of Agriculture, personal communication (1953).
128. Wyatt, G. H., *Analyst,* **71**, 122 (1946).

CHAPTER 5

Standard Solutions
(Microtitration Techniques)

In general, standard solutions for microchemical work are prepared and standardized in the same manner as those for macroanalytical work. Methods for the latter are found in many works on analytical chemistry.[72,155,215] Where macromethods use normal and tenth normal solutions, microanalysis employs one or two hundredth normal solutions.[42,72,83,84,144,145,147,161,168-171,194] The reader is referred to the above-mentioned works if there exist any doubts regarding the preparation of standard solutions required, other than those discussed below.

Fortunately, practically all of the standard solutions are commercially[1] available and have proved satisfactory in the author's laboratory. Where the standards are prepared in the laboratory, freshly boiled, distilled water is generally used.

Standard solutions of hydrochloric acid, sodium hydroxide, potassium biiodate, iodine, sodium thiosulfate, potassium sulfate and barium chloride are the most frequently used. Potassium biiodate (also called biniodate) may be used either as an acid in alkalimetric and acidimetric or in iodometric titrations[53,145,194] as shown by the following equations:

(1) Alkalimetric—Acidimetric

$$KH(IO_3)_2 + NaOH \rightarrow NaIO_3 + KIO_3 + H_2O$$

(2) Iodometric

$$KH(IO_3)_2 + 10KI + 11HCl = 11KCl + 6H_2O + 6I_2$$
$$6I_2 + 12Na_2S_2O_3 = 12NaI + 6Na_2S_4O_6$$

The normality of a solution of potassium biiodate differs depending on whether it is to be used as an acid or in iodometric titrations, the same solution being twelve times as strong for the latter as for the former, as seen from the equations.

Dissolved carbon dioxide must be removed before proper end points can be obtained in acid-base titrations. Consequently, the solutions should be boiled for about 30 seconds before beginning the titration and then again for a few seconds toward the end. (The one exception to this is the titration of the dissolved ammonia in boric acid, obtained in the Kjeldahl determination of nitrogen—see Chapter 8.)

Reagents

STARCH SOLUTION*

Starch solution may be prepared by any of the methods generally used in macroanalysis (see above-noted references), or by that suggested by Elek[53,145,194] which is as follows:

An approximate 1% solution is made by triturating one gram of water-soluble starch with a few milliliters of cold distilled water and adding this mixture to about 95 ml. of boiling 20% aqueous solution of sodium chloride. The resulting mixture is boiled for 5 minutes, cooled and filtered. The solution should be kept in a refrigerator.

An alternate reagent is prepared according to the following which is recommended by the Association of Official Agricultural Chemists[12]:

Two grams of finely powdered potato starch is mixed with a little cold distilled water to form a thin paste. About 200 ml. of boiling distilled water is added with *constant stirring* and the resulting mixture is immediately allowed to cool. One milliliter of mercury is added as a preservative.

THYMOLPHTHALEIN[84]

A 0.1% solution in 95% ethanol is prepared.

SODIUM ALIZARIN SULFONATE[127,199]

A 0.035% solution in water is prepared and stored in a glass-stoppered flask or bottle. The material is also known as Alizarin Red S.

BROMOCRESOL GREEN—METHYL RED[94,109,128,194,226]

Five parts of 0.2% bromocresol green and 1 part of 0.2% methyl red, both in 95% ethanol, are mixed and stored in a dropping bottle.

METHYL RED—METHYLENE BLUE[36,149,190,194,225,226]

Two parts of 0.2% methyl red and 1 part of 0.2% methylene blue, both in 95% ethanol are mixed and stored in a dropping bottle.

METHYL RED[144,145,194]

In a flask are placed 0.15 gram of methyl red powder and 40 ml. of 0.01N sodium hydroxide solution. The mixture is shaken and then filtered and the filtrate stored in a glass-stoppered dropping bottle.

* A 0.2% aqueous solution of amylose,[188] was found to be an excellent indicator by Aluise, Hall, Staats and Becker.[8]

5. Standard Solutions

PHENOLPHTHALEIN[144,145,194]

A 1% solution of phenolphthalein in 95% ethanol is prepared.

TETRAHYDROXYQUINONE (THQ)[4,25,87,150,194,198,210,212]

This is kept in the solid state and used as such. It is used together with the glass filter listed below.

Apparatus

BURETTES

Automatic filling microburettes of the type shown in Figs. 69 and 70, are preferred for storing and titrating standard solutions, although storage in ordinary bottles or volumetric flasks and titrating from the hand-filled type of burette is permissible. The automatic burette[83,84,144,145,168-171,194] shown in Fig. 69 has a storage reservoir of 0.5–1.0 liter. The solution is forced up into the top of the burette when the pressure in the reservoir is increased by pumping the hand-operated bulb. The level of the solution is automatically adjusted to coincide with the 0.00-ml. mark since the delivery tube is at this mark and the tube is so constructed that it acts as a siphon to remove the excess. The flow of solution from the burette is controlled by means of a glass bead inside the rubber tube, a pinch clamp, or a ground glass stopcock. The burettes are usually 10-ml. capacity, graduated to 0.05-ml. The solution in the burette and in the reservoir is protected from carbon dioxide of the air by the Ascarite tubes at the top and between the reservoir and bulb, respectively. The stopcock between the reservoir and the bulb should be closed at all times except when the solution is being pumped up into the burette. Solution which has remained in the burette for more than an hour or two should be discarded. Obviously, the reservoir should be shaken each day to mix any condensate that occurs on standing and which alters the normality of the solution, unless mixed in.

The type[106,194] of burette shown in Fig. 70 is available in sizes of 1, 2, 5, and 10 ml. This type is graduated in 0.01 or 0.05 ml. and has its reservoir at the top. Manipulation of the three-way stopcock allows the solution to flow from the reservoir into the burette. The capacities of the reservoirs are usually approximately 100 ml.

ILLUMINATED TITRATION STAND ASSEMBLY[78,150,194,198]

The titration stand shown in Fig. 71 is used for standardizing $BaCl_2$ solution using tetrahydroxyquinone (THQ) indicator together with the orange-brown glass filter plate described below. The stand is illuminated from below by means of fluorescent lighting. The base is a frosted glass plate. This should be covered with the masking plate or with black paper which has an oblong section cut from it, just large enough for the filter plate (see below) and cuvette (Fig. 72) to set in, side by side. The cuvette which is prepared from

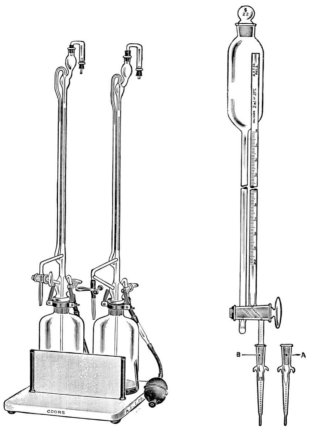

FIG. 69. (*Left*) Automatic burettes.

FIG. 70. (*Right*) Automatic burette with stoppered reservoir and two interchangeable ground tips; for calibration (A) and for delivery of finest drops in tests (B).

glass has the dimensions, 20 × 40 × 60 mm. high, and has graduation marks indicating the 15- and 30-ml. capacities.

ORANGE-BROWN GLASS FILTER PLATE[46,150,194,198,212]

Glass color filter, approximately 2.5 × 26 × 45 mm. having a spectral transmission, uncorrected for reflectance, of 37% at 550 mmµ ± 2 mmµ is required.

5. Standard Solutions

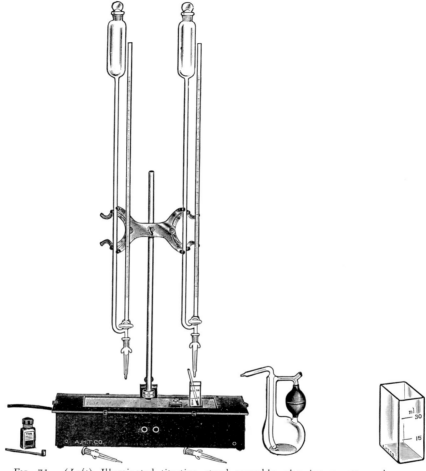

FIG. 71. (*Left*) Illuminated titration stand assembly, showing cuvette and orange-brown filter plate in place.

FIG. 72. (*Far right*) Cuvette.

pH METER

Zeromatic pH meter, No. 9600[16] or comparable instrument is used in the adjustment of the pH of solutions, such as in the standardization of thorium nitrate.

PHOTOELECTRIC FILTER PHOTOMETER[127,132,137,199]

This instrument is of particular value in the titration of fluoride with thorium nitrate using sodium alizarin sulfonate as the indicator, although it undoubtedly can be found to be useful in other titrations in which the color change at the end point is gradual. It also makes possible the handling of larger volumes of solutions of low concentrations. The instrument has a small lamp in the center with a Meyer Trioplan lens (3 inches, f 2.8, provided with adjustable iris diaphragm) and green filter (maximum transmittance, 520 mμ) *on each side*. The light from the lamp, *on each side,* passes through the lens and filter, then through a titrating cell containing solution, and finally impinges onto a

FIG. 73. Photoelectric filter photometer.

photocell (Weston photronic photocell, model 594, Weston Instruments Division of Daystrom Inc., Newark, New Jersey). The two photocells are connected in opposition, a mirror galvanometer being mounted in parallel. The latter is shunted (by Ayrton shunt, Fisher Scientific Company, New York, and Pittsburgh) which permits stepwise variation of the sensitivity over a range. Provision is made for magnetic stirring and titrating. For direct titration, a blank solution and the indicator are placed in the left side titrating cell while the solution to be titrated and the indicator are placed in the right side titrating cell or vice versa. The various stages of operation of the instrument are described below in the standardization of thorium nitrate. Figure 73 shows a commercially available instrument, the type used in the author's laboratories (photoelectric filter photometer, model PFP-1, Marley Products Co., Hyde Park, New York).[132] It is completely encased. The switch on the extreme left controls the galvanometer light, light source and magnetic stirrers. The next control, the second from the left, controls the speed of the magnetic stirrers. The third switch from the left shorts out the galvanometer and photocells when not in use. The switch on the extreme right is a three-stage shunt for varying the sensitivity. At the top are the galvanometer knob for zero adjusting and two diaphragm adjusting wheels which are used to balance the light source to the photocells. Two clear plastic dishes (ordinary refrigerator dishes) are used as titrating cells.

Procedures

It should be remembered that the detecting of an end point is a personal matter, which is more exaggerated when working with one hundredth normal solutions. Therefore, the analyst who is going to perform the particular volumetric procedure should also standardize the solutions, or at least make certain that he sees the same end point as the one who does the standardization.

SODIUM THIOSULFATE, 0.01N

A liter of distilled water is brought to a boil to remove carbon dioxide and cooled while loosely covered. Sodium thiosulfate, $Na_2S_2O_3 \cdot 5H_2O$ (2.48 grams) is dissolved in the water and the solution made up to 1000 ml. It should be transferred to a brown bottle equipped with a rubber stopper. One ml. of chloroform is added and the contents of the bottle shaken for a few minutes. The chloroform acts as a preservative.[93,194] The use of a rubber stopper over a ground glass one is preferred as the former absorbs chloroform, increasing the efficiency of the preservative. This solution is quite stable as long as a pool of chloroform is present, but should be standardized at intervals of about one month. In the absence of chloroform, standardization should be done every few days.[74] Instead of the rubber-stoppered brown bottle, an automatic burette of the type shown in Fig. 69 may be used. If so, it is preferable to use a brown bottle, but if clear glass is used, the solution should be standardized more frequently. Obviously, chloroform must be present at all times, regardless of the setup, or standardization must be done daily.

Standardization:

About 2–3 mg. of solid potassium biiodate is accurately weighed (using a porcelain boat or charging tube—see Chapter 3), transferred to a 125-ml. glass-stoppered Erlenmeyer flask, and dissolved in 5 ml. distilled water. To this is added 1.5 ml. concentrated hydrochloric acid, followed by 1 ml. of *freshly prepared* 4% solution of potassium iodide, and the flask stoppered. After 2 minutes, the solution is diluted with water to 20 ml. and the liberated iodine is titrated with the sodium thiosulfate solution. When the iodine color has almost disappeared (yellow) several drops of the starch solution is added. Thiosulfate is added until the blue color has been converted to a faint pink at most (end point). If it is easier for the analyst to detect the end point by titrating to a colorless solution, such action is permissible, but being consistent is an absolute necessity.

Calculation:

From the equations shown at the beginning of the chapter,

$$\begin{array}{ccccc} 1KH(IO_3)_2 & & 12\,I & & 12Na_2S_2O_3 \\ M.W.\ 389.928 & \rightleftharpoons & M.W.\ 1522.92 & \rightleftharpoons & M.W.\ 1897.368 \end{array}$$

$$\therefore \quad \frac{389.928}{mg.\ KH(IO_3)_2} = \frac{1897.368}{mg.\ Na_2S_2O_3}$$

Now 1 ml. of N $Na_2S_2O_3$ contains 158.114 mg. of $Na_2S_2O_3$

$$\therefore \quad \frac{mg.\ Na_2S_2O_3}{ml.\ Na_2S_2O_3 \times 158.114} = N_{Na_2S_2O_3}$$

$$\therefore \quad \frac{1897.368 \times mg.\ KH(IO_3)_2}{389.928 \times ml.\ Na_2S_2O_3 \times 158.114} = N_{Na_2S_2O_3}$$

$$\therefore \quad \frac{mg.\ KH(IO_3)_2 \times 0.03077}{ml.\ Na_2S_2O_3} = N_{Na_2S_2O_3}$$

IODINE, 0.01N

Approximately 1.27 grams of resublimed iodine crystals are dissolved in a solution of 4 grams of potassium iodide in 4 ml. of water. (Note: Iodine dissolves rather rapidly in a concentrated solution of potassium iodide, but very slowly in a dilute solution.) After about 20–30 minutes (first making certain that solution is complete) the concentrated solution is transferred to a 1-liter volumetric flask and diluted to the mark, with freshly boiled distilled water.

Standardization:

The resulting solution is standardized by titration against the 0.01N thiosulfate (prepared above) using starch indicator.

This solution is not stable and should be standardized daily.

5. Standard Solutions

POTASSIUM SULFATE, 0.01N

Exactly 0.8713 gram of reagent grade potassium sulfate is placed in a 1-liter volumetric flask, dissolved in a little distilled water and the resulting solution diluted to the mark, with distilled water.

BARIUM CHLORIDE, 0.01N

Approximately 1.04 grams of barium chloride is dissolved in water and made up to 1 liter. The solution is standardized against the potassium sulfate, prepared above, using the illuminated titration stand, orange-brown filter plate, cuvette (see above), and tetrahydroxyquinone indicator as follows: Three to 5*[150,194,198] ml. of standard potassium sulfate, 0.01N, is placed in the cuvette, one or two drops of phenolphthalein indicator solution added, and enough 0.1N sodium hydroxide to render the solution alkaline after which it is back-titrated with 0.01N hydrochloric acid just to expel the color. (The titration using THQ indicator must be done in a fairly neutral solution. Although the potassium sulfate solution is already neutral, the above procedure of making alkaline and back-titrating is recommended since this technique is used in the actual titration of sulfate in the determination of sulfur—see Chapter 10—and therefore the $BaCl_2$ solution is standardized under the identical conditions as when used.) Enough water is added to bring the volume to 15 ml. and this is followed by adding 15 ml. of 95% ethanol. One-half scoop (provided with the indicator) of dry THQ indicator is added and dissolved by stirring with a glass rod protected on the end by a rubber sleeve (or policeman) to prevent scratching of the cuvette. The container is then placed on the illuminated titration stand alongside the orange-brown filter plate so that these are illuminated from below and the rest of the stand is blacked out to avoid interference with the visual comparison of the solution with the filter plate. The barium chloride solution is added to the yellow contents of the cuvette, with stirring, until the color gradually shifts towards the red and matches that of the filter plate when viewed from the top, looking through the entire depth of solution.† This is taken as the end point and is very sharp, an additional drop changing the color markedly more toward the red. If the end point has been overstepped the solution cannot be back-titrated. The same solution will have a different normality factor for each filter plate used since the titration would be to a slightly different end point for each. However, this is of no consequence as long as the factor for the particular filter is determined.

* The amount is regulated so that the appearance of solution plus $BaSO_4$ at the end point is about the same as the filter glass. Considerably more or less $BaSO_4$ present does not give a mixture comparable in appearance to the filter.

† Note: Barium salt of indicator is purple-red.[189]

SODIUM HYDROXIDE, 0.01N

This solution may be obtained, commercially,[1] and as such is quite satisfactory. Otherwise, it is prepared from carbonate-free sodium hydroxide as follows: A 50% solution (about 20N)[90,219] of reagent grade of sodium hydroxide is prepared and stored in a stoppered paraffin-lined bottle until the carbonate has settled out, which usually takes some weeks. This is then decanted and diluted with freshly boiled distilled water to about 0.01N and standardized as follows.[72] About 12 to 14 mg. of reference-standard potassium acid phthalate, $KHC_8H_4O_4$, is weighed into a 125-ml. Pyrex Erlenmeyer flask. This is dissolved in about 5 ml. of distilled water, a few drops of phenolphthalein indicator added, and the mixture boiled for about 30 seconds to drive out dissolved carbon dioxide. The hot solution is titrated with the above-prepared sodium hydroxide to a faint pink end point.

$$KHC_8H_4O_4 + NaOH \rightarrow KNaC_8H_4O_4 + H_2O$$

Calculation:

$$\begin{array}{cc} 1NaOH & 1KHC_8H_4O_4 \\ M.W.\ 39.999 & M.W.\ 204.228 \end{array}$$

$$\therefore \frac{mg.\ NaOH}{39.999} = \frac{mg.\ KHC_8H_4O_4}{204.228}$$

or

$$mg.\ NaOH = \frac{mg.\ KHC_8H_4O_4 \times 39.999}{204.228}$$

Now 1 ml. of N NaOH contains 39.999 mg. of NaOH

$$\therefore \frac{mg.\ NaOH}{ml.\ NaOH \times 39.999} = \text{Normality of NaOH}$$

$$\therefore \frac{mg.\ KHC_8H_4O_4 \times 39.999}{204.228 \times ml.\ of\ NaOH \times 39.999} = N_{NaOH}$$

$$\therefore \frac{mg.\ KHC_8H_4O_4}{204.228 \times ml.\ of\ NaOH} = N_{NaOH}$$

Similarly, potassium biiodate[145,194] or benzoic acid[72,194] may be used. For the latter, neutral 95% ethanol (see Chapter 15) is employed as the solvent. For both of these acids, one molecule of acid is equivalent to one molecule of sodium hydroxide:

5. Standard Solutions

$$\begin{array}{cc} 1\text{NaOH} & 1\text{KH}(\text{IO}_3)_2 \\ \text{M.W. } 39.999 & \text{M.W. } 389.928 \end{array}$$

$$\begin{array}{cc} 1\text{NaOH} & 1\text{C}_6\text{H}_5\text{COOH} \\ \text{M.W. } 39.999 & \text{M.W. } 122.125 \end{array}$$

$$\therefore \quad \frac{\text{mg. KH}(\text{IO}_3)_2}{389.928 \times \text{ml. of NaOH}} = N_{\text{NaOH}}$$

and

$$\frac{\text{mg. C}_6\text{H}_5\text{COOH}}{122.125 \times \text{ml. of NaOH}} = N_{\text{NaOH}}$$

HYDROCHLORIC ACID, 0.01N

This solution may be obtained commercially[1] or it may be prepared from more concentrated solution, diluting with freshly boiled distilled water and standardized against standard, 0.01N sodium hydroxide solution, using phenolphthalein as the indicator. Just before the end point is reached, the mixture is boiled for at least 30 seconds to expel carbon dioxide. (Note: If the acid is titrated with the alkali, at the end point, boiling increases the color, so that the end point should be approached only with a hot, well-boiled mixture.)

SODIUM FLUORIDE, 0.01N[127,137,199]

Exactly 0.4200 gram of reagent grade sodium fluoride is placed in a one-liter volumetric flask, dissolved in a small amount of distilled water and the resulting solution diluted to the mark, with distilled water.

THORIUM NITRATE, 0.01N[127,137,199]

Approximately 1.38 grams of thorium nitrate tetrahydrate, $\text{Th}(\text{NO}_3)_4 \cdot 4\text{ H}_2\text{O}$, are dissolved in water and the resulting solution diluted to one liter. This is standardized against the sodium fluoride, prepared above, using sodium alizarin sulfonate as the indicator with the aid of the photoelectric filter photometer described above. (Before performing actual titrations, the proper sensitivity for the instrument must be selected. This is accomplished by doing a blank determination and determining which sensitivity setting gives a deflection of 25 *units* upon the addition of 0.05 ml. of the thorium nitrate when the iris openings are the *same* as during the actual determination, preferably as large as possible.)[132]

One-, 2-, 3-, 4-, 5-, 6-, 7-, 8-, and 10-ml. portions of 0.01N sodium fluoride are titrated with the thorium nitrate so that a plot may be made of ml. of thorium nitrate vs. mg. of fluorine. *Each* of the above-mentioned portions of sodium fluoride are treated in the following manner: The fluoride is transferred to the clear plastic titrating cell and diluted with distilled water to a volume of 450 ml. This solution is adjusted to pH 3.0±0.05 (using 0.1N hydrochloric acid), and 2 ml. of 0.035% sodium alizarin sulfonate indicator is added. A duplicate clear plastic titrating cell containing 450 ml. of distilled water adjusted to pH 3.0±0.05 and 2 ml. of 0.035% sodium alizarin sulfonate indicator is placed in one compartment of the photoelectric filter photometer and the cell containing the fluoride solution in the other. (*Note:* Since sodium alizarin sulfonate is also an acid-base indicator, *all* pH adjustments are most *critical.*) With the *third switch from the left* in the position that both the galvanometer and photocells are *shorted out* (short) and the *lights* and *stirrers on,* the galvanometer is adjusted to the zero mark by rotating the galvanometer adjusting knob at the top of the instrument. (*Note:* The *lights* should be turned on in advance of the time that the instrument is to be used.) The *third switch from the left* is then turned to the "on" position to connect the galvanometer and photocells. The *switch on the extreme right* should be turned to the position of predetermined sensitivity (maximum preferred). The galvanometer is again adjusted to the zero mark *this time by means of the diaphragm adjusting wheels* (usually only one—that on the side of the fluoride solution). The fluoride is then titrated with the thorium nitrate until a deflection of 25 *units*[127,137,199] on the galvanometer is obtained. (*Note:* The color change is from a yellow to a pink.) After a number of different size portions of sodium fluoride have been titrated, a curve is plotted showing the relationship between milliliters of thorium nitrate and milligrams of fluorine titrated. This is preferable to labeling the fluoride as "so-much" normal, since a straight-line function does not exist except possibly over a *small* range of amount of fluorine and then only under certain limited conditions.

ADDITIONAL INFORMATION FOR CHAPTER 5

Although the volumetric procedures described in this chapter as well as throughout this entire book call for the use of burettes of the types shown in Figs. 69, and 70, the analyst encounters problems and procedures elsewhere calling for the use of other volumetric glassware. For the sake of completeness, pieces are shown here for which recommended specifications have been published by the Committee on Microchemical Apparatus of the Division of Analytical Chemistry of the American Chemical Society.[195-197] These include microvolumetric flasks (Fig. 74), pipettes (Fig. 75) to be used with the flasks, and a centrifuge tube (Fig. 61, Chapter 4).

5. Standard Solutions

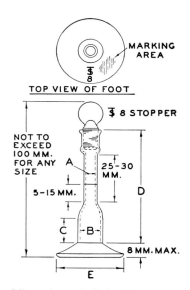

Fig. 74. Microvolumetric flask—details of construction.

Capacity (ml.)	A Inside diameter (mm.)	B Inside diameter (mm.)	C (Approx.) (mm.)	D (Maximum) (mm.)	E (Maximum) (mm.)	Tolerance (ml.)
1	4.2–4.6	8.0–8.5	10	70	37	±0.010
2	5.0–5.4	10.5–11.0	13	70	39	±0.015
3	5.0–5.4	13.25–13.75	14	72	39	±0.015
4	6.2–6.6	13.75–14.25	18	75	39	±0.020
5	6.2–6.6	15.5–16.0	18	75	39	±0.020

To be marked "T.C. (capacity) 20° C."

1-ml. size to weigh less than 19 grams empty (stopper included).

Shape of bases may be either round or hexagonal. Dimensions given in column E are maximum permitted for distance between parallel sides of hexagonal bases and are maximum diameters of round bases.

Additional Information for Chapter 5

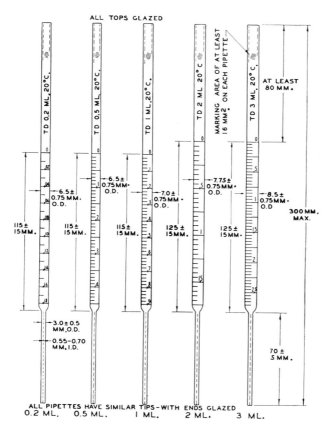

FIG. 75. Micropipette with cylindrical tip—details of construction.

Capacity (ml.)	Subdivision (ml.)	Interval graduated (ml.)	Lining		Number at 0 and each ml.	Tolerance (ml.)
			Ring at each ml.	½ ring at each ml.		
0.2	0.01	0 to 0.18	0.02	0.01	0.02	±0.005
0.5	0.01	0 to 0.45	0.05	0.01	0.1	±0.01
1	0.02	0 to 0.90	0.1	0.02	0.1	±0.02
2	0.05	0 to 1.75	0.25	0.05	0.5	±0.04
3	0.05	0 to 2.70	0.25	0.05	0.5	±0.06

No graduations to appear in tapered portion.
Tip may be tapered at junction with body, but outside diameter at this point may not exceed 4.5 mm.
Tip outlet to be glazed, with least possible constriction.
Calibrated to deliver at 20° C. touching off last drop.

5. Standard Solutions

The volumetric flasks[195] are of a new design which combines conveniency in use with accuracy. The wide base affords greater stability against upset (Fig. 74). In the case of the one-ml. size, the diameter of the base is small enough to permit placement on the microchemical balance pan and the weight is restricted to a maximum of *19 grams* when empty so that when full of a solution of specific gravity of one or less, the capacity of the balance is not exceeded.

In order to increase the usefulness of the above flasks, special measuring pipettes (Fig. 75) were designed.[195] The long narrow delivery stems of the pipettes with cylindrical tip (0.2-, 0.5-, 1-, 2-, and 3-ml. sizes, Fig. 75) reach to the bottom of the flasks, permitting almost complete withdrawal of the contents. Actually, all but a few hundredths of a milliliter can be withdrawn.

The microliter pipettes[196] shown in Figs. 76, and 77 also have dimensions such that they too may be used with the microvolumetric flasks. These types, sizes, and dimensions were recommended after a study was made of the returns from a questionnaire sent out to interested parties. Returns were received from 56 individuals, 48 of whom were users of this type of pipette, and a number of whom have since tested the finished items and found them to be quite satisfactory. The following procedure is recommended for calibration of these microliter pipettes[196]:

"Standard procedure for micropipette calibration consists of filling the pipette with mercury, discharging the mercury into a porcelain dish, weighing the mercury, and making the appropriate weight-temperature-volume calculation. This method has been compared with that of weighing the pipette both empty and mercury filled, and has been found less difficult and equally precise." (*Note:* For calibration purposes where mercury is used, the edge of the mercury meniscus should coincide with the top of the line.)

Micro washout pipettes[197] of the types shown in Figs. 78 and 79, respectively, are used extensively. They are designated as micropipettes instead of as microliter pipettes because of their particular applications and not because of their range. With both, adhering material is transferred by means of wash liquid, being sucked up through the tip in the case of the Folin-type[65] and added to the cuplike top of the other.[84]

The density-type pipettes[2,197] described in Chapter 22, Fig. 217, will also prove useful, particularly because they have ground tips and caps.

Self-filling, self-adjusting polyethylene type of measuring units developed by Sanz are commercially available.[17]

As an aid to pipetting, hypodermic syringes are attached to microliter pipettes by means of plastic tubing. This affords a means of better control.

The burette[9] shown in Fig. 80 may be used as either a micro- or ultramicroburette. It has a 3.5-ml. reservoir through which passes a power driven

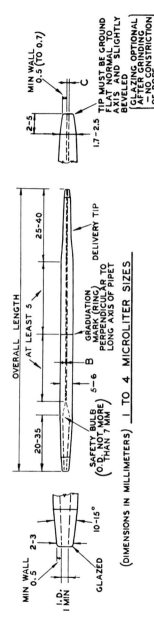

FIG. 76. Microliter pipette for 1- to 4-μl. sizes—details of construction.

Size (Calibd. to contain) (μl.)	Over-all length (mm.)	I.D. tubing (mm.) B	I.D. at end (mm.) C	Min. cap. safety bulb (μl.)	Vol. tol.* (%)
1	140 ± 5	0.12–0.16	0.10–0.20	50	±1
2	140 ± 5	0.16–0.25	0.15–0.25	50	±1
3	140 ± 5	0.20–0.28	0.15–0.25	50	±1
4	140 ± 5	0.24–0.32	0.15–0.25	50	±1

* Closer volumetric standardization must be carried out by user with substances under actual conditions of use.

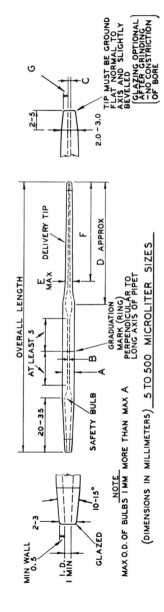

FIG. 77. Microliter pipette for 5- to 500-μl. sizes—details of construction. For dimensions see next page.

FIG. 77. (cont.) Microliter pipette for 5- to 500-μl. sizes—dimensions.

Size (Calibd. to contain) (μl.)	Over-all length (mm.)	O.D. tubing (mm.) A	I.D. tubing (mm.) B	Approx. length (mm.) D	Max. O.D. (mm.) E	Delivery Stem Min. length (mm.) F	Wall at end (mm.) G	I.D. at end (mm.) C	Min. cap. safety bulb (μl.)	Vol. tol.* (%)
5	140 ± 5	5–6	0.18–0.25	65	4	55	0.5–0.7	0.15–0.25	50	±0.5
6	140 ± 5	5–6	0.18–0.25	65	4	55	0.5–0.7	0.15–0.25	50	±0.5
7	140 ± 5	5–6	0.18–0.25	65	4	55	0.5–0.7	0.15–0.25	50	±0.5
8	140 ± 5	5–6	0.18–0.25	65	4	55	0.5–0.7	0.15–0.25	50	±0.5
9	140 ± 5	5–6	0.18–0.25	65	4	55	0.5–0.7	0.15–0.25	50	±0.5
10	140 ± 5	5–6	0.20–0.35	65	4	55	0.5–0.7	0.15–0.25	50	±0.5
15	140 ± 5	5–6	0.25–0.40	65	4	55	0.5–0.7	0.15–0.25	50	±0.5
20	140 ± 5	5–6	0.35–0.50	65	4	55	0.5–0.7	0.25–0.50	50	±0.5
25	140 ± 5	5–6	0.35–0.50	65	4	55	0.5–0.7	0.25–0.50	50	±0.3
35	140 ± 5	5–6	0.35–0.50	65	4	55	0.5–0.7	0.25–0.50	50	±0.3
50	140 ± 5	5–6	0.40–0.55	65	4	55	0.5–0.7	0.30–0.50	50	±0.3
60	140 ± 5	5–6	0.40–0.60	65	4	55	0.5–0.7	0.30–0.50	50	±0.3
75	140 ± 5	5–6	0.50–0.75	65	4	55	0.5–0.7	0.30–0.50	75	±0.3
100	140 ± 5	5–6	0.75–1.00	65	4	55	0.5–0.7	0.40–0.60	75	±0.3
150	140 ± 5	5–6	0.75–1.00	65	4	55	0.5–0.7	0.40–0.60	100	±0.2
200	145 ± 10	5–6	0.75–1.00	65	4	55	0.6–0.8	0.40–0.60	100	±0.2
250	145 ± 10	5–6	0.75–1.00	65	4	55	0.6–0.8	0.40–0.60	100	±0.2
300	145 ± 10	5–6	0.75–1.00	65	4	55	0.6–0.8	0.40–0.70	200	±0.2
400	150 ± 10	6–7	1.00–1.25	70	6	60	0.6–0.8	0.40–0.70	200	±0.2
500	160 ± 10	6–7	1.25–1.50	70	6	60	0.6–0.8	0.40–0.70	200	±0.2

* Closer volumetric standardization must be carried out by user with substances under actual conditions of use.

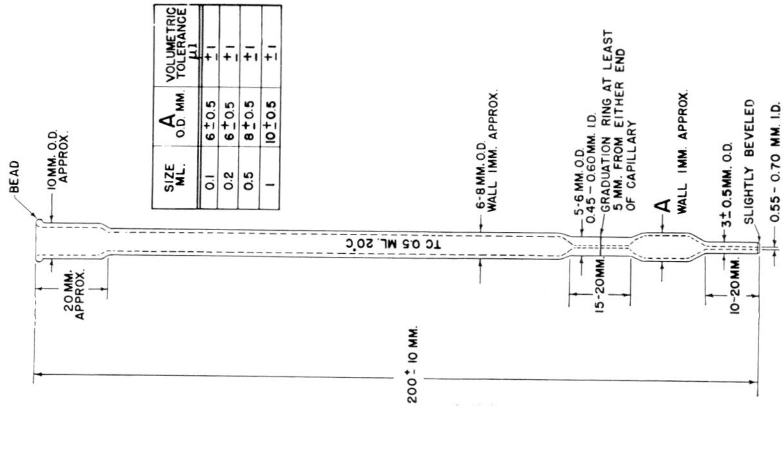

5. Standard Solutions

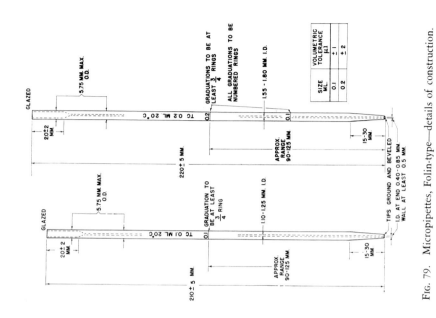

Fig. 79. Micropipettes, Folin-type—details of construction.

Fig. 78. Micro washout pipettes—details of construction.

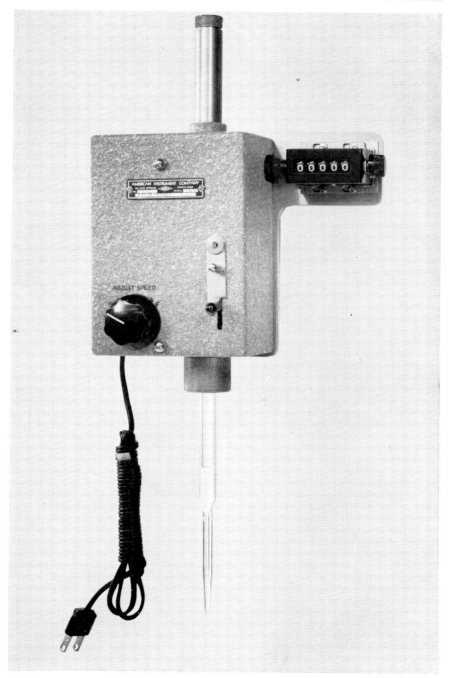

Fig. 80. Aminco automatic power-driven burette.

5. Standard Solutions

Vycor* plunger. A revolution counter is so calibrated that it registers to the nearest 0.001 ml. and a graduated drum mounted on the input shaft to the counter registers 0.0001 ml. per division. Meniscus reading has been eliminated. The delivery type is immersed in the solution to be titrated.

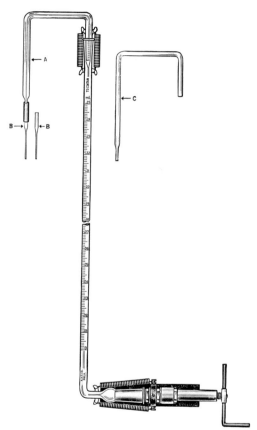

FIG. 81. Rehberg burette. (A) Separate delivery tube with ground joint to facilitate cleaning. (B) Removable glass tip. (C) Gas bubbling tube for insertion in solution being titrated to provide stirring action.

Besides the types of burettes described and shown in the preceding pages, many others have been described, particularly for measuring smaller quantities of solutions. Figures 81 and 82 show two of these capillary burettes[120,163]

* Silica glass (96%) No. 790.[46]

which require micrometer screw type of manipulation. For descriptions of these, the reader is referred to the book by Kirk.[98]

A steaming apparatus of the type shown in Fig. 83 is useful for cleaning flasks in connection with volumetric procedures.[145-147]

For complete information regarding titration curves, equilibrium for acid and for base dissociation, pH, buffer action, the theory and choice of indicators, electrometric titrations, etc., the reader is referred to the treatise by Clark.[44]

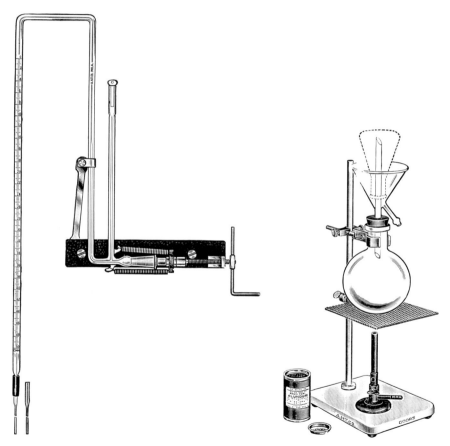

FIG. 82. (*Left*) Linderstrøm-Lang and Holter burette.

FIG. 83. (*Right*) Steaming apparatus, showing method of cleaning flask by flushing with steam.

5. Standard Solutions

TABLE 15

ADDITIONAL INFORMATION ON REFERENCES* RELATED TO CHAPTER 5

For other information the author wishes to call to the attention of the reader, a number of references are given in this table. (See author's statement regarding such information at top of Table 4 of Chapter 1.)

Books
Belcher and Godbert, 22, 23
Clark, E. P., 42
Clark, S. J., 43
Clark, W. M., 44
Grant, 83, 84
Kirk, 98
Knights, McDonald, and Ploompuu, 105
Malissa and Benedetti-Pichler, 130
Milton and Waters, 140, 141
Natelson, 142
Niederl and Niederl, 144, 145
Niederl and Sozzi, 147
Pregl, 161
Roth, 168–171
Schneider, 174
Steyermark, 194

Potentiometric titrations
Alimarin and Petrikova, 5
Cunningham, Kirk, and Brooks, 48
Frediani and Warren, 67
Furman, 70–72
Goddu and Hume, 77
Ingold, 91
Kirk, 98
Kirk and Bentley, 99
Kirsten, Berggren, and Nilsson, 102
Lévy, 119
Lindner and Kirk, 122
Lykken and Rolfson, 126
Schwarz, 180
Simon, 184
Stock, 200–202
Tompkins and Kirk, 214
Wade, 221

Heterometric titration
Bobtelsky and Welwart, 30–33

Automatic titration
Cotlove, Trantham, and Bowman, 47
Simon, 185
Simon, Kováks, Chopard-dit-Jean, and Heilbronner, 186

Conductometric titration
Stock, 203

EDTA titration
Baker, 13
Belcher, Close, and West, 20, 21
Flaschka, 56–59
Flaschka and Abdine, 60
Flaschka and Amin, 61
Flaschka, Barnard, and Broad, 62, 63
Flaschka, ter Haar, and Bazen, 64
Kinnunen and Wennerstrand, 95
Lott and Cheng, 124
Sporek, 192

Amperometric titrations
Bradbury and Hambly, 35
Mader and Frediani, 129
Nikelly and Cooke, 148
Parks, 151
Stock, 204

High-frequency titration
Bien, 26
Blaedel and Malmstadt, 28
Hara and West, 88
Lane, 113
Reilley and McCurdy, 164

Cerimetry
Takahashi, Kimoto, and Sakurai, 211

Nonaqueous titrations
Ballard, Bamford, and Weymouth, 14
Belcher, Berger, and West, 18, 19

* The numbers which appear after each entry in this table refer to the literature citations in the reference list at the end of the chapter.

TABLE 15 (*Continued*)

Nonaqueous titrations (Cont.)
Fritz, 68
Maurmeyer, Margosis, and Ma, 136
Pifer and Wollish, 156, 157
Pifer, Wollish, and Schmall, 158–160
Riddick, 166
Wollish, Colarusso, Pifer, and Schmall, 230
Wollish, Pifer, and Schmall, 231

Spectrometric, colorimetric, etc., equipment
Ellis and Brandt, 54
Kinsey, 96
Mason, 134

Microtitrators
Allen, 7

Burettes, ultramicrotitrations, etc.
Belcher, Berger, and West, 18, 19
Blake, 29
Daimler, 49
Gilmont, 75, 76
Wingo and Johnson, 227
Wise, 229

Burettes, pipettes, general
Alber, 2
Allan, 6
Anderson, 10, 11
Barth and Sze, 15
Beckman Instruments, Inc., 16, 17
Benedetti-Pichler, 24
Birket-Smith, 27
Boguth, 34
British Standards Institution, 37
Buck, Keister, and Zelle, 38
Chinoy, 39
Claff, 40, 41
Dern, Pullman, and Williams, 50
Düsing, 52
Feuer, 55
Flaschka, 56
Folin, 65
Foster and Broeker, 66
Gilmont, 75
Gorbach, 79–81
Gorbach and Haack, 82

Burettes, pipettes, general (Cont.)
Grunbaum and Kirk, 85
Hallett, 86
Harman and Webster, 89
Kirk, 97, 98
Kirk and Bentley, 99
Kirsten, 100, 101
Knights, McDonald, and Ploompuu, 104, 105
Korenman and Rostokin, 107
Koros and Remport-Horvath, 108
Lacourt, 110
Lacourt, Stoffyn, and Timmermans, 111
Lacourt and Timmermans, 112
Langer and Pantages, 114
Lascalzo and Benedetti-Pichler, 115
Lazarow, 116, 117
Levvy, 118
Linderstrøm-Lang and Holter, 120
Llacer and Sozzi, 123
Lundbak, 125
Marsh, 133
Mattenheimer and Borner, 135
Natelson and Zuckerman, 143
Rehberg, 163
Rieman, 167
Saunders, 173
Scholander, 175
Scholander, Edwards, and Irving, 176
Schöniger, 177
Schreiner, 178
Shaffer, Farrington, and Niemann, 182
Sômiya and Kamada, 191
Stock and Fill, 205–209
Thompson, 213
Upson, 217
Willits, 224
Winteringham, 228
Wise, 229
Wyatt, 232

Stirring devices
Claff, 40, 41
Stock and Fill, 205

Calibration
Alber, 2
Buck, Keister, and Zelle, 38

5. Standard Solutions

TABLE 15 (*Continued*)

Calibration (Cont.)
Pećar, 154
Thompson, 213
Upson, 218

Flasks
Thompson, 213

Indicators
Bradbury and Hambly, 35
Cooper, 45
Dorf, 51
Jerie, 92
Kinnunen and Wennerstrand, 95
Manohin, Kakabadse, and Crowder, 131
Preisler and Berger, 162
Schulze, 179
Sendroy and Hastings, 181
Sher, 183
Steinitz, 193

Standard solutions
Underwood, 216

Primary standards
Williams, 223

General, applications, etc.
Albrink, 3
Garschagen, 73

General, applications, etc. (Cont.)
Knights, 103
Lazarow, 116, 117
Pećar, 153, 154
Richter, 165
Simon, Morikofer, and Heilbronner, 187

Titration of sulfate
Fritz and Yamamura, 69
Milner, 139
Smith-New York, 189

Standardization of sodium thiosulfate
Kassner and Kassner, 93
Niederl and Niederl, 144, 145
Peacocke, 152
Pharmacopeia, 155
Walker and Allen, 222

Standardization of hydrochloric acid
Lindner, 121
Niederl and Niederl, 144, 145

Fluoride titration
Ma and Gwirtsman, 127
Mavrodineanu and Gwirtsman, 137
Megregian, 138
Rowley, 172
Venkateswarlu, Ramanthan, and Narayana Rao, 220

REFERENCES

1. Acculate, Anachemia Chemicals, Ltd., Montreal, Quebec and Champlain, New York.
2. Alber, H. K., *Ind. Eng. Chem., Anal. Ed.,* **12**, 764 (1940).
3. Albrink, M. J., *J. Lipid Research,* **1**, 53 (1959).
4. Alicino, J. F., *Anal. Chem.,* **20**, 85 (1948).
5. Alimarin, I. P., and Petrikova, M. N., *J. Anal. Chem., U.S.S.R., (English Translation),* **9**, 137 (1954).
6. Allan, J. C., *S. African J. Med. Sci.,* **11**, 157 (1946).
7. Allen, K. A., *Anal. Chem.,* **28**, 277 (1956).
8. Aluise, V. A., Hall, R. T., Staats, F. C., and Becker, W. W., *Anal. Chem.,* **19**, 347 (1947).
9. American Instrument Co., Silver Springs, Maryland.
10. Anderson, H. H., *Anal. Chem.,* **20**, 1241 (1948).
11. Anderson, H. H., *Anal. Chem.,* **24**, 579 (1952).
12. Association of Official Agricultural Chemists, "Official Methods of Analysis," 8th ed., Association of Official Agricultural Chemists, Washington, D. C., 1955.
13. Baker, J. T., Chemical Co., "The EDTA Titration. Nature and Methods of End Point Detection."
14. Ballard, D.G.H., Bamford, C. H., and Weymouth, F. J., *Analyst,* **81**, 305 (1956).
15. Barth, L. G., and Sze, L. C., *Rev. Sci. Instr.,* **22**, 978 (1951).
16. Beckman Instruments, Inc., Fullerton, California.
17. Beckman Instruments, Inc., Spinco Division, Palo Alto, California.
18. Belcher, R., Berger, J., and West, T. S., *J. Chem. Soc. (London),* p. 2877 (1959).
19. Belcher, R., Berger, J., and West, T. S., *J. Chem. Soc. (London),* p. 2882 (1959).
20. Belcher, R., Close, R. A., and West, T. S., *Chemist Analyst,* **46**, 86 (1957).
21. Belcher, R., Close, R. A., and West, T. S., *Chemist Analyst,* **47**, 2 (1958).
22. Belcher, R., and Godbert, A. L., "Semi-Micro Quantitative Organic Analysis," Longmans, Green, London, and New York, 1945.
23. Belcher, R., and Godbert, A. L., "Semi-micro Quantitative Organic Analysis," 2nd ed., Longmans, Green, London, 1954.
24. Benedetti-Pichler, A. A., "Introduction to the Microtechnique of Inorganic Analysis," p. 234, Wiley, New York, 1942.
25. Betz, W. H., & L. D., Philadelphia, Pennsylvania.
26. Bien, G. S., *Anal. Chem.,* **26**, 909 (1954).
27. Birket-Smith, E., *Scand. J. Clin. & Lab. Invest.,* **3**, 234 (1951).
28. Blaedel, W. J., and Malmstadt, H. V., *Anal. Chem.,* **22**, 734 (1950).
29. Blake, G. G., *Metallurgia,* **41**, 413, 415 (1950).
30. Bobtelsky, M., and Welwart, Y., *Anal. Chim. Acta,* **9**, 281 (1953).
31. Bobtelsky, M., and Welwart, Y., *Anal. Chim. Acta,* **9**, 374 (1953).
32. Bobtelsky, M., and Welwart, Y., *Anal. Chim. Acta,* **10**, 151 (1954).
33. Bobtelsky, M., and Welwart, Y., *Anal. Chim. Acta,* **10**, 464 (1954).
34. Boguth, W., *Z. physiol. Chem. Hoppe-Seyler's,* **285**, 93 (1950).
35. Bradbury, J. H., and Hambly, A. N., *Australian J. Sci. Research,* **5A**, 541 (1952).
36. Brecher, C., *Wien. klin. Wochschr.,* **49**, 1228 (1936).
37. British Standards Institution, *Brit. Standards* **1428**, Pt. D1 (1952), Pt. D2 (1950), Pt. D4 (1954), Pt. D5, and Pt. D6 (1955).
38. Buck, J. B., Keister, M. L., and Zelle, M. R., *Anal. Chim. Acta,* **4**, 130 (1950).
39. Chinoy, J. J., *Current Sci. (India),* **14**, 102 (1945).
40. Claff, C. L., *Science,* **105**, 103 (1947).

5. Standard Solutions

41. Claff, C. L., *Science*, **108**, 67 (1948).
42. Clark, E. P., "Semimicro Quantitative Organic Analysis," Academic Press, New York, 1943.
43. Clark, S. J., "Quantitative Methods of Organic Microanalysis," Butterworths, London, 1956.
44. Clark, W. M., "The Determination of Hydrogen Ions," 2nd ed., Williams & Wilkins, Baltimore, Maryland, 1927.
45. Cooper, S. S., *Ind. Eng. Chem., Anal. Ed.*, **13**, 466 (1941).
46. Corning Glass Works, Corning, New York.
47. Cotlove, E., Trantham, H. V., and Bowman, R. L., *J. Lab. Clin. Med.*, **51**, 461 (1958).
48. Cunningham, B., Kirk, P. L., and Brooks, S. C., *J. Biol. Chem.*, **139**, 11 (1941).
49. Daimler, B. H., *Chem. Ingr. Tech.*, **22**, 104 (1950).
50. Dern, R. J., Pullman, T. N., and Williams, H. R., *J. Lab. Clin. Med.*, **36**, 494 (1950).
51. Dorf, H., *Anal. Chem.*, **25**, 1000 (1953).
52. Düsing, W., *Chem. Fabrik*, p. 313 (1934).
53. Elek, A., and Harte, R. A., *Ind. Eng. Chem., Anal. Ed.*, **8**, 267 (1936).
54. Ellis, G. H., and Brandt, C. S., *Anal. Chem.*, **21**, 1546 (1949).
55. Feuer, I., *Nucleonics*, **13**, 68 (1955).
56. Flaschka, H., *Mikrochemie ver. Mikrochim. Acta*, **36/37**, 269 (1951).
57. Flaschka, H., *Mikrochemie ver. Mikrochim. Acta*, **40**, 21 (1953).
58. Flaschka, H., *Mikrochemie ver. Mikrochim. Acta*, **40**, 42 (1953).
59. Flaschka, H., *Mikrochim. Acta*, p. 55 (1955).
60. Flaschka, H., and Abdine, H., *Mikrochim. Acta*, p. 37 (1955).
61. Flaschka, H., and Amin, A. M., *Mikrochim. Acta*, p. 410 (1953); *Chem. Abstr.* **48**, 1198 (1954).
62. Flaschka, H., Barnard, A. J., and Broad, W. C., *Chemist Analyst*, **46**, 112 (1957).
63. Flaschka, H., Barnard, A. J., and Broad, W. C., *Chemist Analyst*, **47**, 22 (1958).
64. Flaschka, H., Haar, K. ter, and Bazen, J., *Mikrochim. Acta*, p. 345 (1953); *Chem. Abstr.*, **48**, 1198 (1954).
65. Folin, O., *J. Biol. Chem.*, **77**, 421 (1928).
66. Foster, R. H. K., and Broeker, A. G., *J. Lab. Clin. Med.*, **32**, 918 (1947).
67. Frediani, H. A., and Warren, W. B., *Ind. Eng. Chem., Anal. Ed.*, **13**, 646 (1941).
68. Fritz, J. S., "Acid-Base Titrations in Nonaqueous Solvents," G. F. Smith Chemical Co., Columbus, Ohio, 1952.
69. Fritz, J. S., and Yamamura, S. S., *Anal. Chem.*, **27**, 1461 (1955).
70. Furman, N. H., *Anal. Chem.*, **26**, 84 (1954).
71. Furman, N. H., *Ind. Eng. Chem., Anal. Ed.*, **14**, 367 (1942).
72. Furman, N. H., ed., "Scott's Standard Methods of Chemical Analysis," 5th ed., Vol. II, Van Nostrand, New York, 1939.
73. Garschagen, H., *Z. anal. Chem.*, **169**, 49 (1959).
74. Geilmann, W., and Höltje, R., *Z. anorg. Chem.*, **152**, 69 (1926).
75. Gilmont, R., *Anal. Chem.*, **20**, 1109 (1948).
76. Gilmont, R., *Anal. Chem.*, **25**, 1135 (1953).
77. Goddu, R. F., and Hume, D. N., *Anal. Chem.*, **26**, 1740 (1954).
78. Godfrey, P. R., and Shrewsbury, C. L., *J. Assoc. Offic. Agr. Chemists*, **28**, 336 (1945).
79. Gorbach, G., *Chem. Fabrik*, **14**, 390 (1941).

80. Gorbach, G., *Mikrochemie ver. Mikrochim. Acta,* **31**, 109 (1944).
81. Gorbach, G., *Mikrochemie ver. Mikrochim. Acta,* **34**, 183 (1949).
82. Gorbach, G., and Haack, A., *Mikrochim. Acta,* p. 1751 (1956).
83. Grant, J., "Quantitative Organic Microanalysis, Based on the Methods of Fritz Pregl," 4th ed., Blakiston, Philadelphia, Pennsylvania, 1946.
84. Grant, J., "Quantitative Organic Microanalysis," 5th ed., Blakiston, Philadelphia, Pennsylvania, 1951.
85. Grunbaum, B. W., and Kirk, P. L., *Anal. Chem.,* **27**, 333 (1955).
86. Hallett, L. T., *Ind. Eng. Chem., Anal. Ed.,* **14**, 956 (1942).
87. Hallett, L. T., and Kuipers, J. W., *Ind. Eng. Chem., Anal. Ed.,* **12**, 360 (1940).
88. Hara, R., and West, P. W., *Anal. Chim. Acta,* **15**, 193 (1956).
89. Harman, J. W., and Webster, J. H., *Am. J. Clin. Pathol.,* **18**, 750 (1948).
90. Hodgman, C. D., and Lange, N. A., "Handbook of Chemistry and Physics," 9th ed., p. 395, Chemical Rubber, Cleveland, Ohio, 1922.
91. Ingold, W., *Helv. Chim. Acta,* **29**, 1929 (1946).
92. Jerie, H., *Mikrochim. Acta,* **40**, 189 (1953).
93. Kassner, J. L., and Kassner, E. E., *Ind. Eng. Chem., Anal. Ed.,* **12**, 655 (1940).
94. Kaye, I. A., and Weiner, N., *Ind. Eng. Chem., Anal. Ed.,* **17**, 397 (1945).
95. Kinnunen, J., and Wennerstrand, B., *Chemist Analyst,* **46**, 92 (1957).
96. Kinsey, V. E., *Anal. Chem.,* **22**, 362 (1950).
97. Kirk, P. L., *Mikrochemie,* **14**, 1 (1933).
98. Kirk, P. L., "Quantitative Ultramicroanalysis," Wiley, New York, and Chapman & Hall, London, 1950.
99. Kirk, P. L., and Bentley, G. T., *Mikrochemie,* **21**, 250 (1937).
100. Kirsten, W. J., *Anal. Chem.,* **25**, 1137 (1953).
101. Kirsten, W. J., *Anal. Chem.,* **29**, 460 (1957).
102. Kirsten, W. J., Berggren, A., and Nilsson, K., *Anal. Chem.,* **30**, 237 (1958).
103. Knights, E. M., Jr., *J. Am. Med. Assoc.,* **166**, 1175 (1958).
104. Knights, E. M., Jr., McDonald, R. P., and Ploompuu, J., *Am. J. Clin. Pathol.,* **30**, 91 (1958).
105. Knights, E. M., Jr., McDonald, R. P., and Ploompuu, J., "Ultramicro Methods for Clinical Laboratories," Grune & Stratton, New York, and London, 1957.
106. Koch, F. C., *J. Lab. Clin. Med.,* **11**, 774 (1926); **14**, 747 (1929).
107. Korenman, I. M., and Rostokin, A. P., *Zavodskaya Lab.,* **14**, 1391 (1948).
108. Koros, E., and Remport-Horvath, Z., *Chemist Analyst,* **46**, 91 (1957).
109. Kuck, J. A., Kingsley, A., Kinsey, D., Sheehan, F., and Swigert, G. F., *Anal. Chem.,* **22**, 604 (1950).
110. Lacourt, A., *Metallurgia,* **38**, 355 (1948).
111. Lacourt, A., Stoffyn, P., and Timmermans, A. M., *Mikrochemie ver. Mikrochim. Acta,* **33**, 217 (1948).
112. Lacourt, A., and Timmermans, A. M., *Bull. classe sci. Acad. roy. Belg.,* **32**, 52 (1946).
113. Lane, E. S., *Analyst,* **82**, 406 (1957).
114. Langer, S. H., and Pantages, P., *Anal. Chem.,* **30**, 1889 (1958).
115. Lascalzo, A. G., and Benedetti-Pichler, A. A., *Ind. Eng. Chem., Anal. Ed.,* **17**, 187 (1945).
116. Lazarow, A., *J. Lab. Clin. Med.,* **32**, 213 (1947).
117. Lazarow, A., *J. Lab. Clin. Med.,* **35**, 810 (1950).
118. Levvy, G. A., *Chem. & Ind. (London),* p. 4 (1945).

119. Lévy, R., *Bull. soc. chim. France,* p. 497 (1956).
120. Linderstrøm-Lang, K., and Holter, H., *Compt. rend. trav. lab. Carlsberg,* **19**, 1 (1933).
121. Lindner, J., *Z. anal. Chem.,* **91**, 105 (1933); *Mikrochemie, Molisch Festschrift,* p. 301 (1936).
122. Lindner, R., and Kirk, P. L., *Mikrochemie,* **23**, 269 (1938).
123. Llacer, A. J., and Sozzi, J. A., *Anales farm. y bioquím.* (*Buenos Aires*), **16**, 82 (1945).
124. Lott, P. F., and Cheng, K. L., *Chemist Analyst,* **47**, 8 (1958).
125. Lundbak, A., *Kem. Maanedsblad,* **24**, 138 (1943).
126. Lykken, L., and Rolfson, R. B., *Ind. Eng. Chem., Anal. Ed.,* **13**, 653 (1941).
127. Ma, T. S., and Gwirtsman, J., *Anal. Chem.,* **29**, 140 (1957).
128. Ma, T. S., and Zuazaga, G., *Ind. Eng. Chem., Anal. Ed.,* **14**, 280 (1942).
129. Mader, W. J., and Frediani, H. A., *J. Am. Pharm. Assoc. Sci. Ed.,* **40**, 24 (1951).
130. Malissa, H., and Benedetti-Pichler, A. A., "Anorganische qualitative Mikroanalyse," Springer, Wien, 1958.
131. Manohin, B., Kakabadse, G. J., and Crowder, M. M., *Analyst,* **81**, 730 (1956).
132. Marley Products Co., New Hyde Park, New York.
133. Marsh, O., *A.M.A. Arch. Ind. Health,* **12**, 688 (1955).
134. Mason, A. C., *Analyst,* **76**, 172 (1951).
135. Mattenheimer, H., and Borner, K., *Mikrochim. Acta,* p. 916 (1959).
136. Maurmeyer, R. K., Margosis, M., and Ma, T. S., *Mikrochim. Acta,* p. 177 (1959).
137. Mavrodineanu, R., and Gwirtsman, J., *Contribs. Boyce Thompson Inst.,* **18**, 181 (1955).
138. Megregian, S., *Anal. Chem.,* **26**, 1161 (1954).
139. Milner, O. I., *Anal. Chem.,* **24**, 1247 (1952).
140. Milton, R. F., and Waters, W. A., "Methods of Quantitative Microanalysis," Longmans, Green, New York, and Arnold, London, 1949.
141. Milton, R. F., and Waters, W. A., "Methods of Quantitative Microanalysis," 2nd ed., Arnold, London, 1955.
142. Natelson, S., "Microtechniques of Clinical Chemistry for the Routine Laboratory," Thomas, Springfield, Illinois, 1957.
143. Natelson, S., and Zuckerman, J. L., *J. Biol. Chem.,* **170**, 305 (1947).
144. Niederl, J. B., and Niederl, V., "Micromethods of Quantitative Organic Elementary Analysis," Wiley, New York, 1938.
145. Niederl, J. B., and Niederl, V., "Micromethods of Quantitative Organic Analysis," 2nd ed., Wiley, New York, 1942.
146. Niederl, J. B., Niederl, V., and Eitingon, M., *Mikrochemie ver. Mikrochim. Acta,* **25**, 143 (1938).
147. Niederl, J. B., and Sozzi, J. A., *"Microanálisis Elemental Orgánico,"* Calle Arcos, Buenos Aires, 1958.
148. Nikelly, J. B., and Cooke, W. D., *Anal. Chem.,* **28**, 243 (1956).
149. Ogg, C. L., Brand, R. W., and Willits, C. O., *J. Assoc. Offic. Agr. Chemists,* **31**, 663 (1948).
150. Ogg, C. L., Willits, C. O., and Cooper, F. J., *Anal. Chem.,* **20**, 83 (1948).
151. Parks, T. D., *Anal. Chim. Acta,* **6**, 553 (1952).
152. Peacocke, T. A. H., *Chem. & Ind.* (*London*), p. 1245 (1952).
153. Pećar, M., *Microchem. J.,* **3**, 557 (1959).
154. Pećar, M., *Microchem. J.,* **4**, 73 (1960).

References

155. Pharmacopeia of the United States by the Authority of the Pharmacopeial Convention, Vols. 13-16, Mack, Easton, Pennsylvania, 1947, 1950, 1955, 1960.
156. Pifer, C. W., and Wollish, E. G., *Anal. Chem.,* **24**, 300 (1952).
157. Pifer, C. W., and Wollish, E. G., *Anal. Chem.,* **24**, 519 (1952).
158. Pifer, C. W., Wollish, E. G., and Schmall, M., *Anal. Chem.,* **25**, 310 (1953).
159. Pifer, C. W., Wollish, E. G., and Schmall, M., *Anal. Chem.,* **28**, 215 (1954).
160. Pifer, C. W., Wollish, E. G., and Schmall, M., *J. Am. Pharm. Assoc. Sci. Ed.,* **42**, 509 (1953).
161. Pregl, F., "Quantitative Organic Microanalysis," (E. Fyleman, trans. 2nd German ed.), Churchill, London, 1924.
162. Preisler, P. W., and Berger, L., *J. Am. Chem. Soc.,* **64**, 67 (1942).
163. Rehberg, P. B., *Biochem. J.,* **19**, 270 (1925).
164. Reilley, C. N., and McCurdy, W. H., Jr., *Anal. Chem.,* **25**, 86 (1953).
165. Richter, K. M., *Anal. Chem.,* **28**, 2036 (1956).
166. Riddick, J. A., *Anal. Chem.,* **26**, 77 (1954).
167. Rieman, Wm. III, *Ind. Eng. Chem., Anal. Ed.,* **16**, 475 (1944).
168. Roth, H., "Die quantitative organische Mikroanalyse von Fritz Pregl," 4th ed., Springer, Berlin, 1935.
169. Roth, H., "F. Pregl quantitative organische Mikroanalyse," 5th ed., Springer, Wien, 1947.
170. Roth, H., "Pregl-Roth quantitative organische Mikroanalyse," 7th ed., Springer, Wien, 1958.
171. Roth, H., "Quantitative Organic Microanalysis of Fritz Pregl," 3rd ed. (E. B. Daw, trans., 4th German ed., Blakiston, Philadelphia, Pennsylvania, 1937.
172. Rowley, R. J., *Ind. Eng. Chem., Anal. Ed.,* **9**, 551 (1937).
173. Saunders, J. A., *Analyst,* **71**, 528 (1946).
174. Schneider, F., "Qualitative Organic Microanalysis," Wiley, New York, and Chapman & Hall, London, 1946.
175. Scholander, P. F., *Science,* **95**, 177 (1942).
176. Scholander, P. F., Edwards, G. A., and Irving, L., *J. Biol. Chem.,* **148**, 495 (1943).
177. Schöniger, W., *Mikrochemie ver. Mikrochim. Acta,* **40**, 27 (1953).
178. Schreiner, H., *Mikrochemie ver. Mikrochim. Acta,* **38**, 273 (1951).
179. Schulze, H. O., *Science,* **111**, 36 (1950).
180. Schwarz, K., *Mikrochemie,* **13**, 6 (1933).
181. Sendroy, J., and Hastings, A. B., *J. Biol. Chem.,* **98**, 197 (1929).
182. Shaffer, P. A., Jr., Farrington, P. S., and Niemann, C., *Anal. Chem.,* **19**, 492 (1947).
183. Sher, I. H., *Anal. Chem.,* **27**, 831 (1955).
184. Simon, W., *Chimica,* **10**, 286 (1956).
185. Simon, W., *Helv. Chim. Acta,* **39**, 883 (1956).
186. Simon, W., Kováks, E., Chopard-dit-Jean, L. H., and Heilbronner, E., *Helv. Chim. Acta,* **37**, 1872 (1954).
187. Simon, W., Morikofer, A., Heilbronner, E., *Helv. Chim. Acta,* **34**, 1918 (1957).
188. Smith, G. F., Chemical Company, Columbus, Ohio.
189. Smith-New York Company, Inc., "Titration of Sulfate with Tetrahydroxyquinone," Freeport, Long Island, New York, 1948.
190. Sobel, A. E., Yuska, H., and Cohen, J., *J. Biol. Chem.,* **118**, 443 (1937).
191. Sômiya, N., and Kamada, H., *Kagaku no Ryôiki,* **1**, 63 (1947).
192. Sporek, K. F., *Chemist Analyst,* **47**, 12 (1958).
193. Steinitz, K., *Mikrochemie ver. Mikrochim. Acta,* **35**, 176 (1950).

5. Standard Solutions

194. Steyermark, Al, "Quantitative Organic Microanalysis," Blakiston, Philadelphia, Pennsylvania, 1951.
195. Steyermark, Al, Alber, H. K., Aluise, V. A., Huffman, E. W. D., Jolley, E. L., Kuck, J. A., Moran, J. J., and Ogg, C. L., *Anal. Chem.,* **28**, 1993 (1956).
196. Steyermark, Al, Alber, H. K., Aluise, V. A., Huffman, E. W. D., Jolley, E. L., Kuck, J. A., Moran, J. J., and Ogg, C. L., *Anal. Chem.,* **30**, 1702 (1958).
197. Steyermark, Al, Alber, H. K., Aluise, V. A., Huffman, E. W. D., Jolley, E. L., Kuck, J. A., Moran, J. J., Ogg, C. L., and Pietri, C. E., *Anal. Chem.,* **32**, 1045 (1960).
198. Steyermark, Al, Bass, E. A., and Littman, B., *Anal. Chem.,* **20**, 587 (1948).
199. Steyermark, Al, Kaup, R. R., Petras, D. A., and Bass, E. A., *Microchem. J.,* **3**, 523 (1959).
200. Stock, J. T., *Analyst,* **73**, 321 (1948).
201. Stock, J. T., *Metallurgia,* **36**, 51 (1947).
202. Stock, J. T., *Metallurgia,* **37**, 220 (1948).
203. Stock, J. T., *Metallurgia,* **42**, 48 (1950).
204. Stock, J. T., *Microchem. J.,* **3**, 543 (1959).
205. Stock, J. T., and Fill, M. A., *Analyst,* **74**, 318 (1949).
206. Stock, J. T., and Fill, M. A., *Metallurgia,* **31**, 103 (1944).
207. Stock, J. T., and Fill, M. A., *Metallurgia,* **36**, 225 (1947).
208. Stock, J. T., and Fill, M. A., *Metallurgia,* **39**, 335 (1949).
209. Stock, J. T., and Fill, M. A., *Metallurgia,* **41**, 170 (1950).
210. Sundberg, O. E., and Royer, G. L., *Ind. Eng. Chem., Anal. Ed.,* **18**, 719 (1946).
211. Takahashi, T., Kimoto, K., and Sakurai, H., *Rept. Inst. Ind. Sci.,* Tokyo Univ., **5**, 121 (1955); *Chem. Abstr.* **52**, 7028 (1958).
212. Thomas, Arthur H., Company, Philadelphia, Pennsylvania.
213. Thompson, W. R., *Ind. Eng. Chem., Anal. Ed.,* **14**, 268 (1942).
214. Tompkins, E. R., and Kirk, P. L., *J. Biol. Chem.,* **142**, 477 (1942).
215. Treadwell, F. P., and Hall, W. T., "Analytical Chemistry," 6th ed., Vol. II, Wiley, New York, and Chapman & Hall, London, 1924.
216. Underwood, H. G., *J. Assoc. Offic. Agr. Chemists,* **38**, 380 (1955).
217. Upson, U. L., *Anal. Chem.,* **25**, 977 (1953).
218. Upson, U. L., Personal communication (February 10, 1954).
219. Van Slyke, D. D., Folch, J., and Plazin, J., *J. Biol. Chem.,* **136**, 509 (1940).
220. Venkateswarlu, P., Ramanthan, A. N., and Narayana Rao, D., *Indian J. Med. Research,* **41**, 253 (1953).
221. Wade, P., *Analyst,* **76**, 606 (1951).
222. Walker, G. T., and Allen, S. A., *Chem. & Ind. (London),* p. 568 (1953).
223. Williams, M., *Ind. Chem. Mfr.,* **32**, 442 (1956).
224. Willits, C. O., *Anal. Chem.,* **21**, 132 (1949).
225. Willits, C. O., Coe, M. R., and Ogg, C. L., *J. Assoc. Offic. Agr. Chemists,* **32**, 118 (1949).
226. Willits, C. O., and Ogg, C. L., *J. Assoc. Offic. Agr. Chemists,* **33**, 179 (1950).
227. Wingo, W. J., and Johnson, W. H., *Anal. Chem.,* **28**, 1215 (1956).
228. Winteringham, F. P. W., *Analyst,* **70**, 173 (1945).
229. Wise, E. N., *Anal. Chem.,* **23**, 1479 (1951).
230. Wollish, E. G., Colarusso, R. J., Pifer, C. W., and Schmall, M., *Anal. Chem.,* **26**, 1753 (1954).
231. Wollish, E. G., Pifer, C. W., and Schmall, M., *Anal. Chem.,* **26**, 1704 (1954).
232. Wyatt, G. H., *Analyst,* **69**, 180 (1944).

CHAPTER 6

Microdetermination of Metals by the Ashing Technique

A number of metals* in organic compounds may be determined by a simple ashing procedure with or without the addition of sulfuric or nitric acids.[17,18,90,91,171-173,176,184,195-198,223] However, only one metal may be present in the compound, or this method is not applicable and determination must be accomplished by adapting one of the macromethods found throughout the literature.[82,218,235] Before attempting to determine any metal by this method, the reader should consult the literature[107,108,165] to make certain in which state the metal would be present at the end, namely, the oxide, sulfate, or free metal, and if these would or would not be lost by volatilization at the temperature involved. The method is not applicable to compounds containing phosphorus, since this is not completely driven off.[165]

The substance is treated with sulfuric acid† and placed in a muffle in which the acid is driven off and the residue heated at red heat.

The reactions may be represented by the following, depending upon the metal involved.

(1) Organic —Na   Na$_2$SO$_4$
        —K   K$_2$SO$_4$
        —Ca $\xrightarrow{H_2SO_4}$ CaSO$_4$
        —Ba   BaSO$_4$
        etc.   etc.

(2) Organic —Ag   Ag
        —Au $\longrightarrow$ Au
        —Pt   Pt

(3) Organic —Cu   CuO
        —Fe $\xrightarrow{(HNO_2)\ (H_2SO_4)}$ Fe$_2$O$_3$
        etc.   etc.

---

* Determination of arsenic is carried out according to the method given in Chapter 13.
† Sulfuric acid is added in cases where the sulfate is stable at red heat. For silver, gold and platinum, no acid need be added. Where the oxide will be formed at the end of the determination, nitric acid may be added, as for example with copper, iron, etc. Silver may exist as sulfate or chloride in which case the factor would not be 1.000.

Reagents

DILUTED SULFURIC ACID

1 part conc. H₂SO₄ (sp. gr. 1.84)
5 parts distilled water.

DILUTED NITRIC ACID

1 part conc. HNO₃ (sp. gr. 1.42)
1 part distilled water.

Apparatus

PLATINUM BOAT

The platinum boat required is the standard item[5,223-225] described in Chapter 3, Fig. 24.

PLATINUM CYLINDER

The platinum cylinder[1,61] shown in Fig. 84 is used as a sleeve into which the platinum boat is placed. If spattering occurs during the determination the material is deposited on the inner surface of the cylinder, preventing loss. It is made of platinum foil approximately 0.04 mm. thick, is about 3 cm. long and has a diameter of approximately 7.5 mm. To the cylinder is attached a piece of thin platinum wire, the end of which is shaped into a hook. This is used in pushing the cylinder into the micromuffle and pulling it out.

FIG. 84. The Coombs-Alber platinum cylinder (sleeve).

MICROMUFFLE

The micromuffle consists of a tube (Pyrex 1720 glass,* Vycor,* or quartz) which is heated to 680°–700° C. and through which passes a slow current of air. Although the original micromuffle of Pregl[17,18,90,91,171-173,184,195-198] was gas heated, the author prefers electric[172,176,192] units as these provide more constant heat, with equal distribution. This helps to prevent creeping of the material during the early stages of the heating and insures more constant ashing conditions.[172] The apparatus used by the author is shown in Figs. 85 and 86. It is a double unit which permits two determinations to be carried out simultaneously. It consists of two sections of combustion tubing about 225 mm. in length, one tube for each determination. These are portions of standard combustion tubing, Pyrex 1720 glass,* Vycor,* or quartz, approximately 11 mm. O.D. and 8 mm. I.D. [a section of the tube shown in Fig. 98 (Chapter 7) or Fig. 121

---

* Corning Glass Works, Corning, New York.

(Chapter 9)]. The tubes are fastened in a support and pass through the two holes of the movable electric burner* or furnace. At one end, each tube is open and at the other end it is attached to a small bubble counter† which contains about 1 ml. of concentrated H_2SO_4 so that the rate of flow of air through

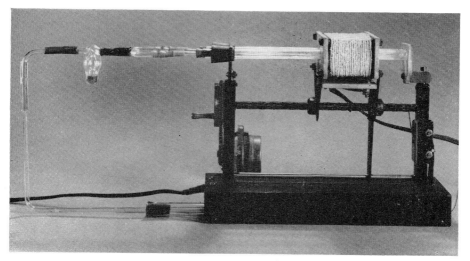

FIG. 85. Micromuffle.

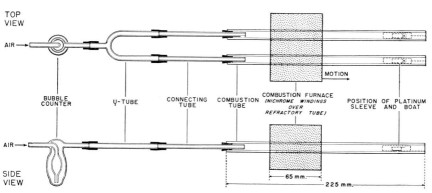

FIG. 86. Micromuffle—details of construction.

the tubes may be observed. The opposite end of the bubble counter is attached to a filter tube containing cotton, which in turn is connected to a compressed air supply.

The movable electric furnace is made from windings of small nichrome

---

* See footnote p. 154, Chapter 7.
† Portion of a broken bubble counter-U-tube (Chapter 9) may be used.

6. Microdetermination of Metals: Ashing Method

wire wound around two sections of alundum tubing approximately 13–14 mm. I.D. and about 65 mm. in length. The temperature is controlled to 680°–700° C. by operating the furnace from a variable auto-transformer. The furnace is attached to a worm drive which is powered by a small electric clock motor,[222] making the drive automatic with a forward speed of approximately 2.5 cm. in 10 minutes.

In many respects, the micromuffle is quite similar to that described by A. R. Norton, G. L. Royer, and R. Koegel.[176]

Procedure

The boat and cylinder are cleaned by immersion in hot dilute nitric acid, washed with distilled water, strongly ignited in the flame of a gas burner, and set in a microdesiccator (Chapter 3, Figs. 43–46) to cool. The dismantled combination is then accurately weighed on the microchemical balance (taking into account the zero reading). About 5 mg. of sample is added to the boat and the combination reweighed. (Note: If the sample is hygroscopic, the weighing pig is used to protect the sample as described in Chapter 3—see Figs. 29, 30, and 32). Three drops of the diluted sulfuric acid* are carefully added to the sample and the boat is then placed in the cylinder using a platinum-tipped forceps (Chapter 3, Fig. 41) to hold each piece during the manipulation. The combination is then placed in the open end of the combustion tubing and pushed far enough in so that all portions, including the wire hook, will be heated eventually. Care should be taken so that the boat is kept upright during the placing. The passage of air through the bubble counter is regulated so that bubbles pass through at the rate of about 1–2 per second. The furnace is placed back so that it is about at the one end of the cylinder opposite to the opening (about 5–6 cm. from the open end of the combustion tube). It is then switched on and the temperature brought up to 680°–700° C. The mechanical drive is set into motion and the hot furnace is passed over the entire length of the cylinder and is allowed to remain in the end position (over the cylinder) for about 15–20 minutes. The combination is then carefully removed from the micromuffle and allowed to cool in a microdesiccator after which it is weighed to obtain the weight of the ash (taking into account the zero reading). It is good practice to reheat the combination in the micromuffle for 5 minutes and reweigh. This insures that constant weight has been attained.

Calculation:

$$\% \text{ Metal} = \frac{\text{Wt. ash} \times \text{factor} \times 100}{\text{Wt. sample}}$$

* See †, p. 133.

*Factors:**

Ash	Element sought	Factor
Na_2SO_4	Na	0.3237
K_2SO_4	K	0.4487
$BaSO_4$	Ba	0.5885
$CaSO_4$	Ca	0.2944
CuO	Cu	0.7988
Fe_2O_3	Fe	0.6994
Pt	Pt	1.000
Au	Au	1.000
Ag	Ag	1.000

Examples:

(a) 5.632 mg. of ash (Na_2SO_4) is obtained from a 7.631-mg. sample

$$\therefore \frac{5.632 \times 0.3237 \times 100}{7.631} = 23.89\% \text{ Na}$$

(b) 2.610 mg. of ash (CuO) is obtained from a 8.001-mg. sample

$$\therefore \frac{2.610 \times 0.7988 \times 100}{8.001} = 26.06\% \text{ Cu}$$

(c) 1.074 mg. of ash (Ag) is obtained from a 7.016-mg. sample

$$\therefore \frac{1.074 \times 1 \times 100}{7.016} = 15.31\% \text{ Ag}$$

ADDITIONAL INFORMATION FOR CHAPTER 6

The micromuffle used by Pregl[17,18,90,91,171-173,184,195-198] is shown in Fig. 87. It consists of a right angle tube and a straight one. The vertical portion of the right angle is heated to supply a hot air current.

Norton, Royer, and Koegel[176] described an automatic electric microfurnace. It provides for the ashing of two samples simultaneously. *The two quartz combustion tubes are drawn through the furnace* in contrast to that shown in Figs. 85 and 86 in which the tubes are stationary and the furnace moves along. The above authors used a standard two-speed governor controlled phonograph motor geared so that the rate of travel is approximately either 1 or 2 inches (2.5 or 5 cm.) in 10 minutes. Platinum heating coils are used. The voltage to these is controlled by means of a variable auto-transformer. The furnace is operated at a temperature of 800° C.

---

* Only a few typical factors are given here. For other cases, the reader is referred to the various handbooks containing gravimetric factors.[59,107,108] Also see Chapter 23.
Also see footnote on page 133.

6. Microdetermination of Metals: Ashing Method

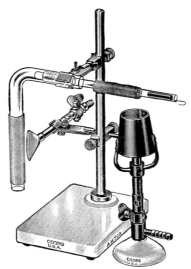

FIG. 87. Pregl micromuffle.

TABLE 16
Additional Information on References* Related to Chapter 6

Although this chapter deals with the determination of metals by the ashing technique, the author wishes to present additional information in the form of a table as in the preceding chapters. (See author's statement at top of Table 4 of Chapter 1.) The references listed in this table refer to the determination of metals by other means as well as the determination of several amphoteric elements not covered in detail in other chapters.

Books
  Belcher and Godbert, 17, 18
  Clark, E. P., 59
  Clark, S. J., 60
  Fritz and Hammond, 81
  Furman, 82
  Grant, 90, 91
  Milton and Waters, 167, 168
  Niederl and Niederl, 171, 172
  Niederl and Sozzi, 173
  Pregl, 184
  Roth, 195–198
  Steyermark, 223

General, miscellaneous
  Belcher, Gibbons, and Sykes, 16
  Belcher, Macdonald, and West, 19
  Cimerman and Selzer, 54, 55

General, miscellaneous (Cont.)
  Duval, 69
  Martin, 158
  Meyrowitz, and Massoni, 166
  Norton, Royer, and Koegel, 176
  Sykes, 231
  Van Etten and Wiele, 239

Reviews
  Belcher, Gibbons, and Sykes, 16
  Sykes, 231

Apparatus
  Norton, Royer, and Koegel, 176

Bombs
  Kondo, 134
  Kuck and Grim, 142

* The numbers which appear after each entry in this table refer to the literature citations in the reference list at the end of the chapter.

TABLE 16 (*Continued*)

Perchloric acid
Smith, 217

Oxygen flask combustion
Belcher, Macdonald, and West, 19
Corner, 62
Southworth, Hodecker, and Fleischer, 221

Kjeldahl flask digestion
Kreshkov, Syavtsillo, and Shemyatenkova, 141

Simultaneous determinations
Klimova and Bereznitskaya, 129, 130
Klimova, Korshun, and Bereznitskaya, 131, 132
Korshun and Chumachenko, 137

General volumetric methods
Belcher and Robinson, 20
Belcher and West, 21
Cimerman and Ariel, 48–50
Cimerman and Bogin, 51
Cimerman and Frenkel, 52
Cimerman and Selzer, 54
Deibner, 65
Gautier and Pellerin, 83
Gusev, Kumov, and Stroganova, 93
Kuck and Grim, 142
Schulitz, 210
Sloviter, McNabb, and Wagner, 215
Strahm and Hawthorne, 226
Sudo, Shimoe, and Miyahara, 230
Venkateswarlu, Ramanthan, and Narayana Rao, 240
Zabrodina and Bagreeva, 248

Heterometric titration
Bobtelsky and Eisenstadter, 24
Bobtelsky and Graus, 25, 26
Bobtelsky and Hapern, 27
Bobtelsky and Jungreis, 28–32
Bobtelsky and Rafailoff, 33, 34
Bobtelsky and Welwart, 35–38

Complexometric, EDTA titration
Amin, 6
Ashby and Roberts, 10

Complexometric, EDTA titration (Cont.)
Flaschka, 74–77
Flaschka and Amin, 78
Flaschka, Amin, and Zaki, 79
Harrison and Harrison, 96
Karrman and Borgstrom, 121
Kimbel, 122
Kinnunen and Merikanto, 125
Parry, 179
Socolar and Salach, 219
Southworth, Hodecker, and Fleischer, 221

Flame photometric, X-ray spectrophotometric, colorimetric, etc., methods
Almassy and Kavai, 4
Arnold and Pray, 9
Ašperger and Murati, 11
Barrett, 14
Beauchene, Berneking, Schrenk, Mitchell, and Silker, 15
Buell, 41
Chatagnon and Chatagnon, 47
Gautier and Pellerin, 83
Gillam, 88
Grogan, Cahnmann, and Lethco, 92
Hegedüs, Fukker, and Dvorszky, 99
Hunter, 112
Ikeda, 114
Ishibashi and Higashi, 115
Kingsley and Schaffert, 124
Kozawa, Tanaka and Sasaki, 139
Kreisky, 140
Lapin and Makarova, 148
Leonard, Sellers, and Swim, 149
Leroux, Maffett, and Monkman, 150
Lewis, 151
Lindstrom, 153
Lipscomb, 154
Marier and Boulet, 157
Martin, 159, 160
Mehlig, 163
Natelson and Penniall, 170
Nonowa, 175
Nozaki, 177
Polley and Miller, 183
Sakuraba, 201

6. Microdetermination of Metals: Ashing Method 140

TABLE 16 (*Continued*)

Flame photometric, X-ray spectrophotometric, colorimetric, etc., methods (*Cont.*)

Saltzman, 202
Schuhknecht, 206
Schuhknecht and Schinkel, 207
Schultz, 211
Solomon, 220
Teeri, 232
Toribara and Sherman, 234
Tunnicliffe, 236
Tutundžić and Mladenović, 237
Umland and Weyer, 238
Zuehlke and Ballard, 250

Potentiometric, polarographic, etc., methods

Carruthers, 45
De Francesco, 64
Heyrovský, 103
Jensen, 116
Kadowaki, Okamoto, and Nakajima, 118
Kuck and Grim, 142
Lambert and Walker, 145
Well, 243

Manometric methods

See Chapter 18
Hoagland, 106

Gravimetric methods, precipitation, etc.

Barber and Kolthoff, 13
Cimerman and Selzer, 55
Fennel and Webb, 73
Gusev, Kumov, and Stroganova, 93
Kondo, 134
Walton and Smith, 242

Electrodeposition

Llacer, Sozzi and Benedetti-Pichler, 155

Chromatographic methods

Lacourt, 144
Lamm, 146

Radioactive methods

Korenman, Sheyanova, and Glazunova, 136

Antimony

Bahr, Bieling, and Thiele, 12
Gellhorn, Krahl, and Fertig, 85
Haight, 94
Jureček and Jenik, 117
Schulek and Wolstadt, 209
Silvert and Kirner, 214

Arsenic

See Chapter 13

Barium

Pungor and Hegedüs, 186
Pungor and Thege, 187

Beryllium

Toribara and Sherman, 234

Bismuth

Silvert and Kirner, 214

Boron

Allen and Tannenbaum, 3
Belcher, Macdonald, and West, 19
Buell, 41
Corner, 62
Gautier and Pignard, 84
Kuck and Grim, 142
Martin, 160
Pflaum and Wenske, 181
Roth, 194
Strahm and Hawthorne, 226

Cadmium

Cimerman and Selzer, 54
Furman, 82
Saltzman, 202
Steyermark, 223

Calcium

Andersch, 7
Ashby and Roberts, 10
Bobtelsky and Eisenstadter, 24
Gilbert, 86, 87
Harrison and Raymond, 95
Harrison and Harrison, 96
Herrmann, 102
Hunter, 112, 113

TABLE 16 (*Continued*)

Calcium (*Cont.*)
Ikeda, 114
Kimbel, 122
Kingsley and Schaffert, 123, 124
Kirk and Tompkins, 127
Lindner and Kirk, 152
Marier and Boulet, 157
McGregor, 161
Natelson and Penniall, 170
Nonowa, 175
Patel, 180
Pungor and Hegedüs, 186
Pungor and Thege, 187
Rappaport and Rappaport, 189
Scholtis, 205
Schuhknecht, 206
Schultz, 211
Socolar and Salach, 219
Steyermark, 223
Teeri, 232
Wilkinson, 246

Chromium
Grogan, Cahnmann, and Lethco, 92
Kozawa, Tanaka, and Sasaki, 139
Lacourt, 144

Cobalt
Bobtelsky and Jungreis, 28
Cimerman and Selzer, 54
Ellis and Gibson, 70
Pohl and Demmel, 182
Sakuraba, 201
Saltzman, 203
Wenger, Cimerman, and Corbaz, 244

Copper
Amin, 6
Beauchene, Berneking, Schrenk, Mitchell, and Silker, 15
Benedetti-Pichler, 22
Bobtelsky and Graus, 26
Bobtelsky and Jungreis, 32
Bobtelsky and Welwart, 38
Carruthers, 45
Chatagnon and Chatagnon, 47
Dezsö and Fulöp, 66
Diehl and Smith, 67
Ellis and Gibson, 71

Copper (*Cont.*)
Emeléus and Haszeldine, 72
Furman, 82
Hecht and Reissner, 97
Hubbard and Spettelm, 111
Lapin and Makarova, 148
Llacer, Sozzi, and Benedetti-Pichler, 155
MacNevin and Bournique, 156
Mehlig, 163
Sakuraba, 201
Smith, 216
Steyermark, 223
Umland and Weyer, 238

Gold
Furman, 82
Konig, Crowell, and Benedetti-Pichler, 135
Onishi, 178
Steyermark, 223

Iron
Almassy and Kavai, 4
Belcher and West, 21
Bobtelsky and Jungreis, 29
Deibner, 65
Flaschka, 77
Furman, 82
Kirk and Bentley, 126
Knop and Kubelkova, 133
Nieuwenburg and Blumendal, 174
Rappaport and Hohenberg, 188
Steyermark, 223
Straub, 227
Tutundžić and Mladenović, 237
Umland and Weyer, 238
Well, 243

Lead
Bobtelsky and Graus, 25
Bobtelsky and Rafailoff, 34
Brantner and Hecht, 40
Cimerman and Ariel, 48–50
Cimerman and Bogin, 51
De Francesco, 64
Furman, 82
Jensen, 116
Kuhn and Schretzmann, 143
Steyermark, 223

TABLE 16 (*Continued*)

Lithium

Hegedüs and Dvorszky, 98
Nozaki, 177
Schuhknecht and Schinkel, 207
Steyermark, 223

Magnesium

Benedetti-Pichler and Schneider, 23
Bobtelsky and Welwart, 37
Butler, 43
Davidson, 63
Gillam, 88
Gusev, Kumov, and Stroganova, 93
Hoagland, 106
Hunter, 112, 113
Ikeda, 114
Karrman and Borgstrom, 121
Steyermark, 223
Strebinger and Reif, 229
Tunnicliffe, 236
Wilkinson, 246
Winkler, 247

Manganese

Flaschka, Amin, and Zaki, 79
Kozawa, Tanaka, and Sasaki, 139
Scott, 212

Mercury

Ašperger and Murati, 11
Barrett, 14
Bobtelsky and Jungreis, 30, 31
Bobtelsky and Rafailoff, 33
Boëtius, 39
Cimerman and Frenkel, 52
Druzhinin and Kislitsin, 68
Emeléus and Haszeldine, 72
Furman, 82
Gautier and Pellerin, 83
Grant, 90, 91
Herd, 100
Hernler, 101
Hirai and Hayatsu, 105
Kadowaki, Okamoto, and Nakajima, 118
Korshun and Chumachenko, 137
Korshun and Lavrovskaya, 138
Leroux, Maffett, and Monkman, 150
Lindstrom, 153

Mercury (Cont.)

Lipscomb, 154
Meixner and Kröcker, 164
Miura, 169
Parry, 179
Polley and Miller, 183
Rauscher, 190
Roth, F. J., 193
Roth, H., 195–198
Rutgers, 199
Sachs, 200
Schulitz, 210
Shukis and Tallman, 213
Sloviter, McNabb, and Wagner, 215
Smith, 216
Southworth, Hodecker, and Fleischer, 221
Steyermark, 223
Sudo, Shimoe, and Miyahara, 230
Verdino, 241
Walton and Smith, 242
Zuehlke and Ballard, 250

Nickel

Benedetti-Pichler, 22
Bobtelsky and Welwart, 35, 36
Furman, 82
Llacer, Sozzi, and Benedetti-Pichler, 155
Steyermark, 223

Palladium

Kinnunen and Merikanto, 125
Konig, Crowell, and Benedetti-Pichler, 135
Steyermark, 223

Platinum

Steyermark, 223

Potassium

Belcher and Robinson, 20
Bullock and Kirk, 42
Chapman, 46
Cimerman and Rzymowska, 53
Cimerman, Wenger, and Rzymowska, 58
Flaschka and Amin, 78
Hegedüs and Dvorszky, 98
Herrmann, 102

TABLE 16 (*Continued*)

Potassium (*Cont.*)

Heyrovský, 103
Kingsley and Schaffert, 123, 124
Klein and Jacobi, 128
Korenman, Sheyanova, and Glazunova, 136
Kreisky, 140
Lewis, 151
Robinson and Hauschildt, 191
Schuhknecht and Schinkel, 207
Steyermark, 223

Selenium

Alber and Harand, 2
Gould, 89
Kahane and Korach, 119
Kan, 120
Kondo, 134
Wernimont and Hopkinson, 245
Zabrodina and Bagreeva, 248

Silicon

Fennell and Webb, 73
Klimova and Bereznitskaya, 129, 130
Klimova, Korshun and Bereznitskaya, 131, 132
Kreshkov, Syavtsillo and Shemyatenkova, 141
McHard, Servais and Clark, 162
Schoklitsch, 204
Thurnwald and Benedetti-Pichler, 233

Silver

Amin, 6
Emeléus and Haszeldine, 72
Flaschka, 75
Foulk and Bawden, 80
Furman, 82
Kuhn and Schretzmann, 143
Lambert and Walker, 145
Schulek, 208
Steyermark, 223

Sodium

Arnold and Pray, 9
Barber and Kolthoff, 13
Caley and Foulk, 44
Flaschka and Amin, 78

Sodium (*Cont.*)

Hegedüs and Dvorszky, 98
Hegedüs, Fukker, and Dvorszky, 99
Herrmann, 102
Holmes and Kirk, 109
Kingsley and Schaffert, 123, 124
Kreisky, 140
Schuhknecht and Schinkel, 207
Solomon, 220
Steyermark, 223

Strontium

Pungor and Hegedüs, 186
Pungor and Thege, 187
Steyermark, 223
Strebinger and Mandl, 228
Zombory, 249

Thallium

Cimerman and Selzer, 55
Flaschka, 76

Thorium

Ishibashi and Higashi, 115
Venkateswarlu, Ramanthan, and Narayana Rao, 240

Tin

Furman, 82
Höltje, 110
Price, 185
Silvert and Kirner, 214

Uranium

Bobtelsky and Hapern, 27
Emeléus and Haszeldine, 72

Zinc

Anderson, 8
Cimerman and Selzer, 54
Cimerman and Wenger, 56, 57
Furman, 82
Hibbard, 104
Lamm, 146
Lang, 147
Martin, 159

Zirconium

Leonard, Sellers, and Swim, 149

REFERENCES

1. Alber, H. K., *Mikrochemie,* **18**, 92 (1935).
2. Alber, H. K., and Harand, J., *J. Franklin Inst.,* **228**, 243 (1939).
3. Allen, H., Jr., and Tannenbaum, S., *Anal. Chem.,* **31**, 265 (1959).
4. Almassy, G., and Kavai, M. Z., *Magyar Kém. Folyóirat,* **61**, 246 (1955).
5. American Society for Testing Materials, *ASTM Designations,* **E124-57T** (1957).
6. Amin, A. M., *Chemist Analyst,* **44**, 17 (1955).
7. Andersch, M. A., *J. Lab. Clin. Med.,* **49**, 486 (1957).
8. Anderson, C. W., *Ind. Eng. Chem., Anal. Ed.,* **13**, 367 (1941).
9. Arnold, E. A., and Pray, A. R., *Ind. Eng. Chem., Anal. Ed.,* **15**, 294 (1943).
10. Ashby, R. O., and Roberts, M., *J. Lab. Clin. Med.,* **49**, 958 (1957)
11. Ašperger, S., and Murati, I., *Anal. Chem.,* **26**, 543 (1954).
12. Bahr, G., Bieling, H., and Thiele, K. H., *Z. anal. Chem.,* **145**, 105 (1955).
13. Barber, H. H., and Kolthoff, I. M., *J. Am. Chem. Soc.,* **50**, 1625 (1928); **51**, 3233 (1929).
14. Barrett, F. R., *Analyst,* **81**, 294 (1956).
15. Beauchene, R. E., Berneking, A. D., Schrenk, W. G., Mitchell, H. L., and Silker, R. E., *J. Biol. Chem.,* **214**, 731 (1955).
16. Belcher, R., Gibbons, D., and Sykes, A., *Mikrochemie ver. Mikrochim. Acta,* **40**, 76 (1952).
17. Belcher, R., and Godbert, A. L., "Semi-Micro Quantitative Organic Analysis," Longmans, Green, London, and New York, 1945.
18. Belcher, R., and Godbert, A. L., "Semi-Micro Quantitative Organic Analysis," 2nd ed., Longmans, Green, London, 1954.
19. Belcher, R., Macdonald, A. M. G., and West, T. S., *Talanta,* **1**, 408 (1958).
20. Belcher, R., and Robinson, J. W., *Mikrochim. Acta,* p. 49 (1954).
21. Belcher, R., and West, T. S., *Anal. Chim. Acta,* **5**, 472 (1951).
22. Benedetti-Pichler, A. A., "Introduction to the Microtechniques of Inorganic Analyses," Wiley, New York, and Chapman & Hall, London, 1942.
23. Benedetti-Pichler, A. A., and Schneider, F., *Mikrochemie, Emich Festschrift,* pp. 1-17 (1930).
24. Bobtelsky, M., and Eisenstadter, J., *Anal. Chim. Acta,* **14**, 89 (1956).
25. Bobtelsky, M., and Graus, B., *Anal. Chim. Acta,* **9**, 163 (1953).
26. Bobtelsky, M., and Graus, B., *Anal. Chim. Acta,* **11**, 253 (1954).
27. Bobtelsky, M., and Hapern, M., *Anal. Chim. Acta,* **11**, 84 (1954).
28. Bobtelsky, M., and Jungreis, E., *Anal. Chim. Acta,* **12**, 248 (1955).
29. Bobtelsky, M., and Jungreis, E., *Anal. Chim. Acta,* **12**, 351 (1955).
30. Bobtelsky, M., and Jungreis, E., *Anal. Chim. Acta,* **12**, 562 (1955).
31. Bobtelsky, M., and Jungreis, E., *Anal. Chim. Acta,* **13**, 72 (1955).
32. Bobtelsky, M., and Jungreis, E., *Anal. Chim. Acta,* **13**, 449 (1955).
33. Bobtelsky, M., and Rafailoff, R., *Anal. Chim. Acta,* **14**, 339 (1956).
34. Bobtelsky, M., and Rafailoff, R., *Anal. Chim. Acta,* **16**, 321 (1957).
35. Bobtelsky, M., and Welwart, Y., *Anal. Chim. Acta,* **9**, 281 (1953).
36. Bobtelsky, M., and Welwart, Y., *Anal. Chim. Acta,* **9**, 374 (1953).
37. Bobtelsky, M., and Welwart, Y., *Anal. Chim. Acta,* **10**, 156 (1954).
38. Bobtelsky, M., and Welwart, Y., *Anal. Chim. Acta,* **10**, 459 (1954).
39. Boëtius, M., *J. prakt. Chem.,* **151**, 279 (1938).
40. Brantner, H., and Hecht, F., *Mikrochemie,* **14**, 30 (1934).
41. Buell, B. E., *Anal. Chem.,* **30**, 1514 (1958).

42. Bullock, B., and Kirk, P. L., *Ind. Eng. Chem., Anal. Ed.,* **7**, 178 (1935).
43. Butler, E. J., *Analyst,* **81**, 615 (1956).
44. Caley, E. R., and Foulk, C. W., *J. Am. Chem. Soc.,* **51**, 1664 (1929).
45. Carruthers, C., *Ind. Eng. Chem., Anal. Ed.,* **17**, 398 (1945).
46. Chapman, G. W., *J. Agr. Sci.,* **37**, 29 (1947).
47. Chatagnon, C., and Chatagnon, P., *Bull. soc. chim. biol.,* **36**, 911 (1954).
48. Cimerman, C., and Ariel, M., *Anal. Chim. Acta,* **12**, 13 (1955).
49. Cimerman, C., and Ariel, M., *Anal. Chim. Acta,* **14**, 48 (1956).
50. Cimerman, C., and Ariel, M., *Anal. Chim. Acta,* **15**, 207 (1956).
51. Cimerman, C., and Bogin, D., *Anal. Chim. Acta,* **12**, 218 (1955).
52. Cimerman, C., and Frenkel, S., *Anal. Chim. Acta,* **16**, 305 (1957).
53. Cimerman, C., and Rzymowska, C. J., *Mikrochemie,* **20**, 129 (1936).
54. Cimerman, C., and Selzer, M., *Anal. Chim. Acta,* **9**, 26 (1953).
55. Cimerman, C., and Selzer, G., *Anal. Chim. Acta,* **15**, 213 (1956).
56. Cimerman, C., and Wenger, P., *Mikrochemie,* **24**, 148 (1938).
57. Cimerman, C., and Wenger, P., *Mikrochemie,* **24**, 153 (1938).
58. Cimerman, C., Wenger, P., and Rzymowska, C. J., *Mikrochemie,* **20**, 1 (1936).
59. Clark, E. P., "Semimicro Quantitative Organic Analysis," Academic Press, New York, 1943.
60. Clark, S. J., "Quantitative Methods of Organic Microanalysis," Butterworths, London, 1956.
61. Coombs, H. I., *Biochem. J.,* **21**, 404 (1927).
62. Corner, M., *Analyst,* **84**, 41 (1959).
63. Davidson, J., *Analyst,* **77**, 263 (1952).
64. De Francesco, F., *Boll. lab. chim. provinciali (Bologna),* **6**, 10 (1955).
65. Deibner, L., *Mikrochemie ver. Mikrochim. Acta,* **35**, 488 (1950).
66. Dezsö, I., and Fülöp, T., *Mikrochim. Acta,* p. 592 (1959).
67. Diehl, H., and Smith, G. F., "The Copper Reagents. Cuproine, Neocuproine, Bathocuproine," G. F. Smith Chemical Co., Columbus, Ohio, 1958.
68. Druzhinin, I. G., and Kislitsin, P. S., *Trudy Inst. Khim. Akad. Nauk Kirgiz. S.S.R.,* p. 21 (1957); *Referat. Zhur. Khim.,* Abstr. No. 57237 (1958); *Anal. Abstr.,* **6**, Abstr. No. 587 (1959).
69. Duval, C., *Chim. anal.,* **34**, 209 (1952).
70. Ellis, K. W., and Gibson, N. A., *Anal. Chim. Acta,* **9**, 275 (1953).
71. Ellis, K. W., and Gibson, N. A., *Anal. Chim. Acta,* **9**, 368 (1953).
72. Emeléus, H. J., and Haszeldine, R. N., *J. Chem. Soc.,* p. 2948 (1949).
73. Fennell, T. R. F. W., and Webb, J. R., *Talanta,* **2**, 389 (1959).
74. Flaschka, H., *Mikrochemie ver. Mikrochim. Acta,* **39**, 38 (1952).
75. Flaschka, H., *Mikrochemie ver. Mikrochim. Acta,* **40**, 21 (1953).
76. Flaschka, H., *Mikrochemie ver. Mikrochim. Acta,* **40**, 42 (1953).
77. Flaschka, H., *Mikrochim. Acta,* p. 361 (1954).
78. Flaschka, H., and Amin, A. M., *Chemist Analyst,* **42**, 78 (1953).
79. Flaschka, H., Amin, A. M., and Zaki, R., *Chemist Analyst,* **43**, 67 (1954).
80. Foulk, C. W., and Bawden, A. T., *J. Am. Chem. Soc.,* **48**, 2045 (1926).
81. Fritz, J. S., and Hammond, G. S., "Quantitative Organic Analysis," Wiley, New York, and Chapman & Hall, London, 1957.
82. Furman, N. H., ed., "Scott's Standard Methods of Chemical Analysis," 5th ed., Vol. II, Van Nostrand, New York, 1939.
83. Gautier, J. A., and Pellerin, F., *Prods. pharm.,* **13**, 149 (1958); *Chem. Abstr.,* **52**, 15338 (1958).

84. Gautier, J. A., and Pignard, P., *Mikrochemie ver. Mikrochim. Acta,* **36/37**, 793 (1951).
85. Gellhorn, A., Krahl, M. E., and Fertig, J. W., *J. Pharmacol. Exptl. Therap.* **87**, 159 (1946).
86. Gilbert, A. B., *Nature,* **183**, 888 (1959).
87. Gilbert, A. B., *Nature,* **183**, 1754 (1959).
88. Gillam, W. S., *Ind. Eng. Chem., Anal. Ed.,* **13**, 499 (1941).
89. Gould, E. S., *Anal. Chem.,* **23**, 1502 (1951).
90. Grant, J., "Quantitative Organic Microanalysis, Based on the Methods of Fritz Pregl," 4th ed., Blakiston, Philadelphia, Pennsylvania, 1946.
91. Grant, J., "Quantitative Organic Microanalysis," 5th ed., Blakiston, Philadelphia, Pennsylvania, 1951.
92. Grogan, C. H., Cahnmann, H. J., and Lethco, E., *Anal. Chem.,* **27**, 983 (1955).
93. Gusev, S. I., Kumov, V. I., and Stroganova, A. M., *Zhur. Anal. Khim.,* **10**, 349 (1955); *Chem. Abstr.,* **50**, 7654 (1956).
94. Haight, G. P., *Anal. Chem.,* **26**, 593 (1954).
95. Harrison, G. E., and Raymond, W. H. A., *Analyst,* **78**, 528 (1953).
96. Harrison, H. E., and Harrison, H. C., *J. Lab. Clin. Med.,* **46**, 662 (1955).
97. Hecht, F., and Reissner, R., *Mikrochemie,* **17**, 127 (1935).
98. Hegedüs, A. J., and Dvorszky, M., *Mikrochim. Acta,* p. 160 (1959).
99. Hegedüs, A., Fukker, F. K., and Dvorszky, M., *Magyar Kém. Folyóirat,* **59**, 334 (1953); *Referat. Zhur. Khim.,* Abstr. No. 36,400 (1954).
100. Herd, R. L., *J. Assoc. Offic. Agr. Chemists,* **38**, 645 (1955).
101. Hernler, F., *Mikrochemie, Pregl Festschrift,* p. 154 (1929).
102. Herrmann, R., *Z. ges. exptl. Med.,* **122**, 84 (1953).
103. Heyrovský, A., *Chem. listy,* **50**, 69 (1956).
104. Hibbard, P. L., *Ind. Eng. Chem., Anal. Ed.,* **6**, 423 (1934).
105. Hirai, M., and Hayatsu, R., *Yakugaku Zasshi,* **70**, 670 (1950).
106. Hoagland, C. L., *J. Biol. Chem.,* **136**, 553 (1940).
107. Hodgman, C. D., "Handbook of Chemistry and Physics," 28th ed., Chemical Rubber, Cleveland, Ohio, 1944.
108. Hodgman, C. D., and Lange, N. A., "Handbook of Chemistry and Physics," 9th ed., Chemical Rubber, Cleveland, Ohio, 1922.
109. Holmes, B., and Kirk, P. L., *J. Biol. Chem.,* **116**, 377 (1936).
110. Höltje, R., *Z. anorg. u. allgem. Chem.,* **198**, 287 (1931).
111. Hubbard, D. M., and Spettelm, E. C., *Anal. Chem.,* **25**, 1245 (1953).
112. Hunter, G., *Analyst,* **83**, 93 (1958).
113. Hunter, G., *Analyst,* **84**, 24 (1959).
114. Ikeda, S., *Nippon Kagaku Zasshi,* **76**, 783 (1955).
115. Ishibashi, M., and Higashi, S., *Bunseki Kagaku,* **4**, 14 (1955).
116. Jensen, R., *Chim. anal.,* **37**, 53 (1955).
117. Jureček, M., and Jenik, J., *Chem. listy,* **51**, 1316 (1957).
118. Kadowaki, H., Okamoto, J., and Nakajima, M., *Yakugaku Zasshi,* **75**, 485 (1955).
119. Kahane, E., and Korach, S., *Mikrochemie ver. Mikrochim. Acta,* **36/37**, 781 (1951).
120. Kan, M., *Takeda Kenkyusho Nempo,* **11**, 54 (1952).
121. Karrman, K. J., and Borgstrom, S., *Svensk Kem. Tidskr.,* **67**, 18 (1955).
122. Kimbel, K. H., *Z. physiol. Chem., Hoppe-Seyler's,* **293**, 272 (1953).
123. Kingsley, G. R., and Schaffert, R. R., *Anal. Chem.,* **25**, 1738 (1953).
124. Kingsley, G. R., and Schaffert, R. R., *J. Biol. Chem.,* **206**, 807 (1954).
125. Kinnunen, J., and Merikanto, B., *Chemist Analyst,* **47**, 11 (1958).

126. Kirk, P. L., and Bentley, G. T., *Mikrochemie*, **21**, 250 (1937).
127. Kirk, P. L., and Tompkins, P. C., *Ind. Eng. Chem., Anal. Ed.*, **13**, 277 (1941).
128. Klein, B., and Jacobi, M., *Ind. Eng. Chem., Anal. Ed.*, **12**, 687 (1940).
129. Klimova, V. A., and Bereznitskaya, E. G., *Zhur. Anal. Khim.*, **11**, 292 (1956).
130. Klimova, V. A., and Bereznitskaya, E. G., *Zhur. Anal. Khim.*, **12**, 424 (1957); *Chem. Abstr.*, **52**, 1853 (1958).
131. Klimova, V. A., Korshun, M. O., and Bereznitskaya, E. G., *Doklady Akad. Nauk S.S.S.R.*, **96**, 81 (1954).
132. Klimova, V. A., Korshun, M. O., and Bereznitskaya, E. G., *Zhur. Anal. Khim.*, **11**, 223 (1956).
133. Knop, J., and Kubelkova, O., *Z. anal. Chem.*, **100**, 161 (1935).
134. Kondo, A., *Bunseki Kagaku*, **6**, 583 (1957); *Chem. Abstr.*, **52**, 15345 (1958).
135. Konig, O., Crowell, W. R., and Benedetti-Pichler, A. A., *Mikrochemie ver. Mikrochim. Acta*, **33**, 281 (1948).
136. Korenman, I. M., Sheyanova, F. R., and Glazunova, Z. I., *Zavodskaya Lab.*, **21**, 774 (1955).
137. Korshun, M. O., and Chumachenko, N. M., *Doklady Akad. Nauk S.S.S.R.*, **99**, 769 (1954).
138. Korshun, M. O., and Lavrovskaya, E. V., *Zhur. Anal. Khim.*, **3**, 322 (1948).
139. Kozawa, A., Tanaka, M., and Sasaki, K., *Bull. Chem. Soc. Japan*, **27**, 345 (1954).
140. Kreisky, F., *Mikrochim. Acta*, p. 242 (1959).
141. Kreshkov, A. P., Syavtsillo, S. V., and Shemyatenkova, V. T., *Zavodskaya Lab.*, **22**, 1425 (1956).
142. Kuck, J. A., and Grim, E. C., *Microchem. J.*, **3**, 35 (1959).
143. Kuhn, R., and Schretzmann, H., *Chem. Ber.*, **90**, 554 (1957).
144. Lacourt, A., *Mikrochim. Acta*, p. 550 (1954).
145. Lambert, R. H., and Walker, R. D., *Ind. Eng. Chem., Anal. Ed.*, **13**, 846 (1941).
146. Lamm, G. G., *Acta Chem. Scand.*, **7**, 1420 (1953).
147. Lang, R., *Z. anal. Chem.*, **93**, 21 (1933).
148. Lapin, L. N., and Makarova, V. P., *Pochvovedenie*, p. 82 (1953); *Referat. Zhur. Khim.*, Abstr. No. 22,146 (1954).
149. Leonard, G. W., Jr., Sellers, D. E., and Swim, L. E., *Anal. Chem.*, **26**, 1621 (1954).
150. Leroux, J., Maffett, P. A., and Monkman, J. L., *Anal. Chem.*, **29**, 1089 (1957).
151. Lewis, R. P., *Analyst*, **80**, 768 (1955).
152. Lindner, R., and Kirk, P. L., *Mikrochemie*, **22**, 291 (1937).
153. Lindstrom, O., *Anal. Chem.*, **31**, 461 (1959).
154. Lipscomb, W. N., *Anal. Chem.*, **25**, 737 (1953).
155. Llacer, A. J., Sozzi, J. A., and Benedetti-Pichler, A. A., *Ind. Eng. Chem., Anal. Ed.*, **13**, 507 (1941).
156. MacNevin, W. M., and Bournique, R. A., *Ind. Eng. Chem., Anal. Ed.*, **12**, 431 (1940).
157. Marier, J. R., and Boulet, M. A., *J. Agr. Food. Chem.*, **4**, 720 (1956).
158. Martin, F., *Mikrochemie ver. Mikrochim. Acta*, **36/37**, 660 (1951).
159. Martin, G., *Bull. soc. chim. biol.*, **34**, 1174 (1952).
160. Martin, G., *Bull. soc. chim. biol.*, **36**, 719 (1954).
161. McGregor, A. J., *Analyst*, **75**, 211 (1950).
162. McHard, J. A., Servais, P. C., and Clark, H. A., *Anal. Chem.*, **20**, 325 (1948).
163. Mehlig, J. P., *Ind. Eng. Chem., Anal. Ed.*, **13**, 533 (1941).
164. Meixner, A., and Kröcker, F., *Mikrochemie*, **5**, 131 (1927).

165. Mellor, J. W., "A Comprehensive Treatise on Inorganic and Theoretical Chemistry," Longmans, Green, New York, and London, 1922–1937.
166. Meyrowitz, R., and Massoni, C. J., *Anal. Chem.,* **27**, 475 (1955).
167. Milton, R. F., and Waters, W. A., "Methods of Quantitative Microanalysis," Longmans, Green, New York, and Arnold, London, 1949.
168. Milton, R. F., and Waters, W. A., "Methods of Quantitative Microanalysis," 2nd ed., Arnold, London, 1955.
169. Miura, H., *Kokumin Eisei,* **25**, 196 (1956); *Chem. Abstr.,* **52**, 18083 (1958).
170. Natelson, S., and Penniall, R., *Anal. Chem.,* **27**, 434 (1955).
171. Niederl, J. B., and Niederl, V., "Micromethods of Quantitative Organic Elementary Analysis," Wiley, New York, 1938.
172. Niederl, J. B., and Niederl, V., "Micromethods of Quantitative Organic Analysis," 2nd ed., Wiley, New York, 1942.
173. Niederl, J. B., and Sozzi, J. A., "Microanálisis Elemental Orgánico," Calle Arcos, Buenos Aires, 1958.
174. Nieuwenburg, C. J. van, and Blumendal, H. B., *Mikrochemie,* **18**, 39 (1935).
175. Nonowa, D. C., *Mikrochim. Acta,* p. 111 (1958).
176. Norton, A. R., Royer, G. L., and Koegel, R., *Ind. Eng. Chem., Anal. Ed.,* **12**, 121 (1940).
177. Nozaki, T., *Nippon Kagaku Zasshi,* **76**, 445 (1955).
178. Onishi, H., *Mikrochim. Acta,* p. 9 (1959).
179. Parry, E. P., *Anal. Chem.,* **29**, 546 (1957).
180. Patel, H. R., *Drug Standards,* **24**, 159 (1956).
181. Pflaum, D. J., and Wenske, H. H., *Ind. Eng. Chem., Anal. Ed.,* **4**, 392 (1932).
182. Pohl, E. A., and Demmel, H., *Anal. Chim. Acta,* **10**, 554 (1954).
183. Polley, D., and Miller, V. L., *Anal. Chem.,* **27**, 1162 (1955).
184. Pregl, F., "Quantitative Organic Microanalysis," (E. Fyleman, trans., 2nd German ed.), p. 136, Churchill, London, 1924.
185. Price, J. W., *Paint Manuf.,* **28**, 147 (1958); *Anal. Abstr.,* **6**, No. 1384 (1959).
186. Pungor, E., and Hegedüs, A. J., *Mikrochim. Acta,* p. 87 (1960).
187. Pungor, E., and Thege, I. K., *Mikrochim. Acta,* p. 712 (1959).
188. Rappaport, F., and Hohenberg, E., *Mikrochemie,* **14**, 119 (1934).
189. Rappaport, F., and Rappaport, D., *Mikrochemie,* **15**, 107 (1934).
190. Rauscher, W. H., *Ind. Eng. Chem., Anal. Ed.,* **10**, 331 (1938).
191. Robinson, R. J., and Hauschildt, J. D., *Ind. Eng. Chem., Anal. Ed.,* **12**, 676 (1940).
192. Rodden, C. J., *Mikrochemie,* **18**, 97 (1935).
193. Roth, F. J., *J. Assoc. Off. Agri. Chemists,* **40**, 302 (1957).
194. Roth, H., *Angew. Chem.,* **50**, 593 (1937).
195. Roth, H., "Die quantitative organische Mikroanalyse von Fritz Pregl," 4th ed., Springer, Berlin, 1935.
196. Roth, H., "F. Pregl quantitative organische Mikroanalyse," 5th ed., Springer, Wien, 1947.
197. Roth, H., "Pregl-Roth quantitative organische Mikroanalyse," 7th ed., Springer, Wien, 1958.
198. Roth, H., "Quantitative Organic Microanalysis of Fritz Pregl," 3rd ed. (E. B. Daw, trans., 4th German ed.), Blakiston, Philadelphia, Pennsylvania, 1937.
199. Rutgers, J. J., *Compt. rend. acad. sci.,* **190**, 746 (1930).
200. Sachs, G., *Analyst,* **78**, 185 (1953).
201. Sakuraba, S., *Bunseki Kagaku,* **4**, 496 (1955).

202. Saltzman, B. E., *Anal. Chem.*, **25**, 493 (1953).
203. Saltzman, B. E., *Anal. Chem.*, **27**, 284 (1955).
204. Schoklitsch, K., *Mikrochemie*, **18**, 144 (1935).
205. Scholtis, K., *Mikrochemie ver. Mikrochim. Acta*, **26**, 150 (1939).
206. Schuhknecht, W., *Z. anal. Chem.*, **157**, 338 (1957).
207. Schuhknecht, W., and Schinkel, H., *Z. anal. Chem.*, **143**, 321 (1954).
208. Schulek, E., *Mikrochemie, Emich Festschrift*, p. 260 (1930).
209. Schulek, E., and Wolstadt, R., *Z. anal. Chem.*, **108**, 400 (1937).
210. Schulitz, P. H., *Arch. Pharm.*, **286**, 506 (1953).
211. Schultz, Y. O., *Schweiz. med. Wochschr.*, **83**, 452 (1953).
212. Scott, F. W., *Chemist Analyst*, **27** (1938).
213. Shukis, A., Jr., and Tallman, R. C., *Ind. Eng. Chem., Anal. Ed.*, **12**, 123 (1940).
214. Silvert, F. C., and Kirner, W. R., *Ind. Eng. Chem., Anal. Ed.*, **8**, 353 (1936).
215. Sloviter, H. A., McNabb, W. M., and Wagner, E. C., *Ind. Eng. Chem., Anal. Ed.*, **13**, 890 (1941).
216. Smith, G. F., Chemical Co., "The Trace Element Determination of Copper and Mercury in Pulp and Paper," Columbus, Ohio.
217. Smith, G. F., Chemical Co., "The Wet Ashing of Organic Matter Employing Hot Concentrated Perchloric Acid. The Liquid Fire Reaction," Columbus, Ohio.
218. Smith, G. McPhail, "A Course of Instruction in Quantitative Chemical Analysis for Beginning Students," rev. ed., Macmillan, New York, 1922.
219. Socolar, S. J., and Salach, J. I., *Anal. Chem.*, **31**, 473 (1959).
220. Solomon, A. K., *Anal. Chem.*, **27**, 1849 (1955).
221. Southworth, B. C., Hodecker, J. H., and Fleischer, K. D., *Anal. Chem.*, **30**, 1152 (1958).
222. Steyermark, Al, *Ind. Eng. Chem., Anal. Ed.*, **17**, 523 (1945).
223. Steyermark, Al, "Quantitative Organic Microanalysis," Blakiston, Philadelphia, Pennsylvania, 1951.
224. Steyermark, Al, Alber, H. K., Aluise, V. A., Huffman, E. W. D., Jolley, E. L., Kuck, J. A., Moran, J. J., Ogg, C. L., and Willits, C. O., *Anal. Chem.*, **26**, 1186 (1954).
225. Steyermark, Al, Alber, H. K., Aluise, V. A., Huffman, E. W. D., Kuck, J. A., Moran, J. J., and Willits, C. O., *Anal. Chem.*, **21**, 1555 (1949).
226. Strahm, R. D., and Hawthorne, M. F., *Anal. Chem.*, **32**, 530 (1960).
227. Straub, J., *Mikrochemie*, **14**, 251 (1934).
228. Strebinger, R., and Mandl, J., *Mikrochemie*, **4**, 168 (1926).
229. Strebinger, R., and Reif, W., *Mikrochemie, Pregl Festschrift*, p. 319 (1929).
230. Sudo, T., Shimoe, D., and Miyahara, F., *Bunseki Kagaku*, **4**, 88 (1955).
231. Sykes, A., *Mikrochim. Acta*, p. 1155 (1956).
232. Teeri, A. E., *Chemist Analyst*, **43**, 43 (1954).
233. Thurnwald, H., and Benedetti-Pichler, A. A., *Mikrochemie*, **11**, 212 (1932).
234. Toribara, T. Y., and Sherman, R. E., *Anal. Chem.*, **25**, 1594 (1953).
235. Treadwell, F. P., and Hall, W. T., "Analytical Chemistry," 6th ed., Vol. II, Wiley, New York, and Chapman & Hall, London, 1924.
236. Tunnicliffe, M. E., *Trans. Inst. Rubber Ind.*, **31**, T141 (1955).
237. Tutundžić, P. S., and Mladenović, S., *Anal. Chim. Acta*, **12**, 390 (1955).
238. Umland, F., and Weyer, F. G., *Klin. Wochschr.*, **33**, 237 (1955).
239. Van Etten, C. H., and Wiele, M. B., *Anal. Chem.*, **25**, 1109 (1953).
240. Venkateswarlu, P., Ramanthan, A. N., and Narayana Rao, D., *Indian J. Med. Research*, **41**, 253 (1953).

6. Microdetermination of Metals: Ashing Method

241. Verdino, A., *Mikrochemie*, **6**, 5 (1928).
242. Walton, H. F., and Smith, A. A., *Anal. Chem.*, **28**, 406 (1956).
243. Well, I. C., *Anal. Chem.*, **23**, 511 (1951).
244. Wenger, P., Cimerman, C., and Corbaz, A., *Mikrochim. Acta*, **2**, 314 (1938); *Mikrochemie ver. Mikrochim. Acta*, **27**, 85 (1939).
245. Wernimont, G., and Hopkinson, F. J., *Ind. Eng. Chem., Anal. Ed.*, **12**, 308 (1940).
246. Wilkinson, R. H., *J. Clin. Pathol.*, **10**, 126 (1957).
247. Winkler, L. W., *Z. anal. Chem.*, **96**, 241 (1934).
248. Zabrodina, A. S., and Bagreeva, M. R., *Vestnik. Moskov. Univ. Ser. Mat. Mekhan., Astron.-Fiz. i Khim.*, **13**, No. 4, 187 (1958); *Chem. Abstr.*, **53**, 12946 (1959).
249. Zombory, L., *Technika (Budapest)*, **10**, 147 (1929).
250. Zuehlke, C. W., and Ballard, A. E., *Anal. Chem.*, **22**, 953 (1950).

CHAPTER 7

Microdetermination of Nitrogen by the Dumas Method

The micro-Dumas,* sometimes called the ultimate or gasometric method for the determination of nitrogen is based on the fact that organic compounds containing nitrogen, when decomposed at red heat with copper oxide yield nitrogen and some oxides of nitrogen. The oxides of nitrogen are reduced to nitrogen by a section of reduced copper. The reactions involved are represented by the following:

$$\text{Organic N} \xrightarrow{\text{CuO}} N_2 + \text{N oxides} + CO_2 + H_2O + Cu$$

$$\text{N oxides} \xrightarrow{\text{Cu}} N_2 + CuO$$

$$O_2 \xrightarrow{\text{Cu}} CuO$$

The entire procedure is carried out in an atmosphere of pure carbon dioxide after which the liberated nitrogen is swept from the tube with carbon dioxide and collected over potassium hydroxide. The volume of nitrogen is read at atmospheric pressure and its weight is calculated from the volume, pressure, and temperature.

This method was believed to successfully handle all types of compounds but experience has taught that many substances require some modification.[11,12,58,59,65,66,93,139,140,162-166,186,211,212] Unless experience has shown that the type to be analyzed gives difficulties, no modification should be used. Pyrimidines often give low results unless copper acetate[66] or potassium chlorate† or both[120,139,140,166,181,186,188] are added along with the sample. Compounds containing N-methyl groups also give low results probably due to formation of methylamine‡ instead of nitrogen.[186,188] The addition of copper acetate and

---
* Please see references 11, 12, 27, 48, 58, 59, 65, 121, 122, 139–141, 158, 163–166, 186, 212.
† The potassium chlorate liberates oxygen to aid in the combustion. The excess oxygen is trapped on the reduced copper.
‡ The methylamine partially dissolves in the KOH. This idea is substantiated by the fact that these same substances when determined by the Kjeldahl method yield methyl amine instead of ammonia.[186,188] The former can be detected by odor on making the distillate akaline—see Chapter 8.

potassium chlorate often causes the combustion to go to completion. Sulfonamides, hydrazines, hydrazones, semicarbazides, and nitrates often give low results and the addition* of potassium dichromate[48,58,59,166,186,188] or vanadium pentoxide to the sample is useful in these cases. However, many compounds are better analyzed by the Kjeldahl method, regular, or one of the two modifications described in Chapter 8. If samples, such as hydrazines, nitro-compounds, nitrates, etc., fail to give good results using the Dumas method, attempts should be made with the Kjeldahl, unless the samples are known to contain nitrogen linked to nitrogen in a ring, such as pyrazolones, 1,2-diazines, 1,2,3,-triazoles, etc. Nitrogen connected to oxygen (ring or chain) and nitrogen connected to nitrogen in a chain may be analyzed by the latter method—see Chapter 8. Substances containing long aliphatic chains often give *very* high results[93,186,188] due to the formation of methane instead of carbon dioxide during the combustion. Potassium chlorate and copper acetate are often useful in correcting this condition.[186,188]

Reagents

POTASSIUM HYDROXIDE SOLUTION, 50%

Reagent grade of potassium hydroxide pellets are dissolved in an equal weight of distilled water and stored in a rubber-stoppered Pyrex flask. This is used in the micronitrometer.[48,58,59,121,122,139-141,158,163-166,186]

MERCURY

About 15 ml. of mercury is used in the lower part of the reservoir of the micronitrometer.

CALOMEL

A few milligrams of calomel[137,157,186] are used in the micronitrometer to prevent gas bubbles from sticking to the mercury surface.

STOPCOCK GREASE

Sisco No. 300 high-vacuum stopcock grease† is the best lubricant for the micronitrometer stopcock. When some of the other greases are used there is considerable frothing of the potassium hydroxide solution which makes reading of the meniscus impossible. With the Sisco No. 300, there is no frothing. It is also used to lubricate the ball and socket joint attaching the needle valve and the micronitrometer—see below.

* The dichromate or varadium pentoxide must be kept in contact with the sample. Therefore, the normal method of mixing the sample with the copper oxide in the tube as described later is dispensed with.
† Swedish Iron and Steel Corp., Westfield, New Jersey.

KROENIG GLASS CEMENT[198]

Kroenig glass cement is used for sealing the joints on the combustion tube to those on the carbon dioxide generator and the needle valve—see below. It is warmed, applied, and allowed to cool to effect the seals.

COPPER TURNINGS (METALLIC)

Clean, degreased copper turnings are used as a section of reduced copper in the combustion tube.

COPPER OXIDE (CuO)

Two sizes of copper oxide are required, namely, coarse and fine powder. Reagent grade of copper oxide (CuO) wire is ground in a mortar and pestle

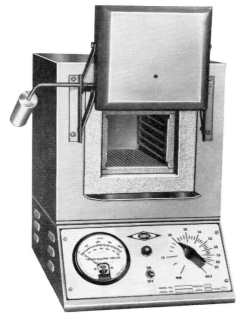

Fig. 88. Muffle furnace.

so that the pieces are about 1–3 mm. in length. The fine powder is obtained by grinding until the material will pass through sieves of 40–100-mesh. Before using, both sizes of copper oxide must be conditioned in the following manner. The material is put into nickel or stainless steel crucibles and then placed in a muffle furnace of the type shown in Fig. 88 which has a small hole opening in the door. Into this hole is inserted a small piece of quartz, Pyrex 1720* glass,

* Corning Glass Works, Corning, New York.

or metal tubing through which is passed a stream of carbon dioxide from a cylinder. The muffle is then heated to about 700° C. and kept at this temperature for 10–15 minutes. The muffle is then allowed to cool to room temperature during which time the stream of carbon dioxide is continued. The copper oxide is then removed from the muffle and stored in screw cap closed bottles. (The two sizes of copper oxide may be mixed for the conditioning and separated by sifting after being removed from the muffle.)

The above treatment destroys any organic material present and drives out pocketed air and replaces it with carbon dioxide. This eliminates any contamination and the replacing of air by the carbon dioxide makes it possible to obtain a much better grade of so-called microbubbles during the determination (see below—Procedure).

The copper oxide (both sizes) is used over and over again. After each determination the temporary filling (see below—Procedure) is removed, treated with carbon dioxide at 700° C., cooled in an atmosphere of carbon dioxide, sifted to separate the two sizes and reused for another determination.

Some of the copper oxide is reduced to metallic copper during the course of each determination. This does not need to be oxidized. In fact, its presence is possibly advantageous as it would act similarly to having additional bands of reduced copper, which is the practice in one large laboratory.[211,213]

DRY ICE

A pure grade of dry ice (solid CO_2) is used as a source of carbon dioxide.

Apparatus

The apparatus consists of a train composed of the following pieces, connected to each other in the order named: (1) source of carbon dioxide, (2) combustion tube containing copper oxide and copper, (3) needle valve for regulating the flow of gas through the train, and (4) the micronitrometer containing potassium hydroxide over which the nitrogen is collected and the carbon dioxide is absorbed. The combustion tube is surrounded by the heating elements of a combustion apparatus, which has two parts, namely, the long stationary furnace* and the short movable or sample furnace.* All of these parts are described in detail in the following pages.

* Common practice has been to refer to these parts as the "long burner" and "short burner" or "sample burner" and the entire outfit as the "combustion furnace." The new Tentative Specifications for the Apparatus for the Microdetermination of Nitrogen by the Dumas Method of the American Society for Testing Materials,[4] however, refer to the two parts as the "long furnace" and the "sample furnace."

DEWAR (THERMOS) FLASK[65,93,165,169,182,186,219]

A 2-liter (1.5 minimum) thermos flask (bottle) of the type shown in Figs. 89, and 94 is used as a carbon dioxide generator. As a safety measure, the flask should be completely wrapped with some form of adhesive tape to prevent shattering in case of breakage. It is filled with powdered dry ice as described below. In the top of the flask is a two-hole rubber stopper. Through one of these holes is a glass tube which connects the flask, through a stopcock, to the combustion tube by means of a 14/35 interchangeable ground joint (female member of the joint on the generator end)—see Figs. 89, and 94. In the other hole of the rubber stopper is a bent glass tube which leads to the bottom of a cylinder of water approximately 37–40 cm. in height (an ordinary 1-liter graduate cylinder is quite satisfactory). By this means, the excess CO_2 escapes through the head of water and maintains the gas under constant pressure. (Instead of the tube extending to the bottom of the water-filled graduate cylinder, the mercury valve shown in Fig. 102 may be used.)

Pure dry ice is crushed and ground to the size of rice grains, or smaller, by means of a mortar and pestle.* This removes air pockets from the material. The powdered dry ice is placed in the Dewar (thermos) flask, the stopper with the tube in the cylinder of water is attached and the system allowed to blow off the excess pressure overnight to rid the system of air before using for a determination. A good 2-liter vacuum bottle will continue to give CO_2 for about 10 days.

COMBUSTION APPARATUS

A unit of the type shown in Fig. 89 (one of those in use in the author's laboratories) is required for the determination. This combustion apparatus consists of a long furnace and a power driven† short or sample furnace. The furnaces are composed of nichrome wire windings in asbestos blocks‡ and are of the split type that can be pushed back away from the combustion tube. The long furnace is 205 mm. in length and can accommodate combustion tubes whose outside diameters are up to 13 mm. The short furnace also accommodates this size tube and is 110 mm. in length. The windings of each are of such resistance, that when operated through a variable auto-transformer, the desired operating temperature of 680° C., in the combustion tube, may

---

* Where a number of units are in operation, a power-driven dry ice pulverizer is recommended. Such units are commercially available, made by Franklin P. Miller & Son, Inc., 36 Meadow Street, East Orange, New Jersey (their model Supreme 16 Pulverizer is used by the author).

† Electric clock motors[185] are quite suitable for this purpose as their r.p.m. is low enough so that they need no reduction gear system. Some of the available commercial furnaces are not mechanized and the above furnishes a rather cheap means of doing so.

‡ Available from C. and H. Menzer Co., Inc., 105 Barclay Street, New York.

be attained. The short furnace travels at the rate of approximately 7 mm. per minute and can travel 75–100 mm. Safety shields* are attached to the furnaces in the vicinity of where the sample is placed, for the protection of the analyst.

Suitable electric combustion apparatus are commercially available,[16,171,198] such as those shown in Figs. 90, 124, and 125 (Chapter 9), minus the heating mortars (Chapter 9) shown in Figs. 124 and 125, which are required

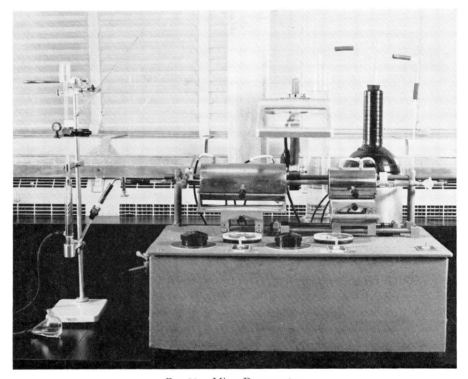

FIG. 89. Micro-Dumas setup.

for the carbon-hydrogen determination. The same type of apparatus (long and short furnaces) may be used for several of the combustion procedures described in the following chapters. The furnaces shown are mechanized and have wide ranges of operating temperatures making them suitable for the various combustion methods. Radiant heating is employed in the Sargent unit.

Recently, tentative specifications for the combustion furnace have been adopted by the American Society for Testing Materials. They are as follows[4]:

* Stanley No. 600 shields, Stanley, New Britain, Connecticut.

"COMBUSTION UNIT

The combustion unit shall consist of a long furnace, a sample furnace, a combustion tube, a combustion tube closure, and a connection tube.

(a) *Long Furnace:*

(1) The long furnace shall have a maximum overall length of 8 in. (203 mm.) with the wall thickness at the ends not to exceed $\frac{1}{4}$ in. (6 mm.). The heating well in the furnace shall accommodate combustion tubes up to 13 mm. outside diameter. Electric heating elements shall be easily replaceable. The furnace shall be mounted firmly on a substantial support.

FIG. 90. Thomas combustion furnace.

(2) The furnace shall be capable of continuous operation at temperatures up to 800° C., as measured inside the combustion tube in the middle of the furnace. The temperature drop from the center to points 1 in. (25 mm.) and 1¾ in. (45 mm.) from either end shall not exceed 15 per cent and 7 per cent, respectively, based on the temperature in the middle of the furnace. Means shall be provided for varying the temperature.

(3) Provision shall be made for rapid cooling of the combustion tube between analyses. The furnace shall be designed so that it may either be moved away from the tube or turned off and cooled rapidly. In the latter case, the time required for cooling and reheating to operating temperature shall not exceed 30 min.

(4) The furnace shall be equipped with some device for continuous or provisional indication of the temperature at the middle of the furnace.

(b) *Sample Furnace:*

(1) The sample furnace shall have an overall length not less than 2½ in. (65 mm.) nor more than 4 in. (102 mm.) with wall thickness at the ends not to exceed ¼ in.

7. Nitrogen: Micro-Dumas Method

(6 mm.). The heating well in the furnace shall accommodate combustion tubes up to 13 mm. outside diameter. Electric heating elements shall be easily replaceable. The furnace shall be mounted firmly on a substantial support.

(2) The furnace shall be capable of continuous operation at temperatures up to 800° C., as measured inside the combustion tube in the middle of the furnace. The temperature drop from the center to points $3/4$ in. (19 mm.) from either end shall not exceed 17 per cent of the temperature in the middle of the furnace. Means shall be provided for varying the temperature.

(3) Items (3) and (4) of Paragraph (*a*) also apply to the sample furnace.

(4) If the furnace is moved mechanically, it shall travel a distance of $6\frac{1}{4}$ in. (159 mm.) min., with provision for manual setting at any point. Whether the furnace has only one or more than one speed of movement over this distance, the rates of travel shall be within the limits of $1/8$ to $5/8$ in. (3 to 16 mm.) per min. An automatic control shall be provided to stop the travel of the sample furnace when it reaches the long furnace."

COMBUSTION TUBE

The combustion tube shown in Fig. 91 is prepared by sealing two male members of interchangeable ground joints to the ends of a piece of Pyrex 1720* glass tubing approximately 520 mm. in length, 11 mm. O.D., 8 mm. I.D.† To the one end is sealed a 7/25 joint and to the other a 14/35 member.‡ The large end connects to the carbon dioxide generator while the small end is attached to the needle valve (see Fig. 92). [Male joints are used on the combustion tube so that the cement used to hold the parts together (see below) does not enter the combustion tube.]

FILLING THE COMBUSTION TUBE (PERMANENT FILLING)—FIG. 91. Only new tubes should be used. No attempt should be made to clean out the old filling from a tube and replace it with new material. With constant use a tube will usually give good results for at least one month. When good determinations are no longer obtained, the filling may be recovered but the tube should be discarded. Most tubes develop cracks or pinholes where the copper oxide fuses through, accounting for the failure. The filling is composed of two parts, the permanent and the temporary. The new tube, before filling, should be dusted with a cotton swab. Enough asbestos fibers (acid-washed) should be placed in the tube and pushed into place with a glass rod to form a plug of 3–4 mm. in length. About a 15-cm. section of the coarse copper oxide wire is then added with tapping to cause the particles to pack solidly. Another asbestos plug of 3–4 mm. in length is added with the aid of the glass rod. Enough metallic copper turnings are added then with the aid of a glass ramrod to make a tightly packed section 4 cm. in length. Another asbestos plug 3–4 mm.

---

* Corning Glass Works, Corning, New York. Vycor (also Corning) or quartz may be used.

† For this tubing use combustion tube with the tip removed—see Fig. 98.

‡ Compare references 56, 121, 122, 210.

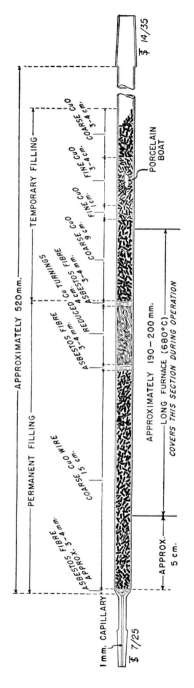

Fig. 91. Dumas tube, permanent and temporary fillings.

in length is added. This comprises the permanent filling which remains in place for the life of the tube (Fig. 91). A new temporary filling is added to the tube for each determination—see below under Procedure.

NEEDLE VALVE[4,70,189,190]

The all-metal needle valve[4,189,190] (Fig. 92), is used to regulate the flow of gas through the train. A female member of a 7/25 interchangeable ground glass joint is attached to the horizontal portion on the right and a $\mathbf{\S}$ 12/1½ standard glass ball joint is attached to the downward slanting tube on the left.

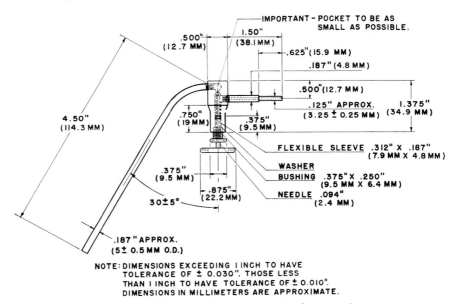

FIG. 92. All-metal needle valve—details of construction.

The glass and metal parts are connected by means of heavy-walled paraffin impregnated rubber tubing of 2-mm. bore using a trace of glycerin as a lubricant (see chapter on carbon-hydrogen determination). The 7/25 joint connects to the smaller joint of the combustion tube while the ball joint fits the socket joint of the micronitrometer (see below). Gas passing through the combustion tube enters the valve through the horizontal capillary tube on the right, passes upward through the conoidal seat and then downward through the slanting tube on the left into the micronitrometer. The flow of gas through the valve is precisely regulated by turning the needle provided with a large knurled disk. The fittings are made gas-tight by means of a flexible sleeve, but a droplet of mercury is used at the point indicated as an additional precaution against leakage. The mercury is introduced through the horizontal capillary. The valve must not leak with working pressures up to and including 10 pounds per square inch.

MICRONITROMETER (MICROAZOTOMETER)[4,48,58,59,65,93,121,122,139-141,158,163-166,186,190,212]

The micronitrometer (Fig. 93) is the reservoir for collecting the nitrogen obtained during the combustion. It consists of (a) a downward slanting

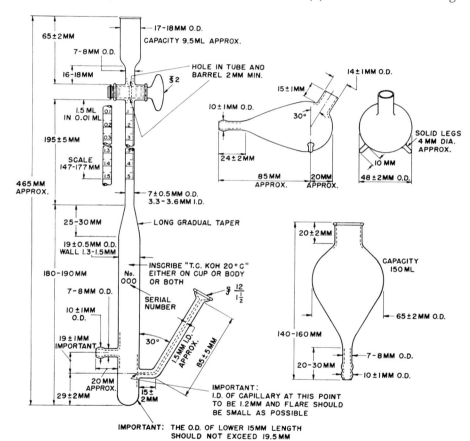

FIG. 93. Micronitrometer, Pregl-type, with leveling bulbs—details of construction.

capillary on the right through which the gas enters, (b) a reservoir to contain 50% KOH for absorbing the carbon dioxide, (c) a leveling bulb* which

---

* Fig. 93 shows two types of leveling bulbs[4,58,59,139-141,158,163-169,186,190]—either may be used. The one at the top rests on the table during the determination while the bottom one must be held in an iron ring on the stand.

7. Nitrogen: Micro-Dumas Method

is connected with rubber tubing to a small side arm on the left and is used for raising and lowering the KOH, (d) a graduated portion for reading the volume of gas, (e) a stopcock and (f) a cylinder above the stopcock.

In the lower portion of the reservoir is placed enough mercury to be above the point where the downward slanting capillary on the right enters the reservoir but below the small side arm on the left connected to the leveling bulb. Gas entering the micronitrometer passes up through the mercury. In order to prevent gas bubbles from sticking to the mercury surface, a few milligrams of calomel (Hg_2Cl_2)[137,157,186] are added.* The leveling bulb is then filled with a 50% solution of pure KOH. [The KOH is replaced about once a week, or sooner if the bubbles (see below) rise more than about 5 cm. before reducing in size.]

Connection to the needle valve is made by means of the ball and socket joint which is lubricated with Sisco No. 300 high-vacuum grease and the two parts of the joint are held together by means of a suitable clamp.†

The graduated portion of the tube has a capacity of 1.5 ml. and is divided into hundredths. Being a precision piece, calibration by the manufacturer is necessary and should be done according to the directions of the American Society for Testing Materials which are as follows:[4,81,133]

"The volume contained by the nitrometer at 20° C shall be certified by the manufacturer to the nearest 0.001 ml. at 6 points. The following procedure, in use by the National Bureau of Standards, is recommended.‡

Calibrate the nitrometer, without the chamber, at the 0.1, 0.3, 0.5, 0.8, 1.1, and 1.5-ml. marks. After cleaning and drying, lubricate the stopcock with a minimum of petroleum jelly. Tare the instrument and mount it in the inverted position with the stopcock closed. Suspend a thermometer beside it and place a transparent screen around the assembly. Pour mercury into a small glass reservoir provided with a stopcock and having a long slender delivery tube. Insert the delivery tube into the tube of the nitrometer down to the stopcock for the initial filling at the 0.1.-ml. mark.

Carefully run in the mercury, raising the delivery stem as the filling progresses until the orifice is just above the surface. Fill nearly to the test mark, remove the stem, and by means of a long slender steel rod, probe around the filled portion of the nitrometer to work out air pockets. Insert the delivery stem nearly to the surface of the mercury and fill to slightly below the mark, agitate the tube to round the meniscus, and, if necessary, add or remove mercury to complete the setting on the line. Round the meniscus for every reading. After observing the temperature, allow the nitrometer to stand for about 3 min, check the setting and temperature, and, if these are satisfactory, weigh the nitrometer. Again clamp the nitrometer in position and, after standing for 5 min, insert the delivery stem, fill to the next test mark, and weigh the instrument as before. Follow the same procedure for the other points. Air pockets at the stopcock seem to be the main source of trouble.

* When the KOH is added, the calomel is converted to the oxide.
† Arthur H. Thomas Company clamp, catalog No. 3241.
‡ See references 81 and 133.

In general practice, the actual volume above 50 per cent potassium hydroxide solution in contact with glass walls is accepted as being 0.001 ml. less than that found by the mercury calibration, based on a determination by the Physikalisch-Technische Reichsanstalt."

Micronitrometers having stopcocks with Teflon* plugs are now commercially available† and indications are that this type will become popular. Obviously, no lubricant should be used with these.

Procedure

Enough sample should be taken to obtain 0.3–0.4 ml. of nitrogen. If the sample is either a solid or a high-boiling liquid, it should be weighed in a clean porcelain boat (if hygroscopic, with the aid of the weighing pig—see Chapter 3). If the sample is a volatile liquid, it should be weighed in a capillary containing potassium chlorate—see Chapter 3.

TEMPORARY FILLING

To the combustion tube containing the permanent filling is added an approximate 9-cm. layer of coarse copper oxide followed by 1 cm. of fine copper oxide powder. The sample is then added (in the boat or capillary) to the combustion tube. Another 2 cm. of fine copper oxide powder is added and the combustion tube rotated with an occasional tilting so that the solid sample becomes mixed with the fine powder. Another 1–2 cm. of fine is added (with tapping) followed by several (3–4) cm. of coarse (see Fig. 91). This completes the temporary filling. (Note: If the sample is a pyrimidine, pyrazine, semicarbazide, or contains N-methyl groups or long aliphatic chains, 20–30 mg. of copper acetate and 10–20 mg. of potassium chlorate[120,139,140,166,181,186] should be added to the sample in the boat before being placed in the combustion tube. If the sample is a sulfonamide or a nitrate it should be covered in the boat by about 50 mg. of powdered potassium dichromate‡ but,[186] in this case, the sample is *not* mixed with the copper oxide by rotating the tube as the dichromate must remain on the sample during the combustion to be effective. Even though the above-mentioned types are not being analyzed, if poor results are obtained and the apparatus is generally giving good results, either or both of the above modifications should be tried. If these fail, vanadium pentoxide should be mixed with the sample as this material has been used successfully with substances difficult to burn.[39])

* E. I. du Pont de Nemours & Co., Inc., Wilmington, Delaware.
† Kimble Glass Company, Vineland, New Jersey and Arthur H. Thomas Company, Philadelphia, Pennsylvania.
‡ Compare Furman,[48] Roth[166] and Grant.[58,59]

ASSEMBLING THE TRAIN
(FIGS. 89, 91, AND 94)

The filled combustion tube is then placed in the cold combustion furnace (the ⚥ 7/25 joint plus 5 cm. of the CuO filling extending beyond the long furnace) and attached to the carbon dioxide generator, using Kroenig glass cement as a sealing agent between the members of the ground joint. (The two members are warmed gently with a flame, a film of the wax applied to the male member, the parts joined, and held together until cool.*) The stopcock on the generator is opened and carbon dioxide is passed through the tube for about 5 minutes before the furnace is heated. This removes the air from the tube and prevents the oxidation of the metallic copper section. The long furnace is then heated to 670°–680° C.[212] (Avoid too high a temperature since the equilibrium between carbon dioxide and carbon monoxide is shifted toward the latter as the temperature is increased and a good grade of microbubbles will not be obtained—see Chapter 14.) During this time, the short movable sample furnace should be on the extreme right away from the long furnace. (If the two furnaces were in contact the conduction of heat might be enough to decompose the sample before desired.) The stopcock on the generator is closed* and the combustion tube is now connected to the needle valve-micronitrometer combination using Kroenig glass cement as a sealing agent between the members of the ground joint (see above).† The stopcock at the top of the micronitrometer is opened then and the leveling bulb containing KOH is raised so that the solution fills the graduated tube and rises about halfway into the cylinder above the stopcock, after which the stopcock on the nitrometer is closed. (KOH must be in the cylinder above the cock at all times or air might suck into the system at this point.)

THE SWEEP OUT

The leveling bulb is then placed on the table (or in a ring near the table top if it is the conventional type). The needle valve is opened, the stopcock on the generator opened, and carbon dioxide allowed to sweep rapidly through until the bubbles in the micronitrometer become so-called "micro" in size. These bubbles are obtained when all of the air has been replaced in the system and only carbon dioxide is coming through. Obviously, the size of the bubble will depend on the purity of the gas. For some generators and with fresh

---

* Otherwise the carbon dioxide pressure is apt to force a channel through the warm wax, giving a poor seal.

† When analyzing fluorine-containing compounds with a high percentage of that element, there is danger of clogging the capillary of the micronitrometer with deposited silica. This may be prevented by including in the setup a small bubble counter (of the type on the bubble counter-U-tube, Fig. 120, Chapter 9) which is filled with water. This is placed between the combustion tube and needle valve.[142]

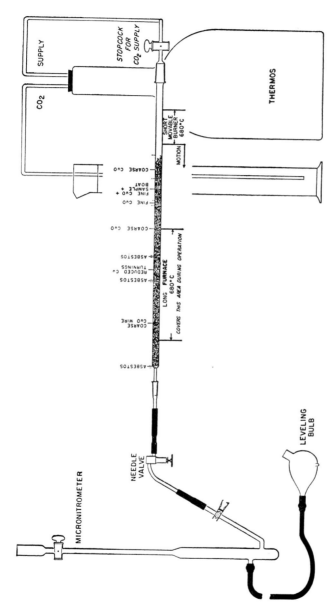

FIG. 94. Diagram of Dumas apparatus. First combustion—stopcock on CO_2 supply closed, needle valve open, short movable furnace (burner) moves over to the long furnace (burner). Second combustion—stopcock on CO_2 supply open, needle valve partly open, short movable furnace (burner) again traverses distance up to the long furnace (burner)—second burning. Sweep out—stopcock on CO_2 supply open, needle valve partly open, short movable furnace (burner) off. For tube filling, see Fig. 91.

KOH in the micronitrometer, the bubbles will almost vanish while with others a much larger bubble will be obtained. However, bubbles are considered to be micro in size, and the gas suitable for use, if their diameter is less than the distance in the graduated tube representing 0.002 ml. (diameter = $\frac{1}{5}$ distance between graduations of 0.01 ml.). If the diameter of these bubbles is assumed to be approximately 0.2 mm. the volume of each bubble would be about 0.004 cu. mm. or 250 would represent a volume of approximately 0.001 ml. Bubbles coming through at the rate of one per second would represent a total volume of about 0.014 ml. per hour.[158]

(Note: If the generator has been giving a good grade of bubble and suddenly does not, the system should be checked for a leak in the following manner. After the bubbles no longer reduce in size, the stopcock on the carbon dioxide generator is closed and the system allowed to stand with the leveling bulb on the table top for 15–20 minutes. This keeps the system under reduced pressure and if a leak is present, air will suck in. The stopcock on the carbon dioxide generator is then opened and if a leak is present, fairly large bubbles will appear in the micronitrometer and a readable quantity of gas will appear in the graduated portion. If there is no leak, the first few bubbles will be slightly larger, but then they will rapidly again reduce in size and no readable amount of gas will collect within several minutes.)

FIRST COMBUSTION

As soon as microbubbles are obtained, the leveling bulb is raised and the stopcock on the micronitrometer is opened to push out any gas present. The stopcock is then closed and the bulb returned to the table top. The stopcock on the carbon dioxide generator is then closed, the needle valve is opened a little more by turning the disc counterclockwise several turns, and the short furnace heated. When the temperature of the sample furnace is 670°–680° C. it is moved over towards the position in the tube near to the sample. This should be done with extreme care as too rapid a combustion leads to very high results.* As the sample begins to burn, bubbles will come through into the micronitrometer. Burning must be done slowly enough so that bubbles come through at a rate of one in 2 seconds or less.[58,59,139,140,158,163-166,186] If burning is continuing at the desired rate, the mechanical drive may be set into motion. The movable furnace is allowed to pass over the entire section of the tube outside of the long furnace. This operation is called the first combustion. When the movable furnace has traversed this distance it is then allowed to remain against the long furnace for several minutes. When the sample has been completely burned, a slight reduction in pressure will be

* Samples containing no nitrogen will give several tenths of a milliliter of gas if burned very rapidly.[186]

noted in the tube as evidenced by the mercury rising in the capillary on the micronitrometer.

SECOND COMBUSTION

The needle valve should then be closed and the short movable furnace moved all of the way back to the right side of the tube. The hot furnace expands the cold gas in the end of the combustion tube and also volatilizes any water condensed in the right portion of the tube. This would cause a rapid flow of gas into the micronitrometer if the needle valve were not closed. The needle valve is slightly opened so that the expanded gas passes into the micronitrometer at the rate of one bubble per second. As soon as the bubbles cease passing through, the needle valve is momentarily closed, the stopcock on the carbon

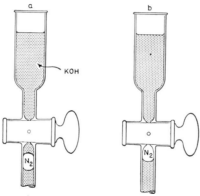

FIG. 95. Cleaning top of micronitrometer. (a) Small amount of KOH sucked into graduated portion during combustions. (b) Appearance after the KOH has been pushed out and N_2 occupies its proper place for reading volume.

dioxide generator is opened and then the needle valve adjusted so that the rate into the micronitrometer continues as above. The hot movable furnace is passed along the tube at such a rate that about 10 minutes is required for it to traverse again the section of the tube up to the long furnace. This is known as the second combustion. It assures against any unburned material remaining. After the movable furnace has passed over the above distance, it is again pulled back to the far right end and allowed to cool. By this time, the bubbles coming through the micronitrometer begin to reduce in size considerably. By means of the needle valve, the rate of flow is increased next to two bubbles per second and finally to about five, as microbubbles are approached.[158] When microbubbles are obtained, the leveling bulb is raised to a position above the cylinder at the top of the micronitrometer and, with the magnifier, the portion of the graduated tube is examined. If KOH has sucked in the top above the gas (Fig. 95a), it is completely pushed out into the bore of the plug by *carefully* turning the stopcock (Fig. 95b). (Caution: Do not allow the gas to enter the stopcock bore or this will represent a loss. If a bubble does enter the bore, it can be recovered by the following. The leveling bulb is lowered to the table top and the stopcock turned quickly so that the bubble plus a little KOH from the cylinder above the stopcock are pulled down into the graduated tube and then the stopcock is closed. The leveling bulb is raised then above the cylinder and the stopcock carefully turned to push the liquid into the bore of the plug.) After standing at least 10 minutes, to insure drainage,[158] the leveling bulb is lowered so that the surface of the liquid in the bulb and in the graduated tube stand at the same height (gas in the micronitrometer at atmospheric pressure) and the volume is read with the aid of the magnifier, estimating the thousandths of a milliliter. The temperature as recorded by a thermometer hanging next to the micronitrometer and at the level of the gas is noted, being read to the nearest half degree.* The atmospheric pressure in the room is recorded to the nearest half millimeter,* using a mercurial barometer of the type shown in Fig. 96.

Calculation of Results

The volume of gas read in the graduated tube must have two corrections applied to it. First, the correction found on calibration by the manufacturer, as shown in the certificate furnished with all precision micronitrometers. The calibration is made at intervals of 0.1 ml. or more (see calibration of micronitrometers under Micronitrometer, above). All micronitrometers are calibrated with mercury in the inverted position. According to the accepted practice[4,81,133] diminishing these values by 0.001 ml. gives the calibration for 50% KOH solution. Some manufacturers furnish a certificate merely showing the correct values at the several intervals over mercury, while others interpret the correction for all points and include the 0.001 ml. deduction in the values. The two types of certificates encountered are shown in the following examples. With the micronitrometer, one manufacturer might furnish certificate *A,* while another might furnish certificate *B* (Table 17).

The second correction for volume is one essentially due to the volume of KOH adhering to the walls of the tube and to the alteration in vapor tension. For these, 2% is deducted from the volume of gas after correction for the calibration.[58,59,158,163-166,186]

A temperature correction for the barometer must also be made.[72] To reduce the readings of a mercurial barometer with a brass scale (type shown in Fig. 96) to 0° C., subtract 3 mm. from the reading at room temperature. The deduction is 3.0 mm. for only a very few combinations of temperature and pressure but

* Compare references 58, 59, 137, 139, 158, 163–166.

for most laboratory conditions the corrections range from 2.5–3.5 mm.[72] and since the barometric pressure is only read to the nearest 0.5 mm., a flat deduction of 3 mm. suffices. Where extreme laboratory conditions exist, the exact deduction should be used (see table in reference cited).

From the temperature, volume, and pressure, the results are calculated using the gas laws. The Nitrogen Reduction Table[139,140] given on pp. 574–583,

TABLE 17

A:

CORRECTION TABLE

Interval (ml.)	Contains (ml.)
0–0.1	0.100
0–0.3	0.300
0–0.5	0.496
0–0.8	0.797
0–1.1	1.097
0–1.5	1.497

(Calibrated with mercury in the inverted position. To obtain the volume of gas over 50% KOH solution deduct 0.001 ml.)

B:

CORRECTION TABLE—MICROPRECISION NITROMETER, PREGL

(Recommended Specifications A.C.S. 1949) Contains at 20° C., KOH
Serial No. 114 Date of Certification January 14, 1950

ml.	0.00	0.01	0.02	0.03	0.04	0.05	0.06	0.07	0.08	0.09
				Corrected volumes in milliliters						
0.0	—	—	—	—	—	0.049	0.059	0.069	0.079	0.089
0.1	0.099	0.109	0.119	0.129	0.139	0.149	0.159	0.169	0.179	0.189
0.2	0.199	0.209	0.219	0.229	0.239	0.249	0.259	0.269	0.279	0.289
0.3	0.299	0.309	0.319	0.329	0.339	0.349	0.359	0.369	0.379	0.389
0.4	0.399	0.409	0.418	0.428	0.438	0.447	0.457	0.466	0.476	0.485
0.5	0.495	0.505	0.515	0.525	0.535	0.545	0.555	0.565	0.575	0.585
0.6	0.595	0.605	0.615	0.625	0.635	0.645	0.655	0.665	0.675	0.685
0.7	0.695	0.705	0.715	0.725	0.735	0.746	0.756	0.766	0.776	0.786
0.8	0.796	0.806	0.816	0.826	0.836	0.846	0.856	0.866	0.876	0.886
0.9	0.896	0.906	0.916	0.926	0.936	0.946	0.956	0.966	0.976	0.986
1.0	0.996	1.006	1.016	1.026	1.036	1.046	1.056	1.066	1.076	1.086
1.1	1.096	1.106	1.116	1.126	1.136	1.146	1.156	1.166	1.176	1.186
1.2	1.196	1.206	1.216	1.226	1.236	1.246	1.256	1.266	1.276	1.286
1.3	1.296	1.306	1.316	1.326	1.336	1.346	1.356	1.366	1.376	1.386
1.4	1.396	1.406	1.416	1.426	1.436	1.446	1.456	1.466	1.476	1.486

7. Nitrogen: Micro-Dumas Method 170

Table 43, Chapter 23, is used in the calculation. One ml. of nitrogen at 0° C. and 760 mm. of pressure weighs 1.2505 mg. The table gives the logarithm of the weight of one ml. of nitrogen at $t°$ C. and p mm. The calculation is shown in the following example. Suppose that during a determination the micro-

FIG. 96. Barometer.

nitrometer whose calibration is shown by the two certificates A and B (Table 17) is used. Also suppose the following:

Wt. sample 3.962 mg.

Vol. of N_2 as read (not corrected) = 0.351 ml.

Temperature recorded by thermometer near micronitrometer = 23° C.

Barometric pressure in laboratory = 761 mm.

Calculation:

Vol. N$_2$ read		0.351 ml.
Vol. correction for calibration	=	—0.001 ml.
Vol. N$_2$ corrected for calibration	=	0.350 ml.
Minus 2% (0.350 × 0.02 = 0.007)	=	0.007 ml.
Vol. of N$_2$ corrected		0.343 ml.
Barometric pressure read	=	761 mm.
Temperature correction		—3
Corrected barometric pressure	=	758 mm.
Log. wt. 1 ml. N$_2$ at 758 mm. and 23° C. (from Table 43, pp. 574–583)	=	0.06076
Plus log of corrected vol. of N$_2$ (0.343 ml.)	=	9.53529 — 10
Log. wt. of N$_2$ obtained	=	9.59605 — 10
Minus log. wt. sample (3.962)	=	0.59791
Log. fraction N	=	8.99814 — 10
Plus log. 100	=	2.00000
Log. % N	=	10.99814 — 10
% N	=	9.96

The allowable error of the method is ± 0.2% (± 0.3% for high percentages of nitrogen).

ADDITIONAL INFORMATION FOR CHAPTER 7

The original micro-Dumas setup of Pregl[58,59,158,163-166] consisted of a Kipp generator similar to the type[4] shown in Fig. 97, a combustion tube similar to that[4] shown in Fig. 98, a three-way stopcock similar to that[4,190] shown in Fig. 99 but without the ball joint, and a micronitrometer (similar to that of Fig. 93 but without the socket joint). (Note: The figures above referred to differ somewhat from those of Pregl[158] since they are those recommended by the Committee for the Standardization of Microchemical Apparatus[190] of the Division of Analytical Chemistry of the American Chemical Society and Subcommittee 29 on Microchemical Apparatus under Committee E1 on Methods of Testing of the American Society for Testing Materials.[4]) The various parts were connected by means of sections of heavy-walled rubber tubing impregnated with paraffin and by one rubber stopper (in the end of the combustion tube). The combustion tube was heated by means of gas burners.

Niederl and Niederl[139,140] describe a setup using two Kipp generators

7. Nitrogen: Micro-Dumas Method

and a gasometer (Fig. 100). The authors[139,140,199,200] use a blank and a 1.1% correction.

The author strongly advocates the use of interchangeable ground joints, although some workers still use rubber stoppers and rubber connectors. For those who so prefer, the American Society for Testing Materials[4] tentative specifications include the connection tube (Z-tube) (Fig. 101) and the one-

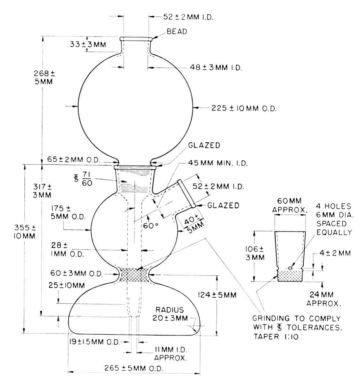

FIG. 97. Kipp generator (2000-ml. capacity)—details of construction.

hole rubber stopper,[4,187] (Fig. 153, Chapter 10). Also for those who wish to use a combustion tube with side arm (Fig. 121, Chapter 9), the above specifications include the combustion tube with tip and side arm and the solid rubber stopper[4,187] (Fig. 165, Chapter 12).

The mercury valve[4,71,186,190] shown in Fig. 102 can be used to maintain the desired pressure of carbon dioxide in the thermos flask and allows the excess pressure to be released. It consists of a center tube which is placed in the hole of a one-hole rubber stopper in the neck of the thermos flask. Connected to this is the mercury reservoir containing a small head of mercury (about 2.5–5 cm.).

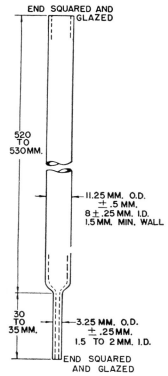

FIG. 98. Standard Dumas combustion tube with tip—details of construction.

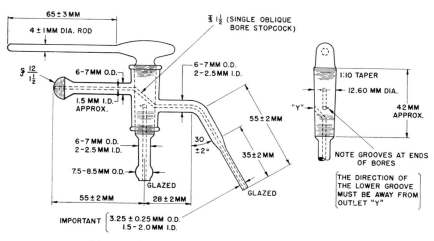

FIG. 99. Precision stopcock—details of construction.

A paper cap between the mercury column and the carbon dioxide generator acts as a bleeder for the excess pressure. The stopcock connects the flask to the combustion tube.

Besides the dry ice-filled thermos bottle as a source of carbon dioxide, Kipp generators (see above) charged with marble and hydrochloric acid are used.[11,12,58,59,121,122,137–139,158,163–166] The marble must be specially treated but,

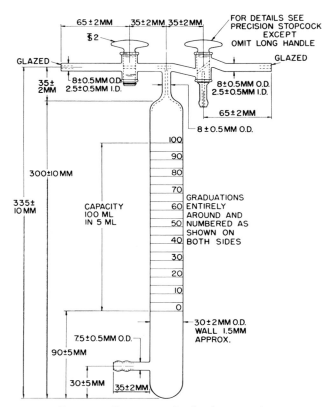

FIG. 100. Gasometer—details of construction.

in spite of this, air pockets down in the body of the pieces are released, in time causing contamination resulting in a poor grade of microbubbles.

Royer, Norton and Foster[169] and Aluise[3] have used carbon dioxide from cylinders but this source is a rather uncertain one, some cylinders supplying an excellent grade of gas while from others the microbubbles are of a poor grade.

To eliminate this uncertainty, Gustin[63] purified the gas by storing the cylinder at —70° C. overnight. The remaining pressure in the frozen cylinder is quickly blown off to the atmosphere. This purge is repeated about four or five times about 5 minutes apart after the cylinder has been removed from the cold storage. The cylinder is then allowed to come to room temperature and is ready for use.

Pomatti[155] introduced an iron wire into the micronitrometer which is floated on the mercury and moved over the surface by means of a magnet. By this means, bubbles are prevented from adhering to the mercury surface.

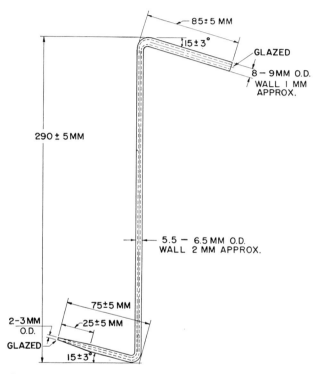

Fig. 101. Connection tube (Z-tube)—details of construction.

To accomplish the same thing, Kirsten and Wallberg-Olausson[103] added a trace of carbon disulfide to the potassium hydroxide to produce a mercury sulfide film. Finely divided copper oxide[58,59,139,140,166] and a trace of amyl alcohol[157] or mercurous oxide[137] have also been used for this purpose. However, in the author's opinion, the use of calomel (yielding mercurous oxide) described in this chapter, provides the simplest and most efficient means.

Barium hydroxide has been added to the potassium hydroxide to prevent foaming.[58,59,158,163–166] The author finds, however, that this is not necessary if reagent grade of potassium hydroxide is used. He has found that when certain types of stopcock grease are used on the stopcock of the micronitrometer,

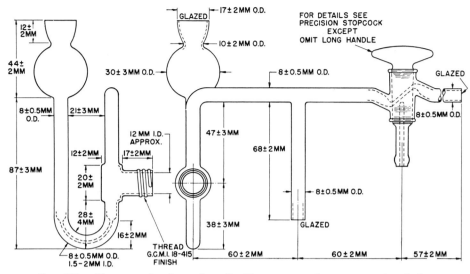

FIG. 102. Mercury valve for carbon dioxide generator (pressure regulator) for use with dry ice—details of construction.

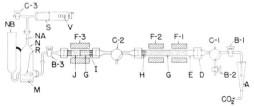

FIG. 103. Diagram of Gustin automatic modification. Combustion and absorption system. (A) Flow meter, 5–300 ml./min. flow range. (B-1, B-2, and B-3) Capillary stopcock, 2 mm. straight bore plug. (C-1 and C-2) Capillary stopcock, three way, 120°, 2 mm. bore plug. (C-3) Capillary stopcock, 2 mm. straight bore plug. (D) Synthetic rubber pressure tubing (neoprene), ¼ inch i.d., ¾ inch o.d. (E) Combustion tube, 96% silica, 7 mm. i.d., 9 mm. o.d. (F-1 and F-2) Sample furnace, movable split type, 2½–3 inches long 12–13 mm. i.d. opening, at 850° C. (F-3) Heater, 4 inches long at 400–500° C. (micro C & H preheater type, A. H. Thomas Co., Philadelphia, Pennsylvania). (G) CuO wire, 20–60 mesh, ignited. (H) Pyrex glass-wool retainer. (I) Constant temperature tube, Pyrex, 8 mm. i.d., 10 mm. o.d., ca. 6 inches long. (J) Copper wire, 20–60 mesh. (M) Bar magnet, plastic covered, ½ inch long. (N) Absorption chamber, 36 mm. o.d., 6 inches high, 3½ inches to capillary column, side opening ca. 15 mm. from base suitable for No. 2 rubber stopper. (NA) Capillary column, 2 mm. i.d., ca. 2½ inches long, having a 1 mm. constriction in the column ca. 10 mm. long, reference line etched around mid-point of the constriction. ⌀ 12/2 socket joint top. (NB) Rigid side arm, 18 mm. o.d., top level with socket joint, 2 mm. i.d. capillary connection to absorption chamber. (R) Natural rubber policeman, ⅛ inch i.d., ⁹⁄₁₆ inch o.d., ca. 1 inch long, with 4–6 mm. slit cut in side, connected to capillary inlet, 2 mm. i.d. with ⌀ 12/2 socket joint. (S) Syringe, precision ground, 5 ml. capacity. (V) Direct reading counter, four digits, attached to syringe to show syringe displacement precisely to 0.001 ml.

frothing is very bad—see Reagents. Niederl and Niederl[139,140] recommended heavy petroleum jelly and lanolin as a stopcock grease.

Instead of placing the sample in a porcelain boat,[25,48] weighed amounts of finely powdered substances have been placed in mixing tubes, fine copper oxide powder added, and the mixture transferred to the combustion tube using a dry wash.[11,12,58,59,139,140,158,163-166] The author feels that this procedure is a

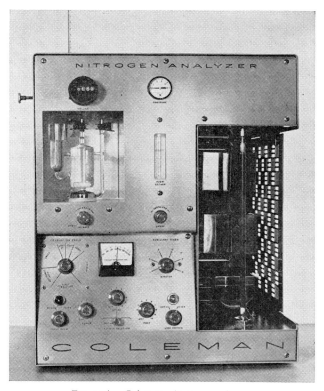

FIG. 104. Coleman nitrogen analyzer.

dangerous one, since materials that are slightly sticky or carry electrostatic charges cannot be always completely transferred.

A completely automatic combustion apparatus has been developed by Gustin[61-63] (and now has become commercially available*[30]). Figures 103 and 104 show the diagram of Gustin's setup and a photograph of the commercial

* As this book goes to press the author has seen only a demonstration of the commercial unit[30] and as yet one has not been delivered to his laboratory. Consequently, he can make no *positive* statement regarding these units.

7. Nitrogen: Micro-Dumas Method 178

model, respectively. Instead of the usual micronitrometer, a precision ground, 5 ml. capacity syringe is used in combination with a direct reading counter which makes it possible to read precisely to 0.001 ml. After the sample-packed combustion tube is attached, the automatic control is started. At the end of the cycle, the increased volume resulting from the combustion is read. With this apparatus, Gustin completed six analyses per hour.

TABLE 18
Additional Information on References* Related to Chapter 7

Following the plan of previous chapters, additional information in the form of the listing of certain references is given in Table 18. (See statement at top of Table 4 of Chapter 1, regarding completeness of this material.)†

General, miscellaneous

Alford, 2
Belcher and Godbert, 11, 12
Belcher and Macdonald, 13
Buchanan, Grimes, Smith, and Heinrich, 19
Bussmann, 20
Clark, E. P., 25
Clark, H. S. and Rees, 26
Clark, S. J., 27
Eder, 41
Etienne and Herrmann, 42
Fritz and Hammond, 46
Garch and Valdener, 50
Grant, 58, 59
Hozumi, 76
Ingram, 84
Iwasaki, 87, 88
Kainz, 89, 90
Kao and Woodland, 91
Kirsten, 93–101
Klimova and Dubinina, 106, 107
Ma, 116
Milton and Waters, 121, 122
Mitsui and Nishimura, 128
Murakami, Miyahara, and Nakai, 134
Murakami, 135
Niederl and Niederl, 139–140
Niederl and Sozzi, 141

General, miscellaneous (Cont.)

Pagel and Oita, 150
Pregl, 158
Roth, 163–166
Royer, Alber et al., 167, 168
Rush, Cruikshank, and Rhodes, 170
Schenck, 172
Steyermark, 186
Večeřa and Šnobl, 207

Collaborative studies

Ogg, 143–145
Steyermark, Alber, Aluise, Huffman, Kuck, Moran, and Willits, 191
Willits and Ogg, 214

Calculations

Belcher and Godbert, 11, 12
Clark, E. P., 25
Clark, S. J., 27
Grant, 58, 59
Hozumi, 75
Niederl and Niederl, 139, 140
Pregl, 158
Roth, 163–166
Steyermark, 186
Trautz, 199
Trautz and Niederl, 200

* The numbers which appear after each entry in this table refer to the literature citations in the reference list at the end of the chapter.

† The large number of articles on the Dumas determination indicate that the determination is by no means "perfect." The author also again wishes to emphasize that *both* the Dumas and Kjeldahl (Chapter 8) methods should be constantly used, the one as a check on the other.

TABLE 18 (*Continued*)

Methods other than Dumas for determination of nitrogen

  (a) Kjeldahl—see Chapter 8
  (b) Manometric—see Chapter 18
  (c) Exchange resins, combination, flame photometry, iodometric, colorimetric, magnesium nitride, hydrogenation, fusion, etc.

Banks, Cuthbertson, and Musgrave, 6
Batt, 7
Beauchene, Berneking, Schrenk, Mitchell, and Silker, 8
Brown and Musgrave, 18
Gel'man and Korshun, 51–53
Herbain, 68
Holowchak, Wear, and Baldeschwieler, 73
Honma and Smith, 74
Ohashi, 146
Ohashi, Takayami, Taki, and Toya, 147
Polley, 154
Remy and Pitiot, 160
Renard and Deschamps, 161
Schöniger, 177
Takagi and Hayashi, 196
Weill and Bedekian, 209
Wust, 218

Ultramicro-, submicro methods

Kirsten, 100

Simultaneous determination of nitrogen and other elements

Arthur, Annino, and Donahoo, 5
Berg and Gromakova, 14
Fedoseev and Ivashova, 44
Frazer, 45
Garch and Valdener, 50
Gel'man, Korshun, Chumachenko, and Larina, 54
Klimova and Anisimova, 104, 105
Schöniger, 173, 174, 178
Terent'ev, Fedoseev, and Ivasheva, 197
Wurzschmitt, 217

Analysis of fluoro-compounds

Ma, 116

Analysis of nitrates

Kieselbach, 92
Leithe, 114
Steyermark, 186

Analysis of nitro-compounds

Genevois, 55
Steyermark, 186

Modifications

Abramson and Laurent, 1
Alford, 2
Arthur, Annino, and Donahoo, 5
Belcher and Bhatty, 10
Belcher and Macdonald, 13
Brancone and Fulmor, 15
Bussmann, 20
Charlton, 22
Childs, Meyers, Johnston, and Mitulski, 23
Clark and Dando, 28
Colson, 31, 32
Corliss, 33
Cropper, Reed, and Rothwell, 35
Dirscherl, Padowetz, and Wagner, 37
Dirscherl and Wagner, 38
Eder, 41
Fedoseev and Ignatenko, 43
Fukuda, 47
Gustin, 61–63
Gysel, 64
Heron, 69
Hozumi and Amako, 77
Hozumi, Imaeda, and Kinoshita, 78
Hozumi and Kinoshita, 79, 80
Inada, 83
Ingram 84, 86
Kainz, 90
Kirsten, 93–101
Kirsten and Grunbaum, 102
Kono, 109
Mangeney, 117
Mitsui, 123–127
Mitsui and Tanaka, 129

7. Nitrogen: Micro-Dumas Method

TABLE 18 (*Continued*)

Modifications (Cont.)

Mizukami and Miyahara, 130
Mizukami, Miyahara and Numoto, 131
Narita and Ishii, 136
Otter, 148
Parks, Bastin, Agazzi, and Brooks, 152
Schöniger, 176
Shah, Pansare, and Mulay, 179
Shelberg, 180
Sternglanz, Thompson, and Savell, 184
Swift, 194
Swift and Morton, 195
Trutnovsky, 201
Unterzaucher, 202, 203
Večeřa, 205, 206
Večeřa and Šynek, 208
Wolfgang and Grunbaum, 216
Zimmermann, 219

Various generators

British Standards Institution, 17
Childs and Moore, 24
Clark, E. P., 25
Dirscherl and Wagner, 38
Furter and Bussmann, 49
Hein, 67
Konovalov, 110
Pagel, 149
Parkin, Fernandez, Braun, and Rietz, 151
Poth, 156
Rauscher, 159
Schöniger, 175
Stock and Fill, 193

Kipp generators

Belcher and Godbert, 11, 12
Dirscherl and Wagner, 38
Grant, 58, 59
Milton and Waters, 121, 122
Nichols, 137
Niederl and Clegg, 138
Niederl and Niederl, 139, 140
Pregl, 158
Roth, 163–166

Pressure regulator for CO_2

Di Pietro, Sassaman, and Merritt, 36

Marble used as source of CO_2

Belcher and Godbert, 11, 12
Grant, 58, 59
Milton and Waters, 121, 122
Nichols, 137
Niederl and Niederl, 139, 140
Pregl, 158
Roth, 163–161

Salts other than marble used as source of CO_2

Dubsky, 40
Govaert, 57
Ide, 82
Kuck, 112
Meixner and Kröcker, 118
Pagel, 149
Rauscher, 159

CO_2 from cylinders

Aluise, 3
Gustin, 62, 63
Royer, Norton, and Foster, 169

Dry ice as source of CO_2

Belcher and Godbert, 11, 12
Milton and Waters, 121, 122
Roth, 163–166
Shelberg, 180
Steyermark, 186

Needle valves

Hershberg and Southworth, 70
Perrine, 153
Vango, 204

Nitrometers

British Standards Institution, 17
Clarke and Winans, 29
Cropper, 34
Guss, 60
Kirsten and Wallberg-Olausson, 103
Koch, Simonson, and Tashinian, 108
Kuck and Altieri, 113
Miller and Latimer, 119
Milner and Sherman, 120
Milton and Waters, 121, 122

TABLE 18 (*Continued*)

Nitrometers (Cont.)
Mitsui, 127
Mitsui and Tanaka, 129
Mizukami and Miyahara, 130
Müller, 132
Pomatti, 155
Stehr, 182, 183
Willits and Ogg, 213
Wise and Roark, 215
Zimmermann, 219

Two micronitrometers
Hallett, 65
Kirsten, 93

Prevention of sticking of bubbles
Grant, 58, 59
Kirsten and Wallberg-Olausson, 103
Nichols, 137
Niederl and Niederl, 139, 140
Pomatti, 155
Power, 157
Roth, 163–166
Steyermark, 186

Prevention of foaming of KOH
Grant, 58, 59
Niederl and Niederl, 139, 140
Pregl, 158
Roth, 163–166
Steyermark, 186

Furnaces
Beazley, 9
Cannon, 21
Clark, E. P., 25
Clark, S. J., 27

Furnaces (Cont.)
Colson, 31
Corliss, 33
Cropper, Reed, and Rothwell, 35
Dirscherl and Wagner, 38
Hallett, 65
Ingram, 84, 85
Kirsten, 93
Korshun and Gel'man, 111
Lévy and Mathieu, 115
Milton and Waters, 121, 122
Mitsui, 123, 125
Roth, 165
Royer, Norton, and Foster, 169
Steyermark, 185, 186
Zimmermann, 219

All glass apparatus (ground joints)
Gonick, Tunnicliff, Peters, Lykken, and Zahn, 56
Steyermark, 186
Weygand, 210

Small ground glass joints
Stock and Fill, 192

Different methods of introducing sample
Belcher and Godbert, 11, 12
Clark, E. P., 25
Furman, 48
Grant, 58, 59
Kirsten, 93
Milton and Waters, 121, 122
Niederl and Niederl, 139, 140
Pregl, 158
Roth, 163–166

REFERENCES

1. Abramson, E., and Laurent, J., *Mikrochim. Acta,* p. 786 (1959).
2. Alford, W. C., *Anal. Chem.,* **24**, 881 (1952).
3. Aluise, V. A., Personal communication.
4. American Society for Testing Materials, *ASTM Designation,* **E148-59T** (1959).
5. Arthur, P., Annino, R., and Donahoo, W. P., *Anal. Chem.,* **29**, 1852 (1957).
6. Banks, R. E., Cuthbertson, F., and Musgrave, W. K. R., *Anal. Chim. Acta,* **13**, 442 (1955).
7. Batt, W. G., *J. Franklin Inst.,* **248**, 451 (1949).

8. Beauchene, R. E., Berneking, A. D., Schrenk, W. G., Mitchell, H. L., and Silker, R. E., *J. Biol. Chem.*, **214**, 731 (1955).
9. Beazley, C. W., *Ind. Eng. Chem., Anal. Ed.*, **10**, 605 (1938).
10. Belcher, R., and Bhatty, M. K., *Fuel*, **37**, 159 (1958); *Anal. Abstr.*, **5**, No. 3802 (1958).
11. Belcher, R., and Godbert, A. L., "Semi-Micro Quantitative Organic Analysis," Longmans, Green, London, New York, and Toronto, 1945.
12. Belcher, R., and Godbert, A. L., "Semi-Micro Quantitative Organic Analysis," 2nd ed., Longmans, Green, London, 1954.
13. Belcher, R., and Macdonald, A. M. G., *Mikrochim. Acta*, p. 1111 (1956).
14. Berg, L. G., and Gromakova, L. M., *Trudy Kazan. Filiala Akad. Nauk S.S.S.R., Ser. Khim. Nauk*, No. 3, 73 (1956); *Referat. Zhur. Khim.*, No. 15875 (1957).
15. Brancone, L. M., and Fulmor, W., *Anal. Chem.*, **21**, 1147 (1949).
16. Brinkmann, C. A., & Co., Inc., Great Neck, Long Island, New York.
17. British Standards Institution, *Brit. Standard* **1428**, Pt. A (1952), and Pt. D3 (1950).
18. Brown, F., and Musgrave, W. K. R., *Anal. Chim. Acta*, **12**, 29 (1955).
19. Buchanan, B. B., Grimes, M. D., Smith, D. E., and Heinrich, B. J., *Proc. Am. Petrol. Inst., Sect. III*, **28**, 64 (1948).
20. Bussmann, G., *Helv. Chim. Acta*, **32**, 995 (1949).
21. Cannon, W. A., *Chemist Analyst*, **39**, 64 (1950).
22. Charlton, F. E., *Analyst*, **82**, 643 (1957).
23. Childs, C. E., Meyers, E. E., Johnston, C. K., and Mitulski, J. D., *Anal. Chem.*, **28**, 1193 (1956).
24. Childs, C. E., and Moore, V. A., *Anal. Chem.*, **25**, 204 (1953).
25. Clark, E. P., "Semimicro Quantitative Organic Analysis," Academic Press, New York, 1943.
26. Clark, H. S., and Rees, O. W., *Illinois State Geol. Survey, Rept. Invest.* **No. 169**, (1954).
27. Clark, S. J., "Quantitative Methods of Organic Microanalysis," Butterworths, London, 1956.
28. Clark, S. J., and Dando, B., *Mikrochim. Acta*, p. 1012 (1955).
29. Clarke, R. G., and Winans, W. R., *Ind. Eng. Chem., Anal. Ed.*, **14**, 522 (1942).
30. Coleman Instruments, Inc., Maywood, Illinois.
31. Colson, A. F., *Analyst*, **75**, 264 (1950).
32. Colson, A. F., "Proceedings of the International Symposium on Microchemistry 1958," p. 105, Pergamon, Oxford, London, New York, and Paris, 1960.
33. Corliss, J. M., *Anal. Chem.*, **29**, 1902 (1957).
34. Cropper, F. R., *Analyst*, **79**, 178 (1954).
35. Cropper, F. R., Reed, R. H., and Rothwell, R., *Mikrochim. Acta*, p. 223 (1954).
36. Di Pietro, C., Sassaman, W. A., and Merritt, C., Jr., *Microchem. J.*, **4**, 97 (1960).
37. Dirscherl, A., Padowetz, W., and Wagner, H., *Mikrochemie ver. Mikrochim. Acta*, **38**, 271 (1951).
38. Dirscherl, A., and Wagner, H., *Mikrochemie ver. Mikrochim. Acta*, **36/37**, 628 (1951).
39. Downing, R., Personal communication.
40. Dubsky, J. V., "Vereinfachte quantitative Mikroelementaranalyse organischer Substanzen," Veit, Leipzig, 1917; *Ber.*, **50**, 710 (1917).
41. Eder, K., *Mikrochim. Acta*, p. 631 (1959).
42. Étienne, A., and Herrmann, J., *Chim. anal.*, **33**, 1 (1951).

43. Fedoseev, P. N., and Ignatenko, L. S., *Izvest. Akad. Nauk Turkmen. S.S.R.,* No. 1, p. 45 (1959) (in Russian); *Chem. Abstr.,* **53**, 12925 (1959).
44. Fedoseev, P. N., and Ivashova, N. P., *Zhur. Anal. Khim.,* **13**, 230 (1958).
45. Frazer, J. W., *U.S. At. Energy Comm.* UCRL-5134, 8 pp. (1958); *Chem. Abstr.,* **52**, 16118 (1958).
46. Fritz, J. S., and Hammond, G. S., "Quantitative Organic Analyses," Wiley, New York, and Chapman & Hall, London, 1957.
47. Fukuda, M., *Yakugaku Zasshi,* **76**, 1041 (1956).
48. Furman, N. H., ed., "Scott's Standard Methods of Chemical Analysis," 5th ed., Vol. II, Van Nostrand, New York, 1939.
49. Furter, M. F., and Bussmann, G., *Helv. Chim. Acta,* **32**, 993 (1949).
50. Garch, J., and Valdener, G., *Chim. anal.,* **36**, 211 (1954).
51. Gel'man, N. E., and Korshun, M. O., *Doklady Acad. Nauk S.S.S.R.,* **72**, 895 (1950).
52. Gel'man, N. E., and Korshun, M. O., *J. Anal. Chem. U.S.S.R. (English Translation),* **12**, 123 (1957); *Chem. Abstr.,* **52**, 7944 (1958).
53. Gel'man, N. E., and Korshun, M. O., *Zhur. Anal. Khim.,* **12**, 128 (1957).
54. Gel'man, N. E., Korshun, M. O., Chumachenko, M. N., and Larina, N. I., *Doklady Akad. Nauk S.S.S.R.,* **123**, 468 (1958); *Chem. Abstr.,* **53**, 3985 (1959).
55. Genevois, L., *Chim. anal.,* **29**, 101 (1947).
56. Gonick, H., Tunnicliff, D. D., Peters, E. D., Lykken, L., and Zahn, V., *Ind. Eng. Chem., Anal. Ed.,* **17**, 677 (1945).
57. Govaert, F., *Mikrochemie,* **9**, 338 (1931).
58. Grant, J., "Quantitative Organic Microanalysis, Based on the Methods of Fritz Pregl," 4th ed., Blakiston, Philadelphia, Pennsylvania, 1946.
59. Grant, J., "Quantitative Organic Microanalysis," 5th ed., Blakiston, Philadelphia, Pennsylvania, 1951.
60. Guss, H. C., *Analyst,* **60**, 401 (1935).
61. Gustin, G. M., *Chem. Eng. News,* **37**, No. 38, 52 (1959).
62. Gustin, G. M., *Microchem. J.,* **1**, 75 (1957).
63. Gustin, G. M., *Microchem. J.,* **4**, 43 (1960).
64. Gysel, H., *Helv. Chim. Acta,* **35**, 802 (1952).
65. Hallett, L. T., *Ind. Eng. Chem., Anal. Ed.,* **10**, 101 (1938); **14**, 956 (1942).
66. Hayman, D. F., and Adler, S., *Ind. Eng. Chem., Anal. Ed.,* **9**, 197 (1937).
67. Hein, F., *Z. angew Chem.,* **40**, 864 (1927).
68. Herbain, M., *Bull. soc. chim. biol.,* **35**, 1233 (1953).
69. Heron, A. E., *Analyst,* **73**, 314 (1948).
70. Hershberg, E. B., and Southworth, L., *Ind. Eng. Chem., Anal. Ed.,* **11**, 404 (1939).
71. Hershberg, E. B., and Wellwood, G. W., *Ind. Eng. Chem., Anal. Ed.,* **9**, 303 (1937).
72. Hodgman, C. D., "Handbook of Chemistry and Physics," 28th ed., pp. 1834–1835, Chemical Rubber, Cleveland, Ohio, 1944.
73. Holowchak, J., Wear, G. E. C., and Baldeschwieler, E. L., *Anal. Chem.,* **24**, 1754 (1952).
74. Honma, M., and Smith, C. L., *Anal. Chem.,* **28**, 458 (1954).
75. Hozumi, K., *Kagaku no Ryôiki, Spec. Ed.* **No. 9** (1955).
76. Hozumi, K., *Kaguku no Ryôiki, Spec. Ed.* **No. 19** (1955).
77. Hozumi, K., and Amako, S., *Mikrochim. Acta,* p. 230 (1959).
78. Hozumi, K., Imaeda, K., and Kinoshita, S., *Yakugaku Zasshi,* **76**, 1161 (1956).
79. Hozumi, K., and Kinoshita, S., *Yakugaku Zasshi,* **76**, 1157 (1956).
80. Hozumi, K., and Kinoshita, S., *Yakugaku Zasshi,* **76**, 1167 (1956).

81. Hughes, J. C., Personal communication. 1959.
82. Ide, W. S., *Ind. Eng. Chem., Anal. Ed.,* **8**, 56 (1936).
83. Inada, M., *Yakugaku Zasshi,* **75**, 151 (1955).
84. Ingram, G., *Chem. & Ind. (London),* p. 103 (1956).
85. Ingram, G., *Metallurgia,* **41**, 54 (1949).
86. Ingram, G., *Mikrochim. Acta,* p. 131 (1953).
87. Iwasaki, K., Information on the Exhibits at the International Symposium on Enzyme Chemistry, Tokyo, Japan, 1957.
88. Iwasaki, K., *J. Am. Med. Assoc.,* **155**, 385 (1954).
89. Kainz, G., *Österr. Chemiker-Ztg.,* **57**, 242 (1956).
90. Kainz, G., *Z. anal. Chem.,* **166**, 427 (1959).
91. Kao, S. S., and Woodland, W. C., *Mikrochemie ver. Mikrochim. Acta,* **38**, 309 (1951).
92. Kieselbach, R., *Ind. Eng. Chem., Anal. Ed.,* **16**, 764 (1944).
93. Kirsten, W., *Anal. Chem.,* **19**, 925 (1947).
94. Kirsten, W., *Anal. Chem.,* **22**, 358 (1950).
95. Kirsten, W. J., *Anal. Chem.,* **29**, 1084 (1957).
96. Kirsten, W. J., *Chim. anal.,* **40**, 253 (1958).
97. Kirsten, W., *Mikrochemie ver. Mikrochim. Acta,* **39**, 245 (1952).
98. Kirsten, W., *Mikrochemie ver. Mikrochim. Acta,* **39**, 389 (1952).
99. Kirsten, W., *Mikrochemie ver. Mikrochim. Acta,* **40**, 121 (1953).
100. Kirsten, W. J., *Microchem. J.,* **2**, 179 (1958).
101. Kirsten, W. J., *Mikrochim. Acta,* p. 836 (1956).
102. Kirsten, W. J., and Grunbaum, B. W., *Anal. Chem.,* **27**, 1806 (1955).
103. Kirsten, W., and Wallberg-Olausson, B., *Anal. Chem.,* **23**, 927 (1951).
104. Klimova, V. A., and Anisimova, G. F., *Bull. Acad. Sci. U.S.S.R., Div. Chem. Sci. S.S.R. (English Translation),* p. 773 (1958).
105. Klimova, V. A., and Anisimova, G. F., *Izvest. Akad. Nauk S.S.S.R. Otdel. Khim. Nauk,* p. 791 (1958); *Chem. Abstr.,* **52**, 16975 (1958).
106. Klimova, V. A., and Dubinina, I. F., *Bull. Acad. Sci. U.S.S.R., Div. Chem. Sci. S.S.R. (English Translation),* p. 123 (1958).
107. Klimova, V. A., and Dubinina, I. F., *Izvest. Akad. Nauk S.S.S.R. Otdel. Khim. Nauk,* p. 129 (1958).
108. Koch, C. W., Simonson, T. R., and Tashinian, V. H., *Anal. Chem.,* **21**, 1133 (1949).
109. Kono, T., *Nippon Nôgei-kagaku Kaishi,* **31**, 622 (1957); *Chem. Abstr.,* **52**, 12662 (1958).
110. Konovalov, A., *Ind. chim. belge,* **16**, 209 (1951).
111. Korshun, M. O., and Gel'man, N. E., *Zavodskaya Lab.,* **12**, 617 (1946).
112. Kuck, J. A., Personal communication.
113. Kuck, J. A., and Altieri, P. L., *Mikrochim. Acta,* p. 17 (1954).
114. Leithe, W., *Mikrochemie ver. Mikrochim. Acta,* **33**, 149 (1948).
115. Lévy, R., and Mathieu, P., *Bull. soc. chim. France,* p. 737 (1952).
116. Ma, T. S., *Microchem. J.,* **2**, 91 (1958).
117. Mangeney, G., *Bull. soc. chim. France,* p. 74 (1950).
118. Meixner, A., and Kröcker, F., *Mikrochemie,* **5**, 125 (1927).
119. Miller, D. M., and Latimer, R. A., *Anal. Chem.,* **31**, 1926 (1959).
120. Milner, R. T., and Sherman, M. S., *Ind. Eng. Chem., Anal. Ed.,* **8**, 331 (1936).
121. Milton, R. F., and Waters, W. A., "Methods of Quantitative Microanalysis," Longmans, Green, New York, and Arnold, London, 1949.

References

122. Milton, R. F., and Waters, W. A., "Methods of Quantitative Microanalysis," 2nd ed., Arnold, London, 1955.
123. Mitsui, T., *Kyoto Daigaku Shokuryô Kagaku Kenkyujo Hôkoku*, No. 11, p. 24 (1953).
124. Mitsui, T., *Kyoto Daigaku Shokuryô Kagaku Kenkyujo Hôkoku*, No. 17, p. 20 (1955).
125. Mitsui, T., *Bunseki Kagaku*, 1, 130 (1952).
126. Mitsui, T., *Bunseki Kagaku*, 2, 117 (1953).
127. Mitsui, T., *Mikrochim. Acta*, p. 150 (1960).
128. Mitsui, T., and Nishimura, A., *Kyoto Daigaku Shokuryô Kagaku Kenkyujo Hôkoku*, No. 17, p. 13 (1955).
129. Mitsui, T., and Tanaka, J., *Nippon Nôgei-kagaku Kaishi*, 24, 188 (1950).
130. Mizukami, S., and Miyahara, K., *Yakugaku Zasshi*, 77, 316 (1957).
131. Mizukami, S., Miyahara, K., and Numoto, K., *Yakugaku Zasshi*, 78, 842 (1958); *Chem. Abstr.*, 52, 18074 (1958).
132. Müller, A., *Mikrochemie ver. Mikrochim. Acta*, 33, 192 (1948).
133. Mulligan, G. C., Personal communication. 1948.
134. Murakami, S., Miyahara, K., and Nakai, H., *Yakugaku Zasshi*, 77, 312 (1957).
135. Murakami, T., *Bunseki Kagaku*, 7, 761 (1958).
136. Narita, K., and Ishii, M., *Yakugaku Zasshi*, 79, 407 (1959); *Chem. Abstr.*, 53, 13870 (1959).
137. Nichols, M. L., *Ind. Eng. Chem., Anal. Ed.*, 5, 149 (1933).
138. Niederl, J. B., and Clegg, D. L., *Mikrochemie ver. Mikrochim Acta*, 35, 132 (1950).
139. Niederl, J. B., and Niederl, V., "Micromethods of Quantitative Organic Elementary Analysis," Wiley, New York, 1938.
140. Niederl, J. B., and Niederl, V., "Micromethods of Quantitative Organic Analysis," 2nd ed., John Wiley, New York, 1942.
141. Niederl, J. B., and Sozzi, J. A., "Microanálisis Elemental Orgánico," Calle Arcos, Buenos Aires, 1958.
142. Nippoldt, B. W., Dis., Symposium on Analysis of Fluorine-containing Compounds, 137th National Meeting of the American Chemical Society, Cleveland, Ohio, April 1960.
143. Ogg, C. L., *J. Assoc. Offic. Agr. Chemists*, 35, 305 (1952).
144. Ogg, C. L., *J. Assoc. Offic. Agr. Chemists*, 36, 344 (1953).
145. Ogg, C. L., *J. Assoc. Offic. Agr. Chemists*, 37, 450 (1954).
146. Ohashi, S., *Bull. Chem. Soc. Japan*, 28, 177 (1955); *Chem. Abstr.*, 52, 15344 (1958).
147. Ohashi, S., Takayama, Y., Taki, S., and Toya, H., *Bunseki Kagaku*, 7, 205 (1958).
148. Otter, I. K. H., *Nature*, 182, 656 (1958).
149. Pagel, H. A., *Ind. Eng. Chem., Anal. Ed.*, 16, 344 (1944).
150. Pagel, H. A., and Oita, I. J., *Anal. Chem.*, 24, 756 (1952).
151. Parkin, B. A., Fernandez, J. B., Braun, J. C., and Rietz, E. G., *Anal. Chem.*, 25, 841 (1953).
152. Parks, T. D., Bastin, E. L., Agazzi, E. J., and Brooks, F. R., *Anal. Chem.*, 26, 229 (1954).
153. Perrine, T. D., *Anal. Chem.*, 28, 286 (1956).
154. Polley, J. R., *Anal. Chem.*, 26, 1523 (1954).
155. Pomatti, R., *Ind. Eng. Chem., Anal. Ed.*, 18, 63 (1946).
156. Poth, E. J., *Ind. Eng. Chem., Anal. Ed.*, 3, 202 (1931).

157. Power, F. W., Personal communication. 1940.
158. Pregl, F., "Quantitative Organic Microanalysis," (E. Fyleman, trans., 2nd German ed.), Churchill, London, 1924.
159. Rauscher, W. H., *Ind. Eng. Chem., Anal. Ed.,* **12**, 694 (1940).
160. Remy, J., and Pitiot, J., *Bull. soc. chim. biol.,* **33**, 405 (1951).
161. Renard, M., and Deschamps, P., *Mikrochemie ver. Mikrochim. Acta,* **36/37**, 665 (1951).
162. Ronzio, A. R., *Ind. Eng. Chem., Anal. Ed.,* **8**, 122 (1936); **12**, 303 (1940).
163. Roth, H., "Die quantitative organische Mikroanalyse von Fritz Pregl," 4th ed., Springer, Berlin, 1935.
164. Roth, H., "F. Pregl quantitative organische Mikroanalyse," 5th ed., Springer, Wien, 1947.
165. Roth, H., "Pregl-Roth quantitative organische Mikroanalyse," 7th ed., Springer, Wien, 1958.
166. Roth, H., "Quantitative Organic Microanalysis of Fritz Pregl," 3rd ed., (E. B. Daw, trans., 4th German ed.), Blakiston, Philadelphia, Pennsylvania, 1937.
167. Royer, G. L., Alber, H. K., Hallett, L. T., and Kuck, J. A., *Ind. Eng. Chem., Anal. Ed.,* **15**, 476 (1943).
168. Royer, G. L., Alber, H. K., Hallett, L. T., Spikes, W. F., and Kuck, J. A., *Ind. Eng. Chem., Anal. Ed.,* **13**, 574 (1941).
169. Royer, G. L., Norton, A. R., and Foster, F. J., *Ind. Eng. Chem., Anal. Ed.,* **14**, 79 (1942).
170. Rush, C. A., Cruikshank, S. S., and Rhodes, E. J. H., *Mikrochim. Acta,* p. 858, (1956).
171. Sargent, E. H., & Company, Chicago, Illinois.
172. Schenck, W. J., *Anal. Chem.,* **26**, 788 (1954).
173. Schöniger, W., *Helv. Chim. Acta,* **39**, 650 (1956).
174. Schöniger, W., Regional Analytical Symposium, Philadelphia, Pennsylvania, February 1957.
175. Schöniger, W., *Mikrochemie ver. Mikrochim. Acta,* **34**, 201 (1949).
176. Schöniger, W., *Mikrochemie ver. Mikrochim. Acta,* **39**, 229 (1952).
177. Schöniger, W., *Mikrochim. Acta,* p. 44 (1955).
178. Schöniger, W., *Mikrochim. Acta,* p. 545 (1957).
179. Shah, G. D., Pansare, V. S., and Mulay, V. V., *Mikrochim. Acta,* p. 1140 (1956).
180. Shelberg, E. F., *Anal. Chem.,* **23**, 1492 (1951).
181. Spies, J. R., and Harris, T. H., *Ind. Eng. Chem., Anal. Ed.,* **9**, 304 (1937).
182. Stehr, E., *Ind. Eng. Chem., Anal. Ed.,* **18**, 513 (1946).
183. Stehr, E., *Mikrochim. Acta,* p. 213 (1954).
184. Sternglanz, P. D., Thompson, R. C., and Savell, W. L., *Anal. Chem.,* **23**, 1027 (1951).
185. Steyermark, Al, *Ind. Eng. Chem., Anal. Ed.,* **17**, 523 (1945).
186. Steyermark, Al, "Quantitative Organic Microanalysis," Blakiston, Philadelphia, Pennsylvania, 1951.
187. Steyermark, Al, *Anal. Chem.,* **22**, 1228 (1950).
188. Steyermark, Al, Unpublished work.
189. Steyermark, Al, Alber, H. K., Aluise, V. A., Huffman, E. W. D., Kuck, J. A., Moran, J. J., and Willits, C. O., *Anal. Chem.,* **21**, 1283 (1949).
190. Steyermark, Al, Alber, H. K., Aluise, V. A., Huffman, E. W. D., Kuck, J. A., Moran, J. J., and Willits, C. O., *Anal. Chem.,* **21**, 1555 (1949).

191. Steyermark, Al, Alber, H. K., Aluise, V. A., Huffman, E. W. D., Kuck, J. A., Moran, J. J., and Willits, C. O., *Anal. Chem.*, **23**, 537 (1951).
192. Stock, J. T., and Fill, M. A., *Metallurgia*, **36**, 54 (1947).
193. Stock, J. T., and Fill, M. A., *Metallurgia*, **38**, 118 (1948).
194. Swift, H., *Analyst*, **79**, 718 (1954).
195. Swift, H., and Morton, E. S., *Analyst*, **77**, 392 (1952).
196. Takagi, T., and Hayashi, N., *Nippon Kagaku Zasshi*, **78**, 445 (1957).
197. Terent'ev, A. P., Fedoseev, P. N., and Ivasheva, N. P., *Zhur. Anal. Khim.*, **13**, 344 (1958); *Chem. Abstr.*, **53**, 117 (1959).
198. Thomas, Arthur H., Company, Philadelphia, Pennsylvania.
199. Trautz, O., *Mikrochemie*, **9**, 300 (1931).
200. Trautz, O., and Niederl, J. B., *Ind. Eng. Chem., Anal. Ed.*, **3**, 151 (1931).
201. Trutnovsky, H., *Mikrochim. Acta*, p. 157 (1960).
202. Unterzaucher, J., *Chem. Ing. Tech.*, **22**, 128 (1950).
203. Unterzaucher, J., *Mikrochemie ver. Mikrochim. Acta*, **36/37**, 706 (1951).
204. Vango, S. P., *Chemist Analyst*, **46**, 72 (1957).
205. Večeřa, M., *Chem. listy*, **47**, 1090 (1953).
206. Večeřa, M., *Mikrochim. Acta*, p. 88 (1955).
207. Večeřa, M., and Šnobl, D., *Chem. listy*, **51**, 1482 (1957).
208. Večeřa, M., and Šynek, L., *Collection Czechoslov. Chem. Comms.*, **24**, 3402 (1959).
209. Weill, C. E., and Bedekian, A., *Microchem. J.*, **1**, 89 (1957).
210. Weygand, C., *Chem. Tech. (Berlin)*, **16**, 15 (1943).
211. Willits, C. O., Personal communication; Metropol. Microchem. Soc., 5th Symposium, New York, March, 1950.
212. Willits, C. O., *Anal. Chem.*, **21**, 134 (1949).
213. Willits, C. O., and Ogg, C. L., Personal communication.
214. Willits, C. O., and Ogg, C. L., *J. Assoc. Offic. Agr. Chemists*, **32**, 561 (1949).
215. Wise, R. W., and Roark, J. N., *Chemist Analyst*, **45**, 56 (1956).
216. Wolfgang, J. K., and Grunbaum, B. J., *Microchem. J.*, **1**, 138 (1957).
217. Wurzschmitt, B., *Chemiker.-Ztg.*, **74**, 419 (1950).
218. Wust, H., *Klin. Wochschr.*, **33**, 185 (1955).
219. Zimmermann, W., *Mikrochemie ver. Mikrochim. Acta*, **31**, 42 (1943).

CHAPTER 8

Microdetermination of Nitrogen by the Kjeldahl Method

The Kjeldahl method* for the determination of nitrogen is applicable to many types of organic compounds, although the method is sometimes referred to as being that for the determination of aminoid nitrogen. It is based upon the fact that on digestion with sulfuric acid and various catalysts, the organic material is destroyed and the nitrogen is converted to ammonium acid sulfate. On making the reaction mixture alkaline, ammonia is liberated which is removed by steam distillation, collected, and titrated. The equations involved are as follows:

(a) Organic N $\xrightarrow[\underset{\text{Catalysts}}{\Delta}]{H_2SO_4}$ $CO_2 + H_2O + NH_4HSO_4$

(b) $NH_4HSO_4 + 2NaOH \rightarrow NH_3 + Na_2SO_4 + 2H_2O$

(c) $NH_3 + HCl \rightarrow NH_4Cl$

A great deal of work has been done to obtain the best catalysts and many have been used such as mercury,[43,73,74,137,138,148,189,218] copper,[12,43,73,74,137,157,163,166] and selenium.[43,73,80,109,137,138] The method described here is largely the outcome of the work done during the collaborative studies on the standardization of microchemical methods for the Association of Official Agricultural Chemists under the direction of Willits and Ogg.[142-146,214-221] This investigation has shown that mercury is the best catalyst, the amount of potassium sulfate is most critical, and above all, the temperature of the digest must be high enough to maintain it at a vigorous boil (340° C.) and for 4 hours. This is in contrast to the many other papers on the subject, many authors recommending 15 to 60 minutes.[40,41,73,130,137,138,153,166]

Where the nitrogen is present in a chain, such as in amines, amides, etc., the digestion proceeds with greater ease than when present as part of a ring, but the collaborative study referred to above proves that the latter are successfully handled. In the author's laboratory,[192,193] nitrogen present in compounds of the folic acid type, with condensed pyrimidine and pyrazine rings in the

---

* Please see references 31, 41, 66, 67, 73, 74, 95, 98, 130, 131, 137, 138, 144, 157, 163–166, 193, 213, 214, 216, 218, 221.

same molecule, can be better analyzed by this method than by the Dumas (refer to Chapter 7). Also compounds having a number of N-methyl groups repeatedly give low values by the Dumas method but give correct results when analyzed by the Kjeldahl procedure.[192,193] (The low figures obtained by the Dumas method probably result from traces of methylamine escaping combustion. This idea is substantiated by the fact that after 4 or more hours of Kjeldahl digestion these compounds yield, in part at least, methylamine which can be detected by odor on making alkaline. In spite of the conversion to methylamine instead of to ammonia, correct results are obtained since, in the titration with standard acid, either of these two require the same amount, mol for mol.)

Compounds containing the N—N, N=N, NO, and NO_2 groupings cannot be analyzed by straight kjeldahlization as described above.* However, with some form of pretreatment the method is reliable for many compounds.[41,49,50,65,66,73,74,116,164,165,193,195,218] Two of these modifications[195] are presented in this chapter in addition to the regular method, and before analyzing a compound by the Kjeldahl method particular note must be made regarding the groupings present and the *particular procedure which applies to the compound in question must be selected*. The reactions involved in the procedures dealing with these two modifications are:

(d) For azo- and nitro-compounds, oximes, isoxazoles, hydrazines, hydrazones, etc.,

$$\text{Organic N (N—N, N=N, NO, NO}_2\text{)} \xrightarrow[\text{(Zn, Fe, HCl)}]{H_2} \text{Organic N (Amine)}$$

(e) For nitrates,

$$-NO_3 \xrightarrow[H_2SO_4]{\text{Salicylic acid}} \text{Nitrosalicylic acid}$$

$$\text{Nitrosalicylic acid} \xrightarrow{Na_2S_2O_3} \text{Aminosalicylic acid}$$

(f) Kjeldahlization of the products of (d) and (e), followed by distillation of ammonia and titration as shown in (a), (b), and (c) above.

The zinc-iron reduction method[143,195] gives good results with many compounds containing N—N, N=N, NO, and NO_2 groupings, where the nitrogen to nitrogen linkage is not part of a ring (nitrogen may be linked to oxygen as part of a ring). Where nitrogen is connected to nitrogen as part of a ring, such as in pyrazolones (aminopyrine), 1,2-diazines, 1,2,3-triazoles, etc., the method is not reliable.

---

* An exception to this is the compounds of the type, $\geqslant N \rightarrow O$, in which the oxygen to nitrogen linkage is so weak that the split takes place at once at the beginning of the digestion and the resulting cyclic nitrogen compound kjeldahlizes in the normal manner *without* need for pretreatment.[192]

8. Nitrogen: Micro-Kjeldahl Method

The zinc-iron reduction method is not reliable for nitrates, but the salicylic acid treatment gives excellent results, and requires only slightly more time than the regular Kjedahl procedure.

When analyzing nitriles by the regular micro-Kjeldahl method, Willits and Ogg[195,221] found that fresh concentrated sulfuric acid, sp. gr. 1.84, must be used. The presence of even the small amounts of water normally absorbed by bottles standing on the shelf produces enough hydrolysis to cause losses. This has been verified in the author's laboratory with many compounds in the past. If a compound to be analyzed contains both nitrile groups as well as nitrogen in other groups, and must be subjected to the zinc-iron reduction before kjeldahlization, the nitrile is reduced before the hydrolysis and subsequent loss in the diluted sulfuric acid (from the hydrochloric acid) occurs.

For the regular micro-Kjeldahl method, 1 ml. of sulfuric acid is used with 0.65 gram of potassium sulfate and 0.016 gram of mercuric oxide. When boiled vigorously, the temperature of the mixture is approximately 340° C. If these same amounts were used, in the presence of the zinc and iron salts there would be a tendency for the hot digestion mixture to become too dry, the temperature rise in considerable excess of the above figure, and the results low due to loss of nitrogen. Consequently, the amount of sulfuric acid is increased to 1.5 ml. which gives a mixture that does not become dry, and which when boiled vigorously, attains a temperature of only about 340° C.

The main disadvantage of the zinc-iron reduction method is the length of time required. However, because a large number of reductions and subsequent digestions can be done simultaneously, the time element become unimportant. The method has the advantages of requiring only the regular Kjeldahl equipment and also, that certain types of compounds—for example, hydrazines, nitro-compounds and others—which are apt to be troublesome when analyzed by the Dumas method are easily handled by the zinc-iron reduction method with accurate results.

Reagents

BORIC ACID[117,118,143,144,168,193,195,214,216,218]

Four per cent aqueous solution of reagent grade of boric acid is used in the receiving flask.

MERCURIC OXIDE[43,73,74,137,138,143,148,189,193,195]

Reagent grade of powdered mercuric oxide is used as a catalyst for the digestion.

POTASSIUM SULFATE[73,74,137,138,143,157,193,195,218]

Reagent grade of powdered potassium sulfate is used in the digestion mixture.

CONCENTRATED SULFURIC ACID

Reagent grade of concentrated sulfuric acid, sp. gr. 1.84, is used for the digestion. *Fresh* acid must be used. Acid which has been standing on the shelf for periods of time absorb enough water from the atmosphere to cause low results with some compounds, particularly nitriles.[195,221]

FORMIC ACID[50,195]

Reagent grade of formic acid (98–100%) is used for solvent purposes in the reduction method.

CONCENTRATED HYDROCHLORIC ACID

Reagent grade of concentrated hydrochloric acid, sp. gr. 1.190, is used in the reduction method for liberating hydrogen.

ZINC DUST[50,195]

Reagent grade (nitrogen-free) zinc dust is used in the reduction method as a means of producing hydrogen.

IRON POWDER[50,195]

Reagent grade of iron powder (prepared by hydrogenation) is also used for reducing purposes.

ETHANOL, 95%

This, too, is used only in the reduction method.

SALICYLIC ACID[66,195]

Reagent grade salicylic acid is used only in the procedure for nitrates, yielding nitrosalicylic acid.

CRYSTALLINE SODIUM THIOSULFATE[66,195]

Reagent grade material is used to reduce the nitrosalicylic acid formed in the procedure for nitrates, aminosalicylic acid being formed.

INDICATOR MIXTURES (a) or (b) (See Chapter 5)

(a) BROMOCRESOL GREEN-METHYL RED.[117,193,195,218] This indicator is prepared by mixing five parts of 0.2% bromocresol green with one part of 0.2% methyl red, both in 95% ethanol.

(b) METHYL RED-METHYLENE BLUE INDICATOR.[144,193,214,216,218] This is prepared by mixing two parts of 0.2% methyl red with one part of 0.2% methylene blue, both in 95% ethanol.

STANDARD SOLUTION OF HYDROCHLORIC ACID, 0.01N

One-hundredth normal hydrochloric acid is used for titrating the distilled ammonia. Either commercially available standard acid[2] may be used or it is standardized by titration against standard, 0.01N sodium hydroxide (see Chapter 5).

SODIUM HYDROXIDE-SODIUM THIOSULFATE MIXTURE[41,143,144,193,195,214,216,218]

This mixture is prepared from 50 grams of sodium hydroxide and 5 grams of sodium thiosulfate ($Na_2S_2O_3 \cdot 5H_2O$), both reagent grades, dissolved in distilled water and diluted to 100 ml. It is used for liberating the ammonia in the distillation flask. The purpose of the thiosulfate is to break up the mercury-ammonia complex formed. Otherwise, low results would be obtained. The equation involved is the following[41]:

$$Hg\begin{matrix}\diagup NH_3 \diagdown \\ \diagdown NH_3 \diagup\end{matrix}SO_4 + Na_2S_2O_3 + H_2O \rightarrow HgS + Na_2SO_4 + (NH_4)_2SO_4$$

(According to Clark,[41] if too much alkali is used in the distillation the HgS decomposes. This is evidenced by the contents of the distilling flask becoming yellow instead of black (HgS), and mercury distilling into the receiver. He states that the contents of the distillation flask should remain black during the distillation.)

Apparatus

The apparatus, with the exception of weighing and titration pieces, consists of digestion flasks, a digestion rack, and a distillation apparatus. Recommended specifications for these were first published by the Committee on Microchemical Apparatus of the Division of Analytical Chemistry of the American Chemical Society.[194] More recently, these were reviewed, changes made where found necessary, and accepted as tentative specifications by the American Society for Testing Materials.[5] The various pieces of apparatus described and shown in this chapter conform to the above tentative specifications.[5]

DIGESTION FLASKS[5,193,194]

Either of the two digestion flasks shown in Figs. 105, and 106 may be used, but the Soltys type is preferable when danger of spattering is great, particularly with the zinc-iron reduction procedure. The flasks should be made of shock-resistant glass with linear coefficient of expansion not greater than 33.5×10^{-7} cm./cm./°C.

DIGESTION RACK[5,193,194]

The digestion rack consists of flask heaters, a flask support, and a fume duct. (The fume duct has been referred to as a manifold in the past.) Suitable units are commercially available, such as those shown in Figs. 107 and 108. In general, they should conform to the following requirements.

HEATERS. The source of heat may be either electric or gas, but the author prefers the former. Each heater should have enough capacity so that it can

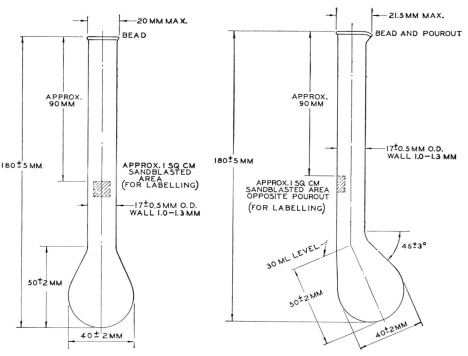

FIG. 105. (*Left*) Kjeldahl digestion flask, 30 ml.—details of construction.

FIG. 106. (*Right*) Kjeldahl digestion flask, 30 ml., Soltys—details of construction.

supply sufficient heat to a 30 ml. flask to cause 15 ml. of water at 25° C. to come to a rolling boil (boiling chips) in not less than 2 or more than 3 minutes. Means should be provided for regulating the amount of heat supplied to each flask so that the above conditions can be met and also should be so adjustable that low-temperature digestions are possible. The switches and controls should be conveniently located and should remain cool even after long periods of operation.

FLASK SUPPORT.[5,193,194] The support on which the bulbs of the flasks rest

8. Nitrogen: Micro-Kjeldahl Method

FIG. 107. Circular type digestion rack with fume duct and digestion flasks in place.

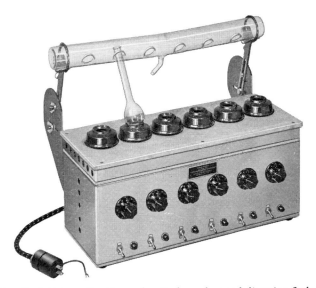

FIG. 108. Straight type digestion rack with fume duct and digestion flask in place.

should shield the necks of the flasks from excessive heating during the digestion and should be provided with circular openings not exceeding 26 mm. in diameter, centered over each heater.

FUME DUCT OR MANIFOLD.[5,193,194] The fume duct should provide means for adequately removing by suction the fumes evolved during a digestion and should support the necks of the flasks at an angle greater than 35° to the

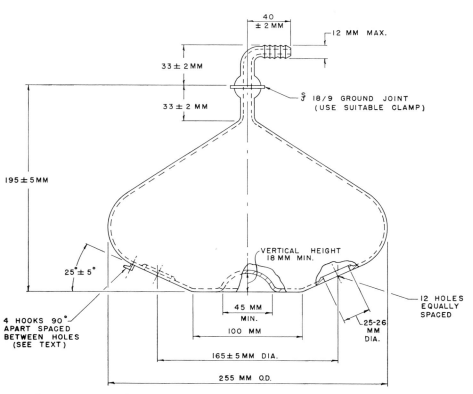

FIG. 109. Fume duct (manifold) for circular type micro-Kjeldahl digestion rack—details of construction.

horizontal. Figures 109 and 110 show details of construction of fume ducts used with setups of the types shown in Figs. 107 and 108, respectively. The fume duct illustrated in Fig. 110 is supported and held in place by suitable clamps mounted at either end of the digestion rack while the duct shown in Fig. 109 rests upon a circular support plate and is held in place by springs from the plate to the four glass hooks. The concave center permits clearance for the nut which fastens the support plate to the center support post.

8. Nitrogen: Micro-Kjeldahl Method

DISTILLATION APPARATUS

Either the one-piece model[5,10,120,178,193,194] or the Pregl (Parnas-Wagner)[5,73,74,137,138,157,163–166,193,194] type distillation apparatus may be used, but the author prefers the use of the former.*

ONE-PIECE DISTILLATION APPARATUS.[5,10,120,178,193,194] Steam is generated in the generator shown in Figs. 111, 112, and 113 which consists of a wide-mouthed commercially available resin-reaction kettle and an immersion heater.

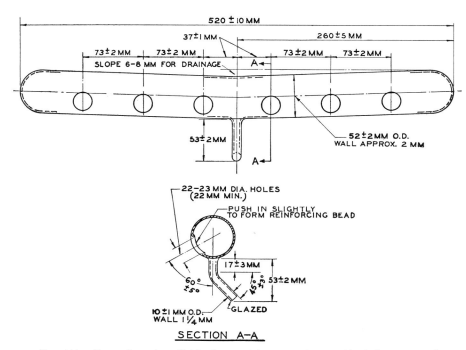

FIG. 110. Fume duct (manifold) for straight type micro-Kjeldahl digestion rack—details of construction.

The interchangeable cover of the kettle, with flat-ground rim and four ground-glass tubulations, is held in place with a suitable clamp. The distillation unit is connected to the center tubulation. Two of the outer three tubulations are used for the leads of the immersion heater assembly, the heater of which should be of the type that does not burn out when not immersed. The remaining tubulation is used for adding water to the kettle. (Note: The addition of boric

---

* The one-piece model was designed by Dr. E. C. Noonan while he was working as a graduate student in the laboratory of Professor H. A. Iddles at the University of New Hampshire in 1935. His work was never published.

acid to the water in the vessel is recommended for reducing blank values.) The rate of steam generation is readily controlled by means of a 7.5 ampere (*minimum* rating) variable transformer. The steam enters the distillation apparatus through the vertical tube extending almost halfway up the outer jacket at the left (Fig. 114), surrounds the distillation flask proper (inner

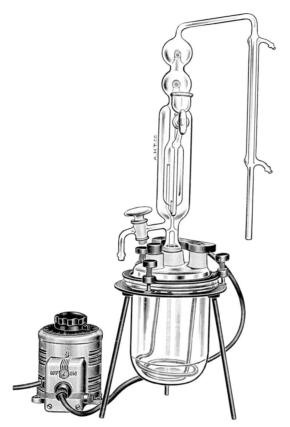

FIG. 111. Micro-Kjeldahl distillation apparatus, one-piece model and steam generator assembly.

chamber), passes into the small bent tube near the top at the right (above the sample funnel inlet), then downward and through the bent portion of the tube at the lower end, and up into the inner chamber or distillation flask. The steam then passes through the reaction mixture, traps, and condenser. The two traps with T-shaped tubes are very efficient for holding back alkali spray. The outer jacket is equipped with a drain tube for removal of the spent reaction mixture.

8. Nitrogen: Micro-Kjeldahl Method

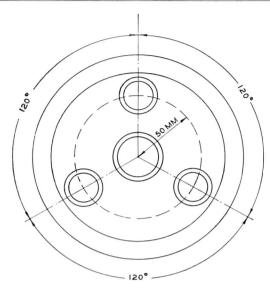
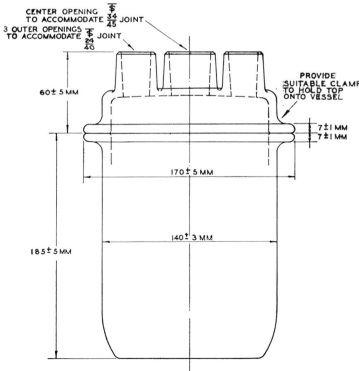

FIG. 112. Steam generator vessel, 2000-ml. capacity—details of construction.

PREGL TYPE (PARNAS-WAGNER) DISTILLATION APPARATUS.[5,73,74,137,138,157,163-6,193,194] The steam generator, consisting of the resin-reaction kettle and immersion heater (Figs. 112, 113, and 115) is also used with the Pregl type distillation apparatus. Besides the generator, the principal parts of the unit are the trap, distillation flask, and condenser. The *trap* is used both to collect

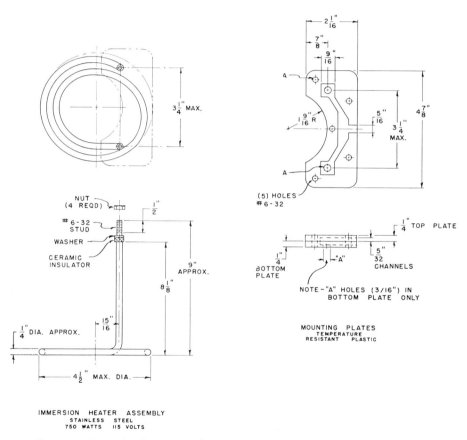

FIG. 113. Immersion heater assembly and mounting plates—details of construction.

condensate before it enters the distillation flask and to prevent the spent reaction mixture from entering the steam generator since, like the one-piece model, the Pregl type is automatically cleaned when steam generation is stopped. The *distillation flask* has entrance tubes for both the steam from the steam trap and for introduction of the digest through the funnel. The flask is equipped with a vacuum jacket to facilitate the heating by the passage of steam. During the distillation, steam enters and passes downward through the inner tube which extends almost to the bottom of the flask. After passing through the reaction mixture it passes through the bulb and condenser. The condenser is of the West type[193,194] for more efficient cooling. Figure 115 shows the assembly and Figs. 115 B, 115 D, 115 E, 115 F, 115 G, 115 H, 115 J, 115 K, 115 L show the details of construction of the various parts shown in the assembly.

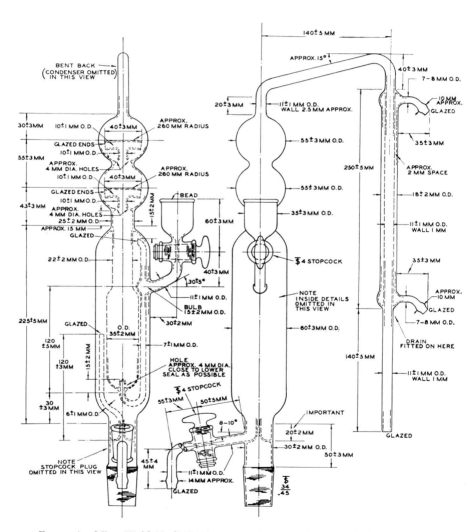

FIG. 114. Micro-Kjeldahl distillation apparatus, one-piece model—two views showing details of construction. Drain, Fig. 115 L, may be fitted in position marked "Drain fitted on here."

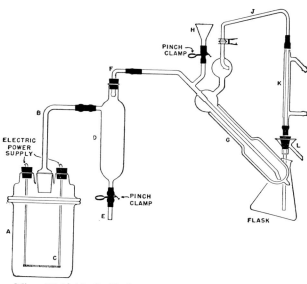

Fig. 115. Micro-Kjeldahl distillation apparatus, Pregl (Parnas-Wagner) type assembly. (A) Steam generator. (B) Steam tube. (C) Immersion heater assembly. (D) Steam trap. (E) Tube. (F) Connecting tube. (G) Distillation flask. (H) Filling funnel. (J) Condenser tube. (K) Condenser jacket. (L) Drain. See details of construction in Figs. 112, 113, and 115B–115L.

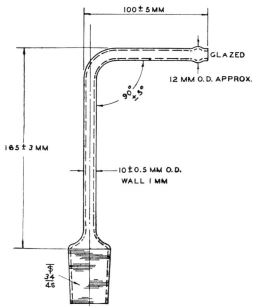

Fig. 115 B. Steam tube—details of construction.

8. Nitrogen: Micro-Kjeldahl Method

The steam generator is connected to the steam trap by means of suitable rubber tubing. The connection between the trap and steam inlet tube on the distillation flask is also made by means of suitable rubber tubing (or, if desired, a ball and socket joint may be used). The funnel, through which the sample

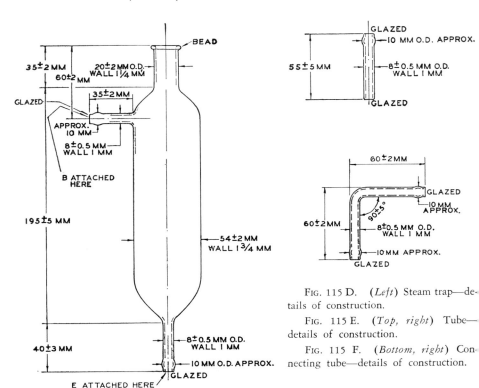

FIG. 115 D. (*Left*) Steam trap—details of construction.

FIG. 115 E. (*Top, right*) Tube—details of construction.

FIG. 115 F. (*Bottom, right*) Connecting tube—details of construction.

and also the sodium hydroxide-thiosulfate mixture is added, is connected to the distillation flask using rubber tubing as above. The trap portion of the distillation flask is joined to the condenser tube by means of a ball and socket joint lubricated with stopcock grease* and held together with a suitable clamp.†

* The author has found Sisco No. 300 grease,[178,193,198] to be very good for this purpose.
† Arthur H. Thomas Company[202] clamps No. 3241 are preferred by the author.

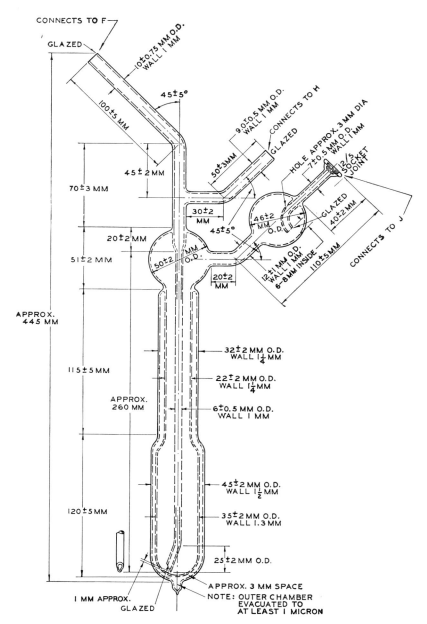

Fig. 115 G. Distillation flask—details of construction.

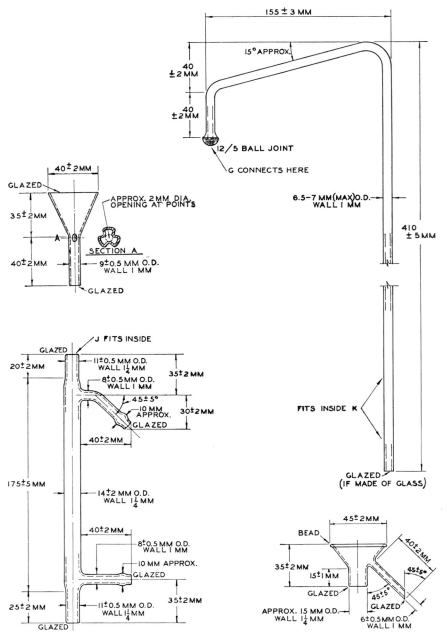

FIG. 115 H. (*Top, left*) Filling funnel—details of construction.
FIG. 115 J. (*Top, right*) Condenser tube (glass or silver, glass preferred)—details of construction.
FIG. 115 K. (*Bottom, left*) Condenser jacket (West-type)—details of construction.
FIG. 115 L. (*Bottom, right*) Drain—details of construction.

Regular Procedure*

About 5 mg. of sample (or enough to require several milliliters of 0.01N hydrochloric acid in the titration of the distilled ammonia) is weighed into a Kjeldahl flask, 30 ml., of the regular (Fig. 105) or the Soltys design (Fig. 106). If the sample is a solid, the weighing should be accomplished using a charging tube (see Figs. 47–49, Chapter 3). If it is a high-boiling liquid, it is weighed in a porcelain boat (Fig. 22, Chapter 3) and the boat dropped into the flask. If this method is used, it might be necessary to use boiling chips to prevent bumping, during digestion, which might cause the boat to crack the flask. [*Caution: Platinum boats or boiling chips (tetrahedra) should never be used.* When platinum is present during the digestion, the nitrogen is *completely lost.*[192,225]] A blank test should be run on the chips to make certain that they do not contain traces of nitrogen. A volatile liquid sample is weighed in a capillary (Fig. 54, Chapter 3) which is broken below the surface of the digestion mixture (see below) with the aid of a glass rod. To the flask are added 0.65 gram† of potassium sulfate, 0.016 gram of mercuric oxide, and 1 ml. of concentrated sulfuric acid sp. gr. 1.84, and the mixture is boiled on the digestion rack for 4 hours† (suitable boiling chips may be added to prevent bumping). The mixture is boiled vigorously enough so that refluxing takes place about one-half way up in the neck of the flask. (Under these conditions tests[192,193,221] show that the temperature of the digestion mixture is approximately 340° C.) After the completion of the digestion, the digest is subjected to distillation as described below under (A) or (B).

(A) USING ONE-PIECE MODEL‡

After digestion of the sample with mercuric oxide, potassium sulfate, and sulfuric acid, as described above, the contents of the Kjeldahl flask are diluted with 2 ml. distilled water and the rim of the flask coated with a thin film of stopcock grease. As a safety measure, to prevent possible suck back, the stopper should be left out of the tubulation (of the kettle cover) through which water is normally added. Then with *both* stopcocks (on sample funnel and on drain-

---

* Not applicable to compounds containing N—N, N=N, NO, NO_2, and NO_3 groups, except $\geqslant N \rightarrow O$.

† In a recent collaborative study,[143] it was found that by using 1.00 gram of potassium sulfate and boiling for only one hour, satisfactory results are obtained. The increased temperature brought about by the increased amount of the salt used affects complete digestion in the shorter period of time. Extreme care should be exercised in the amount of potassium sulfate used because if more than this amount is used, the temperature will be too high and nitrogen will be lost.

‡ Before being used each day, the apparatus should be cleaned by passing steam through it at a rapid rate for about 10 minutes. A new setup should be steamed out for considerably longer.

8. Nitrogen: Micro-Kjeldahl Method

age tube) of the distillation apparatus *open,* the diluted digest is poured through the sample funnel (and the curved tube) into the inner chamber or distillation flask. An additional 10–12 ml. of distilled water, in several 2–3 ml. portions, is then used in quantitatively transferring the contents of the digestion flask to the distillation flask. A 125-ml. Erlenmeyer flask containing 5 ml. of 4% boric acid solution, and four drops of indicator (either bromocresol green-methyl red or methyl red-methylene blue), is placed under the delivery tube of the condenser, in a tilted manner so that the tip is well immersed in the liquid. A *rapid* stream of *cold* water is run through the condenser. Five milliliters of the sodium hydroxide-sodium thiosulfate mixture is carefully run into the distillation flask through the sample funnel. (Since the alkali enters from the bottom, it merely forces the acid layer upward and there is not as much danger of mixing as there is with the Pregl type of apparatus.) The stopper is replaced in the tubulation, both stopcocks are then closed and the steam generator switched on. Steam should pass through at a rather rapid rate but not at such that the strongly alkaline liquid mixture is splashed up into the portion at the top of the flask, below the traps, with the danger of alkaline spray being swept into the condenser. The alkaline mixture in the distillation flask should be steam distilled for 5 minutes from the time that condensation begins in the condenser. The flask containing the boric acid is then lowered so that the tip of the condenser tube is in the neck of the flask, well above the surface of the solution. Steam distillation is continued for 2 to 3 minutes longer to wash out the condenser tip. A few milliliters of distilled water are then used to wash any solution adhering to the outside of the condenser tube into the flask. The boric acid solution of ammonia in the flask is then titrated with $0.01N$ hydrochloric acid to the endpoint. (A second flask containing 5 ml. of boric acid, four drops of indicator, distilled water and an *excess* of $0.01N$ hydrochloric acid should be used for comparison purposes.)

Calculation: See below after description of procedure with Pregl-type apparatus.

The apparatus may be cleaned by either of the following methods:

(1) The steam generator is switched off and both stopcocks are allowed to remain closed. The reduced pressure in the generator, brought about by the cooling, causes the alkaline mixture in the distillation flask to be sucked up through the vertical tube (on the right) and into the outer jacket (reverse of the path taken by the steam) where it can be discarded through the drainage stopcock and tube. Since the vertical tube on the left extends about halfway up the outer jacket, the alkaline mixture does not enter the steam generator. Distilled water is then run into the distillation chamber through the sample funnel. Both stopcocks are again closed and steam generated for about one minute and then shut off. As described above, the rinse water is then auto-

matically sucked out of the distillation flask into the outer jacket and is discarded.

(2)[5] When boiling chips are employed in the digestion the following is particularly satisfactory. The steam generator is switched off while both stopcocks are closed. The reduced pressure in the generator, brought about by the cooling, causes the alkaline mixture in the distillation flask to be sucked into the outer jacket. Both stopcocks are opened to allow the liquid to drain off. Then, while steam is being generated, both stopcocks are closed, the sample inlet funnel is filled with water, and the tip of the condenser is immersed in about 100 ml. of water in a beaker. Steam generation is stopped and the stopcock of the sample inlet funnel opened *slowly* just long enough to admit *almost* all of the water from the funnel, after which it is closed. The resulting reduced pressure in the system causes the water to be sucked from the beaker into the apparatus, thereby washing it. The stopcock on the drainage tube is opened to empty the liquid that has collected in the outer jacket. If necessary, steam generation may be started again and the rinsing repeated several times.

(3) Instead of using the reduced pressure from the cooling to suck the rinse water through the system, a water suction pump may be used, attached to the drainage tube.

(B) USING THE PREGL DISTILLATION APPARATUS (PARNAS-WAGNER TYPE)*

With the distillation flask cool† and the pinch clamps on the filling funnel and on the steam trap open, steam is generated in the generator. With the apparatus so arranged most of the steam passes out through the bottom of the trap and a little passes out through the filling funnel. The diluted digest‡ is poured into the funnel and passes into the distillation flask. About 10–12 ml. of distilled water is then used for quantitatively washing the contents of the Kjeldahl flask into the distillation flask. (The completeness of the rinsing can be tested by adding a drop of methyl orange to the digestion flask after the final rinse.[221]) A 125-ml. Erlenmeyer flask containing 5 ml. of the boric acid solution and four drops of the indicator (either bromocresol green-methyl red or methyl red-methylene blue) is then placed under the delivery tube of the condenser and slightly tilted so that the tip extends well below the liquid

---

* Before the apparatus is used each day, it is thoroughly steamed out as stated above in connection with the one-piece model.

† If the distillation flask is warm, as it cools air is sucked in through the inner tube leading to the bottom of the flask and causes stirring of the acid and alkali when the latter is added. This can cause a possible loss of ammonia, as there should be two distinct layers, acid on top to trap the ammonia, prior to the passage of steam through the mixture.

‡ Diluted with 2 ml. distilled water, as described above under (A) Using One-Piece Model.

8. Nitrogen: Micro-Kjeldahl Method

surface. Five milliliters of the sodium hydroxide-sodium thiosulfate mixture is carefully added through the funnel so that two liquid layers are formed (acid on top, alkali on bottom). In this way the ammonia liberated on contact with alkali passes into the acid layer and is again converted to ammonium acid sulfate. The pinch clamps beneath the filling funnel and steam trap are closed in the respective order which causes steam to enter the tube leading to the bottom of the distillation flask, mixing the two layers. The resulting strongly alkaline mixture is steam distilled for 5 minutes (from the time that condensation begins in the condenser) with the end of the delivery tube below the surface of the boric acid and for an additional 2–3 minutes after the flask has been lowered so that the end of the tube is out of the liquid and up towards the neck of the flask. Steam distillation should be at a rather rapid rate with a *rapid* stream of *cold* water passing through the condenser. However, the rate should not be great enough so that the alkaline mixture splashes up into the bulb at the top of the distillation flask below the Kjeldahl trap and an alkaline spray is carried over into the condenser. The tip of the delivery tube of the condenser is washed with a few milliliters of distilled water to transfer any boric acid adhering to the outside into the flask. The boric acid solution of ammonia is then titrated with 0.01N hydrochloric acid to the end point. For comparison, a second flask containing 5 ml. boric acid, distilled water, four drops of indicator and an *excess* of 0.01N hydrochloric acid should be kept on the titration table.

The apparatus is automatically cleaned when the heat is removed from the steam generator at the completion of the steam distillation, if the two pinch clamps are kept tightly closed. The reduced pressure which results in the *closed* steam generator, on cooling, causes the alkaline mixture to be sucked back into the steam trap which is then drained by opening the pinch clamp beneath it. Distilled water is introduced into the distillation flask through the filling funnel, both pinch clamps closed again, a little steam generated, and the system then allowed to cool, causing the rinse water to be sucked back into the trap as above. An alternate method of washing is the following: With both pinch clamps closed and steam being generated and passing through the system at a rapid rate, the end of the delivery tube of the condenser is submerged in distilled water. Heat is removed from the closed steam generator and the reduced pressure resulting on cooling causes the rinse water to be sucked up into the condenser, into the distillation flask and then into the trap from which it is drained.

Calculation:*

Each ml. of 0.01N hydrochloric acid is equivalent to 0.14008 mg. of nitrogen (see factors, Chapter 23).

* Blank determinations should be made in the absence of a sample, using all reagents, and the value obtained subtracted.

$$\therefore \frac{\text{ml. of } 0.01N \text{ HCl} \times 0.14008 \times 100}{\text{Wt. of sample}} = \% \ N$$

Example:

Suppose that 3.15 ml. of 0.01N hydrochloric acid is required to titrate the ammonia formed from a sample weighing 4.382 mg.

$$\therefore \frac{3.15 \times 0.14008 \times 100}{4.382} = 10.07\% \ N$$

The allowable error of the method is the same as that of the Dumas method, namely, ±0.2%.

Procedure for Determination of Nitrogen in Azo and Nitro Compounds, Oximes, Isoxazoles, Hydrazines, and Hydrazones

From 5 to 8 mg. of sample, 0.2 ml. of 98 to 100% formic acid, and 0.1 ml. of concentrated hydrochloric acid, sp. gr. 1.18, in a 30-ml. micro-Kjeldahl flask (preferably, Soltys type), are heated in a water bath (80 to 85° C.) until the sample is dissolved. (With the few cases where solubility is not affected by formic acid, glacial acetic acid should be used instead.) To this is added 80 mg. of zinc dust (nitrogen-free), and the contents mixed by swirling for about 2 minutes, and heated in the water bath for 5 additional minutes. Then 40 mg. of iron powder is added and the flask is swirled for 2 minutes to effect mixing. To this are added 0.1 ml. of concentrated hydrochloric acid, sp. gr. 1.18, and 0.15 ml. of ethanol, and the mixture is again heated for 5 minutes (water bath). Every 5 minutes a 0.1-ml. portion of concentrated hydrochloric acid is added, with swirling and heating in the water bath, until the iron is dissolved. While the flask is swirled under a hood, 1.0 ml. of concentrated sulfuric acid, sp. gr. 1.84, is added (cautiously), and the contents are warmed until the evolution of gas (hydrogen chloride) has subsided. Then 0.65 gram of potassium sulfate, 0.016 gram of mercuric oxide, and, finally, 0.5 ml. of concentrated sulfuric acid, sp. gr. 1.84, are added. (It is preferable to add the sulfuric acid in this manner rather than all at once, because the final 0.5 ml. can be used to wash down any materials adhering to the walls.)

The mixture is boiled on a micro-Kjeldahl digestion rack for 4 hours, making certain that boiling is vigorous enough so that refluxing takes place almost halfway up in the neck of the flask. The digest is transferred to the micro-Kjeldahl distillation apparatus (preferably the one-piece model), 7.5 ml. of sodium hydroxide-sodium thiosulfate mixture are added, and the ammonia is distilled out into boric acid solution (8 minutes with the delivery tube below

8. Nitrogen: Micro-Kjeldahl Method

the boric acid solution, 2 minutes with it above). The distillate is then titrated with 0.01N hydrochloric acid, using bromocresol green-methyl red (or methyl red-methylene blue) mixture as the indicator. A blank determination should be run using all of the materials, minus the sample, and going through the entire procedure. Any blank value so obtained should be subtracted.

Calculation: Same as above.

Procedure for Determination of Nitrogen in Nitrates[195]

In a 30-ml. micro-Kjeldahl flask are placed 5 to 8 mg. of sample, 35 mg. of salicylic acid and 1 ml. of concentrated sulfuric acid, sp. gr. 1.84. The mixture is first cooled and then allowed to stand at room temperature for 30 minutes, during which time nitration of the salicylic acid takes place. Then 100 mg. of crystalline sodium thiosulfate is added with *cooling* and the mixture is allowed to stand at room temperature for an additional 10 to 15 minutes. Finally 0.65 gram of potassium sulfate and 0.016 gram of mercuric oxide are added, and the mixture is boiled on the digestion rack in the usual manner for 4 hours.

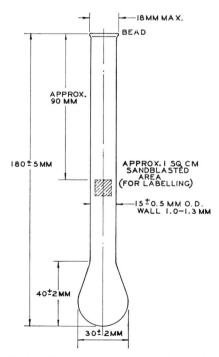

FIG. 116. Kjeldahl digestion flask, 10 ml.—details of construction.

Distillation, titration, and blank determination are done as described under Regular Procedure.

Calculation: Same as above.

ADDITIONAL INFORMATION FOR CHAPTER 8

Figures 116 and 117 show two other sizes of digestion flasks sometimes used, namely the 10-[5,193,194] and 100-[193,194]ml. sizes.

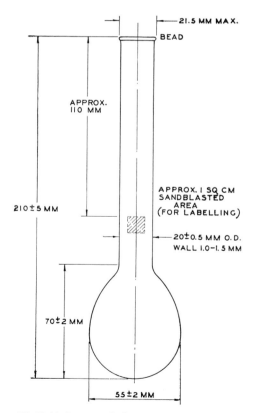

FIG. 117. Kjeldahl digestion flask, 100 ml.—details of construction.

8. Nitrogen: Micro-Kjeldahl Method

TABLE 19

ADDITIONAL INFORMATION ON REFERENCES* RELATED TO CHAPTER 8

In addition to the procedures described in detail in the preceding pages of this chapter, the author wishes to call to the attention of the reader many of the investigations in connection with the microdetermination of nitrogen by the Kjeldahl method. These are listed in Table 19 and the same plan is used here as was used in the preceding chapters. (See statement at top of Table 4 of Chapter 1, regarding completeness of this material.)

Books

Association of Official Agricultural Chemists, 6
Belcher and Godbert, 21, 22
Clark, E. P., 41
Clark, S. J., 42
Furman, 66
Grant, 73, 74
Kirk, 98
Milton and Waters, 130, 131
Niederl and Niederl, 137, 138
Niederl and Sozzi, 139
Pregl, 157
Roth, 163–166
Steyermark, 193

General, miscellaneous, reviews

American Instrument Company, 4
Bode, 26
Bradstreet, 31
Dittrich and deVries, 51
Gómez Vigide, 70
González Carreró, Carballido Ramallo, and Gómez Vigide, 71, 72
Hirai and Hayatsu, 85
Hu, 86
Kainz, 91
Kirk, 95, 98
Levin, Oberholzer, and Whitehead, 112
Machemer and McNabb, 118
McKenzie, Wake and Wallace, 124
Rush, Cruikshank, and Rhodes, 167
Schulek, Burger, and Fehér, 174
Stegemann, 190
Stetten, 191
White and Secor, 211
White, Secor, and Long, 212
Wingo, Davis, and Anderson, 222

Collaborative studies

Acree, 1
Clark, 40
Lake, 107
Ogg, 142, 143
Ogg and Willits, 145, 146
Willits and Ogg, 216–220

Without microchemical balance

Niederl and McBeth, 136

Apparatus

Asami, 7
Ballentine and Gregg, 10
Berck, 24
Brant and Sievers, 34, 35
British Standards Institution, 37
Clark, E. P., 40, 41
Doherty and Ogg, 53
Eder, 55
Furman, 66
Grant, 73, 74
Hallett, 78
Jenden and Taylor, 89
Johanson, 90
Kemmerer and Hallett, 93
Kirk, 96–98
Kirsten, 99
Klingmüller, 101
Leurquin and Delville, 111
Machemer and McNabb, 118
Marino, 120
Miller, 126
Möhlau, 132, 133
Niederl and Niederl, 137, 138
Nógrády, 141
Parnas, 149
Parnas, 150

* The numbers which appear after each entry in this table refer to the literature citations in the reference list at the end of the chapter.

TABLE 19 (*Continued*)

Apparatus (*Cont.*)
Parnas and Wagner, 151
Roth, 162
Scandrett, 169
Schöniger and Haack, 173
Sheers and Cole, 182
Shepard and Jacobs, 183
Sickels and Schultz, 185
Silverstein and Perthel, 186
Skidmore, 187
Willits, John, and Ross, 215

Digestion conditions
Acree, 1
Ballentine and Gregg, 10
Beatty, 12
Beet and Belcher, 16
Belcher and Godbert, 20
Bradstreet, 33
Brüel and Holter, 38
Clark, E. P., 40, 41
Cole and Parks, 43
Dalrymple and King, 44
Elek and Sobotka, 57
Harte, 79
Hartley, 80
Kaye and Weiner, 92
Lauro, 109
Miller, H. S., 127
Niederl and Niederl, 137, 138
Ogg, Brand, and Willits, 144
Osborn, 148
Parnas, 149
Parnas and Wagner, 151
Pepkowitz, Prince, and Bear, 153
Portner, 156
Pregl, 157
Schulek and Vastagh, 176
Steyermark, 193
Stubblefield and Deturk, 197
Vene, 208
White and Secor, 211
Willits, Coe, and Ogg, 214
Zakrzewski and Fuchs, 224

Modifications
Beet, 13–15
Blom and Schwarz, 25

Modifications (*Cont.*)
Bradstreet, 30
Breyhan, 36
Cepciansky and Chromcova, 39
Dermelj and Strauch, 46–48
Edwards, 56
Esafox, 58
Fawcett, 60
Fels and Veatch, 61
Fish and Collier, 64
Hayazu, 82
Hu, 86
Jacobs, 88
Kaye and Weiner, 92
Kirk, 96, 97
Lada and Usiekniewicz, 106
Lang, 108
Lestra and Roux, 110
Marzadro, 121–123
McKenzie and Wallace, 125
Milner and Zahner, 128
Möhlau, 132, 133
Mott and Wilkinson, 135
Noble, 140
Ogg, 143
Ogg and Willits, 145
Pepkowitz, 152
Pepkowitz, Prince, and Bear, 153
Pepkowitz and Shive, 154
Perrin, 155
Schöniger, 172
Silverstein and Perthel, 186
Terent'ev and Luskina, 201
Tourtellotte, Parker, Alving, and De Jong, 204
Zinneke, 225

Sealed tube digestion
Baker, 9
Grunbaum, Kirk, Green, and Koch, 75
Grunbaum, Schaffer and Kirk, 76
Schaffer and Sprecher, 170
White and Long, 210

Indicators
Grant, 73, 74
Kaye and Weiner, 92
Niederl and Niederl, 137, 138

8. Nitrogen: Micro-Kjeldahl Method

TABLE 19 (*Continued*)

Indicators (*Cont.*)
Ogg, Brand, and Willits, 144
Pregl, 157
Roth, 163–166
Sher, 184
Taylor and Smith, 200
Wagner, 209

Catalysts
Baker, 8
Beatty, 12
Bradstreet, 31–33
Dalrymple and King, 44
Fawcett, 60
Fish and Collier, 64
Gómez Vigide, 70
Levin, Oberholzer, and Whitehead, 112
McKenzie and Wallace, 125
Rauen and Buchka, 159
Ribas and López Capont, 160
Ribas and Vázquez-Gesto, 161
Takeda and Senda, 199
White, Secor, and Long, 212

Ultra-, submicro methods
Baudet and Cherbuliez, 11
Belcher, Bhasin and West, 17
Belcher, West, and Williams, 23
Boissonnas and Haselbach, 27
Day, Bernstorf, and Hill, 45
Dixon, 52
Doyle and Omoto, 54
Exley, 59
Gaberman, 68
Grunbaum, Kirk, Green, and Koch, 75
Kirsten, 100
Kuck, Kingsley, Kinsey, Sheehan, and Swigert, 105
Okada and Hanafusa, 147
Schaffer and Sprecher, 170
Scheurer and Smith, 171
Schulek and Fóti, 175
Shaw and Beadle, 181
Sobel, Mayer, and Gottfried, 189
Vallentyne, 205

Fluorine compounds
Fennell and Webb, 62
Ma, 115

N—O, N—N compounds
Baker, 9
Belcher, Bhasin, and West, 17
Belcher and Bhatty, 18
Bradstreet, 28, 29
Clark, E. P., 41
Dickinson, 49, 50
Fish, 63
Friedrich, 65
Grant, 73, 74
Hayazu, 83, 84
Konovalov, 102, 103
Kuch, Kingsley, Kinsey, Sheehan, and Swigert, 105
Ma, Lang, and McKinley, 116
Marzadro, 122
Ogg and Willits, 146
Pepkowitz, 152
Roth, 163–166
Secor, Long, Kilpatrick, and White, 179
Willits and Ogg, 218, 219

Iodometric procedures
Morgulis and Friedman, 134
Rappaport, 158

Omission of distillation
Adams and Spaulding, 3
Belcher and Bhatty, 19
Harvey, 81
Lestra and Roux, 110
Llacer, 113
Marcali and Rieman, 119
Taylor and Smith, 200

Elimination of titration
Taylor and Smith, 200
Also see Chapter 18

Spectrophotometric, colorimetric, chromatographic, etc., methods
Boissonnas and Haselbach, 27
Gardon and Leopold, 69
Gonzáles Carreró, Carballido Ramallo, and Gómez Vigide, 71, 72
Hale, Hale, and Jones, 77
Kotlyarov, 104
Lubochinsky and Zalta, 114

TABLE 19 (*Continued*)

Spectrophotometric, colorimetric, chromatographic, etc., methods (Cont.)

Milner, Zahner, Hepner, and Cowell, 129
Scheurer and Smith, 171
Schwab and Caramanos, 177
Stone, 196
Thompson and Morrison, 203
Wust, 223
Zinneke, 225

Other methods

Dumas, see Chapter 7
Hussey and Maurer, 87
King and Faulconer, 94
Manometric, see Chapter 18
Seligson and Seligson, 180
Slavik and Smetana, 188
Van Slyke, 206
Van Slyke and Kugel, 207

REFERENCES

1. Acree, F., Jr., *J. Assoc. Offic. Agr. Chemists,* **24**, 648 (1941).
2. Acculate, Anachemia Chemicals, Ltd., Montreal, Quebec, and Champlain, New York.
3. Adams, C. I., and Spaulding, G. H., *Anal. Chem.,* **27**, 1003 (1955).
4. American Instrument Co., Inc., "Digestion and Distillation Apparatus for the Micro Kjeldahl Nitrogen Determination," Bull. No. 2271, Silver Springs, Maryland, 1955.
5. American Society for Testing Materials, *ASTM Designation,* **E 147-59 T** (1959).
6. Association of Official Agricultural Chemists, "Official Methods of Analysis," 8th ed., pp. 801–811, Washington, D. C., 1955.
7. Asami, T., *Anal. Chem.,* **31**, 630 (1959).
8. Baker, P. R. W., *Analyst,* **78**, 500 (1953).
9. Baker, P. R. W., *Analyst,* **80**, 481 (1955).
10. Ballentine, R., and Gregg, J. R., *Anal. Chem.,* **19**, 281 (1947).
11. Baudet, P., and Cherbuliez, E., *Helv. Chim. Acta,* **40**, 1612 (1957); *Chem. Abstr.,* **53**, 15328 (1958).
12. Beatty, C., III, *Ind. Eng. Chem., Anal. Ed.,* **15**, 476 (1943).
13. Beet, A. E., *J. Appl. Chem. (London),* **4**, 373 (1954).
14. Beet, A. E., *Nature,* **175**, 513 (1955).
15. Beet, A. E., *Fuel. Soc. J. Sheffield Univ.,* **6**, 12 (1955).
16. Beet, A. E., and Belcher, R., *Mikrochemie,* **24**, 145 (1938).
17. Belcher, R., Bhasin, R. L., and West, T. S., *J. Chem. Soc.,* p. 2585 (1959).
18. Belcher, R., and Bhatty, M. K., *Analyst,* **81**, 124 (1956).
19. Belcher, R., and Bhatty, M. K., *Mikrochim. Acta,* p. 1183, 1956.
20. Belcher, R., and Godbert, A. L., *J. Soc. Chem. Ind. (London),* **60**, 196 (1941).
21. Belcher, R., and Godbert, A. L., "Semi-Micro Quantitative Organic Analysis," Longmans, Green, London, New York, and Toronto, 1945.
22. Belcher, R., and Godbert, A. L., "Semi-Micro Quantitative Organic Analysis," 2nd ed., Longmans, Green, London, 1954.
23. Belcher, R., West, T. S., and Williams, M., *J. Chem. Soc.,* p. 4323 (1957).
24. Berck, B., *Chemist Analyst,* **43**, 52 (1954).
25. Blom, J., and Schwarz, B., *Acta Chem. Scand.,* **3**, 1439 (1949).
26. Bode, F., *Experientia,* **9**, 271 (1953).
27. Boissonnas, R. A., and Haselbach, C. H., *Helv. Chim. Acta,* **36**, 576 (1953).
28. Bradstreet, R. B., *Anal. Chem.,* **26**, 185 (1954); *Anal Abstr.,* **1**, No. 1278 (1954).
29. Bradstreet, R. B., *Anal. Chem.,* **26**, 235 (1954); *Anal. Abstr.,* **1**, No. 1279 (1954).

30. Bradstreet, R. B., *Anal. Chem.*, **32**, 114 (1960).
31. Bradstreet, R. B., *Chem. Revs.*, **27**, 331 (1940).
32. Bradstreet, R. B., *Ind. Eng. Chem., Anal. Ed.*, **10**, 696 (1938).
33. Bradstreet, R. B., *Ind. Eng. Chem., Anal. Ed.*, **12**, 657 (1940).
34. Brant, J. H., and Sievers, D. C., *Ind. Eng. Chem., Anal. Ed.*, **12**, 133 (1940).
35. Brant, J. H., and Sievers, D. C., *Ind. Eng. Chem, Anal. Ed.*, **13**, 133 (1941).
36. Breyhan, T., *Z. anal. Chem.*, **152**, 412 (1956).
37. British Standards Institution, *Brit. Standards,* **1428**, Pt. B1, and Pt. B2 (1953).
38. Brüel, D., and Holter, H., *Compt. rend. trav. lab. Carlsberg. Sér. chim.*, **25**, 289 (1947).
39. Cepciansky, I., and Chromcova, L., *Chem. průmysl,* **9**, 188 (1959); *Anal. Abstr.,* **6**, No. 4802 (1959).
40. Clark, E. P., *J. Assoc. Offic. Agr. Chemists,* **24**, 641 (1941).
41. Clark, E. P., "Semimicro Quantitative Organic Analysis," Academic Press, New York, 1943.
42. Clark, S. J., "Quantitative Methods of Organic Microanalysis," Butterworths, London, 1956.
43. Cole, J. O., and Parks, C. R., *Ind. Eng. Chem., Anal. Ed.*, **18**, 61 (1946).
44. Dalrymple, R. S., and King, G. B., *Ind. Eng. Chem., Anal. Ed.*, **17**, 403 (1945).
45. Day, H. G., Bernstorf, E., and Hill, R. T., *Anal. Chem.*, **21**, 1290 (1949).
46. Dermelj, M., and Strauch, L., *Bull. sci. Conseil acad. RPF Yougoslavie,* **3**, 6 (1956).
47. Dermelj, M., and Strauch, L., *Mikrochim. Acta,* p. 96 (1957).
48. Dermelj, M., and Strauch, L., *Vestnik. Sloven. kemi. društva,* **2**, 77 (1955).
49. Dickinson, W. E., *Anal. Chem.*, **26**, 777 (1954).
50. Dickinson, W. E., *Anal. Chem.*, **30**, 992 (1958).
51. Dittrich, S., and Vries, J. X. de, *pR (Montevideo)*, **5**, 78D (1955); *Chem. Abstr.,* **52**, 169 (1958).
52. Dixon, J. P., *Anal. Chim. Acta,* **13**, 12 (1955).
53. Doherty, D. G., and Ogg, C. L., *Ind. Eng. Chem., Anal. Ed.*, **15**, 751 (1943).
54. Doyle, W. L., and Omoto, J. H., *Anal. Chem.*, **22**, 603 (1950).
55. Eder, K., *Mikrochim. Acta,* p. 227 (1957).
56. Edwards, A. H., *J. Appl. Chem. (London),* **4**, 330 (1954).
57. Elek, A., and Sobotka, H., *J. Am. Chem. Soc.*, **48**, 501 (1926).
58. Esafox, V. I., *Zavodskaya Lab.*, **21**, 1160 (1955).
59. Exley, D., *Biochem. J.,* **63**, 496 (1956).
60. Fawcett, J. K., *J. Med. Lab. Technol.*, **12**, 1 (1954).
61. Fels, G., and Veatch, R., *Anal. Chem.*, **31**, 451 (1959).
62. Fennell, T. R. F. W., and Webb, J. R., *Analyst,* **83**, 694 (1958).
63. Fish, V. B., *Anal. Chem.*, **24**, 760 (1952).
64. Fish, V. B., and Collier, P. R., *Anal. Chem.*, **30**, 151 (1958).
65. Friedrich, A., *Z. physiol. Chem. Hoppe-Seyler's,* **216**, 68 (1933).
66. Furman, N. H., ed., "Scott's Standard Methods of Chemical Analysis," 5th ed., Vol. II, Van Nostrand, New York, 1939.
67. Furter, M. F., Personal communication.
68. Gaberman, V., *Voprosy Med. Khim.*, **3**, 464 (1957); *Chem. Abstr.,* **52**, 12663 (1958).
69. Gardon, J. L., and Leopold, B., *Anal. Chem.*, **30**, 2057 (1958).
70. Gómez Vigide, R. F., *Inform. quím. anal. (Madrid),* **12**, 9 (1958); *Anal. Abstr.,* **5**, No. 3767 (1958).

71. González Carreró, J., Carballido Ramallo, O., and Gómez Vigide, F., *Inform. quím. anal.* (*Madrid*), **10**, 199 (1956).
72. González Carreró, J., Carbillado Ramallo, O., and Gómez Vigide, F., *Proc. Intern. Cong. Pure and Appl. Chem., Anal. Chem. 15th Congr., Lisbon*, **1**, 101 (1956); *Anal. Abstr.*, **6**, No. 1175 (1959).
73. Grant, J., "Quantitative Organic Microanalysis, Based on the Methods of Fritz Pregl," 4th ed., Blakiston, Philadelphia, Pennsylvania, 1946.
74. Grant, J., "Quantitative Organic Microanalysis," 5th ed., Blakiston, Philadelphia, Pennsylvania, 1951.
75. Grunbaum, B. W., Kirk, P. L., Green, L. G., and Koch, C. W., *Anal. Chem.*, **27**, 384 (1955).
76. Grunbaum, B. W., Schaffer, F. L., and Kirk, P. L., *Anal. Chem.*, **24**, 1487 (1952).
77. Hale, C. H., Hale, M. N., and Jones, W. H., *Anal. Chem.*, **21**, 1549 (1949).
78. Hallett, L. T., *Ind. Eng. Chem., Anal. Ed.*, **14**, 956 (1942).
79. Harte, R. A., *Ind. Eng. Chem., Anal. Ed.*, **7**, 432 (1935).
80. Hartley, O., *Ind. Eng. Chem., Anal. Ed.*, **6**, 249 (1934).
81. Harvey, H. W., *Analyst*, **76**, 657 (1951).
82. Hayazu, R., *Yakugaku Zasshi*, **70**, 357 (1950).
83. Hayazu, R., *Yakugaku Zasshi*, **71**, 135 (1951).
84. Hayazu, R., *Yakugaku Zasshi*, **71**, 207 (1951).
85. Hirai, M., and Hayatsu, R., *Yakugaku Zasshi*, **71**, 765 (1951).
86. Hu, C.-C., *Hua Hsüeh Shih Chieh*, p. 131 (1959); *Chem. Abstr.*, **53**, 21396 (1959).
87. Hussey, A. S., and Maurer, J. E., *Anal. Chem.*, **24**, 1642 (1952).
88. Jacobs, M. B., *J. Am. Pharm. Assoc. Sci. Ed.*, **40**, 151 (1951).
89. Jenden, D. J., and Taylor, D. B., *Anal. Chem*, **25**, 685 (1953).
90. Johanson, R., *J. Proc. Australian Chem. Inst.*, **15**, 183 (1948).
91. Kainz, G., *Österr. Chemiker-Ztg.*, **57**, 242 (1956).
92. Kaye, I. A., and Weiner, N., *Ind. Eng. Chem., Anal. Ed.*, **17**, 397 (1945).
93. Kemmerer, G., and Hallett, L. T., *Ind. Eng. Chem.*, **19**, 1295 (1927).
94. King, R. W., and Faulconer, W. B. M., *Anal. Chem.*, **28**, 255 (1956).
95. Kirk, P. L., *Anal. Chem.*, **22**, 354 (1950).
96. Kirk, P. L., *Ind. Eng. Chem., Anal. Ed.*, **8**, 223 (1936).
97. Kirk, P. L., *Mikrochemie*, **16**, 13 (1934).
98. Kirk, P. L., "Quantitative Ultramicroanalysis," Wiley, New York, and Chapman & Hall, London, 1950; "The Chemical Determination of Proteins," *in* "Advances in Protein Chemistry" (M. L. Anson, and J. T. Edsall, eds.), Vol. 3, Academic Press, New York, 1947.
99. Kirsten, W., *Anal. Chem.*, **24**, 1078 (1952).
100. Kirsten, W. J., *Microchem. J.*, **2**, 179 (1958).
101. Klingmüller, V., *Z. anal. Chem.*, **131**, 17 (1950).
102. Konovalov, A., *Ind. chim. belge*, **18**, 329 (1953).
103. Konovalov, A., *Ind. chim. belge*, **24**, 259 (1959); *Anal. Abstr.*, **6**, No. 4803 (1959).
104. Kotlyarov, I. I., *Fiziol. Zhur. S.S.S.R.*, **33**, 123 (1947).
105. Kuck, J. A., Kingsley, A., Kinsey, D., Sheehan, F., and Swigert, G. F., *Anal. Chem.*, **22**, 604 (1950).
106. Lada, Z., and Usiekniewicz, K., *Chem. Anal.* (*Warsaw*), **2**, 351 (1957); *Chem Abstr.*, **52**, 4395 (1958).
107. Lake, G. R., *Anal. Chem.*, **24**, 1806 (1952).
108. Lang, C. A., *Anal. Chem.*, **30**, 1692 (1958).

109. Lauro, M. F., *Ind. Eng. Chem., Anal. Ed.,* **3**, 401 (1931).
110. Lestra, H., and Roux, G., *Compt. rend. acad. sci.,* **233**, 1453 (1951).
111. Leurquin, J., and Delville, J. P., *Experientia,* **6**, 274 (1950).
112. Levin, B., Oberholzer, V. G., and Whitehead, T. P., *Analyst,* **75**, 561 (1950).
113. Llacer, A. J., *Mikrochemie ver. Mikrochim. Acta,* **40**, 173 (1953).
114. Lubochinsky, B., and Zalta, J. P., *Bull. soc. chim. biol.,* **36**, 1363 (1954).
115. Ma, T. S., *Microchem. J.,* **2**, 91 (1958).
116. Ma, T. S., Lang, R. E., and McKinley, J. D., Jr., *Mikrochim. Acta,* p. 368 (1957).
117. Ma, T. S., and Zuazaga, G., *Ind. Eng. Chem., Anal. Ed.,* **14**, 280 (1942).
118. Machemer, P. E., and McNabb, W. M., *Anal. Chim. Acta,* **3**, 428 (1949).
119. Marcali, K., and Rieman, Wm., III, *Ind. Eng. Chem., Anal. Ed.,* **18**, 709 (1946).
120. Marino, S. P., *Chemist Analyst,* **39**, 42 (1950).
121. Marzadro, M., *Mikrochemie ver. Mikrochim. Acta,* **36/37**, 671 (1951).
122. Marzadro, M., *Mikrochemie ver. Mikrochim. Acta,* **38**, 372 (1951).
123. Marzadro, M., *Mikrochemie ver. Mikrochim. Acta,* **40**, 359 (1953).
124. McKenzie, H. A., Wake, R. G., and Wallace, H. S., *Proc. Intern. Congr. Pure and Appl. Chem., Anal. Chem. 15th Congr., Lisbon* **1** (1956); *Anal. Abstr.,* **6**, No. 1175 (1959).
125. McKenzie, H. A., and Wallace, H. S., *Australian J. Chem.,* **7**, 55 (1954).
126. Miller, S. P., *Chemist Analyst,* **43**, 52 (1954).
127. Miller, H. S., *Ind. Eng. Chem., Anal. Ed.,* **8**, 50 (1936).
128. Milner, O. I., and Zahner, R. J., *Anal. Chem.,* **32**, 294 (1960).
129. Milner, O. I., Zahner, R. J., Hepner, L. S., and Cowell, W. H., *Anal. Chem.,* **30**, 1528 (1958).
130. Milton, R. F., and Waters, W. A., "Methods of Quantitative Microanalysis," Longmans, Green, New York, and Arnold, London, 1949.
131. Milton, R. F., and Waters, W. A., "Methods of Quantitative Microanalysis," 2nd ed., Arnold, London, 1955.
132. Möhlau, E., *Pharm. Zentralhalle,* **89**, 105 (1950).
133. Möhlau, E., *Pharm. Zentralhalle,* **89**, 334 (1950).
134. Morgulis, S. M., and Friedman, A. F., *Bull. soc. chim. biol.,* **18**, 1074 (1937).
135. Mott, R. A., and Wilkinson, H. A., *Fuel,* **37**, 151 (1958).
136. Niederl, J. B., and McBeth, C. H., *Mikrochemie ver. Mikrochim. Acta,* **35**, 98 (1950).
137. Niederl, J. B., and Niederl, V., "Micromethods of Quantitative Organic Elementary Analysis," Wiley, New York, 1938.
138. Niederl, J. B., and Niederl, V., "Micromethods of Quantitative Organic Analysis," 2nd ed., Wiley, New York, 1942.
139. Niederl, J. B., and Sozzi, J. A., "Microanálisis Elemental Orgánico," Calle Arcos, Buenos Aires, 1958.
140. Noble, E. D., *Anal. Chem.,* **27**, 1413 (1955).
141. Nógrády, G., *Magyar Kém. Lapja,* **4**, 350 (1949).
142. Ogg, C. L., *J. Assoc. Offic. Agr. Chemists,* **38**, 365 (1955).
143. Ogg, C. L., *J. Assoc. Offic. Agr. Chemists,* **43**, 682, 689 (1960).
144. Ogg, C. L., Brand, R. W., and Willits, C. O., *J. Assoc. Offic. Agr. Chemists,* **31**, 663 (1948).
145. Ogg, C. L., and Willits, C. O., *J. Assoc. Offic. Agr. Chemists,* **33**, 100 (1950).
146. Ogg, C. L., and Willits, C. O., *J. Assoc. Offic. Agr. Chemists,* **35**, 288 (1952).
147. Okada, Y., and Hanafusa, H., *Bull. Chem. Soc. Japan,* **27**, 478 (1954).
148. Osborn, R. A., *J. Assoc. Offic. Agr. Chemists,* **16**, 107 (1953); **18**, 604 (1935).

149. Parnas, J. K., *Z. anal. Chem.*, **114**, 261 (1938).
150. Parnas, Ya. O., *Zhur. Anal. Khim.*, **4**, 54 (1949).
151. Parnas, J. K., and Wagner, R., *Biochem. Z.*, **125**, 253 (1931).
152. Pepkowitz, L. P., *Anal. Chem.*, **24**, 900 (1952).
153. Pepkowitz, L. P., Prince, A. L., and Bear, F. E., *Ind. Eng. Chem., Anal. Ed.*, **14**, 856 (1942).
154. Pepkowitz, L. P., and Shive, W. J., *Ind. Eng. Chem., Anal. Ed.*, **14**, 914 (1942).
155. Perrin, C. H., *Anal. Chem.*, **25**, 968 (1953).
156. Portner, P. E., *Anal. Chem.*, **19**, 502 (1947).
157. Pregl, F., "Quantitative Organic Microanalysis," (E. Fyleman, trans., 2nd German ed.), Churchill, London, 1924.
158. Rappaport, F., *Mikrochemie,* **14**, 49 (1933).
159. Rauen, H. M., and Buchka, M., *Angew. Chem.*, **60A**, 209 (1948).
160. Ribas, I., and López Capont, F., *Anales real soc. españ. fís. y quím. (Madrid)*, **46B**, 581 (1950).
161. Ribas, I., and Vásquez-Gesto, D., *Inform. quím. anal. (Madrid)*, **7**, 29 (1953).
162. Roth, H., *Mikrochemie ver. Mikrochim. Acta,* **31**, 287 (1944).
163. Roth, H., "Die quantitative organische Mikroanalyse von Fritz Pregl," 4th ed., Springer, Berlin, 1935.
164. Roth, H., "F. Pregl quantitative organische Mikroanalyse," 5th ed., Springer, Wien, 1947.
165. Roth, H., "Pregl-Roth quantitative organische Mikroanalyse," 7th ed., Springer, Wien, 1958.
166. Roth, H., "Quantitative Organic Microanalysis of Fritz Pregl," 3rd ed. (E. B. Daw, trans., 4th German ed.), Blakiston, Philadelphia, Pennsylvania, 1937.
167. Rush, C. A., Cruikshank, S. S., and Rhodes, E. J. H., *Mikrochim. Acta,* p. 858 (1956).
168. Scales, F. M., and Harrison, A. P., *Ind. Eng. Chem.*, **12**, 350 (1920).
169. Scandrett, F. J., *Analyst,* **78**, 734 (1953).
170. Schaffer, F. L., and Sprecher, J. C., *Anal. Chem.*, **29**, 437 (1957).
171. Scheurer, P. G., and Smith, F., *Anal. Chem.*, **27**, 1616 (1955).
172. Schöniger, W., *Mikrochim. Acta,* p. 44 (1955).
173. Schöniger, W., and Haack, A., *Mikrochim. Acta,* p. 1369 (1956).
174. Schulek, E., Burger, K., and Fehér, M., *Z. anal. Chem.*, **167**, 28 (1959); *Anal. Abstr.,* **7**, No. 168 (1960).
175. Schulek, E., and Fóti, G., *Anal. Chim. Acta,* **3**, 665 (1949).
176. Schulek, E., and Vastagh, G., *Z. anal. Chem.*, **92**, 352 (1933).
177. Schwab, G. M., and Caramanos, S., *Monatsh. Chem.*, **86**, 341 (1955).
178. Scientific Glass Apparatus Company, Bloomfield, New Jersey.
179. Secor, G. E., Long, M. C., Kilpatrick, M. D., and White, L. M., *J. Assoc. Offic. Agr. Chemists,* **33**, 872 (1950).
180. Seligson, D., and Seligson, H., *J. Lab. Clin. Med.*, **38**, 324 (1951).
181. Shaw, J., and Beadle, L. C., *J. Exptl. Biol.,* **26**, 15 (1949).
182. Sheers, E. H., and Cole, M. S., *Anal. Chem.*, **25**, 535 (1953).
183. Shepard, D. L., and Jacobs, M. B., *J. Am. Pharm. Assoc. Sci. Ed.*, **40**, 154 (1951).
184. Sher, I. H., *Anal. Chem.*, **27**, 831 (1955).
185. Sickels, J. P., and Schultz, H. P., *Quart. J. Florida Acad. Sci.*, **14**, 35 (1951).
186. Silverstein, R. M., and Perthel, R., Jr., *Anal. Chem.*, **22**, 949 (1950).
187. Skidmore, D. W., *Ind. Chemist,* **30**, 386 (1954).
188. Slavik, K., and Smetana, R., *Chem. listy,* **46**, 648 (1952).

8. Nitrogen: Micro-Kjeldahl Method

189. Sobel, A. E., Mayer, A. M., and Gottfried, S. P., *J. Biol. Chem.,* **156**, 355 (1944).
190. Stegemann, H., *Z. physiol. Chem.,* **312**, 255 (1958).
191. Stetten, D., Jr., *Anal. Chem.,* **23**, 1177 (1951).
192. Steyermark, Al, Unpublished work.
193. Steyermark, Al, "Quantitative Organic Microanalysis," Blakiston, Philadelphia, Pennsylvania, 1951.
194. Steyermark, Al, Alber, H. K., Aluise, V. A., Huffman, E. W. D., Kuck, J. A., Moran, J. J., and Willits, C. O., *Anal. Chem.,* **23**, 523 (1951).
195. Steyermark, Al, McGee, B. E., Bass, E. A., and Kaup, R. R., *Anal. Chem.,* **30**, 1561 (1958).
196. Stone, E. W., *Proc. Soc. Exptl. Biol. Med.,* **93**, 589 (1956).
197. Stubblefield, F. M., and Deturk, E. E., *Ind. Eng. Chem., Anal. Ed.,* **12**, 396 (1940).
198. Swedish Iron & Steel Corp., Westfield, New Jersey.
199. Takeda, A., and Senda, J., *Ber. Ōhara Inst. landwirtsch. Biol., Okayama Univ.,* **10**, 241 (1956).
200. Taylor, W. H., and Smith, G. F., *Ind. Eng. Chem., Anal. Ed.,* **14**, 437 (1942).
201. Terent'ev, A. P., and Luskina, B. M., *Zhur. Anal. Khim.,* **14**, 112 (1959).
202. Thomas, Arthur H., Company, Philadelphia, Pennsylvania.
203. Thompson, J. F., and Morrison, G. R., *Anal. Chem.,* **23**, 1153 (1951).
204. Tourtellotte, W. W., Parker, J. A., Alving, R. E., and De Jong, R. N., *Anal. Chem.,* **30**, 1563 (1958).
205. Vallentyne, J. R., *Analyst,* **80**, 841 (1955).
206. Van Slyke, D. D., *J. Biol. Chem.,* **71**, 235 (1926).
207. Van Slyke, D. D., and Kugel, V. H., *J. Biol. Chem.,* **102**, 489 (1933).
208. Vene, J., *Bull. soc. sci. Bretagne,* **15**, 49 (1938).
209. Wagner, E. C., *Ind. Eng. Chem., Anal. Ed.,* **12**, 771 (1940).
210. White, L. M., and Long, M. C., *Anal. Chem.,* **23**, 363 (1951).
211. White, L. M., and Secor, G. E., *Ind., Eng. Chem., Anal. Ed.,* **18**, 457 (1946).
212. White, L. M., Secor, G. E., and Long, M. D. C., *J. Assoc. Offic. Agr. Chemists,* **31**, 657 (1948).
213. Willits, C. O., *Anal. Chem.,* **21**, 132 (1949).
214. Willits, C. O., Coe, M. R., and Ogg, C. L., *J. Assoc. Offic. Agr. Chemists,* **32**, 118 (1949).
215. Willits, C. O., John, H. J., and Ross, L. R., *J. Assoc. Offic. Agr. Chemists,* **31**, 432 (1948).
216. Willits, C. O., and Ogg, C. L., *J. Assoc. Offic. Agr. Chemists,* **31**, 565 (1948).
217. Willits, C. O., and Ogg, C. L., *J. Assoc. Offic. Agr. Chemists,* **32**, 561 (1949).
218. Willits, C. O., and Ogg, C. L., *J. Assoc. Offic. Agr. Chemists,* **33**, 100, 179 (1950).
219. Willits, C. O., and Ogg, C. L., *J. Assoc. Offic. Agr. Chemists,* **34**, 607 (1951).
220. Willits, C. O., and Ogg, C. L., *J. Assoc. Offic. Agr. Chemists,* **34**, 615 (1951).
221. Willits, C. O., and Ogg, C. L., Personal communication.
222. Wingo, W. J., Davis, O. L., and Anderson, L., *Anal. Chem.,* **22**, 1340 (1950).
223. Wust, H., *Klin. Wochschr.,* **32**, 27, 660 (1954).
224. Zakrzewski, Z., and Fuchs, H. J., *Biochem. Z.,* **285**, 390 (1936).
225. Zinneke, F., *Angew. Chem.,* **64**, 220 (1952).

CHAPTER 9

Microdetermination of Carbon and Hydrogen

Since all organic compounds contain carbon, the determination of carbon and hydrogen is the most frequently performed analysis and, therefore, should be considered to be the most important. It also requires the greatest skill and the most exacting of conditions. The determination has been* the subject of much discussion in the past and collaborative studies have been conducted with the hope that from these a more foolproof method will be developed.[383,384] In the opinion of the author, the beginner will be successful with this determination, almost from the beginning, if his laboratory conditions are controlled (see Chapter 1) and if he closely adheres to the directions given in the following pages. It should be understood, however, that few persons achieve acceptable results on their first few tries.

Beginners should not attempt to use some modifications of their own. It is true that out of a group of experienced microanalysts, probably no two will perform the determination in exactly the same manner. However, each variation has resulted from years of experience.

The organic compound is burned at red heat in an atmosphere of oxygen and the carbon and hydrogen are converted into carbon dioxide and water respectively as shown by the following:

$$\text{Organic C} \xrightarrow[O_2]{\Delta} CO_2$$

$$\text{Organic H} \xrightarrow[O_2]{\Delta} H_2O$$

The water is absorbed by magnesium perchlorate (Anhydrone[357] or Dehydrite[357]) and the carbon dioxide is absorbed by sodium hydroxide on asbestos (Ascarite[357]).

Generally, the combustion tube contains sections of the following materials in the order named, starting at the capillary tip: silver, lead peroxide, silver, a mixture of copper oxide and lead chromate, platinum, and finally silver. The sections of silver remove chlorine, bromine, iodine, and oxides of sul-

---

* See Additional Information, pp. 255–264.

9. Carbon and Hydrogen 222

fur.[54,112,113,146,284,285,288,308,317-320,350] The platinum assists in the combustion of condensed ring systems, particularly those containing an angular methyl group which might come off in the form of methane and escape complete combustion.[285] The copper oxide-lead chromate[112,113,284,308,317-320,350] mixture acts as an oxidizing agent. The lead peroxide reacts with and retains oxides of nitrogen which is represented by the following equations[174,175,285,305]:

$$2NO + 2PbO_2 \rightarrow Pb(NO_2)_2 \cdot PbO + \tfrac{1}{2}O_2$$
$$2NO_2 + 2PbO_2 \rightarrow Pb(NO_3)_2 + PbO + \tfrac{1}{2}O_2$$

When fluoro-compounds are being analyzed, *either* of the two following slight modifications in the tube filling is necessary so that fluorine is prevented from passing through:

(a) Two small sections of magnesium oxide are substituted for a portion of the copper oxide-lead chromate mixture,[348] or

(b) A section of aluminum oxide is added between the copper oxide-lead chromate mixture and the platinum.[338,348]

Although both modifications have proven satisfactory, the author prefers the use of the two sections of magnesium oxide, particularly with compounds *also* having a high sulfur content. (When fluoro-compounds are analyzed only *occasionally*, the ordinary tube filling, without the added aluminum or magnesium oxide, may be used *provided* the sample is covered with magnesium oxide.[338,348] This, however, usually gives slightly higher results than when the oxides are *in* the filling, which is to be expected.)

Compounds containing arsenic, antimony, bismuth, and mercury should not be analyzed unless further modifications in the procedure (see Additional Information for this chapter) are made since these elements would ruin the combustion tube filling.[103,112,317-320,350] However, they can be successfully analyzed for carbon by the Van Slyke manometric method[367,368] described in Chapter 18. (Yeh[388a] has obtained excellent results with fluoro-, arsenic, antimony, tin, mercury, and selenium compounds using a special tube filling composed of silver vanadate, zirconium oxide, and magnesium oxide.) Compounds containing the alkali or alkaline earth metals in the absence of sulfur or phosphorus under the conditions of the determination would leave an ash of the carbonate, and the result[112,317-320,350] would be lacking the percentage of one carbon atom. Therefore, potassium dichromate[112,317-320,350] or vanadium pentoxide is added to the sample and during the combustion all of the carbon is converted to carbon dioxide. If, however, enough sulfur or phosphorus is present in the compound to react with the alkali or alkaline earth metal to form the sulfate or phosphate, respectively, no dichromate or vanadium pentoxide need be added.[348,350] If halogens are present in sufficient amounts to form the halides, it should not be necessary to add dichromate or vanadium pentoxide, but in spite of this, the author prefers the addition.[348]

Silver salts, aurates, and platinates give correct results and, in addition, weighing of the residue gives the amount of precious metal (as free metal—compare Chapter 6). Similarly, salts of copper and iron leave as an ash the oxide (see Chapter 6) and offer no interference.

Substances containing phosphorus in the absence of alkali or alkaline earth metals leave an ash of oxides of phosphorus[262,348,350] which holds onto small amounts of carbon and must be ignited *strongly* or low values result.

Reagents

All reagents listed here are commercially available and can be used without further treatment.

SODIUM HYDROXIDE SOLUTION

About 1 liter of a solution of sodium hydroxide (5%) in water is required for the pressure regulators.

CONCENTRATED SULFURIC ACID

About 1 ml. of concentrated sulfuric acid is required for the bubble counter (Fig. 120).

ANHYDRONE[357] OR DEHYDRITE[357] (MAGNESIUM PERCHLORATE)

This material is used in the following pieces for the purpose of absorbing water: U-tube (Fig. 120), both absorption tubes (Fig. 127), and guard tube (Fig. 128).

Neither large pieces nor fine powder should be used—otherwise no special selection of size is necessary.

ASCARITE[357] (SODIUM HYDROXIDE ON ASBESTOS)

The commercially available material of size 8–20 mesh is used in the U-tube (Fig. 120) and in the carbon dioxide absorption tube (Fig. 127).

SILVER WIRE (OR RIBBON)

Fine silver wire (or ribbon) of approximately 34 gauge is used in the combustion tube.

ASBESTOS

Ordinary acid-washed Gooch crucible asbestos is used without further treatment.

9. Carbon and Hydrogen

POTASSIUM DICHROMATE

Reagent grade is melted in a crucible by heating at 400° C.[112,128,317-320,350] The melt is cooled, powdered, and dried in an oven at 120° C. for a short while. It is stored in a glass-stoppered bottle.

PLATINIZED ASBESTOS, 30%[9,87,357]

This material is used in the combustion tube.

COPPER OXIDE WIRE

Ordinary reagent grade of cupric oxide wire is used. No special screening is necessary but neither large pieces nor fine powder should be used (about 1–2 mm. lengths are quite suitable).

LEAD CHROMATE

Special lead chromate for microanalysis is commercially available[87,263,357] and should be used.

COPPER OXIDE-LEAD CHROMATE MIXTURE

This mixture is made of two parts of copper oxide and one part of lead chromate (by weight) and thoroughly mixed.[112,284,308,317-320,350]

LEAD PEROXIDE (LEAD DIOXIDE)

Special grades for microanalysis are commercially available[87,263,282,357] and should be used since they require no further treatment. Otherwise they must be purified.[112,285,317-320]

VANADIUM PENTOXIDE

Reagent grade is used for the same purpose as the potassium dichromate.[348]

ALUMINUM OXIDE

Reagent grade of aluminum oxide (granular) is used in the combustion tube for determinations on fluoro-compounds.[338,348]

MAGNESIUM OXIDE[348]

Reagent grade of precipitated magnesium oxide. This is used either in the combustion tube or on the sample when fluoro-compounds are being analyzed. It must be *free* from moisture and carbonate.

MAGNESIUM OXIDE-ASBESTOS[348]

Equal parts by weight of the magnesium oxide and the asbestos are thoroughly mixed by shaking in a closed bottle. This is used in the combustion tube when dealing with fluoro-compounds.

ABSORBENT COTTON

Ordinary sterile absorbent cotton is used in the absorption tubes.

FIBERGLAS[64]

This material is used to hold the Anhydrone and Ascarite in place in the U-tube.

KROENIG GLASS CEMENT[87,357]

All ground joints are held together with this material. It is prepared from one part white beeswax and four parts of rosin and is commercially available in stick form. It is warmed before using and the parts held in place until the wax sets.

Apparatus

The apparatus consists of a cylinder of oxygen having a reducing valve, a supply of compressed air to which a filter system is attached, two pressure regulators, one each for oxygen and air, one bubble counter-U-tube, one combustion tube, one combustion apparatus, one heating mortar, two absorption tubes, one guard tube, one Mariotte bottle, one graduate cylinder, and two types of rubber tubing. With the exception of the pressure regulators,[106,350] all of the glass parts are those for which recommended specifications have been published by the Committee for the Standardization of Microchemical Apparatus of the Division of Analytical Chemistry of the American Chemical Society.[351]

OXYGEN CYLINDER

A small cylinder of pure oxygen having some sort of reducing valve is required. Any convenient size ranging from the small lecture bottle to the large size may be used. The oxygen should be prepared from liquid air rather than by electrolysis of water since the latter type might contain traces of hydrogen.

COMPRESSED AIR SUPPLY

The ordinary laboratory supply of compressed air may be used but must be filtered through a section of cotton to remove dust, oil, and water. A convenient filter is made by filling a Mariotte bottle* (see below) with cotton. Since the compressed air is very slowly taken from the line, very little oil or water is removed and a filter will last for years without needing replacement.

PRESSURE REGULATORS[106,350]

By means of the pressure regulators, oxygen and air are supplied to the combustion train at constant pressure. Two regulators are required, one for each

---

* Only the bottle is used—not the other parts. Air is led in at the bottom on the side and is taken out at the top.

of the above gases. Each regulator consists of two essential parts: the pressure regulator proper and a leveling tube for obtaining the desired height, connected by means of rubber tubing as shown in Fig. 118. The two regulators

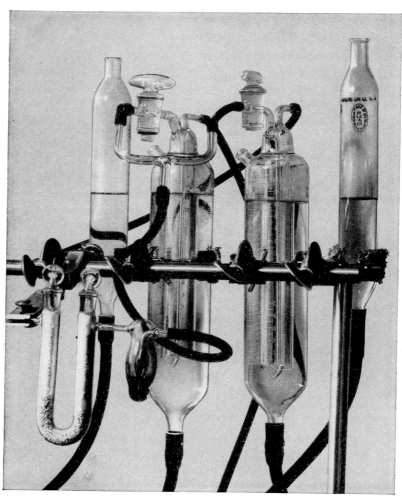

FIG. 118. Pressure regulators—photograph showing connections.

are connected by means of a T- or Y-tube, which in turn is connected to the bubble counter of the combustion train. (See diagram of combustion train—Fig. 130.) Five per cent solution of sodium hydroxide is used in the regulators. Figure 119 shows the details of construction of the unit: a is the leveling tube and b the regulator proper; c is the inlet tube which is connected to the

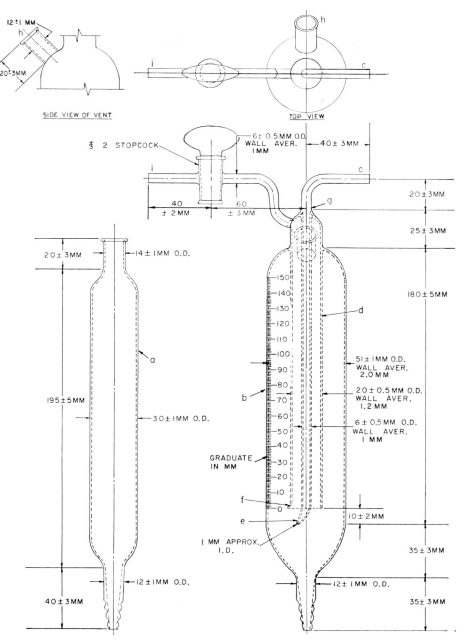

FIG. 119. Pressure regulator—details of construction.

source of oxygen or air. The gas passes vertically down through it, coming out at outlet e and displacing the solution between the vertical part of c and the jacket, d. When the space between c and d is completely filled, the excess gas overflows at f and passes up through the solution and out into the air through h. The position of e should be such that the gas bubbles strike the notch, f. (If, on testing a regulator, it is found that the outlet has been improperly placed, the condition may be corrected by applying the flame of a small blast lamp at point g. When the glass has softened slightly at g, that section of the tube between g and e is allowed to fall into place by gravity.) The height at which the solution stands in b is shown on the graduated scale, this also being the pressure head maintained in d. The gas under the desired pressure passes out through the stopcock and tube i to the bubble counter and the combustion tube.

BUBBLE COUNTER-U-TUBE[350,351]

The bubble counter portion is used as an indicator for determining the rate of flow of oxygen and of air through the system while the U-tube part removes water and carbon dioxide from these gases before they enter the combustion tube. The details of construction are shown in Fig. 120. The gas enters the bubble counter on the left from the pressure regulators (see Fig. 130, diagram of combustion train)* and passes downward through the inner bulb which contains enough concentrated sulfuric acid to immerse the lower 5–7 mm. of the tip of the capillary. The gas then passes through the U-tube into the combustion tube.

FILLING THE BUBBLE COUNTER-U-TUBE.[112,113,284,285,308,317-320,350] Enough Ascarite is placed in the U-tube through the glass ground joints so that the bottom is filled and the absorbent stands at the height of about 6–7 cm. on the right side* (next to the combustion tube) and about 3–4 cm. on the left side* (next to the bubble counter). Small layers (2–3 mm.) of Fiberglas† are put on top of the Ascarite to hold it in place. Enough Anhydrone is then added through each ground joint so that it comes to within 1–2 mm. below the horizontal connecting tubes. (During the addition of both Ascarite and Anhydrone, care must be exercised so that neither material enters the bulb of the bubble counter.) Enough Fiberglas is added to each side to fill the remaining spaces up to the bottoms of the glass stoppers. The ground joints and stoppers are warmed gently with a burner and a film of Kroenig glass cement is applied. The stoppers are put in place and the cement allowed to cool. With the aid of a medicine dropper, enough concentrated sulfuric acid

* In Fig. 130, the direction is reversed.
† Cotton should not be used in the bubble counter-U-tube because of the possibility of perchloric acid ($Mg(ClO_4)_2 + H_2SO_4$) accidentally coming in contact with organic material resulting in an explosion.

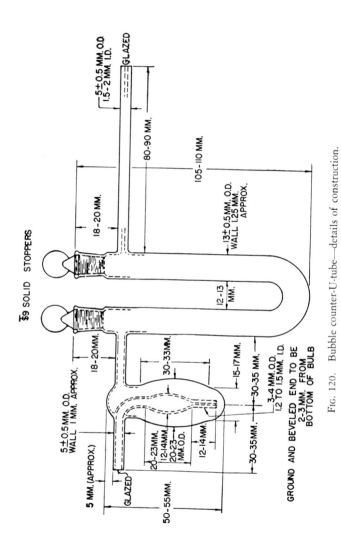

Fig. 120. Bubble counter-U-tube—details of construction.

9. Carbon and Hydrogen

is added, through the horizontal tube on the left, so that the bulb contains enough to immerse the lower 5–7 mm. of the tip of the vertical capillary.

The bubble counter-U-tube can be used until the Anhydrone on the side of the bubble counter appears to be wet. Usually this condition exists after about 75–100 determinations. The apparatus is then cleaned out by washing with water and refilled.

THE COMBUSTION TUBE[10,350,351]

The combustion tube should have the dimensions shown in Fig. 121. Tubes prepared from thinner stock have been shown to be less reliable. Tubes may be prepared of Pyrex No. 1720 glass,[64] Jena Supermax,[337,357] the British hard glass,[357] Vycor[64] (96% silica glass No. 790), or of quartz. (Listed in the order of increasing temperature of softening point.) The author prefers tubes made of Pyrex No. 1720 as these have proved most satisfactory in his laboratory.[349] *Under the conditions of use described in the following pages,* one of these tubes will give good results for about 300 determinations, on the average. (Occasionally, a tube will be good for 500 while others are discarded after 100.)

FILLING THE COMBUSTION TUBE.[112,113,284,308,317–320,350]—(See Fig. 122.) (*A*) *Regular Filling.* [Note: When used for fluoro-compounds, magnesium oxide must be added to the sample[338,348]—see Procedure. However, when fluoro-compounds are being analyzed, it is preferable to use the modified fillings described below under (B) or (C)]. Considerable care must be exercised in the filling of the tubes. The combustible gases passing through it must come in contact with very large surfaces of the various components, but the packing must be loose enough so that the recommended gas velocities may be maintained using limited pressure heads. Fortunately, a rather large variation is possible. The pressure regulator can furnish gas under a constant pressure of from 0.1 to 15.0 cm. of sodium hydroxide solution (see above). The recommended pressure head for bubble counter-U-tube and combustion tube together is between 5 and 15 cm., although tubes have been known to give good results with a pressure head of even less than the lower figure. The tube is so packed that when *in use* with the above conditions 10 ml. of gas pass through it per minute. Other speeds ranging from 3 to 50 or more have been used[27,104,112,284,285,308,317–320,350,385] but the author recommends the above for the beginner. After he has mastered the technique at 10 ml. per minute he may increase the speed slightly, preferably to 12 ml. per minute.

The technique of filling the C-H combustion tube is quite different from that employed for preparing the Dumas one. With the latter, the components could be packed tightly, because CO_2 under a pressure head of about 37–40 cm. of water is used, several times that which is obtainable with the carbon-hydrogen pressure regulator.

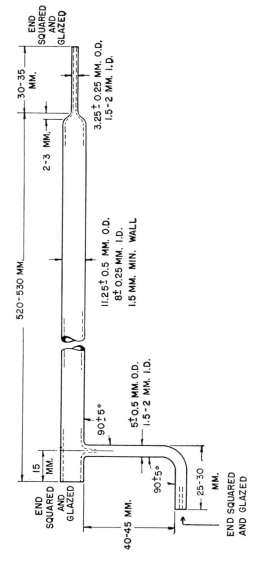

FIG. 121. Combustion tube with tip and side arm—details of construction.

9. Carbon and Hydrogen

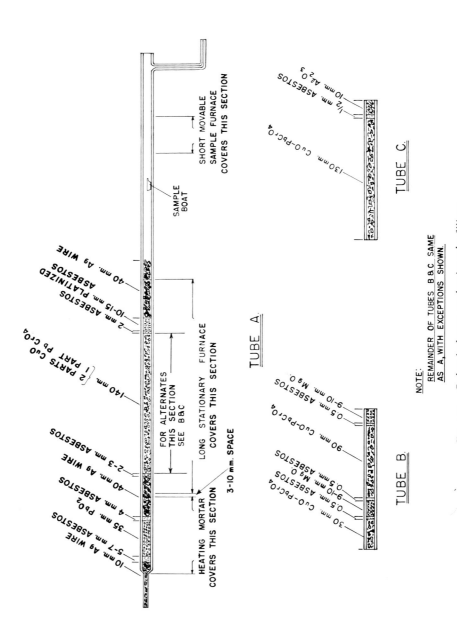

FIG. 122. Carbon-hydrogen combustion tube fillings.

The inside of the tube is wiped free from dust with a cotton swab on the end of a piece of rigid wire whose end is knurled. The author does not believe that elaborate cleaning of the tube or treatment or handling of the components of the filling is necessary, as recommended in the past,[112,284,285,308,317-320] because any organic material present is destroyed during the burning out process (see below).

Two or three strands of *silver wire* (or ribbon) about 100 mm. long are twisted together and inserted in the capillary tip of combustion tube so that the wire is flush with the end and extends well into the main body portion (11 mm. O.D.). Enough silver wire (or ribbon) is then rolled together to form a wad about *10 mm. in length* and whose diameter is approximately that of the I.D. of the combustion tube. This is forced into the tube and all the way down to the tip while the twisted strands are prevented from being pushed out. In this way the wire passing through the capillary tip is locked securely in place. If a portion protrudes past the end of the tip it should be cut off flush with the end, using scissors. The function of the silver at this end and in the tip is to conduct heat so that water will not condense during the combustion.

A section of *asbestos of 5–7 mm.* is next added by *gently* compressing with a glass rod. This is called the *choking plug*[112,284,285,308,317-320,350] and offers resistance to the flow of the gases. It must be kept in mind during preparation that the asbestos expands when heated and gives the effect of a tighter packing.

Enough *lead peroxide* is added, with gentle tapping of the side of the tube to secure good packing, to form a section about *35 mm. in length*. The exact length of the lead peroxide portion will depend upon the size of the heating mortar used as most of the portion of the tube in it contains this material. The inside wall of the tube is wiped clean by means of a cotton swab. Another plug of *asbestos, about 4 mm. in length,* is gently pushed into place and this is followed by a section of *silver wire,* which has been rolled together to form a wad, about *40 mm. long*. When the tube is inserted in the furnace this section should extend 1 cm. into the tube portion in the heating mortar, bridge the gap between this and the long furnace and extend at least 1 cm. into the hot portion of the latter (Figs. 122, and 130). This silver wad acts chiefly as a conductor of heat between the two and prevents condensation of water.

A small section of *asbestos* (*about 2–3 mm.*) is added with gentle pressure. At this point, it is advisable to make certain that the filling is not too tight. This can be done by attaching the capillary tip to the Mariotte bottle. With the drainage tube of the latter depressed about 20°–30° below the horizontal, gas should be sucked through at the rate of about 30–40 ml. per minute.

Enough of a *mixture of copper oxide and lead chromate* (*2 parts CuO :1*

9. Carbon and Hydrogen

part PbCrO₄) is added, with gentle tapping on the side of the tube to insure even packing, to form a section *140 mm.* in length. This is held in place with about *2 mm.* of *asbestos*. The inside wall of the tube is cleaned with a cotton swab. Enough *platinized asbestos* is added to form a very loose, but evenly packed, section about *10–15 mm.* in length. This material is quite fine and if it is more than gently compressed, too much resistance will be offered at this point.

The final portion of the tube filling is *silver wire* (or ribbon) rolled together to form a section about *40 mm.* in length. When the tube is in the combustion apparatus, this should extend *beyond* the end of the long furnace about 10–15 mm. for the purpose of conducting heat, when the two furnaces are in contact. The silver at this point is nearest to the sample and it is here that the chlorine, bromine, iodine, and oxides of sulfur are trapped.[146,350] The filled combustion tube is then stoppered with a clean well-rolled cork and is ready to be connected with the other members of the combustion train. When the tube is heated to 680° C., the contents expand and the filling becomes more compressed. This must be kept in mind so that the filled tube, *in use* (at the above temperature and connected to the bubble counter-U-tube), will allow 10 ml. of air or oxygen per minute to pass through it when supplied at a pressure equal to 5–15 cm. of 5% sodium hydroxide solution.

(B) Filling for Fluoro-Compounds.[348] The filling is done as described under *(A) Regular Filling,* except that *two* sections of *10 mm.* each of the mixture of *magnesium oxide** *and asbestos* (50% each by weight) replace part of the *copper oxide-lead chromate mixture* as shown in Fig. 122 B. After the *silver wire, choking plug, lead peroxide, silver wire* (ribbon) between the heating mortar and long furnace, and 2–3 mm. of *asbestos* are in place (in the order named) only enough of the mixture of *copper oxide and lead chromate* is added to form a section *30 mm.* in length. This is held in place by about $\frac{1}{2}$ *mm.* of *asbestos* and then enough of the *magnesium oxide and asbestos* mixture is added to form a very *loose,* but evenly packed, section of about *9–10 mm.* in length. This is held in place by about $\frac{1}{2}$ *mm.* of asbestos and then another section of evenly packed *copper oxide-lead chromate mixture* about *90 mm.* in length is added, followed by another section of the *magnesium oxide-asbestos mixture* about *9–10 mm.* in length. (The *total* length of the copper oxide-lead chromate and magnesium oxide sections, four in all, should be *140 mm.,* or the exact length of the copper oxide-lead chromate mixture present in the regular filling described above.) Enough *platinized asbestos* is added to form the very loose section about *10–15 mm.* in length as described under *(A) Regular Filling,* followed by the final section of *silver wire,* about

---

* Magnesium aluminate has been used for the retention of fluorine by some workers.[324]

40 mm. in length. Extreme care should be exercised in the filling of this modification as there is a distinct tendency towards packing the tube too tightly.

(C) Alternate Filling for Fluoro-Compounds.[338,348] The modification described under (B), above, is preferred by the author, particularly when large amounts of sulfur also are present, but the following one has been used successfully on a large variety of compounds. The tube is filled as described under *(A) Regular Filling* except that only *130 mm.* of the *copper oxide-lead chromate mixture* is used, followed by enough granular *aluminum oxide** (with an occasional interspersed shred of asbestos to prevent packing too tightly) to form a very loose, but evenly packed, section about *10 mm.* in length. The filling is completed by the addition of the regular amounts of *platinized asbestos* and *silver wire* (or ribbon) (see Fig. 122 C).

COMBUSTION APPARATUS

The combustion apparatus used for the determination of carbon and hydrogen should be electrically[8,59,104,115,116,146,212,285,318,319,349,350,382-385] heated and the voltage supplied to the furnace should be quite constant.[285,348,350] If the regular

Fig. 123. Photograph of carbon-hydrogen apparatus, custom made, author's laboratory.

line voltage in the laboratory varies more than about 2 volts per day, a voltage regulator of some type such as a constant-voltage transformer should be used (refer to Chapter 1). A large percentage of the units in use in the various large laboratories have been constructed in their respective machine shops.[8,59,212,385] Figure 123 shows a photograph of a suitable apparatus, one of those used by the author. This combustion apparatus consists of a long furnace and a power-driven short or sample furnace (see footnote, p. 154, Chapter 7). The

* Magnesium aluminate has been used for the retention of fluorine by some workers.[324]

9. Carbon and Hydrogen

furnaces are composed of nichrome wire windings in asbestos blocks* and are of the split type that can be pushed back away from the combustion tube. The long furnace is 205 mm. in length and can accommodate combustion tubes whose outside diameters are up to 13 mm. The short furnace also accommodates this size tube and is 110 mm. in length. The windings of each are of such resistance, that when operated through a variable auto-transformer, the desired operating temperature of 680° C., in the combustion tube, may be attained. The short furnace travels at the rate of approximately 7 mm. per minute and can travel 75–100 mm. The furnaces used by the author for the carbon-hydrogen determination are the same as those used for the Dumas nitrogen method. Safety shields† are attached to the furnaces in the vicinity of where the sample is placed, for the protection of the analyst.

FIG. 124. Sargent combustion apparatus.

At this writing, no American Society for Testing Materials specifications have been adopted for furnaces to be used for the microdetermination of carbon and hydrogen, but such are in preparation and are expected to be similar to those for the Dumas nitrogen (see Chapter 7).

Several types of furnaces[45,87,129,328,357] are commercially available, some of which are mechanized while some are not. It is the opinion of the author as well as others[8,59,104,116,322,349,350,382-385] that the former should be used. Figures 124 and 125 show two of the commercially available combustion apparatuses, with attached heating mortars (see below). Both are mechanized and have wide ranges of operating temperatures, making them suitable for the various combustion methods. (Also see Additional Information in regard to a new automatic apparatus.[61a])

* Available from C. and H. Menzer Co., Inc., 105 Barclay Street, New York.
† Stanley No. 600 shields, Stanley, New Britain, Connecticut.

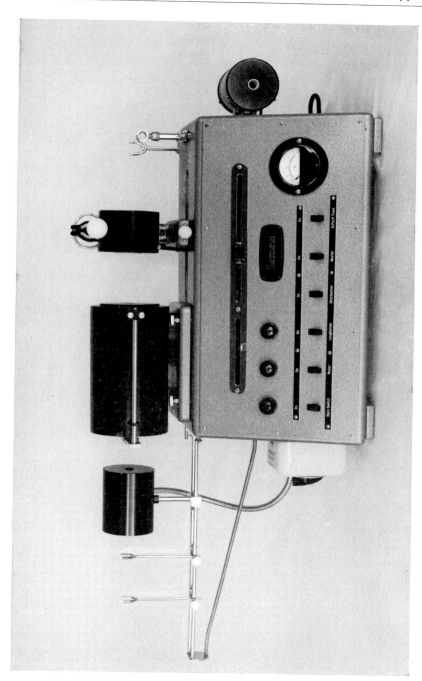

Fig. 125. Heraeus combustion apparatus.

9. Carbon and Hydrogen

HEATING MORTAR[112,113,284,285,308,317-320,350]

The heating mortar can be considered to be almost part of the combustion furnace. It maintains the lead peroxide portion of the tube at a constant temperature, usually to within 1° C., somewhere in the range of 175°–190° C. Electric heating mortars are commercially available and are of two types: (1) those using thermostatic control[87,328,357] and those which are controlled by maintaining constant current.[129] Both types are quite satisfactory. They may be purchased either as part of the combustion apparatus or separately. Figures 123, 124, and 125 show types of heating mortars attached to combustion furnaces and Fig. 126 shows a separate unit.

FIG. 126. Heating mortar.

Electric heating mortars are approximately 80 mm. in length and the opening through which the combustion tube is inserted is approximately 14 mm., which insures a close, but safe, fit with the latter. Attached to the heating mortar is a small metal hook which can be rotated to make contact with the capillary opening of the water absorption tube, when attached, thereby preventing condensation at this point.

The heating mortar is mounted next to the long furnace of the combustion apparatus, usually 3–10 mm. of space separating the two parts.

ABSORPTION TUBES[350,351]

Two absorption tubes (Fig. 127) are required for the determination—one for water and the other for carbon dioxide. They should be made from soda-lime glass in order to reduce accumulation of electrostatic charges. There are two standard lengths and the longer is recommended, if the balance can accommodate it, as more determinations can be done per tube. When Anhydrone and Ascarite are used for absorbing the water and carbon dioxide, respectively, the Ascarite tube clearly shows the extent to which it is exhausted, but the Anhydrone does not until it becomes actually wet at the end. The tubes should be used as a unit and when the Ascarite tube indicates time for cleaning and refilling, the same should be done to the Anhydrone tube. If, however, an accident occurs and one tube is broken long before refilling is indicated, only the one need be replaced. Tubes give good results until the Ascarite tube shows sodium carbonate (lighter in color than the Ascarite and appears to be somewhat fused) to within about 5–10 mm. of the end. (For tube fillings compare Fig. 130.)

(a) THE WATER TUBE. Enough absorbent cotton is forced into the bottom of the absorption tube with the aid of a glass rod, to form a section 4–5 mm. in length. This prevents Anhydrone from passing through the opening. The cotton wad is followed by about 10 mm. of Anhydrone and then another section of cotton, 2–3 mm. long. Anhydrone is then added, with tapping to

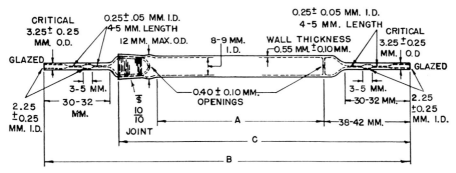

FIG. 127. Absorption tube—details of construction. No markings or etchings on outside.

Size	Approx. length for absorbent (mm.) A	Over-all length with stopper (mm.) B	Length of body (mm.) C
1	80	170 ± 5	135 ± 5
2	100	190 ± 5	155 ± 5

insure even packing, to within 3–5 mm. of the ground joint. (Better packing is obtained if the material is introduced up to the joint, a wad of cotton added, and the tube tapped. The cotton prevents the material from jumping up and down during the process.) Enough cotton is added to fill the tube and be in contact with the ground glass plug so that the packing cannot loosen. Both parts of the ground joint are warmed gently in a burner flame and a little Kroenig glass cement is applied to the plug. Care should be exercised so that the opening in the plug does not become clogged. The two parts are forced together and held thusly until cool. While the cement is still soft and the parts being held together, a cloth should be used to remove the excess material from the rim of the joint. The seal between the two parts must be transparent and show no markings of any kind. Otherwise the joint should be

9. Carbon and Hydrogen

opened and resealed with more cement. The tube should be tested to make certain that the hole in the plug has not become clogged and that gas will pass through. A small piece of cloth should be moistened with either alcohol or benzene and used to clean the rim of the joint free from all traces of remaining cement, which would collect dust and be a source of error. Next, the entire tube should be wiped with a cloth moistened with alcohol to remove adhering material of any kind. During the wiping, care should be exercised so that no alcohol enters the capillaries. A counterpoise tare flask should be prepared for the Anhydrone tube with the aid of lead shot (or glass beads),[212] so that its weight is within 1–2 mg. of the latter and should be kept in the balance case.

(b) CARBON DIOXIDE TUBE. The filling of this tube is accomplished in the same manner as that used for the water tube except that the main filling is Ascarite instead of Anhydrone. About 4–5 mm. of cotton is used to protect the opening. This is followed by a section of Anhydrone about 10 mm. in length. This prevents the escape of water formed by the reaction:

$$2NaOH + CO_2 \rightarrow Na_2CO_3 + H_2O$$

A wad of cotton approximately 2–3 mm. long is next added which is followed by Ascarite to within 3–5 mm. of the ground joint and finally enough cotton to fill the tube. Tapping, sealing, and cleaning are done exactly as described under (a) above. A counterpoise tare flask is prepared, using lead shot (or glass beads), for the Ascarite tube, adjusting the weight to within 1–2 mg. of that of the latter and kept in the balance case.

GUARD TUBE[350,351]

The guard tube shown in Fig. 128 protects the train from moisture in the Mariotte bottle (see below). A layer of cotton about 2–3 mm. in length is placed in the bottom over the small opening to prevent material dropping through. The tube is then filled to within 5–10 mm. of the top with Anhydrone. A layer of cotton completes the filling and holds the desiccant in place with the aid of the glass tube through the rubber stopper.* The stopper and tube must fit snugly so that there is no leak at this point. A single filling of a guard tube lasts for many months unless it is carelessly allowed to remain unstoppered when not in use. (See Figs. 130 and 131b.) †

MARIOTTE BOTTLE[350,351]

The Mariotte bottle shown in Fig. 129 is used to measure the volume of gas passing through the train and to produce a slightly reduced pressure which overcomes the resistance of the absorption tubes. By the latter a

---
* See Fig. 153, Chapter 10.
† The author prefers to use the guard tube in the position shown in these figures, instead of as shown in Fig. 128.

pressure approximately atmospheric is maintained at the two rubber connections between the combustion tube, water, and carbon dioxide absorption tubes. With pressures greater than atmospheric at these connections, loss of carbon dioxide or water could occur while reduced pressures might allow moisture to enter.[112,113,284,285,308,317-320,350]

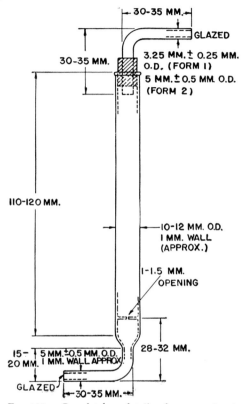

FIG. 128. Guard tube—details of construction.*

In the top of the Mariotte bottle is placed a one-hole rubber stopper with the long piece of glass tubing (connected to the three-way stopcock) in it and extending to within 7–10 mm. of the bottom.

In the side opening of the bottle is placed a cork and drainage tube, the former of which is treated as follows: A large cork is selected, one large enough so that it fits well after considerable softening in a cork roller. A hole is bored in it a few millimeters less than the O.D. of the drainage tube which will eventually be put into it. A round file is selected having a diameter approximately 1–2 mm. less than that of the drainage tube. The file is heated

* The author prefers to use the guard tube in the position shown in Figs. 130 and 131b.

9. Carbon and Hydrogen

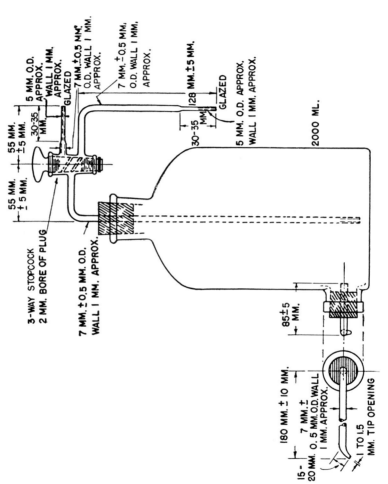

FIG. 129. Mariotte bottle—details of construction.

in a burner flame, almost to the point of redness, and pushed into the hole of the cork burning it almost the size of the drainage tube. The entire cork is placed in the flame and allowed to burn just long enough to obtain a burnt coating over its entire surface. The drainage tube is inserted and the cork forced into the side opening of the bottle. In this manner a stopper is obtained which will not leak and which permits raising and lowering of the drainage tube without further lubrication or danger of breakage.

The Mariotte bottle is filled, each morning, with water. (By using a three-hole rubber stopper in the top of the Mariotte bottle, two additional tubes may be inserted, one attached to the water line and the other for venting during the addition of water, making a more convenient set-up—see Fig. 123.)

GRADUATE CYLINDER

An ordinary 250-ml graduate cylinder is required. For ease in reading, the 75-, 120-, and 210-ml graduations are heavily marked with a china marking pencil.

THIN-WALLED RUBBER TUBING

Ordinary rubber tubing, O.D. approximately 8 mm., I.D. approximately 4 mm., black, red, or para, may be used for connecting the following: (a) oxygen cylinder to the pressure regulator, (b) air supply to the pressure regulator, (c) pressure regulators to bubble counter-U-tube, and (d) guard tube to the Mariotte bottle. The tubing should be cleaned by sweeping a *rapid* stream of oxygen through it for a few minutes before being used. If considerable powder or talcum is present it may be necessary to wash with water and air dry before using the oxygen.[112,113,284,285,308,317-320,350]

HEAVY-WALLED IMPREGNATED CONNECTORS[112,113,284,285,308,317-320,350]

To connect the following: (a) bubble counter-U-tube to the combustion tube, (b) combustion tube to the water absorption tube, (c) water absorption tube to the CO_2 absorption tube, and (d) CO_2 absorption tube to the guard tube, a specially treated heavy-walled rubber tubing must be used. Such tubing is commercially available,[87,357] but only the pliable ones should be selected. Old supplies which have become semirigid should be avoided. If good commercial tubing is not available, the following method produces an excellent product. Rubber tubing having an O.D. of 10–12 mm. and an I.D. of 1–2.5 mm. is cut into lengths of 40–45 mm. and these are placed in a round-bottomed flask containing enough molten paraffin wax (low to medium temperature melting) to completely cover the pieces. The flask is heated on a steam bath and evacuated. At first, considerable frothing occurs caused by the escape of air in the tubing, but after several minutes this ceases and heating should be continued at a pressure of only a few millimeters of mercury for one hour.

9. Carbon and Hydrogen

Several times during the hour of heating, the vacuum should be quickly broken and then the flask re-evacuated. This forces paraffin into the pores of the rubber which previously contained air. Treatment of more than one hour should be avoided as the rubber eventually swells and disintegrates. The pieces should be removed from the flask, the excess paraffin removed with a cloth or towel and then stored in a stoppered bottle.

Before the pieces are put into use, the excess paraffin in the bore is removed by a cotton swab, moistened with glycerin, on the end of the knurled iron wire (see below). The excess glycerin is removed by a dry swab. Certain commercial tubing requires the use of a little benzene before the glycerin swab is used. Otherwise the tube tends to stick to the glassware to which it is attached.

CHAMOIS SKIN

Several pieces of skin, about 6 inches square, treated as described in Chapter 2, are required and are kept in Petri dishes.

FLANNEL OR FELT[112,113,284,285,308,317-320,350]

Two pieces, about 6 inches square, of either clean flannel or thin felt are required. One is used dampened with distilled water while the other is kept dry. Each is kept in a separate Petri dish.

KNURLED IRON WIRE[112,113,284,285,308,317-320,350]

A piece of iron wire, 1 mm. O.D., knurled at the ends, is used with absorbent cotton as a swab to clean the tips of the absorption tubes—see Chapter 3.

FORK OR FORCEPS[24,25,112,113,284,285,308,317-320,350]

Either a fork of the type shown in Fig. 51, Chapter 3, or the special forceps shown in Fig. 42, Chapter 3, is used to handle the absorption tubes.

RACK[24,25,284,285,350]

A rigid wire rack of the type shown in Fig. 50, Chapter 3, is used to hold the absorption tubes when they are not attached to the combustion tube.

TARE FLASKS[102,112,113,284,285,308,317-320,350]

Two tare flasks of the type described in Chapter 3 are needed—one for each of the absorption tubes and clearly marked to avoid mixup.

LEAD SHOT (OR GLASS BEADS)[102,112,113,284,285,308,317-320,350]

Lead shot (or glass beads) is required for increasing the weight of the tare flasks (see Chapter 3).

Assembling the Combustion Train

As shown in the diagram of the combustion train, Fig. 130, oxygen or air passes through the following parts in the order listed:

a. Pressure regulator
b. Bubble counter-U-tube (first through the bubble counter)
c. Combustion tube
d. Anhydrone absorption tube
e. Ascarite absorption tube
f. Guard tube
g. Mariotte bottle

The combustion tube is mounted in the combustion furnaces-heating mortar combination, and the screw clamp which holds the tube tightened so that the latter is rigidly held. The side arm of the tube is moistened with a trace of glycerin—just enough so the fingers glide over it but no liquid film can be seen. A heavy-walled rubber connector is cleaned with glycerin as described above (see description of connector) and the excess removed with a cotton swab on the end of a knurled iron wire. Too much glycerin is to be avoided. However, if not enough is present, the side arm of the tube will be broken off. The side arm is supported with the left hand and one-half the length of the heavy-walled rubber connector is forced over it. (If too much glycerin has been used, the tubing will slide on too easily. It should be removed and wiped with a cotton swab to remove the excess. If, however, not enough has been used, the tubing will stick to the glass. Should this happen, the connector should be cut from the side arm using a sharp razor blade and no attempt made to pull it off or there will be breakage. This rule should always be followed.)

The long capillary side arm of the bubble counter-U-tube is moistened with a trace of glycerin and it is inserted into the rubber connector attached to the side arm of the combustion tube effecting a glass-to-glass contact. (If preferred, the connector may be attached first to the bubble counter-U-tube and this in turn attached to the side arm of the combustion tube.)

A rubber connector (using a trace of glycerin) is attached to the tip of the combustion tube leaving one-half of the rubber tubing extending beyond the tip so that the water absorption tube may be inserted later (Fig. 131a). The rubber connector should not make contact with the side of the heating mortar and it should be left permanently attached to the tip. The extending portion of the rubber connector should be only long enough so that it will eventually cover the capillary of the absorption tube almost up to the small bulb portion (about 18 mm. from the end).

9. Carbon and Hydrogen

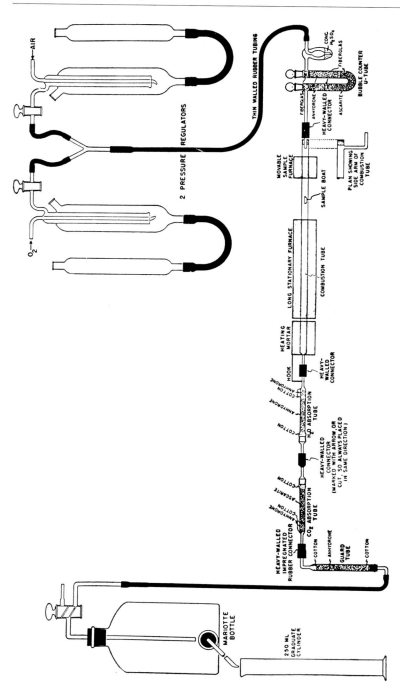

FIG. 130. Diagram of carbon-hydrogen combustion train. Not too clearly illustrated are cotton plugs on *each* side of the Anhydrone layer in both the CO_2 and the H_2O absorption tubes.

The screw clamp holding the combustion tube is loosened so that the tube may freely expand on heating. Oxygen, under a pressure head of several centimeters, is passed through the tube and the heating mortar and both long and short furnaces are switched on. When operating temperatures have been reached (175°–190° C.[24,25,54,112,113,284,285,308,317–320,350] for the heating mortar and 680° C.[104,349,350] for the two furnaces) the tube is pushed up against the heating mortar so that there is no more play, and securely clamped in place.

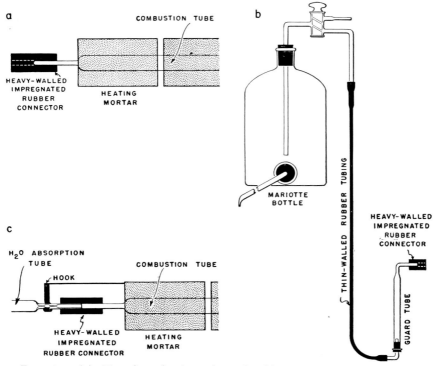

Fig. 131. (a) Tip of combustion tube and rubber connector. (b) Guard tube, Mariotte bottle. (c) Hook on heating mortar making contact with absorption tube capillary.

The hot short furnace is moved the length of the tube between the clamp and the long furnace to remove any combustible material. The short furnace is switched off and oxygen is passed through the hot tube overnight. *Once the long furnace and heating mortar are brought up to working temperatures, they are never switched off for the entire life of the combustion tube.*

After* the tube has been heated overnight with oxygen passing through,

---

* This test cannot be made until the tube has been heating overnight and all combustible material and moisture have been driven out.

9. Carbon and Hydrogen 248

it must be tested for a possible leak by stoppering the rubber connector attached to the tip. Under these conditions, no gas should pass through the bubble counter. If the tube is gas-tight, proceed as below.

A rubber connector is attached, with the aid of a trace of glycerin to the horizontal piece at the top of the filled guard tube leaving one-half of the rubber tubing extending beyond the end of the glass so that the Ascarite absorption tube may be connected here (Fig. 131b). The other end of the guard tube is connected to the vertical tube of the three-way stopcock at the top of the Mariotte bottle, by means of rubber tubing, forming a permanent connection. The Mariotte bottle is filled with water and the graduate cylinder placed in position to collect the water which will come from the former.

PRESSURE REGULATION[112,113,284,285,308,317-320,350]

After the new tube has been burned out overnight, a pressure regulation must be performed to determine the pressure head needed to pass oxygen or air through it at the rate of 10* ml. per minute. Since the bubble counter-U-tube is connected between the pressure regulator and the combustion tube, a new pressure regulation must be done again whenever a new bubble counter-U-tube is attached to an old combustion tube.

The air pressure regulator is adjusted to a height of 5–10 cm. and the stopcocks turned so that only this one is connected to the bubble counter-U-tube and air flows through the system.

A trace of glycerin is added to the tips of the two absorption tubes and they are joined by means of a rubber connector so that the two ground joints *are together* and the capillary tips of the plugs form a glass-to-glass contact. The rubber connector is marked with an arrow, or the end trimmed, so that each end is *always* attached to the same absorption tube. By so doing, the gas will always flow through any exposed portion in the same direction.

The free end of the Anhydrone tube is inserted into the rubber connector on the tip of the combustion tube, effecting a glass-to-glass contact between the two capillary ends (Figs. 130 and 131c).

The free end of the Ascarite tube is connected to the guard tube by means of the rubber connector, effecting a glass-to-glass contact. The drainage tube of the Mariotte bottle is lowered so that it makes an angle of 20°–30° below the horizontal.

With the entire combustion train connected—pressure regulator, bubble counter-U-tube, combustion tube, two absorption tubes, guard tube and Mariotte bottle—the three stopcocks are set as follows: (a) on the oxygen pressure regulator, closed (b) on the air pressure regulator, open (c) on the Mariotte

* Twelve milliliters per minute for the more experienced analyst.

bottle, open so that air can enter from the combustion train and water flow out through the drainage tube. The amount of water dropping into the graduate cylinder is equal to the volume of air passing through the system. The quantity per minute should be noted. The drainage tip should be raised or lowered, as indicated to regulate the flow to a speed of 10* ml. per minute. When this has been attained, the rate at which the air bubbles pass through the water in the Mariotte bottle, via the vertical glass tube is recorded. For example, let us assume that during a 10-second interval, 15 bubbles rise in the Mariotte bottle for a speed of 10* ml. per minute. Therefore, at any later date, we will know that if 15 bubbles rise in the Mariotte bottle in 10 seconds, gas is passing through at the rate of 10* ml. per minute. This is used as the measure rather than observing the rate of flow of the water.

After the flow has been adjusted to a speed of 10* ml. per minute, the rate at which bubbles pass through the sulfuric acid in the bubble counter is recorded. For example, let us assume that 22 bubbles pass through during a 10-second interval, corresponding to 10* ml. per minute.

The two absorption tubes, the guard tube, and Mariotte bottle are then disconnected from the combustion tube. The height of the sodium hydroxide in the air pressure regulator is then adjusted, by means of the leveling tube, to that necessary to cause the gas to pass through the combustion tube at the rate of 10* ml. per minute or, in our example, above, at the rate of 22 bubbles in 10 seconds through the bubble counter. This height of the sodium hydroxide is always maintained throughout the life of the bubble counter-U-tube combination as explained above.

The oxygen pressure regulator is adjusted to the same liquid level as the air regulator. Because one of the regulators might have been incorrectly graduated, it is advisable to make a rapid check on the oxygen one the first time a pair is used.

The data obtained during the pressure regulation should be permanently recorded. For example, if the conditions above were accompanied by a regulation of 9.7 cm. of sodium hydroxide, the record kept for 10* ml. per minute would be:

> Pressure regulator 9.7 cm.
> Bubble counter 22 bubbles/10 seconds
> Mariotte bottle 15 bubbles/10 seconds

The bubble counter is necessary for obtaining the pressure regulation, but the author prefers to use thereafter only the height of the pressure regulator and the rate through the Mariotte bottle for the daily adjustment of the train as explained in the following pages.

* Twelve milliliters per minute for the more experienced analyst.

9. Carbon and Hydrogen

CONDITIONING THE COMBUSTION TUBE[284,285,350]

After hours of heating, the lead peroxide portion of the filling becomes dehydrated and must be resaturated before attempting to perform any determinations. Otherwise, a low hydrogen value would be obtained for the first analysis, since some of the water would be held back. The resaturation is accomplished by burning an unweighed sample of about 6 mg. of benzoic acid* through the tube under exact combustion conditions as described below under Procedure. *This is done, not only with the new tube, but at the beginning of each day's work* since the tube is never allowed to cool, as mentioned above under Assembling the Combustion Train.

Procedure[112,113,308,317-320,350]

The two absorption tubes are connected into the train as described under Pressure Regulation (*ground joints together*,†[112,113,284,285,308,317-320,350] Anhydrone tube connected to the combustion tube, the Ascarite tube to the guard tube and Mariotte bottle, Fig. 130). The height of the two pressure regulators are checked to make certain that they stand at the same figure obtained during the pressure regulation, which in our example above, is 9.7 cm. The stopcock on the air regulator is opened and that on the oxygen regulator closed. The stopcock on the Mariotte bottle is opened so gas enters from the combustion train and water flows out into the graduate cylinder. The drainage tip on the Mariotte bottle is then adjusted so air bubbles rise through the water at the rate equivalent to 10‡ ml. per minute, or, as in our example above, 15 bubbles in 10 seconds. (Note: The readjustment of the Mariotte bottle is necessary even though the pressure regulation of a new tube has just been completed as, in all probability, the pressure regulators do not stand at the same height *after* regulation as they did when the original adjustment was made on the Mariotte bottle in the beginning. Except in the rare case where the regulators are not moved during the process, the Mariotte bottle must be readjusted to have gas pass through at the rate of 10‡ ml. per minute.)

The stopcock on the Mariotte bottle is now closed so that no air will be

---

* Any other organic compound will do but those containing large amounts of nitrogen should be avoided for this purpose because of the effect of oxides of nitrogen on the PbO_2.

† Gas must *always* flow through the absorption tubes in the same direction or error will result due to displacement of water from one tube to the other. If the tubes have been accidentally connected incorrectly and no combustion has been done, they should be reassembled and air swept through for about 10 minutes to re-establish equilibrium before using. If, however, a combustion has been done with a tube reversed, it should be replaced with a new one. If both were reversed, both should be replaced.

‡ Twelve milliliters per minute for the more experienced analyst.

sucked through the tube during the introduction of the sample. The cork is then removed from the end of the combustion tube and the air from the pressure regulator allowed to sweep through the opening. This also insures that no air enters the tube. With as little delay as possible, the previously weighed (4–6 mg., depending on the percentages of carbon and hydrogen present) sample* in the platinum boat or capillary (see Chapter 3) is inserted in the open end and pushed to within about 5 cm. of the long furnace with the aid of a platinum wire hook attached to a glass rod (see Fig. 53, Chapter 3). [Note: If the regular tube filling, (A) above, has been used and a fluoro-compound is to be analyzed, the sample should be covered with magnesium oxide.† If one of the modified fillings, (B) or (C) above, has been used, no magnesium oxide need be added. However, when large amounts of sulfur are present along with fluorine,† the addition of magnesium oxide might prove advantageous even though a modified filling has been used.[348] If the liquid sample has been weighed in a capillary, the latter should be encased in a Coombs-Alber platinum sleeve (Fig. 84, Chapter 6) or placed in a large platinum boat (Fig. 25, Chapter 3) to prevent its fusing to the inside of the combustion tube.] The end of the combustion tube is again stoppered with the cork. (A number of clean, well-rolled corks should be at the setup and they should be rotated in use. This insures a tight fit each time.)

The stopcock on the air regulator is closed and that on the oxygen one opened. The small metal heating hook of the heating mortar is turned so that contact is made with that portion of the capillary of the Anhydrone tube extending beyond the rubber connector which joins it to the combustion tube (Fig. 131c). This prevents the condensation of water in the capillary. The stopcock on the Mariotte bottle is now opened allowing oxygen to pass through the train and water to flow out.

FIRST COMBUSTION

The short movable furnace is moved to the end farthest from the long furnace and the current switched on. As soon as it reaches the working temperature of 680° C., it is brought up to the sample but not over it. As soon as some physical change begins to occur, such as sublimation, melting, darkening, frothing, distillation, or charring, the graduate cylinder is placed under the Mariotte bottle tip and the water flowing from it collected. Simultaneously, the mechanical drive attached to the short movable furnace is put into action and the furnace allowed to travel over the sample and the entire length of the tube up to the long furnace (about 5 cm.) and rest against it

* For salts of alkali and alkaline earth metals, see p. 255.
† Before being used, a test analysis should be made with benzoic acid to which has been added magnesium oxide to make certain that the oxide contains no water or carbonate. Even in the *absence* of fluorine, regardless of the filling, with certain sulfonamides better results are obtained by using magnesium oxide.

9. Carbon and Hydrogen 252

for a couple of minutes. The time required for this to occur is also that for 75 ml. of water to drop into the graduate cylinder. This is termed the *first combustion* and all organic material has been destroyed during it. Phosphorus-containing compounds usually require excessive burning. (Note: If the short movable furnace is not up to about 680° C. when the sample is approached, a sublimate often appears *in back* of the short furnace—side opposite to that of the boat. This must be burned during the first combustion and it is best to do so with a gas burner in back of the sublimate, without disturbing the short movable furnace. This is particularly true if the latter is not of the split type, as any attempt to move back the short furnace would merely force the sublimate further back towards the cork.)

SECOND COMBUSTION

After the short movable furnace has been up against the long furnace for a couple of minutes and 75 ml. of water are in the graduate cylinder, the short furnace is moved back toward the corked end as far as possible. It is then moved again up to the long furnace but more rapidly than during the first combustion. (If the mechanical drive is adjustable, its speed is increased. Otherwise, the furnace is moved along now and then by hand.) The time required for this second burning out, or *second combustion,* as it is called, is also that required for the water to rise in the graduate cylinder from the 75-ml. to the 120-ml. mark, but during most of this time, the short furnace should be up against the long furnace. The moving of the short furnace back and forth is merely a precautionary measure, since all *visible* organic material must be burned during the *first* combustion as explained above.

SWEEP OUT

After 120 ml. of oxygen has passed through the train and the first and second combustions have been done, the short furnace is switched off and moved back towards the corked end. Simultaneously, the stopcock on the oxygen regulator is closed and that on the air regulator opened. Air is passed through the train until the water in the graduate cylinder is at the 210-ml. mark. This part of the determination is called the "sweep out." (Note: The author found that with penicillin derivatives a total volume of 250 ml. is required. This seems to be a peculiarity of this type of compound since no others have ever needed *more* than 210.)

WIPING THE ABSORPTION TUBES[112,113,284,285,308,317–320,350]

The absorption tubes are then disconnected from each other and from the combustion train and removed to the balance room. Any excess glycerin is wiped from the tips with a clean towel or cloth. The insides of the tips are cleaned twice with cotton swabs on the end of a knurled iron wire. *The tubes are then always wiped and weighed in the following manner and in the same order— the Anhydrone tube first, then the Ascarite.* Because much of the success of the determination depends on the consistency of wiping, the procedure should be practiced until the tubes may be reweighed with a reproducibility of 0.01 mg. after being handled. The tube is held by one capillary end with the aid of the cloth or towel used above. The body of the tube and part of the capillaries are wiped with the damp flannel or felt, care being exercised to avoid the ends of the capillaries where water might be absorbed. The entire tube is then wiped with the dry flannel or felt. The tube is next held by one capillary using a chamois skin and the *entire tube and capillaries* wiped with another chamois skin, using lengthwise strokes while the tube is rotated, until there appears to be no resistance to the strokes and the chamois skin seems to glide over the surface. (If soda-lime glass absorption tubes are being used and the relative humidity of the balance room is 40–50%, the electrostatic charge developed during the wiping will be eliminated by the time the tubes are to be weighed— refer to Chapters 1 and 3.) The Anhydrone tube is grounded against a light fixture or pipe and then placed on the rack (Fig. 50) next to the balance while the Ascarite tube is cleaned and wiped in the identical manner as was the Anhydrone tube (including grounding).

Many experienced analysts wipe the absorption tubes only *once* each day,[104,252,322,383-385] that is, *before* they are attached to the train for the conditioning of the combustion tube. During the time required to burn the unweighed sample, the tubes lose their charge. Obviously, each time they are handled after being wiped initially, chamois skins must be used. The tips, however, must be cleaned with a dry cloth and cotton swabs after each determination to remove traces of glycerin. Although this is timesaving and gives equally good results in the hands of the experienced workers, the author recommends wiping the tubes after each analysis for the beginners.

WEIGHING THE ABSORPTION TUBES

As soon as both tubes have been wiped and placed on the rack, a stopwatch is set into motion to record the time interval before weighing, since timing is essential. After about 5 minutes the zero reading of the balance is determined. Weighing of the tubes begins after they have been on the rack 15 minutes. Either of the following methods is acceptable, the author preferring (A) for the beginner and (B) after some practice has been had.

METHOD (A). After 15 minutes have elapsed since the *Ascarite* tube has been placed on the rack, the *Anhydrone* tube is placed on the balance pan hooks with the aid of the fork (Fig. 51) or forceps (Fig. 42) using the previously prepared tare flask as a counterpoise. The tube is weighed in the manner described in Chapter 3, allowing the balance to swing until equilibrium is

9. Carbon and Hydrogen

reached. As soon as the Anhydrone tube has been weighed, it is returned to the rack and the Ascarite tube immediately weighed as above.

METHOD (B). Approximately 12 minutes after the Ascarite tube has been placed on the rack, the Anhydrone tube is placed on the balance pan hooks as described above. The balance is set into swinging immediately and after exactly 15 minues have elapsed since the *Ascarite* tube was placed on the *rack*, the deflection of the balance is taken (three swings to the left and two to the right or two swings to the left and one to the right, whichever is preferred, the slight discrepancy occasionally arising being of no consequence). No attention is paid to whether or not the balance is swinging in equilibrium. The Anhydrone tube is removed and the Ascarite one placed in weighing position. The balance is immediately set into motion and at exactly 20 minutes after the *Ascarite* tube has been placed on the *rack* the deflection is taken as above.

The absorption tubes are then connected to the combustion train and the next determination begun. The sample for the next determination should be weighed while the previous one is being burned. Naturally, weighed samples must be protected in a pig (Figs. 29 or 30, Chapter 3), or desiccator (Fig. 44, Chapter 3) while waiting to be burned.

The determinations are done in succession so that the tubes need be weighed only before the first and after each successive analysis and the weights of the water and carbon dioxide obtained by difference. In this manner, for five determinations, each absorption tube need be weighed only six times.

Calculations:

Factors: (See Chapter 23)

$$\text{For } H_2O, \left(\frac{H_2}{H_2O}\right) = 0.1119$$

$$\text{For } CO_2, \left(\frac{C}{CO_2}\right) = 0.2729$$

$$\therefore \frac{\text{Wt. } H_2O \times 0.1119 \times 100}{\text{Wt. sample}} = \% \text{ H}$$

$$\therefore \frac{\text{Wt. } CO_2 \times 0.2729 \times 100}{\text{Wt. sample}} = \% \text{ C}$$

Example:

Wt. of sample = 4.541 mg.
Increase in weight of the Anhydrone tube = 2.595 mg. (H_2O)
Increase in weight of the Ascarite tube = 10.475 mg. (CO_2)

$$\therefore \frac{2.595 \times 0.1119 \times 100}{4.541} = 6.39\% \text{ H}$$

and

$$\therefore \frac{10.475 \times 0.2729 \times 100}{4.541} = 62.95\% \text{ C}$$

The acceptable accuracy of the determinations is ±0.3% of the theoretical (compare references 97, 285, 320, 350).

SALTS OF ALKALI AND ALKALINE EARTH METALS

Compounds containing alkali or alkaline earth metals which also contain an equivalent amount of sulfur or phosphorus to form the sulfate or phosphate, respectively, may be treated exactly as described above. Phosphates are usually difficult to burn and the ash will be dark (traces of carbon withheld) unless it is strongly ignited. If the salt does not contain either of the above elements, or in insufficient amounts, a carbonate would remain as an ash withholding carbon equivalent to one atom. To prevent this, about 50 mg. of powdered potassium dichromate or vanadium pentoxide is added to the sample in the platinum boat. To prevent spattering of the dichromate or vanadium pentoxide over the inside surface of the combustion tube, the boat is inserted into a Coombs-Alber platinum sleeve (Fig. 84, Chapter 6). Since physical changes in the sample cannot be observed under these conditions, the position of the sample should be estimated and the short movable furnace cautiously brought up to it. By the addition of the dichromate or pentoxide, all of the carbon is converted into carbon dioxide and is absorbed in the Ascarite tube. (Samples which contain halogens should not need dichromate or vanadium pentoxide, but with these the author still prefers the addition.[348])

ADDITIONAL INFORMATION FOR CHAPTER 9

PRESSURE REGULATORS

Pressure regulators of the type used by Pregl[112,113,284,285,308,317-320,350] and for which specifications have been recommended by the Committee for the Standardization of Microchemical Apparatus of the American Chemical Society[351] may be used in place of the ones described. Figure 132 shows the glass parts of this regulator, unassembled. Two are required when air and oxygen are used as described under Procedure. The cylinder on the right is filled with 5% sodium hydroxide and the center piece, held by a metal clamp which is not shown in the drawing, Fig. 132, provides for the pressure head in much the same manner as the regulator shown in Figs. 118 and 119. The three-way stopcock is used to connect the air and oxygen regulators to the bubble counter-U-tube.

Oxygen alone may be used for the determination. For such a setup, one pressure regulator is used. The absorption tubes may be weighed, filled with oxygen, or it may be replaced by air employing an aspirator.[284,285] If the oxygen

9. Carbon and Hydrogen

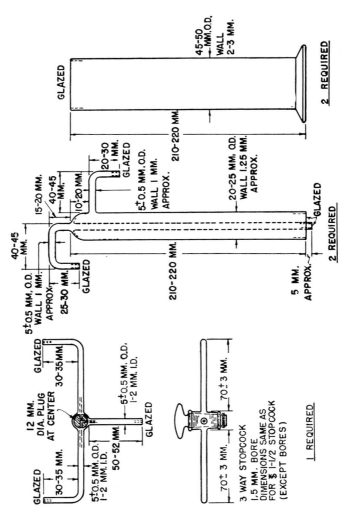

FIG. 132. Glass parts for pressure regulator—details of construction.

is not replaced, the tubes must be rapidly weighed or a special type of absorption tube used—see below.

PREHEATERS

Preheaters of the type shown in Fig. 133 have been used[41,112,113,284,285,317-320,350,351] to purify the air and oxygen if traces of hydrogen from electrolytic oxygen or particles of dust from the inside of the rubber tubing are present.

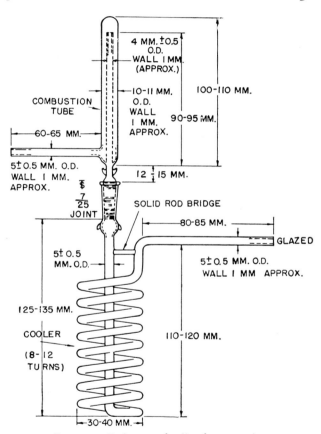

FIG. 133. Preheater—details of construction.

It consists of a section of combustion tube containing copper oxide and heated to about 600° C. The oxygen or air enters from the pressure regulator on the left, passes through the hot copper oxide, then down through the cooling spiral and into the bubble counter-U-tube. With the use of a preheater, an extra drying tube is used between it and the pressure regulator. The author does not recommend the use of preheaters and his opinion is shared by others.[382]

9. Carbon and Hydrogen

HEATING MORTARS

Pregl[308] first used a copper block as a heating mortar and later used a hollow unit containing either a petroleum fraction boiling between 190° and 220° C. or *p*-cymene, boiling at 176° C. The unit was heated by means of a small gas flame and the liquid kept at its boiling point. Specifications for an all-glass unit of this type have been recommended by the Committee for the Standardization of Microchemical Apparatus of the American Chemical So-

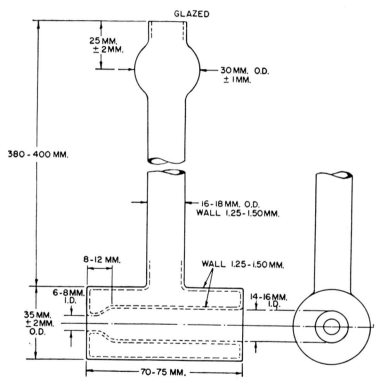

FIG. 134. Heating mortar—details of construction.

ciety[350,351] which is shown in Fig. 134. It is filled with *p*-cymene and heated by either electricity or gas.

ABSORPTION TUBES

Pyrex tubes have been used by MacNevin and Varner.[252] They did not wipe them and, in addition, used an absorption tube as a tare. Control absorption tubes have also been used by Friedrich.[97] The author[349,350] does not recommend the use of Pyrex tubes because they easily become charged with static electricity and often require almost an hour to dissipate their charges even after grounding[285] and exposure to ultraviolet light,[285,316] pitchblende,[120,285] "spent radium tubes" (capsules containing inorganic salts exposed to emanations and previously used for treatment of cancer),[120,285] or, on occasion, the spark of a high-frequency coil.[285,369]

GENERAL

For obvious reasons, the determination of carbon and hydrogen should be considered to be the most important of all the determinations on organic compounds and the large number of articles appearing on the subject substantiate this idea, there always being hopes of improving the method. It has been emphasized that the procedures presented in the preceding pages have proven reliable over the years. There are, however, a number of very reliable procedures which have been developed in recent years and which are being used extensively. The "empty-tube" methods developed by Belcher, Ingram, and Spooner[27,34,148] are being used in the United Kingdom to a greater extent than elsewhere. These use a quartz tube containing baffles and a small amount of silver. The temperatures employed are higher than usual, the tube is held vertically, and high rates of flow of oxygen are used. The so-called Körbl fillings[196-202] are rapidly gaining favor throughout the world. The most common consists of the decomposition products of silver permanganate. The filling is usually shorter than normal and the temperature lower. Körbl's studies have included other oxides. The simultaneous determinations of several elements have obvious great value as does the extremely important submicro- or microgram procedures developed by Unterzaucher[363-365] and by Kirsten[176,177] in Sweden. As this book goes to press, an automatic apparatus has become commercially available.[61a] Its construction and appearance are very similar to that of the nitrogen analyzer (Fig. 104, Chapter 7) except that in place of the nitrometer and counter there are various absorption tubes. On trial in the author's laboratory, good results were obtained on a number of compounds.

For additional material in regard to the subject matter of this chapter, the reader is referred to Table 20.

TABLE 20
ADDITIONAL INFORMATION ON REFERENCES* RELATED TO CHAPTER 9

As in the case of previous chapters, the author presents in the form of Table 20 the additional information which he wishes to call to the attention of the reader. (See statement at top of Table 4 of Chapter 1, regarding completeness of this material.)

Books

Association of Official Agricultural Chemists, 14
Belcher and Godbert, 24, 25
Boëtius, 43
Clark, E. P., 54
Clark, S. J., 60
Furman, 102
Grant, 112, 113
Lindner, 240
Milton and Waters, 264, 265
Niederl and Niederl, 284, 285
Niederl and Sozzi, 288
Pregl, 308
Roth, 317–320
Steyermark, 350

Reviews

Körbl, 198
Renard, 311

Calculator, calculations

Kasagi, 168
Thompson, 358

General, miscellaneous

Apelgot and Mars, 11
Clark and Stillson, 58, 59
Colaitis and Lesbre, 61
Egorova and Zabrodina, 76
Etienne and Herrmann, 79
Fischer, 86
Foreman, 89
Frey, 93
Fujiwara and Kan, 100, 101
Guzman, 114
Horning and Horning, 140
Kainz, 161–164
Kawano, Yamamoto, and Nakayama, 169
Kirsten, 178
Kuck, Altieri, and Towne, 215
Kumpan, 217

General, miscellaneous (Cont.)

Lescher, 224
Malissa, 253–255
Mangeney, 256
Mitsui, 272
Mizukami, Ieki, and Morita, 274
Naughton and Frodyma, 281
Ogg, Willits, Ricciuti, and Connelly, 294
Pepkowitz, 299
Pickhardt, Safranski, and Mitchell, 303
Schöniger, 336

Statistical studies

Fieser and Jackson, 85
Ogg, Willits, Ricciuti, and Connelly, 294
Power, 305
Renoll, Midgley, and Henne, 312
Večeřa and Šnobl, 370
Willits and Ogg, 384

Collaborative studies

Association of Official Agricultural Chemists, 14
Ogg, 293
Willits and Ogg, 383, 384

Test substances

Commission on Microchemical Techniques, IUPAC, 63
Moelants, 276
Sax and Stross, 330
Smith, 342

Apparatus

Abrahamczik, 1
Backeberg, 16
Belcher and Ingram, 27–29
Belcher and Phillips, 31
Belcher and Spooner, 32–34
Böck and Beaucourt, 41
See "Books"
British Standards Institution, 46

* The numbers which appear after each entry in this table refer to the literature citations in the reference list at the end of the chapter.

TABLE 20 (*Continued*)

Apparatus (Cont.)
Caldwell and Barham, 47
Clark and Stillson, 58, 59
Cornwell, 65
Eder, 75
Fischer, 86
Fisher Scientific Company, 87
Flaschenträger, 88
Friedrich, 95–97
Friedrich and Sternberg, 98
Furter, 104
Goulden, 110
Hallett, 115, 116
Hamill, 117
Hoesli, 129
Horeischy and Bühler, 139
Hozumi and Imaeda, 142–144
Hozumi, Imaeda, and Tanaka, 145
Huffman, 146
Ingram, 148–152
Johns, 158
Kemmerer and Hallett, 171
Kirby, 172
Kirsten, 179
Körbl, 196
Kuck, 212
Kuck and Altieri, 213
Kuck, Altieri, and Arnold, 214
Kuck and Arnold, 216
Lacourt, 219
Langer, 222
Lindner, 237
Lindner and Wirth, 246
Lunde, 247
MacNevin and Varner, 252
Mitsui, 266, 267, 269, 270
Müller, 279
Müller and Willenberg, 280
Niederl and Roth, 287
Niemann and Danford, 289
Pickhardt, Safranski, and Mitchell, 303
Prater, 307
Reihlen, 310
Riesenfeld, 313
Robinson and Doan, 315
Royer, Alber, Hallett, Spikes, and Kuck, 321
Royer, Norton, and Sundberg, 322

Apparatus (Cont.)
Sakamoto, 327
Sargent, E. H., Co., 328
Smith and Taylor, 341
Stehr, 345
Steyermark, 349, 350
Steyermark, Alber, Aluise, Huffman, Kuck, Moran, and Willits, 351
Sugiyama and Furuhashi, 352
Thomas, A. H., Co., 357
Vance, 366
Večeřa and Synek, 372
White, Campanile, Agazzi, TeSelle, Tait, Brooks, and Peters, 381
Willits, 382
Willits and Ogg, 385
Zimmermann, 392

Weighing of absorption tubes and problems arising
Boëtius, 43
See "Books"
Evans, Davenport, and Revukas, 81
Friedrich, 97
Hayman, 120
Hernler, 124
Hozumi and Imaeda, 142
Kainz and Hainberger, 165
Lindner, 227–238, 241, 242
Lindner and Figala, 243
Lindner and Hernler, 244, 245
MacNevin and Varner, 252
Niederl and Roth, 287
Power, 305, 306
Rodden, 316
Royer, Norton, and Sundberg, 322
Schmitt and Niederl, 332
Sternberg, 346, 347
Van Straten and Ehret, 369
Weygand, 379
Willits and Ogg, 383, 384

Role of lead dioxide, omission of lead dioxide, substitutes, etc.
Abramson, 2
Abramson and Brochet, 3
Baxter and Hale, 21
Belcher and Phillips, 31

TABLE 20 (*Continued*)

Role of lead dioxide, omission of lead dioxide, substitutes, etc. (*Cont.*)

Berret and Poirier, 39
Canal, 48
Clark and Rees, 57
Corwin, 66, 67
Cropper, 69
Cross and Wright, 70
Dombrowski, 73
Elving and McElroy, 78
Etienne and Mileur, 80
Gränacher, 111
Heron, 125
Horning and Horning, 140
Hussey, Sorenson, and DeFord, 147
Ingram, 150
Kainz, 161, 162
Kainz, Resch, and Schöller, 166
Kirner, 174, 175
Klimova and Korshun, 186
Korshun and Klimova, 207, 208
Kuck, 212
Lévy and Cousin, 225
Macdonald, 251
Mitsui, 268, 271
Mizukami, Ieki, and Morita, 274, 275
Niederl and Niederl, 284, 285
Wagner, 377
Weil, 378
Weygand, 379
Weygand and Hennig, 380
Willits, 382
Wurzschmitt, 387

Metallic, boron, phosphorus, mercury, silicon, etc., compounds

Allen and Tannenbaum, 6
Arthur, Annino, and Donahoo, 13
Belcher, Fildes, and Nutten, 23
Furter, 103
Head and Holley, 122
Kissa, 181
Klimova, Korshun, and Bereznitskaya, 187–190
Körbl and Komers, 201
La Force, Ketchum, and Ballard, 221
Lysyj and Zarembo, 248

Metallic, boron, phosphorus, mercury, silicon, etc., compounds (*Cont.*)

Nunemakes and Shrader, 290
Silbert and Kirner, 339
Sirotenko, 340
Steyermark, 350
Yeh, 388a
Zabrodina and Levina, 389
Zabrodina and Miroshina, 390

Fluoro-compounds

Balis, 19
Belcher and Goulden, 26
Belcher and Macdonald, 30
Bodenheimer and Goldstein, 42
Charlton, 49
Clark, H. S., 56
Clark and Rees, 57
Freier, Nippoldt, Olson, and Weiblen, 92
Gel'man and Korshun, 108
Gel'man, Korshun, and Sheveleva, 109
Korshun, Gel'man, and Glazova, 205
Ma, 249
Mazor, 258
McCoy and Bastin, 259
Morgan and Turnstall, 278
Rush, Cruikshank, and Rhodes, 325, 326
Schwartzkopf, 338
Teston and McKenna, 356
Throckmorton and Hutton, 360
Yeh, 388a
Zimin, Churmanteev, Gubanova, and Verina, 391

Modifications

Backeberg and Istaelstam, 17
Balis, Liebhafsky, and Bronk, 20
Belcher, 22
Belcher and Godbert, 24, 25
Belcher and Ingram, 27, 28
Bennett, 37
Bobranski, 40
Charlton, 49
Christman, 52
Christman, Stuber, and Bothner-By, 53
Clark and Stillson, 58, 59
Cowell and Zahner, 68
Elek, 77

TABLE 20 (*Continued*)

Modifications (Cont.)
Etienne and Mileur, 80
Fedoseev and Ignatenko, 82–84
Fischer, F. O., 86
Friedrich, 95
Garch and Valdener, 107
Goulden, 110
Hammond, 118
Hazenberg, 121
Heinemann, 123
Heron, 125
Hirooka, 126
Holt, 130, 131
Horáček and Körbl, 134–138
Horning and Horning, 140
Ingram, 148–152
Irimescu and Popescu, 154
Israelstam, 155
Kainz and Schöller, 167
Knižáková and Körbl, 193, 194
Körbl, 196–199
Körbl and Blabolil, 200
Körbl and Pribil, 202
Kurihara, 218
Lescher, 224
Lysyj and Zarembo, 248
Ma and Sweeney, 250
Mizukami, Ieki, and Kondo, 273
Mizukami, Ieki, and Morita, 274, 275
Niederl and Niederl, 286
Niemann and Danford, 289
Oerin, Moga, and Bertescu, 292
Okáč and Vrchlabsky, 296
Panicker and Banerjee, 297
Pella, 298
Pickhardt, Safranski, and Mitchell, 303
Robertson, Jett, and Dorfman, 314
Royer, Norton, and Sundberg, 322
Ruf, 323
Šatava and Körbl, 329
Teston and McKenna, 356
Throckmorton and Hutton, 360
Tunnicliff, Peters, Lykken, and Tuemmler, 362
Unterzaucher, 363–365
Večeřa, Šnobl, and Synek, 371
Večeřa and Synek, 372–374
Večeřa, Vojtech, and Synek, 375
Willits and Ogg, 383, 384

Modifications (Cont.)
Wurzschmitt, 388
Yeh, 388a

Ultramicro-submicro methods
Ayers, Belcher, and West, 15
Boos, Jones, and Trenner, 44
Kirsten, 176, 180
Unterzaucher, 363–365

Empty tube technique
Belcher and Goulden, 26
Belcher and Ingram, 27–29
Belcher and Spooner, 33, 34
Clark, S. J., 60
Colson, 62
Hazenberg, 121
Ingram, 148–152
Ingram and Lonsdale, 153
Korshun, 203
Korshun and Klimova, 207
deVries and van Dalen, 376

Wet combustion
See Chapter 18
Adams, 4
Allam and Agiza, 5
Archer, 12
Chen and Lauer, 50
Christensen and Wong, 51
Clark, E. P., 55
Dieterle, 72
Forsblad, 90
Friedmann and Kendall, 94
Hockenhull, 127
Kay, 170
Kirk and Williams, 173
Lieb and Krainick, 226
McCready and Hassid, 260
Nicloux and Boivin, 283
Oberhauser, Rivera, and Etcheverry, 291
Ohashi, 295
Pickhardt, Oemler, and Mitchell, 302
Pollard and Forsee, 304
Schadendorff and Zacherl, 331
Stanek and Nemes, 344
Terent'ev and Luskina, 355
Van Slyke, 367
Van Slyke, Folch and Plazin, 368

9. Carbon and Hydrogen

TABLE 20 (*Continued*)

Simultaneous determination of carbon, hydrogen, and other elements

Allen and Tannenbaum, 6
Belcher and Spooner, 32
Berg and Gromakova, 38
Frazer, 91
Freier, Nippoldt, Olson, and Weiblen, 92
Fujimoto, Utsui, and Ose, 99
Gel'man and Korshun, 108
Klimova and Anisimova, 182, 183
Klimova and Bereznitskaya, 184, 185
Klimova, Korshun, and Bereznitskaya, 188–190
Klimova, Korshun, and Terent'eva, 191
Klimova and Merkulova, 192
Koch, Eckhard, and Malissa, 195
Korshun, Gel'man, and Glazova, 204, 205
Korshun, Gel'man, and Sheveleva, 206
Korshun and Sheveleva, 209, 210
Korshun and Terent'eva, 211
Lacourt and Timmermans, 220
Margolis and Egorova, 257
Mazor, 258
Nunemakes and Shrader, 290
Schöniger, 333–335
Sokolova, 343
Teston and McKenna, 356
Zabrodina and Miroshina, 390
Zimin, Chumanteev, Gubanova, and Verina, 391

Titrimetric, colorimetric, chromatographic, gas chromatographic, etc., procedures

Belcher, Thompson, and West, 35, 36
Delaby, Damiens, and Tsatsas, 71
Duswalt and Brandt, 74
Houghton, 141
Johansson, 157
Juránek, 159

Titrimetric, colorimetric, chromatographic, gas chromatographic, etc., procedures (*Cont.*)

Lee and Meyer, 223
Lindner, 239, 240
Momose, Ueda, and Mukai, 277
Pepkowitz and Proud, 300
Sundberg and Maresh, 353
Unterzaucher, 365

Isotopic C

See Chapter 18
Baertschi and Thürkauf, 18
Christman, 52
Christman, Stuber, and Bothner-By, 53
Harper, Neal, and Rogers, 119
Horáček and Grünberger, 132, 133
Peters and Gutmann, 301
Rabinowitz, 309
Technical News Bulletin, 354
Thorn and Shu, 359
Trenner, Arison, and Walker, 361
Van Slyke, 367
Wilzbach and Sykes, 386

Bombs, oxygen flask methods

Juvet and Chiu, 160
Mehta, 261

Explosive samples

Almstrom, 7
Furter and L'Orange, 105

Hydrogen, beta-ray absorption

Jacobs, Lewis and Piehl, 156

Manometric methods

Holt, 130, 131
See Chapter 18

References

1. Abrahamczik, E., *Mikrochemie,* **22**, 227 (1937).
2. Abramson, E., *Mikrochim. Acta,* p. 770 (1959).
3. Abramson, E., and Brochet, A., *Bull. soc. chim. France,* p. 367 (1954).
4. Adams, J. E., *Ind. Eng. Chem., Anal. Ed.,* **6**, 277 (1934).
5. Allam, F., and Agiza, A. H., *Cairo Univ., Fac. Agr. Bull.* **No. 28**, 16 pp. (1953); *Chem. Abstr.,* **52**, 16975 (1958).
6. Allen, H., Jr., and Tannenbaum, S., *Anal. Chem.,* **31**, 265 (1959).
7. Almström, G. K., *J. prakt. Chem. II,* **95**, 257 (1917).
8. Aluise, V. A., Personal communication.
9. American Platinum Works, Newark, New Jersey.
10. American Society for Testing Materials, *ASTM Designation,* **E 148-59T**.
11. Apelgot, S., and Mars, S., *Bull. soc. chim. biol.,* **35**, 691 (1953).
12. Archer, E. E., *Analyst,* **79**, 30 (1954).
13. Arthur, P., Annino, R., and Donahoo, W. P., *Anal. Chem.,* **29**, 1852 (1957).
14. Association of Official Agricultural Chemists, "Official Methods of Analysis," 8th ed., p. 801, Washington, D. C., 1955.
15. Ayers, C. W., Belcher, R., and West, T. S., *J. Chem. Soc.,* p. 2582 (1959).
16. Backeberg, O. G., *Mikrochemie,* **21**, 135 (1936).
17. Backeberg, O. G., and Israelstam, S. S., *Anal. Chem.,* **24**, 1209 (1952).
18. Baertschi, P., and Thürkauf, M., *Helv. Chim. Acta,* **39**, 79 (1956).
19. Balis, E. W., *Ohio State Univ. Eng. Expt. Sta. News,* **25**, 38 (1953).
20. Balis, E. W., Liebhafsky, H. A., and Bronk, L. B., *Ind. Eng. Chem., Anal. Ed.,* **17**, 56 (1945).
21. Baxter, G. P., and Hale, A. H., *J. Am. Chem. Soc.,* **58**, 510 (1936).
22. Belcher, R., *Chem. & Ind. (London),* **60**, 605 (1941).
23. Belcher, R., Fildes, J. E., and Nutten, A. J., *Anal. Chim. Acta,* **13**, 431 (1955).
24. Belcher, R., and Godbert, A. L., "Semi-Micro Quantitative Organic Analysis," Longmans, Green, London, New York, and Toronto, 1945.
25. Belcher, R., and Godbert, A. L., "Semi-Micro Quantitative Organic Analysis," 2nd ed., Longmans, Green, London, 1954.
26. Belcher, R., and Goulden, R., *Mikrochemie ver. Mikrochim. Acta,* **36/37**, 679 (1951).
27. Belcher, R., and Ingram, G., *Anal. Chim. Acta,* **4**, 118 (1950).
28. Belcher, R., and Ingram, G., *Anal. Chim. Acta,* **4**, 401 (1950).
29. Belcher, R., and Ingram, G., *Anal. Chim. Acta,* **7**, 319 (1952).
30. Belcher, R., and Macdonald, A. M. G., *Mikrochim. Acta,* p. 899 (1956).
31. Belcher, R., and Phillips, D. F., *BIOS* **No. 1606** (British Intelligence Objectives Sub-Committee) (1950).
32. Belcher, R., and Spooner, C. E., *Fuel,* **20**, 130 (1941); *Power Plant Eng.,* **46**, 58 (1942).
33. Belcher, R., and Spooner, C. E., *Ind. Chemist,* **19**, 653 (1943).
34. Belcher, R., and Spooner, C. E., *J. Chem. Soc.,* p. 313 (1943).
35. Belcher, R., Thompson, J. H., and West, T. S., *Anal. Chim. Acta,* **19**, 148 (1958).
36. Belcher, R., Thompson, J. H. and West, T. S., *Anal. Chim. Acta,* **19**, 309 (1958).
37. Bennett, A., *Analyst,* **74**, 188 (1949).
38. Berg, L. G., and Gromakova, L. M., *Trudy Kazan. Filiala Akad. Nauk S.S.S.R., Ser. Khim. Nauk,* **3**, 73 (1956); *Referat. Zhur. Khim,* No. 15875 (1957); *Chem. Abstr.* **53**, 12951 (1959).
39. Berret, R., and Poirier, P., *Bull. soc. chim. France,* p. 724 (1951).

9. Carbon and Hydrogen

40. Bobranski, B., *Mikrochim. Acta,* p. 1735 (1956).
41. Böck, F., and Beaucourt, K., *Mikrochemie,* **6**, 133 (1928).
42. Bodenheimer, W., and Goldstein, M., *Bull. Research Council Israel,* **3**, 53 (1953).
43. Boëtius, M., "Über die Fehlerquellen in der mikroanalytischen Bestimmung des Kohlen- und Wasserstoffes nach der Methode von Fritz Pregl," Verlag Chemie, Berlin, 1931.
44. Boos, R. N., Jones, S. L. and Trenner, N. R., *Anal. Chem.,* **28**, 390 (1956).
45. Brinkmann, C. A., & Co., Inc., Great Neck, Long Island, New York.
46. British Standards Institution, *Brit. Standards* **1428**; Pt. A1 (1950), Pt. A1 (1958), and Pt. A5 (1957).
47. Caldwell, M. J., and Barham, H. N., *Ind. Eng. Chem., Anal. Ed.,* **14**, 485 (1942).
48. Canal, F., *Lab. sci. (Milan),* **6**, 36 (1958); *Chem. Abstr.,* **53**, 21429 (1959).
49. Charlton, F. E., *Analyst,* **81**, 582 (1956).
50. Chen, S. L., and Lauer, K. J. H., *Anal. Chem.,* **29**, 1225 (1957).
51. Christensen, B. E., and Wong, R., *Ind. Eng. Chem., Anal. Ed.,* **13**, 444 (1941).
52. Christman, D. R., *Anal. Chem.,* **27**, 1935 (1955).
53. Christman, D. R., Stuber, J. E., and Bothner-By, A. A., *Anal. Chem.,* **28**, 1345 (1956).
54. Clark, E. P., "Semimicro Quantitative Organic Analysis," Academic Press, New York, 1943.
55. Clark, E. P., *J. Assoc. Offic. Agr. Chemists,* **16**, 255 (1933).
56. Clark, H. S., *Anal. Chem.,* **23**, 659 (1951).
57. Clark, H. S., and Rees, O. W., *Illinois State Geol. Survey, Rept. Invest.* **No. 169** (1954).
58. Clark, R. O., and Stillson, G. H., *Ind. Eng. Chem., Anal. Ed.,* **12**, 494 (1940).
59. Clark, R. O., and Stillson, G. H., *Ind. Eng. Chem., Anal. Ed.,* **19**, 423 (1947).
60. Clark, S. J., "Quantitative Methods of Organic Microanalysis," Butterworths, London, 1956.
61. Colaitis, D., and Lesbre, M., *Bull. soc. chim. France,* p. 1069 (1952).
61a. Coleman Instruments, Inc., Maywood, Illinois.
62. Colson, A. F., *Analyst,* **73**, 541 (1948).
63. Commission on Microchemical Techniques, Section of Analytical Chemistry, International Union of Pure and Applied Chemistry, *Pure and Appl. Chem.,* **1**, 143 (1960).
64. Corning Glass Works, Corning, New York.
65. Cornwell, R. T. K., *Ind. Eng. Chem., Anal. Ed.,* **3**, 4 (1931).
66. Corwin, A. H., Presented at 94th National Meeting of the American Chemical Society, Rochester, New York, September 1937. Microchemical Section, p. L 3.
67. Corwin, A. H., *Mikrochemie,* **24**, 98 (1938).
68. Cowell, W. H., and Zahner, R. J., Thirteenth Metropol. Microchem. Soc. Symposium, New York, March 1958.
69. Cropper, F. R., *Mikrochim. Acta,* p. 25 (1954).
70. Cross, C. K., and Wright, G. F., *Anal. Chem.,* **26**, 886 (1954).
71. Delaby, R., Damiens, R., and Tsatsas, G., *Proc. Intern. Congr. Pure and Appl. Chem., Anal. Chem., 15th Congr., Lisbon,* **1**, 86 (1956); *Anal. Abstr.,* **6**, No. 1175 (1959).
72. Dieterle, H., *Arch. Pharm.,* **262**, 35 (1924).
73. Dombrowski, A., *Mikrochemie ver. Mikrochim. Acta,* **28**, 136 (1940).
74. Duswalt, A. A., and Brandt, W. W., *Anal. Chem.,* **32**, 272 (1960).

75. Eder, K., *Mikrochim. Acta*, p. 224 (1957).
76. Egorova, N. F., and Zabrodina, A. S., *Vestnik. Moskov. Univ.*, p. 235 (1958); *Anal. Abstr.*, **6**, No. 2190 (1959).
77. Elek, A., *Ind. Eng. Chem., Anal. Ed.*, **10**, 51 (1938).
78. Elving, P. J., and McElroy, W. R., *Ind. Eng. Chem., Anal. Ed.*, **13**, 660 (1941).
79. Etienne, A., and Herrmann, J., *Chim. anal.*, **33**, 1 (1951).
80. Etienne, A., and Mileur, R., *Ann. chim. anal.*, **28**, 215 (1946).
81. Evans, R. N., Davenport, J. E., and Revukas, A. J., *Ind. Eng. Chem., Anal. Ed.*, **11**, 553 (1939).
82. Fedoseev, P. N., and Ignatenko, L. S., *Izvest. Akad. Nauk Turkmen. S.S.R.*, p. 24 (1957); *Anal. Abstr.*, **5**, No. 3764 (1958).
83. Fedoseev, P. N., and Ignatenko, L. S., *Izvest. Akad. Nauk Turkmen. S.S.R.*, p. 84 (1957); *Referat. Zhur. Khim.*, No. 70,617 (1958); *Anal. Abstr.*, **6**, No. 1765 (1959).
84. Fedoseev, P. N., and Ignatenko, L. S., *Izvest. Akad. Nauk Turkmen. S.S.R.*, p. 45 (1959); *Chem. Abstr.*, **53**, 12925 (1959).
85. Fieser, L. F., and Jackson, R. P., *J. Am. Chem. Soc.*, **58**, 943 (1936).
86. Fischer, F. O., *Anal. Chem.*, **21**, 827 (1949).
87. Fisher Scientific Company, New York, and Pittsburgh, Pennsylvania.
88. Flaschenträger, B., *Z. angew. Chem.*, **39**, 717 (1926).
89. Foreman, J. K., *Mikrochim. Acta*, p. 1481 (1956).
90. Forsblad, I., *Mikrochim. Acta*, p. 176 (1955).
91. Frazer, J. W., *U. S. At. Energy Comm.* **UCRL-5134**, 8 pp. (1958); *Chem. Abstr.*, **52**, 16118 (1958).
92. Freier, H. E., Nippoldt, B. W., Olson, P. B., and Weiblen, D. G., *Anal. Chem.*, **27**, 146 (1955).
93. Frey, H. M., *Nature*, **183**, 143 (1959).
94. Friedmann, T. E., and Kendall, A. I., *J. Biol. Chem.*, **82**, 45 (1929).
95. Friedrich, A., *Mikrochemie*, **10**, 329 (1932).
96. Friedrich, A., *Mikrochemie*, **10**, 338 (1932).
97. Friedrich, A., *Mikrochemie*, **19**, 23 (1935).
98. Friedrich, A., and Sternberg, H., *Mikrochemie, Molisch Festschrift*, p. 118 (1936).
99. Fujimoto, R., Utsui, Y., and Ose, S., *Yakugaku Zasshi*, **78**, 722 (1958); *Chem. Abstr.*, **52**, 16118 (1958).
100. Fujiwara, H., and Kan, M., *Yakugaku Zasshi*, **75**, 1213 (1956).
101. Fujiwara, H., and Kan, M., *Yakugaku Zasshi*, **76**, 883, 886 (1956).
102. Furman, N. H., ed., "Scott's Standard Methods of Chemical Analysis," 5th ed., Vol. II, Van Nostrand, New York, 1939.
103. Furter, M. F., *Mikrochemie*, **9**, 27 (1931).
104. Furter, M. F., Personal communication.
105. Furter, M. F., and L'Orange, J., *Mikrochemie*, **17**, 38 (1935).
106. Furter, M. F., and Steyermark, Al, *Anal. Chem.*, **20**, 257 (1948).
107. Garch, J., and Valdener, G. *Chim. anal.*, **36**, 211 (1954).
108. Gel'man, N. E., and Korshun, M. O., *Doklady Akad. Nauk S.S.S.R.*, **89**, 685 (1953).
109. Gel'man, N. E., Korshun, M. O., and Sheveleva, N. S., *Zhur. Anal. Khim.*, **12**, 424, 526, 547 (1957); *Chem. Abstr.*, **52**, 1853, 18074 (1958).
110. Goulden, F., *Analyst*, **73**, 320 (1948).
111. Gränacher, C. H., *Helv. Chim. Acta*, **10**, 449 (1927); *Z. anal. Chem.*, **74**, 409 (1928).

9. Carbon and Hydrogen 268

112. Grant, J., "Quantitative Organic Microanalysis, Based on the Methods of Fritz Pregl," 4th ed., Blakiston, Philadelphia, Pennsylvania, 1946.
113. Grant, J., "Quantitative Organic Microanalysis," 5th ed., Blakiston, Philadelphia, Pennsylvania, 1951.
114. Guzman, G. M., *Rev. cienc. apl.* (*Madrid*), **3**, 23 (1949).
115. Hallett, L. T., *Ind. Eng. Chem., Anal. Ed.*, **10**, 101 (1938).
116. Hallett, L. T., *Ind. Eng. Chem., Anal. Ed.*, **14**, 956 (1942).
117. Hamill, W. H., *Ind. Eng. Chem., Anal. Ed.*, **9**, 355 (1937).
118. Hammond, W. A., Drierite Co., Xenia, Ohio.
119. Harper, P. V., Jr., Neal, W. B., Jr., and Rogers, G. R., *J. Lab. Clin. Med.*, **36**, 321 (1950).
120. Hayman, D. F., *Ind. Eng. Chem., Anal. Ed.*, **8**, 342 (1936).
121. Hazenberg, W. M., *Mikrochim. Acta*, p. 709 (1958).
122. Head, E., and Holley, C., *Anal. Chem.*, **28**, 1172 (1956).
123. Heinemann, H., *Oil & Soap*, **23**, 227 (1946).
124. Hernler, F., *Mikrochemie, Pregl Festschrift*, p. 154 (1929).
125. Heron, A. E., *Analyst*, **73**, 314 (1948).
126. Hirooka, S., *Nippon Kagaku Zasshi*, **75**, 236 (1954).
127. Hockenhull, D. J. D., *Biochem. J.*, **46**, 605 (1950).
128. Hodgman, C. D., "Handbook of Chemistry and Physics," 28th ed., Chemical Rubber, Cleveland, Ohio, 1944.
129. Hoesli, H., c/o Obipectin, A.-G., Bischofszell, St. Gallen, Switzerland.
130. Holt, B. D., *Anal. Chem.*, **27**, 1500 (1955).
131. Holt, B. D., *Anal. Chem.*, **28**, 1153 (1956).
132. Horáček, J., and Grünberger, D., *Chem. listy*, **51**, 1944 (1957); *Chem. Abstr.*, **52**, 1853 (1958).
133. Horáček, J., and Grünberger, D., *Collection Czechoslov. Chem. Communs.*, **23**, 1974 (1958).
134. Horáček, J., and Körbl, J., *Chem. & Ind.* (*London*), p. 101 (1958); *Anal. Abstr.* **5**, No. 3358 (1958).
135. Horáček, J., and Körbl, J., *Chem. listy*, **51**, 2132 (1957); *Chem. Abstr.*, **52**, 2661 (1958).
136. Horáček, J., and Körbl, J., *Collection Czechoslov. Chem. Communs.*, **24**, 286 (1959).
137. Horáček, J., and Körbl, J., *Mikrochim. Acta*, p. 303 (1959).
138. Horáček, J., and Körbl, J., *Mikrochim. Acta*, p. 456 (1959).
139. Horeischy, K., and Bühler, F., *Mikrochemie ver. Mikrochim. Acta*, **33**, 231 (1948).
140. Horning, E. C., and Horning, M. G., *Ind. Eng. Chem., Anal. Ed.*, **19**, 688 (1947).
141. Houghton, A. A., *Analyst*, **70**, 118 (1945).
142. Hozumi, K., and Imaeda, K., *Yakugaku Zasshi*, **74**, 565 (1954).
143. Hozumi, K., and Imaeda, K., *Yakugaku Zasshi*, **74**, 570 (1954).
144. Hozumi, K., and Imaeda, K., *Yakugaku Zasshi*, **74**, 574 (1954).
145. Hozumi, K., Imaeda, K., and Tanaka, M., *Yakugaku Zasshi*, **72**, 658 (1952).
146. Huffman, E. W. D., Personal communication.
147. Hussey, A. S., Sorensen, J. H., and DeFord, D. D., *Anal. Chem.*, **27**, 280 (1955).
148. Ingram, G., *Analyst*, **73**, 548 (1948).
149. Ingram, G., *Mikrochemie ver. Mikrochim. Acta*, **36/37**, 690 (1951).
150. Ingram, G., *Mikrochim. Acta*, p. 71 (1953).
151. Ingram, G., *Mikrochim. Acta*, p. 131 (1953).
152. Ingram, G., *Mikrochim. Acta*, p. 877 (1956).

153. Ingram, G., and Lonsdale, M., *Chem. & Ind. (London)*, p. 276 (1956).
154. Irimescu, I., and Popescu, B., *Z. anal. Chem.*, **128**, 185 (1948).
155. Israelstam, S. S., *Anal. Chem.*, **24**, 1207 (1952).
156. Jacobs, R. B., Lewis, L. G., and Piehl, F. J., *Anal. Chem.*, **28**, 324 (1956).
157. Johansson, A., *Anal. Chem.*, **26**, 1183 (1954).
158. Johns, I. B., *Mikrochemie*, **24**, 217 (1938); **25**, 382 (1938).
159. Juránek, J., *Collection of Czechoslov. Chem. Communs.*, **24**, 135 (1959).
160. Juvet, R. S., and Chiu, J., *Anal. Chem.*, **32**, 130 (1960).
161. Kainz, G., *Mikrochemie ver. Mikrochim. Acta*, **35**, 569 (1950).
162. Kainz, G., *Mikrochemie ver. Mikrochim. Acta*, **39**, 166 (1952).
163. Kainz, G., *Österr. Chemiker-Ztg.*, **57**, 216 (1956).
164. Kainz, G., *Z. anal. Chem.*, **166**, 427 (1959).
165. Kainz, G., and Hainberger, L., *Mikrochim. Acta*, p. 870 (1959).
166. Kainz, G., Resch, A., and Schöller, F., *Mikrochim. Acta*, p. 850 (1956).
167. Kainz, G., and Schöller, F., *Z. anal. Chem.*, **148** 6 (1955).
168. Kasagi, M., *Bunseki Kagaku*, **1**, 148 (1952).
169. Kawano, T., Yamamoto, S., and Nakayama, M., *Bunseki Kagaku*, **1**, 126 (1952).
170. Kay, H., *Chem. Ing. Tech.*, **26**, 156 (1954).
171. Kemmerer, G., and Hallett, L. T., *Ind. Eng. Chem.*, **19**, 173 (1927).
172. Kirby, H., *Chem. & Ind. (London)*, **58**, 117 (1939).
173. Kirk, P. L., and Williams, P. A., *Ind. Eng. Chem., Anal. Ed.*, **4**, 403 (1932).
174. Kirner, W. R., *Ind. Eng. Chem., Anal. Ed.*, **5**, 363 (1933); **7**, 366 (1935); **8**, 57 (1936); **9**, 535 (1937); **10**, 342 (1938).
175. Kirner, W. R., *Mikrochemie*, **24**, 98, 219 (1938).
176. Kirsten, W., *Chem. anal.*, p. 253 (1958).
177. Kirsten, W. J., *Microchem. J.*, **2**, 179 (1958).
178. Kirsten, W., *Mikrochemie ver. Mikrochim. Acta*, **35**, 217 (1950).
179. Kirsten, W., *Mikrochim. Acta*, p. 41 (1953).
180. Kirsten, W., *Mikrochim. Acta*, p. 836 (1956).
181. Kissa, E., *Microchem. J.*, **1**, 203 (1957).
182. Klimova, V. A., and Anisimova, G. F., *Bull. Acad. Sci. U.S.S.R., Div. Chem. Sci. S.S.R. (English Translation)*, p. 773 (1958).
183. Klimova, V. A., and Anisimova, G. F., *Izvest. Akad. Nauk S.S.S.R., Otdel. Khim. Nauk*, p. 791 (1958); *Chem. Abstr.*, **52**, 16975 (1958).
184. Klimova, V. A., and Bereznitskaya, E. G., *Zhur. Anal. Khim.*, **11**, 292 (1956).
185. Klimova, V. A., and Bereznitskaya, E. G., *Zhur. Anal. Khim.*, **12**, 424 (1957); *Chem. Abstr.*, **52**, 1853 (1958).
186. Klimova, V. A., and Korshun, M. O., *Zhur. Anal. Khim.*, **6**, 230 (1951).
187. Klimova, V. A., Korshun, M. O., and Bereznitskaya, E. G., *Doklady Akad. Nauk S.S.S.R.*, **84**, 1175 (1952).
188. Klimova, V. A., Korshun, M. O., and Bereznitskaya, E. G., *Doklady Akad. Nauk S.S.S.R.*, **96**, 81 (1954).
189. Klimova, V. A., Korshun, M. O., and Bereznitskaya, E. G., *Doklady Akad. Nauk S.S.S.R.*, **96**, 287 (1954).
190. Klimova, V. A., Korshun, M. O., and Bereznitskaya, E. G., *Zhur. Anal. Khim.*, **11**, 223 (1956).
191. Klimova, V. A., Korshun, M. O., and Terent'eva, E. A., *Zhur. Anal. Khim.* **9**, 275 (1954); *Chem. Abstr.*, **49**, 2942 (1955).
192. Klimova, V. A., and Merkulova, E. N., *Izvest. Akad. Nauk S.S S.R.*, p. 781 (1959); *Anal. Abstr.*, **7**, No. 552 (1960).

193. Knižáková, E., and Körbl, J., *Chem. listy,* **52,** 750 (1958); *Chem. Abstr.,* **52,** 12662 (1958).
194. Knižáková, E., and Körbl, J., *Collection Czechoslov. Chem. Communs.,* **24,** 2420 (1959).
195. Koch, W., Eckhard, S., and Malissa, H., *Arch. Eisenhüttenw.,* **29,** 543 (1958).
196. Körbl, J., *Chem. listy,* **49,** 929 (1955).
197. Körbl, J., *Collection Czechoslov. Chem. Communs.,* **20,** 953 (1955).
198. Körbl, J., *Ind. Chemist,* **34,** 507 (1958).
199. Körbl, J., *Mikrochim. Acta,* p. 1705 (1956).
200. Körbl, J., and Blabolil, K., *Chem. listy,* **49,** 1664 (1955); *Collection Czechoslov. Chem. Communs.,* **21,** 318 (1956) (in German); *Anal. Abstr.,* **3,** No. 3067 (1956).
201. Körbl, J., and Komers, R., *Chem. listy,* **50,** 1120 (1956).
202. Körbl, J., and Pribil, R., *Collection Czechoslov. Chem. Communs.,* **21,** 955 (1956).
203. Korshun, M. O., *Zhur. Anal. Khim.,* **7,** 96 (1952).
204. Korshun, M. O., Gel'man, N. E., and Glazova, K. I., *Doklady Akad. Nauk, S.S.S.R.,* **111,** 1255 (1956); *Referat. Zhur. Khim.,* No. 44,900 (1957).
205. Korshun, M. O., Gel'man, N. E., and Glazova, K. I., *Proc. Acad. Sci., U.S.S.R., Sect. Chem. (English Translation),* **111,** 761 (1956); *Chem. Abstr.,* **52,** 6065 (1958).
206. Korshun, M. O., Gel'man, N. E., and Sheveleva, N. S., *Zhur. Anal. Khim.,* **13,** 695 (1958); *Chem. Abstr.,* **53,** 5958 (1959).
207. Korshun, M. O., and Klimova, V. A, *Zhur. Anal. Khim.,* **2,** 274 (1947).
208. Korshun, M. O., and Klimova, V. A., *Zhur. Anal. Khim.,* **4,** 292 (1949).
209. Korshun, M. O., and Sheveleva, N. S., *Zhur. Anal. Khim.,* **7,** 104 (1952).
210. Korshun, M. O., and Sheveleva, N. S., *Zhur. Anal. Khim.,* **11,** 376 (1956).
211. Korshun, M. O., and Terent'eva, E. A., *Doklady Akad. Nauk S.S.S.R.,* **100,** 707 (1955).
212. Kuck, J. A., Personal communication.
213. Kuck, J. A., and Altieri, P. L., *Mikrochim. Acta,* p. 1550 (1956).
214. Kuck, J. A., Altieri, P. L., and Arnold, M., *Mikrochim. Acta,* p. 1544 (1956).
215. Kuck, J. A., Altieri, P. L., and Towne, A. K., *Mikrochim. Acta,* p. 1 (1954).
216. Kuck, J. A., and Arnold, M., *Mikrochemie ver. Mikrochim. Acta,* **38,** 521 (1951).
217. Kumpan, P., *Glasapparate tech.* **3,** 17, Beil. *Chem. Tech. (Berlin)* **10,** 1958; *Chem. Abstr.,* **53,** 7858 (1959).
218. Kurihara, B., *Yakugaku Zasshi,* **77,** 546 (1957); *Microchem. J.,* **3,** 119 (1959).
219. Lacourt, A., *Metallurgia,* **36,** 289 (1947).
220. Lacourt, A., and Timmermans, A. M., *Anal. Chim. Acta,* **1,** 140 (1947).
221. La Force, J. R., Ketchum, D. F., and Ballard, A. E., *Anal. Chem.,* **21,** 879 (1949).
222. Langer, A., *Ind. Eng. Chem., Anal. Ed.,* **17,** 266 (1945)
223. Lee, T. S., and Meyer, R., *Anal. Chim. Acta,* **13,** 340 (1955).
224. Lescher, V. L., *Anal. Chem.,* **21,** 1246 (1949).
225. Lévy, R., and Cousin, B., *Bull. soc. chim. France,* p. 728 (1952).
226. Lieb, H., and Krainick, H. G., *Mikrochemie,* **10,** 99 (1932).
227. Lindner, J., *Ber.,* **59,** 2561 (1926).
228. Lindner, J., *Ber.,* **59,** 2806 (1926).
229. Lindner, J., *Ber.,* **60,** 124 (1927).
230. Lindner, J., *Ber.,* **63,** 949 (1930).
231. Lindner, J., *Ber.,* **63,** 1123 (1930).
232. Lindner, J., *Ber.,* **63,** 1396 (1930).

233. Lindner, J., *Ber.*, **63**, 1672 (1930).
234. Lindner, J., *Ber.*, **65**, 1696 (1932).
235. Lindner, J., *Ber.*, **70**, 1025 (1937).
236. Lindner, J., *Mikrochemie*, **10**, 321 (1931).
237. Lindner, J., *Mikrochemie*, **20**, 209 (1936).
238. Lindner, J., *Mikrochemie ver. Mikrochim. Acta*, **25**, 197 (1938).
239. Lindner, J., *Mikrochemie ver. Mikrochim. Acta*, **32**, 133 (1944).
240. Lindner, J., "Mikromassanalytische Bestimmung des Kohlenstoffes und Wasserstoffes mit grundlegender Behandlung der Fehlerquellen in der Elementaranalyse," Verlag Chemie, Berlin, 1935.
241. Lindner, J., *Z. anal. Chem.*, **66**, 305 (1925).
242. Lindner, J., *Z. anal. Chem.*, **72**, 135 (1927).
243. Lindner, J., and Figala, N., *Mikrochemie*, **10**, 440 (1932).
244. Lindner, J., and Hernler, F., *Mikrochemie, Emich Festschrift*, p. 191 (1930).
245. Lindner, J., and Hernler, F., *Z. angew. Chem.*, **40**, 462 (1927).
246. Lindner, J., and Wirth, W., *Ber.*, **70**, 1025 (1937).
247. Lunde, G., *Biochem. Z.*, **176**, 157 (1926).
248. Lysyj, I., and Zarembo, J. E., *Microchem. J.*, **2**, 245 (1958).
249. Ma, T. S., *Microchem. J.*, **2**, 91 (1958).
250. Ma, T. S., and Sweeney, R. F., *Mikrochim. Acta*, p. 198 (1956).
251. Macdonald, A. M. G., *Ind. Chemist*, **35**, 193 (1959).
252. MacNevin, W. M., and Varner, J. E., *Ind. Eng. Chem., Anal. Ed.*, **15**, 224 (1943).
253. Malissa, H., *Mikrochim. Acta*, p. 553 (1957).
254. Malissa, H., *Mikrochim. Acta*, p. 127 (1960).
255. Malissa, H., "Proceedings of the International Symposium on Microchemistry 1958," p. 97, Pergamon, Oxford, London, New York, and Paris, 1960.
256. Mangeney, G., *Bull. soc. chim. France*, p. 809 (1951).
257. Margolis, E. I., and Egorova, N. F., *Vestnik Moskov. Univ.*, p. 209 (1958); *Chem. Abstr.*, **53**, 9888 (1959); *Anal. Abstr.*, **6**, No. 2191 (1959).
258. Mazor, L., *Mikrochim. Acta*, p. 114 (1957).
259. McCoy, R. N., and Bastin, E. L., *Anal. Chem.*, **28**, 1776 (1956).
260. McCready, R. M., and Hassid, W. Z., *Ind. Eng. Chem., Anal. Ed.*, **14**, 525 (1942).
261. Mehta, R. K. A., *J. Sci. Ind. Research (India)*, **13B**, 195 (1954).
262. Mellor, J. W., "A Comprehensive Treatise on Inorganic and Theoretical Chemistry," Longmans, Green, London, and New York, 1922-1937.
263. Merck & Co., Inc., Rahway, New Jersey.
264. Milton, R. F., and Waters, W. A., "Methods of Quantitative Microanalysis," Longmans, Green, New York, and Arnold, London, 1949.
265. Milton, R. F., and Waters, W. A., "Methods of Quantitative Microanalysis," 2nd ed., Arnold, London, 1955.
266. Mitsui, T., *Kyoto Daigaku Shokuryô Kagaku Kenkyujo Hôkoku*, **8**, 36 (1952).
267. Mitsui, T., *Kyoto Daigaku Shokuryô Kagaku Kenkyujo Hôkoku*, **11**, 39 (1953).
268. Mitsui, T., *Kyoto Daigaku Shokuryô Kagaku Kenkyujo Hôkoku*, **12**, 1 (1953).
269. Mitsui, T., *Kyoto Daigaku Shokuryô Kagaku Kenkyujo Hôkoku*, **17**, 20 (1955).
270. Mitsui, T., *Bunseki Kagaku*, **1**, 130 (1952).
271. Mitsui, T., *Bunseki Kagaku*, **2**, 3 (1953).
272. Mitsui, T., *Kagaku no Ryôiki*, **5**, 687 (1951).
273. Mizukami, S., Ieki, T., and Kondo, H., *Yakugaku Zasshi*, **77**, 517 (1957); *Microchem. J.*, **3**, 119 (1959).

274. Mizukami, S., Ieki, T., and Morita, N., *Yakugaku Zasshi*, **76**, 553, 559, 601 (1956).
275. Mizukami, S., Ieki, T., and Morita, N., *Yakugaku Zasshi*, **77**, 552 (1957); *Microchem. J.*, **3**, 119 (1959).
276. Moelants, L. J., *Ind. chim. belge*, **21**, 207 (1956).
277. Momose, T., Ueda, Y., and Mukai, Y., *Chem. & Pharm. Bull. (Tokyo)*, **6**, 322 (1958); *Chem. Abstr.*, **52**, 19680 (1958).
278. Morgan, G. T., and Turnstall, R. B., *J. Chem. Soc.*, **125**, 1963 (1924).
279. Müller, R. H., *Ind. Eng. Chem., Anal. Ed.*, **12**, 620 (1940).
280. Müller, E., and Willenberg, H., *J. prakt. Chem.*, **99**, 34 (1919); *Z. anal. Chem.*, **61**, 3 (1922).
281. Naughton, J. J., and Frodyma, M. M., *Anal. Chem.*, **22**, 711 (1950).
282. New York Laboratory Supply Company, New York.
283. Nicloux, M., and Boivin, A., *Compt. rend. acad. sci.*, **184**, 890 (1927).
284. Niederl, J. B., and Niederl, V., "Micromethods of Quantitative Organic Elementary Analysis," Wiley, New York, 1938.
285. Niederl, J. B., and Niederl, V., "Micromethods of Quantitative Organic Analysis," 2nd ed., Wiley, New York, 1942.
286. Niederl, J. B., and Niederl, V., *Mikrochemie ver. Mikrochim. Acta*, **26**, 28 (1939).
287. Niederl, J. B., and Roth, R. T., *Ind. Eng. Chem., Anal. Ed.*, **6**, 272 (1934).
288. Niederl, J. B., and Sozzi, J. A., "Microánalisis Elemental Orgánico," Calle Arcos, Buenos Aires, 1958.
289. Niemann, C., and Danford, V., *Ind. Eng. Chem., Anal. Ed.*, **12**, 563 (1940).
290. Nunemakes, R. B., and Shrader, S. A., *Anal. Chem.*, **28**, 1040 (1956).
291. Oberhauser, B. F., Rivera, R., and Etcheverry, D. H., *Anales fac. filosof. y educ. Univ. Chile, Sección quím.*, **3**, 61 (1946).
292. Oerin, S., Moga, V., and Bertescu, L., *Acad. rep. populare Romîne, Bul. ştinţ., Secţ. ştiinţe tehnice şi chim.*, **4**, 439 (1952).
293. Ogg, C. L., *J. Assoc. Offic. Agr. Chemists*, **38**, 365 (1955).
294. Ogg, C. L., Willits, C. O., Ricciuti, C., and Connelly, J. A., *Anal. Chem.*, **23**, 911 (1951).
295. Ohashi, S., *Bull. Chem. Soc. Japan*, **28**, 585 (1955).
296. Okáč, A., and Vrchlabsky, M., *Chem. listy*, **50**, 2042 (1956).
297. Panicker, A. R., and Banerjee, N. G., *Analyst*, **83**, 296 (1958).
298. Pella, E., *Mikrochim. Acta*, p. 687 (1958).
299. Pepkowitz, L. P., *Anal. Chem.*, **23**, 1716 (1951).
300. Pepkowitz, L. P., and Proud, E. R., *Anal. Chem.*, **21**, 1000 (1949).
301. Peters, J. H., and Gutmann, H. R., *Anal. Chem.*, **25**, 987 (1953).
302. Pickhardt, W. P., Oemler, A. N., and Mitchell, J., Jr., *Anal. Chem.*, **27**, 1784 (1955).
303. Pickhardt, W. P., Safranski, L. W., and Mitchell, J., Jr., *Anal. Chem.*, **30**, 1298 (1958).
304. Pollard, C. B., and Forsee, W. T., *Ind. Eng. Chem., Anal. Ed.*, **7**, 77 (1935).
305. Power, F. W., *Ind. Eng. Chem., Anal. Ed.*, **11**, 660 (1939).
306. Power, F. W., *Mikrochemie*, **22**, 263 (1937).
307. Prater, A. N., *Ind. Eng. Chem., Anal. Ed.*, **12**, 184 (1940).
308. Pregl, F., "Quantitative Organic Microanalysis" (E. Fyleman, trans. 2nd German ed.), Churchill, London, 1924.
309. Rabinowitz, J. L., *Anal. Chem.*, **29**, 982 (1957).
310. Reihlen, H., *Mikrochemie*, **23**, 285 (1938).

311. Renard, M., *Ind. chim. belge,* **13,** 113 (1948).
312. Renoll, M. W., Midgley, T., Jr., and Henne, A. L., *Ind. Eng. Chem., Anal. Ed.,* **9,** 566 (1937).
313. Riesenfeld, E. H., *Chemiker-Ztg.,* **42,** 10 (1918).
314. Robertson, G. I., Jett, L. M., and Dorfman, L., *Anal. Chem.,* **30,** 132 (1958).
315. Robinson, R. J., and Doan, D. J., *Ind. Eng. Chem., Anal. Ed.,* **11,** 406 (1939).
316. Rodden, C. J., *Ind. Eng. Chem., Anal. Ed.,* **12,** 693 (1940).
317. Roth, H., "Die quantitative organische Mikroanalyse von Fritz Pregl," 4th ed., Springer, Berlin, 1935.
318. Roth, H., "F. Pregl quantitative organische Mikroanalyse," 5th ed., Springer, Wien, 1947.
319. Roth, H., "Pregl-Roth quantitative organische Mikroanalyse," 7th ed., Springer, Wien, 1958.
320. Roth, H., "Quantitative Organic Microanalysis of Fritz Pregl," 3rd ed., (E. B. Daw, trans. 4th German ed.), Blakiston, Philadelphia, Pennsylvania, 1937.
321. Royer, G. L., Alber, H. K., Hallett, L. T., Spikes, W. F., and Kuck, J. A., *Ind. Eng. Chem., Anal. Ed.,* **13,** 574 (1941).
322. Royer, G. L., Norton, A. R., and Sundberg, O. E., *Ind. Eng. Chem., Anal. Ed.,* **12,** 688 (1940).
323. Ruf, E., *Z. anal. Chem.,* **163,** 21 (1958).
324. Rush, C. A., Disc., Symposium on Analysis of Fluorine-containing Compounds, 137th National Meeting of the American Chemical Society, Cleveland, Ohio, April 1960.
325. Rush, C. A., Cruikshank, S. S., and Rhodes, E. J. H., *Mikrochim. Acta,* p. 858 (1956).
326. Rush, C. A., Cruikshank, S. S., and Rhodes, E. J. H., *Proc. Intern. Congr. Pure and Appl. Chem., Anal. Chem. 15th Congr. Lisbon,* **1,** 279 (1956); *Anal. Abstr.* **6,** No. 1175 (1959).
327. Sakamoto, S., *Yakugaku Zasshi,* **72,** 509 (1952).
328. Sargent, E. H., Company, Chicago, Illinois.
329. Šatava, V., and Körbl, J., *Collection Czechoslov. Chem. Communs.,* **22,** 1380 (1957); *Chem. Abstr.* **52,** 7929 (1958).
330. Sax, K. J., and Stross, F. H., *Anal. Chem.,* **29,** 1700 (1957).
331. Schadendorff, E., and Zacherl, M. K., *Mikrochemie,* **10,** 99 (1932).
332. Schmitt, R. B., and Niederl, J. B., *Mikrochemie,* **24,** 59 (1938).
333. Schöniger, W., *Helv. Chim. Acta,* **39,** 650 (1956).
334. Schöniger, W., Regional Analytical Symposium, Philadelphia, Pennsylvania, February, 1957.
335. Schöniger, W., *Mikrochim. Acta,* p. 545 (1957).
336. Schöniger, W., "Proceedings of the International Symposium on Microchemistry 1958," p. 86, Pergamon, Oxford, London, New York, and Paris, 1960.
337. Schott u. Gen, Landshut, Bavaria.
338. Schwarzkopf, O., Personal communication, 1955.
339. Silbert, F. C., and Kirner, W. R., *Ind. Eng. Chem., Anal. Ed.,* **8,** 353 (1936).
340. Sirotenko, A. A., *Mikrochemie ver. Mikrochim. Acta,* **40,** 30 (1953).
341. Smith, G. F., and Taylor, W. H., *Ind. Eng. Chem., Anal. Ed.,* **13,** 203 (1941).
342. Smith, W. H., *Anal. Chem.,* **30,** 149 (1958).
343. Sokolova, N. V., *Zhur. Anal. Khim.,* **11,** 728 (1956); *Chem. Abstr.,* **51,** 8582 (1957).
344. Stanek, V., and Nemes, T., *Z. anal. Chem.,* **95,** 244 (1933).

9. Carbon and Hydrogen

345. Stehr, E., *Anal. Chem.*, **31**, 1274 (1959).
346. Sternberg, H., *Mikrochemie*, **22**, 187 (1937).
347. Sternberg, H., *Mikrochemie*, **24**, 65 (1938).
348. Steyermark, Al, Unpublished work.
349. Steyermark, Al, *Ind. Eng. Chem., Anal. Ed.*, **17**, 523 (1945).
350. Steyermark, Al, "Quantitative Organic Microanalysis," Blakiston, Philadelphia, Pennsylvania, 1951.
351. Steyermark, Al, Alber, H. K., Aluise, V. A., Huffman, E. W. D., Kuck, J. A., Moran, J. J., and Willits, C. O., *Anal. Chem.*, **21**, 1555 (1949).
352. Sugiyama, N., and Furuhashi, K., *Nippon Kagaku Zasshi*, **72**, 583 (1951).
353. Sundberg, O. E., and Maresh, C., Presented at the Meeting-in-Miniature of the North Jersey Section of the American Chemical Society, Seton Hall Univ., South Orange, New Jersey, January 1959; *Anal. Chem.*, **32**, 274 (1960).
354. *Tech. News Bull.* **39**, 25 (1955).
355. Terent'ev, A. P., and Luskina, B. M., *Zhur. Anal. Khim.*, **14**, 112 (1959).
356. Teston, R. O'D., and McKenna, F. E., *Ind. Eng. Chem., Anal. Ed.*, **19**, 193 (1947).
357. Thomas, Arthur H., Company, Philadelphia, Pennsylvania.
358. Thompson, R. C., *Anal. Chem.*, **25**, 535 (1953).
359. Thorn, J. A., and Shu, P., *Can. J. Chem.*, **29**, 558 (1951).
360. Throckmorton, W. H., and Hutton, G. H., *Anal. Chem.*, **24**, 2003 (1952).
361. Trenner, N. R., Arison, B. H., and Walker, R. W., *Anal. Chem.*, **28**, 530 (1956).
362. Tunnicliff, D. D., Peters, E. D., Lykken, L., and Tuemmler, F. D., *Ind. Eng. Chem., Anal. Ed.*, **18**, 710 (1946).
363. Unterzaucher, J., *Chem. Ing. Tech.*, **22**, 39 (1950).
364. Unterzaucher, J., *Mikrochemie ver. Mikrochim. Acta*, **36/37**, 706 (1951).
365. Unterzaucher, J., *Mikrochim. Acta*, p. 448 (1957).
366. Vance, J. E., *Ind. Eng. Chem., Anal. Ed.*, **13**, 132 (1941).
367. Van Slyke, D. D., *Anal. Chem.*, **26**, 1706 (1954).
368. Van Slyke, D. D., Folch, J., and Plazin, J., *J. Biol. Chem.*, **136**, 509 (1940).
369. Van Straten, F. W., and Ehret, W. F., *Ind. Eng. Chem., Anal. Ed.*, **9**, 443 (1937).
370. Večeřa, M., and Šnobl, D., *Chem. listy*, **51**, 1482 (1957).
371. Večeřa, M., Šnobl, D., and Synek, L., *Mikrochim. Acta*, p. 9 (1958).
372. Večeřa, M., and Synek, L., *Chem. listy*, **49**, 1891 (1955).
373. Večeřa, M., and Synek, L., *Chem. listy*, **51**, 2266 (1957); *Chem. Abstr.*, **52**, 4393 (1958).
374. Večeřa, M., and Synek, L., *Collection Czechoslov. Chem. Communs.*, **23**, 1202 (1958).
375. Večeřa, M., Vojtech, F., and Synek, L., *Collection Czechoslov. Chem. Communs.*, **25**, 93 (1960).
376. Vries, G. de, and Dalen, E. van, *Anal. Chim. Acta*, **7**, 274 (1952).
377. Wagner, H., *Mikrochemie ver. Mikrochim. Acta*, **36/37**, 634 (1951).
378. Weil, H., *Ber.*, **43**, 149 (1910).
379. Weygand, C., "Quantitative analytische Mikromethoden der organischen Chemie in vergleichender Darstellung," pp. 70–163, Akademische Verlagsges., Leipzig, 1931.
380. Weygand, C., and Hennig, H., *Chem. Fabrik*, **9**, 8 (1936).
381. White, T. T., Campanile, V. A., Agazzi, E. J., TeSelle, L. P., Tait, P. C., Brooks, F. R., and Peters, E. D., *Anal. Chem.*, **30**, 409 (1958).
382. Willits, C. O., *Anal. Chem.*, **21**, 132 (1949).
383. Willits, C. O., and Ogg, C. L., *J. Assoc. Offic. Agr. Chemists*, **32**, 561 (1949).

384. Willits, C. O., and Ogg, C. L., *J. Assoc. Offic. Agr. Chemists,* **34**, 607 (1951).
385. Willits, C. O., and Ogg, C. L., Personal communication.
386. Wilzbach, K. E., and Sykes, W. Y., *Science,* **120**, 494 (1954).
387. Wurzschmitt, B., *Mikrochemie ver. Mikrochim. Acta,* **36/37**, 614 (1951).
388. Wurzschmitt, B., *Mitt. Gebiete Lebensm. u. Hyg.,* **43**, 126 (1952).
388a. Yeh, C. S., Personal communication, 1960; *Anal. Chem.,* in press.
389. Zabrodina, A. S., and Levina, S. Ya., *Vestnik Moskov. Univ., Ser. Mat., Mekhan., Astron., Fiz. i Khim.,* **12**, 181 (1957); *Chem. Abstr.,* **52**, 19680 (1958).
390. Zabrodina, A. S., and Miroshina, V. P., *Vestnik. Moskov. Univ.,* **12**, 195 (1957); *Anal. Abstr.,* **5**, No. 1537 (1958).
391. Zimin, A. V., Churmanteev, S. V., Gubanova, A. V., and Verina, A. D., *Doklady Akad. Nauk S.S.S.R.,* **126**, 784 (1959); *Chem. Abstr.,* **53**, 18736 (1959).
392. Zimmermann, W., *Mikrochemie ver. Mikrochim. Acta,* **31**, 149 (1944).

CHAPTER 10

Microdetermination of Sulfur

The determination of sulfur in organic compounds is based on the combustion of the compound with the subsequent conversion of the sulfur to sulfur trioxide (or sulfate, if alkali or alkaline earth metals are present) and finally to barium sulfate. The organic material can be destroyed by several methods,* but the author prefers either the Carius[37-40,209,214] or the Schöniger.[198-200,213] The Carius method has been proven reliable, over many years, with practically all types of compounds. There have been, however, a few isolated cases in which low unexplained results were obtained. In spite of these, the Carius method still remains the first method of choice of the author. The period required for each combustion is much longer than that for the other methods, but a number of Carius combustions can be carried out simultaneously in one furnace, overnight, which compensates for the time factor. The second method of choice of the author is the Schöniger and although this was introduced only a comparatively short time ago, it has proven to be reliable with a large variety of types of compounds. The author strongly recommends the use of these two methods and *both* should be used in cases of controversy. As a third choice, the Pregl catalytic combustion[11,76,77,161,162,180,186-189,209] should be relied upon since it too has been proven to give good results in the hands of a number of analysts.[11,165-168]

All of the procedures described in this chapter are applicable to fluorine-containing compounds. Phosphorus-containing compounds *must* be analyzed by one of the gravimetric procedures since barium phosphate is *insoluble* in the *neutral* solution required for the titration using tetrahydroxyquinone as the indicator. With the gravimetric procedures, there is no interference.

CARIUS COMBUSTION
VOLUMETRIC CARIUS METHOD[209,214]
(Not Applicable to Phosphorus-containing Compounds)

With the Carius method, the organic material is destroyed by heating with nitric acid in the presence of some alkali metal salt as shown by the following:

* Please see references 5-7, 15, 36-40, 42-44, 51-53, 67, 76, 77, 80-82, 88, 94, 136, 142, 150, 151, 155, 156, 160-162, 169, 173, 174, 180, 186-189, 196-200, 209, 213, 214, 220, 243, 246, 255, 257.

Carius Combustion

$$\text{Organic S} \xrightarrow[\text{(Excess HNO}_3\text{)}]{\text{NaX} \atop [\text{O}]} CO_2 + H_2O + NaHSO_4$$

The acid sulfate is converted to sulfate and titrated with standard barium chloride solution to an end point using tetrahydroxyquinone indicator:[1,5,82,169,205,209,213,214,220]

$$BaCl_2 + Na_2SO_4 \rightarrow 2NaCl + \underline{BaSO_4}$$

(The sodium, calcium, ammonium, and potassium salts of tetrahydroxyquinone

are yellow while the barium salt is red-purple.[205])

Reagents

FUMING NITRIC ACID, REAGENT GRADE[209,213,214]

Reagent grade of fuming nitric acid, sp. gr. 1.49 to 1.50 is used to destroy the organic material. (Caution: This acid must be handled with extreme care.)

PURE SODIUM OR POTASSIUM SALT[160,162,209,213,214]

Any reagent grade of sodium or potassium salt (not containing sulfur), such as oxalate, acid phthalate, chloride, etc., is used for combining with the sulfur trioxide formed during the combustion.

SODIUM HYDROXIDE, APPROX. 0.1N

Approximately 0.1N sodium hydroxide (not standardized) is used to convert the acid sulfate to sulfate previous to the titration.

HYDROCHLORIC ACID, APPROX. 0.01N

Approximately 0.01N hydrochloric acid (not standardized) is used to back-titrate the excess sodium hydroxide referred to above.

PHENOLPHTHALEIN INDICATOR

This solution, prepared as described in Chapter 5, Standard Solutions, is used as an indicator in the titration of the acid sulfate obtained in the combustion.

10. Sulfur

ETHANOL, 95%

This is used so that the titration with barium chloride can be carried out in approximately 50% alcohol.

TETRAHYDROXYQUINONE INDICATOR (THQ)[1,5,26,82,169,205,209,213,214,220]

(Prepared by W. H. & L. D. Betz.[26]) This material is used as a solid—see Chapter 5 on Standard Solutions.

STANDARD POTASSIUM SULFATE, 0.01N[169,209,213,214]

This solution is prepared according to the directions given in Chapter 5 on Standard Solutions.

STANDARD BARIUM CHLORIDE, 0.01N[169,209,213,214]

This solution is prepared and standardized according to the directions given in Chapter 5 on Standard Solutions.

Apparatus

CARIUS COMBUSTION FURNACE[209,211,225]

The combustion furnace used is of the type shown in Fig. 135. It should have at least four wells of approximately 16 mm. inside diameter and 225 mm. long. The wells should be held at a fixed inclined position of approximately 45° or should be adjustable. The furnace should be provided with a device for pushing the combustion tubes from the individual wells.

The furnace must be able to maintain a temperature in the wells of approximately 310° C. and the temperature at any point should not vary more than ± 5° C. from the operating temperature (electrically heated units should be able to perform thusly with voltages as low as 100 volts). The furnace temperature should be adjustable. There should also be a device that shows when the furnace is in operation and a temperature indicator.

The furnace should be equipped with safety devices to confine broken glass in the event of an explosion.

A valuable accessory for the furnace is an automatic time switch (Fig. 136), so that the furnace can be operated at night and be cooled by morning.

CARIUS COMBUSTION TUBES (BOMB TUBES)[211]

Two types of tubes (Fig. 137) have been recommended, namely, heavy-walled and thin-walled, although the former is preferred by the author. Regardless of which is used, the conditions listed in Table 21 *must* be adhered to in order to

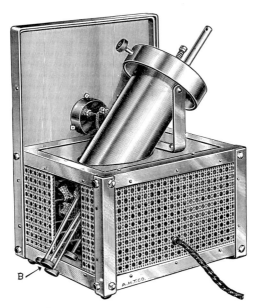

FIG. 135. Micro-Carius furnace. *Top:* Front view. (A) Small adapter tubes for use with undersize Carius tubes. *Bottom:* Rear view. (B) Push rods for removing Carius tubes.

10. Sulfur

minimize the danger of explosion. If directions are followed, the incidence of explosion is extremely small.

The specifications[211] are designed for a maximum operating temperature of 300° C. The length of the sealed tube between the bottom and the start

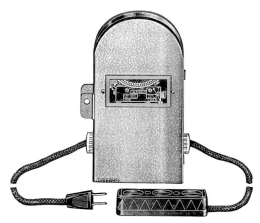

FIG. 136. Automatic time switch.

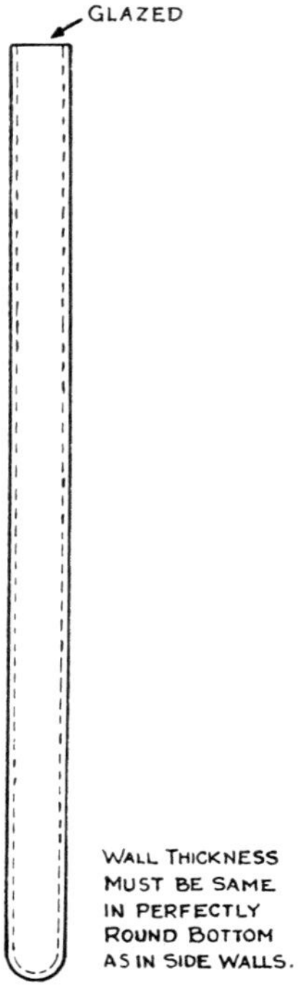

FIG. 137. Carius combustion tube (see Table 21 for details of construction).

of the taper at the shoulder should be 150 to 175 mm. for the heavy-walled tubes and 180 to 210 mm. for the thin-walled type.

TABLE 21
RECOMMENDED SPECIFICATIONS[211] FOR CARIUS COMBUSTION TUBES

Combustion tube	Wall thickness (mm.)	O.D. (mm.)	Length (mm.)	Length of sealed tube between bottom and start of taper at shoulder (mm.)	Volume of HNO_3 (sp. gr. at 60° F., approximately 1.5) (ml.)	Temp. (° C.)
Heavy-walled	2.3 ± 0.3	13 ± 0.8	210 ± 10	150 to 175	More than 0.3 (volume should not exceed 0.7)	250
Thin-walled	1.2 ± 0.2	13 ± 0.7	240 ± 10	180 to 210	0.3 or less	300

The glass should have a coefficient of linear expansion not exceeding 0.0000040 cm. per cm. per 1° C., with a softening point of 820° C. (Corning Pyrex 7740 or equal). Tubes at one end should have a closed round bottom of about the same wall thickness as the side walls and at the other end should be open and glazed. Tubes must be well annealed. The thickness of the wall and the length depend upon the volume of nitric acid used.

BLAST LAMP

Any small blast lamp as, for example, the type shown in Fig. 138, which gives an intensely hot flame when operated with gas and air or oxygen, may be used for sealing the combustion tubes.*

FIG. 138. Blast lamp.

* With beginners, gas and air is preferred, because the glass does not flow as rapidly and the process is better controlled.

10. Sulfur

GRINDING WHEEL

A mechanically driven Carborundum grinding wheel of the type shown in Fig. 139 is used for cutting a groove in the combustion tubes previous to their being cracked open with the aid of a hot rod (see below under Procedure).

FIG. 139. Grinding wheel (glass cutter).

ILLUMINATED TITRATION STAND ASSEMBLY[72,169,209,214,225]

The stand shown in Fig. 71 (Chapter 5—Standard Solutions) is used for the titration of sulfate with barium chloride.

CUVETTE[169,209,214,225]

The cuvette shown in Fig. 72 (Chapter 5—Standard Solutions) is used as a titrating vessel.

ORANGE-BROWN FILTER PLATE[45,169,209,214,225]

The orange-brown filter plate described in connection with the standardization of barum chloride (Chapter 5—Standard Solutions) is used as a comparator for obtaining the end point of the titration of sulfate with barium chloride.

Procedure

The amount of sample used should be enough to require 3 to 5 ml. of $0.01N$ barium chloride in the titration.[169,209,214] If more or less is required the appearance of the titration mixture at the end point does not match that of the filter plate, adding confusion. If considerably more than 5 ml. of barium chloride is required (Ogg, Willits and Cooper[169] allowed *5 ml. of 0.02 N $BaCl_2$*) the determination is best discarded and the same holds if less than 3 ml. is used. In the former case a smaller sample should be used. In the latter case, a larger sample should be used, or if this is not feasible, enough standard potassium sulfate solution is added, just before titrating with barium chloride, so that the recommended quantity is required.

If the sample is a solid, it should be weighed by difference using a charging tube (Figs. 47–49, Chaper 3). Care should be exercised so as not to have sample on the walls of the Carius tube. This is accomplished by holding the charging tube, containing the sample, upright. The empty Carius combustion tube is brought down over it (closed end upward) as far as possible while still holding the charging tube. The combination is quickly inverted and gently tapped so that the sample drops into the Carius tube and no particles adhere to the charging tube rim. If the sample is a high-boiling liquid, it should be weighed in a porcelain boat (Chapter 3) and the boat inserted in the combustion tube. A low-boiling liquid is weighed in a plain capillary (see Chapter 3), the end of which is broken before insertion in the Carius tube. Methylcellulose capsules also may be used. About 15–20 mg.[162,209,214] of some pure sodium or potassium salt (such as oxalate, acid phthalate, chloride, etc.) is added to the combustion tube with gentle tapping to dislodge any material adhering to the walls. (If too much salt is added, a fading end point results in the titration. Consequently, if an organic compound is being analyzed which contains an alkali metal, the amount of sodium or potassium salt added should be proportionately reduced.) Next is added 0.5–0.6 (0.7 *maximum*) ml. fuming nitric acid, sp. gr., 1.49 to 1.50, while rotating the tube so that any adhering material is washed down, and the tube is immediately sealed off as described below. (*Caution!* As seen from Table 21, this quantity of acid is used *only* with the heavy-walled tubes. If the other type is employed, the volume should be *0.3 ml. or less and the dimensions and temperatures mentioned in the following pages must be so adjusted.*) If the sample and acid react at room temperature, the bottom of the tube *and* the acid should first be cooled in dry ice-acetone mixture. As an alternate procedure, the sample may be weighed in a weighing bottle (Fig. 27, Chapter 3), inserted in the *tilted* combustion tube *after* the acid has been added and kept from sliding to the bottom until *after* the tube has been sealed.)

SEALING THE CARIUS COMBUSTION TUBE

(Note: The beginner will do well to practice both the sealing and cutting—see below—Procedure—using empty tubes before attempting these for actual determinations. It is also good practice to employ a safety glass shield between the operator and the blast lamp.)

The filled Carius tube is *held at about a 45° angle to the horizontal.* The section near the open end is gently warmed to evaporate off adhering nitric acid which might otherwise cause the tube to crack. The open tip is then strongly heated with a blast lamp and a section of Pyrex tubing about 150–175 mm. long is sealed on for use as a handle (Fig. 140a). (A section of an old Carius tube may be used for this purpose. Care should be exercised so that the combustion tube containing the sample and nitric acid is not sealed off by

10. Sulfur

the addition of the handle.) The tube is held by one hand at the filled closed end and by the sealed-on handle with the other hand. The hot flame of a blast lamp is played on the section of the tube about 160 to 180* mm. from the closed end (Fig. 140a), while it is slowly rotated, using both hands. The soft glass is allowed to flow downward so that the heated section thickens

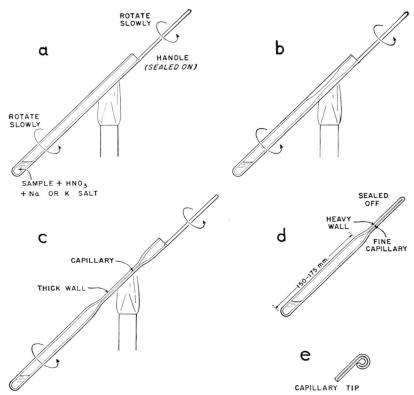

FIG. 140. Sealing Carius tubes. (a) Handle attached to Carius tube and sealing process begun. (b) Walls thickened. (c) Tube drawn out a little and walls again thickened. (d) Tube sealed off. (e) Small hook on end for attaching wire. It is important that length of sealed tube from *rounded bottom* up to *shoulder of taper* be 150–175 mm. (see Table 21).

considerably (Fig. 140b). The tube is slightly drawn out and the side walls of the drawn section again allowed to thicken by the flow of soft glass (Fig. 140c). Finally, when the constricted soft section is attaining capillary dimensions (approximately 0.25 mm. I.D.), the tube is pulled out and the capillary sealed off making a tip of 2–3 cm. in length, about 2–3 mm. O.D. and a very fine inside diameter—a small fraction of a millimeter (Fig. 140d). A well-sealed tube will have the same wall thickness at the constricted portion as it

* These figures apply for the heavy-walled tube only (see Table 21).

does in the main body. This produces one of great strength capable of withstanding the high pressures encountered. Even though very thick-walled capillaries are not needed to withstand the pressures, they are preferred because of less danger of breakage during handling. If there is no means of pulling the tubes out of the furnace, a small hook (Fig. 140e), may be made on the finished capillary by placing in a burner flame and allowing the soft tip to be pulled down by gravity or forcing it with a pair of forceps. A section of wire is then attached to the hook.

The sealed tube is allowed to cool and is then placed in the *cold* combustion furnace in any desired position, ranging from vertical to about 30° to the horizontal. The furnace is slowly heated to 250°* C. and that temperature maintained for 7–8 hours. (*Caution: High pressure.*) (The author charges the furnace at the close of the day, setting a time switch which shuts off the current during the following early morning hours. The furnace is then back to room temperature at the beginning of the next working day.)

REMOVAL OF TUBES FROM THE COMBUSTION FURNACE

Before they can be safely handled, the *cooled*† Carius combustion tubes must have the residual pressure released. The tube is forced or pulled up part of the way out of the furnace so that the tip is exposed and supported thusly. A small amount of nitric acid condensate will always be present in the capillary tip (Fig. 141a). This is forced back into the main body of the tube by waving a flame near it. The tip is then strongly heated until the internal pressure causes a hole to be blown out of the molten capillary and the gases escape with a hiss (Fig. 141b).

The cutting or grinding wheel (Fig. 139), is adjusted so that it will make a cut of not greater than one-half the depth of the wall thickness of the tube.‡ As soon as the pressure has been released in the tube, *it is held at an angle of about 45° to the horizontal against the revolving wheel* and a groove cut, all the way around, at a distance of about 75 mm. from the bottom. Distilled water is poured over the groove to remove grindings as well as to wet it. The end of a piece of the soft glass rod is held in a blast lamp flame until it is molten. It is then quickly pressed against the wet groove on the combustion tube and held there for several seconds (Fig. 142). The tube usually cracks either completely around or almost so. Examination against an illuminated area such as a window will reveal the extent of the crack. If it is not complete, the groove should be rewet and the molten tip of a soft glass rod reapplied to whatever portion has not cracked. This should be repeated until

---

* These figures apply for the heavy-walled tube only (see Table 21).
† Caution: Never handle a warm sealed tube.
‡ The tube should not be completely cut through by the wheel or contamination will result.

10. Sulfur

the crack makes a complete circle and on gently tapping the top of the tube falls off. (If the groove has not been cut deeply enough or the soft glass rod not hot enough, the operation is not successful.) The cut end is then carefully fire polished and the tube set aside to cool. The contents of the tube are then *carefully* diluted with distilled water and then transferred quantitatively to a small beaker (about 50 ml. capacity). The nitric acid is evaporated off on a steam bath. The dry residue of sodium or potassium acid sulfate is dissolved in a few milliliters of distilled water and the solution transferred quantitatively

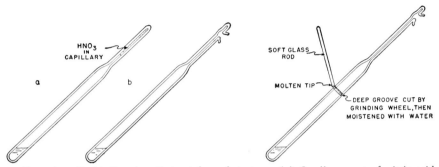

FIG. 141   (*Left*) Opening Carius tubes—first stage. (a) Small amount of nitric acid in capillary of sealed tube, after combustion. (b) Pressure released from tube.

FIG. 142.   (*Right*) Opening Carius tubes—final stage.

to the cuvette (Fig. 72), using less than 15 ml. water in all. A few drops of phenolphthalein indicator are added and the solution is made alkaline with approximately 0.1N sodium hydroxide. It is then back-titrated with approximately 0.01N hydrochloric acid, just to expel the color. The contents are then diluted to 15 ml. with distilled water and then 15 ml. of 95% ethanol added, followed by half a scoop of powdered THQ. The vessel is placed on the illuminated titration stand (Fig. 71—Chapter 5 on Standard Solutions) and is titrated with barium chloride, 0.01N, to the end point (identical in appearance to that of the orange filter plate—see Standardization of 0.01N $BaCl_2$, above-mentioned chapter).

Calculation:

1 ml. of 0.01N $BaCl_2$ is equivalent to 0.1603 mg. of sulfur

$$\therefore \frac{\text{ml. of 0.01N } BaCl_2 \times 0.1603 \times 100}{\text{Wt. sample}} = \% \text{ S}$$

Example:

3.44 ml. of 0.01N $BaCl_2$ is required to titrate the sulfate resulting from the combustion of a 2.960-mg. sample

$$\therefore \frac{3.44 \times 0.1603 \times 100}{2.960} = 18.63\% \text{ S}$$

The allowable error is $\pm 0.3\%$.

GRAVIMETRIC CARIUS METHOD
(Applicable to Phosphorus-containing Compounds)

Instead of the preferred Carius volumetric method given in the preceding pages, the determination may be done gravimetrically.[37–40,42,53,161,162,186–189,209] In the presence of phosphorus, the determination *must*[11] be done gravimetrically, since barium phosphate is insoluble in the *neutral* solution of the volumetric procedure and would interfere. Barium chloride (about 15 mg.) is added to the sample (4–9 mg.) plus nitric acid in the Carius tube, no sodium or potas-

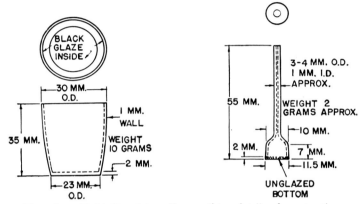

FIG. 143. (*Left*) Porcelain sulfur crucible—details of construction.
FIG. 144. (*Right*) Porcelain filter stick—details of construction.

sium salt being required.[37–40,42,161,186–189,209] The sulfuric acid formed during combustion is immediately converted into barium sulfate:

$$H_2SO_4 + BaCl_2 \rightarrow \underline{BaSO_4} + 2HCl$$

The combustion tube is opened and the contents quantitatively washed into a previously weighed* porcelain crucible (with black interior) (Fig. 143), containing a porcelain filter stick† (Fig. 144).[209,212] The nitric acid is removed, by evaporation, on a steam bath and the dry residue is treated with about 3 ml. of 1:300 hydrochloric acid to redissolve the soluble salts. The filter stick is then attached to the vertical tube of the siphon, receiver and inner container[209,212] shown in Figs. 147 and 148, using a small section of rubber tubing. Vacuum is applied to the side arm of the apparatus and the solution

* The combination should be treated *before weighing* in exactly the same manner as is done after transfer of the precipitate to it, that is, washing, drying, igniting, etc.
† Transfer is best accomplished by a stream of wash water from a wash bottle (Figs. 145[209,212] and 146) into the tilted tube, open end down, the precipitate being carried out with the liquid.

10. Sulfur

of barium nitrate-chloride sucked from the crucible into the inner container and subsequently discarded. The precipitate remains in the crucible and is washed three times with 1-ml. portions of 1:300 hydrochloric acid, sucking as dry as possible between washings. The crucible plus filter stick is then

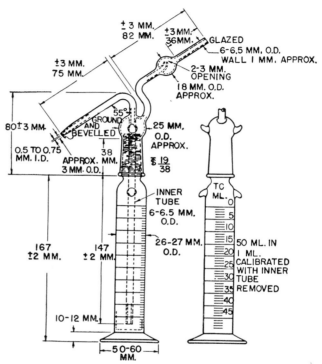

FIG. 145. Two views of a graduated wash bottle—details of construction.

FIG. 146. Wash bottle.

heated in an ordinary laboratory oven at 120° C. for 20–30 minutes to thoroughly dry. Crucible and filter stick are then heated in a small muffle furnace of the type shown in Fig. 88, at 700° C. for 5 minutes. The combina-

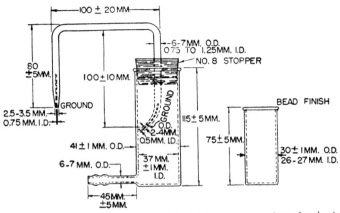

FIG. 147. (*Left*) Siphon, receiver, and (*right*) inner container for barium sulfate filtration—details of construction.

FIG. 148. Siphon, receiver, and inner container for barium sulfate filtration, showing method of use with filter stick and crucible.

tion is then allowed to cool on a metal block or in a metal crucible container with glass dome (desiccator minus the metal cooling block), (Figs. 43, 45, and 46, Chapter 3), rewashed three times with 1-ml. portions of 1:300 hydrochloric acid to remove occluded barium nitrate, redried, reignited at 700° C., cooled for one hour, and weighed (using a tare flask, Figs. 34–36, Chapter 3, as a counterpoise weight). The crucible and filter may be cleaned with concentrated sulfuric acid or it may be used without cleaning for successive determinations.

Unfortunately, much difficulty has been experienced with these filter sticks in the past, tiny holes being present between glazed and unglazed portions which permitted the precipitate to pass through.

Calculation:

Factor:

$$\frac{S}{BaSO_4} = 0.1374$$

$$\therefore \frac{Wt.\ BaSO_4 \times 0.1374 \times 100}{Wt.\ sample} \ \% \ S$$

Example:

6.231 mg. of $BaSO_4$ was obtained from 4.702 mg. of sample

$$\therefore \frac{6.231 \times 0.1374 \times 100}{4.702} = 18.21\% \ S$$

Alternate Gravimetric Carius Procedure
(Applicable to Phosphorus-containing Compounds)

An alternate gravimetric procedure employing the Carius method precipitates the barium sulfate *after* removal from the combustion tube.[160,162] The combustion mixture is the same as given for the THQ titration method, namely, sample, nitric acid, and sodium salt. After removal from the combustion tube and evaporation to dryness, the residue is dissolved in 5 ml. of 1:300 hydrochloric acid and transferred to the crucible (Fig. 143). (The clean, weighed filter stick must be kept separated until filtration or soluble sulfate might pass through the filter and be lost.) The crucible is then heated on a steam bath and the volume reduced, if necessary, so that there will be no danger of loss of precipitate (after precipitation) from creeping to the rim. A total volume of *less* than 10 ml. is preferred. One milliliter of 10% barium chloride solution is added and the mixture digested until the total volume has been reduced to 2–3 ml. Regardless of the initial volume, digesting should be done for *30 minutes*. The crucible is removed from the steam bath, cooled at least 15 minutes. Filtering, washing, etc., of the precipitated barium sulfate is done as described above.

SCHÖNIGER COMBUSTION

VOLUMETRIC SCHÖNIGER METHOD
(Not Applicable to Phosphorus-containing Compounds)

With the Schöniger method,[196,198-200,213] the organic material is destroyed by burning the sample in a special oxygen-filled flask in which the combustion takes place at high temperature, probably around 1200° C. The reaction may be represented by the following:

$$\text{Organic S} \xrightarrow{O_2} CO_2 + H_2O + SO_3 + SO_2$$

The resulting oxides of sulfur are absorbed and finally converted to sulfuric acid according to the following[149]:

$$SO_3 + SO_2 + 3H_2O + Br_2 \longrightarrow 2H_2SO_4 + 2HBr$$

or

$$SO_3 + SO_2 \xrightarrow[H_2O]{HNO_3} H_2SO_4$$

(A number of equations for the conversion of SO_2 into H_2SO_4 by HNO_3 are given by Mellor.[149])

Reagents

SODIUM HYDROXIDE, APPROX. 0.01N

Approximately 0.01N sodium hydroxide (not standardized) is used for absorbing oxides of sulfur in the combustion flask.

SODIUM HYDROXIDE, APPROX. 0.1N
HYDROCHLORIC ACID, APPROX. 0.01N

Same as for the Volumetric Carius Method.

HYDROCHLORIC ACID, APPROX. 0.1N

Approximately 0.1N hydrochloric acid (not standardized) is used for acidifying purposes in the procedure using bromine water to oxidize sulfur to the hexavalent state.

BROMINE WATER

Water saturated with reagent grade of bromine is used in one of the procedures to oxidize sulfur to the hexavalent state.

10. Sulfur

FUMING NITRIC ACID, REAGENT GRADE, SP. GR. 1.49–1.50

This is used to oxidize sulfur to the hexavalent state in one of the procedures.[213]

PHENOLPHTHALEIN INDICATOR
DISTILLED WATER
ETHANOL, 95%
TETRAHYDROXYQUINONE INDICATOR (THQ)
STANDARD POTASSIUM SULFATE, 0.01N
STANDARD BARIUM CHLORIDE, 0.01N

} Same as for the Volumetric Carius Method.

Apparatus

SCHÖNIGER COMBUSTION FLASK[146,198-200,225]

The combustion flask (Fig. 149) consists of a heavy wall, conical flask of borosilicate glass, with a deep bell-shaped flaring lip and elongated interchangeable ground-glass stopper into which has been sealed a heavy platinum

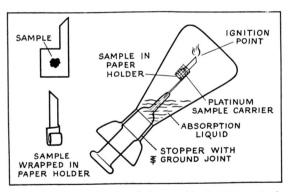

FIG. 149. Schöniger combustion flask assembly showing method of use.

wire gauze sample carrier. Both 300- and 500-ml. flasks are commercially available, but the author prefers the use of the larger size because of the extra available oxygen. *Organic solvents should not be used for cleaning, and stopcock grease should never be used on the ground joints because of the possible fire and explosion hazards from so doing.* Flasks in constant use need not be dried between determinations. (Figure 150 shows a modification of the Schöniger flask which permits electrical ignition and provides shielding.[56a,146])

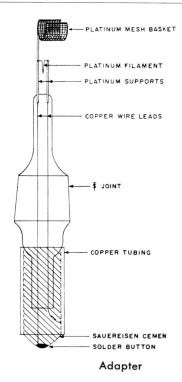

Adapter

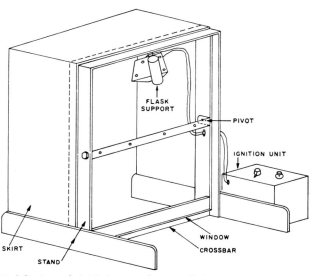

Fig. 150. Modification of Schöniger combustion flask permitting electrical ignition and shielding.

10. Sulfur

FILTER PAPER CARRIERS[198-200,225]

Flag-shaped strips of filter paper are used to hold the sample. They are folded over several times to completely wrap the sample, leaving the small tail sticking out for use as a fuse or ignition point.

METHYLCELLULOSE CAPSULES[225]

These are used for liquid samples (see Chapter 3). The sealed capsule, containing the sample, is wrapped in the filter paper carrier and placed in the platinum gauze basket.

ILLUMINATED TITRATION STAND ASSEMBLY
CUVETTE
ORANGE-BROWN FILTER PLATE

Same as for the Volumetric Carius Method.

Procedure

Enough sample is weighed by difference, using a charging tube (Figs. 47–49) onto the filter paper carrier (Fig. 149) to require 3–5 ml. of 0.01N barium chloride in the final titration (0.48–0.8 mg. S). (Liquid samples are weighed in methylcellulose capsules (Chapter 3) and then placed on the paper carrier.) The paper is folded so as to seal in the sample, but the small tail is left extending for use as the point of ignition (see Fig. 149). The paper is then inserted into the platinum gauze basket attached to the stopper. Ten milliliters of approximately 0.01N sodium hydroxide are added to the combustion flask which is then flushed with oxygen from a cylinder for a few minutes with the tube extending almost to the bottom of the flask. (*Caution:* No grease should be used on the ground joint.) The exposed tail is ignited by means of a burner and the stopper is inserted immediately into the oxygen filled flask. The stoppered flask is held by the stopper in the *inverted* position (see Fig. 149), preferably with the open end of the basket *upward* to prevent the dropping of unburned particles, until the combustion is completed, which takes place at temperatures around 1200° C. and requires a fraction of a minute. By holding the flask in the inverted position, the sodium hydroxide forms a seal around the stopper. (NOTE: *As a safety measure,* the flask should be held behind a safety glass shield, in a hood, while the operator is protected by means of gloves and goggles.) The flask is allowed to cool for about one minute in the inverted position and the contents are shaken vigorously until cloudiness disappears, after which a few ml. of water is placed in the cup surrounding the stopper and the flask is allowed to remain undisturbed for about 15 minutes to insure complete absorption of the oxidation products. The flask is now under a slightly reduced pressure, due to the consumption of some of the oxygen originally present, which makes removal of the stopper sometimes slightly difficult. Gentle manipulation loosens the stopper and the water, which was in the cup surrounding the stopper, is sucked in, washing the ground joint. (In case the stopper cannot be loosened easily, the flask is placed, *momentarily,* on the steam bath in order to increase its internal pressure more *nearly* to that of atmospheric, *but not above it.*) The stopper is removed and washed, collecting the washings in the flask. The contents of the flask are then transferred to a 100 ml. beaker and treated according to *either* (a) *or* (b), the former being preferred by the author:

(a) One milliliter of fuming nitric acid, sp. gr., 1.49–1.50 is added and the solution evaporated to *dryness* on a steam bath. The residue is dissolved in water, transferred to a cuvette, neutralized, etc., and titrated with $0.01N$ barium chloride solution using THQ as the indicator as described under the Volumetric Carius Method.

(b) Bromine water is added dropwise until the color of bromine persists. About 1.1 ml. of approximately $0.1N$ hydrochloric acid is added (to acidify) and the solution evaporated down on a steam bath to expel the bromine. (Although evaporation to dryness is not necessary, it has the advantage of volume control in the cuvette.) The residue (or *small* volume) is dissolved in water, transferred to a cuvette, etc., and titrated as described above under (a).

Calculation:

Same as for Volumetric Carius Method.

PREGL CATALYTIC COMBUSTION

It has been stated previously that the destruction of organic material may be accomplished by other means. During the first few years of the operation of his laboratory, the author used the Pregl catalytic combustion[11,76,77,161,162,180,186-189,209] extensively with good results. It is not the method of choice of the author, but its reliability has been definitely proven through collaborative studies.[11,165-168,247]

The sample is burned in an atmosphere of oxygen at red heat in the presence of platinum. The resulting oxides of sulfur are absorbed either by hydrogen peroxide or by bromine and converted into sulfuric acid according to the following[149]:

$$\text{Organic S} \xrightarrow[O_2 \text{ (Pt)}]{\Delta} SO_2 + SO_3 + CO_2 + H_2O$$

$$SO_2 + SO_3 + H_2O_2 + H_2O \rightarrow 2H_2SO_4$$

or

$$SO_2 + SO_3 + Br_2 + 3H_2O \rightarrow 2H_2SO_4 + 2HBr$$

10. Sulfur

In the absence of nitrogen, halogens, and phosphorus, the sulfuric acid may be titrated with standard alkali[76,77,161,162,180,186-189,209] using methyl red as the indicator:

$$H_2SO_4 + 2NaOH \rightarrow Na_2SO_4 + 2H_2O$$

In the presence of nitrogen and halogens, but in the absence of phosphorus, the volumetric procedure, using barium chloride with THQ as the indicator, may be used *provided* that *bromine* is used in the spiral to absorb the oxides. The use of hydrogen peroxide eliminates the possibility of using THQ indicator, because a fading end point[82,83] results *regardless* of attempts to destroy the peroxide and the sulfuric acid must be determined gravimetrically:

$$H_2SO_4 + BaCl_2 \longrightarrow \underline{BaSO_4} + 2HCl$$

In the presence of phosphorus, the gravimetric procedure must[11] be used as explained under the Carius method.

Reagents

For Gravimetric Procedure

HYDROGEN PEROXIDE, 30% (SUPEROXOL)

Reagent grade of 30% hydrogen peroxide (Superoxol) is used as the oxidizing agent to convert sulfur dioxide into sulfur trioxide. However, for this purpose it is diluted as described below. (*Caution: This material must be handled with extreme care.* It is stored in a refrigerator.)

DILUTE HYDROGEN PEROXIDE

Twenty ml. of 30% hydrogen peroxide is added to 80 ml. of distilled water.

DILUTE HYDROCHLORIC ACID

(One part by volume of reagent grade concentrated acid to 300 parts by volume of water.) This is used for the gravimetric procedure only.

BARIUM CHLORIDE SOLUTION

A 10% solution of reagent grade of barium chloride in distilled water is used in the gravimetric procedure.

For Acidimetric (Direct Neutralization) Procedure

DILUTED HYDROGEN PEROXIDE

See above under Gravimetric Procedure.

STANDARD SODIUM HYDROXIDE, 0.01N

This is prepared and standardized according to the directions given in Chapter 5. It is used only in the absence of nitrogen, halogens, and phosphorus.

METHYL RED INDICATOR

For preparation, see Chapter 5. This is used only in the absence of nitrogen, halogens, and phosphorus.

For Volumetric Procedure

SODIUM HYDROXIDE, APPROX. 0.1N
HYDROCHLORIC ACID, APPROX. 0.01N
PHENOLPHTHALEIN INDICATOR
DISTILLED WATER
ETHANOL, 95%
TETRAHYDROXYQUINONE (THQ) INDICATOR
STANDARD POTASSIUM SULFATE, 0.01N
STANDARD BARIUM CHLORIDE, 0.01N

} Same as for Volumetric Carius Method

BROMINE WATER

Saturated solution of bromine in water. This should be stored in a glass-stoppered bottle.

Apparatus

OXYGEN CYLINDER AND REDUCING VALVE

These are the same as used for the carbon-hydrogen determination (Chapter 9).

PRESSURE REGULATOR

This is the same as used for the carbon-hydrogen determination (Figs. 118 and 119, Chapter 9).

BUBBLE COUNTER-U-TUBE

This is the same as used for the carbon-hydrogen determination (Fig. 120, Chapter 9).

COMBUSTION TUBE WITH INNER SPIRAL[45,76,77,161,162,186-189,209,212]

The combustion tube (Fig. 151),[209,212] is prepared from quartz or Vycor (96% silica glass No. 790).[45] In reality, it serves two purposes, namely, combustion tube and absorber of the combustion product. The plain portion is placed in the combustion furnace. The part containing the spiral is moistened with the absorbent and not heated.

PLATINUM CONTACT STARS[187,188,209,212]

Two platinum contact stars (Fig. 152) are used in the combustion tube.

COMBUSTION APPARATUS

The apparatus* is the same as used in the Dumas determination of nitrogen [Fig. 89, Chapter 7 or Figs. 123–125 (minus the heating mortar), Chapter 9].

TEST TUBE

A standard 8-inch Pyrex[45] test tube is used as a cover for the tip of the combustion tube and for collecting washings.

ADDITIONAL APPARATUS REQUIRED

For the *gravimetric* procedure, the crucible (Fig. 143), filter stick (Fig. 144), wash bottles (Figs. 145 and 146) and siphon, receiver, and inner container (Figs. 147 and 148) are also required.

For the *acidimetric* (direct neutralization) procedure, an automatic burette (Figs. 69 or 70) is required instead of the crucible assembly, above.

For the *volumetric* procedure, the illuminated titration stand assembly (Fig. 71), cuvette (Fig. 72) and orange-brown filter plate (Chapter 5) are required instead of the crucible assembly, above.

Assembly and Gravimetric Procedure
(Applicable to Phosphorus-containing Compounds)

The oxygen cylinder, pressure regulator and bubble counter-U-tube are connected in the order named by means of thin-walled rubber tubing (see Chapter 9). The pressure is then regulated so that approximately 12–15 ml. of oxygen flows through the bubble counter-U-tube per minute. (This is accomplished with the aid of a Mariotte bottle—Fig. 129, Chapter 9—see Pressure Regulation, carbon-hydrogen determination, Chapter 9. The pressure regulation is done without the combustion tube being attached as it offers little added resistance.) The free end of the bubble counter-U-tube is then connected by means of a section of rubber tubing several centimeters in length to a small glass tip inserted in a one-hole rubber stopper† that fits the open end of the combustion tube.

---

* See footnote, p. 154, Chapter 7.
† Universal stopper[9,210] cut to size (see Fig. 153).

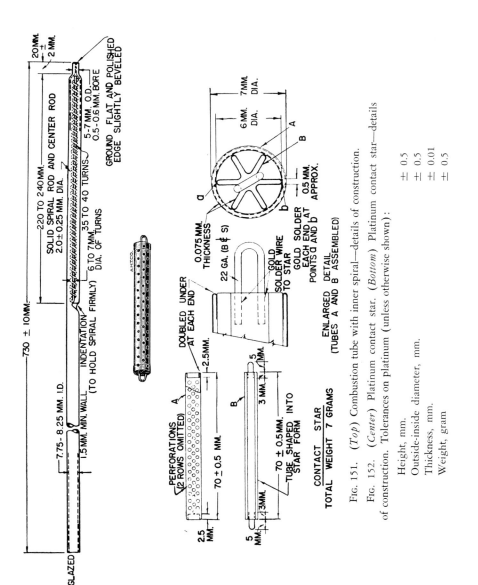

FIG. 151. (*Top*) Combustion tube with inner spiral—details of construction.

FIG. 152. (*Center*) Platinum contact star. (*Bottom*) Platinum contact star—details of construction. Tolerances on platinum (unless otherwise shown):

Height, mm. ± 0.5
Outside-inside diameter, mm. ± 0.5
Thickness, mm. ± 0.01
Weight, gram ± 0.5

10. Sulfur

The combustion tube is securely held vertically in a suitable clamp and stand with the spiral portion downward. The tip is immersed in 4–5 ml. dilute hydrogen peroxide solution contained in the test tube. Gentle suction is applied to the open end of the combustion tube, protecting it from contamination with a cotton-filled air filter of the type shown in Fig. 154, so that the liquid is drawn up covering the entire spiral and a few millimeters beyond. The suction is then removed and the hydrogen peroxide allowed to drain back into the test tube leaving the spiral wet. The liquid is poured from the test tube, leaving the latter moist.

The combustion tube is then placed in the combustion apparatus, the entire spiral portion *extending beyond* the long furnace so that it receives no heat.

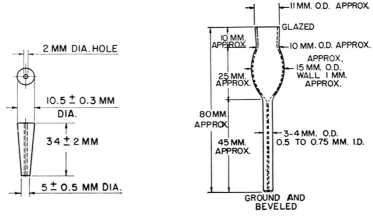

FIG. 153. (*Left*) One-hole rubber (or neoprene) stopper—details of construction.
FIG. 154 (*Right*) Air filter—details of construction.

The test tube, still moist with hydrogen peroxide, is placed over the tip of the combustion tube for protection. A clean platinum contact star is placed in the open end of the combustion tube and pushed with the aid of a platinum wire (Fig. 53, Chapter 3) into the section surrounded by the long furnace and adjacent to the spiral. The end or wire loop of the star should be about 2 cm. from the end of the long furnace (see Fig. 155). The second platinum contact star is then inserted and put into place so that its one wire loop is within a few centimeters of the other contact star and its other loop is about 2 cm. from the end of the long furnace adjacent to the short movable sample furnace (Fig. 155).

The sample (4–9 mg.), previously weighed in a platinum boat or capillary (if the latter, protected with a platinum sleeve, Fig. 84, Chapter 6, or large platinum boat, Fig. 25, Chapter 4) is placed, with the aid of a platinum wire, in the combustion tube at a position about 5 cm. from the end of the long

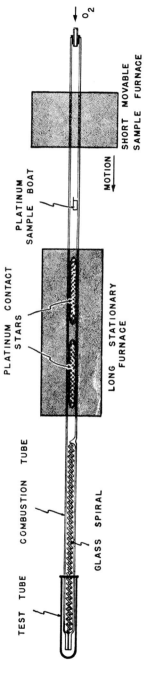

FIG. 155. Diagram of catalytic combustion setup.

10. Sulfur

furnace. The rubber stopper which connects the oxygen supply via the bubble counter-U-tube, is inserted into the open end of the combustion tube and the oxygen allowed to flow through at the rate of 12 to 15 ml. per minute.

The long furnace is heated to 800° C. (at least 750°).[11,244,247] The short furnace is then heated to 800° C. (at least 750°)[11,244,247] and moved up towards the sample cautiously. After combustion has started the short movable sample furnace is *slowly* moved across the sample and up against the long furnace. Too rapid combustion is liable to allow unburned material to pass through into the spiral. Consequently, it is preferred to operate the movable furnace by hand, at least in the early stages of the combustion. After the first combustion is over, it is best to follow with a rapid second to make certain that nothing remains unburned.

The furnaces are then shut off and the tube allowed to cool while oxygen is swept through. When cool, the tube is removed from the furnaces and the test tube removed from the tip. The combustion tube is then mounted vertically, with spiral at the bottom, in the clamp and stand used at the beginning of the determination. A previously weighed porcelain crucible with black interior[209,212] (Fig. 143) (see above under gravimetric Carius determination) is placed immediately under the tip to prevent loss upon dripping. (If nitrogen, phosphorus, and halogens are absent, the titrimetric procedure described later may be used.) About 4 ml. of 1:300 hydrochloric acid is added to the crucible, the tip of the tube immersed and the liquid sucked up (see above) until it is about 2 cm. above the spiral. The suction is removed, the tube raised about 2 cm. and the acid allowed to drain into the crucible. Several small portions of dilute acid are added from the top and the wash liquid allowed to drain into the crucible. The test tube previously used is also rinsed with a small amount of acid and this is added to the contents of the crucible. The test tube is then returned to the position under the spiral for further washing later. One milliliter of 10% barium chloride solution is added to the contents of the crucible to convert the sulfuric acid into barium sulfate. (*Caution:* The crucible should not be nearly full or the precipitate might creep up to the rim.) The crucible is then placed on a steam bath and the contents evaporated down to a few milliliters. In the meantime, the spiral is washed with several small portions of 1:300 hydrochloric acid, catching the washings in the test tube. These are eventually added to the contents of the crucible and the total liquid concentrated to a volume of 2 to 3 ml. The crucible is then allowed to cool. A previously weighed clean filter stick* (Fig. 144—see preceding pages) is attached to siphon, receiver, and inner container[209,212] (Figs. 147 and 148), inserted into the crucible and the liquid sucked off as described above. The

---

* Platinum filter sticks[8,209,212,225] (Fig. 156), having platinum sponge as the filter medium have been used successfully for this particular procedure.[128]

precipitate is then treated identically to that described for the gravimetric Carius determination (see preceding pages), that is, washing, drying, igniting, rewashing, redrying, reigniting, and weighing. The crucible and filter stick must be cleaned between determinations for this procedure (see preceding pages).

Calculation:

Same as for Gravimetric Carius Method.

FIG. 156. Platinum filter stick—details of construction. Tolerances on platinum (unless otherwise shown):

Height, mm.	± 0.5
Outside-inside diameter, mm.	± 0.5
Thickness, mm.	± 0.01
Weight, gram	± 0.5

Acidimetric (Direct Neutralization) Procedure
(Not Applicable to Nitrogen-, Halogen-, or Phosphorus-containing Compounds)

In the absence of nitrogen, phosphorus, and halogens, the spiral is washed with water (instead of hydrochloric acid as described above) catching the washings in a 125-ml. Erlenmeyer flask. Two drops of methyl red indicator (Chapter 5) are added, the liquid boiled for 30 seconds to remove carbon dioxide, and the sulfuric acid titrated with 0.01N sodium hydroxide to the end point (canary yellow for 2 minutes).[161,162,180]

Calculation:

1 ml. of 0.01N NaOH is equivalent to 0.1603 mg. of sulfur

$$\therefore \frac{\text{ml. of } 0.01N \text{ NaOH} \times 0.1603 \times 100}{\text{Wt. sample}} = \% \text{ S}$$

Example:

3.05 ml. of 0.01N NaOH is required to titrate the sulfuric acid formed from a 6.087-mg. sample

$$\therefore \frac{3.05 \times 0.1603 \times 100}{6.087} = 8.03\% \text{ S}$$

Volumetric Procedure

(Not Applicable to Phosphorus-containing Compounds)

In the absence of phosphorus, the THQ titration procedure may be used, *provided* that bromine (and *not* hydrogen peroxide[82,83]) is used in the spiral as the absorbent. The rest of the procedure is the same as for the volumetric Carius or Schöniger ones.

Calculation:

Same as for the Volumetric Carius Method.

ADDITIONAL INFORMATION FOR CHAPTER 10

Instead of the combustion tube with inner spiral, the combustion tube and absorber may be separate and connected by means of a ground joint[11] as shown in Fig. 157. With this system, the joint is heated to 350° C. by means of a third furnace or heater.

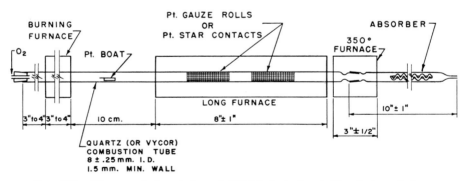

FIG. 157. Diagram of Association of Official Agricultural Chemists' catalytic combustion setup showing details. (Same setup *without* the 350° C. furnace used by that Association for determination of bromine and chlorine—see Chapter 11.)

This assembly was used in the procedures (volumetric and gravimetric) adopted[11,165-168] by the Association of Official Agricultural Chemists following collaborative studies. For the gravimetric, the precipitated barium sulfate was washed with five or six 3-ml portions of 1:300 HCl instead of the rewashing and reigniting before weighing.

TABLE 22

ADDITIONAL INFORMATION ON REFERENCES* RELATED TO CHAPTER 10

The determination of sulfur in organic compounds has been the subject of many investigations due to the tremendous importance of the sulfur compounds, particularly the sulfonamides, and in the first part of this chapter it was emphasized that the use of referee methods is often desirable. Table 22 lists a number of references which the author wishes to call to the attention of the reader. (See statement at top of Table 4 of Chapter 1, regarding completeness of this material.)

Books
Association of Official Agricultural Chemists, 11
Belcher and Godbert, 17, 18
Clark, E. P., 42
Clark, S. J., 43
Friedrich, 63
Furman, 67
Grant, 76, 77
Milton and Waters, 155, 156
Niederl and Niederl, 161, 162
Niederl and Sozzi, 163
Pregl, 180
Roth, 186–189
Steyermark, 209

Reviews
Alicino, 7
Hallett, 81
Horáček, 91
Kainz, 103
Lamo and Doadrio, 132
Thomson, 227
Willits, 246

Collaborative studies
Association of Official Agricultural Chemists, 11
Ogg, 165–168

General, miscellaneous
Batt, 13
Battles, 14
Bussmann, 35
Dixon, 47
Doležil, 48
Gorsuch, 73
Gouverneur and Van Dijk, 74
Horeischy and Bühler, 92

General, miscellaneous (Cont.)
Kent and Whitehouse, 104
Kirsten, 109, 111, 114
Kono, 121
Kress, 127
Lees and Folch, 134
Lincoln, Carney, and Wagner, 136
Malissa, 144
Pepkowitz, 176
Pepkowitz and Shirley, 177
Rieman and Hagen, 182
Rodden, 183
Romyn, 184
Večeřa and Šnobl, 233, 234
Volynskiĭ, 239
Volynskiĭ and Chudakova, 240

Nitrogen compounds
Belcher, Nutten, and Stephen, 21
Lysyj and Zarembo, 138
Steyermark, 209

Fluoro-compounds
Belcher and Macdonald, 20
Ma, 139
Neudorffer, 159
Rush, Cruikshank, and Rhodes, 192
Steyermark, Bass, Johnston, and Dell, 213

Phosphorus compounds
Association of Official Agricultural Chemists, 11
Fischer and Chen, 59
Lysyj and Zarembo, 138
Ogg, 168

Apparatus
Beazley, 15

* The numbers which appear after each entry in this table refer to the literature citations in the reference list at the end of the chapter.

10. Sulfur

TABLE 22 (*Continued*)

Apparatus (Cont.)

British Standards Institution, 33
Clark, E. P., 42
Ingram, 97
Kuck and Griffel, 129
Ma and Benedetti-Pichler, 140
Ma, Kaimowitz, and Benedetti-Pichler, 141
Martin and Deveraux, 146
Peters, Rounds, and Agazzi, 178
Royer, Alber, Hallett, and Kuck, 191

Submicro-, ultramicro-, microgram-methods

Belcher, Bhasin, Shah, and West, 16
Dunicz and Rosenqvist, 49
Granatelli, 75
Holeton and Linch, 89
Jacobs, Braverman, and Hochheiser, 100
Jones and Letham, 102
Kirsten, 108, 109, 113
Larsen, Ross, and Ingber, 133
Stratmann, 217
White, 245
Wilson and Straw, 249

Simultaneous determination of sulfur and other elements

Agazzi, Fredericks, and Brooks, 2
Belcher and Spooner, 22, 23
Boëtius, Gutbier, and Reith, 29
Etienne and Herrmann, 55
Fedoseev and Ivashova, 57
Friedrich, 63
Fujimoto, Utsui, and Ose, 66
Klimova and Bereznitskaya, 118
Korshun and Chumachenko, 124
Korshun and Terent'eva, 126
Margolis and Egorova, 145
Mizukami, Ieki, and Kondo, 157
Oda, Kubo, and Norimasa, 164
Roth, 189

Carius combustion

Carius, 37–40
Clark, E. P., 42
Emich and Donau, 53
Grant, 76, 77

Carius combustion (Cont.)

Horeischy and Bühler, 92
Kuck and Griffel, 129
Milton and Waters, 155, 156
Niederl, Baum, McCoy, and Kuck, 160
Niederl and Niederl, 161, 162
Roth, 186–189
Tanaka, 223
Yagi and Egami, 251

Oxygen flask combustion

Gildenberg, 71
Hempel, 87
Horáček, 91
Lysyj and Zarembo, 137
Mikl and Pech, 152–153
Ottosson and Snellman, 171
Schöniger, 196, 198–200
Soep, 206
Soep and Demoen, 207
Steyermark, Bass, Johnston, and Dell, 213
Thomas, 226

Empty tube technique, oxyhydrogen flame, catalytic combustion, etc.

Beazley, 15
Belcher and Ingram, 19
Etienne and Leger, 56
Grant, 76, 77
Graue and Zöhler, 79
Grote and Krekeler, 80
Hallett and Kuipers, 82
Heine, 85
Hudy and Mair, 93
Huffman, 94
Ingram, 96
Körbl and Pribil, 122
Lévy, 135
Makovetskii and Kholodkovskaya, 143
McChesney and Banks, 148
Niederl and Niederl, 161, 162
Pregl, 180
Roth, 186–189
Sakamoto, Hayazu, and Takenaka, 193
Stragand and Safford, 216
Sundberg and Royer, 219, 220
Večeřa, 230–232

TABLE 22 (*Continued*)

Empty tube technique, oxyhydrogen flame, catalytic combustion, etc. (Cont.)

Večeřa and Šnobl, 234, 235
Večeřa and Synek, 238
Wagner, 241
Walter, 243, 244
Wilson and Straw, 249
Zinneke, 259

Silver gauze absorbent technique

Bladh, Karrman, and Andersson, 27
Etienne and Leger, 56
Kuck and Grim, 130, 131
Stragand and Safford, 216
Sudo, Shimoe, Tsuji, and Soeda, 218
Večeřa and Šnobl, 234, 235
Zinneke, 259

Bombs, fusion

Agazzi, Parks, and Brooks, 3
Alicino, 6
Broekhuysen and Bechet, 34
Callan and Toennies, 36
Colson, 44
Elek and Hill, 52
Furman, 67
Inglis, 95
Kimball and Tufts, 107
Lincoln, Carney, and Wagner, 136
Mahoney and Michell, 142
Niederl and Niederl, 161, 162
Parr, 173
Peel, Clark, and Wagner, 174
Siegfriedt, Wiberley, and Moore, 202
Steyermark and Biava, 215
Wurzschmitt, 250

Potassium, sodium, magnesium, etc., fusion

Bussmann, 35
Dirscherl, 46
Hayazu, 84
Kirsten, 110
Kirsten and Carstens, 115
Klimenko, 117
Monand, 158
Reznik, 181

Potassium, sodium, magnesium, etc., fusion (Cont.)

Schöniger, 197
Večeřa and Spěvák, 236
Zimmermann, 255–258

Perchloric acid combustion, wet digestion with dichromate and nitric acid, chloric acid, etc.

Bethge, 25
McChesney and Banks, 148
Rosenthaler, 185
Szekeres, Fóti, and Pályi, 221
Tanaka, 223
Zdybek, McCann, and Boyle, 254

Hydrogenation methods

Furman, 67
Gel'man, 69
Irimescu and Chirnoaga, 98
Korshun and Gel'man, 125
Meulen, ter, 150
Meulen, ter, and Heslinga, 151
Stratmann, 217
Yudasina and Vysochina, 253

Manometric and gasometric procedures

See Chapter 18
Hoagland, 88
Holter and Løvtrup, 90

Various volumetric procedures

Bladh, Karrman, and Andersson, 27
Boos, 31
Bovee and Robinson, 32
Callan and Toennies, 36
Chalmers and Rigby, 41
Dirscherl, 46
Erdos, 54
Fischer and Chen, 59
Fritz and Freeland, 64
Fritz and Yamamura, 65
Geilmann and Bretschneider, 68
Gildenberg, 71
Hallett and Kuipers, 83
Inglis, 95
Iritani and Tanaka, 99

10. Sulfur

TABLE 22 (*Continued*)

Various volumetric procedures (*Cont.*)

Kirsten, 112, 114
Koch, Eckhard, and Malissa, 119
Kondo, 120
Lysyj and Zarembo, 137
Makovetskii and Kholodkovskaya, 143
Massie, 147
Milner, 154
Ottosson and Snellman, 171
Padowetz, 172
Pepkowitz, 175
Scalamandre and Guerrero, 195
Siegfriedt, Wiberley, and Moore, 202
Sirotenko, 203
Smith and Syme, 204
Soep and Demoen, 207
Soibel'man, 208
Tamiya, 222
Tanaka, 223
Tettweiler and Pilz, 224
Večeřa, 230–232
Večeřa and Šnobl, 233–235
Večeřa and Spěvák, 236, 237
Wagner, 241
Walter, 243
White, 245
Yamaji, 252
Zinneke, 259

EDTA titration

Bather, 12
Belcher, Bhasin, Shah, and West, 16
Belcher and Macdonald, 20
Wilson, Pearson, and Fitzgerald, 248

Indicators—THQ, and others

Abrahamzcik and Blümel, 1
Alicino, 5
Hallett and Kuipers, 82, 83
Ogg, Willits, and Cooper, 169
Smith-New York Company, 205
Sundberg and Royer, 220

Gravimetric procedures

Bladh, Karrman, and Andersson, 27
Bogan, 30

Gravimetric procedures (*Cont.*)

Callan and Toennies, 36
Etienne and Léger, 56
Fischer, 58
Fischer and Sprague, 60
Fiske, 61
Freri, 62
Heller, 86
Klein, 116
Lincoln, Carney, and Wagner, 136
Lysyj and Zarembo, 137
Ma and Benedetti-Pichler, 140
Ma, Kaimowitz, and Benedetti-Pichler, 141
Saschek, 194
Schulek, Pungor, and Guba, 201
Stragand and Safford, 216
Wagner and Miles, 242

Spectrophotometric, colorimetric, X-ray, etc., methods

Ahmed and Lawson, 4
Anderson, 10
Bergamini and Maltagliati, 24
Blanc, Bertrand and Liandier, 28
Broekhuysen and Bechet, 34
Eccleston and Whisman, 50
Gerhard and Johnstone, 70
Grassner, 78
Holeton and Linch, 89
Johnson and Nishita, 101
Jones and Letham, 102
Kiba, Akaza, and Taki, 105
Kiba and Kishi, 106
Koren and Gierlinger, 123
Lysyj and Zarembo, 138
Ory, Warren, and Williams, 170
Philips Electronics, Inc., 179
Roth, 190
Toennies and Bakay, 228
Trifonov, Ivanov, and Pavlov, 229
Večeřa and Spěvák, 237
Wagner, 241
Walter, 243
Zdybek, McCann, and Boyle, 254

REFERENCES

1. Abrahamczik, E., and Blümel, F., *Mikrochim. Acta*, **1**, 354 (1937).
2. Agazzi, E. J., Fredericks, E. M., and Brooks, F. R., *Anal. Chem.*, **30**, 1566 (1958).
3. Agazzi, E. J., Parks, T. D., and Brooks, F. R., *Anal. Chem.*, **23**, 1011 (1951).
4. Ahmed, M. N., and Lawson, G. J., *Talanta*, **1**, 142 (1958).
5. Alicino, J. F., *Anal. Chem.*, **20**, 85 (1948).
6. Alicino, J. F., *Ind. Eng. Chem., Anal. Ed.*, **13**, 506 (1941).
7. Alicino, J. F., *Microchem. J.*, **2**, 83 (1958).
8. American Platinum Works, Newark, New Jersey.
9. American Society for Testing Materials, *ASTM Designation*, **E 148-59T**.
10. Anderson, L., *Acta Chem. Scand.*, **7**, 689 (1953).
11. Association of Official Agricultural Chemists, "Official Methods of Analysis," 8th ed., pp. 801–811, Washington, D. C., 1955.
12. Bather, J. M., *Shirley Inst. Mem.*, **29**, 115 (1956).
13. Batt, W. G., *J. Franklin Inst.*, **248**, 451 (1949).
14. Battles, W., *Petrol. Engr.*, **27**, C-41 (1955).
15. Beazley, C. W., *Ind. Eng. Chem., Anal. Ed.*, **11**, 229 (1939).
16. Belcher, R., Bhasin, R. L., Shah, R. A., and West, T. S., *J. Chem. Soc.*, p. 4054 (1958).
17. Belcher, R., and Godbert, A. L., "Semi-Micro Quantitative Organic Analysis," Longmans, Green, London, New York, and Toronto, 1945.
18. Belcher, R., and Godbert, A. L., "Semi-Micro Quantitative Organic Analysis," 2nd ed., Longmans, Green, London, 1954.
19. Belcher, R., and Ingram, G., *Anal. Chim. Acta*, **7**, 319 (1952).
20. Belcher, R., and Macdonald, A. M. G., *Mikrochim. Acta*, p. 1187 (1956).
21. Belcher, R., Nutten, A. J., and Stephen, W. I., *Mikrochim. Acta*, p. 51 (1953).
22. Belcher, R., and Spooner, C. E., *Fuel*, **20**, 130 (1941).
23. Belcher, R., and Spooner, C. E., *Power Plant Eng.*, **46**, 58 (1942).
24. Bergamini, C., and Maltagliati, M., *Sperimentale, Sez. chim. biol.*, **6**, 65 (1956).
25. Bethge, P. O., *Anal. Chem.*, **28**, 119 (1956).
26. Betz, W. H., & L. D., Philadelphia, Pennsylvania.
27. Bladh, E., Karrman, K. J., and Andersson, O., *Mikrochim. Acta*, p. 60 (1958).
28. Blanc, P., Bertrand, P., and Liandier, L., *Chim. anal.*, **37**, 305 (1955).
29. Boëtius, M., Gutbier, B., and Reith, H., *Mikrochim. Acta*, p. 321 (1958).
30. Bogan, E. J., *Anal. Chem.*, **27**, 1505 (1955).
31. Boos, R. N., *Analyst*, **84**, 633 (1959).
32. Bovee, H. H., and Robinson, R. J., *Anal. Chem.*, **29**, 1353 (1957).
33. British Standards Institution, *Brit. Standards* **1428**, Pt. A3 (1952), Pt. A4 (1953), Pt. A5, and Pt. F1 (1957), and Part I1 (1953).
34. Broekhuysen, J., and Bechet, J., *Anal. Chim. Acta*, **13**, 277 (1955).
35. Bussmann, G., *Helv. Chim. Acta*, **33**, 1566 (1950).
36. Callan, T. P., and Toennies, G., *Ind. Eng. Chem., Anal. Ed.*, **13**, 450 (1941).
37. Carius, L., *Ann.*, **116**, 1 (1860).
38. Carius, L., *Ann.*, **136**, 129 (1865).
39. Carius, L., *Ann.*, **145**, 301 (1868).
40. Carius, L., *Ber.*, **3**, 697 (1870).
41. Chalmers, A., and Rigby, G. W., *Ind. Eng. Chem., Anal. Ed.*, **4**, 162 (1932).
42. Clark, E. P., "Semimicro Quantitative Organic Analysis," Academic Press, New York, 1943.

43. Clark, S. J., "Quantitative Methods of Organic Microanalysis," Butterworths, London, 1956.
44. Colson, A. F., *Analyst,* **67**, 47 (1942).
45. Corning Glass Works, Corning, New York.
46. Dirscherl, A., *Mikrochim. Acta,* p. 421 (1957).
47. Dixon, J. P., *Chem. & Ind. (London),* p. 156 (1959).
48. Doležil, M., *Collection Czechoslov. Chem. Communs.,* **22**, 396 (1957).
49. Dunicz, B. L., and Rosenqvist, T., *Anal. Chem.,* **24**, 404 (1952).
50. Eccleston, B. H., and Whisman, M. L., *Anal. Chem.,* **28**, 545 (1956).
51. Elek, A., and Harte, R. A., *Ind. Eng. Chem., Anal. Ed.,* **9**, 502 (1937).
52. Elek, A., and Hill, D. W., *J. Am. Chem. Soc.,* **55**, 3479 (1933).
53. Emich, F., and Donau, J., *Monatsh. Chem.,* **30**, 754 (1909).
54. Erdos, J., *Mikrochemie ver. Mikrochim. Acta,* **34**, 286 (1949).
55. Étienne, A., and Herrmann, J., *Chim. anal.,* **33**, 1 (1951).
56. Étienne, A., and Léger, J., *Chim. anal.,* **40**, 43 (1958); *Anal. Abstr.,* **5**, No. 3769 (1958); *Chem. Abstr.,* **52**, 8835 (1958).
56a. F. & M. Scientific Corp., New Castle, Delaware.
57. Fedoseev, P. N., and Ivashova, N. P., *Zhur. Anal. Khim.,* **13**, 230 (1958).
58. Fischer, R. B., *Anal. Chem.,* **23**, 1667 (1951).
59. Fischer, R. B., and Chen, W. K., *Anal. Chim. Acta,* **5**, 102 (1951).
60. Fischer, R. B., and Sprague, R. S., *Anal. Chim. Acta,* **5**, 98 (1951).
61. Fiske, C. H., *J. Biol. Chem.,* **47**, 59 (1921).
62. Freri, M., *Gazz. chim. ital.,* **76**, 108 (1946).
63. Friedrich, A., "Die Praxis der quantitativen organischen Mikroanalyse," p. 109, Deuticke, Vienna, and Leipzig, 1933.
64. Fritz, J. S., and Freeland, M. Q., *Anal. Chem.,* **26**, 1593 (1954).
65. Fritz, J. S., and Yamamura, S. S., *Anal. Chem.,* **27**, 1461 (1955).
66. Fujimoto, R., Utsui, Y., and Ose, S., *Yakugaku Zasshi,* **78**, 722 (1958); *Chem. Abstr.,* **52**, 16118 (1958).
67. Furman, N. H., ed., "Scott's Standard Methods of Chemical Analysis," 5th ed., Vol. II, Van Nostrand, New York, 1939.
68. Geilmann, W., and Bretschneider, G., *Z. anal. Chem.,* **139**, 412 (1953); *Anal. Abstr.,* **1**, No. 1248 (1954).
69. Gel'man, N. E., *Zavodskaya Lab.,* **8**, 673 (1939).
70. Gerhard, E. R., and Johnstone, H. F., *Anal. Chem.,* **27**, 702 (1955).
71. Gildenberg, L., *Microchem. J.,* **3**, 167 (1959).
72. Godfrey, P. R., and Shrewsbury, C. L., *J. Assoc. Offic. Agr. Chemists,* **28**, 336 (1945).
73. Gorsuch, T. T., *Analyst,* **81**, 501 (1956).
74. Gouverneur, P., and Van Dijk, H., *Anal. Chim. Acta,* **9**, 59 (1953).
75. Granatelli, L., *Anal. Chem.,* **31**, 434 (1959).
76. Grant, J., "Quantitative Organic Microanalysis, Based on the Methods of Fritz Pregl," 4th ed., Blakiston, Philadelphia, Pennsylvania, 1946.
77. Grant, J., "Quantitative Organic Microanalysis," 5th ed., Blakiston, Philadelphia, Pennsylvania, 1951.
78. Grassner, F., *Z. anal. Chem.,* **135**, 186 (1952).
79. Graue, G., and Zöhler, A., *Angew. Chem.,* **66**, 437 (1954).
80. Grote, W., and Krekeler, H., *Angew. Chem.,* **46**, 103 (1933).
81. Hallett, L. T., *Ind. Eng. Chem., Anal. Ed.,* **14**, 956 (1942).
82. Hallett, L. T., and Kuipers, J. W., *Ind. Eng. Chem., Anal. Ed.,* **12**, 357 (1940).

83. Hallett, L. T., and Kuipers, J. W., *Ind. Eng. Chem., Anal. Ed.,* **12**, 360 (1940).
84. Hayazu, R., *Yakugaku Zasshi,* **70**, 741 (1950).
85. Heine, V. E., *Pharmazie,* **9**, 972 (1954).
86. Heller, K., *Mikrochemie,* **7**, 208 (1929).
87. Hempel, W., *Z. angew. Chem.,* **13**, 393 (1892).
88. Hoagland, C. L., *J. Biol. Chem.,* **136**, 543 (1940).
89. Holeton, R. E., and Linch, A. L., *Anal. Chem.,* **22**, 819 (1950).
90. Holter, H., and Løvtrup, Søren, *Compt. rend. trav. lab. Carlsberg Sér. chim.,* **27**, 72 (1949).
91. Horáček, J., *Chem. listy,* **53**, 6 (1959).
92. Horeischy, K., and Bühler, F., *Mikrochemie ver. Mikrochim. Acta,* **33**, 231 (1948).
93. Hudy, J. A., and Mair, R. D., *Anal. Chem.,* **27**, 802 (1955).
94. Huffman, E. W. D., *Ind. Eng. Chem., Anal. Ed.,* **12**, 53 (1940).
95. Inglis, A. S., *Mikrochim. Acta,* p. 1834 (1956).
96. Ingram, G., *Analyst,* **69**, 265 (1944).
97. Ingram, G., *Mikrochim. Acta,* p. 877 (1956).
98. Irimescu, I., and Chirnoaga, E., *Z. anal. Chem.,* **128**, 71 (1947).
99. Iritani, N., and Tanaka, Y., *Kumamoto Pharm. Bull.,* p. 30 (1955).
100. Jacobs, M. B., Braverman, M. M., and Hochheiser, S., *Anal. Chem.,* **29**, 1349 (1957).
101. Johnson, C. M., and Nishita, H., *Anal. Chem.,* **24**, 736 (1952).
102. Jones, A. S., and Letham, D. S., *Analyst,* **81**, 15 (1956).
103. Kainz, G., *Österr. Chemiker-Ztg.,* **58**, 8 (1957).
104. Kent, P. W., and Whitehouse, M. W., *Analyst,* **80**, 630 (1955).
105. Kiba, T., Akaza (nee Kishi), I., and Taki, S., *Bull. Chem. Soc. Japan,* **30**, 482 (1957); *Chem. Abstr.,* **52**, 2646 (1958).
106. Kiba, T., and Kishi, I., *Bull. Chem. Soc. Japan,* **30**, 44 (1957).
107. Kimball, R. H., and Tufts, L. E., *Anal. Chem.,* **19**, 150 (1947).
108. Kirsten, W., *Chim. anal.,* **40**, 253 (1958).
109. Kirsten, W., *Microchem. J.,* **2**, 179 (1958).
110. Kirsten, W., *Mikrochemie ver. Mikrochim. Acta,* **34**, 151 (1949).
111. Kirsten, W., *Mikrochemie ver. Mikrochim. Acta,* **35**, 1 (1950).
112. Kirsten, W., *Mikrochemie ver. Mikrochim. Acta,* **35**, 174 (1950).
113. Kirsten, W. J., *Mikrochim. Acta,* p. 836 (1956).
114. Kirsten, W. J., "Proceedings of the International Symposium on Microchemistry 1958," p. 132, Pergamon, Oxford, London, New York, and Paris, 1960.
115. Kirsten, W., and Carstens, C., *Anal. Chim. Acta,* **5**, 272 (1951).
116. Klein, B., *Ind. Eng. Chem., Anal. Ed.,* **16**, 536 (1944).
117. Klimenko, V. G., *Biokhimiya,* **14**, 1 (1949).
118. Klimova, V. A., and Bereznitskaya, E. G., *Zhur. Anal. Khim.,* **12**, 424 (1957); *Chem. Abstr.,* **52**, 1853 (1958).
119. Koch, W., Eckhard, S., and Malissa, H., *Arch. Eisenhüttenw.,* **29**, 543 (1958); *Anal. Abstr.,* **6**, No. 2633 (1959).
120. Kondo, A., *Bunseki Kagaku,* **6**, 174 (1957); *Chem. Abstr.,* **52**, 15345 (1958).
121. Kono, T., *Nippon Nôgei-kagaku Kaishi,* **31**, 622 (1957).
122. Körbl, J., and Pribil, R., *Collection Czechoslov. Chem. Communs.,* **21**, 322 (1956).
123. Koren, H., and Gierlinger, W., *Mikrochim. Acta,* p. 220 (1953).
124. Korshun, M. O., and Chumachenko, N. M., *Doklady Akad. Nauk S.S.S.R.,* **99**, 769 (1954).
125. Korshun, M. O., and Gel'man, N. E., *Zavodskaya Lab.,* **12**, 754 (1946).

126. Korshun, M. O., and Terent'eva, Ev. A., *Doklady Akad. Nauk S.S.S.R.,* **100**, 707 (1955).
127. Kress, K. E., *Anal. Chem.,* **27**, 1618 (1955).
128. Kuck, J. A., Personal communication.
129. Kuck, J. A., and Griffel, M., *Ind. Eng. Chem., Anal. Ed.,* **12**, 125 (1940).
130. Kuck, J. A., and Grim, E. C., *Mikrochim. Acta,* p. 201 (1954).
131. Kuck, J. A., and Grim, E. C., *Mikrochim. Acta,* p. 361 (1957).
132. Lamo, M. A. de, and Doadrio, A., *Anales real soc. españ. fís. y quím. (Madrid) Ser. B.,* **44**, 723 (1948).
133. Larsen, R. P., Ross, L. E., and Ingber, N. M., *Anal. Chem.,* **31**, 1596 (1959).
134. Lees, M. B., and Folch, J., *Biochim. et Biophys. Acta,* **31**, 272 (1959).
135. Lévy, R., "Proceedings of the International Symposium on Microchemistry 1958," p. 112, Pergamon, Oxford, London, New York, and Paris, 1960.
136. Lincoln, R. M., Carney, A. S., and Wagner, E. C., *Ind. Eng. Chem., Anal. Ed.,* **13**, 358 (1941).
137. Lysyj, I., and Zarembo, J. E., *Anal. Chem.,* **30**, 428 (1958).
138. Lysyj, I., and Zarembo, J. E., *Microchem. J.,* **3**, 173 (1959).
139. Ma, T. S., *Microchem. J.,* **2**, 91 (1958).
140. Ma, T. S., and Benedetti-Pichler, A. A., *Anal. Chem.,* **25**, 999 (1953).
141. Ma, T. S., Kaimowitz, I., and Benedetti-Pichler, A. A., *Mikrochim. Acta,* p. 648 (1954).
142. Mahoney, J. F., and Michell, J. H., *Ind. Eng. Chem., Anal. Ed.,* **14**, 97 (1942).
143. Makovetskii, P. S., and Kholodkovskaya, K. B., *Khim. i Tekhnol. Topliva i Masel,* **3**, 71 (1958).
144. Malissa, H., *Mikrochim. Acta,* p. 127 (1960).
145. Margolis, E. I., and Egorova, N. F., *Vestnik Moskov. Univ., Ser. Mat., Mekhan., Astron., Fiz. i Khim.,* **13**, 209 (1958).
146. Martin, A. J., and Deveraux, H., *Anal. Chem.,* **31**, 1932 (1959).
147. Massie, W. H. S., *Analyst,* **82**, 352 (1957).
148. McChesney, E. W., and Banks, W. F., Jr., *Anal. Chem.,* **27**, 987 (1955).
149. Mellor, J. W., "Comprehensive Treatise on Inorganic and Theoretical Chemistry," Vol. X, pp. 215, 372, Longmans, Green, London, New York, and Toronto, 1930.
150. Meulen, H. ter, *Rec. trav. chim.,* **41**, 112 (1922).
151. Meulen, H. ter, and Heslinga, J., "Neue Methoden der organisch-chemischen Analyse," Akademische, Leipzig, 1927.
152. Mikl, O., and Pech, J., *Chem. listy,* **46**, 382 (1952); *Chem. Abstr.,* **47**, 70 (1953).
153. Mikl, O., and Pech, J., *Chem. listy,* **47**, 904 (1953); *Chem. Abstr.,* **48**, 3838 (1954).
154. Milner, O. I., *Anal. Chem.,* **24**, 1247 (1952).
155. Milton, R. F., and Waters, W. A., "Methods of Quantitative Microanalysis," Longmans, Green, New York, and Arnold, London, 1949.
156. Milton, R. F., and Waters, W. A., "Methods of Quantitative Microanalysis," 2nd ed., Arnold, London, 1955.
157. Mizukami, S., Ieki, T., and Kondo, H., *Yakugaku Zasshi,* **77**, 520 (1957).
158. Monand, P., *Bull. soc. chim. France,* **20**, 11, 1063 (1953).
159. Neudorffer, J., *Compt. rend. acad. sci,* **230**, 750 (1950).
160. Niederl, J. B., Baum, H., McCoy, J. S., and Kuck, J. A., *Ind. Eng. Chem., Anal. Ed.,* **12**, 428 (1940).
161. Niederl, J. B., and Niederl, V., "Micromethods of Quantitative Organic Elementary Analysis," Wiley, New York, 1938.

162. Niederl, J. B., and Niederl, V., "Micromethods of Quantitative Organic Analysis," 2nd ed., Wiley, New York, 1942.
163. Niederl, J. B., and Sozzi, J. A., "Microanálisis Elemental Orgánico," Calle Arcos, Buenos Aires, 1958.
164. Oda, N., Kubo, M., and Norimasa, K., *Yakugaku Zasshi*, **72**, 1079 (1952).
165. Ogg, C. L., J. *Assoc. Offic. Agr. Chemists*, **35**, 305 (1952).
166. Ogg, C. L., J. *Assoc. Offic. Agr. Chemists*, **36**, 335 (1953).
167. Ogg, C. L., J. *Assoc. Offic. Agr. Chemists*, **38**, 365 (1955).
168. Ogg, C. L., J. *Assoc. Offic. Agr. Chemists*, **38**, 377 (1955).
169. Ogg, C. L., Willits, C. O., and Cooper, F. J., *Anal. Chem.*, **20**, 83 (1948).
170. Ory, H. A., Warren, V. L., and Williams, H. G., *Analyst*, **82**, 189 (1957).
171. Ottosson, R., and Snellman, O., *Acta Chem. Scand.*, **11**, 185 (1957); *Chem. Abstr.*, **52**, 8835 (1958).
172. Padowetz, W., *Mikrochemie ver. Mikrochim. Acta*, **36/37**, 648 (1951).
173. Parr, S. W., *J. Am. Chem. Soc.*, **30**, 764 (1903).
174. Peel, E. W., Clark, R. H., and Wagner, E. C., *Ind. Eng. Chem., Anal. Ed.*, **15**, 149 (1943).
175. Pepkowitz, L. P., *Anal. Chem.*, **20**, 968 (1948).
176. Pepkowitz, L. P., Included in Manhattan Project Classified Report LA-416 (1945).
177. Pepkowitz, L. P., and Shirley, E. L., *Anal. Chem.*, **23**, 1709 (1951).
178. Peters, E. D., Rounds, G. C., and Agazzi, E. J., *Anal. Chem.*, **24**, 710 (1952).
179. Philips Electronics, Inc., Mount Vernon, New York.
180. Pregl, F., "Quantitative Organic Microanalysis" (E. Fyleman, trans., 2nd German ed.), Churchill, London, 1924.
181. Reznik, B. A., *Zavodskaya Lab.*, **16**, 363 (1950).
182. Rieman, Wm., III, and Hagen, G., *Ind. Eng. Chem., Anal. Ed.*, **14**, 150 (1942).
183. Rodden, C. J., ed. "Analytical Chemistry of the Manhattan Project," McGraw-Hill, New York, 1950.
184. Romyn, H. M., *Pharm. Weekblad*, **93**, 929 (1958); *Anal. Abstr.*, **6**, No. 3009 (1959).
185. Rosenthaler, L., *Pharm. Acta Helv.*, **30**, 282 (1955).
186. Roth, H., "Die quantitative organische Mikroanalyse von Fritz Pregl," 4th ed., Springer, Berlin, 1935.
187. Roth, H., "F. Pregl quantitative organische Mikroanalyse," 5th ed., Springer, Wien, 1947.
188. Roth, H., "Pregl-Roth quantitative organische Mikroanalyse," 7th ed., Springer, Wien, 1958.
189. Roth, H., "Quantitative Organic Microanalysis of Fritz Pregl," 3rd ed., (E. B. Daw, trans., 4th German ed.), Blakiston, Philadelphia, Pennsylvania, 1937.
190. Roth, H., *Mikrochemie ver. Mikrochim. Acta*, **36/37**, 379 (1951).
191. Royer, G. L., Alber, H. K., Hallett, L. T., and Kuck, J. A., *Ind. Eng. Chem., Anal. Ed.*, **15**, 230 (1943).
192. Rush, C. A., Cruikshank, S. S., and Rhodes, E. J. H., *Mikrochim. Acta*, p. 858 (1956).
193. Sakamoto, S., Hayazu, R., and Takenaka, R., *Yakugaku Zasshi*, **70**, 287 (1950).
194. Saschek, W., *Ind. Eng. Chem., Anal. Ed.*, **9**, 491 (1937).
195. Scalamandre, A. A., and Guerrero, A. H., *Anales asoc. quím. arg.*, **39**, 28 (1951).
196. Schöniger, W., Regional Analytical Symposium, Philadelphia, Pennsylvania, February 1957.
197. Schöniger, W., *Mikrochim. Acta*, p. 74 (1954).

198. Schöniger, W., *Mikrochim. Acta,* p. 123 (1955).
199. Schöniger, W., *Mikrochim. Acta,* p. 869 (1956).
200. Schöniger, W., "Proceedings of the International Symposium on Microchemistry 1958," p. 93, Pergamon, Oxford, London, New York, and Paris, 1960.
201. Schulek, E., Pungor, E., and Guba, F., *Anal. Chim. Acta,* **10**, 506 (1954).
202. Siegfriedt, R. K., Wiberley, J. S., and Moore, R. W., *Anal. Chem.,* **23**, 1008 (1951).
203. Sirotenko, A. A., *Mikrochim. Acta,* p. 153 (1955).
204. Smith, J., and Syme, A. C., *Analyst,* **81**, 302 (1956).
205. Smith-New York Company, Inc., "Titration of Sulfate with Tetrahydroxyquinone," Freeport, Long Island, New York, 1948.
206. Soep, H., *Mededel. Vlaam. Chem. Ver.,* **21**, 31 (1959); *Chem. Abstr.,* **53**, 16804 (1959).
207. Soep, H., and Demoen, P., *Microchem. J.,* **4**, 77 (1960).
208. Soibel'man, B. I., *Zhur. Anal. Khim.,* **3**, 258 (1948).
209. Steyermark, Al, "Quantitative Organic Microanalysis," Blakiston, Philadelphia, Pennsylvania, 1951.
210. Steyermark, Al, *Anal. Chem.,* **22**, 1228 (1950).
211. Steyermark, Al, Alber, H. K., Aluise, V. A., Huffman, E. W. D., Jolley, E. L., Kuck, J. A., Moran, J. J., and Willits, C. O., *Anal. Chem.,* **23**, 1689 (1951).
212. Steyermark, Al, Alber, H. K., Aluise, V. A., Huffman, E. W. D., Kuck, J. A., Moran, J. J., and Willits, C. O., *Anal. Chem.,* **21**, 1555 (1949).
213. Steyermark, Al, Bass, E. A., Johnston, C. C., and Dell, J. C., *Microchem. J.,* **4**, 55 (1960).
214. Steyermark, Al, Bass, E. A., and Littman, B., *Anal. Chem.,* **20**, 587 (1948).
215. Steyermark, Al, and Biava, F. P., *Anal. Chem.,* **30**, 1579 (1958).
216. Stragand, G. L., and Safford, H. W., *Anal. Chem.,* **21**, 625 (1949).
217. Stratmann, H., *Mikrochim. Acta,* p. 1031 (1956).
218. Sudo, T., Shimoe, D., Tsuji, T., and Soeda, Y., *Bunseki Kagaku,* **8**, 42 (1959); *Anal. Abstr.,* **7**, No. 169 (1960).
219. Sundberg, O. E., and Royer, G. L., *Anal. Chem.,* **24**, 907 (1952).
220. Sundberg, O. E., and Royer, G. L., *Ind. Eng. Chem., Anal. Ed.,* **18**, 719 (1946).
221. Szekeres, L., Fóti, G., and Pályi, E., *Magyar Kém. Folyóirat,* **56**, 377 (1950).
222. Tamiya, N., *Seikagaku,* **22**, 59 (1950).
223. Tanaka, Y., *Yakugaku Zasshi,* **75**, 653 (1955).
224. Tettweiler, K., and Pilz, W., *Naturwissenschaften,* **41**, 332 (1954).
225. Thomas, Arthur H., Company, Philadelphia, Pennsylvania.
226. Thomas, Arthur H., Company, "Technical Service, DU 37-5M-8'58," Philadelphia, Pennsylvania, 1958.
227. Thomson, M. L., *Metallurgia,* **39**, 46 (1948).
228. Toennies, G., and Bakay, V., *Anal. Chem.,* **25**, 160 (1953).
229. Trifonov, A., Ivanov, C. P., and Pavlov, D. N., *Izvest. Khim. Inst. Bulgar. Akad. Nauk,* **3**, 3 (1955).
230. Večeřa, M., *Chem. listy,* **48**, 613 (1954).
231. Večeřa, M., *Chem. listy,* **50**, 308 (1956).
232. Večeřa, M., *Mikrochim. Acta,* p. 88 (1955).
233. Večeřa, M., and Šnobl, D., *Chem. listy,* **51**, 1482 (1957).
234. Večeřa, M., and Šnobl, D., *Collection Czechoslov. Chem. Communs.,* **22**, 986 (1957); *Chem. Abstr.,* **52**, 6056 (1958).
235. Večeřa, M., and Šnobl, D., *Mikrochim. Acta,* p. 28 (1958).

236. Večeřa, M., and Spěvák, A., *Chem. listy,* **50**, 765 (1956).
237. Večeřa, M., and Spěvák, A., *Collection Czechoslov. Chem. Communs.,* **21**, 1278 (1956).
238. Večeřa, M., and Synek, L., *Chem. listy,* **51**, 171 (1957).
239. Volynskiǐ, N., *Zavodskaya Lab.,* **21**, 536 (1955).
240. Volynskiǐ, N. P., and Chudakova, I. K., U.S.S.R. Patent 113,669 (1959); *Chem. Abstr.,* **53**, 2943 (1959).
241. Wagner, H., *Mikrochim. Acta,* p. 19 (1957).
242. Wagner, E. C., and Miles, S. H., *Anal. Chem.,* **19**, 274 (1947).
243. Walter, R. N., *Anal. Chem.,* **22**, 1332 (1950).
244. Walter, R. N., Personal communication. 1951.
245. White, D. C., *Mikrochim. Acta,* p. 254 (1959).
246. Willits, C. O., *Anal. Chem.,* **21**, 132 (1949).
247. Willits, C. O., and Ogg, C. L., Personal communication. 1951.
248. Wilson, H. N., Pearson, R. M., and Fitzgerald, D. M., *J. Appl. Chem. (London),* **4**, 488 (1954).
249. Wilson, H. N., and Straw, H. T., *J. Soc. Chem. Ind. (London),* **69**, 79 (1950).
250. Wurzschmitt, B., *Mikrochemie ver. Mikrochim. Acta,* **36/37**, 769 (1951).
251. Yagi, Y., and Egami, F., *Nippon Kagaku Zasshi,* **67**, 20 (1946).
252. Yamaji, I., *Tokyo Kôgyô Shikensho Hôkoku,* **50**, 203 (1955).
253. Yudasina, A. G., and Vysochina, L. D., *Nauch. Zapiski. Dnepropetrovsk. Gosudarst. Univ.,* **43**, 53 (1953); *Referat. Zhur. Khim.,* No. 15,061 (1954).
254. Zdybek, G., McCann, D. S., and Boyle, A. J., *Anal. Chem.,* **32**, 558 (1960).
255. Zimmermann, W., *Mikrochemie ver. Mikrochim. Acta,* **31**, 15 (1944).
256. Zimmermann, W., *Mikrochemie ver. Mikrochim. Acta,* **33**, 122 (1948).
257. Zimmermann, W., *Mikrochemie ver. Mikrochim. Acta,* **35**, 80 (1950).
258. Zimmermann, W., *Mikrochemie ver. Mikrochim. Acta,* **40**, 162 (1953).
259. Zinneke, F., *Z. anal. Chem.,* **132**, 175 (1951).

CHAPTER II

Microdetermination of Halogens

DETERMINATION OF BROMINE, CHLORINE, AND IODINE

The organic material may be destroyed by the same methods* used for the determination of sulfur in the preceding chapter, but the method of choice, in the opinion of the author, is the Carius.[64-67,379] This method has been used extensively and its reliability is shown by the fact that it has been adopted by the Association of Official Agricultural Chemists after collaborative studies.[10,299,375,376,383,384] However, since the author believes that there should be available another method for referee purposes, such is described. The Pregl[313,379] method given below also proved reliable in collaborative studies.[10,299,383,384] In case of a controversy, *both* methods should be used. In a few isolated cases one method will give better results than the other.

After decomposition and conversion to organic halide, the silver bromide, chloride, or iodide is precipitated and determined gravimetrically. Fluorine is not determined by this method, since silver fluoride is *very*[275] soluble. The presence of fluorine also does not interfere.

CARIUS METHOD

When organic compounds containing bromine, chlorine, or iodine are oxidized with nitric acid, in the presence of silver nitrate, the organic material is destroyed and the halogen is converted to the corresponding silver halide, thusly (compare oxidation of sulfur-containing compounds—Chapter 10):

$$\text{Organic X} \xrightarrow[\substack{[O] \\ (\text{Excess HNO}_3)}]{\text{AgNO}_3} CO_2 + H_2O + \underline{AgX}$$

This is known as the Carius method for halogens,[64-67,97,108,219,292] just as that described in Chapter 10 is the Carius method for sulfur, both being named after

---

* Please see references 5, 26, 61, 64–67, 78, 103, 104, 106, 118, 127, 130, 139, 140, 143, 144, 147, 148, 156, 166, 178, 193, 197, 216, 230, 244, 252, 293, 294, 303, 306, 313, 322–325, 339, 341–344, 358, 379, 394, 395, 398, 399, 401, 402, 418–421, 426, 428.

the discoverer.[64-67] The method is applicable to chlorine, bromine, and iodine, whether ionic or not, and all are determined in the same way.

Reagents

Obviously, all reagents and wash liquids must be halogen-free. It is advisable to check this point qualitatively before attempting determinations.

SILVER NITRATE

Reagent grade of silver nitrate crystals are used in the Carius combustion tubes to react with the halogens to form the silver halides.

FUMING NITRIC ACID

Reagent grade of fuming nitric acid, sp. gr. 1.49 to 1.50 is used to burn the sample. (*Caution:* As stated in Chapter 10, this should be handled with extreme care.)

DILUTE NITRIC ACID

A dilute solution of nitric acid, 1:200, is used for washing the precipitate.

ETHANOL, 95%

POTASSIUM IODIDE SOLUTION, 30%[294]

A 30% solution of potassium iodide is prepared, filtered and kept in a brown glass-stoppered bottle. This is used only in the determinations in which it is known that the precipitate is contaminated with glass. It is a solvent for the precipitate.

Apparatus

CARIUS COMBUSTION FURNACE[380]

The furnace used is that which was described in the previous chapter on the Determination of Sulfur (Chapter 10, Fig. 135).

CARIUS COMBUSTION TUBES[380]

The combustion tubes are the same as are used for the Determination of Sulfur (Chapter 10, Fig. 137).

FILTER TUBE[139,140,293,294,313,322-325,379,381]

The filter tube used for collecting and weighing the precipitate of silver halide is shown in Fig. 158.[379,381] It is constructed of glass and has a sintered filter plate of the same material. Tests have shown that this plate should have an average pore diameter of 15–25 μ[379,381] The tubes of this type made in America, at present, are all of Pyrex glass.[85] These are the exception to the rule stated in Chapter 3, that the glass objects weighed on the balance should be constructed of soda-lime glass. Tubes of the latter type are manufactured in Europe, and are quite satisfactory.

As an alternate to the sintered glass type of filter, soda-lime glass ones (Fig. 159), may be used with an asbestos matting to hold the precipitate.[313,379] These tubes are made essentially like the sintered plate type, except that there

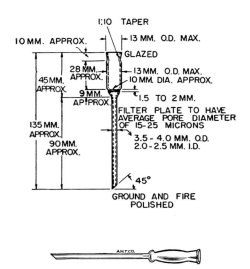

FIG. 158. (*Top*) Filter tube—details of construction.
FIG. 159. (*Bottom*) Filter tube without sintered plate.

is a constriction into which is placed a small wad of glass wool or a glass bead as a support for the asbestos. Ordinary acid-washed Gooch asbestos is then packed down to form a mat several millimeters thick. However, the sinter plate type is so much more convenient that the asbestos type should only be used in case the former are not available.

Before using a new filter, it should be cleaned by passing 1:1 nitric acid through it. (Chromic-sulfuric acid cleaning solution or strong caustic should never be used.) It is then good practice to filter a few milligrams of silver halide onto the filter to make certain that the precipitate will not pass through. The filter tube is then cleaned with 30% potassium iodide, washed well with water, 1:200 nitric acid, and ethanol, in this order and dried in an oven at 120° C. for 30 minutes. If the Pyrex variety is used, the tube should be wiped with a chamois skin (Chapter 3) *before* drying so that the electrostatic charge will be lost by the time the tube is placed on the balance. After the tube is removed from the oven it is allowed to cool for 45 minutes in the balance room and then weighed using a marked tare flask (Figs. 34–36) as a counterpoise weight.

FILTRATION ASSEMBLY[379,381]

The filtration assembly shown in Fig. 160 is used in the transfer of the precipitate from the Carius combustion tube to the filter tube. (Figure 160 shows the setup with a test tube, which is used in connection with the Pregl catalytic

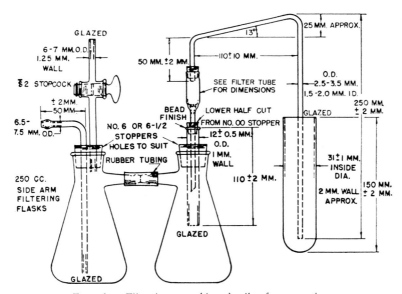

FIG. 160. Filtration assembly—details of construction.

combustion. With the Carius combustion, the opened combustion tube is used instead of the test tube, since the precipitate is transferred *directly* from the combustion tube to the filter.) It consists of one suction flask into which fits the filter tube and a second suction flask which acts as a trap between the former and the source of vacuum. A bent tube attached to the top of the filter tube by means of a one-hole rubber stopper (preferably the universal type,[7,374,379] Fig. 153) acts as the connecting link to the Carius (or test) tube containing the precipitate. When suction is applied the silver halide is drawn over into the filter tube, affecting a quantitative transfer. The precipitate is likewise washed by the addition of the liquids to the Carius (or test) tube instead of directly to the filter tube.

11. Halogens

BLACK PAPER

Black paper is used as a cover for the filter tube during the filtration of silver chloride or bromide, to protect these from the light. Silver iodide precipitates need no such protection. The paper should be wrapped around the tube and held in place by a rubber band or clamp.

OVEN

Any suitable electric laboratory oven is satisfactory for drying the silver halide precipitate at 120° C.

Procedure

Five to 8 mg. of sample is added to the Carius combustion tube exactly as described in the determination of sulfur [Chapter 10 (see also Chapter 3 for weighing the sample)]. About 20 mg. of silver nitrate crystals is added to the Carius tube and the latter gently tapped to cause adhering material to drop to the bottom. Electrified particles often stick to the sides, but grounding against a water pipe corrects this condition. Five- to six-tenths (seven-tenths *maximum*) milliliters (see Table 21, Chapter 10) of fuming nitric acid, sp. gr. 1.49–1.50, is added while the tube is rotated to wash down any remaining adhering particles. (If the sample and acid react at room temperature, the bottom of the tube *and* the acid should first be cooled in dry ice-acetone mixture. As an alternate procedure, the sample may be weighed in a weighing bottle (Fig. 27, Chapter 3), inserted in the *tilted* combustion tube *after* the acid has been added and kept from sliding to the bottom until *after* the tube has been sealed). The Carius tube is immediately sealed in the manner described in detail in the determination of sulfur (Chapter 10). The closed tube is then heated in the Carius furnace for 7–8 hours at 250° C. after which it is allowed to cool, the pressure released, and then opened exactly as described in the above-mentioned chapter. However, the Carius tubes for the halogen determination are cut as near to the top as possible, instead of about 75 mm. from the bottom as was recommended in the sulfur determination. (The extra length is desirable for the halogens, since the contents are diluted with considerably more water than in the sulfur determination. Also whereas the contents of the tube are poured out in the determination of sulfur, they are removed in the halogen determination by inserting the tube connected to the filter which in turn is attached to a vacuum system.) A few drops of water are added to dilute the nitric acid and this is followed, with *cautious mixing*, by enough (water) to fill the opened Carius tube to within about 3–4 cm. of the top. It is then placed in a hot water or steam bath, protected from the light,* for about 20 minutes,

---

* Silver iodide does not need to be protected from the light.

if the precipitate is either chloride or bromide. This period of digestion is more than sufficient, since the precipitate is already coagulated during the combustion. In the case of iodide, a longer digestion is usually required to break up the eutectic mixture of silver nitrate-silver iodide formed during the combustion. The eutectic mixture melts below the temperature of the steam bath and persists as a heavy yellow oil on the bottom of the tube. Stirring with a glass rod speeds up the solution of the silver nitrate and greatly reduces the time of digestion, which must be continued until the precipitate is in the form of a fine powder. [If excessive amounts of silver nitrate have been used in the Carius combustion tubes, the solutions must be diluted to a greater volume than recommended above to obtain complete precipitation of the silver halide (compare Mellor[275]). This is particularly true in the determination of iodine. Consequently, after the precipitate has coagulated, a small amount of the supernatant liquid should be quantitatively transferred with the aid of a pipette into a test tube containing a few milliliters of water. If there is no additional precipitation on dilution, the small removed portion may be discarded and the contents of the Carius tube treated as below. Otherwise, the entire contents of the Carius tube should be quantitatively transferred to an 8-inch test tube, about 10 ml. (or more, if necessary) of distilled water added, and the contents heated on the water bath until the newly formed precipitate coagulates.]

The precipitate is then transferred to the previously weighed filter tube in the following manner. The filter tube* is placed in the one-hole rubber stopper of the suction flask. The short side of the long bent connecting tube (or siphon) is placed in the top of the filter tube with the aid of a one-hole rubber stopper. With the stopcock opened, a source of vacuum is attached to the horizontal tube on the trap flask. The Carius combustion tube, containing the diluted nitric acid and coagulated precipitate (after digestion in the hot-water or steam bath), is placed under the tip of the long side of the bent connecting tube (or siphon) and then raised and supported so that the tip extends to the *bottom*† and surrounds the precipitate. (Position somewhat similar to that occupied by the test tube in Fig. 160. The test tube is used when the sample is subjected to catalytic combustion—see following pages.) The stopcock is then partly closed so that there is just enough suction to draw the precipitate (first) and supernatant liquid up the siphon tube, around, and down into the filter tube at the rate of about two drops per second. The rate is controlled, by means of the stopcock, so that the filter tube never becomes more than about one-half full of liquid. Should this occur, the Carius tube is lowered, for a moment or two, to stop the passage of liquid upward in

---

* Covered with dark paper if filtering chloride or bromide.
† By so doing, the precipitate is first drawn over followed by the liquid. This ensures quantitative transfer with a minimum of wash liquids.

11. Halogens

the siphon tube. As soon as all of the precipitate and the liquid have been transferred from the Carius to the filter tube, a few milliliters of 1:200 nitric acid are added to the Carius tube, using a wash bottle of the type shown in Fig. 145[379,381] or Fig. 146. This is sucked over onto the filter and a few milliliters of ethanol are added in the same manner. The washing, alternately with 1:200 nitric acid and ethanol is repeated several times to ensure complete transfer of the precipitate. (Alternating the liquids is more efficient than the use of either singly.) The stopper and siphon tube are removed from the filter tube and washed first with a stream of 1:200 nitric acid and then with one of ethanol, catching the washings in the filter tube. The precipitate is finally washed with a little ethanol and sucked dry.

The filter tube is then carefully removed from the filtration assembly and the tip cleaned with a cotton swab on the end of a knurled iron wire. The entire tube is then gently, but completely, wiped with a chamois skin (see Chapter 3), care being exercised so that the precipitate does not "jump" up during the wiping and be lost. The tube is grounded and then placed in an oven and the precipitate dried at 120° C. for about 30 minutes. Then the filter is removed to the balance room, handling it only with forceps or chamois skin, and allowed to cool (protected from light, if chloride or bromide). It is weighed after 45 minutes. If the precipitate has become contaminated with glass, the silver halide is dissolved by filling the filter tube with 30% KI solution, letting stand awhile, sucking off, repeating, washing with water, 1:200 nitric acid, and ethanol (in this order), drying at 120° C. for 30 minutes, and reweighing. The difference in weight between before and after cleaning represents the silver halide. (If a number of determinations are being done, the filter need not be cleaned, before filtering the next precipitate, provided adequate precautions are taken. Under these conditions, naturally, the *gain* in weight from the *previous* determination is used in the calculation.)

Calculation:

Factors:

$$\frac{Cl}{AgCl} = 0.2474$$

$$\frac{Br}{AgBr} = 0.4255$$

$$\frac{I}{AgI} = 0.5405$$

$$\therefore \frac{\text{Wt. precipitate} \times \text{factor} \times 100}{\text{Wt. sample}} = \% \text{ Halogen}$$

Examples:

(a) 5.782 mg. of AgCl was obtained from an 8.623-mg. sample

$$\therefore \frac{5.782 \times 0.2474 \times 100}{8.623} = 16.59\% \text{ Cl}$$

(b) 5.202 mg. of AgBr was obtained from a 7.301-mg. sample

$$\therefore \frac{5.202 \times 0.4255 \times 100}{7.301} = 30.32\% \text{ Br}$$

(c) 6.001 mg. of AgI was obtained from a 5.906-mg. sample

$$\therefore \frac{6.001 \times 0.5405 \times 100}{5.906} = 54.92\% \text{ I}$$

The allowable error of the determination is ± 0.3%.

PREGL CATALYTIC COMBUSTION METHOD

In the beginning of this chapter it was stated that the destruction of the organic material may be accomplished by other methods, but that the author prefers the Carius (compare Chapter 10). Foremost among these is the Pregl catalytic combustion (see Chapter 10) which the author used during the first few years of the operation of his laboratory and which gave good results (for bromine and chlorine) in a collaborative study.[10,299,383,384] The sample is burned in an atmosphere of oxygen at red heat, in the presence of platinum, and the resulting halogen absorbed in an alkaline-reducing medium. The reducing agent is oxidized, the resulting solution acidified, the silver halide precipitated, and determined gravimetrically. The reactions represented are the following:

$$\text{Organic X} \xrightarrow[\text{O}_2 \text{ (Pt)}]{\Delta} X_2 + CO_2 + H_2O$$
$$(X = Cl, Br, I)$$

$$X_2 \xrightarrow[\text{(Hydrazine)}]{Na_2CO_3} 2NaX \quad [4]$$

$$2NaX + 2AgNO_3 \rightarrow 2NaNO_3 + \underline{2AgX}$$

Reagents

As stated in the first part of this chapter, all reagents must be halogen-free.

SODIUM CARBONATE[293,294,313,379]

Twenty-five grams of reagent grade sodium carbonate is dissolved in 100 ml. of water. This is used as the absorbing agent for the liberated halogens.

HYDRAZINE SULFATE[4,103,104]

A saturated solution of pure hydrazine sulfate in water is used as the reducing agent.

DILUTED NITRIC ACID SOLUTION

Concentrated nitric acid, sp. gr. 1.42, is diluted, 1:200, with water. This is used as a wash liquid.

HYDROGEN PEROXIDE, 30% (SUPEROXOL)

Reagent grade of 30% hydrogen peroxide (Superoxol) is used as an oxidizing agent. This substance is stored in a refrigerator. *It must be handled with extreme care.*

CONCENTRATED NITRIC ACID

Reagent grade of concentrated nitric acid, sp. gr. 1.42 is used to ensure an excess of acid previous to precipitation.

SILVER NITRATE

A 5% solution of reagent grade of silver nitrate in water is used to precipitate the halide. This solution is stored in a brown glass-stoppered bottle.

ETHANOL, 95%

Apparatus

COMBUSTION TRAIN[139,140,293,294,313,322-325,379]

This is identical with that used for the Pregl catalytic combustion method for the determination of sulfur (Chapter 10). It consists of the following:

Thin-walled rubber tubing	Combustion tube with inner spiral
Oxygen cylinder	Two platinum contact stars
Pressure regulator	Combustion apparatus
Bubble counter-U-tube	Mariotte bottle

TEST TUBE

A standard 8-inch Pyrex[85] test tube is used for collecting the washings and for the precipitation of the silver halide.

FILTER TUBE[379,381]

The same filter tube is used that was described in the early part of this chapter.

FILTRATION ASSEMBLY[379,381]

The same assembly is used that was described in the early part of this chapter.

OVEN

Any standard laboratory oven may be used for drying the precipitate as described previously in this chapter.

Assembly and Procedure

The assembly[139,140,293,294,313,322–325,379] of the combustion train is done exactly as described for the Pregl catalytic combustion method for the determination of sulfur (Chapter 10).

Approximately 4–5 ml. of sodium carbonate solution and four to five drops of hydrazine sulfate[4,103,104] solution are mixed in a standard 8-inch Pyrex test tube. The combustion tube is held in a vertical position, by a suitable clamp and stand, and the tip immersed in the sodium carbonate-hydrazine sulfate mixture. Mild suction is applied to the end of the combustion tube, while protecting it from contamination with a cotton-filled filter tube of the type shown in Fig. 154. The alkaline mixture (sodium carbonate-hydrazine sulfate) is carefully sucked up so that the liquid covers the entire spiral and extends about 5–7 mm. above it. The suction is then removed and the liquid allowed to drain back into the test tube, leaving the spiral wet. The alkaline mixture in the test tube is poured off and discarded, but the test tube is not washed since it is to be used for the precipitation following the combustion. The tube is then placed in the combustion furnace, platinum contact stars put in place, the sample (5–9 mg.) added and burned identically as described for the catalytic combustion method for the determination of sulfur.

Chlorine and bromine present no difficulties, but iodine crystals collect, after the combustion, in the tube just beyond the long furnace and must be driven, by very careful heating, into the alkaline mixture on the spiral. This is accomplished by the use of a small gas flame or by *slowly* moving the combustion tube in the direction that brings the end of the spiral closer to the long furnace. (The spiral should not be brought into the red hot area as it is not made of combustion tube type of glass and may melt.)

The absorbed halogen is washed quantitatively into the test tube with distilled water by sucking water up above the spiral as well as washing down from the top in the manner described for the catalytic combustion method for the determination of sulfur (see Chapter 10). The combined washings in the test tube are treated with two to three drops of 30% hydrogen peroxide (Superoxol) and then heated on a steam bath for 5 minutes. The test tube contents are then cooled under the cold water tap and 2 ml. of concentrated nitric acid, sp. gr. 1.42, is added, followed by 2 ml.. of 5% silver nitrate solution. The test tube is again heated on the steam bath (protecting from the light if chloride or bromide are present) until the precipitate is completely coagulated. This usually requires at least one hour. The contents of the tube are again cooled and the precipitate transferred to a filter tube (Fig. 158 or 159) washed, dried, and weighed as described under the Carius method.

Calculation:

Same as for the Carius method.

SIMULTANEOUS DETERMINATION OF CHLORINE AND BROMINE

Chlorine and bromine may be determined simultaneously by the method of Moser and Miksch.[284,294,325] The two silver halides are collected in a microporcelain filter (Fig. 166, Chapter 13) using the crucible filter assembly (Fig. 167, Chapter 13). For the description of these and the method of using, see determination of arsenic, gravimetric method (Chapter 13). The mixed halide precipitate is weighed and then treated with about six–seven times its weight of ammonium iodide (or bromide). The mixture is heated to 300° C. and the mixture is converted to silver iodide (or bromide) and reweighed. To ensure complete conversion, the treatment with ammonium iodide (or bromide) is repeated and the precipitate reweighed (no gain in weight after last treatment means complete conversion). The increase in weight is then used in the calculation of the percentages of chlorine and bromine.

DETERMINATION OF FLUORINE

The determination of fluorine is based on the destruction of organic material at high temperature in an atmosphere of oxygen and conversion of the organic fluorine to hydrofluoric (or fluosilicic) acid and subsequent titration with thorium nitrate using sodium alizarin sulfonate as the indicator:

$$\text{Organic F} \xrightarrow{\underset{(H_2O)}{O_2}} CO_2 + H_2O + HF$$

and

$$HF \text{ (or } H_2SiF_6) \xrightarrow{Th(NO_3)_4} ThF_4$$

No interference is observed from nitrogen, bromine, chlorine, iodine, or sulfur; and in the absence of arsenic, mercury, and phosphorus, no distillation of the resulting acid is necessary.[385] In the presence of arsenic, mercury, or phosphorus, distillation of *fluosilicic* acid previous to titration with thorium nitrate is necessary.[385]

Judging from the number of papers which have been listed in recent reviews,[246,247,378] it would seem that no method for the microdetermination of fluorine in organic compounds has received wide acceptance. However, the method described below has given excellent results in the author's laboratories on many different types of fluoro-compounds, including mono- and perfluoro-compounds, in the hands of a number of analysts. In addition, in a recent collaborative study[377] it was noted that the majority of the collaborators who obtained good results used procedures which included the features of this method. Consequently, in the 1960 collaborative study, this method was used and, as this book goes to press, the returns were such that this method has been adopted as an Association of Official Agricultural Chemists' "First Action" procedure.[377a]

The success of the determination depends upon a close control of the conditions and, consequently, there should be no deviations.[385] Of prime importance is the adjustment of the pH since the indicator, sodium alizarin sulfonate, is also an acid-base indicator. The relationship between milliliters of $0.01N$ thorium nitrate and the weight of fluorine is not a straight line function except possibly under very limited conditions and over a *small* range of amount of fluorine. Although titration may be accomplished without the instrument described, in the opinion of the author, better results are obtained with it than by visual means.

Reagents

SODIUM PEROXIDE
Reagent grade is required.

DISTILLED WATER

SODIUM HYDROXIDE, APPROX. 0.1N
This is used in the adjustment of the pH previous to titration.

SODIUM HYDROXIDE, APPROX. 1N
This, too, is used in the adjustment of the pH.

HYDROCHLORIC ACID, APPROX. 0.1N
For pH adjustment purposes.

HYDROCHLORIC ACID, APPROX. 1N
Same as above.

SODIUM ALIZARIN SULFONATE INDICATOR (ALIZARIN RED S)
This solution is prepared as described in Chapter 5.

STANDARD SODIUM FLUORIDE, 0.01N
This solution is prepared according to the directions given in Chapter 5.

STANDARD THORIUM NITRATE, 0.01N
This solution is prepared and standardized according to the directions given in Chapter 5.

11. Halogens

PERCHLORIC ACID (70–72%), REAGENT GRADE

This is used in the distillation procedure when arsenic, mercury, or phosphorus is present. (*Caution:* All precautions normally used with this acid should be observed.)

SILVER PERCHLORATE SOLUTION, 25%

This is used only in the distillation procedure with perchloric acid.

Apparatus

SCHÖNIGER COMBUSTION FLASK[395]

Same as used for the determination of sulfur (Fig. 149, Chapter 10).

FILTER PAPER CARRIERS

See Chapter 10.

OXYGEN CYLINDER AND REDUCING VALVE

AUTOMATIC BURETTES

See Chapter 5.

pH METER

The Zeromatic pH meter, No. 9600 (Beckman Instruments, Inc.[27]) or comparable instrument is used for determining the pH of the solution previous to titration.

PHOTOELECTRIC FILTER PHOTOMETER[248,260,268]

The instrument described in Chapter 5 is a definite aid in obtaining the end point of the titration of fluoride with thorium nitrate using sodium alizarin sulfonate as the indicator.

DISTILLATION APPARATUS

The distillation apparatus is that described by Ma and Gwirtsman[248] and shown in Figs. 161 and 162. This is attached to the steam generator (Figs. 112 and 113) used in connection with the Kjeldahl distillation apparatus (Chapter 8). It has some points of similarity to that used in the above-mentioned determination. It is composed of a distillation flask and condenser. Steam enters through the joint, J_1 (Fig. 162), passes through the two concentric tubes, IT_1 and ET_1, and enters the distillation flask, D, through the two openings. The vapors enter the condenser, C, which consists of three concentric tubes. In IT_2 and ET_2, the vapors are condensed, and in ET_3, the cooling water is circulated. The distillate drains off on the right through the descending tube. The ground joint, J_2, serves both as an opening for the introduction of the solution and as a seat for the thermometer which records the temperature of the liquid, L. An electric heating jacket, H, surrounds the section of the distillation flask which contains the liquid, L, inasmuch as the steam distilla-

FIG. 161. Apparatus for distillation of fluosilicic acid.

11. Halogens

tion from the perchloric acid solution is done at a temperature of 135° C. The jacket is prepared from nichrome wire, W (resistance, 2.120 ohms/foot), 600 cm. in length. Of this length, 420 cm. forms the resistance spiral. This is wound on a glass cylinder, 48 mm. diameter, previously covered with aluminum foil and asbestos. The construction is completed by layers of insulating cement, asbestos and another layer of cement. The heating jacket is

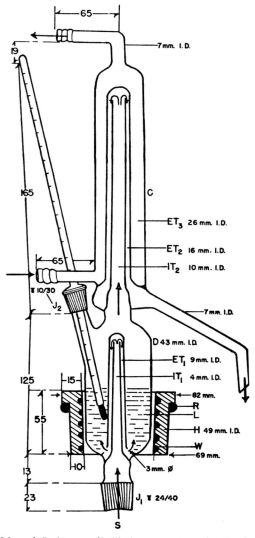

FIG. 162. Ma and Gwirtsman distillation apparatus—details of construction.

held in place by the ring, R. The temperature is readily controlled by means of a 7.5 ampere variable transformer.

Procedure[385]

IN THE ABSENCE OF ARSENIC, MERCURY, AND PHOSPHORUS[385]

Enough sample to contain about 0.5–0.7 mg. (preferably not more) of fluorine is placed on the filter paper carrier used in connection with the Schöniger oxygen flask (see Chapter 10). [Liquid samples are weighed in gelatin or methylcellulose capsules (Chapter 3), and the closed capsule is placed on the filter paper.] Approximately 15–20 mg. of sodium peroxide is added and the mixture wrapped in the filter paper and placed in the platinum basket carrier in the stopper of a 500-ml. Schöniger flask. Twenty milliliters of distilled water is placed in the flask, oxygen introduced for several minutes, the sample ignited, and immediately inserted into the flask. After combustion is complete, the contents are shaken vigorously until cloudiness disappears, after which the flask is allowed to remain undisturbed for about 15 minutes to insure complete absorption of the oxidation products. (If iodine is present in quantities great enough to impart a yellow color, the solution should be warmed on a steam bath to dispel this color.) The contents of the combustion flask are washed into the clear plastic titrating cell of the photoelectric filter photometer and brought up to a volume of 450 ml. with distilled water. The solution is adjusted to pH 3.0 ± 0.05 with the aid of $1N$ and $0.1N$ hydrochloric acid (and sodium hydroxide) using the pH meter and 2 ml. of 0.035% sodium alizarin sulfonate indicator is added. A duplicate clear plastic titrating cell containing 450 ml. of distilled water adjusted to pH 3.0 ± 0.05 and 2 ml. of 0.035% sodium alizarin sulfonate indicator is placed in one compartment of the photoelectric filter photometer and the cell containing the fluoride solution in the other compartment. (Note: Since sodium alizarin sulfonate is also an acid-base indicator, the pH adjustment of *both* are *critical*.) The unknown is then titrated with $0.01N$ thorium nitrate following the directions given in Chapter 5 in connection with the standardization of thorium nitrate.

Calculation:

$$\frac{\text{mg. of F} \times 100}{\text{Wt. sample}} = \% \text{ F}$$

Example:

4.105 mg. of sample required 5.52 ml. of thorium nitrate

From the plot of ml. of thorium nitrate vs. mg. of fluorine obtained during the standardization, this volume is equivalent to 0.528 mg. F

$$\therefore \frac{0.528 \times 100}{4.105} = 12.86\% \text{ F}$$

IN THE PRESENCE OF ARSENIC, MERCURY, OR PHOSPHORUS[248,385]

After combustion of the sample as described above, the contents of the Schöniger flask are transferred to the distillation apparatus (Fig. 162) through the joint, J$_2$, *using as little water as possible*. Twenty milliliters of perchloric acid (70–72%) and one ml. of silver perchlorate solution (25%) are added together with about a dozen small glass beads (for added surface contact). The reaction mixture is heated by means of the electric heating jacket, H, and as the temperature rises, steam generation is started. The temperature of the mixture is maintained at 135 ± 2° C. and the distillate collected in a 250 ml. volumetric flask. (Actually, the distillate is collected from a temperature of about 95° C., but is raised as *quickly* as possible to 135 ± 2° C. by changing the setting on the transformer. Practice is required for successful manipulation but the beginner will have earlier success if the volume of liquid in the flask is kept at a *minimum* and the steam generation is kept constant so that there will be no temporary reduction in internal pressure resulting in the sucking back of the reaction mixture. The addition of a *small* amount of alkali and phenolphthalein indicator to the steam generator provides a means of determining whether suck back has occurred at any time since even a small amount of the strong perchloric acid solution would make the water in the generator acid.) After 250 ml. of distillate has been collected, it is transferred to the clear plastic titrating cell and diluted with water to a volume of 450 ml. This is adjusted to pH 3.0 ± 0.05, 2 ml. of 0.035% sodium alizarin sulfonate added and the fluosilicic acid titrated with thorium nitrate, all as described above.

The distillation apparatus is easily cleaned between determinations by replacing the steam generator with some type of bottle connected to a water aspirator and immersing the distillate delivery tube in fluorine-free distilled water, which is then sucked through the entire system.

Calculations:

Same as described above.

ADDITIONAL INFORMATION FOR CHAPTER 11

Automatic titration of chloride has been described by Cotlove, Trantham, and Bowman.[86] A commercial[6] adaptation of their instrument is shown in Fig. 163. A constant direct current is passed between a pair of silver generator electrodes, causing electrochemical oxidation of the generator anode to the silver ions which are continuously released at a steady rate into the titration solution. When all the chloride has been combined with silver ions, the appearance of free silver ions causes an abrupt increase in current between a pair of silver indicator electrodes. When the indicator current reaches a preselected magnitude, the sensitive meter-relay circuitry is actuated which removes the potential from the generator electrodes and the timer which runs concurrently with

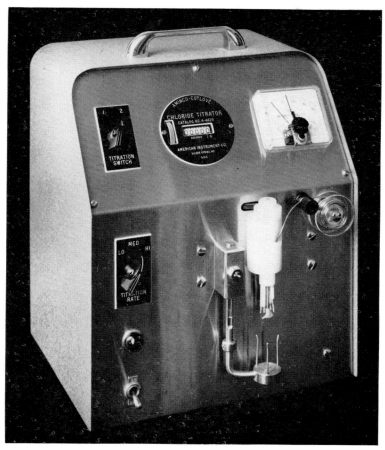

FIG. 163. Automatic chloride titrator.

the generation of silver ions. Since silver ions are generated at a constant rate, the amount used to precipitate the chloride ions is proportional to the elapsed time. Hence, the chloride content of the titration solution can be determined. Obviously, organic chloride must first be converted to the chloride ions.[6]

For those workers who prefer the alkali metal fusion method of converting organic halogen, particularly fluorine, to inorganic halide, the micro-Parr

11. Halogens

bomb assembly[382] shown in Fig. 164 will prove very useful, since it stands up under continued use when fusions are made at 700° C.

Figure 164 shows the details of construction of the nut and bolt clamping device (which should be machined from Type 303 free machining stainless steel), dimensions of the copper gaskets,[32,39,45,247,248,382] and the assembly showing a micro-Parr bomb[23,304] in position. A stainless steel block may be drilled out as a support for the assembly. For a tight seal, before use, the gaskets should be annealed at 700° C. (preferably in an atmosphere of carbon dioxide or nitrogen), and any oxide coating removed with dilute nitric acid.

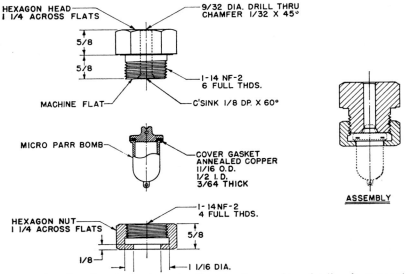

Fig. 164. Clamping device for micro-Parr bomb assembly—details of construction. All dimensions in inches. Use Type 303 free machining stainless steel.

Annealed copper gaskets of the correct dimensions are commercially available (Parr Instrument Company, Moline, Illinois, No. 7MB copper, or Arthur H. Thomas Company,[395] No. 2196-E). Figure 164 shows the dimension of both the hexagon head and hexagon nut to be $1\frac{1}{4}$ inches across the flats. For greater durability this may be increased to $1\frac{1}{2}$ inches, all other dimensions being left as shown.

To prevent leakage around the gasket during fusion, the cup and cover should be cleaned after each determination with fine emery cloth, while being turned in a lathe. For closing, the nut portion of the unit should be held in a vise while the bolt portion is tightened using a long-handled (20 inches or more) socket wrench. After fusion the cup and lid may be separated by holding the former in a vise while tapping the edge of the latter with hammer and chisel.

TABLE 23
Additional Information on References* Related to Chapter 11

Besides the procedures described in the preceding pages of this chapter, the author wishes to call to the attention of the reader many of the articles which have been published on the determination of the various halides. Presentation is the same as used for the previous chapters. (See statement at top of Table 4 of Chapter 1, regarding completeness of this material.) The importance of the determination of these elements is proven by the extremely large number of articles which have been published, obviously with the aim of improvement of existing methods in one way or another. The Table is divided into two parts, A and B, bromine, chlorine, and iodine being grouped in the former and fluorine in the latter because of the different problems connected with these.

A. Determination of Bromine, Chlorine, and Iodine

Books

Association of Official Agricultural Chemists, 10
Belcher and Godbert, 34, 35
Clark, E. P., 78
Clark, S. J., 81
Friedrich, 127
Furman, 130
Grant, 139, 140
Milton and Waters, 279, 280
Niederl and Niederl, 293, 294
Niederl and Sozzi, 295
Pregl, 313
Roth, 322–325
Steyermark, 379

Reviews

Beaucourt, 25
Fellenberg, 118
Hallett, 148
Hernler and Pfenningberger, 156
Kainz, 183
Leipert, 230
Lunde, Gloss, and Böe, 244
Willits, 426

Collaborative studies

Ogg, 299
Steyermark, 375, 376
Steyermark and Faulkner, 383
Steyermark and Garner, 384

Apparatus

Belcher and Ingram, 37
British Standards Institution, 56

Apparatus (Cont.)

Dean and Hawley, 92
Étienne and Herrmann, 112
Ingram, 163
Kimball and Tufts, 190
Kirsten, 191
Kuck and Griffel, 219
Maurmeyer and Ma, 265
Northrop, 297
Royer, Alber, Hallett, and Kuck, 328
Schall and Williamson, 337
Sundberg and Royer, 388
Thomas, Shinn, Wiseman, and Moore, 398

General, miscellaneous

Analytical Methods Committee, Pesticides Residues in Foodstuffs Subcommittee, 8
Barakat and El-Wahab, 20
Belcher and Goulden, 36
Belcher, Shah, and West, 42
Bennett and Debbrecht, 46
Brown and Musgrave, 57
Ellis and Duncan, 105
Elving and Ligett, 106
Gouverneur and Van Dijk, 136
Hasselmann and Laustriat, 153
Itai and Nakashima, 167
Jeník, Patek, and Jureček, 171
Jungreis and Gedalia, 177
Jurany, 180
Kaplan and Schnerb, 189
Kirsten, 192, 194, 196

* The numbers which appear after each entry in this table refer to the literature citations in the reference list at the end of the chapter.

TABLE 23 (*Continued*)

General, miscellaneous (Cont.)

Kirsten and Alperowitz, 198
Klein, 200
Kolthoff and Yutzy, 204
Konovalov, 207
Lachiver, 223
Lacourt and Timmermans, 224
Lane, 226
Lemp and Broderson, 232
Lohr, Bonstein, and Frauenfelder, 242
Malmstadt and Winefordner, 259
Menschenfreund, 276
MonteBovi, Halpern, Koretsky, and Dunne, 283
Naylor, 287
Németh and Menkyna, 288
Nutten, 298
Pirt and Chain, 310
Rao and Shah, 316
Rogina and Dubravcic, 321
Rush, Cruikshank, and Rhodes, 330
Servigne, 351
Sokolova, Orestova, and Nikolayeva, 365
Spitzy and Lieb, 366
Spitzy, Reese, and Skrube, 367
Spitzy and Skrube, 368
Steinitz, 372
Terent'ev and Luskina, 392
Urrutia, Ramirez, and Aguayo, 400
Večeřa, 403
Večeřa and Šnobl, 410
Volynskiĭ and Chudakova, 415
Zacherl and Stöckl, 429

Simultaneous determination of bromine, chlorine, iodine, as well as one with other elements

Alicino, Crickenberger, and Reynolds, 5
Belcher and Spooner, 43, 44
Boëtius, Gutbier, and Reith, 51
Chateau, 69
Fedoseev and Ivashova, 114
Fedoseev and Sobko, 116, 117
Friedrich, 127
Fujimoto, Utsui, and Ose, 128
Gysel, 146
Intonti and Gargiulo, 164

Simultaneous determination of bromine, chlorine, iodine, as well as one with other elements (Cont.)

Klimova and Bereznitskaya, 201
Konovalov, 208
Korshun and Bondarevskaya, 209
Korshun and Chumachenko, 210
Korshun, Gel'man, and Sheveleva, 212
Korshun and Sheveleva, 214
Lacourt and Timmermans, 224
Lévy, 237
Merz, 277
Michel and Deltour, 278
Niederl and Niederl, 294
Roth, 322–325
Teston and McKenna, 394
Wagner and Bühler, 417

Ultramicro-, submicro-methods

Ballczo and Hainberger, 13
Bather, 21
Cannon, 63
Custer and Natelson, 89
Farlow, 113
Judah, 175
Jungreis and Gedalia, 177
Kirsten, 192
Kuck, Daugherty, and Batdorf, 218
Sanz, Brechbühler, and Green, 334
Staemmler, 370
Suter and Hadorn, 389
Viswanathan, 414

Determination of bromine, chlorine, and iodine in fluoro-compounds

Belcher, Macdonald, and Nutten, 41
Rush, Cruikshank, and Rhodes, 330

Carius combustion

Carius, 64–67
Clark, 78
Doering, 97
Kuck and Griffel, 219
Makineni, McCorkindale, and Syme, 258
Niederl, Baum, McCoy, and Kuck, 292
Pirt and Chain, 311
Weygand and Werner, 419
White and Kilpatrick, 420
White and Secor, 421

TABLE 23 (*Continued*)

Chromic acid combustion

Grant, 139, 140
Kunori, 221
Leithe, 231
Niederl and Niederl, 294
Rao and Shah, 316
Roth, 322–325
Shakrokh and Chesbro, 354
Thomas, Shinn, Wiseman, and Moore, 398
Zacherl and Krainick, 428

Catalytic combustion, empty tube technique, etc.

Agazzi, Fredericks, and Brooks, 2
Beazley, 26
Belcher and Ingram, 37
Clark, E. P., 78
Clark and Rees, 80
Estevan and Serra, 111
Fedoseev and Sobko, 115–117
Fildes, 120
Friedrich, 127
Grote and Krekeler, 143, 144
Hallett, 147, 148
Heine, 154
Kainz, 184
Kirsten, 193, 194, 197
Konovalov, 207
Krekeler, 216
Mitsui and Sato, 281
Nicksic and Farley, 291
Otter, 301
Pregl, 313
Schöberl, 339
Sokolova and Loseva, 364
Sundberg and Royer, 388
Teston and McKenna, 394
Večeřa and Bulušek, 404–409
Večeřa and Spevak, 411, 412

Oxygen flask combustion

Bennewitz, 47
Cheng, 72
Corner, 84
Erdey, Mázor, and Meisel, 109
Johnson and Vickers, 174

Oxygen flask combustion (*Cont.*)

Konovalov, 207
Praeger and Fürst, 312
Schöniger, 341–344
Soep, 363
Thomas, 396

Oxy-hydrogen flame

Ehrenberger, 102
Lévy, 240

Bombs, fusion with various agents

Agazzi, Parks, and Brooks, 3
Batt, 22
Beamish, 23, 24
Belcher, Caldas, Clark, and Macdonald, 32
Belcher, Macdonald, and Nutten, 41
Brown and Musgrave, 57
Bürger, 61
Chiang, 73
Clark, E. P., 78
Crespi and Cevolani, 87
Daudel, Flon, and Herczeg, 91
Elek and Harte, 103
Elek and Hill, 104
Elving and Ligett, 106
Grodsky, 141
Haslam and Hall, 152
Inglis, 162
Kainz, 181, 182
Kainz and Resch, 185, 186
Kimball and Tufts, 190
Kondo, 205, 206
Korshun and Chumachenko, 210
Lévy, 236, 238, 239
Lohr, Bonstein, and Frauenfelder, 242
MacNevin and Baxley, 252
MacNevin and Brown, 253
Martin, 262
Parr, 303, 304
Peel, Clark, and Wagner, 306
Pringsheim, 314
Schöniger, 340
Terent'ev, Obtemperanskaya, and Ermolenko, 393
Weber, 418

TABLE 23 (*Continued*)

Decomposition by disodium biphenyl, alkali metals in organic solvents, etc.

Bennett and Debbrecht, 46
Blinn, 49
Clark, E. P., 78
Ionescu and Goia, 165
Irimescu and Chirnoaga, 166
Johncock, Musgrave, and Wiper, 172
Pecherer, Gambrill, and Wilcox, 305
Rauscher, 317
Sezerat and Laniece, 352
Sisido and Yagi, 358

Gravimetric procedures

Fedoseev and Sobko, 115
Intonti and Gargiulo, 164
Kuhn and Schretzmann, 220
Lohr, Bonstein, and Frauenfelder, 242
Pirt and Chain, 311
Safford and Stragand, 331

Volumetric procedures

Alicino, Crickenberger, and Reynolds, 5
Archer, 9
Bather, 21
Belcher, 28
Belcher, Fildes, and Macdonald, 33
Belcher and Goulden, 36
Belcher, Macdonald, and Nutten, 41
Caldwell and Moyer, 62
Chateau and Hervier, 70
Cheng, 72
Christensen, 74
Čihalík and Voráček, 75–77
Dean, Wiser, Martin, and Barnum, 93
Dubouloz, Monge-Hedde, and Fondarai, 98
Ehrenberger, 102
Erdey, Mazor, and Meisel, 109
Grangaud, 137
Grodsky, 141
Harlay, 149
Haslam and Hall, 152
Hasselmann and Laustriat, 153
Inglis, 162
Intonti and Gargiulo, 164
Ionescu and Goia, 165

Volumetric procedures (*Cont.*)

Jureček and Večeřa, 179
Kainz, 182
Kainz and Resch, 185
Kimball and Tufts, 190
Kirsten, 193, 197
Kirsten and Ehrlich-Rogozinsky, 199
Lachiver, 223
Lacourt and Timmermans, 224
Lapin and Zamanov, 227
Lein and Schwartz, 229
Lévy, 238
Ma and Nystrom, 250
Makineni, McCorkindale, and Syme, 258
Milton and Waters, 279, 280
Pääbo and Rottenberg, 302
Pilz, 309
Pirt and Chain, 311
Praeger and Fürst, 312
Schöniger, 340–344
Schulek, 345
Schulek and Pungor, 346
Shakrokh, 353
Smirk, 359
Smith, G. McP., 361
Stempel, 373
Suter and Hadorn, 389
Van Winkle and Smith, 401
Večeřa and Bulušek, 404–409
Večeřa and Spevak, 411, 412
White and Kilpatrick, 420
White and Secor, 421

Spectrophotometric, colorimetric optical methods

Bode and Waldschmidt, 50
Bosch and Rubia Pacheco, 52
Brandt and Dahlenborg, 54
Brandt and Duswalt, 55
Custer and Natelson, 89
Desassis and Macheboeuf, 96
Dubouloz, Monge-Hedde, and Fondarai, 98
Gatterer, 131
Gordon, 135
Gross, Wood, and McHargue, 142
Hunter, 160
Hunter and Goldspink, 161

TABLE 23 (*Continued*)*

Spectrophotometric, colorimetric optical methods (Cont.)
Iwasaki, Utsumi, and Ozawa, 168
Iwasaki, Utsumi, Ozawa, and Hasegewa, 169
Lancaster and Hodgson, 225
Lysyj, 245
Marsh, 261
Maruyama and Seno, 263
Schall and Williamson, 337
Shakrokh and Chesbro, 354
Staemmler, 370
Večeřa and Spevak, 411, 412

X-ray fluorescence
Kokotailo and Damon, 203
Leroux, Maffett, and Monkman, 233
Natelson, 286
Philips Electronics, 308

Neutron absorption
Kusaka, 222

Microdiffusion
Cheek, 71
Pirt and Chain, 310

Chromatographic, ion exchange methods
Banks, Cuthbertson, and Musgrave, 19
Gerbaulet and Maurer, 134
Hashmi, 151
Hülsen, 159
Kondo, 206

Indicators
Brandt and Dahlenborg, 54
Bullock and Kirk, 59
Dubouloz, Monge-Hedde, and Fondarai, 98
Furman, 130
Jureček and Večeřa, 179

Indicators (Cont.)
Sakaguchi, 332
Steinitz, 372
Thomas, J. F., 397
Treadwell and Hall, 399

Polarography, potentiometric, amperometric, high-frequency titrations
Blaedel, Lewis, and Thomas, 48
Cogbill and Kirkland, 83
Duxbury, 99
Inglis, 162
Kirsten, 194, 195
Kramer, Moore, and Ballinger, 215
Kuck, Daugherty, and Batdorf, 218
Lefferts, 228
Lévy, 234, 235, 237–239
Lingane and Small, 241
Mader and Frediani, 257
Monand, 282
Pecherer, Gambrill, and Wilcox, 305
Simon and Zyka, 356
Tělupilová-Krestýnová and Šantavý, 391
Wade, 416

Nonaqueous titration
Lefferts, 228

Automatic titration
Cotlove, Trantham, and Bowman, 86
Juliard, 176
Sundberg, Craig, and Parsons, 387

EDTA titration
Flaschka, 123

Determination of radioactive iodine
Nesh and Peacock, 289
Raben, 315

* See p. 340 for *B. Determination of Fluorine.*

11. Halogens

TABLE 23 (*Continued*)

B. *Determination of Fluorine*

Books

Association of Official Agricultural
  Chemists, 10
Belcher and Godbert, 35
Clark, S. J., 81
Milton and Waters, 280
Rodden, 319
Roth, 324
Simons, 357

Reviews

Kainz, 183
Ma, 246, 247
Macdonald, 251
McKenna, 272
Simons, 357

Collaborative studies

Steyermark, 377, 377a

Apparatus

Belcher and Tatlow, 45
Gwirtsman, Mavrodineanu, and Coe, 145
Jakl, 170
Kimball and Tufts, 190
Ma and Gwirtsman, 248
Marley Products, 260
Mavrodineanu and Coe, 266
Mavrodineanu and Gwirtsman, 267, 268
Parr Instrument Company, 304
Samachson, Slovik, and Sobel, 333
Schall and Williamson, 337
Stacey, Tatlow, and Massingham, 369
Steyermark and Biava, 382
Sweetser, 390
Wickbold, 422

General, miscellaneous

Abrahamczik and Merz, 1
Ballczo and Weisz, 18
Belcher, 29
Brown and Musgrave, 57
Chapman, Heap, and Saunders, 68
Debal, Lévy, and Moureau, 94, 95
Eger and Lipke, 100
Eger and Yarden, 101
Ehrenberger, 102

General, miscellaneous (*Cont.*)

Elving and Ligett, 106
Funasaka, Kawane, Kojima, and
  Ishihara, 129
Ma, 247
Mázor, 271
Rush, Cruikshank, and Rhodes, 330
Saunders, 335
Schloemer, 338
Schwarzkopf, 348
Schwarzkopf and Henlein, 349
Stegemann and Jung, 371
Zimmerman, Hitchcock, and Gwirtsman, 431

Simultaneous determination of fluorine with other elements

Freier, Nippoldt, Olson, and
  Weiblen, 126
Gel'man and Korshun, 132
Gel'man, Korshun, Chumachenko, and
  Larina, 133
Korshun, Gel'man, and Glazova, 211
Teston and McKenna, 394
Zimin, Churmanteev, Gubanova, and
  Verina, 430

Ultramicro-, submicro-methods

Baker and Morrison, 11
Ballczo and Kaufmann, 14–16
Harms and Jander, 150
MacNulty and Hunter, 254
Murty, Viswanathan, and Ramakrishna, 285
Nielsen, 296
Samachson, Slovik, and Sobel, 333
Williams, 425

Determination in presence of phosphorus, arsenic, mercury, etc.

Belcher and Macdonald, 38, 39
Chapman, Heap, and Saunders, 68
Eger and Lipke, 100

Catalytic combustion

Clark, H. S., 79
Clark and Rees, 80

TABLE 23 (*Continued*)

Catalytic combustion (Cont.)
Kojima, Nagase, and Muramatsu, 202
Mázor, 271
Peregud and Boĭkina, 307
Rickard, Ball, and Harris, 318
Schumb and Radimer, 347

Oxygen flask combustion
Francis, 125
Rogers and Yasuda, 320
Schöniger, 342
Senkowski, Wollish and Shafer, 350
Thomas, 396

Oxy-hydrogen flame
Ehrenberger, 102
Lévy, 240
Sweetser, 390
Wickbold, 422

Bombs, fusion
Belcher, 29
Belcher, Caldas, and Clark, 31
Belcher, Caldas, Clark, and Macdonald, 32
Belcher and Macdonald, 38–40
Belcher, Macdonald, and Nutten, 41
Belcher and Tatlow, 45
Brown and Musgrave, 57
Hennart and Merlin, 155
Kainz and Schöller, 187, 188
Kimball and Tufts, 190
Korshun, Klimova, and Chumachenko, 213
Ma and Gwirtsman, 248
Mavrodińeanu and Gwirtsman, 269
Nichols and Olson, 290
Roth, 326
Savchenko, 336
Schwarzkopf, 348
Schwarzkopf and Henlein, 349
Silvey and Cady, 355
Steyermark and Biava, 382

Decomposition by disodium biphenyl, alkali metals in organic solvents, etc.
Bennett and Debbrecht, 46
Horiuchi, 157
Johncock, Musgrave, and Wiper, 172
Strahm, 386
Vaughn and Nieuwland, 402

Gravimetric procedures
Ballczo and Schiffner, 17
Belcher, 29, 30
Belcher and Macdonald, 40
Kimball and Tufts, 190
Ma and Mangravite, 249
Mázor, 271
Ruff, 329
Schwarzkopf, 348
Schwarzkopf and Henlein, 349

Volumetric procedures
Ballczo and Kaufmann, 14–16
Belcher, 29, 30
Belcher, Caldas, and Clark, 31
Belcher and Macdonald, 38, 39
Brandt and Duswalt, 55
Brunisholz and Michod, 58
Clifford, 82
Dahle, Bonnar, and Wichmann, 90
Debal, Lévy, and Moureau, 95
Ehrenberger, 102
Erdey, Mázor and Pápay, 110
Fichera, 119
Fine and Wynne, 121
Fisher Scientific Company, 122
Harms and Jander, 150
Hennart and Merlin, 155
Hubbard and Henne, 158
Johnson, 173
Kainz and Schöller, 187, 188
Kojima, Nagase, and Muramatsu, 202
Ma and Gwirtsman, 248
MacNulty, Hunter, and Barrett, 255
MacNulty and Woollard, 256
Matuszak and Brown, 264
Mavrodineanu and Gwirtsman, 268
Megregian, 273, 274
Rowley and Churchill, 327
Savchenko, 336
Schöniger, 342
Senkowski, Wollish, and Shafer, 350
Smith and Gardner, 360
Venkateswarlu, Ramanthan, and Narayana Rao, 413
Willard and Horton, 423

Spectrophotometric, colorimetric methods
Bumsted and Wells, 60

TABLE 23 (*Continued*)

Spectrophotometric, colorimetric methods (*Cont.*)

Currey and Mellon, 88
Emi and Hayami, 107
Fine and Wynne, 121
Lothe, 243
Ma and Gwirtsman, 248
MacNulty, Hunter, and Barrett, 255
Mavrodineanu and Gwirtsman, 268
Mavrodineanu, Sanford, and Hitchcock, 270
Megregian, 273
Samachson, Slovik, and Sobel, 333
Schall and Williamson, 337
Senkowski, Wollish and Shafer, 350
Yasuda and Lambert, 427

Nuclear magnetic resonance

Brame, 53
Foster, 124

Ion exchange

Banks, Cuthbertson, and Musgrave, 19
Eger and Lipke, 100

Amperometric, high-frequency, oscillometric, conductometric, etc., titrations

Grant and Haendler, 138
Kubota and Surak, 217
Megregian, 274
Monand, 282
Olson and Elving, 300

EDTA Titration

Hennart and Merlin, 155

Distillation of fluosilicic acid

Ballczo, Doppler, and Lanik, 12
Horiuchi, 157
Jakl, 170
Kimball and Tufts, 190
Ma and Gwirtsman, 248
Mavrodineanu and Gwirtsman, 267
Murty, Viswanathan, and Ramakrishna, 285
Samachson, Slovik, and Sobel, 333
Schall and Williamson, 337
Smith and Parks, 362
Willard and Winter, 424

REFERENCES

1. Abrahamczik, E., and Merz, W., *Mikrochim. Acta*, p. 445 (1959).
2. Agazzi, E. J., Fredericks, E. M., and Brooks, F. R., *Anal. Chem.*, **30**, 1566 (1958).
3. Agazzi, E. J., Parks, T. D., and Brooks, F. R., *Anal. Chem.*, **23**, 1011 (1951).
4. Alicino, J. F., Personal communication.
5. Alicino, J. F., Crickenberger, A., and Reynolds, B., *Anal. Chem.*, **21**, 755 (1949).
6. American Instrument Co., Silver Springs, Maryland.
7. American Society for Testing Materials, *ASTM Designation,* **E 148-59T**.
8. Analytical Methods Committee, Report Prepared by the Pesticides Residues in Foodstuffs Subcommittee, *Analyst*, **82**, 378 (1957).
9. Archer, E. E., *Analyst*, **83**, 571 (1958).
10. Association of Official Agricultural Chemists, "Official Methods of Analysis," 8th ed., pp. 801–811, Washington, D. C., 1955.
11. Baker, B. B., and Morrison, J. D., *Anal. Chem.*, **27**, 1306 (1955).
12. Ballczo, H., Doppler, G., and Lanik, A., *Mikrochim. Acta*, p. 809 (1957).
13. Ballczo, H., and Hainberger, L., *Mikrochim. Acta*, p. 466 (1959).
14. Ballczo, H., and Kaufmann, O., *Mikrochemie ver. Mikrochim. Acta*, **38**, 237 (1951).
15. Ballczo, H., and Kaufmann, O., *Mikrochemie ver. Mikrochim. Acta*, **39**, 9 (1952).
16. Ballczo, H., and Kaufmann, O., *Mikrochemie ver. Mikrochim. Acta*, **39**, 13 (1952).
17. Ballczo, H., and Schiffner, H., *Z. anal. Chem.*, **152**, 3 (1956).
18. Ballczo, H., and Weisz, H., *Mikrochim. Acta*, p. 751 (1957).

19. Banks, R. E., Cuthbertson, F., and Musgrave, W. K. R., *Anal. Chim. Acta,* **13**, 442 (1955).
20. Barakat, M. Z., and El-Wahab, M. R., *Anal. Chem.,* **26**, 1973 (1954).
21. Bather, J. M., *Analyst,* **81**, 536 (1956).
22. Batt, W. G., *J. Franklin Inst.,* **248**, 451 (1949).
23. Beamish, F. E., *Ind. Eng. Chem., Anal. Ed.,* **5**, 348 (1933).
24. Beamish, F. E., *Ind. Eng. Chem., Anal. Ed.,* **6**, 352 (1934).
25. Beaucourt, J. H., *Metallurgia,* **38**, 353 (1948).
26. Beazley, C. W., *Ind. Eng. Chem., Anal. Ed.,* **11**, 229 (1939).
27. Beckman Instruments, Inc., Fullerton, California.
28. Belcher, R., Personal communication, 1959.
29. Belcher, R., *Chim. anal.,* **36**, 65 (1954); *Anal. Abstr.,* **1**, No. 1280 (1954).
30. Belcher, R., *Österr. Chemiker-Ztg.,* **55**, 158 (1954).
31. Belcher, R., Caldas, E. F., and Clark, S. J., *Analyst,* **77**, 602 (1952).
32. Belcher, R., Caldas, E. F., Clark, S. J., and Macdonald, A., *Mikrochim. Acta,* p. 283 (1953).
33. Belcher, R., Fildes, J. E., and Macdonald, A. G., *Chem. & Ind. (London),* p. 1402 (1955).
34. Belcher, R., and Godbert, A. L., "Semi-Micro Quantitative Organic Analysis," Longmans, Green, London, New York, and Toronto, 1945.
35. Belcher, R., and Godbert, A. L., "Semi-Micro Quantitative Organic Analysis," 2nd ed., Longmans, Green, London, 1954.
36. Belcher, R., and Goulden, R., *Mikrochim. Acta,* p. 290 (1953).
37. Belcher, R., and Ingram, G., *Anal. Chim. Acta,* **7**, 319 (1952).
38. Belcher, R., and Macdonald, A. M. G., *Mikrochim. Acta,* p. 243 (1955).
39. Belcher, R., and Macdonald, A. M. G., *Mikrochim. Acta,* pp. 899, 1187 (1956).
40. Belcher, R., and Macdonald, A. M. G., *Mikrochim. Acta,* p. 510 (1957).
41. Belcher, R., Macdonald, A. M. G., and Nutten, A. J., *Mikrochim. Acta,* p. 104 (1954).
42. Belcher, R., Shah,, R. A., and West, T. S., *J. Chem. Soc.,* p. 2998 (1958).
43. Belcher, R., and Spooner, C. E., *Fuel,* **20**, 130 (1941).
44. Belcher, R., and Spooner, C. E., *Power Plant Eng.,* **46**, 58 (1942).
45. Belcher, R., and Tatlow, J. C., *Analyst,* **76**, 593 (1951).
46. Bennett, C. E., and Debbrecht, F. J., 131st National Meeting of the American Chemical Society, Miami, Florida, April 1957, *Abstracts,* p. 24B.
47. Bennewitz, R., *Mikrochim. Acta,* p. 54 (1960).
48. Blaedel, W. J., Lewis, W. B., and Thomas, J. W., *Anal. Chem.,* **24**, 509 (1952).
49. Blinn, R. C., *Anal. Chem.,* **32**, 292 (1960).
50. Bode, E., and Waldschmidt, M., *Hoppe-Seyler's Z. physiol. Chem.,* **308**, 204 (1957).
51. Boëtius, M., Gutbier, B., and Reith, H., *Mikrochim. Acta,* p. 321 (1958).
52. Bosch, F. de A., and Rubia Pacheco, J. de la, *Anales real soc. españ. fís. y chím. (Madrid),* **47B**, 263 (1951).
53. Brame, E. G., Jr., Symposium on Analysis of Fluorine-Containing Compounds, 137th National Meeting of the American Chemical Society, Cleveland, Ohio, April 1960, p. 22B.
54. Brandt, K., and Dahlenborg, H., *Acta Chem. Scand.,* **4**, 582 (1950).
55. Brandt, W. W., and Duswalt, A. A., Jr., *Anal. Chem.,* **30**, 1120 (1958).
56. British Standards Institution, *Brit. Standards,* **1428**, Pt. A3 (1952), Pt. A4 (1953), Pt. A5, and Pt. F1 (1957).

57. Brown, F., and Musgrave, W. K. R., *Anal. Chim. Acta,* **12**, 29 (1955).
58. Brunisholz, G., and Michod, J., *Helv. Chim. Acta,* **37**, 598 (1954).
59. Bullock, B., and Kirk, P. L., *Ind. Eng. Chem., Anal. Ed.,* **7**, 178 (1935).
60. Bumsted, H. E., and Wells, J. C., *Anal. Chem.,* **24**, 1595 (1952).
61. Bürger, K., *Chemie, Die,* **55**, 245 (1942).
62. Caldwell, J. R., and Moyer, H. V., *Ind. Eng. Chem., Anal. Ed.,* **7**, 38 (1935).
63. Cannon, J. H., *J. Assoc. Offic. Agr. Chemists,* **41**, 428 (1958).
64. Carius, L., *Ann.,* **116**, 1 (1860).
65. Carius, L., *Ann.,* **136**, 129 (1865).
66. Carius, L., *Ann.,* **145**, 301 (1868).
67. Carius, L., *Ber.,* **3**, 697 (1870).
68. Chapman, N. B., Heap, R., and Saunders, B. C., *Analyst,* **73**, 434 (1948).
69. Chateau, H., *Sci. et inds. phot.,* **26**, 41 (1955).
70. Chateau, H., and Hervier, B., *Sci. et inds. phot.,* **38**, 270 (1957).
71. Cheek, D. B., *J. Appl. Physiol.,* **5**, 639 (1953).
72. Cheng, W., *Microchem. J.,* **3**, 537 (1959).
73. Chiang, Fan-Tih, *Chemistry (Taiwan),* p. 96 (1957); *Chem. Abstr.,* **53**, 120 (1959).
74. Christensen, B. G., *Ugeskrift Laeger,* **105**, 866 (1943).
75. Číhalík, J., and Voráček, J., *Chem. listy,* **52**, 1075 (1958).
76. Číhalík, J., and Voráček, J., *Chem. listy,* **52**, 1269 (1958).
77. Číhalík, J., and Voráček, J., *Collection Czechoslov. Chem. Communs.,* **24**, 1643 (1959).
78. Clark, E. P., "Semimicro Quantitative Organic Analysis," Academic Press, New York, 1943.
79. Clark, H. S., *Anal. Chem.,* **23**, 659 (1951).
80. Clark, H. S., and Rees, O. W., *Illinois State Geol. Survey, Rept. Invest.,* No. **169** (1954).
81. Clark, S. J., "Quantitative Methods of Organic Microanalysis," Butterworths, London, 1956.
82. Clifford, P. A., *J. Assoc. Offic. Agr. Chemists,* **23**, 303 (1940).
83. Cogbill, E. C., and Kirkland, J. J., *Anal. Chem.,* **27**, 1611 (1955).
84. Corner, M., *Analyst,* **84**, 41 (1959).
85. Corning Glass Works, Corning, New York.
86. Cotlove, E., Trantham, H. V., and Bowman, R. L., *J. Lab. Clin. Med.,* **51**, 461 (1958).
87. Crespi, V., and Cevolani, F., *Chim. e ind. (Milan),* **38**, 583 (1956).
88. Currey, R. P., and Mellon, M. G., *Anal. Chem.,* **29**, 1632 (1957).
89. Custer, J. J., and Natelson, S., *Anal. Chem.,* **21**, 1005 (1949).
90. Dahle, D., Bonnar, R. U., and Wichmann, H. I., *J. Assoc. Offic. Agr. Chemists,* **21**, 459 (1938).
91. Daudel, P., Flon, M., and Herczeg, C., *Compt. rend. acad. sci.,* **228**, 1059 (1949).
92. Dean, R. B., and Hawley, R. L., *Anal. Chem.,* **19**, 841 (1947).
93. Dean, R. B., Wiser, W. C., Martin, G. E., and Barnum, D. W., *Anal. Chem.,* **24**, 1638 (1952).
94. Debal, E., Lévy, R., and Moureau, H., *Proc. Intern. Congr. Pure and Appl. Chem., Anal. Chem. 15th Congr. Lisbon,* **1**, 85 (1956); *Anal. Abstracts,* **6**, No. 1175 (1959) (in French).
95. Debal, M., Lévy, R., and Moureau, H., *Mikrochim. Acta,* p. 396 (1957).
96. Desassis, A., and Macheboeuf, M., *Ann. inst. Pasteur,* **76**, 6 (1949).

97. Doering, H., *Ber.,* **70**, 1887 (1937).
98. Dubouloz, P., Monge-Hedde, M. F., and Fondarai, J., *Bull. soc. chim. France,* p. 898 (1947).
99. Duxbury, McD., *Analyst,* **75**, 679 (1950).
100. Eger, C., and Lipke, J., *Anal. Chim. Acta,* **20**, 548 (1959).
101. Eger, C., and Yarden, A., *Anal. Chem.,* **28**, 512 (1956).
102. Ehrenberger, F., *Mikrochim. Acta,* p. 192 (1959).
103. Elek, A., and Harte, R. A., *Ind. Eng. Chem., Anal. Ed.,* **9**, 502 (1937).
104. Elek, A., and Hill, D. W., *J. Am. Chem. Soc.,* **55**, 2550 (1933).
105. Ellis, G. H., and Duncan, G. D., *Anal. Chem.,* **25**, 1558 (1953).
106. Elving, P. J., and Ligett, W. B., *Ind. Eng. Chem., Anal. Ed.,* **14**, 449 (1942).
107. Emi, K., and Hayami, T., *Nippon Kagaku Zasshi,* **76**, 1291 (1955).
108. Emich, F., and Donau, J., *Monatsh. Chem.,* **30**, 745 (1909).
109. Erdey, L., Mázor, L., and Meisel, T., *Mikrochim. Acta,* p. 140 (1958).
110. Erdey, L., Mázor, L., and Pápay, M., *Mikrochim. Acta,* p. 482 (1958).
111. Estevan, J., and Serra, J., *Anales real soc. españ. fís. y chím. (Madrid),* **53**, 233 (1957).
112. Etienne, A., and Herrmann, J., *Chim. anal.,* **33**, 1 (1951).
113. Farlow, N. H., *Anal. Chem.,* **29**, 883 (1957).
114. Fedoseev, P. N., and Ivashova, N. P., *Zhur. Anal. Khim.,* **13**, 230 (1958).
115. Fedoseev, P. N., and Sobko, M. Ya., *Zhur. Anal. Khim.,* **13**, 595 (1958); *Anal. Abstr.,* **6**, No. 1347 (1959).
116. Fedoseev, P. N., and Sobko, M. Ya., *Zhur. Anal. Khim.,* **13**, 702 (1958); *Anal. Abstr.,* **6**, No. 2630 (1959).
117. Fedoseev, P. N., and Sobko, M. Ya., *Zhur. Anal. Khim.,* **14**, 118 (1959); *Anal. Abstr.,* **6**, No. 4006 (1959).
118. Fellenberg, Th. v., *Mikrochemie,* **7**, 242 (1929).
119. Fichera, G., *Boll. sedute accad. Gioenia sci. nat. Catania,* **1**, 599 (1952).
120. Fildes, J. E., *Revs. Pure and Appl. Chem. (Australia),* **9**, 117 (1959); *Chem. Abstr.,* **53**, 17773 (1959).
121. Fine, L., and Wynne, E. A., *Microchem. J.,* **3**, 515 (1959).
122. Fisher Scientific Company, "Technical Data on Lanthanum Chloranilate," New York, and Pittsburgh, Pennsylvania.
123. Flaschka, H., *Mikrochemie ver. Mikrochim. Acta,* **40**, 21 (1953).
124. Foster, H., Analytical Group, North Jersey Section of the American Chemical Society, South Orange, New Jersey, 1957.
125. Francis, H. J., Jr., Personal communication, 1956.
126. Freier, H. E., Nippoldt, B. W., Olson, P. B., and Weiblen, D. G., *Anal. Chem.,* **27**, 146 (1955).
127. Friedrich, A., "Die Praxis der quantitativen organischen Mikroanalyse," pp. 102, 109, Deuticke, Vienna, and Leipzig, 1933.
128. Fujimoto, R., Utsui, Y., and Ose, S., *Yakugaku Zasshi,* **78**, 722 (1958); *Chem. Abstr.,* **52**, 16118 (1958).
129. Funasaka, W., Kawane, M., Kojima, T., and Ishihara, K., *Bunseki Kagaku,* **4**, 607 (1955).
130. Furman, N. H., ed., "Scott's Standard Methods of Chemical Analysis," 5th ed., Vol. II, Van Nostrand, New York, 1939.
131. Gatterer, A., *Mikrochemie ver. Mikrochim. Acta,* **36/37**, 476 (1951).
132. Gel'man, N. E., and Korshun, M. O., *Doklady Akad. Nauk S.S.S.R.,* **89**, 685 (1953).

133. Gel'man, N. E., Korshun, M. O., Chumachenko, M. N., and Larina, N. I., *Doklady Akad. Nauk S.S.S.R.,* **123**, 468 (1958); *Chem. Abstr.,* **53**, 3985 (1959).
134. Gerbaulet, K., and Maurer, W., Radioaktive Isotope in Klinik und Forschung (K. Fellinger, and H. Vetter, eds.), Sonderbände zur *Strahlentherapie,* **33**, 116.
135. Gordon, H. T., *Anal. Chem.,* **24**, 857 (1952).
136. Gouverneur, P., and Van Dijk, H., *Anal. Chim. Acta,* **9**, 59 (1953).
137. Grangaud, R., *Ann. pharm. franç.,* **6**, 212 (1948).
138. Grant, C. L., and Haendler, H. M., *Anal. Chem.,* **28**, 415 (1956).
139. Grant, J., "Quantitative Organic Microanalysis, Based on the Methods of Fritz Pregl," 4th ed., Blakiston, Philadelphia, Pennsylvania, 1946.
140. Grant, J., "Quantitative Organic Microanalysis," 5th ed., Blakiston, Philadelphia, Pennsylvania, 1951.
141. Grodsky, J., *Anal. Chem.,* **21**, 1551 (1949).
142. Gross, W. G., Wood, L. K., and McHargue, J. S., *Anal. Chem.,* **20**, 900 (1948).
143. Grote, W., and Krekeler, H., *Z. anal. Chem.,* **98**, 463 (1934); **114**, 321 (1938).
144. Grote, W., and Krekeler, H., *Angew. Chem.,* **46**, 106 (1933).
145. Gwirtsman, J., Mavrodineanu, R., and Coe, R. R., *Anal. Chem.,* **29**, 887 (1957).
146. Gysel, H., *Helv. Chim. Acta,* **24**, 128 (1941).
147. Hallett, L. T., *Ind. Eng. Chem., Anal. Ed.,* **10**, 111 (1938).
148. Hallett, L. T., *Ind. Eng. Chem., Anal. Ed.,* **14**, 956 (1942).
149. Harlay, V., *Ann. pharm. franç.,* **5**, 81 (1947).
150. Harms, J., and Jander, G., *Z. Elektrochem.,* **42**, 315 (1936).
151. Hashmi, H. H., *Anal. Chim. Acta,* **17**, 291 (1957).
152. Haslam, J., and Hall, J. I., *Analyst,* **83**, 196 (1958); *Anal. Abstr.,* **5**, No. 3061 (1958); *Chem. Abstr.,* **52**, 13522 (1958).
153. Hasselmann, M., and Laustriat, G., *Compt. rend. acad. sci.,* **234**, 625 (1952).
154. Heine, V. E., *Pharmazie,* **9**, 972 (1954).
155. Hennart, C., and Merlin, E., *Anal. Chim. Acta,* **17**, 463 (1957).
156. Hernler, F., and Pfenningberger, R., *Mikrochemie ver. Mikrochim. Acta,* **25**, 267 (1938).
157. Horiuchi, N., *Ann. Rept. Takamine Lab.,* **7**, 209 (1955).
158. Hubbard, D. M., and Henne, A. L., *J. Am. Chem. Soc.,* **56**, 1078 (1934).
159. Hülsen, W., *J. Chromatog.,* **1**, 91 (1958).
160. Hunter, G., *Biochem. J.,* **60**, 261 (1955).
161. Hunter, G., and Goldspink, A. A., *Analyst,* **79**, 467 (1954).
162. Inglis, A. S., *Mikrochim. Acta,* p. 934 (1955); p. 1488 (1956).
163. Ingram, G., *Mikrochim. Acta,* p. 877 (1956).
164. Intonti, R., and Gargiulo, M., *Rend. ist. super. sanità,* **11**, 1071 (1948).
165. Ionescu, M., and Goia, I., *Acad. rep. populare Romîne, Filiala Cluj. Studii cercetări chim.,* **8**, 173 (1957).
166. Irimescu, I., and Chirnoaga, E., *Z. anal. Chem.,* **125**, 32 (1942).
167. Itai, T., and Nakashima, T., *Eisei Shikenjo Hôkoku,* **76**, 17 (1958); *Chem. Abstr.* **53**, 15871 (1959).
168. Iwasaki, I., Utsumi, S., and Ozawa, T., *Nippon Kagaku Zasshi,* **78**, 474 (1957).
169. Iwasaki, I., Utsumi, S., Ozawa, T., and Hasegewa, R., *Nippon Kagaku Zasshi,* **78**, 468 (1957).
170. Jakl, F., *Mikrochemie ver. Mikrochim. Acta,* **32**, 195 (1944).
171. Jeník, J., Patek, V., and Jureček, M., *Collection Czechoslov. Chem. Communs.,* **24**, 4040 (1959).
172. Johncock, P., Musgrave, W. K. R., and Wiper, A., *Analyst,* **84**, 245 (1959).

173. Johnson, B. H., *Dissertation Abstr.*, **14**, 763 (1954).
174. Johnson, C. A., and Vickers, C., *J. Pharm. and Pharmacol.*, **11**, 218T (1959).
175. Judah, J. D., *Biochem. J.*, **45**, 60 (1949).
176. Juliard, A. L., *Anal. Chem.*, **30**, 136 (1958).
177. Jungreis, E., and Gedalia, I., *Mikrochim. Acta*, p. 145 (1960).
178. Jureček, M., *Collection Czechoslov. Chem. Communs.*, **12**, 455 (1947).
179. Jureček, M., and Večeřa, M., *Chem. listy*, **46**, 620 (1952).
180. Jurány, H., *Mikrochim. Acta*, p. 134 (1955).
181. Kainz, G., *Mikrochemie ver. Mikrochim. Acta*, **35**, 466 (1950).
182. Kainz, G., *Mikrochemie ver. Mikrochim. Acta*, **38**, 124 (1951).
183. Kainz, G., *Österr. Chemiker-Ztg.*, **58**, 8 (1957).
184. Kainz, G., *Z. anal. Chem.*, **166**, 427 (1959).
185. Kainz, G., and Resch, A., *Mikrochemie ver. Mikrochim. Acta*, **39**, 1 (1952).
186. Kainz, G., and Resch, A., *Mikrochemie ver. Mikrochim. Acta*, **39**, 292 (1952).
187. Kainz, G., and Schöller, F., *Mikrochim. Acta*, p. 211 (1955).
188. Kainz, G., and Schöller, F., *Mikrochim. Acta*, p. 843 (1956).
189. Kaplan, D., and Schnerb, I., *Anal. Chem.*, **30**, 1703 (1958).
190. Kimball, R. H., and Tufts, L. E., *Anal. Chem.*, **19**, 150 (1947).
191. Kirsten, W. J., *Anal. Chem.*, **25**, 805 (1953).
192. Kirsten, W. J., *Microchem. J.*, **2**, 179 (1958).
193. Kirsten, W. J., *Mikrochemie ver. Mikrochim. Acta*, **34**, 149 (1949).
194. Kirsten, W. J., *Mikrochemie ver. Mikrochim. Acta*, **40**, 170 (1953).
195. Kirsten, W. J., *Mikrochim. Acta*, p. 289 (1957).
196. Kirsten, W. J., "Proceedings of the International Symposium on Microchemistry 1958," p. 132, Pergamon, Oxford, London, New York, and Paris, 1960.
197. Kirsten, W. J., *Särtryck ur Svensk Kem. Tidskr.*, **57**, 69 (1945).
198. Kirsten, W. J., and Alperowicz, I., *Mikrochemie ver. Mikrochim. Acta*, **39**, 234 (1952).
199. Kirsten, W. J., and Ehrlich-Rogozinsky, S., *Chemist Analyst*, **47**, 58 (1958).
200. Klein, E., *Biochem. Z.*, **322**, 388 (1952).
201. Klimova, V. A., and Bereznitskaya, E. G., *Zhur. Anal. Khim.*, **11**, 292 (1956).
202. Kojima, R., Nagase, S., and Muramatsu, H., *Bunseki Kagaku*, **4**, 518 (1955).
203. Kokotailo, G. T., and Damon, G. F., *Anal. Chem.*, **25**, 1185 (1953).
204. Kolthoff, I. M., and Yutzy, H. C., *Ind. Eng. Chem., Anal. Ed.*, **9**, 75 (1937).
205. Kondo, A., *Bunseki Kagaku*, **6**, 238 (1957); *Chem. Abstr.*, **52**, 15345 (1958).
206. Kondo, A., *Bunseki Kagaku*, **7**, 232 (1958); *Anal. Abstr.*, **6**, No. 1349 (1959).
207. Konovalov, A., *Ind. chim. belge*, **23**, 19 (1958); *Chem. Abstr.*, **52**, 7944 (1958).
208. Konovalov, A., *Ind. chim belge*, **24**, 259 (1959); *Anal Abstr.*, **6**, No. 4803 (1959).
209. Korshun, M. O., and Bondarevskaya, E. A., *Proc. Acad. Sci. U.S.S.R., Sect. Chem. (English Translation)*, **110**, 553 (1956); *Chem. Abstr.*, **52**, 4393 (1958).
210. Korshun, M. O., and Chumachenko, N. M., *Doklady Akad. Nauk S.S.S.R.*, **99**, 769 (1954).
211. Korshun, M. O., Gel'man, N. E., and Glazova, K. I., *Proc. Acad. Sci. U.S.S.R., Sect. Chem. (English Translation)*, **111**, 761 (1956); *Chem. Abstr.*, **52**, 6065 (1958).
212. Korshun, M. O., Gel'man, N. E., and Sheveleva, N. S., *Zhur. Anal. Khim.*, **13**, 695 (1958).
213. Korshun, M. O., Klimova, V. A., and Chumachenko, M. N., *Zhur. Anal. Khim.*, **10**, 358 (1955).
214. Korshun, M. O., and Sheveleva, N. S., *Zhur. Anal. Khim.*, **11**, 376 (1956).

215. Kramer, H. P., Moore, W. A., and Ballinger, D. G., *Anal. Chem.*, **24**, 1892 (1952).
216. Krekeler, H., *Angew. Chem.*, **50**, 337 (1937).
217. Kubota, H., and Surak, J. G., *Anal. Chem.*, **31**, 283 (1959).
218. Kuck, J. A., Daugherty, M., and Batdorf, D. K., *Mikrochim. Acta*, p. 297 (1954).
219. Kuck, J. A., and Griffel, M., *Ind. Eng. Chem., Anal. Ed.*, **12**, 125 (1940).
220. Kuhn, R., and Schretzmann, H., *Ber.*, **90**, 554 (1957).
221. Kunori, M., *Nippon Kagaku Zasshi*, **79**, 773 (1958); *Anal. Abstr.*, **6**, No. 1348 (1959).
222. Kusaka, Y., *Nippon Kagaku Zasshi*, **79**, 1266 (1958); *Anal. Abstr.*, **6**, No. 2629 (1959).
223. Lachiver, F., *Ann. chim. (Paris)*, **10**, 92 (1955).
224. Lacourt, A., and Timmermans, A. M., *Anal. Chim. Acta*, **1**, 140 (1947).
225. Lancaster, J. E., and Hodgson, W. G., Metropol. Microchem. Soc., New York, February 1960.
226. Lane, M., *Ind. Eng. Chem., Anal. Ed.*, **14**, 149 (1942).
227. Lapin, L. N., and Zamanov, R. K., *Zhur. Anal. Khim.*, **10**, 364 (1955).
228. Lefferts, D. T., *Microchem. J.*, **2**, 257 (1958).
229. Lein, A., and Schwartz, N., *Anal. Chem.*, **23**, 1507 (1951).
230. Leipert, T., *Mikrochim. Acta*, **3**, 73, 147 (1938).
231. Leithe, W., *Mikrochemie ver. Mikrochim. Acta*, **33**, 167 (1948).
232. Lemp, J. F., and Broderson, H. H., *J. Am. Chem. Soc.*, **39**, 2069 (1917).
233. Leroux, J., Maffett, P. A., and Monkman, J. L., *Anal. Chem.*, **29**, 1089 (1957).
234. Lévy, R., *Bull. soc. chim. France*, p. 497 (1956).
235. Lévy, R., *Bull. soc. chim. France*, p. 507 (1956).
236. Lévy, R., *Compt. rend. acad. sci.*, **230**, 1958 (1950).
237. Lévy, R., *Compt. rend. acad. sci.*, **235**, 882 (1952).
238. Lévy, R., *Mikrochemie ver. Mikrochim. Acta*, **36/37**, 741 (1951).
239. Lévy, R., *Mikrochim. Acta*, p. 906 (1956).
240. Lévy, R., "Proceedings of the International Symposium on Microchemistry 1958," p. 112, Pergamon Press, Oxford, London, New York, and Paris, 1960.
241. Lingane, J. J., and Small, L. A., *Anal. Chem.*, **21**, 1119 (1949).
242. Lohr, L. J., Bonstein, T. E., and Frauenfelder, L. J., *Anal. Chem.*, **25**, 1115 (1953).
243. Lothe, J. J., *Anal. Chem.*, **28**, 949 (1956).
244. Lunde, G., Gloss, K., and Böe, J., *Mikrochemie, Pregl Festschrift*, p. 272 (1929).
245. Lysyj, I., *Microchem. J.*, **3**, 529 (1959).
246. Ma, T. S., *Anal. Chem.*, **30**, 1557 (1958).
247. Ma, T. S., *Microchem. J.*, **2**, 91 (1958).
248. Ma, T. S., and Gwirtsman, J., *Anal. Chem.*, **29**, 140 (1957).
249. Ma, T. S., and Mangravite, R., Unpublished work.
250. Ma, T. S., and Nystrom, R. F., Unpublished work.
251. Macdonald, A. M. G., *Ind. Chemist*, **33**, 310 (1957).
252. MacNevin, W. M., and Baxley, W. H., *Ind. Eng. Chem., Anal. Ed.*, **12**, 299 (1940).
253. MacNevin, W. M., and Brown, G. H., *Ind. Eng. Chem., Anal. Ed.*, **14**, 908 (1942).
254. MacNulty, B. J., and Hunter, G. J., *Anal. Chim. Acta*, **9**, 425 (1953).
255. MacNulty, B. J., Hunter, G. J., and Barrett, D. G., *Anal. Chim. Acta*, **14**, 368 (1956).
256. MacNulty, B. J., and Woollard, L. D., *Anal. Chim. Acta*, **14**, 452 (1956).
257. Mader, W. J., and Frediani, H. A., *J. Am. Pharm. Assoc. Sci. Sect.*, **40**, 24 (1951).

258. Makineni, S., McCorkindale, W., and Syme, A. C., *J. Appl. Chem. (London)*, **8**, 310 (1958); *Anal. Abstr.*, **6**, No. 584 (1959).
259. Malmstadt, H. V., and Winefordner, J. D., *Anal. Chem.*, **32**, 281 (1960).
260. Marley Products Company, New Hyde Park, New York.
261. Marsh, G. E., *Appl. Spectroscopy*, **12**, 113 (1958); *Anal. Abstr.*, **6**, No. 2627 (1959).
262. Martin, F., *Mikrochemie ver. Mikrochim. Acta*, **36/37**, 653 (1951).
263. Maruyama, M., and Seno, S., *Bull. Chem. Soc. Japan*, **32**, 486 (1959).
264. Matuszak, M. P., and Brown, D. R., *Ind. Eng. Chem., Anal. Ed.*, **17**, 100 (1945).
265. Maurmeyer, R. K., and Ma, T. S., *Mikrochim. Acta*, p. 563 (1957).
266. Mavrodineanu, R., and Coe, R. R., *Contribs. Boyce Thompson Inst.*, **18**, 173 (1955).
267. Mavrodineanu, R., and Gwirtsman, J., *Contribs. Boyce Thompson Inst.*, **17**, 489 (1954).
268. Mavrodineanu, R., and Gwirtsman, J., *Contribs. Boyce Thompson Inst.*, **18**, 181 (1955).
269. Mavrodineanu, R., and Gwirtsman, J., *Contribs. Boyce Thompson Inst.*, **18**, 419 (1956).
270. Mavrodineanu, R., Sanford, W. W., and Hitchcock, A. E., *Contribs. Boyce Thompson Inst.*, **18**, 167 (1955).
271. Mázor, L., *Mikrochim. Acta*, p. 113 (1957).
272. McKenna, F. E., *Nucleonics*, **8**, 24 (1951); **9**, 40, 51 (1951).
273. Megregian, S., *Anal. Chem.*, **26**, 1161 (1954).
274. Megregian, S., *Anal. Chem.*, **29**, 1063 (1957).
275. Mellor, J. W., "A Comprehensive Treatise on Inorganic and Theoretical Chemistry," Vol. II, Longmans, Green, London, 1928.
276. Menschenfreund, D., *J. Assoc. Offic. Agr. Chemists*, **39**, 523 (1956).
277. Merz, W., *Mikrochim. Acta*, p. 456 (1959).
278. Michel, O., and Deltour, G., *Bull. soc. chim. biol.*, **31**, 1125 (1949).
279. Milton, R. F., and Waters, W. A., "Methods of Quantitative Microanalysis," Longmans, Green, New York, and Arnold, London, 1949.
280. Milton, R. F., and Waters, W. A., "Methods of Quantitative Microanalysis," 2nd ed., Arnold, London, 1955.
281. Mitsui, T., and Sato, H., *Mikrochim. Acta*, p. 1603 (1956).
282. Monand, P., *Bull. soc. chim. France*, p. 704 (1956).
283. MonteBovi, A. J., Halpern, A., Koretsky, H., and Dunne, T., *Am. J. Pharm.*, **124**, 12 (1952).
284. Moser, L., and Miksch, R., *Mikrochemie, Pregl Festschrift*, p. 293 (1929).
285. Murty, G. V. L. N., Viswanathan, T. S., and Ramakrishna, V., *Anal. Chim. Acta*, **16**, 213 (1957).
286. Natelson, S., Personal communication, 1959.
287. Naylor, H., *Shirley Inst. Mem.*, **28**, 123 (1955).
288. Németh, Š., and Menkyna, M., *Mikrochim. Acta*, p. 510 (1958).
289. Nesh, F., and Peacock, W. C., *Anal. Chem.*, **22**, 1573 (1950).
290. Nichols, M. L., and Olsen, J. S., *Ind. Eng. Chem., Anal. Ed.*, **15**, 342 (1943).
291. Nicksic, S. W., and Farley, L. L., *Anal. Chem.*, **30**, 1802 (1958).
292. Niederl, J. B., Baum, H., McCoy, J. S., and Kuck, J. A., *Ind. Eng. Chem., Anal. Ed.*, **12**, 428 (1940).
293. Niederl, J. B., and Niederl, V., "Micromethods of Quantitative Organic Elementary Analysis," Wiley, New York, 1938.

294. Niederl, J. B., and Niederl, V., "Micromethods of Quantitative Organic Analysis," 2nd ed., Wiley, New York, 1942.
295. Niederl, J. B., and Sozzi, J. A., "Microanálisis Elemental Orgánico," Calle Arcos, Buenos Aires, 1958.
296. Nielsen, H. M., *Anal. Chem.*, **30**, 1009 (1958).
297. Northrop, J. H., *J. Gen. Physiol.*, **31**, 213 (1948).
298. Nutten, A. J., *Mikrochemie ver. Mikrochim. Acta*, **39**, 355 (1952).
299. Ogg, C. L., *J. Assoc. Offic. Agr. Chemists*, **38**, 365 (1955).
300. Olson, E. C., and Elving, P. J., *Anal. Chem.*, **26**, 1747 (1954).
301. Otter, I. K. H., *Nature*, **182**, 393 (1958).
302. Pääbo, K., and Rottenberg, M., *Acta Chem. Scand.*, **3**, 1444 (1949).
303. Parr, S. W., *J. Am. Chem. Soc.*, **30**, 764 (1908).
304. Parr Instrument Company, Parr Manual No. 121, pp. 37, 47, Moline, Illinois, 1950.
305. Pecherer, B., Gambrill, C. M., and Wilcox, C. W., *Anal. Chem.*, **22**, 311 (1950).
306. Peel, E. W., Clark, R. H., and Wagner, E. C., *Ind. Eng. Chem., Anal. Ed.*, **15**, 149 (1943).
307. Peregud, E. A., and Boĭkina, B. S., *Zavodskaya Lab.*, **22**, 287 (1956).
308. Philips Electronics, Inc., Mount Vernon, New York.
309. Pilz, W., *Z. anal. Chem.*, **155**, 423 (1957).
310. Pirt, S. J., and Chain, E. B., *Biochem. J.*, **50**, 716 (1952).
311. Pirt, S. J., and Chain, E. B., *Rend. ist. super. sanità.*, **16**, 363 (1953).
312. Praeger, K., and Fürst, H., *Chem. Tech. (Berlin)*, **10**, 537 (1958); *Anal. Abstr.*, **6**, No. 2196 (1959).
313. Pregl, F., "Quantitative Organic Microanalysis," (E. Fyleman, trans. 2nd German ed.), Churchill, London, 1924.
314. Pringsheim, H., *Ber.*, **36**, 4244 (1903).
315. Raben, M. S., *Anal. Chem.*, **22**, 480 (1950).
316. Rao, D. S., and Shah, G. D., *Mikrochemie ver. Mikcrohim. Acta*, **40**, 254 (1953).
317. Rauscher, W. H., *Ind. Eng. Chem., Anal. Ed.*, **9**, 296 (1937).
318. Rickard, R. R., Ball, F. L., and Harris, W. W., *Anal. Chem.*, **23**, 919 (1951).
319. Rodden, C. J., ed., "Analytical Chemistry of the Manhattan Project," McGraw-Hill, New York, 1950.
320. Rogers, R. N., and Yasuda, S. K., *Anal. Chem.*, **31**, 616 (1959).
321. Rogina, B., and Dubravcic, M., *Analyst*, **78**, 594 (1953).
322. Roth, H., "Die quantitative organische Mikroanalyse von Fritz Pregl," 4th ed., Springer, Berlin, 1935.
323. Roth, H., "F. Pregl quantitative organische Mikroanalyse," 5th ed., Springer, Wien, 1947.
324. Roth, H., "Pregl-Roth quantitative organische Mikroanalyse," 7th ed., Springer, Wien, 1958.
325. Roth, H., "Quantitative Organic Microanalysis of Fritz Pregl," 3rd ed. (E. B. Daw, trans. 4th German ed.), Blakiston, Philadelphia, Pennsylvania, 1937.
326. Roth, H., *in* Houben-Weyl-Müller, "Methoden der organischen Chemie," 4th ed., "Methoden der organischen Chemie" (Houben-Weyl), Müller, E., ed., Vol. 2, Analytische Methoden, p. 137, Thieme, Stuttgart, 1953.
327. Rowley, R. J., and Churchill, H. V., *Ind. Eng. Chem., Anal. Ed.*, **9**, 551 (1937).
328. Royer, G. L., Alber, H. K., Hallett, L. T., and Kuck, J. A., *Ind. Eng. Chem., Anal. Ed.*, **15**, 230 (1943).
329. Ruff, O., *Ber.*, **69**, 299 (1936).

330. Rush, C. A., Cruikshank, S. S., and Rhodes, E. J. H., *Mikrochim. Acta,* p. 858 (1956).
331. Safford, H. W., and Stragand, G. L., *Anal. Chem.,* **23**, 520 (1951).
332. Sakaguchi, T., *Yakugaku Zasshi,* **62**, 404 (1942).
333. Samachson, J., Slovik, N., and Sobel, A. E., *Anal. Chem.,* **29**, 1888 (1957).
334. Sanz, M. C., Brechbühler, T., and Green, I. J., *Proc. Intern. Congr. Pure and Appl. Chem., Anal. Chem. 15th Congr. Lisbon,* **1**, 288 (1956) (in French); *Anal. Abstr.,* **6**, No. 1175 (1959).
335. Saunders, B. C., "Some Aspects of the Chemistry and Toxic Action of Organic Compounds Containing Phosphorus and Fluorine," Cambridge Univ. Press, London, and New York, 1957.
336. Savchenko, A. Ya., *Zhur. Anal. Khim.,* **10**, 355 (1955).
337. Schall, E. D., and Williamson, H. G., *J. Assoc. Offic. Agr. Chemists,* **38**, 452 (1955).
338. Schloemer, A., *Mikrochemie ver. Mikrochim. Acta,* **31**, 123 (1944).
339. Schöberl, A., *Angew Chem.,* **50**, 334 (1937).
340. Schöniger, W., *Mikrochim. Acta,* p. 74 (1954).
341. Schöniger, W., *Mikrochim. Acta,* p. 123 (1955).
342. Schöniger, W., *Mikrochim. Acta,* p. 869 (1956).
343. Schöniger, W., Regional Analytical Symposium, Philadelphia, Pennsylvania, February 1957.
344. Schöniger, W., "Proceedings of the International Symposium on Microchemistry 1958," p. 93, Pergamon, Oxford, London, New York, and Paris, 1960.
345. Schulek, E., *Anal. Chim. Acta,* **2**, 74 (1948).
346. Schulek, E., and Pungor, E., *Anal. Chim. Acta,* **4**, 109 (1950).
347. Schumb, W. C., and Radimer, K. J., *Anal. Chem.,* **20**, 871 (1948).
348. Schwarzkopf, O., Personal communication. 1956.
349. Schwarzkopf, O., and Henlein, R., *Proc. Intern. Congr. Pure and Appl. Chem., Anal. Chem. 15th Congr. Lisbon,* **1**, 301 (1956); *Anal. Abstr.,* **6**, No. 1175 (1959).
350. Senkowski, B., Wollish, E. G., and Shafer, E. G. E., *Anal. Chem.,* **31**, 1574 (1959).
351. Servigne, Y., *Proc. Intern. Congr. Pure and Appl. Chem., Anal Chem. 15th Congr. Lisbon,* **1**, 343 (1956); *Anal Abstr.,* **6**, No. 1175 (1959).
352. Sezerat, A., and Laniece, M., *Ann. pharm. franç.,* **13**, 745 (1955).
353. Shakrokh, B. K., *J. Biol. Chem.,* **154**, 517 (1944).
354. Shakrokh, B. K., and Chesbro, R. M., *Anal. Chem.,* **21**, 1003 (1949).
355. Silvey, G. A., and Cady, G. H., *J. Am. Chem. Soc.,* **72**, 3624 (1950).
356. Simon, V., and Zyka, J., *Pharmazie,* **10**, 648 (1955).
357. Simons, J. H., "Fluorine Chemistry," Vol. II, pp. 51–211, Academic Press, New York, 1954.
358. Sisido, K., and Yagi, H., *Anal. Chem.,* **20**, 677 (1948).
359. Smirk, F., *Biochem. J.,* **21**, 31 (1927).
360. Smith, F. A., and Gardner, D. E., *Am. Ind. Hyg. Assoc. Quart.,* **16**, 215 (1955).
361. Smith, G. McP., "Quantitative Chemical Analysis," p. 168, Macmillan, New York, 1942.
362. Smith, O. D., and Parks, T. D., *Anal. Chem.,* **27**, 998 (1955).
363. Soep, H., *Mededel. Vlaam. Chem. Ver.,* **21**, 31 (1959); *Chem. Abstr.,* **53**, 16804 (1959).
364. Sokolova, N. V., and Loseva, K. T., *Zhur. Anal. Khim.,* **13**, 349 (1958); *Chem. Abstr.,* **53**, 117 (1959).

11. Halogens

365. Sokolova, N. V., Orestova, V. A., and Nikolayeva, N. A., *Zhur. Anal. Khim.,* **16**, 472 (1959).
366. Spitzy, H., and Lieb, H., *Mikrochim. Acta,* p. 273 (1956).
367. Spitzy, H., Reese, M., and Skrube, H., *Mikrochim. Acta,* p. 488 (1958).
368. Spitzy, H., and Skrube, H., *Biochem. Z.,* **324**, 60 (1953).
369. Stacey, M., Tatlow, J. C., and Massingham, W. E., Brit. Patent. Appl. **3631/51** (1951).
370. Staemmler, H. J., *Biochem. Z.,* **323**, 74 (1952).
371. Stegemann, H., and Jung, G. F., *Z. physiol. Chem.,* **315**, 222 (1959).
372. Steinitz, K., *Mikrochemie ver. Mikrochim. Acta,* **35**, 176 (1950).
373. Stempel, B., *Z. anal. Chem.,* **141**, 101 (1954); Anal. Abstr., 1, No. 1258 (1954).
374. Steyermark, Al, Chairman, Committee for the Standardization of Microchemical Apparatus, Division of Analytical Chemistry, American Chemical Society, *Anal. Chem.,* **22**, 1228 (1950).
375. Steyermark, Al, *J. Assoc. Offic. Agr. Chemists,* **40**, 381 (1957).
376. Steyermark, Al, *J. Assoc. Offic. Agr. Chemists,* **41**, 297 (1958).
377. Steyermark, Al, *J. Assoc. Offic. Agr. Chemists,* **43**, 683 (1960).
377a. Steyermark, Al, *J. Assoc. Offic. Agr. Chemists,* **44**, in press (1961).
378. Steyermark, Al, *Microchem. J.,* **3**, 399 (1959).
379. Steyermark, Al, "Quantitative Organic Microanalysis," Blakiston, Philadelphia, Pennsylvania, 1951.
380. Steyermark, Al, Alber, H. K., Aluise, V. A., Huffman, E. W. D., Jolley, E. L., Kuck, J. A., Moran, J. J., and Willits, C. O., *Anal. Chem.,* **23**, 1689 (1951).
381. Steyermark, Al, Alber, H. K., Aluise, V. A., Huffman, E. W. D., Kuck, J. A., Moran, J. J., and Willits, C. O., *Anal. Chem.,* **21**, 1555 (1949).
382. Steyermark, Al, and Biava, F. P., *Anal. Chem.,* **30**, 1579 (1958).
383. Steyermark, Al, and Faulkner, M. B., *J. Assoc. Offic. Agr. Chemists,* **35**, 291 (1952).
384. Steyermark, Al, and Garner, M. W., *J. Assoc. Offic. Agr. Chemists,* **36**, 319 (1953).
385. Steyermark, Al, Kaup, R. R., Petras, D. A., and Bass, E. A., *Microchem. J.* **3**, 523 (1959).
386. Strahm, R. D., *Anal. Chem.,* **31**, 615 (1959).
387. Sundberg, O. E., Craig, H. C., and Parsons, J. S., *Anal. Chem.,* **30**, 1842 (1958).
388. Sundberg, O. E., and Royer, G. L., *Ind. Eng. Chem., Anal. Ed.,* **18**, 720 (1946).
389. Suter, H., and Hadorn, H., *Z. anal. Chem.,* **160**, 335 (1958); *Chem. Abstr.,* **53**, 13538 (1958).
390. Sweetser, P. B., *Anal. Chem.,* **23**, 1766 (1956).
391. Tělupilová-Krestýnová, O., and Šantavý, F., *Mikrochim. Acta,* p. 64 (1954).
392. Terent'ev, A. P., and Luskina, B. M., *Zhur. Anal. Khim.,* **14**, 112 (1959).
393. Terent'ev, A. P., Obtemperanskaya, S. I., and Ermolenko, N. V., *Nauch. Doklady Vyssheĭ Shkoly Khim. i Khim. Tekhnol.,* p. 83 (1958); *Anal Abstr.,* No. 2195 (1959).
394. Teston, R. O'D., and McKenna, F. E., Anal. Chem., **19**, 193 (1947).
395. Thomas, Arthur H., Company, Philadelphia, Pennsylvania.
396. Thomas, Arthur H., Company, "Technical Service, DU 37-5M-8'58," Philadelphia, Pennsylvania, 1958.
397. Thomas, J. F., *J. Am. Water Works Assoc.,* **46**, 257 (1954).
398. Thomas, J. W., Shinn, L. A., Wiseman, H. G., and Moore, L. A., *Anal. Chem.,* **22**, 726 (1950).

399. Treadwell, F. P., and Hall, W. T., "Analytical Chemistry," Vol. II, p. 603, Wiley, New York, 1924.
400. Urrutia, H., Ramirez, A., and Aguayo, N., *Bol. soc. chilena quím.,* **8**, 9 (1958); *Chem. Abstr.,* **53**, 21378 (1959).
401. Van Winkle, W. A., and Smith, G. McP., *J. Am. Chem. Soc.,* **42**, 333 (1920).
402. Vaughn, T. H., and Nieuwland, J. A., *Ind. Eng. Chem., Anal. Ed.,* **3**, 274 (1931).
403. Večeřa, M., *Mikrochim. Acta,* p. 88 (1955).
404. Večeřa, M., and Bulušek, J., *Chem. listy,* **51**, 1475 (1957).
405. Večeřa, M., and Bulušek, J., *Chem. listy,* **52**, 1520 (1958); *Chem. Abstr.,* **52**, 19712 (1958).
406. Večeřa, M., and Bulušek, J., *Chem. listy,* **52**, 1526 (1958); *Anal. Abstr.,* **6**, No. 2628 (1959).
407. Večeřa, M., and Bulušek, J., *Collection Czechoslov. Chem. Communs.,* **23**, 257 (1958).
408. Večeřa, M., and Bulušek, J., *Collection Czechoslov. Chem. Communs.,* **24**, 1630 (1959).
409. Večeřa, M., and Bulušek, J., *Mikrochim. Acta,* p. 41 (1958).
410. Večeřa, M., and Šnobl, D., *Chem. listy,* **51**, 1482 (1957).
411. Večeřa, M., and Spevak, A., *Chem. listy,* **51**, 2037 (1957); *Chem. Abstr.,* **52**, 2648 (1958).
412. Večeřa, M., and Spevak, A., *Collection Czechoslov. Chem. Communs.,* **23**, 1197 (1958).
413. Venkateswarlu, P., Ramanthan, A. N., and Narayana Rao, D., *Indian J. Med. Research,* **41**, 253 (1953).
414. Viswanathan, R., *Biochem. J.,* **48**, 239 (1951).
415. Volynskiĭ, N. P., and Chudakova, I. K., U.S.S.R. Patent 113,669 (1959); *Chem. Abstr.,* **53**, 2943 (1959).
416. Wade, P., *Analyst,* **76**, 606 (1951).
417. Wagner, H., and Bühler, F., *Mikrochemie ver. Mikrochim. Acta,* **36/37**, 641 (1951).
418. Weber, A. P., *Rec. trav. chim.,* **59**, 1104 (1940).
419. Weygand, C., and Werner, A., *Mikrochemie,* **26**, 177 (1939).
420. White, L. M., and Kilpatrick, M. D., *Anal. Chem.,* **22**, 1049 (1950).
421. White, L. M., and Secor, G. E., *Anal. Chem.,* **22**, 1047 (1950).
422. Wickbold, R., *Angew. Chem.,* **66**, 175 (1954).
423. Willard, H. H., and Horton, C. A., *Anal. Chem.,* **22**, 1190 (1950).
424. Willard, H. H., and Winter, O. B., *Ind. Eng. Chem., Anal. Ed.,* **5**, 7 (1933).
425. Williams, H. A., *Analyst,* **75**, 510 (1950).
426. Willits, C. O., *Anal. Chem.,* **21**, 132 (1949).
427. Yasuda, S. K., and Lambert, J. L., *Anal. Chem.,* **30**, 1485 (1958).
428. Zacherl, M. K., and Krainick, H. G., *Mikrochemie,* **11**, 61 (1932).
429. Zacherl, M. K., and Stöckl, W., *Mikrochemie ver. Mikrochim. Acta,* **38**, 278 (1951).
430. Zimin, A. V., Churmanteev, S. V., Gubanova, A. V., and Verina, A. D., *Doklady Akad. Nauk S.S.S.R.,* **126**, 784 (1959); *Chem. Abstr.,* **53**, 18736 (1959).
431. Zimmerman, P. W., Hitchcock, A. E., and Gwirtsman, J., *Contribs. Boyce Thompson Inst.,* **19**, 49 (1957).

CHAPTER 12

Microdetermination of Phosphorus

The determination of phosphorus in organic compounds is based on the destruction of organic material, conversion of the phosphorus, in the process to phosphoric acid, and the subsequent conversion of this to ammonium phosphomolybdate according to the following reactions:[66,67,74,89]

$$\text{Organic P} \xrightarrow[\text{(HNO}_3 \text{ or HNO}_3 + \text{H}_2\text{SO}_4)]{\text{Oxidation}} \text{H}_2\text{O} + \text{CO}_2 + \text{H}_3\text{PO}_4$$

and

$$\text{H}_3\text{PO}_4 \xrightarrow[\text{HNO}_3]{\text{(NH}_4)_2 \text{MoO}_4} (\text{NH}_4)_3\text{PO}_4 \cdot 14\text{MoO}_3$$

The formula for the ammonium phosphomolybdate precipitate is empirical, based on the work of Pregl,[74] and is quite different from that accepted in macroanalysis, namely, $(\text{NH}_4)_3\text{PO}_4 \cdot 12\text{MoO}_3 \cdot 2\text{HNO}_3 \cdot \text{H}_2\text{O}$, which on being heated at 160°–180° C is converted into $(\text{NH}_4)_3\text{PO}_4 \cdot 12\text{MoO}_3$.[95]

Two procedures of decomposing the sample are given below, namely, one employing a Kjeldahl-like type of digestion and the other employing Carius combustion. The Kjeldahl-like type of digestion is not applicable to fluorine-containing compounds, but is generally proven to be reliable with other substances. This digestion, together with the rest of the procedure described below, was adopted by the Association of Official Agricultural Chemists after being subjected to collaborative study.[69,70] However, this method of destroying organic material is not as vigorous as employing Carius combustion so that it is quite conceivable that certain compounds which resist oxidation might be better treated by the latter. Therefore, it is recommended that both be employed where there is doubt. The Carius procedure described below is applicable to fluorine-containing compounds.[81]

KJELDAHL GRAVIMETRIC METHOD
(Not Applicable to Fluorine-containing Compounds)*

Reagents

CONCENTRATED SULFURIC ACID, SP. GR. 1.84
Reagent grade of concentrated sulfuric acid, sp. gr. 1.84 is used as part of the combustion mixture.

---
* A white precipitate is present at the end of the digestion. (Compare Fennel, Roberts, and Webb,[30] and Furman and State.[37])

CONCENTRATED NITRIC ACID, SP. GR. 1.42

Reagent grade of concentrated nitric acid is used along with the sulfuric acid, above, for oxidizing the organic material and converting the phosphorus to phosphoric acid.

NITRIC-SULFURIC ACID MIXTURE[66,67,89]

Four hundred and twenty milliliters of concentrated nitric acid, sp. gr. 1.42, is slowly poured into 580 ml. of distilled water. To this is added, slowly, 30 ml. of concentrated sulfuric acid, sp. gr. 1.84.

AMMONIUM NITRATE SOLUTION, 2%

A 2% solution of reagent grade of ammonium nitrate in distilled water is prepared and made slightly acid by the addition of two drops of concentrated nitric acid, sp. gr. 1.42, per liter. This solution, which is used as a wash liquid for the precipitate, is stored in a glass-stoppered bottle. This solution must be filtered immediately before being used, as it often develops a flaky precipitate.

MOLYBDATE REAGENT[66,67,89]

One hundred and fifty grams of reagent grade of powdered ammonium molybdate is dissolved in 400 ml. of boiling distilled water and the resulting solution, which is usually cloudy, is cooled to room temperature under a tap. Fifty grams of reagent grade of ammonium sulfate is placed in a 1-liter volumetric flask and dissolved in a mixture of 105 ml. of distilled water and 395 ml. of concentrated nitric acid, sp. gr. 1.42. The volumetric flask containing the nitric acid solution of ammonium sulfate is cooled under the tap while the cooled solution of ammonium molybdate (even though cloudy) is added to it, slowly, in a thin stream, and with constant stirring. At no time should the mixture in the volumetric flask be allowed to become warm. The resulting solution which should be almost clear is diluted with distilled water to 1000 ml. and stored in a refrigerator for 3 days. It is then filtered through an ordinary filter paper and stored in a paraffin-lined*[9,89] glass-stoppered brown reagent bottle in the refrigerator. It is filtered again immediately before being used. Since this reagent is stable for only some months, it should be repeatedly checked by performing a phosphorus determination with it using some known substance. As the reagent ages, low results are obtained even though duplicate determinations will give excellent checks on each other.

ETHANOL, 95%

Ethanol, 95%, is used as a wash liquid for the precipitate.

---

* The paraffin-lined bottle is prepared by adding molten paraffin (melted on a steam bath) to the dry bottle and rotating until the wax has completely solidified.

12. Phosphorus

ACETONE
Pure acetone is used as a wash liquid for the precipitate.

Apparatus

KJELDAHL DIGESTION FLASK, 30 ML.
The regular straight type of Kjeldahl flask, 30 ml. (Fig. 105, Chapter 8) is used for both oxidation of the organic material and the precipitation of the ammonium phosphomolybdate.

DIGESTION RACK AND MANIFOLD
The Kjeldahl flasks are held during digestion of the sample in the Kjeldahl digestion rack and manifold described in connection with the Kjeldahl determination of nitrogen (Figs. 107–110, Chapter 8).

FILTER TUBE[41,42,66,67,74,77-80,89,91]
The filter tubes described in the chapter on the halogen determination are used for collecting the precipitate (see Figs. 158 and 159, Chapter 11). They should be tested to make certain that the filter will retain the phosphomolybdate precipitate. The filter tubes may be used without cleaning for successive determinations done the same day but not at longer periods. The filter tubes are cleaned between determinations by treatment with $0.1N$ NaOH to dissolve the precipitate followed immediately by water, dilute nitric acid, ammonium nitrate, ethanol, and acetone exactly as done during a determination (see below).

FILTRATION ASSEMBLY[89,91]
The filtration assembly described in the above-mentioned chapter is used in transferring the precipitate from the Kjeldahl flask to the filter tube (see Fig. 160, Chapter 11).

WASH BOTTLES
Two wash bottles of the type shown in Figs. 145 or 146, Chapter 10 (one each for the ammonium nitrate and the ethanol) are used in transferring the precipitate from the Kjeldahl flask to the filter tube.

RUBBER OR NEOPRENE STOPPERS
Two or three small *solid* rubber or neoprene stoppers are used in removing the precipitate from the walls of the Kjeldahl flask during its transfer to the filter tube. These are dropped into the bulb of the flask and shaken around in the presence of a small amount of the wash liquids. They strike the walls

of the flask and act in the same manner as does the ordinary rubber policeman, generally used in macroanalysis. The smaller ends of the universal solid stopper[1,88,89] shown in Fig. 165, serve the purpose very well.

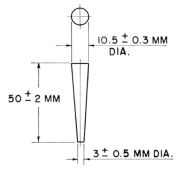

FIG. 165.   Solid rubber (or neoprene) stopper—details of construction.

VACUUM DESICCATOR

An ordinary vacuum desiccator, without desiccant, is used in drying the precipitate.

VACUUM PUMP

An ordinary electric motor-driven vacuum pump, capable of producing a pressure of a fraction of a millimeter of mercury, is used in the drying of the precipitate.

Procedure[66,67,89]

About 5 mg. of sample is weighed and transferred into a Kjeldahl flask, 30 ml., by one of the methods described in Chapter 3, that of using the charging tube (Figs. 47–49, Chapter 3) being preferred to the others for this purpose, since a precipitation also is done in this flask. (In the event that a porcelain boat must be used, it is removed after the digestion and washed with the nitric-sulfuric acid mixture and with the water required to dilute the digest. If the sample must be weighed in a glass capillary, this too must be removed before precipitation either by the use of a platinum wire or by filtration.) The size of the sample should be governed by the percentage of phosphorus present so that a precipitate of not more than 40–50 mg. is obtained. (The weight of the precipitate is roughly sixty-nine times that of the phosphorus present.[74]) One-half milliliter of concentrated sulfuric acid is added to the sample in the Kjeldahl flask followed by four to five drops of concentrated nitric acid. The flask is placed on the digestion rack and the mixture heated, cautiously at first and then strongly, until no more brown fumes of oxides of nitrogen

12. Phosphorus

escape and only white ones of sulfur trioxide are in the neck of the flask. The flask is then cooled under the tap and four to five drops of concentrated nitric acid again added and the above-described heating repeated, cooled, four to five drops of nitric acid added, and the contents heated a third time until white fumes of sulfur trioxide appear. It is then cooled and 2 ml. of the nitric-sulfuric acid mixture added, followed by 12.5 ml. of distilled water. The contents of the Kjeldahl flask are then heated on a steam bath for 15 minutes to insure complete conversion of the phosphorus to orthophosphoric acid. The Kjeldahl flask is then removed from the steam bath, held vertically, and 15 ml. of freshly filtered molybdate reagent added immediately by means of a pipette. The reagent should be added so that it drops into the center of the liquid in the Kjeldahl flask and none flows down the sides. (Note: The mixture should not be heated after the addition of the molybdate reagent.) The resulting yellow solution soon becomes cloudy and after 2–3 minutes the flask is swirled gently to thoroughly mix the contents but not violently enough to cause splashing on the sides. It is then covered so as to prevent contamination with dust and set aside in the dark overnight.

The next morning the precipitate is filtered into a previously weighed filter tube with the aid of the filtration assembly using the technique described for the halogen determination (Chapter 11). After the bulk of the precipitate and the supernatant liquor is transferred, a small amount of precipitate clings to the inner walls of the Kjeldahl flask. Small amounts of ammonium nitrate and ethanol, respectively, are alternately used to complete the transfer of the ammonium phosphomolybdate. If after a few alternate washings, some precipitate still clings to the inner walls, two or three clean tiny solid rubber stoppers are dropped into the flask, followed by a little wash liquid and the stoppers made to bounce around by shaking. This removes the last traces of precipitate from the walls and after several more washings alternately with ammonium nitrate and ethanol, respectively, the transfer is quantitative. (The rubber stoppers are kept in the flask during the transfer and if they have been selected small enough, do not interfere with the siphon tube.) The siphon tube is then disconnected from the filter tube, the stopper which had joined them washed with ammonium nitrate and then with ethanol, catching the washings in the filter tube. The precipitate is then washed with a little more ammonium nitrate, then with ethanol, and finally with acetone and sucked dry. The filter tube is then wiped with a chamois skin (see Chapter 3)* and placed in a vertical position in a vacuum desiccator which contains no desiccant. The desiccator is then evacuated to a pressure of about one mm. Hg for a period of 30 minutes, the pump being kept in continuous operation. The vacuum is then released and the filter tube *immediately* weighed to the nearest

---

* Ground tube after wiping.

0.1 mg. The precipitate is extremely hygroscopic so that it is impossible to attempt to weigh in the normal manner. Unless there is undue delay in weighing the absorbed moisture does not affect the results since the factor is so low.

As explained above, the filter may be used for successive determinations on the same day but not for longer periods.

Calculation:

Factor: 0.014524*[36,42,74,77-79,89]

$$\therefore \frac{\text{Wt. of precipitate} \times 0.014524 \times 100}{\text{Wt. sample}} = \% \text{ P}$$

Example:

33.1 mg. of ammonium phosphomolybdate is obtained from a 5.316-mg. sample

$$\therefore \frac{33.1 \times 0.014524 \times 100}{5.316} = 9.04\% \text{ P}$$

The accuracy of the determination is about $\pm 0.3\%$.

CARIUS GRAVIMETRIC METHOD
(Applicable to Fluorine-containing Compounds)

Although the Kjeldahl gravimetric method has been used with reliability in the author's laboratories for more than twenty years, it is not applicable to fluorine-containing compounds. As stated earlier in this chapter, by substituting Carius combustion for the Kjeldahl digestion these compounds yield correct results. The rest of the method, namely, precipitation as ammonium phosphomolybdate, is the same as for the Kjeldahl. The method is not restricted to fluorine-containing compounds, but may be used on all types of substances. The only disadvantage is the time factor in regard to the combustion, but as pointed out in Chapters 10 and 11, this becomes unimportant. In addition, due to the vigorous treatment during the combustion, certain compounds which resist oxidation are better handled by this method.

---

* Pregl[74] established the factor empirically as 0.014524 (log = 16209). Roth[77-79] also gave these figures. However, the English translations of Roth-Pregl by Daw[80] and by Grant[41] and also in the first and second editions of the book by Niederl and Niederl,[66,67] the factor has been mistakenly given as 0.01454. Niederl and Niederl, and Daw list the correct logarithm for 0.014524, namely, 16209, but Grant has listed the log as 16249 which is neither correct for 0.014524 nor for 0.01454.

Reagents

FUMING NITRIC ACID, REAGENT GRADE

Reagent grade of acid, sp. gr. 1.49–1.50 is used for destroying the organic material and converting the phosphorus to phosphoric acid. (*Caution!* See Chapter 10.)

CONCENTRATED SULFURIC ACID, SP. GR. 1.84
NITRIC-SULFURIC ACID MIXTURE
AMMONIUM NITRATE SOLUTION, 2%
MOLYBDATE REAGENT
ETHANOL, 95%
ACETONE

Same as for Kjeldahl gravimetric method.

Apparatus

CARIUS COMBUSTION FURNACE[90]

See Chapter 10, Fig. 135.

CARIUS COMBUSTION TUBE[90]

See Chapter 10, Fig. 137.

KJELDAHL DIGESTION FLASK, 30 ML.
FILTER TUBE
FILTRATION ASSEMBLY
WASH BOTTLES
RUBBER STOPPERS
VACUUM DESICCATOR
VACUUM PUMP

Same as for Kjeldahl gravimetric method.

Procedure

About 5 mg. of sample is weighed and transferred into a Carius combustion tube by one of the methods described in Chapter 3 (also see Chapter 10). (If a porcelain boat, capillary, or weighing bottle is used, it must be removed previous to precipitation.) The size of the sample should be governed by the percentage of phosphorus present so that not more than 40–50 mg. of ammonium phosphomolybdate precipitate is obtained. (The weight of the precipitate is roughly sixty-nine times that of the phosphorus present.[74]) One-half milliliter of fuming nitric acid is added and the Carius tube sealed. (*Caution!* See Chapter 10 regarding *all* precautions as to quantities of acid, length of sealed tube, temperatures, opening of tubes after combustion, etc.[90])

The sealed tube is heated at 250° C.* for 8 hours in the Carius combustion furnace, after which the tubes are cooled, opened, the contents transferred to a 50 ml. beaker and evaporated to *dryness* on a steam bath. To the residue is added 0.5 ml. concentrated sulfuric acid, followed by 2 ml. of the nitric-sulfuric acid mixture. The resulting solution is transferred to a 30 ml. Kjeldahl digestion flask and diluted with 12.5 ml. of distilled water, which is used in the transfer. The contents of the flask are then heated on a steam bath for 15 minutes to insure complete conversion of the phosphorus to orthophosphoric acid. The flask is then removed from the steam bath, held vertically, and 15 ml. of freshly filtered molybdate reagent added immediately, by means of a pipette, *exactly* as described in the early part of this chapter, in connection with the Kjeldahl gravimetric method. From this point on the procedure is *exactly* the same as that of the above, that is, the mixing, allowing to stand overnight, filtering, washing, drying, and weighing.

Calculation:

Same as for the Kjeldahl Gravimetric Method.

SIMULTANEOUS DETERMINATION OF BARIUM AND PHOSPHORUS

Barium and phosphorus may be determined simultaneously. After Carius combustion, the contents of the tube are washed with distilled water into a previously weighed porcelain sulfur crucible (Fig. 143, Chapter 10), and the contents evaporated to dryness on a steam bath. The residue is treated with 2–3 ml. of distilled water and about one ml. of 10% sulfuric acid, and the contents digested on a steam bath to coagulate the precipitate. (If Kjeldahl digestion[79,80] has been used instead of Carius combustion, the contents of the Kjeldahl flask are washed into the porcelain sulfur crucible. In this case, no additional sulfuric acid should be added. However, since barium phosphate is *soluble* in dilute acid solution, the Carius procedure is to be preferred, because, with the Kjeldahl procedure, the precipitated barium sulfate is difficult to transfer to the porcelain sulfur crucible.) The filtrate is sucked off *and the washings* added to the main filtrate, keeping the volume to about 12 ml. The precipitate ($BaSO_4$) is dried, ignited, and weighed (see Chapter 10). The filtrate and washings are transferred to a 30 ml. Kjeldahl flask and *if* Carius combustion was used, 0.4 ml. of concentrated sulfuric acid is added. (However, if Kjeldahl digestion was used *no additional concentrated sulfuric acid should be added.*) Then, 2 ml. of the nitric acid-sulfuric acid mixture is added,

---

* See Table 21, Chapter 10.

12. Phosphorus

followed by heating on the steam bath for 15 minutes, addition of 15 ml. of molybdate reagent, standing overnight, filtration, drying, and weighing of the ammonium phosphomolybdate as described previously in this chapter.

TABLE 24
ADDITIONAL INFORMATION ON REFERENCES* RELATED TO CHAPTER 12

As with the previous chapters, the author wishes to call to the attention of the reader additional material in regard to the determination of phosphorus. This is shown in Table 24. (See statement at top of Table 4 of Chapter 1, regarding completeness of this material.)

Books
Association of Official Agricultural
 Chemists, 2
Belcher and Godbert, 6, 7
Clark, E. P., 19
Clark, S. J., 20
Furman, 36
Grant, 41, 42
Milton and Waters, 63
Niederl and Niederl, 66, 67
Niederl and Sozzi, 68
Pregl, 74
Roth, 77–80
Steyermark, 89

Collaborative studies
Association of Official Agricultural
 Chemists, 2
Ogg, 69, 70

Ultramicro-, submicro-methods
Harvey, 43
Kirsten, 51, 52
Lindner and Kirk, 59
Nakamura, 64
Schaffer, Fong, and Kirk, 84
Welch and West, 99

Simultaneous determination of phosphorus and other elements
Bartels and Hoyme, 4
Eger and Lipke, 28
Klimova, Korshun, and Terent'eva, 54
Korshun and Terent'eva, 55
Merz, 62
Roth, 79, 80

Fluoro-compounds
Belcher and Macdonald, 8
Fennell, Roberts, and Webb, 30
Fennell and Webb, 31
Furman and State, 37
Rush, Cruikshank, and Rhodes, 81

General, miscellaneous
Bachofer and Wagner, 3
Bartels and Hoyme, 4
Chen, Toribarb, and Warner, 18
Fennell and Webb, 32
Harvey, 43
Kirsten and Carlsson, 53
Ma and McKinley, 61
Tunnicliffe, 97

Carius combustion
Olivier, 71
Rush, Cruikshank, and Rhodes, 81

Kjeldahl-type digestion (sulfuric or perchloric plus nitric acids)
Association of Official Agricultural
 Chemists, 2
Chalmers and Thomson, 17
Clark, S. J., 20
Cogbill, White, and Susano, 21
Kirsten and Carlsson, 53
Lévy, 56
Sakamoto and Hayazu, 82
Sterges, Hardin, and MacIntire, 86
Steyermark, 89
Welch and West, 99

* The numbers which appear after each entry in this table refer to the literature citations in the reference list at the end of the chapter.

TABLE 24 (*Continued*)

Bombs and fusion in general

Batt, 5
Burton and Riley, 15
Clark, E. P., 19
Clark, S. J., 20
Niederl and Niederl, 66, 67
Parr, 72

Oxygen flask combustion

Belcher and Macdonald, 8
Bennewitz and Tänzer, 10
Cohen and Czech, 22
Corner, 23
Fleischer, Southworth, Hodecker, and Tuckerman, 34
Kirsten and Carlsson, 53
Thomas, 94

Gravimetric procedures

Clark, E. P., 19
Clark, S. J., 20
Fennell and Webb, 31
Furman, 36
Gheorghiu and Radulescu, 39
Gottschalk, 40
Heller, 45
Lorenz, 60
Niederl and Niederl, 66, 67
Pregl, 74
Roth, 77–80
Rush, Cruikshank, and Rhodes, 81
Schuhecker, 85
Steyermark, 89
Tsuzuki, Miwa, and Kobayashi, 96
Woy, 100

Volumetric procedures

Belcher and Macdonald, 8
Diemair and Baier, 26
Fleischer, Southworth, Hodecker, and Tuckerman, 34
Lindner and Kirk, 59
Niederl and Niederl, 67
Roth, 79, 80
Sakamoto and Hayazu, 82

Quinoline phosphomolybdate (8-quinolinol phosphomolybdate)

Belcher and Macdonald, 8
Fennell and Webb, 31
Gottschalk, 40

Strychnine phosphomolybdate methods

Embden and Schmidt, 29
Hegedüs and Dvorszky, 44

Spectrophotometric, colorimetric, optical methods

Batt, 5
Bernhart and Wreath, 11
Bruno and Belluco, 14
Burton and Riley, 15
Carles, 16
Chalmers and Thomson, 17
Di Bacco, 24
Dickman and Bray, 25
Fiske and Subbarow, 33
Fleischer, Southworth, Hodecker, and Tuckerman, 34
Fontaine, 35
Furman, 36
Gates, 38
Harvey, 43
Hegedüs and Dvorszky, 44
Jureček and Jeník, 47, 48
King, 49
Kirsten and Carlsson, 53
Nakamura, 64
Rhodes, 75
Roth, 76, 79
Sass, Ludemann, Witten, Fischer, Sisti, and Miller, 83
Schaffer, Fong, and Kirk, 84
Taussky and Shorr, 92
Taylor, Miller, and Roth, 93

X-ray fluorescence

Natelson, 65
Philips Electronics, 73

Flame photometry

Brite, 13
Dippel, 27

TABLE 24 (*Continued*)

Amperometric, potentiometric, polarographic methods
Boos and Conn, 12
Cogbill, White, and Susano, 21
Lévy, 56–58
Stern, 87

EDTA, complexometric procedures
Bennewitz and Tänzer, 10
Fleischer, Southworth, Hodecker, and Tuckerman, 34

Manometric, gasometric methods
See Chapter 18
Hoagland, 46
Kirk, 50
Van Slyke, Page, and Kirk, 98

Cation-exchange, chromatographic methods
Bruno and Belluco, 14
Cogbill, White, and Susano, 21
Eger and Lipke, 28

REFERENCES

1. American Society for Testing Materials, *ASTM Designation*, **E 148-59T**.
2. Association of Official Agricultural Chemists, "Official Methods of Analysis," 8th ed., pp. 801–811, Washington, D. C., 1955.
3. Bachofer, M. D., and Wagner, E. C., *Ind. Eng. Chem., Anal. Ed.*, **15**, 601 (1943).
4. Bartels, U., and Hoyme, H., *Chem. Tech. (Berlin)*, **11**, 156 (1959); *Chem. Abstr.*, **53**, 15863 (1959).
5. Batt, W. G., *J. Franklin Inst.*, **248**, 451 (1949).
6. Belcher, R., and Godbert, A. L., "Semi-Micro Quantitative Organic Analysis," Longmans, Green, London, New York, and Toronto, 1945.
7. Belcher, R., and Godbert, A. L., "Semi-Micro Quantitative Organic Analysis," 2nd ed., Longmans, Green, London, 1954.
8. Belcher, R., and Macdonald, A., *Talanta*, **1**, 185 (1958).
9. Bell, R. D., and Doisy, E. A., *J. Biol. Chem.*, **44**, 55 (1920).
10. Bennewitz, R., and Tänzer, I., *Mikrochim. Acta*, p. 835 (1959).
11. Bernhart, D. N., and Wreath, A. R., *Anal. Chem.*, **27**, 440 (1955).
12. Boos, R. N., and Conn, J. B., *Anal. Chem.*, **23**, 674 (1951).
13. Brite, D. W., *Anal. Chem.*, **27**, 1815 (1955).
14. Bruno, M., and Belluco, U., *Ricerca sci.*, **26**, 3337 (1956).
15. Burton, J. D., and Riley, J. D., *Analyst*, **80**, 391 (1955).
16. Carles, J., *Bull. soc. chim. biol.*, **38**, 255 (1956).
17. Chalmers, R. A., and Thomson, D. A., *Anal. Chim. Acta*, **18**, 575 (1958).
18. Chen, P. S., Toribara, T. Y., and Warner, H., *Anal. Chem.*, **28**, 1756 (1956).
19. Clark, E. P., "Semimicro Quantitative Organic Analysis," Academic Press, New York, 1943.
20. Clark, S. J., "Quantitative Methods of Organic Microanalysis," Butterworths, London, 1956.
21. Cogbill, E. G., White, J. C., and Susano, C. D., *Anal. Chem.*, **27**, 455 (1955).
22. Cohen, L. E., and Czech, F. W., *Chemist Analyst*, **47**, 86 (1958).
23. Corner, M., *Analyst*, **84**, 41 (1959).
24. Di Bacco, G., *Boll. chim. farm.*, **93**, 43, 88 (1954).
25. Dickman, S. R., and Bray, R. H., *Ind. Eng. Chem., Anal. Ed.*, **12**, 665 (1940).
26. Diemair, W., and Baier, R., *Z. anal. Chem.*, **133**, 7 (1951).
27. Dippel, W. A., *Anal. Chem.*, **26**, 553 (1954).
28. Eger, C., and Lipke, J., *Anal. Chim. Acta*, **20**, 548 (1959).

29. Embden, G., and Schmidt, G., *in* "Handbuch der biologischen Arbeitsmethoden" (E. Abderhalden, ed.), Abt. V, Tl. 5A, pp. 11, 1548, Urban & Schwarzenberg, Berlin, 1920–1936.
30. Fennell, T. R. F. W., Roberts, M. W., and Webb, J. R., *Analyst,* **82**, 639 (1957).
31. Fennell, T. R. F. W., and Webb, J. R., *Talanta,* **2**, 105 (1959).
32. Fennell, T. R. F. W., and Webb, J. R., *Talanta,* **2**, 389 (1959).
33. Fiske, C. H., and Subbarow, Y., *J. Biol. Chem.,* **66**, 375 (1925).
34. Fleischer, K. D., Southworth, B. C., Hodecker, J. H., and Tuckerman, M. M., *Anal. Chem.,* **30**, 152 (1958).
35. Fontaine, T. D., *Ind. Eng. Chem., Anal. Ed.,* **14**, 77 (1942).
36. Furman, N. H., ed., "Scott's Standard Methods of Chemical Analysis," 5th ed., Vol. II, Van Nostrand, New York, 1939.
37. Furman, N. H., and State, H. M., *Ind. Eng. Chem., Anal. Ed.,* **8**, 420 (1936).
38. Gates, O. R., *Anal. Chem.,* **26**, 730 (1954).
39. Gheorghiu, C., and Radulescu, E., *Rev. chim. (Bucharest),* **8**, 779 (1957); *Chem. Abstr.,* **52**, 7013 (1958).
40. Gottschalk, G., *Z. anal. Chem.,* **159**, 257 (1958).
41. Grant, J., "Quantitative Organic Microanalysis, Based on the Methods of Fritz Pregl," 4th ed., Blakiston, Philadelphia, Pennsylvania, 1946.
42. Grant, J., "Quantitative Organic Microanalysis," 5th ed., Blakiston, Philadelphia, Pennsylvania, 1951.
43. Harvey, H. W., *Analyst,* **78**, 110 (1953).
44. Hegedüs, A. J., and Dvorszky, M., *Mikrochim. Acta,* p. 141 (1959).
45. Heller, K., *Mikrochemie,* **7**, 208 (1929); *Chem. Abstr.,* **24**, 1818 (1930).
46. Hoagland, C. L., *J. Biol. Chem.,* **136**, 543 (1940).
47. Jureček, M., and Jeník, J., *Chem. listy,* **51**, 1312 (1957).
48. Jureček, M., and Jeník, J., *Collection Czechoslov. Chem. Communs.,* **23**, 447 (1958).
49. King, E. J., *Biochem. J.,* **26**, 292 (1932).
50. Kirk, E., *J. Biol. Chem.,* **106**, 191 (1934).
51. Kirsten, W., *Chim. anal.,* **40**, 253 (1958).
52. Kirsten, W. J., *Microchem. J.,* **2**, 179 (1958).
53. Kirsten, W. J., and Carlsson, M. E., *Microchem. J.,* **4**, 3 (1960).
54. Klimova, V. A., Korshun, M. O., and Terent'eva, E. A., *Zhur. Anal. Khim.,* **9**, 275 (1954); *Chem. Abstr.,* **49**, 2942 (1955).
55. Korshun, M. O., and Terent'eva, E. A., *Doklady Akad. Nauk S.S.S.R.,* **100**, 707 (1955).
56. Lévy, R., *Bull. soc. chim. France,* p. 517 (1956).
57. Lévy, R., *Compt. rend. acad. sci.,* **236**, 1781 (1953).
58. Lévy, R., *Compt. rend. acad. sci.,* **238**, 2320 (1954).
59. Lindner, R., and Kirk, P. L., *Mikrochemie,* **22**, 300 (1937).
60. Lorenz, N., *Landwirtsch. Vers.-Sta.,* **55**, 183 (1901).
61. Ma, T. S., and McKinley, J. D., Jr., *Mikrochim. Acta,* p. 4 (1953).
62. Merz, W., *Mikrochim. Acta,* p. 456 (1959).
63. Milton, R. F., and Waters, W. A., "Methods of Quantitative Microanalysis," 2nd ed., Arnold, London, 1955.
64. Nakamura, G. R., *Anal. Chem.,* **24**, 1372 (1952).
65. Natelson, S., Personal communication. 1959.
66. Niederl, J. B., and Niederl, V., "Micromethods of Quantitative Organic Elementary Analysis," Wiley, New York, 1938.
67. Niederl, J. B., and Niederl, V., "Micromethods of Quantitative Organic Analysis," 2nd ed., Wiley, New York, 1942.

68. Niederl, J. B., and Sozzi, J. A., "Microanálisis Elemental Orgánico," Calle Arcos, Buenos Aires, 1958.
69. Ogg, C. L., *J. Assoc. Offic. Agr. Chemists,* **39**, 408 (1956).
70. Ogg, C. L., *J. Assoc. Offic. Agr. Chemists,* **40**, 386 (1957).
71. Olivier, S. C., *Rec. trav. chim.,* **59**, 872 (1940).
72. Parr, S. W., *J. Am. Chem. Soc.,* **30**, 764 (1908).
73. Philips Electronics, Inc., Mount Vernon, New York.
74. Pregl, F., "Quantitative Organic Microanalysis" (E. Fyleman, trans. 2nd German ed.), Churchill, London, 1924.
75. Rhodes, D. N., *Nature,* **176**, 215 (1955).
76. Roth, H., *Mikrochemie ver. Mikrochim. Acta,* **31**, 290 (1944).
77. Roth, H., "Die quantitative organische Mikroanalyse von Fritz Pregl," 4th ed., Springer, Berlin, 1935.
78. Roth, H., "F. Pregl. quantitative organische Mikroanalyse," 5th ed., Springer, Wien, 1947.
79. Roth, H., "Pregl-Roth quantitative organische Mikroanalyse," 7th ed., Springer, Wien, 1958.
80. Roth, H., "Quantitative Organic Microanalysis of Fritz Pregl," 3rd ed. (E. B. Daw, trans. 4th German ed.), Blakiston, Philadelphia, Pennsylvania, 1937.
81. Rush, C. A., Cruikshank, S. S., and Rhodes, E. J. H., *Mikrochim. Acta,* p. 858 (1956).
82. Sakamoto, S., and Hayazu, R., *Yakugaku Zasshi,* **70**, 698 (1950).
83. Sass, S., Ludemann, W. D., Witten, B., Fischer, V., Sisti, A. J., and Miller, J. I., *Anal. Chem.,* **29**, 1346 (1957).
84. Schaffer, F. L., Fong, J., and Kirk, P. L., *Anal. Chem.,* **25**, 343 (1953).
85. Schuhecker, K., *Z. anal. Chem.,* **116**, 14 (1939).
86. Sterges, A. J., Hardin, L. J., and MacIntire, W. H., *J. Assoc. Offic. Agr. Chemists,* **33**, 114 (1950).
87. Stern, A., *Ind. Eng. Chem., Anal. Ed.,* **14**, 74 (1942).
88. Steyermark, Al, Chairman, Committee for the Standardization of Microchemical Apparatus, Division of Analytical Chemistry, American Chemical Society, *Anal. Chem.,* **22**, 1228 (1950).
89. Steyermark, Al, "Quantitative Organic Microanalysis," Blakiston, Philadelphia, Pennsylvania, 1951.
90. Steyermark, Al, Alber, H. K., Aluise, V. A., Huffman, E. W. D., Jolley, E. L., Kuck, J. A., Moran, J. J., and Willits, C. O., *Anal. Chem.,* **23**, 1689 (1951).
91. Steyermark, Al, Alber, H. K., Aluise, V. A., Huffman, E. W. D., Kuck, J. A., Moran, J. J., and Willits, C. O., *Anal. Chem.,* **21**, 1555 (1949).
92. Taussky, H. H., and Shorr, E., *J. Biol. Chem.,* **202**, 675 (1953).
93. Taylor, A. E., Miller, C. W., and Roth, H., *Mikrochemie ver. Mikrochim. Acta,* **31**, 292 (1944).
94. Thomas, Arthur H., Company, "Technical Service, DU 37-5M-8'58," Philadelphia, Pennsylvania (1958).
95. Treadwell, F. P., and Hall, W. T., "Analytical Chemistry," Vol. II, Wiley, New York, 1924.
96. Tsuzuki, Y., Miwa, M., and Kobayashi, E., *Anal. Chem.,* **23**, 1179 (1951).
97. Tunnicliffe, M. E., *Trans. Inst. Rubber Ind.,* **31**, T141 (1955).
98. Van Slyke, D. D., Page, I. H., and Kirk, E., *J. Biol. Chem.,* **102**, 635 (1933).
99. Welch, C. M., and West, P. W., *Anal. Chem.,* **29**, 874 (1957).
100. Woy, R., *Chemiker-Ztg.,* **21**, 441 (1897).

CHAPTER 13

Microdetermination of Arsenic

CARIUS VOLUMETRIC METHOD

The determination of arsenic in organic compounds is based on the oxidation of the organic material to carbon dioxide and water and the arsenic to arsenic acid, simultaneously.[22-24,43-46,50-53,57,61,65] The oxidation may be effected by several means as well as the subsequent determination of the arsenic acid.[5,6,11-14, 22-24,43-46,50-53,57,61,65] For these, the author prefers the Carius method[11-14,57] of oxidation (compare Chapters 10 and 11) and the iodometric procedure[45,50-53, 57,65] for the arsenic acid. Although other methods[23,24,45,46,50-53] have been used in the author's laboratory for both stages, best results were obtained with these. The reactions are as follows[22,45,50-53,57,65]:

(a) Oxidation:

$$\text{Organic As} \xrightarrow[\text{[O]}]{\text{HNO}_3} \text{CO}_2 + \text{H}_2\text{O} + \text{H}_3\text{AsO}_4$$

(b) $\text{H}_3\text{AsO}_4 + 2\text{HI} \rightarrow \text{H}_3\text{AsO}_3 + \text{H}_2\text{O} + \text{I}_2$

(c) $\text{I}_2 + 2\text{Na}_2\text{S}_2\text{O}_3 \rightarrow 2\text{NaI} + \text{Na}_2\text{S}_4\text{O}_6$

With this procedure, there is no interference from halogens (bromine, chlorine, fluorine, or iodine), sulfur, or phosphorus. With other procedures, interference has been reported.[22]

Reagents

FUMING NITRIC ACID, SP. GR., 1.49–1.50

Reagent grade of fuming nitric acid, sp. gr. 1.49–1.50, is used to oxidize the organic material to carbon dioxide and water and the arsenic to arsenic acid. (*Caution:* see Chapter 10.)

CONCENTRATED HYDROCHLORIC ACID, SP. GR., 1.190

Reagent grade of concentrated hydrochloric acid, sp. gr. 1.190, is treated just before being used, as follows[45,53,57,65]: About 20–25 ml. of acid is placed in a 125-ml. ground glass-stoppered flask with the stopper removed. The acid is boiled gently for 2–3 minutes to drive out any free chlorine present. The flame

13. Arsenic

is removed, the ground glass stopper loosely inserted, and the contents of the flask cooled under the tap. (Note: If the stopper cannot be removed after cooling, the closed flask is placed under a warm water faucet for a few seconds, after which the stopper can be removed.)

POTASSIUM IODIDE SOLUTION, 4%[45,50-53,57]

This solution is prepared immediately before being used and must be colorless. It is converted into hydriodic acid during the determination of the arsenic acid.

STANDARD SODIUM THIOSULFATE, 0.01N

This solution is prepared and standardized according to the directions given in Chapter 5. It is used to titrate the iodine liberated by the reaction between the hydriodic acid (from the KI) and the arsenic acid.

STARCH INDICATOR

This is prepared according to the directions given in Chapter 5.

DISTILLED WATER

This is freshly boiled before the determination.

Apparatus

CARIUS COMBUSTION TUBE[58]

The combustion tube described in Chapter 10 (see Fig. 137) is used in the oxidation.

CARIUS COMBUSTION FURNACE[30,58,60]

The furnace described in Chapter 10 (see Fig. 135) is used for heating the Carius tube.

BLAST LAMP

The blast lamp described in Chapter 10 (see Fig. 138) is used for sealing the Carius tubes.

GRINDING OR CUTTING WHEEL

The motor-driven Carborundum grinding wheel described in Chapter 10 (see Fig. 139) is used for cutting a groove around the Carius tubes after combustion.

BURETTE

An automatic microburette of the types described in Chapter 5 (Figs. 69, and 70) is used in the titration of the liberated iodine (resulting from the reaction between the HI and H_3AsO_4).

Procedure[45,50-53,57,65]

Five to 10 mg. of sample (or enough to require about 5 ml. of thiosulfate to titrate the liberated iodine—see below) is weighed* as described in Chapter 3 and placed in a clean, dry Carius combustion tube (refer to Chapter 10). Five to six tenths† (seven tenths *maximum*)[58] of a ml. of fuming nitric acid, sp. gr. 1.49–1.50, is added to the sample and the combustion tube sealed according to the method described in Chapter 10. The sealed tube is then heated in the Carius combustion furnace at a temperature of 250° C.† for 7–8 hours. The tube is then opened (*refer to method described in Chapter 10*) and the contents quantitatively rinsed into a 30-ml. beaker. The beaker is then placed on a steam bath and the contents evaporated down to dryness. One ml. of the freshly boiled distilled water is then added to dissolve the residue and the resulting solution transferred to a 125-ml. ground glass-stoppered flask. The beaker is then washed with 5 ml. of freshly boiled concentrated hydrochloric acid (in five portions of one ml. each to insure quantitative transfer) and the acid washings added to the water solution of arsenic acid in the ground glass-stoppered flask. To the acid solution is added 2 ml. of 4% potassium iodide (freshly prepared and colorless). The flask is immediately stoppered and the mixture allowed to stand for 10 minutes. The liberated iodine is then titrated with standard $0.01N$ sodium thiosulfate until the solution is light yellow in color. Twenty ml. of recently boiled, cold distilled water is added, followed by three to four drops of starch indicator and the titration with thiosulfate completed (refer to Chapter 5). [Note: A faint pink tint to the solution is considered to be the end point. On standing for several minutes, the blue coloration reappears. In the event that the end point is overstepped, accidentally, a small measured amount of standard iodine solution (see Chapter 5) may be added, followed by thiosulfate to the end point but this should all be accomplished with as little delay as possible or high results will be obtained.]

Calculation:

Factor:

1 ml. of $0.01N$ $Na_2S_2O_3$ is equivalent to 0.3748 mg. of arsenic

$$\therefore \frac{\text{ml. of } 0.01N \text{ } Na_2S_2O_3 \times 0.3748 \times 100}{\text{Wt. sample}} = \% \text{ As}$$

Example:

4.75 ml. of $0.01N$ thiosulfate is required to titrate the iodine liberated in the analysis of a 6.319-mg. sample.

$$\therefore \frac{4.75 \times 0.3748 \times 100}{6.319} = 28.17\% \text{ As}$$

The accuracy of the method is $\pm$ 0.2–0.3%.

* If the sample is weighed into a porcelain boat, the boat is put into the Carius tube.
† See Table 21, Chapter 10.

CARIUS GRAVIMETRIC METHOD

Instead of the preferred iodometric method described previously, the determination may be carried out gravimetrically.[22-24,45,46,50-53,57] The author used this latter method for approximately nine years before changing to the iodometric. Although the gravimetric procedure gives good results there are considerably more manipulations involved besides the necessity of using the small porcelain filter crucible[23,45,46,50-53,57,59] (Neubauer) (Fig. 166) with the attending danger of loss of precipitate during filtration in the hands of a beginner.

The organic material is destroyed by Carius combustion as described for the iodometric method. The resulting arsenic acid is treated with magnesia mixture yielding magnesium ammonium arsenate which in turn is converted into magnesium pyroarsenate according to the following reactions[22,57,61]:

$$H_3AsO_4 \xrightarrow[\substack{NH_4Cl \\ NH_4OH}]{MgCl_2} MgNH_4AsO_4$$

then

$$2MgNH_4AsO_4 \xrightarrow{\Delta\; 1000°\;C.} H_2O + 2NH_3 + Mg_2As_2O_7$$

Reagents

FUMING NITRIC ACID, SP. GR. 1.49–1.50

AMMONIUM HYDROXIDE[45,46,57]

A 2N solution of ammonium hydroxide is prepared by diluting reagent grade of concentrated material (sp. gr. 0.90, 28%[29] NH_3, 15N[29]) with water.

MAGNESIA MIXTURE[22,23,45,46,50-53,57]

Five and one-half grams of crystalline magnesium chloride and 10.5 grams of ammonium chloride, both reagent grade, are dissolved in 100 ml. of distilled water.

ETHANOL, 95%

Apparatus

CARIUS COMBUSTION TUBE
CARIUS COMBUSTION FURNACE
BLAST LAMP
GRINDING OR CUTTING WHEEL

See above.

PORCELAIN FILTER CRUCIBLE AND COVER[57,59]

The porcelain filter crucible and cover shown in Fig. 166, are used for collecting and igniting the precipitate. There is an unglazed porous porcelain bottom, which is the filter medium. The crucible is attached to the filter assembly (see below) and cleaned with dilute hydrochloric acid, rinsed with water,

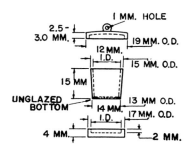

FIG. 166. Porcelain filter crucible and cover (Neubauer)—details of construction.

$2N$ ammonium hydroxide, ethanol, and again with $2N$ ammonium hydroxide, in the order named. It is then dried in an oven for 20 minutes at 120° C., after which it and the cover are heated to 1000° C. for 10 minutes in a muffle furnace, cooled for one hour on a metal block in a desiccator (Figs. 43–46), Chapter 3), and weighed.

CRUCIBLE FILTER ASSEMBLY[57,59]

A suction flask assembly similar to that used in Chapter 11 (Fig. 160) is required. The tube shown in Fig. 167, which fits into the suction flask, takes the place of that used with the above mentioned assembly. The crucible is held securely on top of this tube by means of a rubber sleeve made of thin-walled rubber tubing. This simple assembly is recommended in preference to the more complicated Wintersteiner assembly[57,59,64] which is difficult to control and seems to be very little used at present.

OVEN

An ordinary laboratory type oven (Chapter 4) is used to dry the precipitate.

MUFFLE FURNACE[60]

A small muffle furnace of the type shown in Fig. 88, capable of attaining a temperature of 1000° C., is used to convert the magnesium ammonium arsenate to magnesium pyroarsenate.

13. Arsenic

Procedure[22,23,45,46,50-53,57]

Five to 10 mg. of sample is decomposed in a Carius tube and the resulting nitric-arsenic acid solution evaporated to dryness in a 30-ml. beaker exactly as described for the iodometric method in the preceding pages.

The residue is then dissolved in 4 ml. of 2N ammonium hydroxide. To the resulting solution is added 1 ml. of magnesia mixture and the beaker is then placed in the freezing compartment of a refrigerator overnight. The precipitate which is at first amorphous becomes crystalline. The next morning

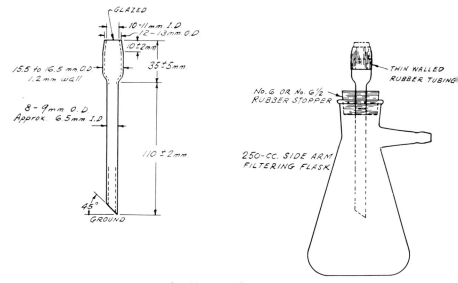

FIG. 167. Crucible filter assembly—details of construction.

the contents are allowed to melt and the precipitate transferred to the porcelain filter crucible with the aid of a medicine dropper.[45,46,54,57] The precipitate and supernatant liquid are sucked up into the dropper and slowly deposited onto the filter while mild suction is applied to the flask. (Or, if preferred, a conventional glass rod covered by a rubber sleeve or policeman may be used.) After the main bulk of the precipitate has been transferred to the filter, the remaining precipitate in the beaker is transferred with the aid of small portions of, alternately, 2N ammonium hydroxide and ethanol. The precipitate is then washed with 3 ml. of 2N ammonium hydroxide and the crucible placed in an oven at 120° C. for 20 minutes. The crucible cover is then put in place, the covered crucible removed to a muffle furnace, and heated at 1000° C. for 10 minutes after which it is allowed to cool for one hour on a metal block in a desiccator (Figs. 43–46, Chapter 3) and weighed.

Calculation:

Factor:

$$\frac{2As}{Mg_2As_2O_7} = 0.4826$$

$$\therefore \frac{Wt.\ precipitate \times 0.4826 \times 100}{Wt.\ sample} = \%\ As$$

Example:

6.053 mg. of $Mg_2As_2O_7$ is obtained from a 7.801-mg. sample

$$\therefore \frac{6.053 \times 0.4826 \times 100}{7.801} = 37.45\%\ As$$

The method has an accuracy of about $\pm 0.3\%$.

TABLE 25

ADDITIONAL INFORMATION ON REFERENCES* RELATED TO CHAPTER 13

In addition to the methods described in detail in the preceding pages of this chapter, the author wishes to call to the attention of the reader the references listed in Table 25. (See statement at top of Table 4 of Chapter 1, regarding completeness of this material.)

Books
Belcher and Godbert, 5, 6
Clark, S. J., 17
Furman, 22
Grant, 23, 24
Milton and Waters, 43, 44
Niederl and Niederl, 45, 46
Pregl, 48
Roth, 50–53
Steyermark, 57

General, miscellaneous, and review material
Heller, 28
How, 31
Jacobs and Nagler, 32
Roth, 49

Ultramicro-, submicro-methods
Kingsley and Schaffert, 35

Simultaneous determination of arsenic and other elements
Grant, 24
Roth, 52
Schulek and Wolstadt, 55

Carius combustion
Steyermark, 57

Kjeldahl-type digestion (nitric and sulfuric acid, sulfuric acid and hydrogen peroxide, chloric acid, etc.)
Belcher and Godbert, 5
Furman, 22
Lévy, 39, 40
Niederl and Niederl, 45
Roth, 51–53
Tuckerman, Hodecker, Southworth, and Fleischer, 62

Oxygen flask combustion
Corner, 18
Merz, 42

Bomb methods and fusion in general
Beamish and Collins, 4
Jureček and Jeník, 33, 34

Volumetric, iodometric methods
Bahr, Bieling, and Thiele, 1, 2
Furman, 22
Jureček and Jeník, 33, 34
Sloviter, McNabb, and Wagner, 56

* The numbers which appear after each entry in this table refer to the literature citations in the reference list at the end of the chapter.

TABLE 25 (*Continued*)

Ceric sulfate titration
Kolthoff and Amdur, 36

Photometric, colorimetric methods
Bricker and Sweetser, 8
Bruno and Belluco, 9
Chaney and Magnuson, 16
Crawford, Palmer, and Wood, 19
Di Bacco, 20
Jacobs and Nagler, 32
Oliver and Funnell, 47
Tuckerman, Hodecker, Southworth, and Fleischer, 62

Electrolytic, potentiometric, coulometric methods
Grant, 23, 24
Lévy, 39–41
Tutundžić and Mladenović, 63
Yoshimura, 66

Gravimetric methods
Heller, 27
Saschek, 54
Wintersteiner, 64

Marsh, Gutzeit,[21,22,25] arsine, etc., methods
Bodnar, Szep, and Cieleszky, 7
Bystrov and Parshikov, 10
Cassil, 15
Furman, 22
Grant, 23, 24
Haight, 26
How, 31
Jureček and Jeník, 33, 34
Lachele, 37
Levvy, G. A., 38
Milton and Waters, 43
Yoshimura, 66

Distillation of trichloride
Bang, 3
Milton and Waters, 43

Chromatographic method
Bruno and Belluco, 9

REFERENCES

1. Bähr, G., Bieling, H., and Thiele, K. H., *Z. anal. Chem.*, **143**, 103 (1954).
2. Bähr, G., Bieling, H., and Thiele, K. H., *Z. anal. Chem.*, **145**, 105 (1955).
3. Bang, I., *Biochem. Z.*, **161**, 195 (1925).
4. Beamish, F. E., and Collins, H. L., *Ind. Eng. Chem., Anal. Ed.*, **6**, 379 (1934).
5. Belcher, R., and Godbert, A. L., "Semi-Micro Quantitative Organic Analysis," Longmans, Green, London, New York, and Toronto, 1945.
6. Belcher, R., and Godbert, A. L., "Semi-Micro Quantitative Organic Analysis," 2nd ed., Longmans, Green, London, 1954.
7. Bodnar, J., Szep, Ö., and Cieleszky, V., *Z. physiol. Chem. Hoppe-Seyler's*, **264**, 1 (1940).
8. Bricker, C. E., and Sweetser, P. B., *Anal. Chem.*, **24**, 409 (1952).
9. Bruno, M., and Belluco, U., *Ricerca sci.*, **26**, 3337 (1956).
10. Bystrov, S. P., and Parshikov, Y. I., *Aptechnoe Delo*, **6**, 38 (1957); *Chem. Abstr.*, **52**, 6059 (1958).
11. Carius, L., *Ann.*, **116**, 1 (1860).
12. Carius, L., *Ann.*, **136**, 129 (1865).
13. Carius, L., *Ann.*, **145**, 301 (1868).
14. Carius, L., *Ber.*, **3**, 697 (1870).
15. Cassil, C. C., *J. Assoc. Offic. Agr. Chemists*, **24**, 196 (1941).
16. Chaney, A. L., and Magnuson, H. J., *Ind. Eng. Chem., Anal. Ed.*, **12**, 691 (1940).

17. Clark, S. J., "Quantitative Methods of Organic Microanalysis," Butterworths, London, 1956.
18. Corner, M., *Analyst*, **84**, 41 (1959).
19. Crawford, A., Palmer, J. G., and Wood, J. H., *Mikrochim. Acta*, p. 277 (1958).
20. Di Bacco, G., *Boll. chim. farm.*, **93**, 43, 88 (1954).
21. Feigl, F., "Spot Tests" (R. E. Oesper, trans.), 4th ed., Vol. I, p. 97, Elsevier, Amsterdam, Houston, London, and New York, 1954.
22. Furman, N. H., ed., "Scott's Standard Methods of Chemical Analysis," 5th ed., Vol. II, Van Nostrand, New York, 1939.
23. Grant, J., "Quantitative Organic Microanalysis, Based on the Methods of Fritz Pregl," 4th ed., Blakiston, Philadelphia, Pennsylvania, 1946.
24. Grant, J., "Quantitative Organic Microanalysis," 5th ed., Blakiston, Philadelphia, Pennsylvania, 1951.
25. Gutzeit, M., *Pharm. Ztg.*, **24**, 263 (1879).
26. Haight, G. P., *Anal. Chem.*, **26**, 593 (1954).
27. Heller, K., *Mikrochemie*, **7**, 208 (1929).
28. Heller, K., *Mikrochemie*, **14**, 369 (1934).
29. Hodgman, C. D., and Lange, N. A., "Handbook of Chemistry and Physics," 9th ed., p. 393, Chemical Rubber, Cleveland, Ohio, 1922.
30. Hoesli, H., c/o Obipectin A.-G., Bischofszell, St. Gallen, Switzerland.
31. How, A. E., *Ind. Eng. Chem., Anal. Ed.*, **10**, 226 (1938).
32. Jacobs, M. B., and Nagler, J., *Ind. Eng. Chem., Anal. Ed.*, **14**, 442 (1942).
33. Jureček, M., and Jeník, J., *Chem. listy*, **50**, 84 (1956).
34. Jureček, M., and Jeník, J., *Collection Czechoslov. Chem. Communs.*, **21**, 890 (1956).
35. Kingsley, G. R., and Schaffert, R. R., *Anal. Chem.*, **23**, 914 (1951).
36. Kolthoff, I. M., and Amdur, E., *Ind. Eng. Chem., Anal. Ed.*, **12**, 177 (1940).
37. Lachele, C. E., *Ind. Eng. Chem., Anal. Ed.*, **6**, 256 (1934).
38. Levvy, G. A., *Biochem. J.*, **37**, 598 (1943).
39. Lévy, R., *Bull. soc. chim. France*, p. 517 (1956).
40. Lévy, R., *Compt. rend. acad. sci.*, **236**, 1781 (1953).
41. Lévy, R., *Compt. rend. acad. sci.*, **238**, 2320 (1954).
42. Merz, W., *Mikrochim. Acta*, p. 640 (1959).
43. Milton, R. F., and Waters, W. A., "Methods of Quantitative Microanalysis," Longmans, Green, New York, and Arnold, London, 1949.
44. Milton, R. F., and Waters, W. A., "Methods of Quantitative Microanalysis," 2nd ed., Arnold, London, 1955.
45. Niederl, J. B., and Niederl, V., "Micromethods of Quantitative Organic Elementary Analysis," Wiley, New York, 1938.
46. Niederl, J. B., and Niederl, V., "Micromethods of Quantitative Organic Analysis," 2nd ed., Wiley, New York, 1942.
47. Oliver, W. T., and Funnell, H. S., *Anal. Chem.*, **31**, 259 (1959).
48. Pregl, F., "Quantitative Organic Microanalysis" (E. Fyleman, trans. 2nd German ed.), Churchill, London, 1924.
49. Roth, H., *Angew. Chem.*, **53**, 441 (1940).
50. Roth, H., "Die quantitative organische Mikroanalyse von Fritz Pregl," 4th ed., Springer, Berlin, 1935.
51. Roth, H., "F. Pregl quantitative organische Mikroanalyse," 5th ed., Springer, Wien, 1947.
52. Roth, H., "Pregl-Roth quantitative organische Mikroanalyse," 7th ed., Springer, Wien, 1958.

53. Roth, H., "Quantitative Organic Microanalysis of Fritz Pregl," 3rd ed. (E. B. Daw, trans., 4th German ed.), Blakiston, Philadelphia, Pennsylvania, 1937.
54. Saschek, W., *Ind. Eng. Chem., Anal. Ed.,* **9**, 491 (1937).
55. Schulek, E., and Wolstadt, R., *Z. anal. Chem.,* **108**, 400 (1937).
56. Sloviter, H. A., McNabb, W. M., and Wagner, E. C., *Ind. Eng. Chem., Anal. Ed.,* **14**, 516 (1942).
57. Steyermark, Al, "Quantitative Organic Microanalysis," Blakiston, Philadelphia, Pennsylvania, 1951.
58. Steyermark, Al, Alber, H. K., Aluise, V. A., Huffman, E. W. D., Jolley, E. L., Kuck, J. A., Moran, J. J., and Willits, C. O., *Anal. Chem.,* **23**, 1689 (1951).
59. Steyermark, Al, Alber, H. K., Aluise, V. A., Huffman, E. W. D., Kuck, J. A., Moran, J. J., and Willits, C. O., *Anal. Chem.,* **21**, 1555 (1949).
60. Thomas, Arthur H., Company, Philadelphia, Pennsylvania.
61. Treadwell, F. P., and Hall, W. T., "Analytical Chemistry," Vol. II, Wiley, New York, 1924.
62. Tuckerman, M. M., Hodecker, J. H., Southworth, B. C., and Fleischer, K. D., *Anal. Chim. Acta,* **21**, 463 (1959).
63. Tutundžić, P. S., and Mladenović, S., *Anal. Chim. Acta,* **12**, 390 (1955).
64. Wintersteiner, O., *Mikrochemie,* **2**, 14 (1924).
65. Wintersteiner, O., *Mikrochemie,* **4**, 155 (1926).
66. Yoshimura, C., *Nippon Kagaku Zasshi,* **78**, 1586 (1957); *Chem. Abstr.,* **52**, 16120 (1958).

CHAPTER 14

Microdetermination of Oxygen

The determination of oxygen in organic compounds is of utmost importance because of its wide occurrence. Although methods for its direct determination were extensively investigated in the past, none proved wholly satisfactory. Therefore, its value was obtained commonly by difference, subtracting the sum of the other determined elements from 100%. Elving and Ligett[24] presented an excellent historical review of the above methods to which the reader is referred.

In recent years a direct method which is becoming very popular and appears[101] to be a satisfactory one was first developed by Schütze,[79,80]* examined by Korshun,[51] modified by Zimmermann[104] and by Unterzaucher[95] and the latter modification thoroughly investigated in this country by Aluise et al.[5]† Modifications have been made more recently by Dundy and Stehr,[22] Holowchak and Wear,[38] Oita and Conway,[65] and by Oliver,[66] all with the purpose of eliminating errors due to the presence of sulfur and hydrogen. In a recent collaborative study,[86,87] conducted by the Association of Official Agricultural Chemists excellent results were reported by each of the collaborators regardless of which modification was used—gravimetric, volumetric (iodometric), or manometric finish.

The number of laboratories in which these procedures are being used is still quite small, but it is safe to conclude that the determination of oxygen will be done within the near future as routinely as are those of the other elements.

Two methods are described here, one gravimetric and one volumetric, those with which the author has obtained good results over the years, although the author definitely prefers the gravimetric. Both are modifications of the Unterzaucher[94-98] which was investigated by Aluise et al.,[5] with the additions recommended by Oita and Conway,[65] Canales and Parks,[12] and Campanile et al.[11] Neither procedure, without further modification is suitable for compounds containing metals, fluorine, or phosphorus. (With fluorine-containing compounds, Rush[73] uses a platinum thermal decomposition tube. With this, there is no interference. He has found that, with the quartz tubes containing magnesium nitride,[60] oxygen is retained as magnesium oxide, if water is formed in the decomposition. This obviously results in low values.)

* Compare Kirsten.[48]
† Compare Roth.[71,72]

14. Oxygen

GRAVIMETRIC METHOD

The determination is based on the thermal decomposition of the sample in an inert atmosphere and passage of the decomposition vapors over carbon at 1120° C., at which temperature the equilibrium between carbon dioxide and carbon monoxide is shifted entirely to monoxide as shown below. The carbon monoxide is then converted to carbon dioxide which is determined gravimetrically. The reactions are represented by the following[88,89]:

$$\text{Organic O} \xrightarrow[C]{1120°\ C.} CO$$

(NOTE: At 1120° C., the reactions,

$$H_2O + C = H_2 + CO$$

and

$$CO_2 + C = 2CO$$

proceed quantitatively to the right.)

Then,[89,102]

$$CO + CuO \xrightarrow{670-680°\ C.} CO_2 + Cu$$

and

$$CO_2 + 2NaOH \longrightarrow Na_2CO_3 + H_2O$$

Reagents

COPPER OXIDE

Reagent grade of copper oxide wire, 1–3 mm. in length, is used to convert the purified carbon monoxide to carbon dioxide.

ASCARITE[93]

Same as used in Chapter 9 (8–20 mesh). This is used in the carbon monoxide scrubber to remove any acidic substances resulting from the presence of sulfur or halogens in the sample. It is also used in the carbon dioxide absorption tube (see Chapter 9 for filling).

ANHYDRONE OR DEHYDRITE (MAGNESIUM PERCHLORATE)[83,93]

(See Chapter 9). This is used in the various tubes for absorbing water.

CONCENTRATED SULFURIC ACID

This is used in the bubble counter-U-tube (see Chapter 9).

PHOSPHORUS PENTOXIDE

This may be used as a drying agent in the purification train.

REDUCED COPPER TURNINGS

Same as in Chapter 7. This is used for removing traces of oxygen from the nitrogen (or helium) in the purification train and for removing[12,15,65,66] sulfur compounds leaving the thermal decomposition tube:

$$4\ Cu + CS_2 \rightarrow 2\ Cu_2S + C$$

and

$$2\ Cu + COS \rightarrow Cu_2S + CO$$

NITROGEN (OR HELIUM)[1,58]

High purity nitrogen or helium is used as the inert atmosphere in the system. Either gas must be purified further by passing through several scrubbing bottles, in series, containing different solid desiccants, such as Anhydrone, calcium chloride, or phosphorus pentoxide, after which the gas must be passed through a tube containing closely packed reduced copper turnings at 600° C., then through Anhydrone and finally through the bubble counter-U-tube.[89]

CARBON

Wyex Compact Black, J. M. Huber, Inc.,[39] available through Arthur H. Thomas Company,[93] in the form of small pellets (passing through a 30-mesh but retained on an 80-mesh sieve) is required for the thermal decomposition tube filling. It is digested with concentrated hydrochloric acid after which it is washed thoroughly with water in the following manner. The carbon and acid are stirred mechanically with a large amount of water and then the carbon allowed to settle. The wash water is decanted off, more water added, stirred, allowed to settle, decanted, etc., until the washings no longer show a test for chloride.* The carbon is then dried, placed in a quartz tube and heated in a slow stream of nitrogen, the temperature being increased gradually to 550° C. and kept thusly for a period of several hours. This treatment removes volatile material and sinters the carbon, thereby preventing channeling when it is packed later in the thermal decomposition tube.

QUARTZ WOOL AND QUARTZ CHIPS[93]

These are used in the thermal decomposition tube along with the carbon. The chips should be washed with hydrofluoric acid, then with water, and dried in an oven.

---

* This treatment with hydrochloric acid is not described in the paper by Aluise *et al.*,[5] but is now being used by those authors.[3]

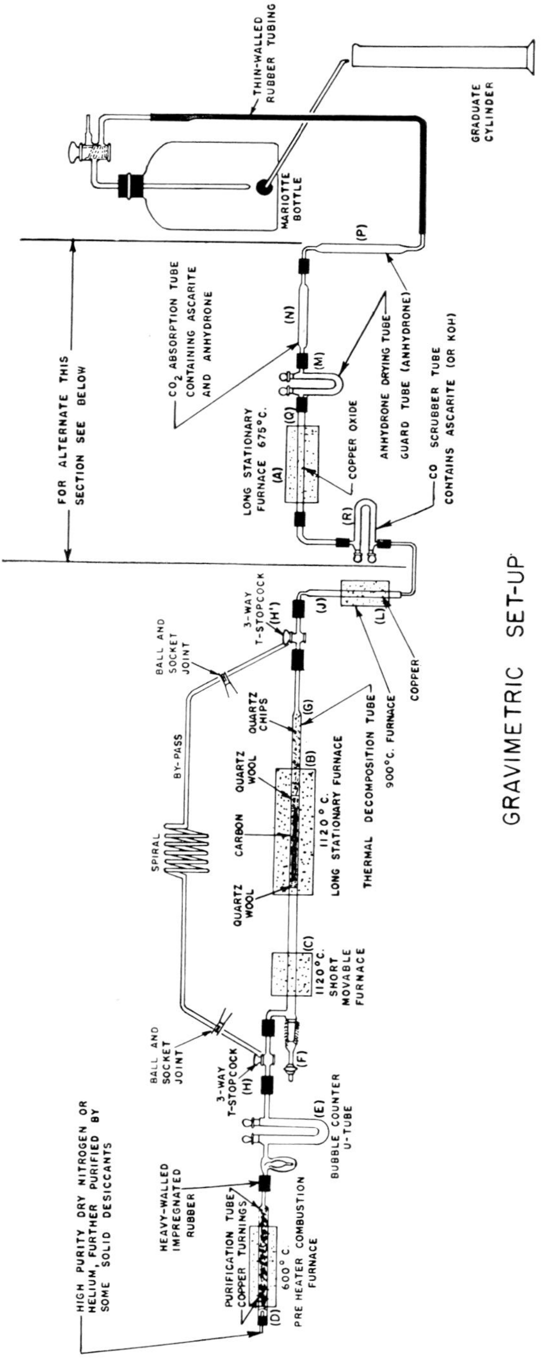

FIG. 168, Part 1.

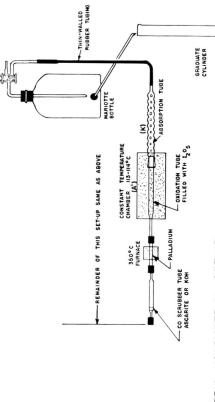

Fig. 168. Diagram of apparatus for determination of oxygen.

Apparatus

The pieces of apparatus required are shown in Figs. 168 and 169 described below. There are many points of similarity in this combustion train and that used for the determination of carbon and hydrogen, there being a bubble counter-U-tube, E (Fig. 168), combustion apparatus (short movable sample furnace, C, and long furnace, B,* extra long furnace, A (or constant temperature chamber, A'), absorption tubes, guard tube, and Mariotte bottle.

FIG. 169. Apparatus for determination of oxygen.

COMBUSTION APPARATUS

The apparatus consists of a long furnace and a short movable sample furnace. In addition, some means, as described below, should be provided for heating to a high temperature the portion of the tube passing through the insulated walls of the furnaces.

In the past, most of the laboratories where the determination was being performed had furnaces custom-built in their respective machine shops. One of the chief problems connected with this determination has been the securing of suitable furnaces that stand up under the high temperature required. Since

---
* See footnote, p. 154, Chapter 7.

Gravimetric Method

no recommended or tentative specifications have been published, it seems in order to describe the features (and some details) of those units which appeared to perform satisfactorily and for which details of construction have been published.

Walton, McCulloch, and Smith[100] give complete details for the construction of their furnace in which the heating element is 90% platinum-10% rhodium wire. Aluise *et al.*,[5] described a long furnace[57] which rides on a track and can be moved forward about 5 cm. to permit heating of that part of the reaction tube passing through the insulating walls. (Note: Where such provision is not made, a ring burner[93] of the type shown in Fig. 170 accomplishes the same effect. This directs flame points radially forward and is placed between the long furnace and the movable sample furnace or burner.) A suitable commercially available gas sample burner[2,93] is shown in Fig. 171.

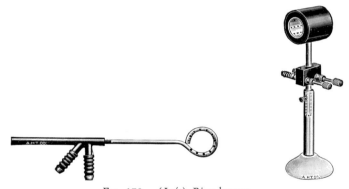

FIG. 170. (*Left*) Ring burner.
FIG. 171. (*Right*) Aluise gas sample burner.

The type of apparatus in use in the author's laboratories[88,89,91] for the past ten years is shown in Fig. 169 and the details of construction are shown in Figs. 172a, 172b, 172c, and 172d. The photograph is that for a gravimetric procedure, although there is but slight modification for the volumetric (iodometric). The furnaces stand up under continuous heating at 1120° C. for many months. (Note: The size pipe shown in the drawing (3/8 inch) is too small in internal diameter to accommodate tubes of the dimensions shown in Fig. 175. In addition, the length of the apparatus is too great for the above. The tubes used by the author, which are custom made, are 9 mm. O.D., 6 mm. I.D., and have an over-all length of approximately 107 cm., including the cap and stopcock.)

The heating elements are composed of Inconel pipe (nickel-chromium-iron-alloy, International Nickel Co., New York). This alloy is heat- and corrosion-

14. Oxygen

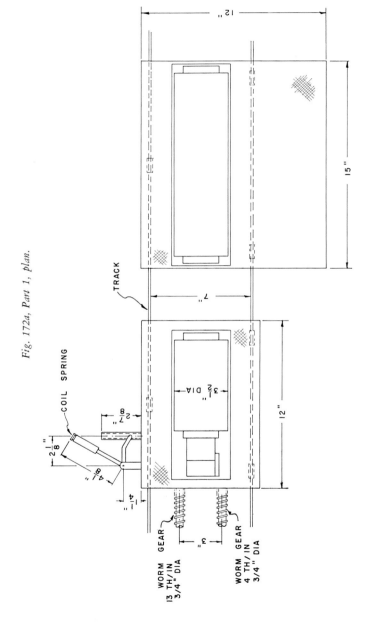

Fig. 1724, Part 1, plan.

385 Gravimetric Method

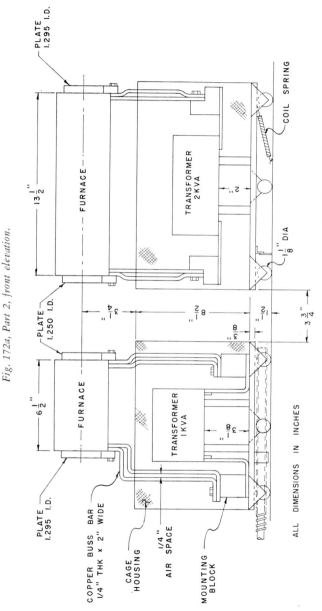

Fig. 172a, Part 2, front elevation.

FIG. 172a. Assembly for thermal decomposition apparatus for determination of oxygen—details of construction.

14. Oxygen

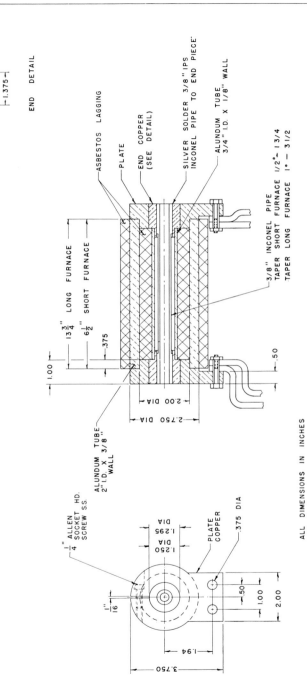

FIG. 172b. Heating elements for oxygen apparatus—details of construction.

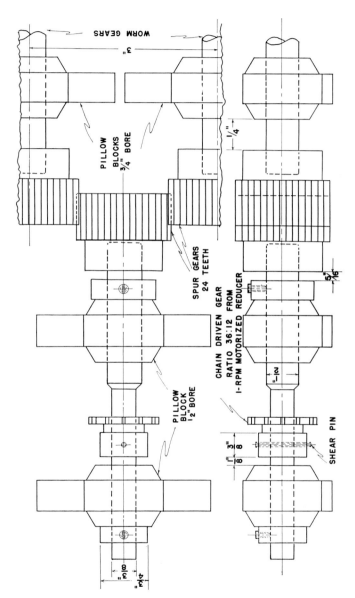

Fig. 172c. Driving mechanism for oxygen apparatus—details of construction.

14. Oxygen

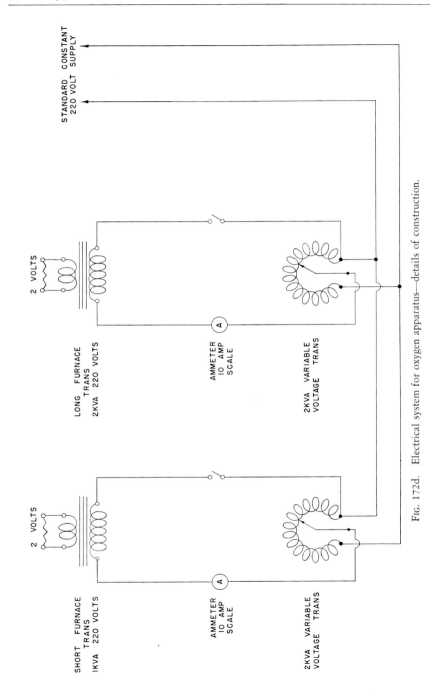

Fig. 172d. Electrical system for oxygen apparatus—details of construction.

resisting, nonmagnetic, and resistant to progressive oxidation and to intracrystalline attack.

Constant voltage is supplied to the furnaces (long and short furnaces) to obtain the constant temperature necessary during each determination.[3,4] After several months of use, slightly higher voltages must be supplied through manipulation of the variable transformers in order to produce the same current consumption as when new. This indicates that under the conditions of operation, the resistance gradually increases. However, the constancy of temperature in the short furnace is of no great importance.

Details of Construction[12a,91]

The long furnace is made from a 2 kv.-amp. transformer with a 220-volt primary winding. This transformer was purchased without a secondary winding, so that one of the desired size could be built. The secondary winding is made from 0.5×1.5 inch copper bus bar bent to fit the primary core. At some distance from the heating element connection is made to two $2 \times 2 \times 3$ inch brass blocks through sections of $\frac{1}{4} \times 2$ inch copper bus bar. The blocks tie the element to the transformer. Heat radiation is sufficient to prevent damage to the transformer, although air or water cooling may be employed. Expansion of the Inconel pipe on heating is permitted by the spring-like action of the connecting pieces ($\frac{1}{4} \times 2$ inch bus bars).

The heating elements are made from sections of $\frac{3}{8}$ inch I.P.S. Inconel pipe, silver soldered to two copper ends bored out to fit the pipe. The copper ends are attached to copper plates using squeeze bolts. As the units operate under low voltage and high amperage, extreme care must be taken in making the connection of each joint. These ends and plates are also machined to hold concentric refractory tubes with air spaces between them, which act as insulators for the Inconel pipe. The outside of the unit is wrapped with asbestos tape. The element (Inconel pipe) of the long furnace is tapered from the inside of the copper end to within 3 inches of the center on both ends. This changes the resistance of the sections of the element so that the same temperature throughout may be maintained, and strengthens the center of the element. The taper runs from nothing in the center to 1° in 3.5 inches on each end. The short furnace is similar to the long, except that a one kv.-amp. transformer is used. During operation at 1120° C., the short furnace draws 620 amperes at 0.8 volt, and the long furnace, 480 amperes at 1.45 volts. (Keep the ammeters outside the influence of the magnetic field set up by the furnace circuits.)

Both furnaces are mounted on tracks with wheels. The long furnace is held in position by means of a tension spring, which permits it to be moved forward to the right several inches. The short furnace is mechanized. It travels forward 6 inches in 25 minutes and returns the same distance in 8 minutes.

When the short furnace comes in contact with the long furnace, it pushes the latter forward to the right about 2 inches before its motion is reversed. This permits heating of that part of the reaction tube passing through the insulating walls. When the short furnace then travels to the left, the tension spring causes the long furnace to return to its original position. The mechanization is accomplished by means of two screws fastened to the bottom of the short furnace, one for the forward motion and one for the backward motion. A toggle bolt screw action arrangement is used to change from one screw to the other at the extreme ends of travel. Power is supplied from a fractional horsepower, one r.p.m. gear head motor (Model SG-25 Flexo action motor) made by the Merkle-Korff Gear Co., 213 North Morgan Street, Chicago, Illinois.

Commercially available furnaces[9,93] are shown in Figs. 90 and 125 (Chapters 7 and 9), the latter requiring a special long furnace unit.

PREHEATER TYPE OF MICROCOMBUSTION FURNACE[93]

A small electric preheater type of microcombustion furnace (Fig. 173) is used in the nitrogen purification train. It maintains a section of copper turn-

FIG. 173. Preheater type of microbustion furnace.

ings at 500°–600° C., thereby converting any oxygen present to copper oxide compare Chapter 7, Tube Filling).

ADDITIONAL COMBUSTION FURNACES

For the gravimetric procedure, an additional combustion furnace (long stationary furnace, only), of the type used for the Dumas determination, is required for maintaining the CuO tube at a temperature of 670–680° C.[89,102] Another short sample furnace, L, capable of a temperature of 900° C., is required to heat the reduced copper turning used to absorb sulfur compounds (CS_2 and COS).[12,15,65,66]

HELIUM OR NITROGEN CYLINDER AND REDUCING VALVE

See above under Reagents. A safety valve of some type, which will blow off at a set pressure, should be used in addition to the reducing valve to prevent accidents.

FRIEDRICH BOTTLE[93]

This is used as a pressure regulator for the system. This type of bottle (Fig. 174) is designed to act as either a pressure or partial vacuum regulator depending upon which of the set of tubes at the top are selected. For positive pressure, the *two tubes* which are *connected* to the vertically downward tube

FIG. 174. Friedrich bottle.

are used, the excess gas bubbles through the mercury in the body and leaves through the other set. [If constant *partial vacuum* is desired, the opposite set of tubes are used, the reduced pressure being kept constant by air (or gas) which is sucked downward through the vertical tube through the mercury.]

BY-PASS TUBE AND STOPCOCKS

The by-pass tube and stopcocks (Fig. 168) are used for backwash purposes, causing a stream of nitrogen to flow backward through the tube in order to exclude air when a sample is being placed in the decomposition tube. The by-pass tube may contain a coil section to give it flexibility to prevent breakage caused by strain.

THERMAL DECOMPOSITION (COMBUSTION OR REACTION) TUBE[93]

The thermal decomposition (combustion or reaction) tube (Figs. 168 and 175) is made of clear fused quartz, Solar radiation grade.[3] Attached to the main body of the tube is a heavy-walled capillary tip. At the other end is a ⚭

14. Oxygen

14/35 interchangeable ground joint over which fits a cap provided with a stopcock. Near to the end, having the ground joint, is a side arm for connecting the nitrogen or helium supply. [Note: The tube used by Aluise et al.,[5] had a body length of 600 mm. and a 150-mm. tip. The tubes commercially available are shorter, having a body length of 550 mm. and a 100-mm. tip. The author has obtained excellent results with tubes having an over-all length of approximately 107 cm. (including the cap and stopcock), approximately 9 mm. O.D., and approximately 6 mm. I.D.] Before being used, the tube should be cleaned by rinsing with hydrofluoric acid and then with water and dried.[3]

The tube is filled as shown in Fig. 175, with repeated tapping to avoid channeling during operation. (This figure shows the filling used by Aluise *et al.* If, however, a different size furnace is used, the carbon layer should

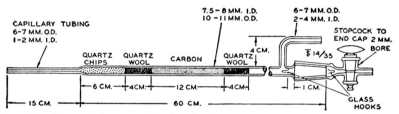

FIG. 175. Thermal decomposition (combustion or reaction) tube, showing filling and details of construction.

be the same, but the amount of quartz chips and wool should be changed in the proper proportions.) It is extremely important that the carbon be in the section of the furnace at 1120° C. at *all times*.

REDUCED COPPER PURIFICATION TUBES

One nitrogen purification tube (Figs. 168 and 169) consists of a section of combustion tubing (Pyrex[16] No. 1720 glass) into which is packed about 10 cm. of reduced copper turnings (compare Chapter 7). This purification tube is mounted in the preheater type of furnace, and the copper turnings heated at 500–600° C., and placed before the bubble counter-U-tube in front of the thermal decomposition tube. A second tube, *J*, but prepared from *quartz*, is placed *after* the thermal decomposition tube and is heated to 900° C. Its function is to remove carbon disulfide and carbonyl sulfide.[12,15,65,66]

COPPER OXIDE OXIDATION TUBE

A section of the standard Dumas combustion tube with tip (Fig. 98, Chapter 7) approximately 30–40 cm. in length, exclusive of the tip, is filled with about 25 cm. of copper oxide wire, similarly to the manner of filling the Dumas tube, except that no reduced copper is used. This is placed in a long stationary furnace *allowing 5 cm. to extend beyond the furnace* as explained in Chapter 7. The tube is maintained at 670–680° C. and is used to oxidize the carbon monoxide to carbon dioxide.[89,102] (Fig. 168, Q.)

BUBBLE COUNTER-U-TUBE[88,90,93]

A bubble counter-U-tube is used in the purification train. It is identical to that used for the carbon-hydrogen determination[88,90] (Fig. 120, Chapter 9) or it may have the ball member of a $ 12/2 interchangeable ball and socket joint on both side arms.[93] The unit is filled exactly as described in Chapter 9.

CARBON MONOXIDE SCRUBBER TUBE

The carbon monoxide scrubber tube (Fig. 168, R) is used for removing the halogens and sulfur present. Any type of drying tube may be used for this purpose and it is filled with the crushed potassium hydroxide pellets or with Ascarite—see above under Reagents.

ANHYDRONE ABSORPTION TUBE

A regulation absorption tube[88,90] (Fig. 127, Chapter 9) is filled with Anhydrone exactly as the water absorption tube of the carbon-hydrogen determination. (As an alternate, a U-tube filled with Anhydrone may be used—Fig. 168, M.) It is placed after the copper oxide tube, Q, before the carbon dioxide absorption tube, N, the positions of the Anhydrone, Ascarite, and guard tubes being the same as in the carbon-hydrogen determination. The Anhydrone tube, however, is a *permanent* part of the oxygen setup.

CARBON DIOXIDE ABSORPTION TUBE

A regulation absorption tube[88,90] (Fig. 127, Chapter 9) is filled with Ascarite and Anhydrone, *exactly* as the carbon dioxide tube of the carbon-hydrogen determination (Fig. 168, N).

MARIOTTE BOTTLE[88,90]

Same as used in Chapter 9.

GUARD TUBE[88,90]

Same as in Chapter 9 (see Fig. 168, P).

GRADUATE CYLINDER

An ordinary 1000-ml. graduate cylinder is used for measuring the volume of gas passing through the system.

Assembling the Apparatus[88,89]

The various parts of the apparatus are connected either by means of paraffin-impregnated heavy-walled tubing, identical to that used in connection with the determination of carbon and hydrogen (see Chapter 9), or by means of the various interchangeable ground joints. Assembly is begun at the connection to the source of nitrogen or helium and allowed to proceed connecting the various parts in their respective order—purification train, bubble counter-U-tube, thermal decomposition tube, by-pass, reduced copper tube, carbon monoxide scrubber, oxidation tube (CuO), and drying tube. The carbon section of the thermal decomposition tube must be in the 1120° C. zone *at all times.* A slow stream of nitrogen (or helium) is passed through the system for about one hour with *all* units at *room temperature,* after which all sections are heated to their respective temperatures (purification train copper at 600° C., thermal decomposition tube at 1120° C., copper at 900° C., and copper oxide at 670–680° C.). The inert gas is passed through for at *least* 2 days at the rate of about 10 ml. per minute (see Chapter 9 for method of determining rate by means of bubble counter). The carbon dioxide absorption tube is then attached to the system and to the guard tube (connected to the Mariotte bottle). With the arm of the Mariotte bottle slightly below the horizontal, the height of the mercury in the pressure regulator (Friedrich bottle) is adjusted so that 10 ml. of gas per minute pass through the system (see Chapter 9 for determining volume) with all parts at operating temperatures. The rates at which the bubbles pass through the bubble counter and Mariotte bottle are recorded for reference just as done in connection with the carbon-hydrogen determination.

Best results are obtained by keeping the furnaces at all times at operating temperatures, even when not in use (compare Chapter 9).

Procedure[88,89]

The rate of flow of nitrogen or helium through the system at operating temperatures (long furnace at 1120° C., copper combustion furnace at 900° C., copper oxide combustion furnace at 670–680° C., but short movable furnace at room temperature) is adjusted to 10 ml. per minute. The three-way T-stopcocks, H, and H' (Fig. 168) are turned* so that nitrogen or helium is passed in the reverse direction through the thermal decomposition tube (reaction tube), G, and out through the stopcock of the cap, F. Enough sample is weighed into a platinum boat to contain 1.0–1.3 mg. of oxygen. (If the sample

* The stopcock, H, is closed to the reaction tube, G, and connects the source of nitrogen or helium with the by-pass tube. The stopcock, H', is closed to the copper tube and connects the by-pass tube with the reaction tube, G.

is a volatile liquid, it is weighed in a capillary tube, preferably quartz, and *inserted in a Coombs-Alber platinum sleeve or long platinum boat.*) The cap, F, is removed and the platinum boat, containing the sample, is inserted and pushed to within about 8 cm. of the long furnace, B, with the aid of a platinum hook on the end of a glass rod. The cap, F, is replaced immediately, its stopcock left open and the reverse flow of nitrogen or helium through the thermal decomposition tube (reaction tube), G, and out through the above stopcock, is continued for 20 minutes in order to expel all of the air that entered the system during the insertion of the sample.

The previously weighed Ascarite absorption tube is attached to the Anhydrone absorption tube and the guard tube (connected to the Mariotte bottle). The stopcock of the cap, F, is closed to the atmosphere and the three-way T-stopcocks, H and H' are turned* so that the nitrogen (or helium) will enter the thermal decomposition (reaction) tube, G, through its side arm. The stopcock on the Mariotte bottle is opened and the nitrogen (or helium) allowed to flow through the entire system at the rate of 10 ml. per minute (making any necessary adjustment of the drainage tip of the Mariotte bottle—see Chapter 9).

The movable sample furnace is heated to a temperature of about 1120° C., then brought up within about 4 cm. of the sample and the automatic drive put into operation. The sample is pyrolyzed *gradually* and the decomposition products are swept into the hot carbon in the portion of the tube inside of the long furnace. Approximately 25 to 30 minutes are required for the movable sample furnace to traverse the distance up to the long furnace. The long furnace is then moved forward to expose that portion of the tube protected previously by the insulating walls of the furnaces. [Note: A ring burner (Fig. 170) may also be used for the same purpose.] The short furnace is moved over and this portion is heated for about 5–10 minutes. The movable furnace is brought back and the long furnace returned to its original position. This insures complete pyrolysis of any material which may have condensed in the cooler portion of the tube. The movable furnace is allowed to cool, after which the sweeping out with nitrogen or helium is continued until approximately 700 ml. of gas† passes through the system during the course of the determination from the beginning of the pyrolysis.

The Ascarite absorption tube is then removed, wiped, and weighed *exactly* as described in Chapter 9 on the determination of carbon-hydrogen.

A blank determination must be made by introducing an empty platinum boat into the reaction tube and duplicating the above procedure using 700 ml.

---

* The stopcock, H, is closed to the by-pass and connects the source of nitrogen or helium to the side arm of the reaction tube, G. The stopcock, H', is closed also to the by-pass tube and connects the reaction and copper tubes.

† *Exactly* the same amount is used for *both* the determination and the blank.

14. Oxygen

of nitrogen or helium.* The weight of carbon dioxide obtained from the blank determination is subtracted from that obtained during a regular determination. A well-functioning setup gives a zero blank.

Calculation:

Factor:

$$\frac{O}{CO_2} = 0.3635$$

$$\therefore \frac{\text{Wt. of } CO_2 \times 0.3635 \times 100}{\text{Wt. sample}} = \% \text{ O}$$

Example:

A 5.683-mg. sample gave 4.893 mg. of carbon dioxide

$$\therefore \frac{4.893 \times 0.3635 \times 100}{5.683} = 31.30\% \text{ O}$$

CLEANING DEPOSITED CARBON FROM REACTION TUBE[88]

After a number of determinations, a deposit of carbon forms on the inside wall of the thermal decomposition tube between the long and short furnaces. This makes insertion and removal of the boat difficult due to poor visibility. The deposit should be removed occasionally by the following procedure. With the entire setup assembled (absorption tube and Mariotte bottle attached) (Fig. 168) the stopcock, H, is closed to *both* the reaction tube, G, and the by-pass tube. The stopcock H' is closed to the by-pass and connects the reaction tube to the copper tube. The stopcock on the cap, F, is opened to the atmosphere, the arm of the Mariotte bottle lowered, and air is sucked through the reaction tube. The short movable furnace is heated to 1120° C. and moved over against the long furnace. The carbon deposit is burned off thusly in a matter of minutes. The stopcock on the cap, F, is then closed and the stopcock, H, turned to connect the reaction tube, G, with the supply of helium or nitrogen. The inert gas is passed through the system overnight to free it of air after which it is again ready for use.

VOLUMETRIC (IODOMETRIC) METHOD

The volumetric (iodometric) method, too, is based on the thermal decomposition of the sample in an inert gas and passage of the decomposition vapors over carbon at 1120° C., at which temperature the equilibrium between carbon dioxide and carbon monoxide is shifted completely to carbon monoxide (see equations above under Gravimetric Method). The carbon monoxide is then converted to carbon dioxide by means of iodine pentoxide, iodine being formed

---

* *Exactly* the same amount is used for *both* the determination and the blank.

in the reaction which is titrated with sodium thiosulfate. The reactions involved are the following[5,88]:

$$\text{Organic O} \xrightarrow[C]{\overset{\Delta}{1120°\ C.}} CO$$

$$5CO + I_2O_5 = 5CO_2 + I_2$$
$$I_2 + Br_2 = 2\,IBr$$
$$2\,IBr + 4Br_2 + 6H_2O = 2HIO_3 + 10HBr$$
$$2HIO_3 + 10HI = 6\,I_2 + 6H_2O$$
$$6\,I_2 + 12Na_2S_2O_3 = 12NaI + 6Na_2S_4O_6$$

Reagents

IODINE PENTOXIDE

Reagent grade of iodine pentoxide is required for the conversion of carbon monoxide to carbon dioxide with the simultaneous liberation of iodine. The particles used should not pass through a 100-mesh sieve.

SODIUM HYDROXIDE

A 20% solution of reagent grade of sodium hydroxide in distilled water is used as the absorbent for the liberated iodine.

REDUCED COPPER TURNINGS

Same as used for Gravimetric Method.

POTASSIUM HYDROXIDE OR ASCARITE[93]

Reagent grade of potassium hydroxide pellets are crushed to 8- to 20-mesh size. They are tightly packed into the carbon monoxide scrubber tube for removal of acidic gaseous products formed during the thermal decomposition of substances containing halogens and sulfur (see below). Ascarite may be used in place of the potassium hydroxide.

ANHYDRONE[93]

This, together with Ascarite, is used in the bubble counter-U-tube (see Chapter 9).

NITROGEN OR HELIUM

Same as for Gravimetric Method.

CARBON

Same as for Gravimetric Method.

14. Oxygen

QUARTZ WOOL AND QUARTZ CHIPS

Same as for Gravimetric Method.

PALLADIUM ON ASBESTOS

Palladium on asbestos (10%) is used for absorbing hydrogen. A palladium thimble[11,12] may be used in place.

BROMINE

Reagent grade of bromine is used to oxidize the iodine to iodate previous to titration.

POTASSIUM ACETATE–GLACIAL ACETIC ACID SOLUTION

A 10% solution of reagent grade of potassium acetate in glacial acetic acid is required.

SODIUM ACETATE SOLUTION, 20%

A 20% solution of reagent grade of sodium acetate in distilled water is required.

FORMIC ACID

Reagent grade of formic acid, 90%, is used to destroy the excess of bromine present after conversion of the iodine to iodate.

POTASSIUM IODIDE

Reagent grade of potassium iodide, crystals, is used to furnish the hydriodic acid needed for liberation of the iodine previous to titration with thiosulfate.

DILUTE SULFURIC ACID, 10%

Dilute sulfuric acid, 10%, is used to convert the potassium iodide to hydriodic acid.

STANDARD SOLUTION OF SODIUM THIOSULFATE, 0.01N

This solution is prepared and standardized according to the directions given in Chapter 5.

STARCH INDICATOR

This is prepared according to the directions given in Chapter 5. (Aluise, Hall, Staats, and Becker[5] suggest the use of an 0.2% aqueous solution of amylose, G. Frederick Smith Chemical Company[83] as the indicator.)

Apparatus

COMBUSTION APPARATUS

Same as for Gravimetric Method.

PREHEATER TYPE OF MICROCOMBUSTION FURNACE

Same as for Gravimetric Method.

ADDITIONAL COMBUSTION FURNACES

For the volumetric (iodometric) method, an additional small combustion furnace similar to the preheater type (Fig. 173) but controlled at a temperature of 350° C. is required for heating the palladium on asbestos (or palladium) tube.[11,12] Another short furnace, identical to that used for the gravimetric method is required to heat the reduced copper filling to 900° C. (for removal of CS_2 and COS).[12,65,66]

CONSTANT TEMPERATURE CHAMBER[93]

An electrically heated, thermostatically controlled, constant temperature chamber (Figs. 168 and 176) is used for maintaining the oxidation tube (I_2O_5) at 113°–114° C. for the purpose of driving out the liberated iodine. This is similar to the heating mortar used in the determination of carbon and hydrogen (Chapter 9).

FRIEDRICH BOTTLE
NITROGEN (OR HELIUM) CYLINDER AND REDUCING VALVE
REDUCED COPPER PURIFICATION TUBES
BY-PASS TUBE AND STOPCOCKS
THERMAL DECOMPOSITION (COMBUSTION OR REACTION) TUBE
CARBON MONOXIDE SCRUBBER TUBE
BUBBLE COUNTER-U-TUBE

} Same as for Gravimetric Method.

PALLADIUM TUBE[11,12,89]

This tube is used for absorption of hydrogen. It is placed between the carbon monoxide scrubber tube and the iodine pentoxide oxidation tube. It is prepared from a section of combustion tubing (Fig. 98, Chapter 7). The palladium on asbestos is loosely packed in the tube for a distance of about 10 cm.[89] The tube is heated in the small combustion furnace to a temperature of 350° C. The palladium on asbestos is a cheap substitute for the palladium thimble

14. Oxygen

described by Campanile et al.,[11] and Canales and Parks[12], but if preferred, the latter should be used.

OXIDATION TUBE[93]

The oxidation tube (Fig. 177) is made of Pyrex[16] glass. It has a heavy-walled capillary tip attached to the main body of the tube and at the opposite end is a ⚜ 14/35 interchangeable ground joint for attaching it to the absorption tube, Vigreux type (see below). The oxidation tube is filled with iodine pent-

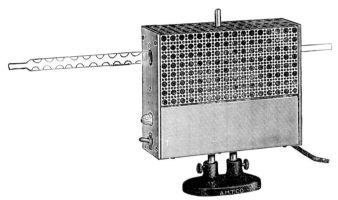

FIG. 176. Constant temperature chamber.

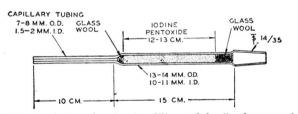

FIG. 177. Oxidation tube, showing filling and details of construction.

oxide, being held in place at both ends by plugs of glass wool. The packing of the iodine pentoxide must be done tightly to avoid channeling during operation. The oxidation tube is surrounded during operation by the constant temperature chamber. After being filled, a new tube should be heated for some days at 113°–114° C. while nitrogen or helium is swept through to condition it.

ABSORPTION TUBE, VIGREUX TYPE[93]

The absorption tube (Fig. 178) is made of Pyrex[16] glass. The main body has indented walls so that a large surface is provided. Attached to the one end of the main body is a heavy-walled capillary tip and at the other end is a

$ 14/35 interchangeable ground joint which fits that on the end of the oxidation tube.

MARIOTTE BOTTLE[88,90]

This is the same as used for the carbon-hydrogen determination (Chapter 9, Fig. 129).

GRADUATE CYLINDER

An ordinary 1000-ml. graduate cylinder is used for measuring the gas flow through the system.

FIG. 178. Absorption tube (Vigreux type), *left,* showing attachment to oxidation tube, *right.*

AUTOMATIC BURETTE

An automatic burette of the type shown in Figs. 69 and 70 is used for titration of the iodine with thiosulfate.

GROUND GLASS-STOPPERED ERLENMEYER FLASK

An ordinary Pyrex 250-ml. ground glass-stoppered Erlenmeyer flask is used for the titration.

Assembling the Apparatus[88,89]

The various parts of the apparatus are connected either by means of paraffin impregnated heavy-walled tubing, identical to that used in connection with the determination of carbon and hydrogen (see Chapter 9), or by means of the various interchangeable ground joints. Assembly is begun at the connection to the source of nitrogen or helium and allowed to proceed connecting the various parts in their respective order (see Gravimetric Method), thermal decomposition tube, by-pass, reduced copper tube, carbon monoxide scrubber tube, palladium tube, oxidation tube (I_2O_5), and absorption tube, Vigreux type.*
The carbon section must be in the 1120° C. zone *at all times.* A slow stream of nitrogen or helium is passed through the system for several hours with all the furnaces at room temperature. The preheater is heated to 500°–600° C., the long furnace is gradually heated to 1120° C., combustion furnace for reduced copper tube to 900° C., the furnace for the palladium tube to 350° C., and the constant temperature chamber is heated to 113°–114° C. Nitrogen or

---

* The ground joint connecting the absorption tube, Vigreux Type, *K,* and the oxidation tube containing the iodine pentoxide, must be in the heated portion of the constant temperature chamber.

14. Oxygen

helium is passed through the system for some days. Then the Mariotte bottle is connected to the absorption tube, Vigreux type, which in turn is attached to the oxidation tube. (Note: No lubricant is used on this joint.) The pressure of the nitrogen or helium is adjusted carefully so that 10 ml. of gas per minute flows through the system when up to operating temperature. The rates at which bubbles pass through the bubble counter and Mariotte bottle under these conditions are recorded so that the same rate of flow may be obtained for all determinations (compare Chapter 9).

Best results are obtained when the long furnace is kept at a temperature of 1120° C. overnight rather than at some lower temperature. Consequently, except when long periods of idleness are anticipated, the operation temperature should be maintained at all times (compare Chapter 9).

Procedure[5,88,89]

The rate of flow of nitrogen or helium through the system at operating temperature (long furnace at 1120° C., constant temperature chamber at 113°–114° C.,* but short movable furnace at room temperature) is adjusted to 10 ml. per minute. The three-way T-stopcocks, H, and H', (Fig. 168) are turned† so that nitrogen or helium is passed in the reverse direction through the thermal decomposition tube (reaction tube), G, and out through the stopcock of the cap, F. Enough sample is weighed into a platinum boat to contain 1.0–1.3 mg. of oxygen. (If the sample is a volatile liquid, it is weighed in a capillary tube, preferably quartz, and *inserted in a Coombs-Alber platinum sleeve* or long platinum boat.) The cap, F, is removed and the platinum boat, containing the sample, is inserted and pushed to within about 8 cm. of the long furnace, B, with the aid of a platinum hook on the end of a glass rod. The cap, F, is replaced immediately, its stopcock left open and the reverse flow of nitrogen or helium through the thermal decomposition tube (reaction tube), G, and out through the above stopcock, is continued for 20 minutes in order to expel all of the air that entered the system during the insertion of the sample.

The tip of the absorption tube, Vigreux type, K, is immersed in the 20% solution of sodium hydroxide and suction applied to fill the tube almost to the ground joint. The alkali is allowed to drain off and the moistened absorption tube attached to the oxidation tube (containing the iodine pentoxide), using a rotary motion so that the two joints are firmly sealed. (No lubricant is used.) This joint must be in the heated portion of the constant temperature chamber, A, so that all of the liberated iodine will be driven into the portion of the absorption tube moistened with sodium hydroxide. The Mariotte bottle

---

* Aluise[3,4] has obtained good results with a temperature of 80° C.

† The stopcock, H, is closed to reaction tube, G, and connects the source of nitrogen or helium with the by-pass tube. The stopcock, H', is closed to the copper tube and connects the by-pass tube with the reaction tube, G.

is attached again to the absorption tube. The stopcock of the cap, F, is closed to the atmosphere and the three-way T-stopcocks, H and H' are turned* so that the nitrogen or helium will enter the thermal decomposition (reaction) tube, G, through its side arm. The stopcock on the Mariotte bottle is opened and nitrogen or helium allowed to flow through the reaction tube, G, in the normal manner at the rate of 10 ml. per second.

The movable sample furnace is heated to a temperature of about 1120° C., then brought up within about 4 cm. of the sample and the automatic drive put into operation. The sample is pyrolyzed *gradually* and the decomposition products are swept into the hot carbon in the portion of the tube inside of the long furnace. Approximately 25–30 minutes are required for the movable sample furnace to traverse the distance up to the long furnace. The long furnace is then moved forward to expose that portion of the tube protected previously by the insulating walls of the furnaces. [Note: A ring burner (Fig. 170) may also be used for the same purpose.] The short furnace is moved over and this portion is heated for about 5–10 minutes. The movable furnace is brought back and the long furnace returned to its original position. This insures complete pyrolysis of any material which may have condensed in the cooler portion of the tube. The movable furnace is allowed to cool, after which the sweeping out with nitrogen or helium is continued until approximately 700 ml. of gas† passes through the system during the course of the determination from the beginning of the pyrolysis.

The absorption tube, Vigreux tube, K, is disconnected from the Mariotte bottle and the oxidation tube in the constant temperature chamber. The contents of the absorption tube are rinsed with about 125 ml. of distilled water into a 250-ml. Pyrex ground glass-stoppered Erlenmeyer flask containing 10 drops of bromine and 10 ml. of the 10% solution of potassium acetate in glacial acetic acid. The contents are stirred and then 10 ml. of the 20% solution of sodium acetate in water is added. The excess bromine is destroyed by adding slowly with shaking 15–20 drops of 90% formic acid using only what is needed to decolorize. The mixture is allowed to stand for about 5 minutes and then 0.3 gram of potassium iodide and 5 ml. of 10% sulfuric acid are added. The contents of the flask are mixed by swirling and then titrated immediately with standard 0.01N sodium thiosulfate using starch indicator.

A blank determination must be made by introducing an empty platinum boat into the reaction tube and duplicating the above procedure using 700 ml. of nitrogen or helium.† The volume of thiosulfate required for this blank

---

* The stopcock, H, is closed to the by-pass and connects the source of nitrogen or helium to the side arm of the reaction tube, G. The stopcock, H', is closed also to the by-pass tube and connects the reaction and copper tubes.

† *Exactly* the same amount is used for *both* the determination and the blank.

determination is the correction which is subtracted from the volume obtained in a regular one.

Calculation:

Factor:

1 ml. of 0.01N sodium thiosulfate is equivalent to 0.06667 mg. of oxygen

$$\therefore \frac{\text{ml. of } 0.01N \text{ Na}_2\text{S}_2\text{O}_3 \text{ (corrected)} \times 0.06667 \times 100}{\text{Wt. of sample}} = \% \text{ O}$$

where ml. of 0.01N $Na_2S_2O_3$ (corrected) = ml. for a determination — ml. for a blank.

Example:

3.518 mg. of sample required 17.00 ml. of 0.01N thiosulfate to titrate the liberated iodine. A blank determination required 0.30 ml. of 0.01N thiosulfate

$$\therefore \text{ ml. of } 0.01N \text{ Na}_2\text{S}_2\text{O}_3 \text{ (corrected)} = 17.00 - 0.30$$
$$= 16.70$$

$$\therefore \frac{16.70 \times 0.06667 \times 100}{3.518} = 31.65\% \text{ O}$$

CLEANING DEPOSITED CARBON FROM REACTION TUBE

See Gravimetric Method.

TABLE 26

ADDITIONAL INFORMATION ON REFERENCES* RELATED TO CHAPTER 14

In addition to the procedures described in detail in the preceding pages of this chapter, the author wishes to call to the attention of the reader the material listed in Table 26. (See statement at top of Table 4 of Chapter 1, regarding completeness of this material.)

Books
    Belcher and Godbert, 7
    Clark, S. J., 13
    Grant, 31
    Milton and Waters, 63
    Roth, 71, 72
    Steyermark, 88

Reviews
    Elving and Ligett, 24
    Fort, 26, 27
    Kainz, 45
    Unterzaucher, 94, 97

Collaborative studies
    Jones, 44
    Steyermark, 86, 87

General, miscellaneous, errors, etc.
    Aluise, Alber, Conway, Harris, Jones, and Smith, 4
    Foreman, 25
    Gouverneur, Schreuders, and Degens, 29
    Grube and Spiedel, 33
    Hinkel and Raymond, 35
    Kirsten, 48
    Kono, 49
    Kono, Sato, Suzuki, and Isobe, 50
    Maylott and Lewis, 59
    Moelants and Wesenbeek, 64
    Otting, 67
    Radmacher and Hoverath, 69
    Schöniger, 77
    Unterzaucher, 95–97

* The numbers which appear after each entry in this table refer to the literature citations in the reference list at the end of the chapter.

TABLE 26 (*Continued*)

Traces
Walton, 99
Walton, McCulloch, and Smith, 100

Simultaneous determination of oxygen and other elements
Korshun and Bondarevskaya, 52

Apparatus
Aluise, Hall, Staats, and Becker, 5
Campanile, Badley, Peters, Agazzi, and Brooks, 11
Canales and Parks, 12
Clark, S. J., 13
Dundy and Stehr, 22
Garch and Valdener, 28
Gouverneur, Schreuders, and Degens, 29
Hinkel and Raymond, 35
Hintermaier and Grützner, 36
Holowchak and Wear, 38
Kuck, 54
Maylott and Lewis, 59
Oita and Conway, 65
Oliver, 66
Siniramed and Manci, 82
Steyermark, 88
Steyermark, McNally, Wiseman, Nivens, and Biava, 91
Unterzaucher, 94, 95, 97
Willits and Ogg, 102
Zimmermann, 103, 104

Purification of carrier gas and gas used (He, N_2, H_2)
Bürger, 10
Canales and Parks, 12
Imaeda, 41
Steyermark, 88

Thermal decomposition tube fillings
Campanile, Badley, Peters, Agazzi, and Brooks, 11
Colson, 14
Gouverneur, Schreuders, and Degens, 29
Hinkel and Raymond, 35
Imaeda, 43
King, 46
Korshun and Bondarevskaya, 53
Maylott and Lewis, 59

Thermal decomposition tube fillings (*Cont.*)
Oita and Conway, 65
Oliver, 66
Schöniger, 77
Unterzaucher, 94

Fluoro-compounds
Mázor, 60
Rush, 73
Rush, Cruikshank, and Rhodes, 74

Thermal decomposition tube temperatures
Colson, 14
Deinum and Schouten, 18
Gouverneur, Schreuders, and Degens, 29
King, 46
Korshun and Bondarevskaya, 53
Oita and Conway, 65
Oliver, 66
Steyermark, 88

Removal of sulfur compounds, interference of sulfur compounds
Canales and Parks, 12
Conway, 15
Dixon, 20
Kono, Sato, Suzuki, and Isobe, 50
Maylott and Lewis, 59
Oita and Conway, 65
Oliver, 66
Radmacher and Hoverath, 69

Interference from hydrogen
Campanile, Badley, Peters, Agazzi, and Brooks, 11
Canales and Parks, 12
Dixon, 20
Jones, 44

Various iodine pentoxide oxidation tube fillings
Campanile, Badley, Peters, Agazzi, and Brooks, 11
Canales and Parks, 12
Colson, 14
Gouverneur, Schreuders, and Degens, 29

TABLE 26 (*Continued*)

Various iodine pentoxide oxidation tube fillings (*Cont.*)
Grant, Katz, and Haines, 30
Imaeda, 42
Jones, 44
King, 46
Kono, Sato, Suzuki, and Isobe, 50
Oita and Conway, 65
Oliver, 66
Steyermark, 88
Unterzaucher, 94, 95, 98

Copper oxide oxidation tube filling
Dixon, 20
Korshun and Bondarevskaya, 53
Willits and Ogg, 102

Use of hopcalite, MnO_2, in oxidation tube
Deinum and Schouten, 18
Imaeda, 43
Spooner, 85

Use of mercuric oxide in oxidation tube
Deinum and Schouten, 18
King, 46

Gravimetric procedures
Berret and Poirier, 8
Dixon, 20
Dundy and Stehr, 22
Jones, 44
Korshun, 51
Oita and Conway, 65
Radmacher and Hoverath, 69
Schütze, 78–80
Steyermark, 88
Willits and Ogg, 102
Zimmermann, 104

Volumetric procedures
Bürger, 10
Campanile, Badley, Peters, Agazzi, and Brooks, 11
Canales and Parks, 12
Deinum and Schouten, 18
Hinkel and Raymond, 35
Jones, 44

Volumetric procedures (*Cont.*)
King, 46
Lacourt, 55
Lee and Meyer, 56
Radmacher and Hoverath, 69
Renard and Jadot, 70
Steyermark, 88
Unterzaucher, 94, 96

Manometric, gasometric procedure
Holowchak and Wear, 38

Physical chemical technique
Harris, Smith, and Mitchell, 34

Amperometric, potentiometric, conductometric methods, automatic titration
Canales and Parks, 12
Drekopf and Braukman, 21
Ehrenberger, Gorbach, and Mann, 23

Determination in presence of metals
Hoekstra and Katz, 37
Huber, 40
Takahashi, Kawane, and Mitsui, 92

Absorption, chromatographic method
Pietsch, 68

Colorimetric method
Walton, McCulloch, and Smith, 100

Fusion methods
Lee and Meyer, 56
Sheft and Katz, 81

Isotopic oxygen
Anbar, Dostrovsky, Klein, and Samuel, 6
Dahn and Moll, 17
Grosse and Kirshenbaum, 32

Hydrogenation methods
Dinnerstein and Klipp, 19
Kirner, 47
Maylott and Lewis, 59
Meulen, ter, 61
Meulen, ter and Heslinga, 62
Russell and Fulton, 75
Russell and Marks, 76
Smith, Duffield, Pierotti, and Mooi, 84

REFERENCES

1. Air Reduction Company, New York.
2. Aluise, V. A., *Anal. Chem.*, **21**, 746 (1949).
3. Aluise, V. A., Personal communication. 1950.
4. Aluise, V. A., Alber, H. K., Conway, H. S., Harris, C. C., Jones, W. H., and Smith, W. H., *Anal. Chem.*, **23**, 530 (1951).
5. Aluise, V. A., Hall, R. T., Staats, F. C., and Becker, W. W., *Anal. Chem.*, **19**, 347 (1947).
6. Anbar, M., Dostrovsky, I., Klein, F., and Samuel, D., *J. Chem. Soc.*, p. 155 (1955).
7. Belcher, R., and Godbert, A. L., "Semi-Micro Quantitative Organic Analysis," 2nd ed., Longmans, Green, London, 1954.
8. Berret, R., and Poirier, P., *Bull. soc. chim. France*, p. D539 (1949).
9. Brinkmann, C. A., & Co., Inc., Great Neck, Long Island, New York.
10. Bürger, K., *Mikrochim. Acta*, p. 313 (1957).
11. Campanile, V. A., Badley, J. H., Peters, E. D., Agazzi, E. J., and Brooks, F. R., *Anal. Chem.*, **23**, 1421 (1951).
12. Canales, A. M., and Parks, T. D., *Anal. Chim. Acta*, **15**, 25 (1956).
12a. Chesapeake Scientific Apparatus Company, Roseland, New Jersey.
13. Clark, S. J., "Quantitative Methods of Organic Microanalysis," Butterworths, London, 1956.
14. Colson, A. F., *Analyst*, **79**, 784 (1954).
15. Conway, H. S., *Mikrochim. Acta*, p. 849 (1957).
16. Corning Glass Works, Corning, New York.
17. Dahn, H., and Moll, H., *Chem. & Ind. (London)*, p. 399 (1959); *Chem. Abstr.*, **53**, 17754 (1959).
18. Deinum, H. W., and Schouten, A., *Anal. Chim. Acta*, **4**, 286 (1950).
19. Dinnerstein, R. A., and Klipp, R. W., *Anal. Chem.*, **21**, 545 (1949).
20. Dixon, J. P., *Anal. Chim. Acta*, **19**, 141 (1958).
21. Drekopf, K., and Braukman, B., *Brennstoff-Chem.*, **36**, 203 (1955).
22. Dundy, M., and Stehr, E., *Anal. Chem.*, **23**, 1408 (1951).
23. Ehrenberger, F., Gorbach, S., and Mann, W., *Mikrochim. Acta*, p. 778 (1958).
24. Elving, P. J., and Ligett, W. B., *Chem. Revs.*, **34**, 129 (1944).
25. Foreman, J. K., *Mikrochim. Acta*, p. 1481 (1956).
26. Fort, R., *Chim. anal.*, **39**, 319 (1957); *Chem. Abstr.*, **52**, 972 (1958).
27. Fort, R., *Chim. anal.*, **39**, 366 (1957); *Chem. Abstr.*, **52**, 4397 (1958).
28. Garch, J., and Valdener, G., *Chim. anal.*, **36**, 211 (1954).
29. Gouverneur, P., Schreuders, M. A., and Degens, P. N., Jr., *Anal. Chim. Acta*, **5**, 293 (1951).
30. Grant, G. A., Katz, M., and Haines, R. L., *Can. J. Technol.*, **29**, 43 (1951).
31. Grant, J., "Quantitative Organic Microanalysis," 5th ed., Blakiston, Philadelphia, Pennsylvania, 1951.
32. Grosse, A. V., and Kirshenbaum, A. D., *Anal. Chem.*, **24**, 584 (1952).
33. Grube, G., and Spiedel, H., *Z. Elektrochem.*, **53**, 339 (1949).
34. Harris, C. C., Smith, D. M., and Mitchell, J., Jr., *Anal. Chem.*, **22**, 1297 (1950).
35. Hinkel, R. D., and Raymond, R., *Anal. Chem.*, **25**, 470 (1953).
36. Hintermaier, A., and Grützner, R., *Mikrochim. Acta*, p. 944 (1955).
37. Hoekstra, H. R., and Katz, J. J., *Anal. Chem.*, **25**, 1608 (1953).
38. Holowchak, J., and Wear, G. E. C., *Anal. Chem.*, **23**, 1404 (1951).
39. Huber, J. M., Inc., New York.
40. Huber, W., *Mikrochim. Acta*, p. 751 (1959).

41. Imaeda, K., *Yakugaku Zasshi,* **77,** 1192 (1957); *Chem. Abstr.,* **52,** 4395 (1958).
42. Imaeda, K., *Yakugaku Zasshi,* **78,** 386 (1958); *Chem. Abstr.,* **52,** 11653 (1958).
43. Imaeda, K., *Yakugaku Zasshi,* **78,** 30 (1958); *Chem. Abstr.,* **52,** 6065 (1958); *Anal. Abstr.,* **6,** No. 3546 (1959).
44. Jones, W. H., *Anal. Chem.,* **25,** 1449 (1953).
45. Kainz, G., *Österr. Chemiker-Ztg.,* **57,** 216 (1956).
46. King, J. G., *Coke and Gas,* **18,** 31 (1956).
47. Kirner, W. R., *Ind. Eng. Chem., Anal. Ed.,* **9,** 535 (1937).
48. Kirsten, W., *Mikrochemie ver. Mikrochim. Acta,* **34,** 152 (1959).
49. Kono, T., *Nippon Nôgei-kagaku Kaishi,* **31,** 622 (1957).
50. Kono, T., Sato, K., Suzuki, M., and Isobe, I., *Nippon Nôgei-kagaku Kaishi,* **31,** 587 (1957); *Chem. Abstr.,* **52,** 12662 (1958).
51. Korshun, M. O., *Zavodskaya Lab.,* **10,** 241 (1941).
52. Korshun, M. O., and Bondarevskaya, E. A., *Proc. Acad. Sci. U.S.S.R., Sect. Chem., (English Translation),* **110,** 553 (1956); *Chem. Abstr.,* **52,** 4393 (1958).
53. Korshun, M. O., and Bondarevskaya, E. A., *Zhur. Anal. Khim.,* **14,** 123 (1959); *Chem. Abstr.,* **53,** 9888 (1959); *Anal. Abstr.,* **6,** No. 4004 (1959).
54. Kuck, J. A., Personal communication, 1946.
55. Lacourt, A., *Mikrochim. Acta,* p. 735 (1954).
56. Lee, T. S., and Meyer, R., *Anal. Chim. Acta,* **13,** 340 (1955).
57. Lindberg Engineering Company, Chicago, Illinois.
58. Linde Air Products Company, New York.
59. Maylott, A. O., and Lewis, J. B., *Anal. Chem.,* **22,** 1051 (1950).
60. Mázor, L., *Mikrochim. Acta,* p. 1757 (1956).
61. Meulen, H. ter, *Rec. trav. chim.,* **41,** 112 (1922).
62. Meulen, H. ter, and Heslinga, J., "Neue Methoden der organisch-chemischen Analyse," Akademische Verlagsges., Leipzig, 1927.
63. Milton, R. F., and Waters, W. A., "Methods of Quantitative Microanalysis," 2nd ed., Arnold, London, 1955.
64. Moelants, L. J., and Wesenbeek, W., *Mikrochim. Acta,* p. 738 (1954).
65. Oita, I. J., and Conway, H. S., *Anal. Chem.,* **26,** 600 (1954).
66. Oliver, F. H., *Analyst,* **80,** 593 (1955).
67. Otting, W., *Mikrochemie ver. Mikrochim. Acta,* **38,** 551 (1951).
68. Pietsch, H., *Erdöl u. Kohle,* **11,** 157 (1958); *Anal. Abstr.,* **6,** No. 200 (1959).
69. Radmacher, W., and Hoverath, A., *Z. anal. Chem.,* **167,** 336 (1959).
70. Renard, M., and Jadot, J., *Mededel. Vlaam. Chem. Ver.,* **12,** 48 (1950).
71. Roth, H., "F. Pregl quantitative organische Mikroanalyse," 5th ed., Springer, Wien, 1947.
72. Roth, H., "Pregl-Roth quantitative organische Mikroanalyse," 7th ed., Springer, Wien, 1958.
73. Rush, C. A., Disc., Symposium on Analysis of Fluorine-containing Compounds, 137th National Meeting of the American Chemical Society, pp. 21B-23B, Cleveland, Ohio, April 1960.
74. Rush, C. A., Cruikshank, S. S., and Rhodes, E. H. J., *Mikrochim. Acta,* p. 858 (1956).
75. Russell, W. W., and Fulton, J. W., *Ind. Eng. Chem., Anal. Ed.,* **5,** 384 (1933).
76. Russell, W. W., and Marks, M. E., *Ind. Eng. Chem., Anal. Ed.,* **6,** 381 (1934).
77. Schöniger, W., *Mikrochim. Acta,* p. 863 (1956).
78. Schütze, M., *Ber.,* **77,** 484 (1944).
79. Schütze, M., *Z. anal. Chem.,* **118,** 241 (1939).

80. Schütze, M., *Z. anal. Chem.,* **118**, 245 (1939).
81. Sheft, I., and Katz, J. J., *Anal. Chem.,* **29**, 1322 (1957).
82. Siniramed, C., and Manci, C., *Riv. combustibili,* **11**, 366 (1957).
83. Smith, G. Frederick, Chemical Company, Columbus, Ohio.
84. Smith, R. N., Duffield, J., Pierotti, R. A., and Mooi, J., *Anal. Chem.,* **28**, 1161 (1956).
85. Spooner, C. E., *Fuel,* **26**, 15 (1947).
86. Steyermark, Al, *J. Assoc. Offic. Agr. Chemists,* **41**, 299 (1958).
87. Steyermark, Al, *J. Assoc. Offic. Agr. Chemists,* **42**, 319 (1959).
88. Steyermark, Al, "Quantitative Organic Microanalysis," Blakiston, Philadelphia, Pennsylvania, 1951.
89. Steyermark, Al, Unpublished work.
90. Steyermark, Al, Alber, H. K., Aluise, V. A., Huffman, E. W. D., Kuck, J. A., Moran, J. J., and Willits, C. O., *Anal. Chem.,* **21**, 1555 (1949).
91. Steyermark, Al, McNally, M. J., Wiseman, W. A., Nivens, R., and Biava, F. P., *Anal. Chem.,* **24**, 589 (1952).
92. Takahashi, M., Kawane, M., and Mitsui, T., *Mikrochim. Acta,* p. 647 (1957).
93. Thomas, Arthur H., Company, Philadelphia, Pennsylvania.
94. Unterzaucher, J., *Analyst,* **77**, 584 (1952).
95. Unterzaucher, J., *Ber.,* **73B**, 391 (1940).
96. Unterzaucher, J., *Bull. soc. chim. France,* p. C71 (1953).
97. Unterzaucher, J., *Mikrochemie ver. Mikrochim. Acta,* **36/37**, 706 (1951).
98. Unterzaucher, J., *Mikrochim. Acta,* p. 822 (1956).
99. Walton, W. W., Personal communication, 1950.
100. Walton, W. W., McCulloch, F. W., and Smith, W. H., *J. Research Natl. Bur. Standards,* **40**, 443 (1948).
101. Willits, C. O., *Anal. Chem.,* **21**, 132 (1949).
102. Willits, C. O., and Ogg, C. L., personal communication, 1950.
103. Zimmermann, W., *Mikrochim. Acta,* p. 888 (1955).
104. Zimmermann, W., *Z. anal. Chem.,* **118**, 258 (1939).

CHAPTER 15

Microdetermination of Neutralization Equivalent, Ionic Hydrogen, or Carboxyl Groups

The number of grams of substance required to neutralize one liter of normal alkali is known as the neutralization equivalent.[32,86,120,121,162,163] It serves both as a determination of the acidic groups and of the molecular weight. The neutralization equivalent is equivalent to the molecular weight divided by the number of acid groups present. For example, the value for benzoic acid would be equal to the molecular weight, while that for phthalic acid would be one-half the molecular weight. The reactions involved may be represented by the following equations:

$$R'COOH + NaOH = R'COONa + H_2O$$

or

$$R''(COOH)_2 + 2NaOH = R''(COONa)_2 + 2H_2O$$

According to Kamm,[86] the determination gives good results for most carboxylic acids. One aromatic amino group in the molecule does not interfere appreciably, but more than one aromatic or one aliphatic amino group does interfere. Hydroxyl groups and even a single phenolic group, as in salicylic acid, do not interfere and give neutralization equivalents equal to the molecular weight. However, in general, the weakly acidic groups, like phenols, amides, and imides, give abnormally high neutralization equivalents. A strongly acidic phenol like *s*-tribromophenol may be titrated quantitatively in alcoholic solution using phenolphthalein as the indicator. Clark[32] has found that with hydroxyl groups the end point fades and the titration is best carried out rapidly. He also preferred to add an excess of alkali and back-titrate when dealing with lactones.

It must be borne in mind that *any* acid group may affect the determination. In fact, the same procedure may be used to determine halogens present as the hydrohalides, sulfur as the acid sulfate or sulfate, phosphorus as the phosphates, etc.

Besides phenolphthalein, thymolphthalein and thymol blue have been used as indicators. The general rules regarding the relationship of the pKa of the indicator and that of the substance to be titrated apply.[34,74] For example, where other titratable groups are present, the pKa of the *carboxyl* might be so low

410

that an indicator such as bromocresol green may be required for titration of the *carboxyl* alone without interference of the other acidic groups present. Organic compounds having several acid groups are comparable to inorganic di- or tribasic acids, such as phosphoric acid.

In the absence of pKa data, type compounds are often helpful for test purpose to help the analyst choose an indicator.

Reagents

STANDARD SODIUM HYDROXIDE, 0.01N[1,50]

This is prepared and standardized according to the directions given in Chapter 5.

STANDARD HYDROCHLORIC ACID, 0.01N[1]

This is prepared and standardized according to the directions given in Chapter 5. It is used only for such cases as lactones where it is advisable to add an excess of alkali and back-titrate or if by accident the end point has been overstepped.

PHENOLPHTHALEIN INDICATOR

This is prepared according to the directions given in Chapter 5.

NEUTRAL ETHANOL, 95%

Several drops of phenolphthalein indicator are added to 200–300 ml. of 95% ethanol in a 1-liter Erlenmeyer flask. The contents of the flask are boiled for 30 seconds on a steam bath and then enough $0.1N$ or $0.01N$ sodium hydroxide is added to produce a *faint* pink coloration. The ethanol is cooled and stored in a ground-glass stoppered bottle. More $0.01N$ alkali must be added from time to time to keep it neutral. This is used only for substances which are not soluble in water.

DISTILLED WATER

After being boiled for 30 seconds, 10 ml. of this should require one drop of $0.01N$ alkali to give an end point with phenolphthalein. Otherwise, it must be neutralized as described above for ethanol.

Apparatus

MICROBURETTES[90,91,132,163]

Two microburettes of the types described in Chapter 5 are required (see Figs. 69 and 70).

15. Neutralization Equivalent

Procedure*[162,163]

Five to 10 mg., or preferably enough sample to require about 5 ml. of 0.01N alkali, is placed in a 125-ml. Erlenmeyer flask. If the substance is water-soluble, 10 ml. of water is added (otherwise, 10 ml. of neutral ethanol is used), warming if necessary to bring about solution. One or two drops of phenolphthalein indicator is added and the contents of the flask boiled for 30 seconds (on a steam bath, if neutral ethanol is used). The *hot* solution is then titrated with 0.01N sodium hydroxide to a pink end point which persists for one minute. (Note: It is good practice to heat the solution once or twice during the titration to make certain that it is carbon dioxide free.) If, by accident, the end point is overstepped, a measured amount of standard 0.01N hydrochloric acid is added and then standard alkali again to the proper end point. Likewise, certain samples are best handled by back-titrating. In these cases, a measured amount (excess) of standard alkali is added and then the hot solution back-titrated with standard acid to just expel the color. Enough standard alkali (one or two drops, at most, should be needed) is then added to produce a pink coloration which lasts for one minute.

Calculations:

$$\text{Neut. Equiv.} = \frac{\text{Wt. of sample in grams} \times 1000}{\text{No. ml. of } N \text{ alkali used}}$$

Since the number of grams $\times$ 1000 is equal to the number of milligrams, the formula becomes

$$\text{Neut. Equiv.} = \frac{\text{Wt. sample in mg.}}{\text{No. ml. of } N \text{ alkali used}}$$

or

$$\text{Neut. Equiv.} = \frac{\text{Wt. sample in mg.} \times 100}{\text{No. ml. of } 0.01N \text{ alkali used}}$$

Example:

4.85 ml. of 0.01N NaOH was required for a 6.025-mg. sample

$$\therefore \text{Neut. Equiv.} = \frac{6.025 \times 100}{4.85} = 124$$

Allowable error $\pm 2.0\%$.

The results may also be calculated as per cent of carboxyl, COOH, if this group is known to be present:

Factor:

1 ml. of 0.01N NaOH is equivalent to 0.4502 mg. of COOH

$$\therefore \frac{\text{ml. of } 0.01N \text{ NaOH} \times 0.4502 \times 100}{\text{Wt. sample}} = \% \text{ COOH}$$

* Compare Clark,[32] Grant,[61,62] Niederl and Niederl,[120,121] and Roth.[136-139]

Example:
Calculating the above example as COOH, instead of as Neutralization Equivalent, it becomes

$$\frac{4.85 \times 0.4502 \times 100}{6.025} = 36.2\% \text{ COOH}$$

The results may be calculated as per cent chlorine, bromine, sulfur, etc., if these are known to be present as the hydrochloride, hydrobromide, sulfate, etc., and *carboxyl* is known to be *absent*. Obviously, the factors would then be as follows:

Factors:
1 ml. of 0.01N NaOH is equivalent to:

0.3546 mg. of Cl
0.7992 mg. of Br
0.1603 mg. of S
etc.

$$\therefore \frac{\text{ml. of } 0.01N \text{ NaOH} \times \text{Factor} \times 100}{\text{Wt. sample}} = \% \text{ Element}$$

Example:
3.00 ml. of 0.01N NaOH was required for a 6.000-mg. sample of a hydrochloride (*no* carboxyl present)

$$\therefore \frac{3.00 \times 0.3546 \times 100}{6.000} = 17.73\% \text{ Cl}$$

TABLE 27

ADDITIONAL INFORMATION ON REFERENCES* RELATED TO CHAPTER 15

In addition to the material presented in the preceding pages of this chapter, the author wishes to call to the attention of the reader the numerous references listed in Table 27. (See statement at top of Table 4 of Chapter 1, regarding completeness of this material.)

Books
  Belcher and Godbert, 9, 10
  Block and Bolling, 18
  Clark, E. P., 32
  Clark, S. J., 33
  Clark, W. M., 34
  Friedrich, 45
  Fritz and Hammond, 46
  Furman, 50
  Grant, 61, 62
  Milton and Waters, 114, 115
  Niederl and Niederl, 120, 121
  Pregl, 132
  Roth, 136–139

Books (Cont.)
  Siggia, 149
  Steyermark, 163

Reviews
  Dunn, 35
  Hallett, 67
  Hammond, 68
  Kirsten, 88
  Lacourt, 98
  Ma, 104
  Mitchell, Montague, and Kinsey, 116
  Stelt, 160
  Steyermark, 162
  Willits, 191

* The numbers which appear after each entry in this table refer to the literature citations in the reference list at the end of the chapter.

TABLE 27 (*Continued*)

Submicro- ultramicro-methods

Giri, Radhakrishnan, and Vaidyana-Bergold and Pister, 12
Bonting, 20
Gordon, 60
Grant, W. M., 64
Grunbaum, Schaffer, and Kirk, 66
Hullin and Noble, 76
Kirsten, 88
Koepsell and Sharpe, 92
Lowry, Lopez, and Bessey, 103
Mannelli, 105
Tsao, Baumann, and Wark, 175
Tsao and Brown, 176
Wellington, 187
West, 188

Direct neutralization (acidimetric methods)

Black, 15
Ellenbogen and Brand, 39
Friedrich, 45
Gorbach, 57, 58
Grassmann and Heyde, 65
Hurka, 78
Jerie, 81
Kul'berg, 97
Owens and Maute, 127
Pregl, 132
Roth, 136–139
Schmidt-Nielson, 144
Schneider and Foulke, 145
Smith, Mitchell, and Billmeyer, 155
Stetten and Grail, 161
Steyermark, 162, 163
Tous and Pizarro, 173
West, 188

Chromatographic methods

Bergmann and Segal, 11
Blackburn and Robson, 16
Block, 17
Bryant and O'Connor, 22
Cavallini and Frontali, 27
Claborn and Patterson, 31
Eastroe, 37
Federico and Ciucani, 42
Fromageot and Colas, 47

Chromatographic methods (Cont.)

Giri, Radhakrishnan, and Vaidyanathan, 54
James and Martin, 80
Jureček, Churáček, and Cervinka, 82
Klatzkin, 89
Löffler and Reichl, 102
Martin and Mittlemann, 106
McFarren and Mills, 111
Nair, 118
Nijkamp, 122, 123
Overell, 126
Pereira and Serra, 128
Pfeil and Goldbach, 130
Ramsey and Patterson, 135
Seligson and Shapiro, 146
Sjöquist, 152
Van de Kamer, Gerritsma, and Wansink, 178
Wellington, 187
Wieland and Feld, 189

Spectrophotometric, colorimetric methods

Bergold and Pister, 12
Bobtelsky and Graus, 19
Bonting, 20
Breusch and Tulus, 21
Calkins, 25
Cherkin, Wolkowitz, and Dunn, 28
Ciaranfi and Fonnesu, 29
Federico and Ciucani, 42
Fonnesu, 43
Forziati, Rowen, and Plyler, 44
Furman, Morrison, and Wagner, 51
Gey, 53
Gordon, 60
Grant, 63, 64
Herb and Riemenschneider, 71
Herrington, 72
Hill, 73
Koepsell and Sharpe, 92
McArdle, 108
McKinney and Reynolds, 112
Nekhorocheff and Wajzer, 119
Perlman, Lardy, and Johnson, 129
Pratt and Corbitt, 131
Pucher, 133
Pucher, Sherman, and Vickery, 134

TABLE 27 (*Continued*)

Spectrophotometric, colorimetric methods (*Cont.*)
Schmall, Pifer, and Wollish, 142
Schmall, Pifer, Wollish, Duschinsky, and Gainer, 143
Sjöquist, 152
Szalkowski and Mader, 167
Taufel and Ruttloff, 169
Tsao, Baumann and Wark, 175
Tsao and Brown, 176
Wagner and Schröpl, 185
Weil-Malherbe and Bone, 186

Polarographic, conductometric, potentiometric, high-frequency, electrolytic methods
Butler and Czepiel, 24
Carson and Ko, 26
Elving and Van Atta, 40
Epstein, Sober, and Silver, 41
Furter and Gubser, 52
Grunbaum, Schaffer, and Kirk, 66
Hara and West, 69
Harlow and Wyld, 70
Hurka, 78
Ingold, 79
Karrman and Johansson, 87
Kolthoff, 93
Martin and Mittlemann, 106
Maurmeyer, Margosis, and Ma, 107
Oelsen and Graue, 124
Van Meurs and Dahmen, 179
Yakubik, Safranski, and Mitchell, 192
Yamamura, 193

Iodometric methods
Alicino, 3
Elek and Harte, 38
Hurka, 78
Kometiani and Sturua, 94
Shimosawa, 148

Oxidation methods
Buffa, 23
Linhardt and Reichold, 101
Pucher, 133
Roth, 140
Shimosawa, 148

Oxidation methods (*Cont.*)
Smith, 156
Weil-Malherbe and Bone, 186

Potassium hydrosulfide methods
Fuchs, 48, 49
Hunter and Edwards, 77
Tsurumi and Sasaki, 177

Manometric methods
See Chapter 18
Tracey, 174

Methods for amino acids
See Chapter 18
Baudet and Cherbuliez, 6
Bettzieche, 14
Blackburn and Robson, 16
Block and Bolling, 18
Bryant and O'Connor, 22
Cherkin, Wolkowitz, and Dunn, 28
Eastroe, 37
Fromageot and Colas, 47
Furman, Morrison, and Wagner, 51
Giri, Radhakrishnan, and Vaidyanathan, 54
Gorbach, 59
Grassmann and Heyde, 65
Kainz and Huber, 83
Kainz and Kasler, 84
Klatzkin, 89
Lacourt, Sommereyns, Francotte, and Delande, 99
Martin and Mittlemann, 106
McCaldin, 109
McFarren and Mills, 111
Merck, 113
Moubasher and Awad, 117
Pereira and Serra, 128
Pfeil and Goldbach, 130
Sjöquist, 152
Smith and Agiza, 153, 154
Spier and Pascher, 157
Steele, Sfortunato, and Ottolenghi, 158
Steers and Sevag, 159
Van Slyke and Dillon, 180
Van Slyke, Dillon, MacFadyen, and Hamilton, 181
Wellington, 187
Zeile and Oetzel, 194

TABLE 27 (*Continued*)

General, miscellaneous
Albrink, 2
Baker, 5
Beroza, 13
Block and Bolling, 18
Buffa, 23
Cimerman and Selzer, 30
Dyer, 36
Glagoleva-Malikova, 55
Goiffon and Couchoud, 56
Hopton, 75
Kaiser and Kagan, 85
Kometiani and Sturua, 94
Kottmeyer, 95, 96
Orekhovich and Tustanovskii, 125
Siggia and Floramo, 150
Stöhr and Scheibl, 164
Strong, Feeney, and Earle, 165
Sudo, Shimoe, and Tsujii, 166
Taufel, Pohloudek-Fabini, and Behnke, 168
Taussky, 170
Taylor, 171
Thomis and Kotionis, 172
Tous and Pizarro, 173
Volpi, 184
Wiese and Hansen, 190

Microdiffusion, microdistillation
Lang and Pfleger, 100
Ryan, 141

Complexometric methods
Ayers, 4
Spier and Pascher, 157

Nonaqueous titration
Maurmeyer, Margosis, and Ma, 107
Sensabaugh, Cundiff, and Markunas, 147
Van Meurs and Dahmen, 179
Yakubik, Safranski, and Mitchell, 192
Yamamura, 193

Determination of apparent dissociation constants
Simon, 151

Fluoro-compounds
Bergmann and Segal, 11

Apparatus
Hunter and Edwards, 77
McClendon, 110
Smith, Mitchell, and Billmeyer, 155
Stetten and Grail, 161
Sudo, Shimoe, and Tsujii, 166
Van Slyke, Folch, and Plazin, 182
Van Slyke and Neill, 183
Yakubik, Safranski, and Mitchell, 192

REFERENCES

1. Acculate, Anachemia Chemicals, Ltd., Montreal, Quebec and Champlain, New York.
2. Albrink, M. J., *J. Lipid Research,* **1,** 53 (1959).
3. Alicino, J. F., *Ind. Eng. Chem., Anal. Ed.,* **15,** 764 (1943).
4. Ayers, C. W., *Anal. Chim. Acta,* **15,** 77 (1956).
5. Baker, P. R. W., *Analyst,* **79,** 289 (1954).
6. Baudet, P., and Cherbuliez, E., *Helv. Chim. Acta,* **38,** 841 (1955).
7. Belcher, R., Berger, J., and West, T. S., *J. Chem. Soc.,* p. 2877 (1959).
8. Belcher, R., Berger, J., and West, T. S., *J. Chem. Soc.,* p. 2882 (1959).
9. Belcher, R., and Godbert, A. L., "Semi-Micro Quantitative Organic Analysis," Longmans, Green, London, New York, and Toronto, 1945.
10. Belcher, R., and Godbert, A. L., "Semi-Micro Quantitative Organic Analysis," 2nd ed., Longmans, Green, London, 1954.
11. Bergmann, G., and Segal, F., *Biochem. J.,* **62,** 542 (1956).
12. Bergold, G., and Pister, L., *Z. Naturforsch.,* **3b,** 406 (1948); *Chem. Abstr.,* **43,** 8420 (1949).
13. Beroza, M., *Anal. Chem.,* **25,** 177 (1953).
14. Bettzieche, F., *Z. physiol. Chem. Hoppe-Seyler's,* **161,** 178 (1926); *Chem. Abstr.,* **21,** 721 (1927).

15. Black, S., *Arch. Biochem.*, **23**, 347 (1949).
16. Blackburn, S., and Robson, A., *Chem. & Ind. (London)*, p. 614 (1950).
17. Block, R. J., *Anal. Chem.*, **22**, 1327 (1950).
18. Block, R. J., and Bolling, D., "The Determination of the Amino Acids," Burgess, Minneapolis, Minnesota, 1940.
19. Bobtelsky, M., and Graus, B., *Anal. Chim. Acta*, **9**, 163 (1953).
20. Bonting, S. L., *Arch. Biochem. Biophys.*, **56**, 307 (1955).
21. Breusch, F. L., and Tulus, R., *Biochim. et Biophys. Acta*, **1**, 77 (1947); *Chem. Abstr.*, **41**, 4405 (1947).
22. Bryant, F., and O'Connor, D. J., *Australian J. Sci.*, **13**, 111 (1951); *Chem. Abstr.*, **45**, 4767 (1951).
23. Buffa, A., *Ann. chim. (Rome)*, **40**, 617 (1950); *Chem. Abstr.*, **45**, 6126 (1951).
24. Butler, J. P., and Czepiel, T. P., *Anal. Chem.*, **28**, 1468 (1956).
25. Calkins, V. P., *Anal. Chem.*, **15**, 762 (1943).
26. Carson, W. N., Jr., and Ko, R., *Anal. Chem.*, **23**, 1019 (1951).
27. Cavallini, D., and Frontali, N., *Ricerca sci.*, **23**, 807 (1953); *Chem. Abstr.*, **47**, 9863 (1953).
28. Cherkin, A., Wolkowitz, H., and Dunn, M. S., *Anal. Chem.*, **28**, 895 (1956).
29. Ciaranfi, E., and Fonnesu, A., *Biochem. J.*, **50**, 698 (1952).
30. Cimerman, C., and Selzer, M., *Anal. Chim. Acta*, **9**, 26 (1953).
31. Claborn, H. V., and Patterson, W. I., *J. Assoc. Offic. Agr. Chem.*, **31**, 134 (1948).
32. Clark, E. P., "Semimicro Quantitative Organic Analysis," Academic Press, New York, 1943.
33. Clark, S. J., "Quantitative Methods of Organic Microanalysis," Butterworths, London, 1956.
34. Clark, W. M., "The Determination of Hydrogen Ions," 2nd ed., Williams & Wilkins, Baltimore, Maryland, 1927.
35. Dunn, M. S., *Advances in Chem. Ser.*, **3**, 13 (1950); *Chem. Abstr.*, **45**, 68 (1951).
36. Dyer, D. C., *J. Biol. Chem.*, **28**, 445 (1917).
37. Eastroe, J. E., *Biochem. J.*, **74**, 8P (1960).
38. Elek, A., and Harte, R. A., *Ind. Eng. Chem., Anal. Ed.*, **8**, 267 (1936).
39. Ellenbogen, E., and Brand, E., *Anal. Chem.*, **27**, 2007 (1955).
40. Elving, P. J., and Van Atta, R. E., *Anal. Chem.*, **26**, 295 (1954).
41. Epstein, J., Sober, H. A., and Silver, S. D., *Anal. Chem.*, **19**, 675 (1947).
42. Federico, K., and Ciucani, M., *Chim. e ind. (Milan)*, **36**, 598 (1954).
43. Fonnesu, A., *Biochem. Z.*, **324**, 512 (1953).
44. Forziati, F. H., Rowen, J. W., and Plyler, E. K., *J. Research Natl. Bur. Standards*, **46**, 288 (1951).
45. Friedrich, A., "Die Praxis der quantitativen organischen Mikroanalyse," p. 173, Deuticke, Leipzig, and Vienna, 1933.
46. Fritz, J. S., and Hammond, G. S., "Quantitative Organic Analysis," Wiley, New York, 1957.
47. Fromageot, C., and Colas, R., *Biochim. et Biophys. Acta*, **3**, 417 (1949) (in French).
48. Fuchs, F., *Monatsh. Chem.*, **9**, 1132, 1143 (1888).
49. Fuchs, F., *Monatsh. Chem.*, **11**, 363 (1890).
50. Furman, N. H., ed., "Scott's Standard Methods of Chemical Analysis," 5th ed., Vol. II, Van Nostrand, New York, 1939.
51. Furman, N. H., Morrison, G. H., and Wagner, A. F., *Anal. Chem.*, **22**, 1561 (1950).

52. Furter, M. F., and Gubser, H., *Helv. Chim. Acta,* **21**, 1725 (1938).
53. Gey, K. F., *Intern. Rev. Vitamin Research,* **25**, 21 (1953).
54. Giri, K. V., Radhakrishnan, A. N., and Vaidyanathan, C. S., *Anal. Chem.,* **24**, 1677 (1952).
55. Glagoleva-Malikova, E. M., *Latvijas PSR Zinātņu Akad. Vēstis,* pp. 121, 124 (1949); *Chem. Abstr.,* **47**, 12122 (1953).
56. Goiffon, R., and Couchoud, M., *Ann. biol. clin. (Paris),* **11**, 327 (1953); *Chem. Abstr.,* **47**, 12129 (1953).
57. Gorbach, G., *Fette u. Seifen,* **49**, 553 (1942); *Chem. Abstr.,* **37**, 5605 (1943).
58. Gorbach, G., *Fette u. Seifen,* **49**, 625 (1942); *Chem. Abstr.,* **38**, 1387 (1944).
59. Gorbach, G., *Mikrochemie ver. Mikrochim. Acta,* **36/37**, 902 (1951).
60. Gordon, H. T., *Anal. Chem.,* **23**, 1853 (1951).
61. Grant, J., "Quantitative Organic Microanalysis, Based on the Methods of Fritz Pregl," 4th ed., Blakiston, Philadelphia, Pennsylvania, 1946.
62. Grant, J., "Quantitative Organic Microanalysis," 5th ed., Blakiston, Philadelphia, Pennsylvania, 1951.
63. Grant, W. M., *Anal. Chem.,* **19**, 206 (1947).
64. Grant, W. M., *Anal. Chem.,* **20**, 267 (1948).
65. Grassmann, W., and Heyde, W., *Z. physiol. Chem. Hoppe-Seyler's,* **183**, 32 (1929).
66. Grunbaum, B. W., Schaffer, F. L., and Kirk, P. L., *Anal. Chem.,* **25**, 480 (1953).
67. Hallett, L. T., *Ind. Eng. Chem., Anal. Ed.,* **14**, 956 (1942).
68. Hammond, C. W., *in* "Organic Analysis" (J. Mitchell, Jr., I. M. Kolthoff, E. S. Proskauer, and A. Weissberger, eds.), Vol. III, p. 97, Interscience, New York, 1956.
69. Hara, R., and West, P. W., *Anal. Chim. Acta,* **15**, 193 (1956).
70. Harlow, G. A., and Wyld, G. E. A., *Anal. Chem.,* **30**, 73 (1958).
71. Herb, S. F., and Riemenschneider, R. W., *Anal. Chem.,* **25**, 953 (1953).
72. Herrington, E. F. G., *Analyst,* **78**, 174 (1953).
73. Hill, U. T., *Ind. Eng. Chem., Anal. Ed.,* **18**, 317 (1946).
74. Hodgman, C. D., and Lange, N. A., "Handbook of Chemistry and Physics," 9th ed., p. 395, Chemical Rubber, Cleveland, Ohio, 1922.
75. Hopton, J. W., *Anal. Chim. Acta,* **8**, 429 (1953).
76. Hullin, R. P., and Noble, R. L., *Biochem. J.,* **55**, 289 (1953).
77. Hunter, W. H., and Edwards, J. D., *J. Am. Chem. Soc.,* **35**, 452 (1913).
78. Hurka, W., *Mikrochemie ver. Mikrochim. Acta,* **31**, 5 (1943).
79. Ingold, W., *Mikrochemie ver. Mikrochim. Acta,* **36/37**, 276 (1951).
80. James, A. T., and Martin, A. J. P., *Biochem. J.,* **50**, 679 (1952).
81. Jerie, H., *Mikrochemie ver. Mikrochim. Acta,* **40**, 189 (1953).
82. Jureček, M., Churáček, J., and Cervinka, V., *Mikrochim. Acta,* p. 102 (1960).
83. Kainz, G., and Huber, H., *Mikrochim. Acta,* p. 38 (1960).
84. Kainz, G., and Kasler, F., *Mikrochim. Acta,* p. 62 (1960).
85. Kaiser, E., and Kagan, B. M., *Anal. Chem.,* **23**, 1879 (1951).
86. Kamm, O., "Qualitative Organic Analysis," p. 138, Wiley, New York, 1923.
87. Karrman, K. J., and Johansson, G., *Mikrochim. Acta,* p. 1573 (1956).
88. Kirsten, W. J., *Microchem. J.,* **2**, 179 (1958).
89. Klatzkin, C., *Nature,* **169**, 422 (1952).
90. Koch, F. C., *J. Lab. Clin. Med.,* **11**, 774 (1926).
91. Koch, F. C., *J. Lab. Clin. Med.,* **14**, 747 (1929).
92. Koepsell, H. J., and Sharpe, E. S., *Arch. Biochem. Biophys.,* **38**, 443 (1952).
93. Kolthoff, I. M., "Die Massanalyse," Springer, Berlin, 1928.

94. Kometiani, P. A., and Sturua, G. G., *Biokhimiya,* **13,** 23 (1948); *Chem. Abstr.,* **42,** 7659 (1948).
95. Kottmeyer, G., *Biochem. Z.,* **322,** 304 (1952).
96. Kottmeyer, G., *Biochem. Z.,* **324,** 160 (1953).
97. Kul'berg, L. M., *Zavodskaya Lab.,* **7,** 417 (1938); *Chem. Abstr.,* **33,** 88 (1939).
98. Lacourt, A., *Bull. soc. chim. Belg.,* **45,** 313 (1936); *Chem. Abstr.,* **30,** 7479 (1936).
99. Lacourt, A., Sommereyns, G., Francotte, C., and Delande, N., *J. pharm. Belg.,* p. 535 (1953); *Mikrochim. Acta,* p. 305 (1953).
100. Lang, K., and Pfleger, K., *Mikrochemie ver. Mikrochim. Acta,* **36/37,** 1174 (1951).
101. Linhardt, K., and Reichold, E., *Biochem. Z.,* **320,** 241 (1950).
102. Löffler, J. E., and Reichl, E. H., *Mikrochim. Acta,* p. 79 (1953).
103. Lowry, O. H., Lopez, J. A., and Bessey, O. A., *J. Biol. Chem.,* **160,** 609 (1945).
104. Ma, T. S., *Microchem. J.,* **3,** 415 (1959).
105. Mannelli, G., *Mikrochemie ver. Mikrochim. Acta,* **35,** 29 (1950).
106. Martin, A. J. P., and Mittlemann, R., *Biochem. J.,* **43,** 353 (1948).
107. Maurmeyer, R. K., Margosis, M., and Ma, T. S., *Mikrochim. Acta,* p. 177 (1959).
108. McArdle, B., *Biochem. J.,* **60,** 647 (1955).
109. McCaldin, D. J., *Chem. Revs.,* **60,** 39 (1960).
110. McClendon, J. F., *J. Biol. Chem.,* **154,** 357 (1944).
111. McFarren, E. F., and Mills, J. A., *Anal. Chem.,* **24,** 650 (1952).
112. McKinney, R. W., and Reynolds, C. A., *Talanta,* **1,** 46 (1958).
113. Merck & Co., Inc., "The Story of the Amino Acids," Rahway, New Jersey, 1940.
114. Milton, R. F., and Waters, W. A., "Methods of Quantitative Microanalysis," Longmans, Green, New York, and Arnold, London, 1949.
115. Milton, R. F., and Waters, W. A., "Methods of Quantitative Microanalysis," 2nd ed., Arnold, London, 1955.
116. Mitchell, J., Jr., Montague, B. A., and Kinsey, R. H., *in* "Organic Analysis" (J. Mitchell, Jr., I. M. Kolthoff, E. S. Proskauer, and A. Weissberger, eds.), Vol. III, p. 1, Interscience, New York, 1956.
117. Moubasher, R., and Awad, W. I., *J. Biol. Chem.,* **179,** 915 (1949).
118. Nair, J. H., III, *Anal. Chem.,* **25,** 1912 (1953).
119. Nekhorocheff, J., and Wajzer, J., *Bull. soc. chim. biol.,* **35,** 695 (1953).
120. Niederl, J. B., and Niederl, V., "Micromethods of Quantitative Organic Elementary Analysis," Wiley, New York, 1938.
121. Niederl, J. B., and Niederl, V., "Micromethods of Quantitative Organic Analysis," 2nd ed., Wiley, New York, 1942.
122. Nijkamp, H. J., *Anal. Chim. Acta,* **5,** 325 (1951).
123. Nijkamp, H. J., *Nature,* **172,** 1102 (1953).
124. Oelsen, W., and Graue, G., *Angew. Chem.,* **64,** 24 (1952).
125. Orekhovich, V. N., and Tustanovskii, A. A., *Doklady Akad. Nauk S.S.S.R.,* **67,** 333 (1949).
126. Overell, B. T., *Australian J. Sci.,* **15,** 28 (1952); *Chem. Abstr.,* **46,** 10039 (1952).
127. Owens, M. L., Jr., and Maute, R. L., *Anal. Chem.,* **27,** 1177 (1955).
128. Pereira, A., and Serra, J. A., *Science,* **113,** 387 (1951).
129. Perlman, D., Lardy, H. A., and Johnson, M. J., *Anal. Chem.,* **16,** 515 (1944).
130. Pfeil, E., and Goldbach, H. J., *Klin. Wochschr.,* **34,** 194 (1956).
131. Pratt, E. L., and Corbitt, H. B., *Anal. Chem.,* **24,** 1665 (1952).
132. Pregl, F., "Quantitative Organic Microanalysis" (E. Fyleman, trans., 2nd German ed.), Churchill, London, 1924.
133. Pucher, G. W., *J. Biol. Chem.,* **153,** 133 (1944).

134. Pucher, G. W., Sherman, C. C., and Vickery, H. B., *J. Biol. Chem.*, **113**, 235 (1936).
135. Ramsey, L. L., and Patterson, W. I., *J. Assoc. Offic. Agr. Chem.*, **31**, 139 (1948).
136. Roth, H., "Die quantitative organische Mikroanalyse von Fritz Pregl," 4th ed., Springer, Berlin, 1935.
137. Roth, H., "F. Pregl quantitative organische Mikroanalyse," 5th ed., Springer, Wien, 1947.
138. Roth, H., "Pregl-Roth quantitative organische Mikroanalyse," 7th ed., Springer, Wien, 1958.
139. Roth, H., "Quantitative Organic Microanalysis of Fritz Pregl," 3rd ed. (E. B. Daw, trans., 4th German ed.), Blakiston, Philadelphia, Pennsylvania, 1937.
140. Roth, H., *Mikrochim. Acta*, p. 766 (1958).
141. Ryan, H., *Analyst*, **83**, 528 (1958); *Anal. Abstracts*, **6**, No. 962 (1959).
142. Schmall, M., Pifer, C. W., and Wollish, E. G., *Anal. Chem.*, **25**, 1486 (1953).
143. Schmall, M., Pifer, C. W., Wollish, E. G., Duschinsky, R. C., and Gainer, H., *Anal. Chem.*, **26**, 521 (1954).
144. Schmidt-Nielson, K., *Compt. rend. trav. lab. Carlsberg. Sér. chim.*, **24**, 233 (1942); *Chem. Abstr.*, **38**, 2225 (1944).
145. Schneider, F., and Foulke, D. G., *Anal. Chem.*, **11**, 111 (1939).
146. Seligson, D., and Shapiro, B., *Anal. Chem.*, **24**, 754 (1952).
147. Sensabaugh, A. J., Cundiff, R. D., and Markunas, P. C., *Anal. Chem.*, **30**, 1145 (1958).
148. Shimosawa, T., *Yakugaku Zasshi*, **63**, 462 (1943); *Chem. Abstr.*, **44**, 6772 (1950).
149. Siggia, S., "Quantitative Organic Analysis Via Functional Groups," 2nd ed., Wiley, New York, and Chapman & Hall, London, 1954.
150. Siggia, S., and Floramo, N. A., *Anal. Chem.*, **25**, 797 (1953).
151. Simon, W., *Helv. Chim. Acta*, **41**, 1835 (1958).
152. Sjöquist, J., *Arkiv. Kemi*, **11**, 129 (1957); *Chem. Abstr.*, **53**, 983 (1958).
153. Smith, A. M., and Agiza, A. H., *Analyst*, **76**, 619 (1951).
154. Smith, A. M., and Agiza, A. H., *Analyst*, **76**, 623 (1951).
155. Smith, D. M., Mitchell, J., Jr., and Billmeyer, A. M., *Anal. Chem.*, **24**, 1847 (1952).
156. Smith, R. J., *Anal. Chem.*, **25**, 505 (1953).
157. Spier, H. W., and Pascher, G., *Z. physiol. Chem. Hoppe-Seyler's*, **296**, 147 (1954).
158. Steele, R., Sfortunato, T., and Ottolenghi, L., *J. Biol. Chem.*, **177**, 231 (1949).
159. Steers, E., and Sevag, M. G., *Anal. Chem.*, **21**, 641 (1949).
160. Stelt, C. van der, *Pharm. Weekblad*, **86**, 69 (1951); *Chem. Abstr.*, **45**, 5573 (1951).
161. Stetten, D., Jr., and Grail, G. F., *Ind. Eng. Chem., Anal. Ed.*, **15**, 300 (1943).
162. Steyermark, Al, *in* "Organic Analysis" (J. Mitchell, Jr., I. M. Kolthoff, E. S. Proskauer, and A. Weissberger, eds.), Vol. II, pp. 1–18, Interscience, New York, 1954.
163. Steyermark, Al, "Quantitative Organic Microanalysis," Blakiston, Philadelphia, Pennsylvania, 1951.
164. Stöhr, R., and Scheibl, F., *Mikrochemie ver. Mikrochim. Acta*, **36/37**, 362 (1951).
165. Strong, F. M., Feeney, R. E., and Earle, A., *Ind. Eng. Chem., Anal. Ed.*, **13**, 566 (1941).
166. Sudo, T., Shimoe, D., and Tsujii, T., *Bunseki Kagaku*, **6**, 494 (1957).
167. Szalkowski, C. R., and Mader, W. J., *Anal. Chem.*, **24**, 1602 (1952).
168. Taufel, K., Pohloudek-Fabini, R., and Behnke, U., *Z. anal. Chem.*, **146**, 244 (1955).

169. Taufel, K., and Ruttloff, H., *Ernährungsforschung,* **1**, 693 (1956).
170. Taussky, H. H., *J. Biol. Chem.,* **181**, 195 (1949).
171. Taylor, T. G., *Biochem. J.,* **54**, 48 (1953).
172. Thomis, G. N., and Kotionis, A. Z., *Anal. Chim. Acta,* **14**, 457 (1956).
173. Tous, J. G., and Pizarro, A. V., *Anales real soc. españ. fís. y quím. (Madrid),* **46B**, 105 (1950); *Chem. Abstr.,* **44**, 10601 (1950).
174. Tracey, M. V., *Biochem. J.,* **43**, 185 (1948).
175. Tsao, M. U., Baumann, M. L., and Wark, S., *Anal. Chem.,* **24**, 722 (1952).
176. Tsao, M. U., and Brown, S., *J. Lab. Clin. Med.,* **35**, 320 (1950).
177. Tsurumi, S., and Sasaki, Y., *Sci. Repts. Tôhoku Univ., First Ser.,* **19**, 681 (1930); *Chem. Abstr.,* **25**, 2077 (1931).
178. Van de Kamer, J. H., Gerritsma, K. W., and Wansink, E. H., *Biochem. J.,* **61**, 174 (1955).
179. Van Meurs, N., and Dahmen, E. A. M. F., *Anal. Chim. Acta,* **21**, 443 (1959).
180. Van Slyke, D. D., and Dillon, R. T., *Compt. rend. trav. lab. Carlsberg. Sér. chim.,* **22**, 480 (1938); *Chem. Abstr.,* **32**, 7950 (1938).
181. Van Slyke, D. D., Dillon, R. T., MacFadyen, D. A., and Hamilton, P., *J. Biol. Chem.,* **141**, 627 (1941).
182. Van Slyke, D. D., Folch, J., and Plazin, J., *J. Biol. Chem.,* **136**, 509 (1940).
183. Van Slyke, D. D., and Neill, J. M., *J. Biol. Chem.,* **61**, 523 (1924).
184. Volpi, A., *Ann. chim. appl.,* **37**, 442 (1947).
185. Wagner, G., and Schröpl, E., *Pharmazie,* **14**, 605 (1959).
186. Weil-Malherbe, H., and Bone, A. D., *Biochem. J.,* **45**, 377 (1949).
187. Wellington, E. F., *Can. J. Chem.,* **30**, 581 (1952).
188. West, T. S., *Analyst,* **82**, 459 (1957).
189. Wieland, T., and Feld, U., *Angew. Chem.,* **63**, 258 (1951).
190. Wiese, H. F., and Hansen, A. E., *J. Biol. Chem.,* **202**, 417 (1953).
191. Willits, C. O., *Anal. Chem.,* **21**, 132 (1949).
192. Yakubik, M. G., Safranski, L. W., and Mitchell, J., Jr., *Anal. Chem.,* **30**, 1741 (1958).
193. Yamamura, S. S., *Dissertation Abstr.,* **17**, 2787 (1957).
194. Zeile, K., and Oetzel, M., *Z. physiol. Chem. Hoppe-Seyler's,* **284**, 1 (1949).

CHAPTER 16

Microdetermination of Alkoxyl Groups (Methoxyl and Ethoxyl)

The determination of alkoxyl groups (methoxyl and ethoxyl) in the form of either ethers or esters is accomplished by treatment of the organic compound with boiling hydriodic acid.[12,13,28,33,44,45,50,51,82-84,86-89,97-100] By this treatment the alkoxyl is split off and converted to methyl or ethyl iodide. Where the alkoxyl is present in the form of an ester, the splitting off is rather rapid and in fact often first yields the alcohol which then is converted to the iodide. Where the alkoxyl is present in the form of an ether, the action takes place more slowly.

The alkyl iodide (methyl or ethyl) is then determined iodometrically or gravimetrically. For the former, a modification of the method of Vieböch and Brecher[108] is used, the alkyl iodide being oxidized with bromine to iodic acid. This is treated with an excess of potassium iodide in acid solution yielding iodine, which is determined by titrating with thiosulfate (compare standardization of thiosulfate, Chapter 5). For the gravimetric procedure, the alkyl iodide (methyl or ethyl) is reacted with alcoholic silver nitrate yielding a double salt, $AgI \cdot AgNO_3$,[50,51,86-89,100] which in turn is split on the addition of water and nitric acid. The resulting silver iodide is then determined gravimetrically. The various reactions for both procedures are shown below.

(a) For esters:

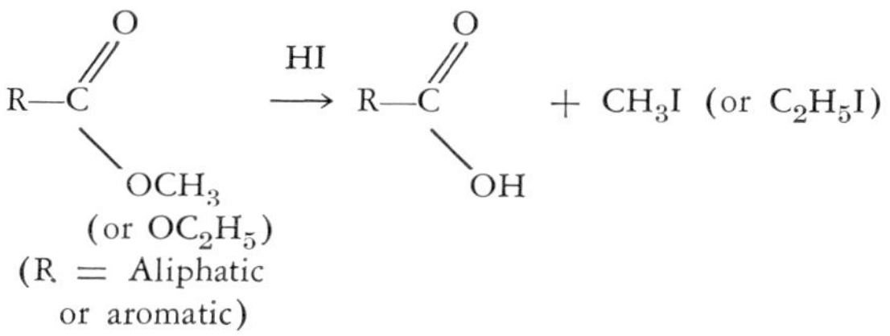

$$R-C(=O)-OCH_3 \text{ (or } OC_2H_5) \xrightarrow{HI} R-C(=O)-OH + CH_3I \text{ (or } C_2H_5I)$$

(R = Aliphatic or aromatic)

(b) For ethers:

$$R-OCH_3 \text{ (or } OC_2H_5) \xrightarrow{HI} R-OH + CH_3I \text{ (or } C_2H_5I)$$

(R = Aliphatic or aromatic)

(c) Then volumetric (iodometric):

$$CH_3I + Br_2 \rightarrow CH_3Br + IBr$$
$$(\text{or } C_2H_5I) \quad (\text{or } C_2H_5Br)$$
$$IBr + 3H_2O + 2Br_2 \rightarrow HIO_3 + 5HBr$$
$$HIO_3 + 5HI \rightarrow 3\,I_2 + 3H_2O$$
$$3\,I_2 + 6Na_2S_2O_3 \rightarrow 6NaI + 3Na_2S_4O_6$$

or gravimetric:

$$CH_3I \xrightarrow{AgNO_3} AgI.AgNO_3$$
$$(\text{or } C_2H_5I) \quad \downarrow\ H_2O$$
$$\phantom{(\text{or } C_2H_5I)} \quad \downarrow\ HNO_3$$
$$\underline{AgI} + AgNO_3$$

Two pieces of apparatus are described in the following pages, *both* of which may be used with a volumetric or gravimetric procedure, and with *either* give excellent results. The first method described is that employing the modified Clark[99,101] apparatus developed by the Committee on Microchemical Apparatus of the Division of Analytical Chemistry of the American Chemical Society and the method adopted by the Association of Official Agricultural Chemists following a collaborative study in which this apparatus was used. This adopted procedure is a volumetric (iodometric) one, but the same apparatus may be employed in the gravimetric procedure. Likewise, the procedure described using the apparatus developed by the author[97,100] is a gravimetric one, but the same apparatus may be used with the volumetric (iodometric) procedure. Each piece of apparatus has an advantage over the other. The Steyermark apparatus, due to its *two* reaction flasks, often gives better results with volatile substances but it is more fragile than the modified Clark apparatus. The latter is suitable also for semimicro-work or where relatively large amounts of material must be taken in order to analyze for small percentages of alkoxyl groups.

Regardless of the method employed, there is no interference from fluorine.

VOLUMETRIC (IODOMETRIC) METHOD

Reagents

PHENOL[33,50,51,82-84,86-89,97-100]

Pure crystalline phenol is used as a solvent for the sample. (*Caution:* This must be handled with care.)

ACETIC ACID-POTASSIUM ACETATE-BROMINE SOLUTION[99]

Ten grams of potassium acetate is dissolved in glacial acetic acid and diluted with the acid to 100 ml., and 3 ml. of bromine is added to complete the reagent. Thus must be freshly prepared.

SODIUM ACETATE SOLUTION[54,99]

Twenty-five grams of sodium acetate, $NaC_2H_3O_2.3H_2O$ is dissolved in distilled water and diluted to a volume of 100 ml. This is used in the scrubber[54] as well as in the titration mixture.

STARCH INDICATOR

Same as described in Chapter 5.

STANDARD SODIUM THIOSULFATE, 0.01N

This is prepared and standardized as described in Chapter 5.

FORMIC ACID

Reagent grade formic acid is used to destroy the excess bromine before titrating.

POTASSIUM IODIDE CRYSTALS

Reagent grade, stored in brown bottle.

DILUTE SULFURIC ACID, 10%

HYDRIODIC ACID,
SP. GR., 1.7[50,51,82,86–89,97,99,100]

The hydriodic acid generally purchased is unsuitable for use without purification, whether reagent grades or those designated as special for the microdetermination of methoxyl are used. They often contain hydrogen sulfide, phosphine, and possibly alkyl iodides formed from the reaction of the acid vapors on the material used to cover the bottle stoppers. These give large blanks, sometimes the amount is equal to or greater than that obtained in actual determinations.

TREATMENT OF HYDRIODIC ACID.[64,97,99,100] One pound of reagent grade hydriodic acid, sp. gr. 1.7, is placed in a round-bottomed flask which, in turn, is connected by means of a ground joint to an air condenser. The acid is heated to gentle boiling for about 2 hours, during which a slow stream of carbon dioxide or nitrogen is bubbled through by means of a glass tube extending to the bottom. At no time should the acid vapors be allowed to come in contact with organic material, which would cause recontamination. When heating is stopped the flow of gas likewise should be discontinued, as fuming acid is formed by passing the gas through at room temperature.[76,97,100] A blank methoxyl determination should then be carried out to determine the quality of the acid. If even a trace is obtained, the refluxing with carbon dioxide or nitrogen should be repeated, as it is possible to obtain a reagent which gives a perfect blank. The acid is stored in a brown glass-stoppered bottle at laboratory temperature and gives good results for a number of weeks. (The color of the product is of no importance. In fact, the presence of dissolved iodine, causing the dark color, is advantageous since it converts organic bond sulfur to the elementary form, preventing its interference.[33,97,100]) Test samples should be run at frequent intervals to make certain that the reagent is still efficient.

Apparatus

CARBON DIOXIDE SOURCE

A small carbon dioxide cylinder, provided with a suitable reducing valve, is used as a source of carbon dioxide. Before entering the alkoxyl apparatus,

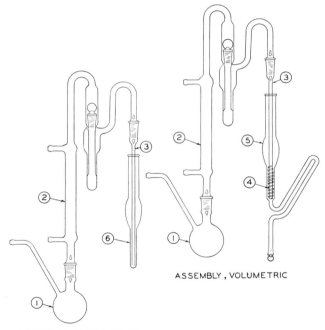

FIG. 179. Modified Clark alkoxyl apparatus, assembly (gravimetric and volumetric).

the gas should be passed through some type of wash bottle or scrubbing tower which contains a concentrated solution of sodium carbonate to remove any acid vapors present. A bubble counter containing concentrated sulfuric acid may be used following the wash bottle or scrubbing tower as an aid to the subsequent regulation of the flow of carbon dioxide through the system. [Either the broken section of a bubble counter-U-tube (see Fig. 120, Chapter 9) may be used for this purpose or a complete unit without filling, except the H_2SO_4.]

MODIFIED CLARK ALKOXYL APPARATUS[99,101]

The apparatus used, shown in Fig. 179, is the modified Clark apparatus with which either volumetric or gravimetric procedures may be used depending upon the type of receiver. For the volumetric procedure, the apparatus consists of the reaction flask with side arm, (1), condenser with scrubber, (2), inlet tube, (3), spiral, (4), and volumetric receiver, (5), and is shown as-

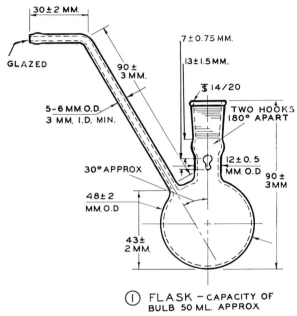

FIG. 179(1). Flask for modified Clark alkoxyl apparatus—details of construction.

sembled. For the gravimetric procedure, the spiral, (4), and volumetric receiver, (5), are replaced by the gravimetric receiver, (6), and this is also shown assembled.

The dimensions for the side arm of the flask, (1), were arrived at after a number of experiments. Capillary tubes, with and without bulbs, were unsatisfactory because of condensation in the tube. The recommended length of the side arm is necessary to minimize contact of acid with the gas connection.

The condenser with scrubber, (2), has an enlarged section between the two parts to prevent suck back of liquid from scrubber into condenser at the end of a determination. (Several types of scrubbers were tested, including one constructed of two compartments connected by a capillary tube. The one selected operated more efficiently than all others tried.)

Volumetric (Iodometric) Method

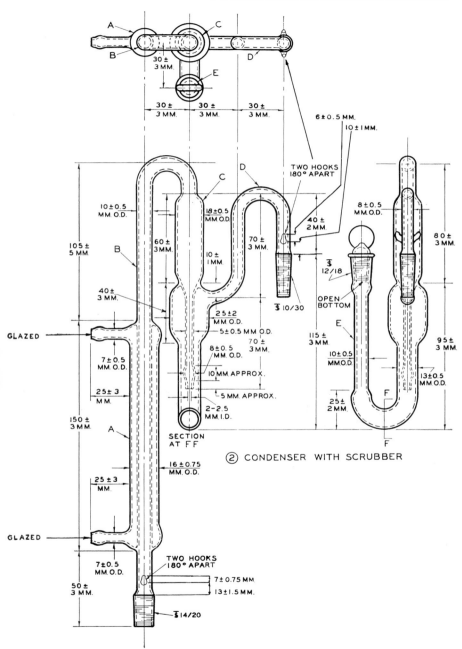

FIG. 179(2). Condenser with scrubber for modified Clark alkoxyl apparatus—details of construction.

16. Alkoxyl Groups

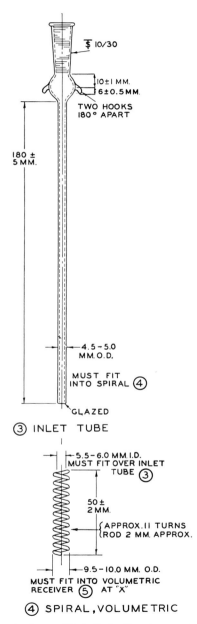

FIG. 179(3). Inlet tube for modified Clark alkoxyl apparatus—details of construction.

FIG. 179(4). Spiral, volumetric, for modified Clark alkoxyl apparatus—details of construction.

Volumetric (Iodometric) Method

The section between the scrubber and the inlet tube, (3), was designed to prevent liquid being carried into the receiver.

Use of the spiral, (4), in the receiver, (5), is optional in the volumetric procedure. Extensive tests have shown that equally good results are obtained without the spiral.[101]

BURETTE

An automatic burette of the type shown in Figs. 69 or 70 (Chapter 5) is used for the standard thiosulfate solution.

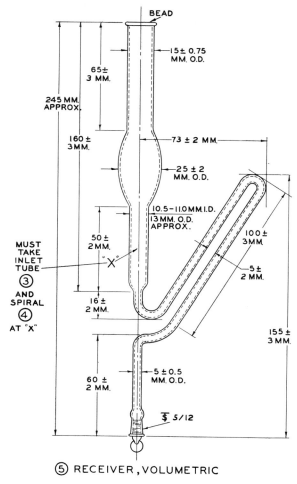

FIG. 179(5). Receiver, volumetric, for modified Clark alkoxyl apparatus—details of construction.

ELECTRIC HEATER

A small electric heater[53] of the type used in the construction of the Kjeldahl digestion rack (Fig. 180) is particularly suitable for heating the contents of the reaction flask. Gas microburners of the type shown in Fig. 181 are also suitable, but the author prefers the use of the electric type since, with these, boiling is more easily controlled.

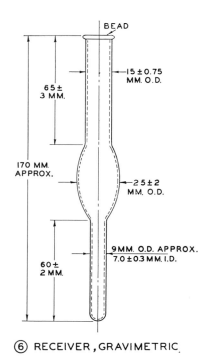

(6) RECEIVER, GRAVIMETRIC

FIG. 179(6). Receiver, gravimetric, for modified Clark alkoxyl apparatus—details of construction.

FIG. 180. (*Left*) Hankes heater.
FIG. 181. (*Right*) Gas microburner.

Procedure

USING THE MODIFIED CLARK APPARATUS (FIG. 179)[99,101]

The scrubber, (2), is filled halfway with the sodium acetate solution and the volumetric receiver, (5), is filled two-thirds full with *freshly prepared* acetic acid-potassium acetate-bromine solution. (If desired, the spiral, (4), may be inserted into the receiver, but extensive tests have shown that equally good results are obtained without it.[101]) Enough sample is added to the reaction flask, (1), to require about 8 ml. of 0.01N sodium thiosulfate in the determination. Solid samples are weighed in a platinum boat (Fig. 24, Chapter 3), while volatile liquids are weighed in the customary capillaries (Fig. 54, Chapter 3), but, *in addition,* an empty platinum boat (or tetrahedra) is added as a means of preventing bumping. Two and one-half ml. of melted phenol is added to the flask, followed by 5 ml. of hydriodic acid. The reaction flask is *immediately* connected to the condenser with scrubber, (2). The source of carbon dioxide is attached to the side arm of the reaction flask and carbon dioxide passed through the apparatus at the rate of about 15 ml. per minute. Cold water is circulated through the condenser and the reaction mixture is allowed to remain at *room temperature for 30 minutes.* The mixture is then heated to boiling by means of the heater (or a microburner) and boiled at such a rate that the vapors of the boiling liquid rise into the condenser, but not more than halfway. Boiling is continued with *water circulating* in the condenser for *one-half hour,* after which the water is *drained* from the condenser and the *boiling continued* for an *additional one-half hour.*

The receiver and inlet tube are disconnected, the stopper removed from the end of the siphon, the unit tilted, and the contents siphoned into a 125-ml. ground glass-stoppered Erlenmeyer flask which contains 5 ml. of the sodium acetate solution. The receiver and inlet tube are washed with enough water so that the washings when added to the Erlenmeyer flask bring the total volume in it to about 50 ml. Formic acid is added, dropwise, with swirling of the flask, until the excess of bromine has been destroyed. Any remaining bromine vapors should be removed by blowing air over the liquid. To the contents of the flask are added 0.5 gram of potassium iodide and 5 ml. of the 10% sulfuric acid. The flask is stoppered and swirled to dissolve the potassium iodide crystals and to mix the contents. The liberated iodine is titrated with 0.01N sodium thiosulfate using starch as the indicator, as described under the standardization of thiosulfate (see Chapter 5). (Any blank value obtained by carrying out the determination in the absence of a sample should be subtracted from the volume of thiosulfate obtained with the sample. However, if the hydriodic acid has been properly treated and if all the reagents are of reagent grade, there should be absolutely no blank.)

16. Alkoxyl Groups

Calculations:

Factors:

1 ml. of 0.01N sodium thiosulfate is equivalent to 0.05173 mg. of methoxyl (OCH_3) or 0.07510 mg. of ethoxyl (OC_2H_5)

$$\therefore \frac{\text{ml. of } 0.01N \ Na_2S_2O_3 \times 0.05173 \times 100}{\text{Wt. sample}} = \text{per cent } OCH_3$$

and

$$\frac{\text{ml. of } 0.01N \ Na_2S_2O_3 \times 0.07510 \times 100}{\text{Wt. sample}} = \text{per cent } OC_2H_5$$

Examples:

a. 6.82 ml. of 0.01N thiosulfate is required to titrate the iodine liberated in the analysis of a 3.348-mg. sample, containing methoxyl groups

$$\therefore \frac{6.82 \times 0.05173 \times 100}{3.348} = 10.54\% \ OCH_3$$

b. 6.03 ml. of 0.01N thiosulfate is required to titrate the iodine liberated in the analysis of a 4.012-mg. sample containing ethoxyl groups

$$\therefore \frac{6.03 \times 0.07510 \times 100}{4.012} = 11.29\% \ OC_2H_5$$

The allowable error is ±0.3%.

USING THE STEYERMARK APPARATUS[97,100]

The volumetric (iodometric) method may be performed with the Steyermark apparatus (Fig. 182) described below after substituting the test tube, *g*, with the volumetric receiver [Fig. 179, (5)]. All reagents are the same as used for the volumetric method described above (using the modified Clark alkoxyl apparatus). With the Steyermark apparatus and volumetric procedure, the scrubber, *d*, is filled halfway with the sodium acetate solution, and the volumetric receiver [Fig. 179, (5)] is two-thirds filled with the acetic acid-potassium acetate-bromine solution. The rest of the manipulation is as described below and the reader is advised to refer to this, a *summary* of which is as follows. Water is run through the condenser (Fig. 182, *c*). The reaction mixture in the flask, *a*, is allowed to remain at *room temperature* for *one-half hour* while the hydriodic acid in the flask, *b*, is *boiled* and a stream of carbon dioxide is passed through the setup. The mixture in the reaction flask, *a*, is then brought to a boil and the contents of *both* flasks, *a* and *b*, boiled for an additional one-half hour with the water running through the condenser, *c*, and then for *at least* another one-half hour after the water has been drained from the condenser, *c*. The volumetric receiver is then removed, emptied, etc., as described above and the liberated iodine titrated. The calculations are the same as above.

GRAVIMETRIC METHOD

This method gives good results for volatile substances as well as those containing more than one group although both of these types are known to present difficulties when analyzed by other procedures.[33,44,45,97,100]

During the course of the determination, a small amount of silver sulfide is formed resulting from the decomposition of the thiosulfate used in the scrubber. In addition, there is formed some reduced silver caused by allowing

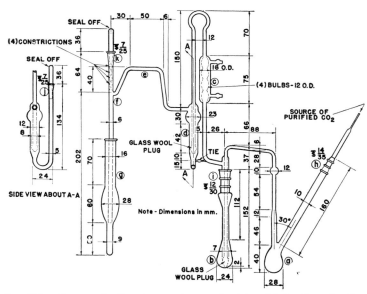

FIG. 182. Steyermark alkoxyl apparatus, gravimetric—details of construction.

the acid alcoholic silver nitrate solution to stand too long before filtration of the silver iodide. Both of these are removed by treating the precipitate with cold concentrated nitric acid[97,100] until the former is yellow in color.

Reagents

PHENOL[33,50,51,82-84,86-89,97-100]

Pure crystalline phenol is used as a solvent for the sample. (*Caution:* This must be handled with care.)

PROPIONIC ANHYDRIDE[33,83,97,100]

Pure propionic anhydride is used also as a solvent for the sample.

16. Alkoxyl Groups

TIN FOIL[50,51,82,84,86-89,97,100]

Thin, pure tin foil is cut into pieces about 2.5 sq. cm. in area and then rolled so that they will pass through the tube into the reaction flask.

CADMIUM SULFATE, 5%[33,50,51,82-84,86-89,97,100]

A 5% solution of cadmium sulfate in distilled water is prepared using reagent grade of material. This is used in the scrubber to remove hydrogen sulfide which might be formed in the determination.

SODIUM THIOSULFATE, 5%[50,51,86-89,97,100]

A 5% solution of sodium thiosulfate in distilled water is prepared using reagent grade of material. It is added to the scrubber to trap any iodine in the reaction flask.

ALCOHOLIC SILVER NITRATE, 3.85%[50,51,82,86-89,97,100]

Four grams of reagent grade of silver nitrate crystals is dissolved in 100 grams of 95% ethanol. The resulting solution is refluxed for 4 hours using a flask and condenser having ground glass joints. The flask is then disconnected from the condenser, closed with a ground glass stopper, and set aside in the dark for one week before using. The solution is then carefully decanted from the separated silver and stored in a brown glass-stoppered bottle. The solution may be used for as long as five to six months but not longer or low results will be obtained.

HYDRIODIC ACID, SP. GR., 1.7

Same as for the Volumetric (Iodometric) Method.

DILUTE NITRIC ACID, 1:200

This is prepared from pure concentrated nitric acid, sp. gr. 1.42, by diluting 1:200 with distilled water. It is used in transferring and in washing the precipitate.

CONCENTRATED NITRIC ACID, SP. GR., 1.42

Concentrated nitric acid, sp. gr. 1.42, is used for washing the precipitate to remove silver sulfide and reduced silver.[97,100]

ETHANOL, 95%

Ethanol, 95%, is used in transferring and washing the precipitate.

Apparatus

CARBON DIOXIDE SOURCE

This is identical to that described in connection with the volumetric procedure and includes: Carbon dioxide cylinder equipped with reducing valve, wash bottle or scrubbing tower containing sodium carbonate solution, and bubble counter containing concentrated sulfuric acid.

STEYERMARK ALKOXYL APPARATUS[97,100]

Fig. 182 gives details of construction. It combines the advantageous features of two others which likewise proved satisfactory, namely, those of Elek[33] and of Furter.[44,45]

The apparatus consists of a recation flask, *a,* which has an inlet tube for the passage of carbon dioxide and a distilling head connected to a tube passing to the bottom of a second reaction flask, *b.* Glass wool or sintered glass is used at the point shown to insure small bubbles. The flask, *b,* is connected to the reflux condenser, *c,* which in turn is connected to the scrubber, *d.* Glass wool or sintered glass again is used as shown for breaking up bubbles. The scrubber, *d,* is connected by means of a side arm, *e,* to the tube, *f,* which passes into the test tube (receiver), *g,* used for collecting the precipitate. Ground joints are present at the points, *h, i, j,* and *k.*

ELECTRIC HEATERS[53] OR MICROBURNERS

Same as for the Volumetric (Iodometric) Method. Two are required, one for each reaction flask.

FILTER TUBE

Same as used in Chapter 11 (Fig. 158).

FILTRATION ASSEMBLY

Same as used in Chapter 11 (Fig. 160).

Procedure[97,100]

USING THE STEYERMARK APPARATUS[97,100]

The sample (5–9 mg.), if solid or a high-boiling liquid, is weighed in a platinum boat and placed in the reaction flask (Fig. 182, *a*). [Volatile samples are weighed in the customary capillaries (Chapter 3) and inserted along with a platinum boat, or tetraheda, for prevention of bumping.] Several crystals of phenol and five to six drops of propionic anhydride are added and the mixture warmed (unless the sample is volatile) to dissolve the sample. Hydriodic acid is added to the second reaction flask, *b,* so that the bulb portion is about one-half full and the flask is then put into place, using a drop of the acid to seal the joint. Equal parts of 5% solutions of cadmium sulfate and sodium thiosulfate are added to the scrubber, *d,* to fill it halfway. Water is run into the tube, *f,* through the ground joint, *k,* and the stopper is attached quickly so as to have the constrictions filled with water. The source of carbon dioxide is attached at the ground joint, *h,* and the gas is bubbled through the apparatus until the excess water in the tube, *f,* has been forced out (usually a minute or less). Two milliliters of 3.85% alcoholic silver nitrate is placed in the test tube (receiver), *g,* and this is put into place so that the delivery tube, *f,* goes to the very bottom. The ground joint, *h,* is opened and a piece of the tin foil (2.5 sq. cm. in area, then rolled to pass through the opening) is added to the flask, *a,* followed by enough hydriodic acid to fill the bulb part of the flask approximately half full. The source of carbon dioxide is *immediately* attached, using a drop of the acid to seal the joint.

The flow of gas is regulated so that the bubbles pass through the silver nitrate in the test tube (receiver), *g,* at the rate of about one bubble per second. Water is run through the condenser, *c,* and a heater, or microburner, is used to heat the acid in the flask, *b,* to gentle boiling. (*Caution:* Fuming acid is formed upon passing gas through at room temperature.[76,97,100]) In this condition, the apparatus is allowed to remain for one-half hour. The reaction mixture in the flask, *a,* is then heated to boiling by means of a heater, or microburner, and the boiling of the contents of the two flasks, *a* and *b,* is continued for one-half hour.

At the end of this period, the water is drained from the condenser, but the boiling of the contents of *both* flasks is *continued* for *at least* an additional one-half hour. The stopper of the ground joint, *k,* is then opened, the test tube (receiver), *g,* lowered, and the precipitate washed into the test tube, *g,* alternately and by means of 1:200 nitric acid and ethanol. Then 1:200 nitric acid is added to the test tube until it is about four-fifths full. Five drops of concentrated nitric acid, sp. gr. 1.42, is added and the contents of the tube are brought quickly to the boiling point on a steam bath and then *immediately* cooled *and* filtered into a filter tube (Fig. 158, Chapter 11) with the aid of the filtration assembly (Fig. 160, Chapter 11). The silver iodide precipitate is washed with 1:200 nitric acid and then with cold concentrated nitric acid, sp. gr. 1.42, by filling the filter tube with the acid, allowing it to soak through for several minutes and then sucking dry. (*Caution:* Silver iodide is insoluble in nitric acid, sp. gr. 1.42, but is quite soluble in fuming acid, sp. gr. 1.49–1.50.[97,100]) The washing should be repeated several times until the precipitate is *yellow.* The precipitate is then washed with 1:200 nitric acid followed by ethanol and dried in an oven at 120° C., after which it is weighed (refer to Chapter 11).

Gravimetric Method

Calculations:

An empirical correction of 0.120 mg. of silver iodide (0.06 mg. per ml. of silver nitrate used) must be added to the weight of the precipitate.[37-39,50,51,82,86,87,89,100]

Factors:

For OCH_3, 0.1322
OC_2H_5, 0.1919

$$\therefore \frac{\text{Wt. of precipitate (corrected)} \times \text{factor} \times 100}{\text{Wt. sample}} = \text{per cent } OCH_3, OC_2H_5$$

Examples:

a. 9.063 mg. of AgI, corrected (8.943 mg., actual, plus 0.120 mg. correction), is obtained from a 7.801-mg. sample containing methoxyl groups

$$\therefore \frac{9.063 \times 0.1322 \times 100}{7.801} = 15.36\% \; OCH_3$$

b. 8.137 mg. of AgI, corrected (8.017 mg., actual, plus 0.120 mg. correction), is obtained from a 6.802-mg. sample containing ethoxyl groups

$$\therefore \frac{8.137 \times 0.1919 \times 100}{6.802} = 22.96\% \; OC_2H_5$$

The allowable error is $\pm 0.3\%$.

CLEANING THE APPARATUS

The apparatus is best cleaned, after draining off the solutions from the flasks and scrubber, by immersing the delivery tip, *f,* in water and applying suction to the flask, *a.* After the apparatus is cleaned by the passage of water it may be dried either by placing in an oven or by passing acetone through it followed by a stream of warm air.

USING THE MODIFIED CLARK APPARATUS[99,101]

The gravimetric method may be performed with the modified Clark apparatus (Fig. 179 described above) using the gravimetric receiver, (6). All reagents are the same as used for the gravimetric method described above (using the Steyermark alkoxyl apparatus). With the modified Clark apparatus and gravimetric procedure, cadmium sulfate and sodium thiosulfate are used in the scrubber and alcoholic silver nitrate is used in the gravimetric receiver. The silver iodide is treated as above, including washing and weighing.

Calculation:

Same as for Gravimetric Method above.

16. Alkoxyl Groups

TABLE 28
Additional Information on References* Related to Chapter 16

In addition to the procedures described in the preceding pages of this chapter, the author wishes to call to the attention of the reader the material presented in the various references listed in Table 28. (See statement at top of Table 4 of Chapter 1, regarding completeness of this material.) This includes work on the determination of higher homologs, simultaneous determination of ethoxyl and methoxyl, and the determination of alkyl groups attached to sulfur instead of to oxygen. Repeated statements in the literature that the hydriodic acid must be fresh and colorless are in error.[97,100] The use of various scrubber solutions, particularly the thiosulfate has been attacked, chiefly by Heron and co-workers.[54] This led to the use of sodium acetate in the volumetric procedure which was adopted by the Association of Official Agricultural Chemists.[99]

Books
  Association of Official Agricultural Chemists, 4
  Belcher and Godbert, 12, 13
  Clark, E. P., 28
  Clark, S. J., 29
  Friedrich, 37
  Furman, 43
  Grant, 50, 51
  Milton and Waters, 77, 78
  Niederl and Niederl, 82, 83
  Pregl, 84
  Roth, 86–89
  Siggia, 94
  Steyermark, 100

Reviews
  Elek, 34

Collaborative studies
  Clark, 26, 27
  Steyermark, 98, 99
  Steyermark and Loeschauer, 102

Ultramicro-, submicro-methods
  Belcher, Bhatty, and West, 9, 10
  Bhatty, 15
  Kirsten and Ehrlich-Rogozinsky, 63
  Mathers and Pro, 75

Simultaneous determination of ethoxyl and methoxyl
  Gran, 49
  Grant, 50, 51

Simultaneous determination of ethoxyl and methoxyl (Cont.)
  Küster and Maag, 69
  Makens, Lothringer, and Donia, 74
  Roth, 86–89

Simultaneous determination of alkoxyl and alkimide
  Belcher, Bhatty, and West, 10
  Steyermark, 100

Determination of S-alkyl groups
  Baernstein, 5–7
  Kuhn, Birkofer, and Quackenbush, 68
  Roth, 86–89
  Vieböck and Brecher, 108

Boron, silicon, etc., compounds
  Alexander, 1
  Nessonova and Pogosyants, 80
  Syavtsillo and Bondarevskaya, 103

General, miscellaneous, etc.
  Anderson and Duncan, 2
  Bailey, 8
  Billitzer, 16
  Bournique, 17
  Christensen, Friedman, and Sato, 22
  Easterbrook and Hamilton, 32
  Fukuda and Sai, 42
  Gran, 48
  Gysel, 52
  Houghton, 56
  Inglis, 59

* The numbers which appear after each entry in this table refer to the literature citations in the reference list at the end of the chapter.

TABLE 28 (*Continued*)

General, miscellaneous, etc. (Cont.)

Kolka and Vogt, 65
Ma, 73
Samsel and McHard, 92
Slotta and Haberland, 95
Stephen, 96
White and Wright, 111
Zeisel, 114, 115
Zeisel and Fanto, 116

Higher homologs

Campbell and Chettleburgh, 20
Ditrych, Rejhova, and Ulbrich, 31
Kirsten and Ehrlich-Rogozinsky, 63
Kuck, 67
Roth, 88
Shaw, 93
Večeřa and Spěvák, 105, 106

Glycols, polyhydric alcohols, hydroxalkyl, etc., groups

Lortz, 72
Morgan, 79
Rudloff, 90

Volumetric procedures

Belcher, Fildes, and Nutten, 11
Bürger and Balaz, 19
Elek, 33, 34
Filipovič and Štefanac, 35
Kinsman and Noller, 61
Kirpal and Bühn, 62
Kirsten and Ehrlich-Rogozinsky, 63
Makens, Lothringer, and Donia, 74
Niederl and Niederl, 83
Roth, 86–89
Steyermark, 98–100
Syavtsillo and Bondarevskaya, 103
Vieböch and Brecher, 108
Vieböch and Schwappach, 109

Gravimetric procedures

Fukuda, 40
Fukuda and Sai, 41
Steyermark, 97, 100

False results in absence of alkoxyl groups

Huang and Morsingh, 58

Preparation and/or treatment of hydriodic acid

Belcher and Godbert, 12, 13
Bethge and Carlson, 14
Clark, E. P., 24, 28
Grant, 50, 51
Knoll, 64
Niederl and Niederl, 82, 83
Pregl, 84
Roth, 86–89
Steyermark, 97–100

Scrubber solutions

Bethge and Carlson, 14
Franzen, Disse, and Eysell, 36
Heron, Reed, Stagg, and Watson, 54
Kirsten and Ehrlich-Rogozinsky, 63
Steyermark, 97–100
Syavtsillo and Bondarevskaya, 103
White, E. P., 110

Solvents

Elek, 33, 34
Inglis, 59
Kirsten and Ehrlich-Rogozinsky, 63
Steyermark, 97–100

Gas-liquid chromatographic methods

Kratzl and Gruber, 66
Vertalier and Martin, 107

Spectrophotometric method

Mathers and Pro, 75

Sulfuric acid cleavage

Langejan, 70

Combustion—silver gauze method

Fukuda, 40
Fukuda and Sai, 41

Nitrometry

Takiura, Takino, and Harada, 104

TABLE 28 (*Continued*)

Apparatus	Apparatus (Cont.)
Arlt, 3	Hoffman and Wolfrom, 55
Belcher and Godbert, 12, 13	Houghton and Wilson, 57
Bethge and Carlson, 14	Kahovec, 60
Billitzer, 16	Kirsten and Ehrlich-Rogozinsky, 63
British Standards Institution, 18	Lieff, Marks, and Wright, 71
Chinoy, 21	Neumann, 81
Christensen, Friedman, and Sato, 22	Niederl and Niederl, 82, 83
Christensen and King, 23	Pregl, 84
Clark, E. P., 25–28	Rigakos, 85
Colson, 30	Roth, 86–89
Elek, 33, 34	Saccardi, 91
Furman, 43	Shaw, 93
Furter, 44, 45	Steyermark, 97–100
Gettler, Niederl, and Benedetti-Pichler, 46	Steyermark, Alber, Aluise, Huffman, Jolley, Kuck, Moran, and Ogg, 101
Gran, 47	Vieböck and Brecher, 108
Grant, 50, 51	Vieböck and Schwappach, 109
Hankes, 53	White, T., 112
Heron, Reed, Stagg, and Watson, 54	Zacherl and Krainick, 113

REFERENCES

1. Alexander, A. P., *Anal. Chem.,* **27**, 105 (1955).
2. Anderson, D. M. W., and Duncan, J. L., *Chem. & Ind. (London),* p. 457 (1959).
3. Arlt, H. G., Jr., *Anal. Chem.,* **28**, 1502 (1956).
4. Association of Official Agricultural Chemists, "Official Methods of Analysis," 8th ed., pp. 801–811, Washington, D. C., 1955.
5. Baernstein, H. D., *J. Biol. Chem.,* **97**, 663 (1932).
6. Baernstein, H. D., *J. Biol. Chem.,* **106**, 451 (1934).
7. Baernstein, H. D., *J. Biol. Chem.,* **115**, 25 (1936).
8. Bailey, A. J., *Ind. Eng. Chem., Anal. Ed.,* **14**, 181 (1942).
9. Belcher, R., Bhatty, M., and West, T. S., *J. Chem. Soc.,* p. 4480 (1957).
10. Belcher, R., Bhatty, M., and West, T. S., *J. Chem. Soc.,* p. 2393 (1958).
11. Belcher, R., Fildes, J. E., and Nutten, A. J., *Anal. Chim. Acta,* **13**, 16 (1955).
12. Belcher, R., and Godbert, A. L., "Semi-Micro Quantitative Organic Analysis," Longmans, Green, London, New York, and Toronto, 1945.
13. Belcher, R., and Godbert, A. L., "Semi-Micro Quantitative Organic Analysis," 2nd ed., Longmans, Green, London, 1954.
14. Bethge, P. O., and Carlson, O. T., *Anal. Chim. Acta,* **15**, 279 (1956).
15. Bhatty, M. K., *Analyst,* **82**, 458 (1957).
16. Billitzer, A. W., *Lab. Practice,* **7**, 289 (1958); *Anal. Abstr.,* **6**, No. 588 (1959).
17. Bournique, R. A., *Chemist Analyst,* **43**, 40 (1954).
18. British Standards Institution, *Brit. Standards* **1428**, Pt. C1 (1954).
19. Bürger, K., and Balaz, F., *Angew. Chem.,* **54**, 58 (1941).
20. Campbell, A. D., and Chettleburgh, V. J., *Analyst,* **84**, 190 (1959).
21. Chinoy, J. J., *Analyst,* **61**, 602 (1936).
22. Christensen, B. E., Friedman, L., and Sato, Y., *Ind. Eng. Chem., Anal. Ed.,* **13**, 276 (1941).

23. Christensen, B. E., and King, A., *Ind. Eng. Chem., Anal. Ed.,* **8**, 194 (1936).
24. Clark, E. P., *Ind. Eng. Chem., Anal. Ed.,* **10**, 677 (1938).
25. Clark, E. P., *J. Am. Chem. Soc.,* **51**, 1479 (1929).
26. Clark, E. P., *J. Assoc. Offic. Agr. Chemists,* **15**, 136 (1932).
27. Clark, E. P., *J. Assoc. Offic. Agr. Chemists,* **22**, 622 (1939).
28. Clark, E. P., "Semimicro Quantitative Organic Analysis," Academic Press, New York, 1943.
29. Clark, S. J., "Quantitative Methods of Organic Microanalysis," Butterworths, London, 1956.
30. Colson, A. F., *Analyst,* **58**, 594 (1933).
31. Ditrych, Z., Rejhova, H., and Ulbrich, V., *Chem. listy,* **49**, 869 (1955).
32. Easterbrook, W. C., and Hamilton, J. B., *Analyst,* **78**, 551 (1953).
33. Elek, A., *Ind. Eng. Chem., Anal. Ed.,* **11**, 174 (1939).
34. Elek, A., *in* "Organic Analysis" (J. Mitchell, Jr., I. M. Kolthoff, E. S. Proskauer, and A. Weissberger, eds.), Vol. I, p. 67, Interscience, New York, 1953.
35. Filipovič, L., and Štefanac, Z., *Croat. Chem. Acta,* **30**, 149 (1958); *Anal. Abstr.,* **6**, No. 1773 (1959).
36. Franzen, F., Disse, W., and Eysell, K., *Mikrochim. Acta,* p. 44 (1953).
37. Friedrich, A., "Die Praxis der quantitativen organischen Mikroanalyse," Deuticke, Leipzig, and Vienna, 1933.
38. Friedrich, A., *Mikrochemie,* **7**, 185, 195 (1929).
39. Friedrich, A., *Z. physiol. Chem. Hoppe-Seyler's,* **163**, 141 (1927).
40. Fukuda, M., *Yakugaku Zasshi,* **77**, 934 (1957).
41. Fukuda, M., and Sai, T., *Yakugaku Zasshi,* **78**, 83 (1958); *Chem. Abstr.,* **53**, 6066 (1958); *Anal. Abstr.,* **6**, No. 3555 (1959).
42. Fukuda, M., and Sai, T., *Yakugaku Zasshi,* **78**, 101 (1958); *Chem. Abstr.,* **53**, 6066 (1958); *Anal. Abstr.,* **6**, No. 3555 (1959).
43. Furman, N. H., ed., "Scott's Standard Methods of Chemical Analysis," 5th ed., Vol. II, Van Nostrand, New York, 1939.
44. Furter, M. F., *Helv. Chim. Acta,* **21**, 1144 (1938).
45. Furter, M. F., *Helv. Chim. Acta,* **21**, 1151 (1938).
46. Gettler, A. O., Niederl, J. B., and Benedetti-Pichler, A. A., *J. Am. Chem. Soc.,* **54**, 1476 (1932).
47. Gran, G., *Svensk Papperstidn.,* **55**, 255-257, 287-290 (1952).
48. Gran, G., *Svensk Papperstidn.,* **56**, 179-180, 202-203 (1953).
49. Gran, G., *Svensk Papperstidn.,* **57**, 702 (1954).
50. Grant, J., "Quantitative Organic Microanalysis, Based on the Methods of Fritz Pregl," 4th ed., Blakiston, Philadelphia, Pennsylvania, 1946.
51. Grant, J., "Quantitative Organic Microanalysis," 5th ed., Blakiston, Philadelphia, Pennsylvania, 1951.
52. Gysel, H., *Mikrochim. Acta,* p. 743 (1954).
53. Hankes, L. H., *Anal. Chem.,* **27**, 166 (1955).
54. Heron, A. E., Reed, R. H., Stagg, H. E., and Watson, H., *Analyst,* **79**, 671 (1954).
55. Hoffman, D. O., and Wolfrom, M. L., *Anal. Chem.,* **19**, 225 (1947).
56. Houghton, A. A., *Analyst,* **70**, 19 (1945).
57. Houghton, A. A., and Wilson, H. A. B., *Analyst,* **69**, 363 (1944).
58. Huang, R. L., and Morsingh, F., *Anal. Chem.,* **24**, 1359 (1952).
59. Inglis, A. S., *Mikrochim. Acta,* p. 677 (1957).
60. Kahovec, L., *Mikrochemie,* **14**, 341 (1934).
61. Kinsman, S., and Noller, C. R., *Ind. Eng. Chem., Anal. Ed.,* **10**, 424 (1938).

16. Alkoxyl Groups

62. Kirpal, A., and Bühn, T., *Monatsh. Chem.*, **36**, 853 (1915).
63. Kirsten, W. J., and Ehrlich-Rogozinsky, S., *Mikrochim. Acta*, p. 787 (1955).
64. Knoll, A., Personal communication, 1945.
65. Kolka, A. G., and Vogt, R. R., *J. Am. Chem. Soc.*, **61**, 1464 (1939).
66. Kratzl, K., and Gruber, K., *Monatsh. Chem.*, **89**, 618 (1958).
67. Kuck, J. A., *Mikrochemie ver. Mikrochim. Acta*, **36/37**, 65 (1951).
68. Kuhn, R., Birkofer, L., and Quackenbush, F. W., *Ber.*, **72B**, 407 (1939).
69. Küster, W., and Maag, W., *Z. physiol. Chem. Hoppe-Seyler's*, **127**, 190 (1923).
70. Langejan, M., *Pharm. Weekblad*, **92**, 667 (1957).
71. Lieff, M., Marks, C., and Wright, G. F., *Can. J. Research*, **15**, 529 (1937).
72. Lortz, H. J., *Anal. Chem.*, **28**, 892 (1956).
73. Ma, T. S., "Proceedings of the International Symposium on Microchemistry 1958," p. 151, Pergamon, Oxford, London, New York, and Paris, 1960.
74. Makens, R. F., Lothringer, R. L., and Donia, R. A., *Anal. Chem.*, **31**, 1265 (1959).
75. Mathers, A. P., and Pro, M. J., *Anal. Chem.*, **27**, 1662 (1955).
76. Mellor, J. W., "Comprehensive Treatise on Inorganic and Theoretical Chemistry," Vol. 2, p. 190, Longmans, Green, London, 1927.
77. Milton, R. F., and Waters, W. A., "Methods of Quantitative Microanalysis," Longmans, Green, New York, and Arnold, London, 1949.
78. Milton, R. F., and Waters, W. A., "Methods of Quantitative Microanalysis," 2nd ed., Arnold, London, 1955.
79. Morgan, P. W., *Ind. Eng. Chem., Anal. Ed.*, **18**, 500 (1946).
80. Nessonova, G. D., and Pogosyants, E. K., *Zavodskaya Lab.*, **24**, 953 (1958); *Anal. Abstr.*, **6**, No. 1772 (1959).
81. Neumann, F., *Ber.*, **70**, 734 (1937).
82. Niederl, J. B., and Niederl, V., "Micromethods of Quantitative Organic Elementary Analysis," Wiley, New York, 1938.
83. Niederl, J. B., and Niederl, V., "Micromethods of Quantitative Organic Analysis," 2nd ed., Wiley, New York, 1942.
84. Pregl, F., "Quantitative Organic Microanalysis," (E. Fyleman, trans., 2nd German ed.), Churchill, London, 1924.
85. Rigakos, D. R., *J. Am. Chem. Soc.*, **53**, 3903 (1931).
86. Roth, H., "Die quantitative organische Mikroanalyse von Fritz Pregl," 4th ed., Springer, Berlin, 1935.
87. Roth, H., "F. Pregl quantitative organische Mikroanalyse," 5th ed., Springer, Wien, 1947.
88. Roth, H., "Pregl-Roth quantitative organische Mikroanalyse," 7th ed., Springer, Wien, 1958.
89. Roth, H., "Quantitative Organic Microanalysis of Fritz Pregl," 3rd ed., E. B. Daw, trans., 4th German ed.), Blakiston, Philadelphia, Pennsylvania, 1937.
90. Rudloff, E. von, *Anal. Chim. Acta*, **16**, 294 (1957).
91. Saccardi, P., *Ann. chim. appl.*, **34**, 18 (1944).
92. Samsel, E. P., and McHard, J. A., *Ind. Eng. Chem., Anal. Ed.*, **14**, 750 (1942).
93. Shaw, B. W., *J. Soc. Chem. Ind. (London)*, **66**, 147 (1947).
94. Siggia, S., "Quantitative Organic Analysis via Functional Groups," 2nd ed., Wiley, New York, and Chapman & Hall, London, 1954.
95. Slotta, K. H., and Haberland, G., *Ber.*, **65B**, 127 (1932).
96. Stephen, W. I., "Proceedings of the International Symposium on Microchemistry 1958," p. 163, Pergamon, Oxford, London, New York, and Paris, 1960.
97. Steyermark, Al, *Anal. Chem.*, **20**, 368 (1948).

98. Steyermark, Al, *J. Assoc. Offic. Agr. Chemists,* **38**, 367 (1955).
99. Steyermark, Al, *J. Assoc. Offic. Agr. Chemists,* **39**, 401 (1956).
100. Steyermark, Al, "Quantitative Organic Microanalysis," Blakiston, Philadelphia, Pennsylvania, 1951.
101. Steyermark, Al, Alber, H. K., Aluise, V. A., Huffman, E. W. D., Jolley, E. L., Kuck, J. A., Moran, J. J., and Ogg, C. L., *Anal. Chem.,* **28**, 112 (1956).
102. Steyermark, Al, and Loeschauer, E. E., *J. Assoc. Offic. Agr. Chemists,* **37**, 433 (1954).
103. Syavtsillo, S. V., and Bondarevskaya, E. A., *Zhur. Anal. Khim.,* **11**, 613 (1956).
104. Takiura, K., Takino, Y., and Harada, S., *Yakugaku Zasshi,* **76**, 1328 (1956).
105. Večeřa, M., and Spěvák, A., *Chem. listy,* **52**, 1520 (1958); *Anal. Abstr.,* **6**, No. 2199 (1959).
106. Večeřa, M., and Spěvák, A., *Collection Czechoslov. Chem. Communs.,* **24**, 413 (1959).
107. Vertalier, S., and Martin, F., *Chim. anal.,* **40**, 80 (1958).
108. Vieböch, F., and Brecher, C., *Ber.,* **63**, 3207 (1930).
109. Vieböch, F., and Schwappach, A., *Ber.,* **63**, 2818 (1930).
110. White, E. P., *Ind. Eng. Chem., Anal. Ed.,* **16**, 207 (1944).
111. White, E. V., and Wright, G. F., *Can. J. Research,* **14B**, 427 (1936).
112. White, T., *Analyst,* **68**, 366 (1943).
113. Zacherl, M. K., and Krainick, H. G., *Mikrochemie,* **11**, 61 (1932).
114. Zeisel, S., *Monatsh. Chem.,* **6**, 989 (1885).
115. Zeisel, S., *Monatsh. Chem.,* **7**, 406 (1886).
116. Zeisel, S., and Fanto, R., *Z. landwirtsch. Versuchsw.,* **5**, 729 (1902).

CHAPTER 17

Microdetermination of Acyl Groups (Acetyl and Formyl)

The determination of acetyl or formyl groups attached to either oxygen or to nitrogen (as in esters, amides, etc.) is carried out according to the method of Elek and Harte.[2,19,44] It is based on the hydrolysis of the compound by means of p-toluenesulfonic acid and treatment of the resulting acetic or formic acid with an excess of potassium iodide and iodate. The liberated iodine (proportional to the acetic or formic acid) is determined by titration with standard thiosulfate. The reactions are shown by the following:

I. CH_3COOR (or $HCOOR$)
   or
   $CH_3CONR'R''$ (or $HCONR'R''$) $\xrightarrow{CH_3C_6H_4SO_3H}$ CH_3COOH (or $HCOOH$)

   where R,R',R'' = Aliphatic or aromatic

II. $6CH_3COOH + 5KI + KIO_3 \rightarrow 6CH_3COOK + 3H_2O + 3I_2$
    (or $6HCOOH$) (or $6HCOOK$)

III. $3I_2 + 6Na_2S_2O_3 \rightarrow 6NaI + 3Na_2S_4O_6$

During the course of the hydrolysis, small amounts of sulfur dioxide are often formed. This reacts with iodine, according to the following equation, so that correction for it must be made.

IV. $2H_2O + SO_2 + I_2 \rightarrow H_2SO_4 + 2HI$

The presence of *labile* fluorine interferes with the determination. The same applies to any other acidic grouping present, or formed as a result of decomposition during the hydrolysis, that would steam distill at reduced pressure, *other than* chlorine, bromine, and iodine (which do *not* interfere since they are retained in the reaction mixture as the silver halides).

Reagents

SILVER SULFATE POWDER

Reagent grade of silver sulfate powder is used in the reaction flask to trap any halogens present either in the sample or as an impurity in the p-toluenesulfonic acid.

p-TOLUENESULFONIC ACID[18]

A 25% solution in water is prepared from a pure grade of p-toluenesulfonic acid.

IODINE SOLUTION, 0.01N

This solution is prepared according to the directions given in Chapter 5. It does not need to be standardized—this is used in the receiving flask.

POTASSIUM IODIDE CRYSTALS

Reagent grade of potassium iodide crystals are used along with the 0.01N iodine in the receiving flask.

POTASSIUM IODATE SOLUTION, 4%

A 4% solution of reagent grade of potassium iodate is prepared using freshly boiled distilled water. It is stored in a ground glass-stoppered bottle.

STANDARD SODIUM THIOSULFATE SOLUTION, 0.01N

This is prepared and standardized as described in Chapter 5.

STARCH INDICATOR

This is prepared as described in Chapter 5.

Apparatus[19,33,34,40,44]

ACYL APPARATUS

The apparatus shown in Fig. 183 consists of a reaction flask, a, an introduction funnel, b, condenser, c, and receiving flask, d. A side arm with a Kjeldahl-type trap is attached to the neck of the reaction flask. Flow of liquid through the introduction funnel is controlled by means of a stopcock (water seal). The delivery end of the condenser tube has a sealed-in sintered glass plate for providing efficient absorption of the distillate in the receiving flask solution. A side arm with stopcock is attached to the receiving flask, d, through which the system may be evacuated. The reaction flask and funnel are joined by means of a ground joint using a water seal. Likewise, the side arm of the flask and the top tube of the condenser are connected by means of a ground joint (water seal), the male member of which is on the flask and the delivery tube extends beyond the male member grinding so that no loss of distillate will occur at this point. (The flask and condenser may also be joined by means of rubber tubing, effecting a glass-to-glass connection but the above-mentioned method is better.) The reaction flask is provided with a constant level type of water bath and the receiving flask with an ice bath.

17. Acyl Groups

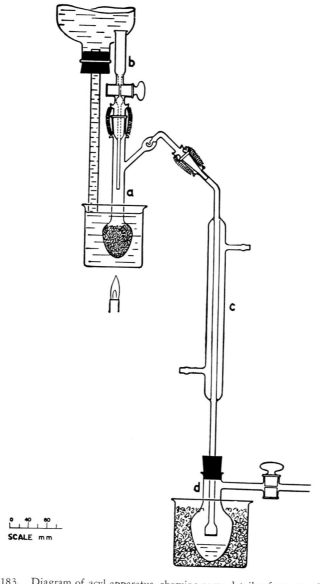

Fig. 183. Diagram of acyl apparatus, showing some details of construction.

SOURCE OF VACUUM

A vacuum pump attached to a trap and a manometer is used as a source of vacuum. The trap should be provided with a "bleeder"* of some type so that a pressure of 50–60 mm. of mercury may be maintained in the setup.

BURETTES

Two automatic burettes of the type shown in Fig. 69 or 70, Chapter 5, are needed, one each for the 0.01N iodine and the 0.01N thiosulfate.

PYREX GLASS ROD[16]

Small fire-polished Pyrex glass rods (about 2×5 mm.) are used in the reaction flask. These must be Pyrex and must have fire-polished edges or low results are obtained due to their becoming porous.

ASCARITE-FILLED DRYING TUBE

Any type of drying tube is filled with Ascarite[50] and equipped with a one-hole stopper that fits the neck of the receiver flask.

Procedure[19,33,34,44]

A weighed sample is transferred to the reaction flask, *a*. (The size of the sample is governed by the amount of acetyl or formyl present. Enough should be taken so as to require at least 4 ml. of 0.01N thiosulfate in the titration.) Onto the sample is added enough dry sections of Pyrex glass rod to three-quarter fill the bulb of the flask. This is followed by 10 to 15 mg. of silver sulfate powder onto the sections of glass rod. The introduction funnel, *b*, is inserted into the flask, *a*, which in turn is attached to the condenser, *c*, using water to lubricate the stopcock on the funnel and to seal the two ground joints (between the funnel and the flask, and between the flask and the condenser). Cold water is run through the condenser. Into the receiving flask, *d*, is placed 1.5 grams of potassium iodide crystals and 5.00 ml. of 0.01N iodine. This flask, *d*, is attached immediately to the condenser, by means of the rubber stopper, with the sintered glass plate on the end of the delivery tube about one cm. above the liquid. Adjustment of the height above the liquid is made easy by wetting the outer wall of the delivery tube and the rubber stopper, after which the latter can be slid up or down to any desired position. The stopcock on the receiving flask is closed and the flask cooled by packing cracked ice all around it.

* An all-metal needle valve [3,44–46,50] (Fig. 92, Chapter 7) is most satisfactory for this purpose.

Two ml. of 25% *p*-toluenesulfonic acid solution is placed in the introduction funnel, the stopcock on it opened and the liquid allowed to flow into the flask. (If the solution does not drain down due to airlock in the closed system, the funnel is momentarily raised to separate the two members of the ground joint. Then after drainage is complete the water seal between the funnel and flask is remade.) The stopcock on the funnel is closed and the *funnel filled with water*.* The water bath surrounding the reaction flask is heated to boiling and maintained so for one hour if the acyl is linked to oxygen and for 3 hours if it is linked to nitrogen.

The boiling water in the bath surrounding the reaction flask is then replaced with cold water. The stopcock on the receiving flask is opened and connection made to the vacuum system. The pressure is adjusted then to 50–60 mm. of mercury. A little water, for lubrication, is placed on top of the rubber stopper holding the receiving flask to the condenser. The receiving flask and stopper are then slid up on the delivery tube until the sintered plate on the tip rests at the bottom of the liquid. No bubbles should escape through the liquid if the system is air-tight. If bubbles do escape, water should be sprayed on the stopcock and on the two ground joints to effect water seals. The water bath surrounding the reaction flask, *a*, is heated to boiling and maintained thusly during the entire subsequent vacuum distillation. (Note: As the temperature of the bath rises, bubbles will escape through the liquid in the receiving flask for a short time until equilibrium is reached again and then cease). After the liquid has distilled from the reaction flask, *a*, and a dry residue remains, 1–2 ml. of water is carefully added through the stopcock on the introduction funnel. (If the addition is too rapid, some *p*-toluenesulfonic acid solution may be mechanically carried from the flask into the condenser and subsequently the receiver, spoiling the determination.) Vacuum distillation is continued again until a dry residue remains in the flask. This procedure is repeated about five times to effect a complete transfer of the acetic or formic acid from the reaction flask to the receiver. Each time that the contents of the reaction flask go to dryness, liquid rises from the receiver in the delivery tube well up into the water-cooled portion of the condenser. Addition of water to the reaction flask each time causes the solution to return to the receiver flask. After the fifth portion of water has been distilled out of the flask, *a*, the receiving flask, *d*, and rubber stopper are slid down so that the sintered plate is several centimeters above the liquid. Air is carefully admitted *through the stopcock on the introduction funnel* and the vacuum source then removed from the receiver flask. The reaction flask is disconnected from the condenser and several 1- to 2-ml. portions of water added to the top of the condenser and allowed to drain into the receiver.

* This insures a water seal during the subsequent vacuum distillation.

The receiver flask is disconnected from the condenser and the iodine solution titrated with standard 0.01N thiosulfate to the end point using two drops of starch indicator. (Note: The starch is not added until the solution is light yellow in color.) A faint pink coloration is considered to be the end point. The above is known as the *first titration* and the volume of thiosulfate required is accurately recorded. Without delay, 2 ml. of 4% potassium iodate is added to the almost colorless contents of the receiver flask (after the first titration). Immediately, the receiver is stoppered with an Ascarite tube (stopcock on the side arm also closed) and then placed in a water bath at 35° C. for 20 minutes. After this period, the intensely blue iodine solution (caused by Equation II, p. 444) is titrated with standard 0.01N sodium thiosulfate to the end point. This is known as the *second titration*.

In order to correct for the effect of the sulfur dioxide formed (Equation IV, p. 444), the following titration must be made. Into a 125-ml. Erlenmeyer flask is placed 1.5 grams of potassium iodide crystals, 5.00 ml. of the same 0.01N iodine used above, and one drop of dilute acetic acid (containing about 2 mg. of CH_3COOH per drop). The mixture is allowed to stand for at least 5 minutes and then titrated to the end point with 0.01N sodium thiosulfate using starch indicator. This is known as the *third titration*. If sulfur dioxide does form in the reaction, the volume of thiosulfate required for the third titration is greater than that for the first, since some of the iodine is converted already to hydriodic acid by the sulfur dioxide (Equation IV, p. 444). If the difference is more than 0.05 ml. the analysis should be repeated.[19] The difference, multiplied by two, to account for the dibasicity of the sulfuric acid formed, is deducted from the second titration as a correction for the acid introduced as sulfur dioxide.

Calculation:

Factors:

1 ml. of 0.01N sodium thiosulfate is equivalent to 0.4305 mg. of acetyl (CH_3CO) or 0.2902 mg. of formyl (HCO).

$$\therefore \frac{\text{ml. of 0.01N } Na_2S_2O_3 \text{ (corr.)} \times 0.4305 \times 100}{\text{Wt. sample}} = \% \, CH_3CO$$

or

$$\frac{\text{ml. of 0.01N } Na_2S_2O_3 \text{ (corr.)} \times 0.2902 \times 100}{\text{Wt. sample}} = \% \, HCO$$

Examples:

a. A 5.694-mg. sample containing an acetyl group required,

    5.05 ml. of 0.01N $Na_2S_2O_3$ for the first titration

    4.80 ml. of 0.01N $Na_2S_2O_3$ for the second titration

    5.10 ml. of 0.01N $Na_2S_2O_3$ for the third titration

    (third titration — first titration) × 2 = correction for SO_2

17. Acyl Groups 450

$\therefore (5.10 - 5.05) \times 2 = 0.10$ ml. of $0.01N$ $Na_2S_2O_3$ = correction for SO_2

$\therefore 4.80 - 0.10 = 4.70$ ml. of $0.01N$ $Na_2S_2O_3$ actually used for the sample

$\therefore \dfrac{4.70 \times 0.4305 \times 100}{5.694} = 35.53\%$ CH_3CO

b. A 6.317-mg. sample containing a formyl group required,

$\quad$ 5.00 ml. of $0.01N$ $Na_2S_2O_3$ for the first titration
$\quad$ 6.15 ml. of $0.01N$ $Na_2S_2O_3$ for the second titration
$\quad$ 5.05 ml. of $0.01N$ $Na_2S_2O_3$ for the third titration
$\quad$ (third titration — first titration) $\times 2$ = correction for SO_2

$\therefore (5.05 - 5.00) \times 2 = 0.10$ ml. of $0.01N$ $Na_2S_2O_3$ = correction for SO_2

$\therefore 6.15 - 0.10 = 6.05$ ml. of $0.01N$ $Na_2S_2O_3$ actually used for the sample

$\therefore \dfrac{6.05 \times 0.2902 \times 100}{6.317} = 27.79\%$ HCO

The accuracy of the determination is ± 0.3–0.5%.

TABLE 29
Additional Information on References* Related to Chapter 17

In addition to the procedure presented in the preceding pages of this chapter, the author wishes to call to the attention of the reader the references listed in Table 29. (See statement at top of Table 4 of Chapter 1, regarding completeness of this material.)

Books
- Belcher and Godbert, 5, 6
- Clark, E. P., 13
- Clark, S. J., 14
- Grant, 20, 21
- Milton and Waters, 30, 31
- Niederl and Niederl, 33, 34
- Roth, 35–38
- Siggia, 41
- Steyermark, 44

Review
- Hall and Shaefer, 22

Collaborative study
- Steyermark and Loeschauer, 47

Apparatus
- British Standards Institution, 8
- Buděšínský, 9

Apparatus (Cont.)
- Chaney and Wolfrom, 10
- Elek and Harte, 19
- Kainz, 24
- Kuhn and Roth, 25
- Niederl and Niederl, 33, 34
- Roth, 35–38
- Schöniger, Lieb, and El Din Ibrahim, 39
- Steyermark, 44
- Wiesenberger, 51–53

General, miscellaneous
- Inglis, 23
- Matchett and Levine, 27
- Mázor and Meisel, 28
- Sudo, Shimoe, and Tsujii, 48

O-Acetyl compounds
- Alicino, 1
- Chaney and Wolfrom, 10

* The numbers which appear after each entry in this table refer to the literature citations in the reference list at the end of the chapter.

TABLE 29 (*Continued*)

O-Acetyl compounds (*Cont.*)
Clark, 13
Cramer, Gardner, and Purves, 17
Elek and Harte, 19
Kunz and Hudson, 26
Mázor and Meisel, 28
Steyermark, 44
Wolfrom, Konigsberg, and Soltzberg, 54

N-Acetyl compounds
Chaney and Wolfrom, 10
Clark, 13
Elek and Harte, 19
Steyermark, 44

Formyl groups
Alicino, 2
Spingler and Markert, 43
Steyermark, 44

Benzoyl groups
Grant, 20, 21
Roth, 35–38
Tani and Nara, 49

Types of compounds which interfere
Clark, E. P., 13
Späth and Gruber, 42

Distillation
Buděšínský, 9
Clark, 11–13
Elek and Harte, 19
Matchett and Levine, 27

Distillation (*Cont.*)
Mizukami, Ieki, and Koyama, 32
Schöniger, Lieb, and El Din Ibrahim, 39
Steyermark, 44
Sudo, Shimoe, and Tsujii, 48

Iodometric procedures
Bradbury, 7
Elek and Harte, 19
Inglis, 23
Mizukami, Ieki, and Koyama, 32
Steyermark, 44

Titration with alkali
Alicino, 1
Clarke and Christensen, 15
Inglis, 23
Kuhn and Roth, 25
Mázor and Meisel, 28
Niederl and Niederl, 33, 34
Roth, 35–38
Schöniger, Lieb, and El Din Ibrahim, 39

Spectrophotometric, colorimetric methods
Bayer and Reuther, 4
McComb and McCready, 29

Gas-chromatographic method
Spingler and Markert, 43

Cation-exchange
Tani and Nara, 49

REFERENCES

1. Alicino, J., *Anal. Chem.*, **20**, 590 (1948).
2. Alicino, J. F., *Ind. Eng. Chem., Anal. Ed.*, **15**, 764 (1943).
3. American Society for Testing Materials, *ASTM Designations*, E 148-59T.
4. Bayer, E., and Reuther, K. H., *Ber.*, **89**, 254 (1956).
5. Belcher, R., and Godbert, A. L., "Semi-Micro Quantitative Organic Analysis," Longmans, Green, London, New York, and Toronto, 1945.
6. Belcher, R., and Godbert, A. L., "Semi-Micro Quantitative Organic Analysis," 2nd ed., Longmans,Green, London, 1954.
7. Bradbury, R. B., *Anal. Chem.*, **21**, 1139 (1949).
8. British Standards Institution, *Brit. Standards*, **1428**, Pt. C2 (1954).

9. Buděšínský, B., *Chem. listy,* **50**, 1936 (1956).
10. Chaney, A., and Wolfrom, M. L., *Anal. Chem.,* **28**, 1614 (1956).
11. Clark, E. P., *Ind. Eng. Chem., Anal. Ed.,* **8**, 487 (1936).
12. Clark, E. P., *Ind. Eng. Chem., Anal. Ed.,* **9**, 539 (1937).
13. Clark, E. P., "Semimicro Quantitative Organic Analysis," Academic Press, New York, 1943.
14. Clark, S. J., "Quantitative Methods of Organic Microanalysis," Butterworths, London, 1956.
15. Clarke, R., and Christensen, B. E., *Ind. Eng. Chem., Anal. Ed.,* **17**, 334 (1945).
16. Corning Glass Works, Corning, New York.
17. Cramer, F. B., Gardner, T. S., and Purves, C. B., *Ind. Eng. Chem., Anal. Ed.,* **15**, 319 (1943).
18. Eastman Kodak Company, Rochester, New York.
19. Elek, A., and Harte, R. A., *Ind. Eng. Chem., Anal. Ed.,* **8**, 267 (1936).
20. Grant, J., "Quantitative Organic Microanalysis, Based on the Methods of Fritz Pregl," 4th ed., Blakiston, Philadelphia, Pennsylvania, 1946 .
21. Grant, J., "Quantitative Organic Microanalysis," 5th ed., Blakiston, Philadelphia, Pennsylvania, 1951.
22. Hall, R. T., and Shaefer, W. E., *in* "Organic Analysis" (J. Mitchell, Jr., I. M. Kolthoff, E. S. Proskauer, and A. Weissberger, eds.), Vol. II, p. 19, Interscience, New York, 1954.
23. Inglis, A. S., *Mikrochim. Acta,* p. 228 (1958).
24. Kainz, G., *Mikrochemie ver. Mikrochim. Acta,* **35**, 89 (1950).
25. Kuhn, R., and Roth, H., *Ber.,* **66**, 1274 (1933).
26. Kunz, A., and Hudson, C. S., *J. Am. Chem. Soc.,* **48**, 1982 (1926).
27. Matchett, J. R., and Levine, J., *Ind. Eng. Chem., Anal. Ed.,* **13**, 98 (1941).
28. Mázor, L., and Meisel, T., *Anal. Chim. Acta,* **20**, 130 (1959).
29. McComb, E. A., and McCready, R. M., *Anal. Chem.,* **29**, 819 (1957).
30. Milton, R. F., and Waters, W. A., "Methods of Quantitative Microanalysis," Longmans, Green, New York, and Arnold, London, 1949.
31. Milton, R. F., and Waters, W. A., "Methods of Quantitative Microanalysis," 2nd ed., Arnold, London, 1955.
32. Mizukami, S., Ieki, T., and Koyama, C., *Yakugaku Zasshi,* **76**, 465 (1956).
33. Niederl, J. B., and Niederl, V., "Micromethods of Quantitative Organic Elementary Analysis," Wiley, New York, 1938.
34. Niederl, J. B., and Niederl, V., "Micromethods of Quantitative Organic Analysis," 2nd ed., Wiley, New York, 1942.
35. Roth, H., "Die quantitative organische Mikroanalyse von Fritz Pregl," 4th ed., Springer, Berlin, 1935.
36. Roth, H., "F. Pregl quantitative organische Mikroanalyse," 5th ed., Springer, Wien, 1947.
37. Roth, H., "Pregl-Roth quantitative organische Mikroanalyse," 7th ed., Springer, Wien, 1958.
38. Roth, H., "Quantitative Organic Microanalysis of Fritz Pregl," 3rd ed. (E. B. Daw, trans., 4th German ed.), Blakiston, Philadelphia, Pennsylvania, 1937.
39. Schöniger, W., Lieb, H., and El Din Ibrahim, M. G., *Mikrochim. Acta,* p. 96 (1954).
40. Scientific Glass Apparatus Company, Bloomfield, New Jersey.
41. Siggia, S., "Quantitative Organic Analysis via Functional Groups," 2nd ed., Wiley, New York, and Chapman & Hall, London, 1954.

42. Späth, E., and Gruber, W., *Ber.,* **71**, 106 (1938).
43. Spingler, H., and Markert, F., *Mikrochim. Acta,* p. 122 (1959).
44. Steyermark, Al, "Quantitative Organic Microanalysis," Blakiston, Philadelphia, Pennsylvania, 1951.
45. Steyermark, Al, Alber, H. K., Aluise, V. A., Huffman, E. W. D., Kuck, J. A., Moran, J. J., and Willits, C. O., *Anal. Chem.,* **21**, 1283 (1949).
46. Steyermark, Al, Alber, H. K., Aluise, V. A., Huffman, E. W. D., Kuck, J. A., Moran, J. J., and Willits, C. O., *Anal. Chem.,* **21**, 1555 (1949).
47. Steyermark, Al, and Loeschauer, E. E., *J. Assoc. Offic. Agr. Chemists,* **37**, 433 (1954).
48. Sudo, T., Shimoe, D., and Tsujii, T., *Bunseki Kagaku,* **6**, 494 (1957).
49. Tani, H., and Nara, A., *Yakugaku Zasshi,* **74**, 1399 (1954).
50. Thomas, Arthur H., Company, Philadelphia, Pennsylvania.
51. Wiesenberger, E., *Mikrochemie ver. Mikrochim. Acta,* **30**, 241 (1942).
52. Wiesenberger, E., *Mikrochemie ver. Mikrochim. Acta,* **33**, 51 (1948).
53. Wiesenberger, E., *Mikrochim. Acta,* p. 127 (1954).
54. Wolfrom, M. L., Konigsberg, M., and Soltzberg, S., *J. Am. Chem. Soc.,* **58**, 490 (1936).

CHAPTER *18*

Microdeterminations Carried Out on the Van Slyke Manometric Apparatus

    I. MANOMETRIC CARBON DETERMINATION
    II. MANOMETRIC DETERMINATION OF PRIMARY AMINO NITROGEN IN ALIPHATIC α-AMINO ACIDS
    III. OTHER DETERMINATIONS USING THIS APPARATUS

The Van Slyke manometric blood gas apparatus shown in Figs. 184 and 185 was first used, as the name implies, for the determination of the gases in blood.[3,24,97] Since that time it has been applied to a number of microchemical determinations as shown in the following pages of this chapter.

I. MANOMETRIC CARBON DETERMINATION[92]

The manometric carbon determination is based on the oxidation of the organic compound to carbon dioxide and the measuring of the pressure exerted by this gas at a constant volume and at room temperature.

    The oxidation is effected by a wet combustion mixture consisting of fuming sulfuric, chromic, phosphoric and iodic acids. The resulting carbon dioxide and halogens, if present, are absorbed in alkaline hydrazine solution (the hydrazine being present to reduce the halogens—see Chapter 11) and the unabsorbed gases (oxygen and nitrogen) are ejected. The absorbed carbon dioxide is liberated by means of lactic acid and the pressure exerted by it is measured. (The other acidic components being stronger acids than lactic remain as salts in the alkaline solution.) The liberated carbon dioxide is reabsorbed and the residual pressure in the apparatus noted. The difference between these readings is the pressure due to the carbon dioxide obtained by combustion of the sample. Elements which interfere with, or require modification of, the dry combustion (Chapter 9) have no effect on the manometric method. The reactions are represented by the following:

(a) Organic compounds $\xrightarrow{\text{Oxidation}}$ CO_2 + H_2O + Halogens + etc.
    (C, H, O, Halogen, etc.)

(b) $CO_2 + \text{Halogens} \xrightarrow[NH_2.NH_2]{NaOH} Na_2CO_3 + NaX$ (X = Halogens)

(c) $Na_2CO_3 \xrightarrow{\text{Lactic acid}} CO_2$

(d) $CO_2 \xrightarrow{NaOH} Na_2CO_3$

This method, using the same manometric apparatus, is applicable for samples containing anywhere from 0.3–15 mg. of carbon. However, for handling the larger amounts some of the accessory pieces are modified as are the technique and the quantities of reagents used. This chapter deals with the analysis of samples containing carbon in the amounts of from 0.3–0.7 mg. and from 2–3.5 mg., which Van Slyke termed for the sake of convenience, "submicroanalysis" and "microanalysis," respectively. All of the pieces of apparatus for these ranges are the same and there are practically no differences in either the amounts of reagents used or the technique. For the analysis of the amounts of carbon in the range of 8–15 mg. which the above author termed "macroanalysis," the reader is referred to the original publication.[92]

The procedure outlined below appears to be given in great detail. However, this is necessary since it is one in which the right stopcock must be turned to its correct position at the right time or the determination is ruined. In spite of this, the average person is able to do precision work after one or two days' practice and most chemists find the technique to be a fascinating one. The determination has the advantages (a) that it may be completed in about 20 minutes; (b) that there appear to be no interfering elements and (c) that "submicro" quantities may be determined. On the other hand, it gives only the carbon values instead of both carbon and hydrogen as does the dry combustion method (Chapter 9).

Reagents

POTASSIUM IODATE

Reagent grade of potassium iodate powder is used in the combustion tube to furnish the iodic acid.

COMBUSTION FLUID
(Caution: Highly Corrosive)

In a 1-liter Pyrex[15] Erlenmeyer ground glass-stoppered flask are placed 25 grams of anhydrous chromic acid (CrO_3) and 5 grams of potassium iodate, both reagent grade, followed by 167 ml. of pure syrupy phosphoric acid (sp. gr., 1.7), and 333 ml. of fuming sulfuric acid (20% excess SO_3). The stopper is left off and the contents of the flask heated slowly to 150° C., with occasional rotation to dissolve the chromic acid and to aid the escape of the carbon dioxide formed by the oxidation of traces of organic material present. The flask is covered by an inverted beaker to prevent contamination and allowed to cool to room temperature. Then the stopper of the flask is inserted and covered by the beaker as an added means of protection from dust. This flask should be used for storage purposes and should be opened to the atmosphere only when used to fill a small bottle from which the mixture is poured during the course of the determination. This small bottle should be of the type shown in Fig. 186, C. It is equipped with both a ground stopper and a stopper cover, the latter being joined to the bottle below the neck by means of ground surfaces. In this manner the combustion fluid is allowed to absorb a minimum of moisture from the atmosphere. The fluid in this small bottle should be replaced quite frequently, discarding the mixture after several days' use. This is done as a precaution, since absorbed water vapor leads to low carbon values.*

ALKALINE HYDRAZINE

A $0.5N$ sodium hydroxide plus $0.3M$ hydrazine solution, approximately carbon dioxide free, is prepared as follows: A concentrated solution of sodium hy-

* In personal communications (June 14, 1949 and November 10, 1950), Doctor Van Slyke advised the author that he is now replacing the combustion fluid by the following:

A. For Carbohydrate and Polyhydroxl Alcohols: SOLID REAGENT. One part of potassium dichromate to ten parts by weight of potassium iodate. These are ground together. (See below.)

LIQUID REAGENT. Fifty ml. reagent grade of concentrated sulfuric acid (sp. gr. 1.84), 50 ml. of reagent grade of syrupy phosphoric acid (sp. gr. 1.70–1.72) and 1.5 grams of potassium iodate are heated at 160°–190° C. until the solid is dissolved (the mixture is saturated with potassium iodate). The heating destroys any organic material which might be present as a contaminant.

B. Combustion Reagents for General Use (Substances Other than Carbohydrates and Polyhydroxyl Alcohols): SOLID REAGENT. One part of potassium dichromate to two parts of potassium iodate by weight. The two substances are ground together. (See below.)

LIQUID REAGENT. Sixty-seven ml. reagent grade of fuming sulfuric acid (20% excess SO_3), 33 ml. of reagent grade of syrupy phosphoric acid (sp. gr. 1.70–1.72), and one gram of potassium iodate. The mixture is heated at 160°–190° C. until the solid is dissolved. (See above.)

(Note: Since so-called fuming grades of sulfuric acid may vary considerably in the excess SO_3 content, Van Slyke advocates standardization by titration with alkali of fuming acids of 10 and 30% excess SO_3 and then mixing them in the proper proportion to obtain 20%.)

For the procedure, either of the potassium dichromate–potassium iodate mixtures is added to the dry sample in the following proportions:

0.15 gram for the "submicro" combustions (0.1–0.7 mg. of carbon).

0.30 gram for "micro" combustions (0.7–3.5 mg. of carbon). The liquid reagent is added in the same volume and manner described under Procedure.

droxide (18–20N) is prepared, stoppered, and then stored until the carbonate has been allowed to settle out.* A little over a liter of distilled water is acidified with a few drops of dilute acid and boiled for a few minutes to expel any dissolved carbon dioxide. The water is allowed to cool under the protection of an Ascarite tube to prevent the reabsorption of the gas. Enough concentrated, carbonate-free, sodium hydroxide (18–20N) is decanted into a portion of the boiled-out distilled water to make a solution of slightly greater concentration than 0.815N. An aliquot portion of this is titrated with standard acid to determine the true normality. Then the bulk of the solution is diluted with whatever amount of boiled-out distilled water that may be necessary to bring the concentration to 0.815 ± 0.005N. This solution is then stored in a paraffin-lined aspirator bottle protected by a stopper into which is inserted an Ascarite[83] drying tube. (Note: Standard 0.815N sodium hydroxide solution, approximately carbon dioxide free, is now commercially available.[26])

Five grams of hydrazine sulfate crystals are placed into a 250-ml. volumetric flask (or if it is desirable to prepare some other volume of solution, 2 grams of hydrazine sulfate is added for each 100-ml. capacity of the flask used). The aspirator tip of the paraffin-lined bottle containing the 0.815N sodium hydroxide is washed with distilled water and then a few milliliters of the alkali is run out and discarded as precautions against collecting carbonate. *The aspirator tip is inserted down into the neck of the 250-ml. volumetric flask* and 0.815N alkali added, while the flask is whirled to dissolve the crystals. The tip is kept in place until the liquid is added up to the 250-ml. mark and is then removed. The resulting 0.5N sodium hydroxide plus 0.3M hydrazine solution is stoppered and mixed well. The stopper is removed from the 250-ml. volumetric flask and the solution *sucked up* immediately into a rubber-tipped stopcock cylinder of the type shown in Fig. 186, D, protected by an Ascarite† tube (see description below). For this purpose, an extra length of clean rubber tubing is attached over the rubber tip on the stopcock cylinder and the free end is inserted to the bottom of the flask so that solution is withdrawn at this point. Since the top layers of the alkali might have absorbed carbon dioxide again during the manipulation, about one-fifth of the solution is left in the volumetric flask and discarded. After removal of the tubing, the rubber tip of the stopcock cylinder is immersed in mercury, covered by normal sulfuric acid, as shown in Fig. 186, D. The alkaline hydrazine solution should be dated and used for a maximum of one month, after which it should be discarded. The solution in the stopcock cylinder is always kept alongside the manometric apparatus so that the temperature of the solution is the same as that of the water-jacketed extraction chamber— see below.

* This solution is now commercially available.[1,26]
† Or soda-lime.

LACTIC ACID, 2N

One volume of reagent grade concentrated lactic acid, sp. gr. 1.20, is diluted with four volumes of water. An aliquot portion of this is titrated with standard alkali using phenolphthalein as the indicator, after which either more water or lactic acid is added to the bulk of the solution to make it 2N. (Note: 2N lactic acid solution is now commercially available.[26])

SODIUM HYDROXIDE SOLUTION, APPROXIMATELY 5N

A sodium hydroxide solution, approximately 5N, is used to reabsorb the liberated carbon dioxide. This need not be carbonate free. It may be stored in a stopcock cylinder of the type shown in Fig. 186, D but *minus* the rubber tip on the delivery tube.

SYRUPY PHOSPHORIC ACID, SP. GR., 1.7

This material is used for lubricating the stopcock on the dropping funnel (Figs. 187 and 188, F), and the ground joint between the dropping funnel and the combustion tube (Figs. 187 and 188, T).

MERCURY

About 20 pounds of triple distilled mercury is needed for filling the manometric system. Note: A trace of glycerol or diethylene glycol should be added through the stopcock at the top of the manometer to absorb water.

STOPCOCK GREASE

A special stopcock grease for lubricating all of the stopcocks on the manometric apparatus (three on the manometer, one on the chamber) and on the dropping cylinder is commercially available.[18,83] In addition, the author has found Sisco No. 300 stopcock grease[71,82] to be equally good for this purpose.

Apparatus[18,51,70,71,80,83,92]

The manometric apparatus used is shown in Figs. 184, 185, 187, and 188, while the accessory apparatus are shown in Figs. 186 and 189.

MANOMETRIC APPARATUS

The manometric apparatus consists of a water-jacketed extraction chamber, C, manometer and leveling bulb, L (Figs. 187 and 188).

The extraction chamber is water-jacketed as a means of maintaining more constant temperature. Sealed to the upper portion of the chamber is an inlet cup, E. Between the chamber and the cup is a three-way stopcock, b. One lead of the three-way stopcock connects the chamber to the cup and the other joins the chamber when desired to the combustion tube by means of the connecting tube, Q. The chamber proper is composed of three connected bulbs, graduated in size, the smallest at the top and the largest at the bottom. Graduations appear above the smallest bulb, between the first and second, and the second and third bulbs, and below the third, being marked respectively 0.5, 2.0, 10.0, and 50.0 ml. An electric shaking (or magnetic stirring) device

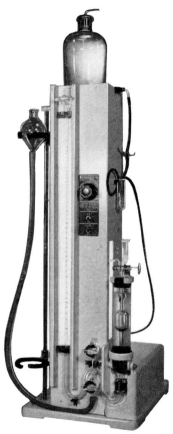

FIG. 184. Sargent model of Van Slyke manometric blood gas apparatus, shaker type.

is attached. By this means, the contents of the chamber may be shaken (or stirred) violently. A heavy-walled rubber tube, approximately 25 mm. O.D., 5 mm. I.D., (or ball and socket joint) connects the extraction chamber to the manometer.

The manometer is equipped with three stopcocks, two of which are used to fill it with mercury and the third, a (Figs. 187 and 188) is used to control the flow of mercury which, in turn, controls the volume of gas in the chamber.

18. Manometric Determinations

Below the stopcock, *a,* is attached a section of heavy-walled rubber tubing whose opposite end is connected to the leveling bulb, *L,* which is used as a reservoir for the mercury.

The assembled manometric apparatus filled with mercury should not be allowed to stand indefinitely with all of the stopcocks closed, or breakage

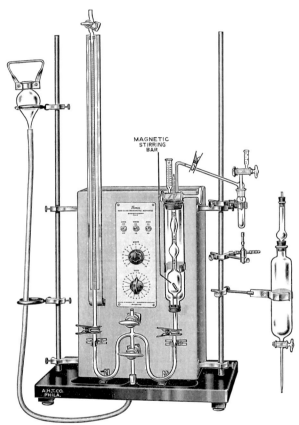

FIG. 185. Thomas model of Van Slyke manometric blood gas apparatus, magnetic stirring.

is apt to occur due to expansion of the mercury if the temperature in the room rises. The stopcock, *a,* should be left open to prevent this.

STAND FOR MANOMETRIC APPARATUS

The extraction chamber, *C,* manometer, and leveling bulb, *L,* are mounted on a stand, made either of wood, plastic, or metal. At the top is a small platform (or other means) for mounting the distilled water reservoir used for washing the extraction chamber between analyses. On the left side of the stand are mounted two cut-away rings for holding the leveling bulb, L. The upper ring holds the bulb *above* the top of the cup, E, attached to the extraction chamber, and the lower ring holds the bulb so that the mercury surface in it stands at about the level of the 50-ml. mark on the extraction chamber. The stand is provided with an illuminated background for that

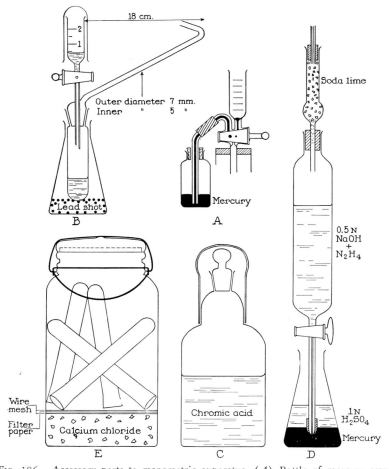

FIG. 186. Accessory parts to manometric apparatus. (A) Bottle of mercury arranged for sealing capillary of stopcock, b, on extraction chamber (Fig. 187). (B) 100-ml. flask weighted with lead shot serving as a stand for combustion tube after analysis. (C) 100-ml. bottle for storage of combustion fluid, provided with cover and ground joint to prevent access of moisture. (D) Stopcock cylinder containing alkaline hydrazine solution with tip protected from atmospheric carbon dioxide by sulfuric acid and mercury. (E) Fruit jar for storing clean combustion tubes.

portion of the manometer which is graduated. The illumination should be on for just long enough to make a reading or the mercury will expand from the heat. For this reason, only fluorescent (or internal reflection) lighting should be used. With some models (Fig. 184) the water jacket surrounding the extraction chamber is held in a clamp which in turn is attached to an electric motor-driven eccentric. This provides the means of shaking the chamber

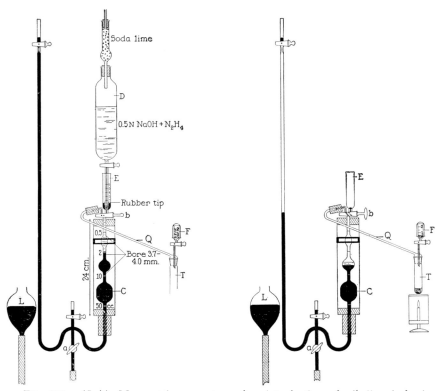

FIG. 187. (*Left*) Manometric apparatus—after introduction of alkaline hydrazine into chamber and before start of combustion.

FIG. 188. (*Right*) Manometric apparatus—at start of combustion but before evolution of carbon dioxide and oxygen has begun.

vigorously to and fro at a tempo of 250 to 300 times per minute. With other models (Fig. 185) the same rate of stirring is accomplished by means of a magnetic stirrer with concealed magnetic stirring apparatus.

CONNECTING TUBE AND CUP

The connecting tube, Q (Figs. 187 and 188), and 2-ml. graduated cup, F (plus stopcock) (also see B, Fig. 186), are sealed together through an interchangeable ground joint which in turn fits the top of the combustion tube. The tube, Q, connects via heavy-walled rubber tubing, approximately 25 mm. O.D., 4 mm. I.D. (or, if desired, an interchangeable ground joint) the combustion tube to the extraction chamber, while the cup, F, is used for the addition of combustion fluid, via the stopcock to the sample. The cup is provided with a cover made from the bottom section of a test tube. A 100-

FIG. 189. Stopcock pipette, showing attached rubber tip and method of introducing liquid into extraction chamber.

ml. Pyrex[15] flask, weighted with lead shot, serves as a stand for the combination when not in use (see B, Fig. 186).

COMBUSTION TUBE

The combustion tube is made of Pyrex[15] glass and is equipped with an interchangeable ground joint at the top which fits the corresponding member of the connecting tube and cup (see Figs. 187 and 188, T).

SMALL MERCURY BOTTLE WITH CAPILLARY

A small bottle, having a loosely fitted capillary inserted through a stopper, is filled with mercury as shown in *A*, Fig. 186. This is used for making mercury seals in the capillary of the stopcock above the chamber. Connection is made to the chamber through the same heavy-walled rubber tubing used when the combustion tube is attached via the tube, *Q*. (Note: If an interchangeable ground joint is used instead of the heavy-walled rubber tubing, the capillary with the small mercury bottle must also be provided with a ground joint to fit.)

STOPCOCK CYLINDER

The stopcock cylinder shown in Figs. 186 and 187, *D,* is used as a reservoir for the alkaline hydrazine and for transferring this solution to the extraction chamber. The delivery end is fitted with a special rubber tip (see below) by means of which a gas-tight connection is made with the extraction chamber.

RUBBER TIPS

Rubber tips are used on the delivery ends of the stopcock cylinder (above) and on the stopcock pipettes (see below) to effect a gas-tight connection between them and the extraction chamber. They are commercially available from the sources supplying the manometric apparatus. They may be made by cutting a ring about one cm. wide from a pliable rubber tube with a bore of 1–2 mm. and walls of 3–4 mm. thick. The ring is then tapered at one end by grinding down the outer edge with sandpaper or a grinding wheel so that it will fit into the bottom of the cup, *E,* above the extraction chamber. Considerate care must be exercised in the grinding so that the taper is smooth and gives a good gas-tight fit when used.

STOPCOCK PIPETTE[97]

A one-ml. capacity stopcock pipette of the type shown in Fig. 189 is required for introducing the lactic acid into the extraction chamber. It is graduated with two marks, the volume between being equal to one ml. The delivery end is fitted with a rubber tip (see above).

MERCURY DROPPING BOTTLE

A small bottle from which fractions of a milliliter of mercury may be poured is used in making mercury seals through the cup, *E,* at the top of the extraction chamber. A one-ounce bottle supplied with a one-hole rubber stopper containing a section of glass capillary tubing (1-mm. bore) is quite satisfactory.

HEAVY-WALLED RUBBER TUBING

Heavy-walled pliable rubber tubing, approximately 25 mm. O.D. and 4 mm. I.D. is used for connecting the extraction chamber to the manometer and to the combustion tube via the glass connecting tube, Q, unless interchangeable ground joints are present.

DISTILLED WATER RESERVOIR

A distilled water reservoir of 1- or 2-liter capacity is used for washing the manometric apparatus. A Mariotte bottle (Fig. 129, Chapter 9) is quite satisfactory for this purpose. It is placed on a platform at the top of the manometer stand so that the water will flow by gravity into the cup, E (Fig. 187) when needed.

WATER SUCTION SYSTEM

A water suction pump, protected by a trap of 1-liter capacity, is used for sucking off the solutions and wash water from the cup, E. The tubing attached to this is fitted with a rubber-tipped glass capillary to facilitate the washing.

ELECTRIC HEATER OR MICROBURNER

Same as used in Chapter 16 (see Figs. 180 and 181), the electric one being preferred.[23,83]

Procedure[80,92]

A synopsis of the procedure may be given by the following:

1. Connection of combustion tube and sample to the manometric apparatus
2. Evacuation of the combustion tube
3. Ejection of air from the combustion tube
4. Introduction of alkaline hydrazine followed by mercury seal
5. Combustion
6. Absorption of carbon dioxide
7. Disconnection of the combustion tube followed by mercury seal
8. Ejection of unabsorbed gases followed by mercury seal
9. Extraction of carbon dioxide (mercury seal)
10. Reading of the p_1 value
11. Reabsorption of the carbon dioxide (mercury seal)
12. Reading of the p_2 value.

CONNECTING COMBUSTION TUBE WITH EXTRACTION CHAMBER

The sample is weighed in a platinum boat (see Chapter 3) and placed in a clean dry combustion tube, T (Figs. 187 and 188). [The size of the sample is governed by the percentage of carbon since not more than 3.5 mg. of this element may be present or the manometer cannot be read. The platinum boat is useful in preventing bumping. If the sample is weighed by difference using a charging tube (see Chapter 3), a few platinum tetrahedra should be added for the above purpose.] For a "microanalysis" (2–3.5 mg. C), 200 mg. of potassium iodate* is added to the sample. (For a "submicroanalysis," 0.3– 0.7 mg. C, only 100 mg. of potassium iodate is added.) The combustion tube is held in an almost horizontal position and a ring of syrupy phosphoric acid is drawn around the upper part of the ground joint with the aid of a glass rod or medicine dropper. The connecting tube, Q, and cup, F, combination is inserted into the combustion tube. (The *stopcock* of the above combination is also lubricated with syrupy phosphoric acid and is *left in the closed position*.) The leveling bulb, L (Figs. 187 and 188), is placed in the upper ring of the stand, the stopcock, a, is opened and the stopcock, b, is turned so that mercury completely fills the extraction chamber, C, and several milliliters pass into the cup, E. The stopcock, b, is then closed and the stopcock, a, is left open. The combustion tube-connecting tube-cup combination (T-Q-F) is joined to the extraction chamber, C, by means of a section of the heavy-walled tubing, effecting a glass-to-glass connection—(or if preferred, an interchangeable ground joint may be used at this point, as is the preference of the author). Slightly more than 2 ml. of fresh combustion fluid is placed in the cup, F, which then is covered to protect the fluid from moisture.

PRELIMINARY EJECTION OF AIR

The leveling bulb, L, is placed in the lower ring on the stand. Then the stopcock, b, at the top of the extraction chamber, C, is turned so as to connect the chamber with the combustion tube, T. The mercury level in the chamber falls immediately since air is drawn from the combustion tube. The leveling bulb, L, then is removed from the lower ring, held in the left hand, and lowered below the table top so that mercury is drawn out of the chamber. When the mercury level in the chamber, C, has reached the 50-ml. mark, the *stopcock, a, is closed,* and the *leveling bulb, L,* is returned to the *upper ring* of the stand. The *stopcock, b, is closed and then stopcock, a, is opened.* This admits mercury to the chamber, C, and compresses the trapped air. Then the stopcock, b, is turned so that the trapped air is forced out through the cup, E, along with several milliliters of mercury. Then the stopcock, b, is closed. [Ejection of the air diminishes to insignificance the part of the blank (see below) that is due to atmospheric carbon dioxide present in the apparatus when the combustion is begun. It also removes the inert gases thereby increasing the efficiency of each subsequent transfer (see below) since there are fewer interfering molecules present].

* See footnote, p. 456, for Combustion Fluid regarding the use of potassium dichromate at this point.

MEASUREMENT OF ALKALINE
HYDRAZINE SOLUTION INTO EXTRACTION CHAMBER

The *leveling bulb*, *L*, is placed in the *lower ring* of the stand. The rubber-tipped stopcock cylinder, *D* (Figs. 186 and 187), containing the alkaline hydrazine solution is placed so that the delivery end makes a gas-tight contact with the bottom of the cup, *E*, which contains a few milliliters of mercury for sealing purposes (see Fig. 187). The stopcock on the cylinder is opened. (No solution flows thusly since the rubber-tipped end is pressed tightly against the bottom of the cup, *E*.) The stopcock, *b*, is opened slightly so that alkaline hydrazine solution *slowly* enters the *extraction chamber*. When the alkaline solution is about one mm. above the 2-ml. mark in the extraction chamber, the stopcock, *b*, is closed and so is that on the cylinder, after which the latter is replaced in its holder with the tip protected from atmospheric carbon dioxide—*D* (Fig. 186). The stopcock, *b*, is opened and closed quickly so that both the alkali in the capillary of the cock and enough mercury for a seal* are drawn into the chamber. The alkali which was contained in the capillary is enough to bring the volume of solution in the chamber to exactly 2 ml. The cup, *E*, is washed with several portions of water, each of which is removed by the water suction pump. (At no time should alkali be allowed to remain in the cup where it would absorb carbon dioxide and be carried back into the chamber.)

COMBUSTION

The *leveling bulb*, *L*, *is removed* from the lower ring and held in the left-hand *below* the table top. Mercury is drawn out of the chamber, *C*, and the manometer until the mercury level in the *manometer* stands a little below the level of the 2-ml. mark on the *chamber*, *C*. Then the *stopcock*, *a*, *is closed* and the leveling bulb, *L*, is placed in the *lower ring* of the stand and left in this position until the combustion is finished. (This position brings the mercury surface in the bulb, *L*, to about the level of the 50-ml. mark on the chamber, *C*.) The stopcock, *b*, is turned to connect the chamber, *C*, with the combustion tube, *T* (see Fig. 188). The stopcock below the cup, *F*, is *carefully* opened to allow 2† ml. of fresh combustion fluid to *slowly* run into the combustion tube, *T*, making certain that no air is allowed to enter the system. For this reason, slightly more than the required amount of combustion fluid was added to the cup, *F*, and the slight excess is allowed to remain in the capillary above the stopcock below the cup.

* At no time should the manometric apparatus be allowed to remain under reduced pressure (leveling bulb, *L*, in the lower ring) unless the capillaries of the stopcock, *b*, are filled with mercury as a seal.

† Two ml. for a "microanalysis" (2–3.5 mg. of carbon) or 1.5 ml. for a "sub-microanalysis" (0.3–0.7 mg. carbon).

An electric heater or microburner is used to heat the contents of the combustion tube. The heat is applied cautiously at first so that the foam collar which forms in the tube is not more than 2 cm. high on the liquid. Within one or two minutes the fluid is brought to boiling. As carbon dioxide and oxygen are evolved, the mercury falls in the extraction chamber, C, and rises in the manometer (Fig. 188). During this stage the stopcock, a, is opened slightly and closed every few seconds to admit mercury from the leveling bulb, L, into the chamber, C, and to keep the gas space in the extraction chamber, C, at about 10 ml. Within about one minute from the beginning of the heating, enough gas has been evolved to press the mercury in the manometer up to its top and to permit the stopcock, a, to be opened completely without causing a backflow of the alkaline solution from the chamber, C, to the combustion tube, T. Then the stopcock, a, is left fully open during the rest of the combustion with the system under a pressure of about 600 mm. of mercury. Vigorous boiling* is continued at this pressure for 1.5 minutes to complete the combustion. During this interval, foam should fill $\frac{1}{3}$ to $\frac{1}{2}$ of the tube, T.) This completes the combustion.

ABSORPTION OF CARBON DIOXIDE BY THE ALKALINE HYDRAZINE IN THE EXTRACTION CHAMBER

After the completion of the combustion above, the heater or flame is left in place under the combustion tube and the leveling bulb, L, is lowered below the table top. This causes mercury to fall in the chamber, C, and the bulb is held in this position until the mercury has reached the 50-ml. mark. The combustion fluid boils vigorously under the reduced pressure, almost the entire tube being filled with foam. The leveling bulb is quickly raised which causes mercury to return to the chamber, C. During this compressive stage, the combustion fluid does not boil. As soon as the gas space in the chamber, C, has been compressed to about 7 ml., the leveling bulb is quickly lowered below the table top until the mercury in the chamber falls again to the 50-ml. mark. This process of alternate expansion and compression should be repeated rapidly 25 times. *The stopcock, a, is closed then while the gas still is expanded (mercury at the 50-ml. mark)*, after which the leveling bulb, L, is returned to the lower ring. The total time required for the 25 excursions should be about 4 minutes.†

* Gentle boiling may yield low results.

† When many determinations are carried out in succession, the alternate lowering and raising of the bulb may be substituted by alternately applying and releasing of suction at the top of the leveling bulb, L, while it is in the lower ring of the stand. For this purpose, a T- or Y-tube is inserted by means of a one-hole stopper into the neck of the leveling bulb (Fig. 190). One of the extending tubes is attached directly to a pump without an intermediate trap. The suction is applied and the finger is used to close the free tube. Mercury is drawn from the chamber into the bulb but rapidly returns when

The heater or flame is then removed from under the combustion tube, *T*, the *stopcock, b, is closed*, and the connecting tube, *Q*, is disconnected from the chamber, *C* (after releasing the vacuum in the combustion tube by opening the stopcock below the cup, *F*). (The combustion tube, *T*, connecting tube, *Q*, and cup, *F*, should be washed and *thoroughly* dried in an oven at 120° C. to prepare them for the next analysis. The practice of using the connecting tube and cup for the next determination is a dangerous one since a drop of combustion fluid is apt to fall upon the sample during the preliminary evacuation.) Immediately after removal of the connecting tube, *Q*, the *curved* capillary of the stopcock, *b*, must be filled with mercury. This is done by attaching the small bottle of mercury, equipped with a loose-fitting capillary,

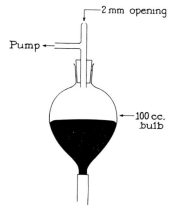

FIG. 190. Leveling bulb attachment—when suction pump is used to raise and lower mercury.

as shown in Fig. 186, *A*, and *carefully* opening the stopcock, *b*, so that mercury fills the cock and a few drops fall into the extraction chamber. The bottle is removed and the excess mercury in the capillary is drawn into the chamber leaving just enough mercury to fill the capillary about one cm. above the stopcock, *b*, which is then closed. (If the excess mercury is not drawn in, it will be thrown on the table top when subsequent shaking is done. The small volume of air admitted to the chamber in the process of making the mercury seal at this point is of no consequence since it will be ejected with the unabsorbed gases as described below.) The *stopcock, a, is then opened* and the *leveling bulb, L, is placed in the upper ring* putting the gases in the extraction chamber under positive pressure.

---

the finger is removed from the tube and the suction is broken. When this method is used it is desirable to have a bulb, *L*, of not over 100-ml capacity and to have enough mercury present in the system so that the bulb is nearly filled when all of the mercury is withdrawn from the extraction chamber.

18. Manometric Determinations

EJECTION OF UNABSORBED GASES

The stopcock, b, is opened carefully allowing the gases to escape through the mercury pool in the cup, E, until the alkaline solution *just reaches* the *bottom* of the stopcock, b. The stopcock, b, is then closed and the leveling bulb, L, placed in the lower ring of the stand. The stopcock, b, is quickly opened and closed so as to make a mercury seal in the capillary and a few drops of mercury fall into the chamber. During this process the small amount of unabsorbed gases which was in the capillary of the stopcock is drawn back into the extraction chamber. This has no influence on the determination since it is carbon dioxide free. If most of the mercury in the cup, E, was drawn into the chamber during the above process, about 1 or 2 ml. more is added so that a good seal is obtained in the subsequent operation.

EXTRACTION OF CARBON DIOXIDE AND THE READING OF p_1

A 1-ml. stopcock pipette fitted with a rubber tip is filled to the top mark with $2N$ lactic acid. The rubber tip is pressed against the bottom of the cup, E, and the stopcock on the pipette is opened. (The liquid in the pipette does not fall since the tip is against the closed-off capillary in the bottom of the cup.) The stopcock, b, is opened carefully to admit exactly one ml. of lactic acid and then closed. The stopcock pipette is removed. The stopcock, b, is opened and closed quickly to effect another mercury seal. The leveling bulb, L, is lowered beneath the level of the table top, drawing mercury out of the extraction chamber, C. When the mercury has fallen to the 50-ml. mark, the stopcock, a, is closed and the leveling bulb is returned to the lower ring. The extraction chamber is shaken (or stirred) for 20–30 seconds.* During this interval, the extracted carbon dioxide forces the mercury down below the 50-ml. mark in the chamber and the stopcock, a, is occasionally opened momentarily to keep the mercury in the extraction chamber at the 50-ml. mark. This is accomplished by the admission of mercury each time that the stopcock, a, is opened. Then the extraction chamber is shaken or stirred vigorously for 1.5 minutes to complete the extraction of the carbon dioxide from the solution. The *leveling bulb*, L, is placed in the *upper ring* and the stopcock, a, is opened which forces mercury into the chamber, C, reducing the volume of gas to the desired mark—see below. (This reduction in volume should be accomplished *rapidly* within 30 or 40 seconds but without setting the mercury in the chamber and manometer to oscillating by jerky opening or closing of the stopcock, a.) The volume of gas in the chamber is reduced to exactly 10 ml. for "microcombustions" with 2–3.5 mg. of carbon (or to

---

* A cork is placed in the top of cup, E, during the shaking so that the mercury is not thrown out.

2 ml. for "submicrocombustions" with about one-fifth as much carbon) and the stopcock, a, is closed. Obviously, the final adjustment of the volume must be done slowly and with extreme care and it is suggested that a reading lens be used as an aid in bringing the meniscus of the aqueous solution exactly to the desired mark on the extraction chamber. The manometer reading, p_1, is taken, making use of the illuminated background and a reading lens. [If it is desired to check this reading, the mercury in the chamber should be lowered again to the 50-ml. mark, the chamber shaken or stirred for one minute and the gas volume reduced again to 10 ml. (or 2 ml.) before the duplicate reading of p_1 is made.]

REABSORPTION OF CARBON DIOXIDE AND THE READING OF p_2

Now the *leveling bulb, L,* is placed in the *lower* ring. The stopcock, a, is opened and the gas is left under a slightly reduced pressure. About one ml. of $5N$ sodium hydroxide is added on top of the mercury in the graduated cup, E. The stopcock, b, is carefully opened to admit the mercury in the cup, E, into the extraction chamber. When the mercury level has fallen into the capillary above the stopcock, b, note is made of the height of the alkali in the graduated cup. The stopcock, b, is allowed to remain slightly open until *exactly 0.5 ml.* of alkali (measured by difference) has entered the extraction chamber.* About 2 ml. of mercury is added to the cup, E, and a mercury seal is made, using more than the ordinary amount of mercury to dislodge any viscous alkali that may be stuck in the tube at the top of the chamber below the stopcock, b. The excess sodium hydroxide is removed from the cup, E, by suction and then it is washed several times with distilled water to rid it of traces of alkali. The contents of the extraction chamber are mixed by lowering the leveling bulb, L, below the table top to bring the surface of the solution down to a point a little below he 10-ml. mark (*not* to the bottom) and then raising the leveling bulb to a point above the ring so that the pressure in the chamber is about atmospheric, as shown on the manometer. This process of lowering and raising the leveling bulb, L, is repeated three times. Then the leveling bulb is lowered below the table top to bring the solution meniscus in the chamber down a little *below* the 10-ml. mark (or the 2-ml. mark, at whichever point p_1 was read), and then the stopcock, a, is closed. The leveling bulb is returned again to the upper ring. The system is allowed to stand for one minute to allow drainage of the alkali from the walls to be complete. The stopcock, a, is carefully opened to raise the meniscus to exactly the 10-ml. (or 2-ml.) mark and the p_2 reading is made on the manometer. The temperature of the water jacket is noted.

* Care must be exercised so that no trace of air enters the chamber during the admission of alkali or mercury. There must be no air bubble between the two layers.

18. Manometric Determinations

The pressure of the carbon dioxide obtained by the combustion of the sample is given by

$$P_{CO_2} = p_1 - p_2 - c,$$

where $c = p_1 - p_2$, obtained from a blank analysis at the same temperature as that of the regular determination. *Consequently, the entire procedure must be repeated, in the absence of a sample.* However, a blank analysis does not have to be performed before or after each determination. Since the value is a check on the purity of the reagents used, it need be done only occasionally if care is exercised in handling these so that they do not become contaminated from dust (in the combustion fluid) or carbon dioxide from the air (in the alkaline hydrazine or lactic acid).

WASHING THE EXTRACTION CHAMBER

It is very important that the extraction chamber be properly washed between analyses. The stopcock, a, is opened and the leveling bulb, L, is placed in the upper ring. The tubing connected to the suction device (supplied with the trap) is placed in the cup, E. The stopcock, b, is opened and the rising mercury forces the solution from the chamber into the cup, E, where it is removed immediately by the suction. When mercury is being sucked out of the cup, E, the stopcock, b, is closed. The leveling bulb is placed in the lower ring. Water is added from the reservoir at the top of the stand and the stopcock, b, is opened carefully, admitting water to the chamber. The cup, E, is kept full of water coming from the reservoir at the same rate as it is drawn into the chamber being careful not to admit air. When no more water will enter the system (chamber, C, full) the water tube is removed and replaced by the suction one. Simultaneously, the leveling bulb is placed in the upper ring. As soon as all of the water has been forced out of the chamber together with a little mercury, the stopcock, b, is closed, the leveling bulb, L, returned to the lower ring, water admitted, etc., as above and the process repeated twice more (three washings in all). A few minutes after the final washing, any water which collects above the mercury surface at the top of the mercury-filled chamber should be forced out into the cup, E. The apparatus is then ready for the next analysis.

Calculation:

The amount of carbon present in the sample is found by multiplying the P_{CO_2} by the factor given in Table 30.

$$\text{mg. C} = P_{CO_2} \times \text{factor}$$

where

$$P_{CO_2} = p_1 - p_2 - c$$

$(c = p_1 - p_2$ for a blank determination$)$

$$\therefore \frac{P_{CO_2} \times \text{factor} \times 100}{\text{Wt. sample}} = \% \text{ C}$$

Example No. 1 "Microanalysis":

In the analysis of a 3.539-mg sample, the following pressures are obtained on compressing the gas to a volume of 10 ml. at 24° C.:

$$\text{Volume} = 10.0 \text{ ml.}$$
$$\text{Temperature} = 24.0° \text{ C.}$$
$$P_1 = 468.0 \text{ mm.}$$
$$P_2 = 29.0 \text{ mm.}$$
$$\left.\begin{array}{l} P_1 \text{ (blank)} = 32.0 \text{ mm.} \\ P_2 \text{ (blank)} = 28.0 \text{ mm.} \end{array}\right\} P_1 - P_2 = c = 4.0 \text{ mm.}$$
$$\therefore P_{CO_2} = 468.0 - 29.0 - 4.0 = 435.0 \text{ mm.}$$

From Table 30 the factor for a volume of 10 ml. at 24.0° C. is 0.006829

$$\therefore \frac{435.0 \times 0.006829 \times 100}{3.539} = 83.94\% \text{ C}$$

Example No. 2 "Submicroanalysis":

In the analysis of a 1.425-mg. sample the following pressure is obtained when the gas was compressed to a volume of 2 ml. at 26.0° C.:

$$\text{Volume} = 2.0 \text{ ml.}$$
$$\text{Temperature} = 26.0° \text{ C.}$$
$$P_1 = 580.4 \text{ mm.}$$
$$P_2 = 106.5 \text{ mm.}$$
$$\left.\begin{array}{l} P_1 \text{ (blank)} = 125.5 \text{ mm.} \\ P_2 \text{ (blank)} = 105.00 \text{ mm.} \end{array}\right\} P_1 - P_2 = c = 20.5 \text{ mm.}$$
$$\therefore P_{CO_2} = 580.4 - 106.5 - 20.5 = 453.4 \text{ mm.}$$

From Table 30, the factor for a volume of 2 ml. at 26.0° C. is 0.001366

$$\therefore \frac{453.4 \times 0.001366 \times 100}{1.425} = 43.46\% \text{ C}$$

The accuracy of the determination is equal to that of the dry combustion method (Chapter 9). Consequently, the allowable error is ±0.3%.

In general, with apparatus supplied by a reliable manufacturer, no correction need be applied for most work. If, however, the calibration chart shows considerable deviation from the amounts marked on the extraction chamber, correction may be made by multiplying the factors given below by the ratio of the two volumes[92] that is, by

$$\frac{\text{Volume shown on correction chart}}{\text{Volume marked on extraction chamber}}$$

Also for most analyses the factors given in Table 30 (determined by the use of an empirical correction)[92] are satisfactory. If extreme accuracy is desired, a whole series of analyses on a known pure compound is made and each factor of Table 30 is multiplied by the ratio[92]:

$$\frac{\text{Theoretical value for carbon}}{\text{Average of the found values for carbon}}$$

TABLE 30
FACTORS FOR CARBON CALCULATION

Temperature (°C.)	Factors	
	"Submicroanalysis" Vol. = 2.000 ml.	"Microanalysis" Vol. = 10.00 ml.
10	0.001474	0.007304
11	0.001466	0.007266
12	0.001458	0.007229
13	0.001451	0.007193
14	0.001444	0.007158
15	0.001437	0.007123
16	0.001430	0.007089
17	0.001424	0.007055
18	0.001417	0.007021
19	0.001410	0.006988
20	0.001403	0.006955
21	0.001397	0.006923
22	0.001390	0.006891
23	0.001384	0.006860
24	0.001378	0.006829
25	0.001372	0.006799
26	0.001366	0.006770
27	0.001360	0.006741
28	0.001354	0.006712
29	0.001349	0.006684
30	0.001343	0.006656
31	0.001337	0.006629
32	0.001332	0.006602
33	0.001327	0.006576
34	0.001321	0.006550
35	0.001316	0.006524

II. MANOMETRIC DETERMINATION OF PRIMARY AMINO NITROGEN IN THE ALIPHATIC α-AMINO ACIDS

The aliphatic compounds containing primary amino nitrogen react with nitrous acid to yield nitrogen according to the equation:

$$RNH_2 + HNO_2 \rightarrow ROH + H_2O + N_2$$

Van Slyke[85,88] found that the NH_2 groups in the α-*amino acids* react quantitatively in 3–4 minutes while the NH_2 groups in other types of substances require much longer periods, sometimes many hours. The group in ammonia requires about 1.5–2 hours[85] (25% completion after approximately 4 minutes[86]); in methylamine, 1.5–2[85]; in urea, 8[85] (6–7% completion after 4 minutes[48]); and in purines and pyrimidines from 2–5 hours for quantitative reactions to occur. Therefore, the α-amino groups in the aliphatic acids may be determined in the presence of other forms of nitrogen. For example, with tryptophan a value for nitrogen is obtained equal to one-half of the total present while with histidine a value is obtained equal to one-third of the total, since there is but one α-amino group present in each, but a total of two and three nitrogen atoms, respectively, in these molecules.

This method, which requires about 15 minutes to complete, once the operator has mastered the technique, gives excellent results, and like the manometric carbon determination is a fascinating one to perform.

Reagents

SODIUM NITRITE SOLUTION

Eight hundred grams of reagent grade of sodium nitrite is dissolved in one liter of water. This is used as the source of nitrous acid.

GLACIAL ACETIC ACID

Reagent grade of glacial acetic acid is used to liberate the nitrous acid in the reaction. About 600 ml. should be set aside and used only for this purpose to avoid the necessity of running frequent blank determinations (see below).

ALKALINE PERMANGANATE

Fifty grams of potassium permanganate is added to one liter of 10% solution of sodium hydroxide, both substances being reagent grade. The mixture is shaken until the alkali is saturated with the permanganate. This is used for absorbing the nitric oxide formed by spontaneous decomposition of nitrous acid.*[53]

CAPRYLIC ALCOHOL

Caprylic alcohol is used, when necessary, to prevent foaming of viscous solutions such as in the presence of protein.

* $3HNO_2 = 2NO + HNO_3 + H_2O$. The original procedure of Van Slyke made use of this reaction by using the nitric oxide to displace air.[85] (The equation given by Van Slyke is unbalanced and incorrect.[53])

Apparatus[18,51,70,71,80,83,88,92]

MANOMETRIC APPARATUS

This is the same as used for the manometric carbon determination. The accessory parts, however, are different.

STOPCOCK PIPETTES

Three stopcock pipettes are required for the determination, one each of the sizes 1, 2, and 5 ml. These are all fitted with rubber tips at the delivery ends.

HEMPEL PIPETTE

One Hempel pipette of the type shown in Figs. 192, 193, and 194, is used for washing the reaction gases with alkaline permanganate to remove the nitric oxide. It consists of two reservoir bulbs and a delivery capillary. Between the one bulb and the capillary is a three-way stopcock which can in turn connect the (1) capillary to the cup above the stopcock, (2) the cup above the stopcock to the bulb, and (3) the delivery capillary to the bulbs. The delivery capillary is fitted with a rubber tip. Between the two bulbs is an outlet which is fitted with a small section of rubber tubing closed off by a screw lamp. This outlet is used for drainage purposes.

The Hempel pipette is filled with alkaline permanganate as follows: The stopcock on the pipette is turned so as to connect the cup above it with the bulbs. The alkaline solution is added through a small funnel to the upper bulb until it is one-half full while the lower bulb is full and the liquid fills most of the capillary under the stopcock. The screw clamp is opened momentarily and a few milliliters of solution drained off to remove any trapped air at this point. The pipette is tilted so that the upper bulb is *above* the stopcock. This forces alkali up into the cup. The stopcock is quickly turned to connect the delivery capillary with either the bulbs or the cup long enough for the delivery tip to be filled with permanganate. The stopcock is closed and the Hempel pipette is ready to be used. (The upper bulb should be one-half full, the cup should contain 1–2 ml., and the rest of the pipette should be *filled* with solution.)

Procedure[80,88]

The manipulation of the manometric apparatus, namely, the technique of making mercury seals, expanding and compressing the gases, etc., is similar to that described under I. Manometric Determination of Carbon. In this part, it is assumed that the reader is already familiar with that section and consequently comparatively brief directions along these lines will be given here.

A synopsis of the steps in the analysis is as follows[80]:

1. Addition of the sample
2. Addition of glacial acetic acid
3. Mercury seal
4. Extraction of air from the above substances
5. Ejection of extracted air
6. Addition of sodium nitrite
7. Mercury seal
8. Decomposition of amino groups
9. Transfer of gases to Hempel pipette; absorption of nitric oxide
10. Washing of extraction chamber
11. Preparation of air-free water (making necessary mercury seal)
12. Return of nitrogen from the Hempel pipette to the extraction chamber (mercury seal)
13. Reading of p_1
14. Ejection of nitrogen (mercury seal)
15. Reading of p_0.

REMOVAL OF AIR FROM THE MIXED SOLUTION OF AMINE AND ACETIC ACID

The maximum amount of nitrogen that can be determined is about 0.6 mg. and the minimum is about 0.0004 mg. The sample is dissolved in water (or very dilute acid) in a small volumetric flask and the solution diluted to the mark. The extraction chamber, C, of the manometric apparatus is filled completely with mercury as are *both* capillaries of the stopcock, b (Fig. 187), and 1–2 ml. in the cup, E. This is done by raising the leveling bulb, L, to the upper ring and opening the stopcock, b, first one way and then the other. The leveling bulb, L, is then placed in the *lower ring* and a 5.0-ml. aliquot portion of the aqueous solution of the sample is added to the extraction chamber from a rubber-tipped 5-ml. stopcock pipette through manipulation of the stopcock, b. This is followed by exactly one ml. of glacial acetic acid from a rubber-tipped one-ml. stopcock pipette.* A mercury seal is made so as to bring down into the chamber the acid left in the stopcock capillary and to make the latter gas-tight. The leveling bulb, L, is lowered below the table top and mercury drawn out of the chamber, C, to the 50-ml. mark after which the stopcock, a (Fig. 187), is closed and the evacuated chamber shaken† (or stirred) for 2 minutes at the tempo of 250 to 300 times per minute to remove air. The leveling bulb, L, is placed in the upper ring, the stopcock, a, opened

---

* If protein or other content of the amine solution makes the latter likely to form troublesome foam, a drop of caprylic alcohol is also added with the acetic acid.

† Insert cork stopper in the top of the cup, E, to prevent the spilling of mercury.

and the gases compressed. The stopcock, *b,* is opened very carefully to eject the trapped air and to allow the solution from the chamber to *rise just high enough to fill the capillary above the stopcock, b, but not into the cup, E.* Then the stopcock, *b,* is closed. The operation must be done exactly as described or low values will result from loss of amine solution in the cup, *E,* or high values will result from readmission of trapped air in the subsequent operation.

DECOMPOSITION OF α-AMINO GROUPS

The leveling bulb, *L,* is placed in the lower ring. Two ml. of sodium nitrite solution is added to the extraction chamber from a rubber-tipped 2-ml. stopcock pipette which immediately is followed by a mercury seal. The evolution of nitrogen and nitric oxide begins at once. The leveling bulb, *L,* is lowered below the table top and mercury drawn out of the chamber down to about 1–2 cm. above the 50-ml. mark. The stopcock, *a,* is closed and the mixture

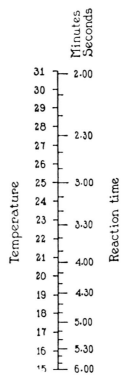

FIG. 191. Scale indicating reaction period required for complete decomposition of α-amino acids, and 0.07 decomposition of urea, when total volume of reacting solution is 8 ml. (Van Slyke).

allowed to stand thusly until within one minute of the end of the reaction time given on the accompanying scale (Fig. 191). During this period if the mercury level is forced down below the 50-ml. mark by the pressure of the evolved gases, the stopcock, *a*, is opened momentarily to bring the mercury level back to the above mark. During the last minute of the reaction time, the mixture is shaken to complete the evolution of nitrogen and, to keep the mercury level from dropping, the stopcock, *a*, is occasionally opened as above.

TRANSFER OF THE NITROGEN AND NITRIC OXIDE MIXTURE TO THE HEMPEL PIPETTE AND THE ABSORPTION OF THE NITRIC OXIDE

The leveling bulb, *L*, is placed in the upper ring and the stopcock, *a* (Fig. 187), is opened. A few milliliters of water are placed in the cup, *E*, above the small pool of mercury and the rubber-tipped capillary of the Hempel pipette is held firmly with the left hand against the bottom of the cup, *E*. The stopcock on the Hempel pipette is turned so as to connect the rubber-tipped delivery capillary with the bulbs, as shown in Fig. 192. The stopcock, *b*, is opened so as to force the mixture of gases slowly over into the Hempel pipette. The aqueous solution will continue to evolve gas (NO) as long as it remains in the chamber but no attempt should be made to collect this. Consequently, the transfer should be stopped as soon as a trace of the liquid has passed the bore of the stopcock on the Hempel pipette. The stopcock, *b*, is closed and so is the stopcock on the Hempel pipette, the latter being turned *counterclockwise* but not far enough to connect the permanganate-filled cup with the bulbs. Then the Hempel pipette is gently shaken for a few seconds with a *swirling horizontal motion,* being careful not to entrap air in the exposed bulb. The stopcock on the Hempel pipette is turned quickly to admit a small amount of the permanganate from the cup into the bulb, forcing the gas and solution in the capillary between these parts into the bulb. Then the stopcock on the Hempel pipette is quickly closed to avoid loss of gas through it. The pipette is shaken gently again horizontally and then set aside while the chamber, *C*, is being washed. Since nitric oxide continues to form in the chamber, *C*, this must be thoroughly cleaned before the next operation. The stopcock, *b*, is opened and the aqueous solution which is forced up into the cup, *E*, is removed by insertion of the suction tube (see description under Manometric Carbon Determination). The stopcock, *b*, is closed and the leveling bulb, *L*, is placed in the lower ring. Water is admitted through the cup, *E*, and the stopcock, *b*, until the chamber, *C*, is full. The stopcock, *b*, is closed, the leveling bulb, *L*, is placed in the upper ring, the stopcock, *b*, is opened and the water which is forced out into the cup, *E*, is again removed by suction. The washing is done three times just as in the case of the carbon determination.

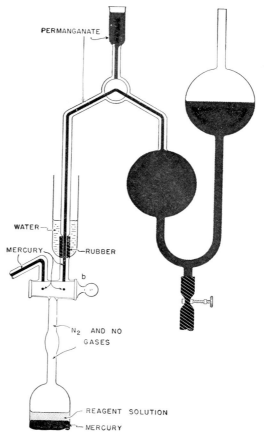

FIG. 192. Hempel pipette—ready for transfer of nitrogen and nitric oxide from extraction chamber. Permanganate in capillaries of Hempel pipette, cup, and bulbs.

RETURN OF PURIFIED NITROGEN TO THE MANOMETRIC APPARATUS

The leveling bulb, L, is in the upper ring and mercury fills the chamber, C, the stopcock, b, and a few milliliters of it is in the cup, E, following the washings. Then the leveling bulb, L, is placed in the lower ring. Ten milliliters of distilled water, in two 5-ml. portions from the cup, E, is admitted to the extraction chamber. A mercury seal is made and the chamber evacuated drawing the mercury level down to the 50-ml. mark, in the usual manner already described several times in this chapter. The evacuated chamber is shaken for one minute to remove the greater part of the air from the water. Then the *leveling bulb, L,* is placed in the *upper ring, the stopcock, a, opened* and the extracted air ejected through the stopcock, *b*. Then one milliliter of the air-free water is forced up into the cup, *E*, and the stopcock, *b*, closed. The

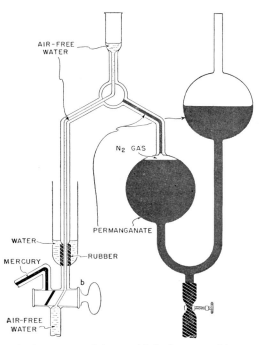

FIG. 193. Hempel pipette—containing purified nitrogen, with cup and one capillary filled with air-free water.

stopcock on the Hempel pipette is turned *quickly counterclockwise* to *connect* the *cup* above it with the *delivery capillary* and the alkaline permanganate present in the cup is allowed to drain off. The stopcock is left in this position and the cup and capillary thoroughly washed with distilled water and allowed to drain. The Hempel pipette is held in the left hand and the rubber-tipped delivery capillary is pressed against the bottom of the cup, *E*. The stopcock, *b*, is opened and air-free water from the extraction chamber is forced up to fill the empty delivery capillary of the Hempel pipette and about one ml. passed into the cup above its stopcock, as shown in Fig. 193. The stopcock, *b*, is closed and the stopcock on the Hempel pipette turned *clockwise* to admit a little air-free water to wash the capillary, between this stopcock and the bulb, free from permanganate. This stopcock is turned again *clockwise* to connect the Hempel bulb with the delivery capillary as shown in Fig. 194. The *leveling bulb, L,* is placed in the *lower ring* and the stopcock, b, is opened to admit the nitrogen from the Hempel pipette to the extraction

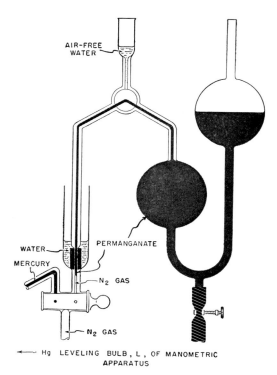

FIG. 194. Hempel pipette—after return of purified nitrogen to extraction chamber. Small quantity of nitrogen in capillary has not yet been drawn into extraction chamber.

chamber. The stopcock, b, is closed to stop the transfer when the column of permanganate following the gas has entered the capillary above the stopcock, b, as shown in Fig. 194. The stopcock on the Hempel pipette is closed and the pipette removed to its stand. The cup, E, above the extraction chamber is washed with several portions of water, to remove as much permanganate as possible. About one ml. of mercury is placed in the cup, E, and the stopcock, b, opened and closed quickly to admit mercury to force the trapped nitrogen and permanganate from the capillary into the extraction chamber and to make a mercury seal.

MEASURING THE NITROGEN

The *leveling bulb, L, is placed below the table top* and the level of the water in the chamber lowered until the water meniscus is at either the 0.5 ml. or the 2.0-ml. mark, depending upon the amount of nitrogen present. (If this exerts less than 100 mm. of pressure at the 2-ml. mark, it is preferable to read the pressure when the gas volume is 0.5 ml.) The meniscus is brought exactly to the desired volume by manipulation of the stopcock, *a* (Fig. 187), making use of a reading lens. With the aid of the illuminated background for the manometer and a reading lens, the manometer reading, p_1, is recorded.

The *leveling bulb, L, is placed* in the *upper ring,* the stopcocks, *a,* and then, *b,* opened (in this order), and the gas ejected. The stopcock, *b,* is closed, the *leveling bulb, L, lowered below the table top* (mercury seal) and the level of the water in the chamber again brought down until the water meniscus is at either the 0.5- or 2.0-ml. mark, at whichever it was for the p_1 reading, and the stopcock, *a,* closed. Then the manometer reading, p_0, is recorded.

BLANK ANALYSIS

Just as was the case with the manometric carbon determination, a blank analysis must be done as a control of the reagents. The exact procedure described above for the determination of α-amino nitrogen must be repeated, substituting 5 ml. of water for the 5 ml. of amine solution. Obviously, the blank analysis must be done at the same volume as was the sample. For a few weeks after the preparation of new sodium nitrite solution the correction continues to change slightly (diminishes), eventually becomes constant, and then need be determined only occasionally. When the analysis yields enough nitrogen to give a pressure of more than 100 mm. at a volume of 2.0 ml., variations in the correction with ordinary changes in room temperature may be neglected. However, if the gas measurements must be made at a volume of 0.5 ml., the correction should be determined at a temperature almost the same as that of the analysis. Corrections usually are equal to about 5–7 mm. with the gas at a volume of 2.0 ml. and about 20–30 mm. at 0.5 ml.

Calculation:

The pressure exerted by the α-amino nitrogen is calculated from the equation:

$$P_{N_2} = p_1 - p_0 - c$$

where $c = p_1 - p_0$ obtained in the blank analysis.

The weight of α-amino nitrogen is equal to the pressure multiplied by the factor shown in Table 31, that is,

$$\text{mg. of α-amino N} = P_{N_2} \times \text{factor}$$

$$\therefore \frac{P_{N_2} \times \text{factor} \times 100}{\text{Wt. of sample in the 5-ml. aliquot}} = \%\ \alpha\text{-amino nitrogen}$$

Example No. 1:

6.031 mg. of sample (in 5-ml. aliquot of solution) gave the following data:

Volume = 2.0 ml.
Temperature = 24° C.
p_1 = 204.5 mm. analysis of sample
p_0 = 49.5 mm. analysis of sample
p_1 (blank) = 54.5 mm.
p_0 (blank) = 49.5 mm. } $p_1 - p_0 = c = 5.0$ mm.
$P_{N_2} = p_1 - p_0 - c$
 = 204.5 − 49.5 − 5.0 = 150.0 mm.

$$\therefore \frac{150.0 \times 0.001511 \times 100}{6.031} = 3.76\%\ \alpha\text{-amino nitrogen}$$

TABLE 31
FACTORS FOR CALCULATION OF α-AMINO NITROGEN

Temperature (°C.)	Factors by which mm. P_{N_2} are multiplied to give mg. α-amino N in sample analyzed	
	Vol. = 0.5 ml.	Vol. = 2.0 ml.
15	0.000390	0.001561
16	0.000389	0.001555
17	0.000387	0.001549
18	0.000386	0.001544
19	0.000385	0.001538
20	0.000383	0.001533
21	0.000382	0.001527
22	0.000380	0.001522
23	0.000379	0.001516
24	0.000378	0.001511
25	0.000376	0.001506
26	0.000375	0.001500
27	0.000374	0.001495
28	0.000372	0.001490
29	0.000371	0.001485
30	0.000370	0.001480
31	0.000368	0.001474
32	0.000367	0.001469
33	0.000366	0.001464
34	0.000365	0.001459

Example No. 2:

1.010 mg. of sample (in 5-ml. aliquot of solution) gave the following data:

Volume = 0.5 ml.
Temperature = 25° C.
P_1 = 497.5 mm. analysis of sample
P_0 = 172.5 mm. analysis of sample
P_1 (blank) = 191.5 mm.
P_0 (blank) = 171.5 mm. } $P_1 - P_0 = c = 20.0$ mm.
$P_{N_2} = P_1 - P_0 - c$
= 497.5 — 172.5 — 20.0 = 305.0 mm.

$$\therefore \frac{305.0 \times 0.000376 \times 100}{1.010} = 11.35\% \text{ α-amino nitrogen}$$

This method permits the measurement of the α-amino nitrogen in 5 ml. of solution to within ± 0.0005 mg.[88]

III. OTHER DETERMINATIONS CARRIED OUT ON THE MANOMETRIC APPARATUS

AMPUL AND VIAL TESTING WITH MANOMETRIC APPARATUS

The above method for the analysis of carbon has been applied to the study of the aging of solutions in sealed ampuls[78,80] and vials,[81] where this process is accompanied by the evolution of small amounts of carbon dioxide. The material under observation is subjected to testing both before and after storage at slightly elevated temperatures, such as, for example, 45° C. The method has been used extensively in the author's laboratory as well as by others.[62] Through the use of the apparatus described below, together with the "submicroanalytical" technique, fractions of a milligram of carbon dioxide may be determined.

Ampuls[78,80]

Apparatus

The apparatus is the same as above with the exception that the combustion tube, T, cup, F, and connecting tube, Q, are replaced by a specially constructed flask since the carbon dioxide is already present and does not need generation. Figure 195 shows the flask and its relation to the rest of the system. It consists of a 100-ml. round-bottomed flask (a, Fig. 196) having a ground joint to which is attached a distilling tube, b, that is bent so that it may be connected onto the Van Slyke-Folch extraction chamber, C (Figs. 187 and 188).

18. Manometric Determinations

Attached to the side of the flask is a compartment, d, for holding the ampul, which varies with the size of the article to be tested, different flasks being required for almost each size of ampul. This compartment is similar in shape to an inverted ampul, having a very thick-walled neck, e. To the upper half of the compartment is attached a heavy glass rod plunger, f, which passes through the ground joint, g, and rubber tubing, h, which permits motion.

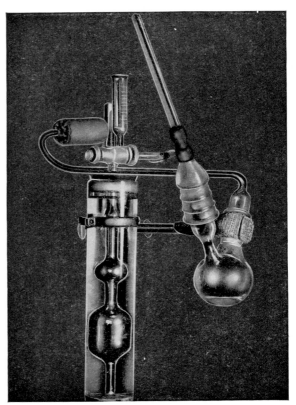

FIG. 195. Ampul testing attachment for manometric apparatus, showing connection to extraction chamber.

Procedure[78,80]

To ensure the proper pH for complete extraction of carbon dioxide, several milliliters of dilute hydrochloric acid are added to the flask, except where addition of acid causes decomposition of the substance in the ampul with evolution of carbon dioxide. The exact quantity of acid added will depend upon the pH of the material under test. The tip of the ampul is scratched

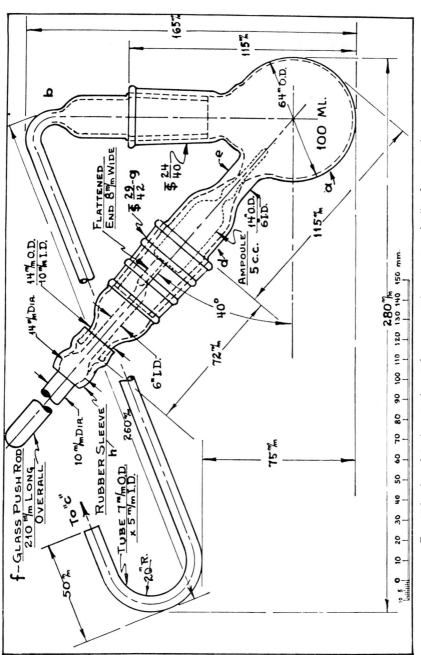

Fig. 196. Ampul testing attachment for manometric apparatus—details of construction.

with a file, care being taken not to produce a crack through which gas might escape. The ampul is then placed in the compartment, neck down, the plunger is attached, and the flask is connected to the extraction chamber as shown in Fig. 195. The entire system is evacuated and the air removed from it in the usual manner. The alkaline hydrazine solution is added to the extraction chamber and the ampul is broken by depression of the plunger. With the system under reduced pressure, the contents are sucked from the ampul into the bottom of the flask and mixed with the hydrochloric acid, if used. The carbon dioxide is then absorbed by twenty-five excursions, after which its pressure is measured in the way described above. The blank values are obtained by repeating the procedure without having an ampul present.

Calculation:
The calculations are the same as above using the factors given in Table 30.

Vials[81]

Apparatus

The vial testing attachment to the manometric apparatus is shown in Fig. 197. Depression of the plunger causes a hypodermic needle to pierce the rubber cap of the sealed vial, thus connecting its contents with the manometric system. Otherwise, the procedure is the same as that used for ampul testing, including calculations.

MANOMETRIC MICRO-KJELDAHL DETERMINATION OF NITROGEN

The Kjeldahl determination may be done manometrically as follows: The sample is digested in the normal manner. The digest is neutralized and washed into the manometric apparatus where it is treated with hypobromite. The following reaction occurs and the liberated nitrogen is measured manometrically:

$$2NH_3 + 3NaBrO \rightarrow 3H_2O + 3NaBr + N_2$$

By this method the distillation and use of standard solutions are eliminated. Van Slyke[87] states that the results are reproducible to within one per cent.

BLOOD AND URINE ANALYSIS

Carbon monoxide, carbon dioxide, oxygen, and nitrogen may be determined in either blood or plasma as indicated in the beginning of this chapter.[3,24,61,72,94,97]

Likewise, other determinations on blood or urine may be performed, such as for the amount of amino nitrogen[61,88,95,96] or reducing sugars present.[61,93]

Phosphate or sulfate present in plasma may be determined as described below under the headings of phosphorus and sulfur.[30] Magnesium in blood and urine filtrates is accomplished by the methods listed below.[31]

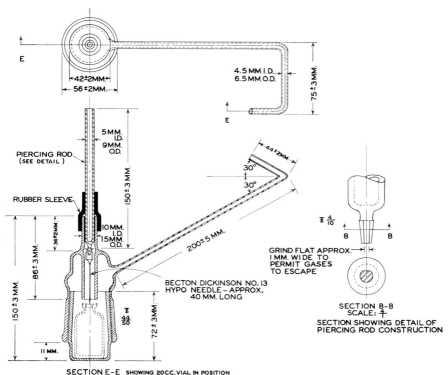

FIG. 197. Vial testing attachment for manometric apparatus—details of construction.

GAS SOLUBILITY STUDIES

The solubilities of such gases as carbon dioxide, nitrogen, oxygen, carbon monoxide, air, and helium in water or in olive oil may be determined with the manometric apparatus.[89]

INDIRECT METHODS BASED ON MANOMETRIC COMBUSTION OF ORGANIC PRECIPITATES

Determination of Phosphorus

Phosphorus, in organic compounds, may be determined manometrically. The substance is subjected to wet combustion and the resulting phosphoric acid precipitated with strychnine molybdate. The pure precipitate is oxidized in the manometric apparatus[30,46,99] The phosphorus content is calculated from the carbon in the precipitated strychnine phosphomolybdate.

Determination of Sulfur

Sulfur may be determined manometrically in the absence of phosphorus by the method of Hoagland.[30] The sulfur is converted to sulfate which in turn is precipitated as benzidine sulfate. This is oxidized and the resulting carbon dioxide measured manometrically. The sulfur content is calculated from the carbon in the precipitate,

$$C_{12}H_8(NH_2)_2 \cdot H_2SO_4$$

which contains twelve atoms of carbon for one of sulfur. Hoagland states that as little as 0.07 mg. of sulfur may be determined with an average error of one part in 200 and that as little as 0.02 mg. may be determined without greatly increasing the error.

Determination of Magnesium

To determine magnesium manometrically,[31] the element is first precipitated as magnesium-8-hydroxyquinoline according to the reaction,

$$Mg^{++} + 2C_9H_6NOH \rightarrow Mg(C_9H_6NO)_2 + 2H^+$$

and the precipitate oxidized in the manometric apparatus. The precipitate contains eighteen carbon atoms to one magnesium and Hoagland[31] states that as little as 0.03 mg. of magnesium may be determined accurately. Obviously, the magnesium is calculated from the amount of carbon in the magnesium-8-hydroxyquinoline.

TABLE 32
ADDITIONAL INFORMATION ON REFERENCES* RELATED TO CHAPTER 18

In addition to the procedures presented in the preceding pages of this chapter, the author wishes to call to the attention of the reader the additional references listed in Table 32. (See statement at top of Table 4 of Chapter 1, regarding completeness of this material.)

Books
Grant, 21, 22
Milton and Waters, 55, 56
Peters and Van Slyke, 61
Roth, 65–68
Steyermark, 80

Reviews
Hillenbrand and Pentz, 29
McCaldin, 52
Van Slyke, 84

Ultramicro-, submicro-methods
Battley, 6

* The numbers which appear after each entry in this table refer to the literature citations in the reference list at the end of the chapter.

TABLE 32 (*Continued*)

Ultramicro-, submicro-methods (Cont.)

Sandkuhle, Cunningham, and Kirk, 69
Steyermark, 80
Van Slyke, Folch, and Plazin, 92

Modifications, general, miscellaneous, etc.

Hurka, 34
Hussey and Maurer, 35
Kainz, 36
Kainz and Huber, 37–39
Kainz and Kasler, 41
Kainz and Schöller, 42–44
Van Slyke and Kirk, 95

Carbon

Austin, 3
Battley, 6
Farrington, Niemann, and Swift, 17
Gabourel, Baker, and Koch, 19
Heald, 27
Hockenhull, 32
Holt, 33
Rappaport, Eichhorn, and Nutman, 63, 64
See Chapter 9

Isotopic carbon

Boos, Jones, and Trenner, 8
Burr, 11
Gabourel, Baker, and Koch, 19
Gora and Hickey, 20
Harper, Neal, and Rogers, 25
Lindenbaum, Schubert, and Armstrong, 50
Neville, 59
See Chapter 9

Nitrogen

Kainz, 36
Kainz and Schöller, 42–44
Merck, 54
Peter, 60
Sandkuhle, Cunningham, and Kirk, 69
Shimoe, 73–76
See Chapters 7 and 8

Oxygen

Sendroy and Liu, 72
See Chapter 14

Gasometric method for amino nitrogen

Carson, 12
Grant, 21, 22
Kainz, Huber, and Kasler, 40
Kainz and Schöller, 43
Peter, 60
Roth, 65–68
Shimoe, 73, 76
Van Slyke, 85, 86
Van Slyke and Kirk, 95
See Chapter 7

Gas chromatographic method

Kainz and Huber, 37

Paper chromatographic method

Woiwod, 103, 104

Ninhydrin and similar reactions

McCaldin, 52
Moubasher and Awad, 58
Van Slyke and Dillon, 90
Van Slyke, Dillon, MacFadyen, and Hamilton, 91
Virtanen and Rautanen, 100

Solubilities of gases in liquids

Van Slyke, 89

Other methods for amino acids

Aquist, 2
Bhatty, 7
Borsook and Dubnoff, 9
Hershenson and Hume, 28
Keen and Fritz, 45
Merck, 54
Sisco, Cunningham, and Kirk, 77
Steyermark, 79
West, 102
Zeile and Oetzel, 105
See Chapters 7 and 8

TABLE 32 (*Continued*)

Apparatus	Use of other types of apparatus (Warburg, Neville, micronitrometers, etc.)
Barcroft, 4	Battley, 6
Barcroft and Haldane, 5	Braganca, Quastel, and Sucher, 10
Charest, Koch, and Gagnon, 14	Burr, 11
Doherty and Ogg, 16	Chapon, 13
Konikov, 47	Hussey and Maurer, 35
Milton and Waters, 55, 56	Kainz and Huber, 38
Möhle, 57	Kainz, Huber, and Kasler, 40
Rappaport, Eichhorn, and Nutman, 63, 64	Ley, J. de, 49
Shimoe, 73	Neville, 59
Steyermark, 78, 80	See Chapter 7
Steyermark and Kaup, 81	
Van Slyke and Neill, 98	
Warburg, 101	

REFERENCES

1. Acculate, Anachemia Chemicals, Ltd., Montreal, Quebec and Champlain, New York.
2. Aquist, S. E. G., *Acta Physiol. Scand.*, **13**, 297 (1947).
3. Austin, J. H., *J. Biol. Chem.*, **61**, 345 (1924).
4. Barcroft, J., *J. Physiol. (London)*, **37**, 12 (1908).
5. Barcroft, J., and Haldane, J. S., *J. Physiol. (London)*, **28**, 232 (1902).
6. Battley, E. H., *J. Biol. Chem.*, **226**, 237 (1957).
7. Bhatty, M. K., *Analyst*, **82**, 458 (1957).
8. Boos, R. N., Jones, S. L., and Trenner, N. R., *Anal. Chem.*, **28**, 390 (1956).
9. Borsook, H., and Dubnoff, J. W., *J. Biol. Chem.*, **131**, 163 (1939).
10. Braganca, B. M., Quastel, J. H., and Sucher, R., *Arch. Biochem. Biophys.*, **52**, 18 (1954).
11. Burr, J. G., Jr., *Anal. Chem.*, **26**, 1395 (1954).
12. Carson, W. N., Jr., *Anal. Chem.*, **23**, 1016 (1951).
13. Chapon, L., *Bull. soc. chim. biol.*, **37**, 171 (1955); *Anal. Abstr.*, **2**, No. 2821 (1955).
14. Charest, M. P., Koch, P., and Gagnon, A., *Rev. can. biol.*, **9**, 422 (1951).
15. Corning Glass Works, Corning, New York.
16. Doherty, D. G., and Ogg, C. L., *Ind. Eng. Chem., Anal. Ed.*, **15**, 751 (1943).
17. Farrington, P. S., Niemann, C., and Swift, E. H., *Anal. Chem.*, **21**, 1423 (1949).
18. Fisher Scientific Company, New York, and Pittsburgh, Pennsylvania.
19. Gabourel, J. D., Baker, M. J., and Koch, C. W., *Anal. Chem.*, **27**, 795 (1955).
20. Gora, E. K., and Hickey, F. C., *Anal. Chem.*, **26**, 1158 (1954).
21. Grant, J., "Quantitative Organic Microanalysis, Based on the Methods of Fritz Pregl," 4th ed., Blakiston, Philadelphia, Pennsylvania, 1946.
22. Grant, J., "Quantitative Organic Microanalysis," 5th ed., Blakiston, Philadelphia, Pennsylvania, 1951.
23. Hankes, L. H., *Anal. Chem.*, **27**, 166 (1955).
24. Harington, C. R., and Van Slyke, D. D., *J. Biol. Chem.*, **61**, 575 (1924).
25. Harper, P. V., Jr., Neal, W. B., Jr., and Rogers, G. R., *J. Lab. Clin. Med.*, **36**, 321 (1950).
26. Hartman-Leddon Company, Philadelphia, Pennsylvania.

27. Heald, P. J., *Biochem. J.*, **49**, 684 (1951).
28. Hershenson, H. M., and Hume, D. N., *Anal. Chem.*, **29**, 16 (1957).
29. Hillenbrand, E. F., Jr., and Pentz, C. A., in "Organic Analysis" (J. Mitchell, Jr., I. M. Kolthoff, E. S. Proskauer, and A. Weissberger, eds.), Vol. III, p. 129, Interscience, New York, 1956.
30. Hoagland, C. L., *J. Biol. Chem.*, **136**, 543 (1940).
31. Hoagland, C. L., *J. Biol. Chem.*, **136**, 553 (1940).
32. Hockenhull, D. J. D., *Biochem. J.*, **46**, 605 (1950).
33. Holt, B. D., *Anal. Chem.*, **27**, 1500 (1955).
34. Hurka, W., *Mikrochemie ver. Mikrochim. Acta*, **31**, 83 (1944).
35. Hussey, A. S., and Maurer, J. E., *Anal. Chem.*, **24**, 1642 (1952).
36. Kainz, G., *Mikrochim. Acta*, p. 349 (1953).
37. Kainz, G., and Huber, H., *Mikrochim. Acta*, p. 51 (1959).
38. Kainz, G., and Huber, H., *Mikrochim. Acta*, p. 337 (1959).
39. Kainz, G., and Huber, H., *Mikrochim. Acta*, p. 38 (1960).
40. Kainz, G., Huber, H., and Kasler, F., *Mikrochim. Acta*, p. 744 (1957).
41. Kainz, G., and Kasler, F., *Mikrochim. Acta*, p. 62 (1960).
42. Kainz, G., and Schöller, F., *Biochem. Z.*, **327**, 292 (1955).
43. Kainz, G., and Schöller, F., *Naturwissenchaften*, **42**, 209 (1955).
44. Kainz, G., and Schöller, F., *Z. physiol. Chem. Hoppe-Seyler's*, **301**, 259 (1955).
45. Keen, R. T., and Fritz, J. S., *Anal. Chem.*, **24**, 564 (1952).
46. Kirk, E., *J. Biol. Chem.*, **106**, 191 (1934).
47. Konikov, A. P., *Biokhimiya*, **12**, 221 (1947).
48. Levene, P. A., and Van Slyke, D. D., *J. Biol. Chem.*, **12**, 301 (1912).
49. Ley, J. de, *Mededel. Vlaam Chem. Ver.*, **12**, 27 (1950).
50. Lindenbaum, A., Schubert, J., and Armstrong, W. D., *Anal. Chem.*, **20**, 1120 (1948).
51. Machlett, E., and Son, New York.
52. McCaldin, D. J., *Chem. Revs.*, **60**, 39 (1960).
53. Mellor, J. W., "A Comprehensive Treatise on Inorganic and Theoretical Chemistry," Vol. 7, p. 459, Longmans, Green, New York, and London, 1928.
54. Merck & Co., Inc., "The Story of the Amino Acids," Rahway, New Jersey, 1940.
55. Milton, R. F., and Waters, W. A., "Methods of Quantitative Microanalysis," Longmans, Green, New York, and Arnold, London, 1949.
56. Milton, R. F., and Waters, W. A., "Methods of Quantitative Microanalysis," 2nd ed., Arnold, London, 1955.
57. Möhle, W., *Chem. Ingr. Tech.*, **22**, 416 (1950).
58. Moubasher, R., and Awad, W. I., *J. Biol. Chem.*, **179**, 915 (1949).
59. Neville, O. K., *J. Am. Chem. Soc.*, **70**, 3499 (1948).
60. Peter, F., *Magyar Kémi. Folyóirat*, **63**, 289 (1957).
61. Peters, J. P., and Van Slyke, D. D., "Quantitative Clinical Chemistry," Vol. II, Williams & Wilkins, Baltimore, Maryland, 1943.
62. Private communications, 1946, 1947.
63. Rappaport, F., Eichhorn, F., and Nutman, M., *Clin. Chim. Acta*, **1**, 305 (1956).
64. Rappaport, F., Eichhorn, F., and Nutman, M., *J. Clin. Pathol.*, **9**, 166 (1956).
65. Roth, H., "Die quantitative organische Mikroanalyse von Fritz Pregl," 4th ed., Springer, Berlin, 1935.
66. Roth, H., "F. Pregl quantitative organische Mikroanalyse," 5th ed., Springer, Wien, 1947.

67. Roth, H., "Pregl-Roth quantitative organische Mikroanalyse," 7th ed., Springer, Wien, 1958.
68. Roth, H., "Quantitative Organic Microanalysis of Fritz Pregl," 3rd ed. (E. B. Daw, trans., 4th German ed.), Blakiston, Philadelphia, Pennsylvania, 1937.
69. Sandkuhle, J., Cunningham, B., and Kirk, P. L., *J. Biol. Chem.*, **146**, 427 (1942).
70. Sargent, E. H., & Co., Chicago, 30, Illinois.
71. Scientific Glass Apparatus Company, Bloomfield, New Jersey.
72. Sendroy, J., Jr., and Liu, S. H., *J. Biol. Chem.*, **89**, 133 (1930).
73. Shimoe, D., *Bunseki Kagaku*, **5**, 517 (1956).
74. Shimoe, D., *Bunseki Kagaku*, **5**, 547 (1956).
75. Shimoe, D., *Bunseki Kagaku*, **5**, 550 (1956).
76. Shimoe, D., *Bunseki Kagaku*, **5**, 617 (1956).
77. Sisco, R. C., Cunningham, B., and Kirk, P. L., *J. Biol. Chem.*, **139**, 1 (1941).
78. Steyermark, Al, *Ind. Eng. Chem., Anal. Ed.*, **17**, 191 (1945).
79. Steyermark, Al, *in* "Organic Analysis" (J. Mitchell, Jr., I. M. Kolthoff, E. S. Proskauer, and A. Weissberger, eds.), Vol. II, p. 1, Interscience, New York, 1954.
80. Steyermark, Al, "Quantitative Organic Microanalysis," Blakiston, Philadelphia, Pennsylvania, 1951.
81. Steyermark, Al, and Kaup, R. R., *Anal. Chem.*, **28**, 924 (1956).
82. Swedish Iron & Steel Corp., Westfield, New Jersey.
83. Thomas, Arthur H., Company, Philadelphia, Pennsylvania.
84. Van Slyke, D. D., *Anal. Chem.*, **26**, 1706 (1954).
85. Van Slyke, D. D., *J. Biol. Chem.*, **9**, 185 (1911).
86. Van Slyke, D. D., *J. Biol. Chem.*, **12**, 275 (1912).
87. Van Slyke, D. D., *J. Biol. Chem.*, **71**, 235 (1927).
88. Van Slyke, D. D., *J. Biol. Chem.*, **83**, 425 (1929).
89. Van Slyke, D. D., *J. Biol. Chem.*, **130**, 545 (1939).
90. Van Slyke, D. D., and Dillon, R. T., *Compt. rend. trav. lab. Carlsberg. Sér. chim.*, **22**, 480 (1936); *Chem. Abstr.*, **32**, 7950 (1938).
91. Van Slyke, D. D., Dillon, R. T., MacFadyen, D. A., and Hamilton, P., *J. Biol. Chem.*, **141**, 627 (1941).
92. Van Slyke, D. D., Folch, J., and Plazin, J., *J. Biol. Chem.*, **136**, 509 (1940).
93. Van Slyke, D. D., and Hawkins, J. A., *J. Biol. Chem.*, **79**, 739 (1928).
94. Van Slyke, D. D., and Hiller, A., *J. Biol. Chem.*, **78**, 807 (1928).
95. Van Slyke, D. D., and Kirk, E., *J. Biol. Chem.*, **102**, 651 (1933).
96. Van Slyke, D. D., and Kugel, V. H., *J. Biol. Chem.*, **102**, 489 (1933).
97. Van Slyke, D. D., and Neill, J. M., *J. Biol. Chem.*, **61**, 523 (1924).
98. Van Slyke, D. D., and Neill, J. M., Catalog No. S-7325, E. H. Sargent & Co., Chicago, 30, Illinois.
99. Van Slyke, D. D., Page, I. H., and Kirk, E., *J. Biol. Chem.*, **102**, 635 (1933).
100. Virtanen, A. I., and Rautanen, N., *Suomen Kemistilehti*, **19B**, 56 (1946).
101. Warburg, O., *Biochem. Z.*, **142**, 317 (1923).
102. West, T. S., *Analyst*, **82**, 459 (1957).
103. Woiwod, A. J., *Biochem. J.*, **45**, 412 (1949).
104. Woiwod, A. J., *Nature*, **161**, 169 (1948).
105. Zeile, K., and Oetzel, M., *Z. physiol. Chem. Hoppe-Seyler's*, **284**, 1 (1949).

CHAPTER 19

Microdetermination of Unsaturation (Double Bonds)—Hydrogen Number

The determination of the amount of unsaturation (double bonds) in an organic compound is accomplished by catalytic hydrogenation carried out at approximately atmospheric pressure using platinic or palladous oxide as the catalyst. The results may be expressed in various ways, e.g., mols of hydrogen consumed per 100 grams of sample, grams of hydrogen consumed per 100 grams of sample or *grams of sample reacting with one mol of hydrogen* (hydrogen number). The theoretical hydrogen number is equal to the molecular weight divided by the number of double bonds per molecule (compare Chapter 15). The reaction may be represented by the following:

$$\ce{\underset{}{\overset{}{C}}=\underset{}{\overset{}{C}} \xrightarrow{H_2} \underset{}{\overset{}{CH}}-\underset{}{\overset{}{CH}}}$$

This procedure, obviously, is applicable only to those compounds which can be hydrogenated at atmospheric pressure and, therefore, the reader is strongly advised to refer to the literature[2,12] on the subject of hydrogenation before attempting a determination on a specific compound.

Quantitative hydrogenation as a means of determining the amount of unsaturation has advantages over the halogenation methods (addition of bromine or iodine to the double bond) which are seldom used because of errors.[32] It is not subject to the errors resulting from the presence of two or more conjugated double bonds or from complex molecular structure.[32] The high ratio of catalyst to sample aids in obtaining complete hydrogenation and in overcoming the effects of poisoning of the catalyst by small amounts of impurities in the sample being analyzed.

Reagents

GLACIAL ACETIC ACID

Reagent grade of glacial acetic acid is used as the solvent, of course, provided that the sample is soluble in it. Otherwise, some other suitable pure solvent must be used, such as ethanol, methanol, etc. With glacial acetic acid, there is less fire hazard than with ethanol and of the above-mentioned three, methanol presents the greatest hazard.

19. Unsaturation (Double Bonds)

CATALYSTS

Platinic oxide,[1,9] palladous oxide,[47] or palladium on carbon (10%),[32] may be used as the catalyst depending upon the nature of the sample. The author *strongly* advises these being purchased, since they are commercially available,* in preference to being prepared by the analyst. At all times, the catalyst should be handled with care due to the fire hazard.

Apparatus

Note: Wherever interchangeable ground joints are not used, connections should be made with plastic tubing—*not rubber.*

HYDROGEN CYLINDER AND REDUCING VALVE

Prepurified hydrogen is required.

NITROGEN CYLINDER AND REDUCING VALVE

Prepurified nitrogen is *desirable, but not necessary.* It is good practice to sweep the apparatus with nitrogen, following the determination, before opening the reaction flask containing catalyst and solvent to the air to reduce fire hazard (see photograph of author's setup, Fig. 198).

FRIEDRICH BOTTLES[51]

These are used as pressure regulators (see Fig. 198). They are the same as described in Chapter 14 (see Fig. 174). The overflow tubes should be vented to an exhaust hood or to the outdoors.

SCRUBBING TOWERS

Scrubbing towers filled with both Ascarite[51] and Anhydrone[51] are used in the purification train (see Fig. 198).

PREHEATER TYPE MICROCOMBUSTION FURNACE[51]

A preheater type of microcombustion furnace (see Fig. 173, Chapter 14) is used in the purification train (see Figs. 198 and 199).

PURIFICATION TUBE

A section of combustion tube with ground joints attached and containing either platinized asbestos (Chapter 9) or platinum contact stars (Fig. 152, Chapter 10) heated to 650° C. in the preheater type of microcombustion furnace is used as part of the purification train.

* American Platinum Works, Newark, New Jersey.

BUBBLE COUNTER-U-TUBE

This is identical with that used for the carbon-hydrogen determination (Chapter 9, Fig. 120). It should, however, have interchangeable ground joints attached to both side arms as shown in Fig. 199. The filling (Ascarite, Anhydrone, sulfuric acid) is identical to that described in Chapter 9.

HYDROGENATION APPARATUS[32,51]

The apparatus used is shown in Figs. 198 and 199. The purification train is attached to a tube, F, for saturating the hydrogen with the solvent, the gas

FIG. 198. Microhydrogenation apparatus. Photograph showing units of purification train and attachment of nitrogen source for sweeping purposes to reduce fire hazard.

passing through a sintered glass plate immersed to a depth of about 2.5 cm. in the solvent. This tube is connected by means of an interchangeable ground joint to the burette-manometer combination, at the top of which are two three-way stopcocks, one above the other, 1 and 2, respectively. All connections, other than by means of interchangeable ground joints are affected by means of plastic tubing. The burette, G, and the manometer, H, have parallel and co-

19. Unsaturation (Double Bonds)

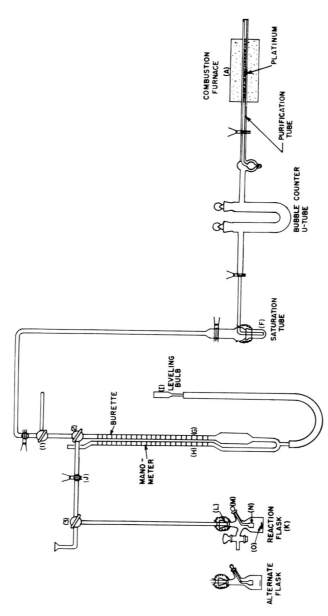

FIG. 199. Diagram of microhydrogenation apparatus. [Combustion furnace (A) shown is preheater type of microcombustion furnace.]

inciding graduations, with 0.02-ml. intervals from 0 to 7 ml. The mercury in the burette, *G,* and the manometer, *H,* is leveled by raising or lowering the leveling bulb, *I,* attached to a rack and pinion device, as shown in Fig. 198 (also compare Fig. 202, Chapter 20). The reaction unit (flask and connecting tube-stopcock) and burette, *G,* are connected by means of ball and socket joint ($ 12/2), *J,* held by a suitable clamp. The reaction flask (20 ml. capacity), *K,* and connecting tube-stopcock, *3,* are connected by means of an interchangeable ground joint (⚜ 14/20), *L.* The side arm and stopper, *M,* are prepared from members of an interchangeable ground joint (⚜ 10/12). The stopper, which extends into the neck of the flask, has a groove near the tip and perpendicular to its long axis. [As an optional attachment, the flask may be equipped with a No. 2 stopcock attached to the neck (Fig. 199), so that tubing may be connected to carry off hydrogen used during the sweep out period]. The sample cup, *N,* is hung on the stopper, *M,* and is dropped to the bottom of the flask by turning the stopper. The sample cup is prepared from aluminum rod and is 8 mm. in length and 6 mm. in diameter with a volume of approximately 0.1 ml. A nichrome handle is attached to the cup through two holes drilled on opposite sides. A glass covered magnetic stirring bar, *O,* is used in the reaction flask. The cup attached to the stopcock, *3,* facilitates cleaning the tube leading to the reaction flask. *All* ground joints should be thoroughly greased with Sisco No. 300 stopcock grease (see Chapter 7).

[*Note:* If nitrogen is used for sweeping purposes at the end of the determination, the stopcock, *1,* is attached to the source of purified nitrogen (see Fig. 198); otherwise, this stopcock is used for venting hydrogen to the atmosphere during initial sweep outs.]

THERMOMETER

Anschütz, room temperature range.

Procedure[32]

Four milliliters of glacial acetic acid (or other suitable solvent), about 20 mg. of the catalyst, and the stirring bar are placed in the reaction flask, *K* (Fig. 199). (*Note:* Whatever solvent is used must also be placed in the vessel which saturates the hydrogen with the solvent.) The purification train is purged by allowing the hydrogen to vent through the side arm of the stopcock, *1.* (Venting should be through a plastic tube either into an exhaust hood or to the outdoors.) Then the stopcock, *1,* is turned so to be open to both the purification train and the stopcock, *2,* but *closed* to the atmosphere. Enough sample is weighed in the cup, *N,* to absorb 3–5 ml. of hydrogen in the reaction. The cup is then *suspended* in the flask by means of the groove in the stopper, M. The reaction flask is then attached to the apparatus and held in place by means of springs. The position of the stopcock, 3, should be such that it is open to both the stopcock, 2, and the reaction flask, K, but closed to the cup above the stopcock, 3. The burette, G, is filled with mercury by raising the leveling bulb, I, and by manipulation, if necessary, of the stopcock, 2. (The burette should be filled up to the bore of the stopcock, but no mercury should pass into the bore.) The stopcock, 2, is then closed to the burette and opened to *both* the reaction flask (through the stopcock, 3) and the stopcock, 1. The stopper, M, is loosened and hydrogen allowed to sweep through the flask at the rate of 25–35 ml. per minute for about 15 minutes. (If the reaction flask is provided with a stopcock, such as shown in Figs. 198 and 199, the stopper, M, is kept in place and the sweeping is done through this stopcock instead. The presence of this stopcock makes it possible to sweep out into an exhaust.) The stopper, M (or stopcock on the reaction flask) is closed making certain that the sample does not drop into the flask. The stopcock, 2, is turned to connect the burette, G, *and* the reaction flask to the purification train and the burette is filled with hydrogen by lowering the leveling bulb, I. The stopcock, 1, is then turned slightly (about one-eight turn) to close all of its openings, and the stopcock, 2, is turned so that the burette, G, is connected only to the reaction flask unit (the stopcock, 2, is open to the burette and reaction flask, but closed to the stopcock, 1). The hydrogen in the burette-reaction flask unit is placed under a few millimeters of positive pressure by slightly raising the leveling bulb, I, and then the excess pressure is released by *momentarily* turning the stopcock, 3, so that it connects the burette-reaction flask unit with the atmosphere (the stopcock, 3, open to the stopcock, 2, the reaction flask *and* the cup above the stopcock, 3). The stopcock, 3, is returned to its original position—open to both the reaction flask and the burette, but closed to the cup above the stopcock, 3. The magnetic stirrer is placed under the reaction flask and the acetic acid and catalyst stirred until there is no more diminution of the gas volume in the burette and then stirred for 5 minutes longer to make certain that the absorption of hydrogen by the catalyst is complete. (If palladium on carbon is used as the catalyst, the diminution is practically nothing, but if either oxide, platinic or palladous, is used, this must first be reduced to the metal, and the uptake of hydrogen will depend upon the amount of catalyst in the flask. The diminution might be so great that it is necessary to again fill the burette with hydrogen from the purification train as described above, before proceeding with the analysis.) The magnetic stirrer then is *removed* from under the reaction flask and the apparatus allowed to stand undisturbed for 15 minutes. (The stirrer must be removed since this becomes warm during operation.) After the 15-minute period has expired, the levels of the mercury in the burette, G, and in the manometer, H, are made equal (by manipulation of the leveling bulb, I), bringing the hydrogen in the closed system to atmospheric pressure. The volume in the burette, the temperature adjacent to the apparatus, and the atmospheric pressure are recorded.

The sample cup, N, is then dropped into the reaction mixture by turning the stopper, M. The stirrer is replaced under the flask and the contents stirred vigorously until the absorption of hydrogen seems to be complete, and then the stirring is continued for another 15 minutes. During the *entire* procedure, the mercury levels in the burette and in the manometer should not differ by more than one cm. This is accomplished by manipulation of the leveling bulb.

After the reaction is complete, the magnetic stirrer is again removed and the apparatus allowed to equilibrate for 15 minutes. The levels of the mercury in the burette and in the manometer are again made equal (gas at atmospheric pressure) and the volume in the burette, temperature, and atmospheric pressure recorded. (See below for determination of *total volume* of flask and parts.)

Calculation:

Several corrections must be made. The barometer reading should be corrected as dicussed in Chapter 7 (subtracting 3 mm.). Since a solvent (acetic acetic) is in the closed system, its vapor pressure at the recorded temperature must also be subtracted from the barometric reading because the pressure exerted by the *hydrogen* is the difference. Since the volume of the reaction flask, connecting parts, *and* portion of the burette *above* the zero mark are also affected by changes in temperature and pressure, this combined volume must be considered *if* a change of temperature and/or atmospheric pressure occur(s) during the analysis. (For obvious reasons, a well air-conditioned room is the best environment for the apparatus.) This volume is determined in the following manner. After the final readings of the temperature, atmospheric pressure, and burette have been made, the leveling bulb is raised (or lowered, if desired) so as to cause a difference in the levels of the mercury in the burette and the manometer of 20–30 mm. (*Note:* The graduations on the burette and manometer coincide with each other, *but* these graduations are *milliliters* on the burette and only by coincidence would be in *millimeters*. Therefore, a millimeter rule must be used to obtain the difference in levels.) The new burette reading is recorded. Where

V_i = burette reading, before reaction of sample (mercury in burette and manometer at *same* level)

T_i = temperature, before reaction of sample

P_i = barometer reading minus 3 mm., before reaction of sample

P_{si} = vapor pressure of solvent at temperature T_i (see below)

V_f = burette reading, after reaction of sample (mercury in burette and manometer at *same* level)

19. Unsaturation (Double Bonds)

T_f = temperature after reaction of sample
P_f = barometer reading minus 3 mm., after reaction of sample
P_{sf} = vapor pressure of solvent at temperature T_f (see below)
V^1 = burette reading, after reaction of sample with mercury in the burette and manometer at *different* levels
p^1 = difference in mercury levels of the burette and manometer to obtain V^1
V_t = total *gas* volume of the reaction flask unit plus the volume of the burette above the zero mark
V_i^0 = volume of H_2 at 0° C., 760 mm. before reaction of sample
V_f^0 = volume of H_2 at 0° C., 760 mm. after reaction of sample

Then

$$\frac{P_f(V_t + V_f)}{T_f} = \frac{(P_f + P^1)(V_t + V^1)}{T_f}$$

or

$$V_t = \frac{P_f V_f - P_f V^1 - P^1 V^1}{P^1}$$

and

$$V_i^0 = \frac{(P_i - P_{si}) \times (V_i + V_t) \times 273.16}{T_i \times 760}$$

and

$$V_f^0 = \frac{(P_f - P_{sf}) \times (V_f + V_t) \times 273.16}{T_f \times 760}$$

and

$\therefore V_f^0 - V_i^0$ = volume of H_2 at 0° C., 760 mm., absorbed by the sample

The amount of hydrogen required per molecule of a substance having one double bond is given by:

M.W. of substance : M.W. of H_2 : : Wt. of sample in grams : Wt. of H_2 used in grams

or M.W. of substance : 2.016 : : Wt. of sample in grams : Wt. of H_2 used in grams

or M.W. of substance : 22415 : : Wt. of sample in grams : Vol. of H_2 used in ml.

or $\quad$ M.W. of substance $= \dfrac{\text{Wt. of substance in grams} \times 22415}{\text{Vol. } H_2 \text{ used in ml.}}$

or* $\quad$ Hydrogen number $= \dfrac{\text{Wt. of substance in mg.} \times 22.415}{\text{Vol. } H_2 \text{ used in ml.}}$

where Hydrogen number = Wt. of sample that will react with 1 mol of H_2

or $\quad$ Hydrogen number $= \dfrac{\text{Wt. sample in mg.} \times 22.415}{V_i^0 - V_f^0}$

* See first paragraph of this chapter.

TABLE 33
Vapor Pressure of Glacial Acetic Acid[a,42]

Temperature (°C.)	Pressure (mm. Hg.)
0	3.3
10	6.4
17	9.2
20	11.7
25	15.2
30	20.6
35	26.6
40	34.8
45	44.7
50	56.6

[a] For obtaining intermediate points, a plot of temperature vs. pressure should be made and the correct value taken from the curve, such as Fig. 200.

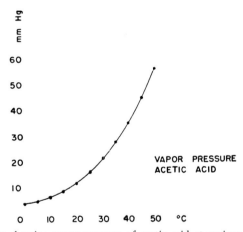

Fig. 200. Curve showing vapor pressure of acetic acid at various room temperatures.

Example:

11.981 mg. of sample in acetic acid was used. The following conditions existed:

V_i = burette reading (mercury levels in burette and manometer equal) before reaction of sample = 5.54 ml.

T_i = temperature before reaction of sample = 24.1° C. or 297.26° K

P_i = barometric reading before reaction of sample = 766.0 mm. uncorrected or 763.0 mm. corrected

P_{si} = vapor pressure of the solvent at 24.1° C. = $\approx$ 15 mm. (from Table 33) (also from Fig. 200)

V_f = burette reading (mercury levels in the burette and manometer equal) after reaction of sample = 3.00 ml.

19. Unsaturation (Double Bonds)

T_f = temperature after reaction of sample = 24.1° C. or 297.26° K

P_f = barometric reading after reaction of sample = 766.0 mm. uncorrected or 763.0 mm. corrected

P_{sf} = vapor pressure of the solvent at 24.1° C. = $\approx$ 15 mm. (from Table 33) (also from Fig. 200)

V^1 = burette reading, after reaction of sample with mercury in the burette and manometer at different levels = 2.00 ml.

P^1 = difference in mercury levels of the burette and manometer to obtain V^1 = 29.0 mm.

$$\therefore V_t = \frac{763.0 \times 3.00 - 763.0 \times 2.00 - 29.0 \times 2.00}{29.0}$$

$$= 24.31 \text{ ml.}$$

$$\therefore V_i^0 = \frac{(763.0 - 15) \times (5.54 + 24.31) \times 273.16}{297.26 \times 760.0}$$

$$= \frac{748.0 \times 29.85 \times 273.16}{297.26 \times 760.0} = 27.00 \text{ ml.}$$

and

$$V_f^0 = \frac{(763.0 - 15) \times (3.00 + 24.31) \times 273.16}{297.26 \times 760.0}$$

$$= \frac{748.0 \times 27.31 \times 273.16}{297.26 \times 760.0} = 24.70 \text{ ml.}$$

$$\therefore V_f^0 - V_i^0 = 2.30 \text{ ml.}$$

and

$$\therefore \text{Hydrogen number} = \frac{11.981 \times 22.415}{2.30} = 116.8$$

TABLE 34
Additional Information on References* Related to Chapter 19

In addition to the procedure presented in the preceding pages of this chapter, the author wishes to call to the attention of the reader the references listed in Table 34. (See statement at top of Table 4 of Chapter 1, regarding completeness of this material.)

Books

Adkins, 2
Ellis, 12
Fritz and Hammond, 16
Grant, 18, 19
Milton and Waters, 29, 30
Roth, 38–41
Siggia, 48
Steyermark, 50

Ultramicro-, submicro- methods

Grunbaum and Kirk, 20

General, miscellaneous

Chaphekar and Gore, 10
Ellis, 12
Polgár and Jungnickel, 35
Steyermark, 50

* The numbers which appear after each entry in this table refer to the literature citations in the reference list at the end of the chapter.

TABLE 34 (*Continued*)

Apparatus

Buděšínský, 7
Colson, 11
Erdos, 14
Flaschka and Hochenegger, 15
Johns and Seiferle, 21
Kuhn and Möller, 23
Mead and Howton, 26
Miller and DeFord, 28
Noller and Barusch, 31
Parrette, 33
Prater and Haagen-Smit, 36
Savacool and Ullyot, 44
Schöniger, 45
Seaman, 46
Vandenheuvel, 53
Weygand and Werner, 57
Zaugg and Lauer, 59

Use of Barcroft-Warburg apparatus

Barcroft, 3
Barcroft and Haldane, 4
Milton and Waters, 29
Warburg, 56

Volatile compounds

Engelbrecht, 13

Spectrophotometric end point

Miller and DeFord, 27

Karl Fischer end point

Seaman, 46

Bromination, halogenation, iodometric methods (iodine value)

Brignoni, 6
Buděšínský, and Vaníčkova, 8
Giraut-Erler and Grimberg, 17
Grunbaum and Kirk, 20
Komori, 22
Kuhn and Roth, 24
Miller and DeFord, 27
Phillips and Wake, 34
Rossmann, 37
Ruziczka, 43
Smits, 49
Unger, 52
Viollier, 54

Coulometric methods

Leisey and Grutsch, 25
Walisch and Ashworth, 55

REFERENCES

1. Adams, R., and Shriner, R. L., *J. Am. Chem. Soc.*, **45**, 2171 (1923).
2. Adkins, H., "Reactions of Hydrogen with Organic Compounds over Copper-Chromium Oxide and Nickel Catalysts," Univ. Wisconsin Press, Madison, Wisconsin, 1937.
3. Barcroft, J., *J. Physiol. (London)*, **37**, 12 (1908).
4. Barcroft, J., and Haldane, J. S., *J. Physiol. (London)*, **28**, 232 (1902).
5. Brancone, L. M., Personal communication, 1947.
6. Brignoní, J. M., *Quím. ind. (Montevideo)*, **1**, 13 (1946).
7. Buděšínský, B., *Collection Czechoslov. Chem. Communs.*, **24**, 2948 (1959).
8. Buděšínský, B., and Vaníčková, E., *Československ. farm.*, **6**, 305 (1957).
9. Carothers, W. H., and Adams, R., *J. Am. Chem. Soc.*, **45**, 1071 (1923).
10. Chaphekar, M. R., and Gore, T. S., *Mikrochim. Acta*, p. 664 (1959).
11. Colson, A. F., *Analyst*, **79**, 298 (1954).
12. Ellis, C., "Hydrogenation of Organic Substances, Including Fats and Fuels," 3rd ed., Van Nostrand, New York, 1930.
13. Engelbrecht, R. M., *Anal. Chem.*, **29**, 1556 (1957).
14. Erdos, J., *Mikrochemie ver. Mikrochim. Acta*, **35**, 236 (1950).
15. Flaschka, H., and Hochenegger, M., *Mikrochim. Acta*, p. 586 (1957).
16. Fritz, J. S., and Hammond, G. S., "Qualitative Organic Analysis," Wiley, New York,

19. Unsaturation (Double Bonds)

and Chapman & Hall, London, 1957.
17. Giraut-Erler, L., and Grimberg, J., *Ann. biol. clin. (Paris)*, **7**, 127 (1949).
18. Grant, J., "Quantitative Organic Microanalysis, Based on the Methods of Fritz Pregl," 4th ed., Blakiston, Philadelphia, Pennsylvania, 1946.
19. Grant, J., "Quantitative Organic Microanalysis," 5th ed., Blakiston, Philadelphia, Pennsylvania, 1951.
20. Grunbaum, B. W., and Kirk, P. L., *Mikrochemie ver. Mikrochim. Acta,* **39**, 268 (1952).
21. Johns, I. B., and Seiferle, E. J., *Ind. Eng. Chem., Anal. Ed.,* **13**, 841 (1941).
22. Komori, S., *Kôgyô Kagaku Zasshi,* **51**, 120 (1948).
23. Kuhn, R., and Möller, E. F., *Angew. Chem.,* **47**, 145 (1934).
24. Kuhn, R., and Roth, H., *Ber.,* **65**, 1285 (1932).
25. Leisey, F. A., and Grutsch, J. F., *Anal. Chem.,* **28**, 1553 (1956).
26. Mead, J. F., and Howton, D. R., *Anal. Chem.,* **22**, 1204 (1950).
27. Miller, J. W., and DeFord, D. D., *Anal. Chem.,* **29**, 475 (1957).
28. Miller, J. W., and DeFord, D. D., *Anal. Chem.,* **30**, 295 (1958).
29. Milton, R. F., and Waters, W. A., "Methods of Quantitative Microanalysis," Longmans, Green, New York, and Arnold, London, 1949.
30. Milton, R. F., and Waters, W. A., "Methods of Quantitative Microanalysis," 2nd ed., Arnold, London, 1955.
31. Noller, C. R., and Barusch, M. R., *Ind. Eng. Chem., Anal. Ed.,* **14**, 907 (1942).
32. Ogg, C. L., and Cooper, F. J., *Anal. Chem.,* **21**, 1400 (1949).
33. Parrette, R. L., *Anal. Chem.,* **26**, 237 (1954).
34. Phillips, W. M., and Wake, W. C., *Analyst,* **74**, 306 (1949).
35. Polgár, A., and Jungnickel, J. L., in "Organic Analysis" (J. Mitchell, Jr., I. M. Kolthoff, E. S. Proskauer, and A. Weissberger, eds.), Vol. III, p. 203, Interscience, 1956.
36. Prater, A. N., and Haagen-Smit, A. J., *Ind. Eng. Chem., Anal. Ed.,* **12**, 705 (1940).
37. Rossmann, E., *Ber.,* **65B**, 1847 (1932).
38. Roth, H., "Die quantitative organische Mikroanalyse von Fritz Pregl," 4th ed., Springer, Berlin, 1935.
39. Roth, H., "F. Pregl quantitative organische Mikroanalyse," 5th ed., Springer, Wien, 1947.
40. Roth, H., "Pregl-Roth quantitative organische Mikroanalyse," 7th ed., Springer, Wien, 1958.
41. Roth, H., "Quantitative Organic Microanalysis of Fritz Pregl," 3rd ed., (E. B. Daw, trans. 4th German ed.), Blakiston, Philadelphia, Pennsylvania, 1937.
42. Roth, W. A., and Scheel, K., "Landolt-Börnstein, Physikalisch-Chemische Tabellen," 5th ed., p. 1365, Springer, Berlin, 1923.
43. Ruziczka, W., *Mikrochemie ver. Mikrochim. Acta,* **36/37**, 924 (1951).
44. Savacool, R. V., and Ullyot, G. E., *Anal. Chem.,* **24**, 714 (1952).
45. Schöniger, W., *Mikrochemie ver. Mikrochim. Acta,* **38**, 132 (1951).
46. Seaman, W., *Anal. Chem.,* **30**, 1840 (1958).
47. Shriner, R. L. and Adams, R., *J. Am. Chem. Soc.,* **46**, 1683 (1924).
48. Siggia, S., "Quantitative Organic Analysis via Functional Groups," 2nd ed., Wiley, New York, and Chapman & Hall, London, 1954.
49. Smits, R., *Rec. trav. chim.,* **78**, 713 (1959).
50. Steyermark, Al, "Quantitative Organic Microanalysis," Blakiston, Philadelphia, Pennsylvania, 1951.
51. Thomas, Arthur H., Company, Philadelphia, Pennsylvania.

52. Unger, E. H., *Anal. Chem.,* **30**, 375 (1958).
53. Vandenheuvel, F. A., *Anal. Chem.,* **24**, 847 (1952).
54. Viollier, G., *Helv. Physiol. et Pharmacol. Acta,* **7**, C26 (1949).
55. Walisch, W., and Ashworth, M. R. F., *Mikrochim. Acta,* p. 497 (1959).
56. Warburg, O., *Biochem. Z.,* **142**, 317 (1923).
57. Weygand, C., and Werner, A., *J. prakt. Chem.,* **149**, 330 (1937).
58. Willits, C. O., and Ogg, C. L., Personal communication, 1949.
59. Zaugg, H. E., and Lauer, W. M., *Anal. Chem.,* **20**, 1022 (1948).

CHAPTER 20

Microdetermination of Other Groups

For the sake of completeness, a number of group determinations less frequently used will be discussed briefly in this chapter. Where a detailed description is required, the reader is referred to the literature cited. These determinations include both those suitable for specific substances and also a few which the author considers to be apt to present difficulties to the beginner. It is true that some of the latter group are being used on a large scale with excellent results by a few microanalysts who have had a great deal of experience with the particular determinations and who are familiar with "all of the tricks" necessary. However, the author prefers to treat these, too, briefly here.

DETERMINATION OF ALKIMIDE GROUPS (N-METHYL AND N-ETHYL)

The determination of N-methyl and N-ethyl groups (secondary, tertiary amines, etc.)[79,80,173,174,183,189-192,234] is based on the formation of the corresponding quaternary alkyl ammonium iodide when treated with hydriodic acid. The quaternary ammonium compound is decomposed by heating to yield the corresponding alkyl iodide which in turn is determined either gravimetrically or iodometrically as described for the determination of methoxyl and ethoxyl (Chapter 16). The reactions are represented by the following:

$$>NR + HI \rightarrow >NR.HI$$

$$>NR.HI \xrightarrow[360° C.]{\Delta} RI$$

Then:

  Gravimetrically

$$RI \xrightarrow{AgNO_3} \underline{AgI}$$

or

  Iodometrically

$$RI + Br_2 \rightarrow RBr + IBr$$
$$IBr + 3H_2O + 2Br_2 \rightarrow HIO_3 + 5HBr$$
$$HIO_3 + 5HI \rightarrow 3I_2 + 3H_2O$$
$$3I_2 + 6Na_2S_2O_3 \rightarrow 6NaI + 3Na_2S_4O_6$$

The determination works quite well for some substances, such as, atropine,[183] cocaine hydrochloride,[183] and theobromine.[183] However, its success depends upon the splitting off, quantitatively, of the alkyl iodide when the quaternary ammonium compound is subjected to pyrolysis, and many substances do not act thusly. The author has had experience with a number of such examples, the purity of the compounds under test having been established by elementary analyses.[234,235] Consequently, in his opinion, this possibility must always be considered when performing this determination.

Alkoxyl and alkimide groups may be determined simultaneously, the alkoxyl groups first, during the initial stages previous to pyrolysis. However, the possibility of alkimide groups splitting off at *low* temperatures cannot be ignored. Under such conditions, an alkimide group would be mistaken for an alkoxyl group (compare Chapter 16).

Reagents

The reagents are the same as required for the alkoxyl determination (Chapter 16) plus the two following:

AQUEOUS SOLUTION OF GOLD CHLORIDE, 5%

This acts catalytically to accelerate the splitting off of the alkyl group.[183,234] It is added to the contents of the reaction flask.

AMMONIUM IODIDE CRYSTALS

Reagent grade of ammonium iodide crystals are added to the reaction flask.

Apparatus

The apparatus[65,66,68,79,80,173,174,183,189-192,234] generally used is that designed by Friedrich[65,66,68] (Fig. 201). The apparatus consists of a reaction flask, condenser tube, AA', scrubber and delivery tube. [For the iodometric procedure, however, it is better to replace the test tube receiver with the volumetric receiver (compare Chapter 16).] The three-way stopcock, BB', makes it possible to suck the condensed hydriodic acid back into the reaction flask for repeated pyrolysis.

The alkoxyl apparatus[233,234] (Fig. 182, Chapter 16) may be used in place[234] of the Friedrich setup. The second reaction flask, b (Fig. 182), serves the same purpose as that of the condenser tube, AA' (Fig. 201). The function of the three-way stopcock, BB' (Fig. 201), is imitated by lowering the second reaction flask, b (Fig. 182), just enough to break the seal so that air may enter through the top during the sucking back of the hydriodic acid.

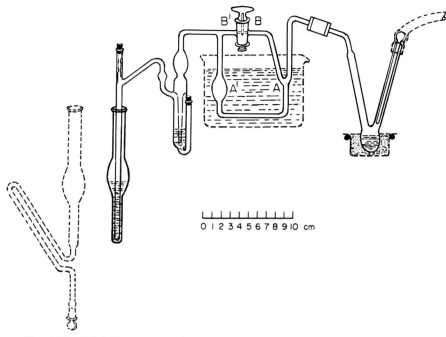

Fig. 201. Friedrich apparatus for determination of alkimide groups (N-ethyl and N-methyl).

The reaction flask is heated by means of a suitable bath so that its contents may be raised eventually to a temperature of 360° C.

Procedure

In the procedure[79,80,173,174,183,189–192,234] the scrubber and receiver are charged exactly as described for the alkoxyl determination. The reaction flask is charged also as for the above determination and in addition there are added one to two drops of gold chloride solution and ammonium iodide crystals equal to about twenty times the weight of the sample. If the Friedrich apparatus is used, the condenser tube, AA', is immersed in a bath of hot water (90° C.). (If the Steyermark apparatus[233,234] is used, proceed exactly as described in Chapter 16.)

A slow stream of carbon dioxide is passed through exactly as described in Chapter 16. If alkoxyl groups are present in addition to the alkimide, the mixture in the reaction flask is allowed to remain at room temperature for 30 minutes after which it is boiled gently for at least one hour (see Chapter 16). After this time, the alkyl iodide which passed over into the receiver is determined (either weighed as silver iodide or the liberated iodine titrated with thiosulfate). (Note: It is advisable to *assume* that alkoxyl is always present, otherwise these groups may be mistaken for alkimide.)

After the alkoxyl groups have been removed, the temperature of the bath under the reaction flask is slowly raised to 360° C. During this period, the hydriodic acid is distilled over into the condenser tube, AA' (Fig 201) [or the second reaction flask, b (Fig. 182, Chapter 16)]. The dry residue in the reaction flask is heated at 360° C. for a few minutes after which it is allowed to cool to room temperature, while the carbon dioxide is passed through the system. The receiver is removed and the alkyl iodide obtained during the first distillation is determined either gravimetrically or iodometrically.

A new receiver is placed under the delivery tube. The three-way stopcock, BB' (Fig. 201), is turned so that the bulb A' is connected to the air—the vertical bore faces the scrubber. (If the Steyermark apparatus is used, the second reaction flask, b (Fig. 182), is lowered just enough to break the seal.) The source of carbon dioxide is disconnected and mild suction applied instead, whereby the hydriodic acid is sucked back into the reaction flask onto the dry residue. About 0.5 ml. of fresh hydriodic acid is run in through the bore of the stopcock (or the flask, b, Fig. 182) and then sucked into the reaction flask as a wash. The stopcock, BB', is returned to its original position as shown in Fig. 201, connecting the reaction flask and delivery tube via the scrubber. The source of carbon dioxide is reconnected and the contents of the reaction flask reheated to 360° C. and the alkyl iodide from the second distillation determined. A third, fourth, etc., distillation is done, if necessary, until there is no further precipitate of silver iodide or iodine liberated. (*Note:* The gravimetric correction of 0.06 mg. silver iodide per ml. of silver nitrate used is applied—see Chapter 16.[65,66,68,173,234])

Calculations:

Factors:

Gravimetric (see Chapter 23)

for CH_3, 0.06404
for C_2H_5, 0.1238

Iodometric

for CH_3, 1 ml. of $0.01N$ $Na_2S_2O_3$ is equivalent to 0.02506 mg. of CH_3
for C_2H_5, 1 ml. of $0.01N$ $Na_2S_2O_3$ is equivalent to 0.04844 mg. of C_2H_5

Gravimetric

$$\frac{\text{Wt. of AgI} \times \text{factor} \times 100}{\text{Wt. sample}} = \% \ CH_3, C_2H_5$$

Iodometric

$$\frac{\text{ml. of } 0.01N \ Na_2S_2O_3 \times \text{factor} \times 100}{\text{Wt. sample}} = \% \ CH_3, C_2H_5$$

Examples:

a. 9.087 mg. of silver iodide, corrected, is obtained from a 5.603-mg. sample containing N-CH₃ groups.

$$\therefore \frac{9.087 \times 0.06404 \times 100}{5.603} = 10.39\% \text{ CH}_3$$

b. 6.608 mg. of silver iodide, corrected, is obtained from a 6.311-mg. sample containing N-C₂H₅ groups.

$$\therefore \frac{6.608 \times 0.1238 \times 100}{6.311} = 12.96\% \text{ C}_2\text{H}_5$$

c. 10.00 ml. of 0.01N Na₂S₂O₃ is required to titrate the iodine liberated on analyzing a 4.630-mg. sample containing N-CH₃ groups.

$$\therefore \frac{10.00 \times 0.02506 \times 100}{4.630} = 5.41\% \text{ CH}_3$$

d. 6.37 ml. of 0.01N Na₂S₂O₃ is required to titrate the iodine liberated on analyzing a 4.002-mg. sample containing N-C₂H₅ groups.

$$\therefore \frac{6.37 \times 0.04844 \times 100}{4.002} = 7.71\% \text{ C}_2\text{H}_5$$

DETERMINATION OF HYDROXYL GROUPS
MICROMETHOD[180,234]

The microdetermination of hydroxyl groups in organic compounds is based on the esterification with acetic anhydride-pyridine mixture.[176,180,236] Following esterification the excess acetic anhydride is hydrolyzed and the resulting acetic acid titrated with standard alkali.

$$\text{ROH} \xrightarrow[\substack{\text{CH}_3\text{CO} \\ \text{CH}_3\text{CO}}\text{O}]{\text{Excess}} \text{R—O—COCH}_3$$

Then

$$\substack{\text{CH}_3\text{CO} \\ \text{CH}_3\text{CO}}\text{O} + \text{H}_2\text{O} \rightarrow 2\text{CH}_3\text{COOH}$$
Excess

and

$$2\text{CH}_3\text{COOH} + 2\text{NaOH} \rightarrow 2\text{CH}_3\text{COONa} + 2\text{H}_2\text{O}$$

Reagents

ACETIC ANHYDRIDE
This should be redistilled and be acetate free.

PYRIDINE
This must be redistilled and water free.

STANDARD SODIUM HYDROXIDE, 0.04N

This should be carbonate free, made from concentrated alkali (see Chapter 18) and standardized with potassium acid phthalate (compare Chapter 5).

PHENOLPHTHALEIN INDICATOR

See Chapter 5.

Apparatus[180,234]

CAPILLARY TUBES

Capillary tubes, sealed at one end, 3 mm. in diameter and 6 cm. in length are prepared from glass tubing.

GLASS PLUNGERS

Small glass rods, 1 mm. O.D. and 5 mm. long are used for stirring the contents of the reaction tubes after sealing.

CENTRIFUGE

A microcentrifuge is required for inserting liquid samples and the acetic anhydride and pyridine (compare Chapter 3).

GRADED GLASS TUBES AND PLUNGER COMBINATION

(See Rast tubes, Chapter 21.) These are used for introducing solid samples.

Procedure[180]

Two to 10 mg. of sample is used. Liquid samples are introduced into the tubes by the technique described in Chapter 3 in connection with volatile liquid samples. Solids are introduced by the technique described for the Rast method for the determination of molecular weight (Chapter 21). After introducing the sample and obtaining its weight, 20–25 mg. ($>$ 100% excess) of acetic anhydride is added to the capillary which is then reweighed. Four to six drops of pyridine are added and a small glass plunger (1 $\times$ 5 mm.) for stirring purposes. The tube is sealed, shaken, and set aside for 24 hours. A blank is run simultaneously.

After 24 hours the tube is placed in a 50-ml. Erlenmeyer flask, covered with 5 ml. of distilled water, and broken by means of a heavy glass rod. The released acetic acid is titrated with standard alkali (0.04N).

Calculation:

$$\frac{(\text{m.e. of anhydride used} - \text{m.e. of acid found}) \times 1700}{\text{mg. of sample}} = \% \text{ OH}$$

where m.e. of acid found = ml. × normality, m.e. of anhydride used = mg. of anhydride × ratio × normality (where ratio = milliliters of base required to neutralize acid derived from one mg. of anhydride).

SEMIMICROMETHOD[176]

Ogg, Porter, and Willits[176] described a semimicromethod for the determination of hydroxyl groups. They used one volume of acetic anhydride and three volumes of pyridine as the acetylating agent. Their apparatus consists of a glass-stoppered, pear-shaped flask which is designed so that small volumes of the sample and the reagent are held in the conical tip, insuring complete mixing. The reagent is measured from an S-shaped capillary burette. The alkali used for the titration is $0.1N$. These authors use a mixed indicator solution[125] composed of one part of 0.1% aqueous solution of cresol red neutralized with sodium hydroxide and three parts of 0.1% thymol blue neutralized with sodium hydroxide. They also correct for any free acid present in the sample.

MICRODETERMINATION OF ACTIVE HYDROGEN

The determination of active hydrogen may be accomplished by reaction with a Grignard reagent[187] resulting in the evolution of methane which is measured.[59,93,94,158,162,173,174,190-193,225,248,265-269] The reaction may be represented by the following:

$$RH \xrightarrow{CH_3MgI} CH_4$$

Many groups that react with the Grignard reagent, however, do not yield methane. Therefore, the reader is referred to the vast amount of literature on the subject of the Grignard reaction before attempting these determinations. The method[225] is adaptable also to those groups which do not yield methane and for these a known quantity of the reagent is used after which the excess is treated with a known amount of aniline which reacts to yield methane. The apparatus used is shown in Fig. 202. It is available commercially.[247] The microhydrogenation apparatus (Chapter 19) is a modification of this piece.

The determination of active hydrogen may be accomplished also by treatment of the compound with heavy water, D_2O, by the method of Rachele and Melville[186] (compare Clarke, Johnson, and Robinson[46]). A transfer occurs and the deuterium replaces the active hydrogen. The excess heavy water now also containing the water from the active hydrogen is removed by distillation *in vacuo*. The compound containing active deuterium is treated with ordinary water and again a transfer occurs, ordinary hydrogen replacing the active deuterium. The resulting water containing the heavy water from the active deuterium is distilled off *in vacuo* and the amount of heavy water present determined by means of the falling drop[47] method (compare K. Fenger-Eriksen, A. Krogh, and H. Ussing[58]).

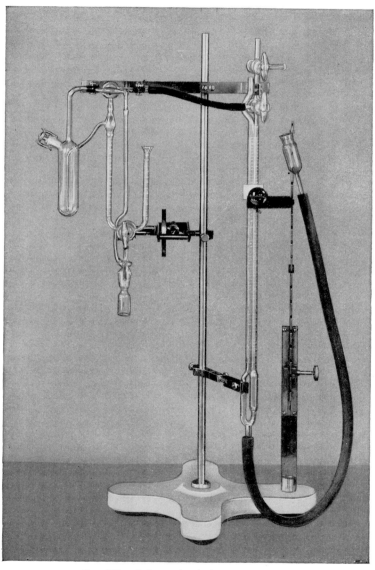

FIG. 202. Soltys active hydrogen apparatus.

20. Other Groups

This second method has the advantage that the substance is recovered quantitatively unchanged. For a less accurate determination the substance containing active deuterium may be weighed and the amount of active hydrogen determined by the difference in the weights of the compound before and after treatment with heavy water.[259] W. H. Hamill[92] described a modification of this method which distinguishes active from labile hydrogen.

Lithium aluminum hydride is becoming quite popular as a reagent for the determination of active hydrogen. Table 35 lists numerous references to articles describing the use of this substance.

MISCELLANEOUS MICRODETERMINATIONS

For the numerous determinations listed below, the reader is referred to the literature in Table 35. The seventh German edition of the Roth book[191] devotes considerable space to the following determinations.

Acetylene linkage
Acid amides
Amines (secondary and tertiary)
Carbonyl (aldehydes and ketones)
Dialkyl sulfide
Disulfide
Dithiocarbamate
Gas analysis
Isocyanate
Isopropylidene
Isothiocyanate
Mercapto
Methyl groups attached to carbon
Nitrates
Nitro
Nitroso
Peroxide
Saponification number
Thiuramdisulfide
Vinyl ether

TABLE 35

ADDITIONAL INFORMATION ON REFERENCES* RELATED TO CHAPTER 20

In addition to the procedures described in the preceding pages of this chapter, the author wishes to call to the attention of the reader a number of other references on these subjects as well as on other determinations less frequently performed. All of these are listed in Table 35. (See statement at top of Table 4 of Chapter 1, regarding completeness of this material.)

Books

Belcher and Godbert, 12
Cheronis, 39
Cheronis and Entrikin, 40
Clark, E. P., 43
Clark, S. J., 44
Fritz and Hammond, 69
Grant, 79, 80
Milton and Waters, 163
Niederl and Niederl, 173, 174
Pregl, 183
Roth, 189–192
Siggia, 214
Steyermark, 234

Reviews

Becker, 6
Kirsten, 123
Ma, 148, 149
Olleman, 177
Prévost and Souchay, 185
Sobotka, 223
Veibel, 252

Ultramicro-, submicro-methods

Belcher, Bhatty, and West, 11
Bhatty, 21
Garbers, Schmid, and Karrer, 71
Gey and Schön, 73
Hack, 88
Holter and Løvtrup, 100
Kirsten, 123
Kolb and Toennies, 129
Rosenberg, Perrone, and Kirk, 188
Vonesch and Guagnini, 253, 255

Nitrates

Becker and Schaefer, 7

Nitroso compounds

Anger, 1
Becker and Schaefer, 7
Jucker, 108
Juvet, Twickler, and Afremow, 111
Ma and Earley, 151

Nitriles

Berinzaghi, 18

Nitro-compounds

Becker and Schaefer, 7
Ehlers and Sease, 54
Hörmann, Lamberts, and Fries, 101
Jucker, 108
Juvet, Twickler and Afremow, 111
Ma and Earley, 151
Perpar, Tišler, and Vrbaški, 179
Suzuki, Muramoto, Ueno, and
   Sugano, 242

Azo compounds

Juvet, Twickler, and Afremow, 111

Hydroxamic acids, oximes

Vonesch and Guagnini, 253–255

Alkimide (N-methyl, N-ethyl)

Belcher, Bhatty, and West, 11
Bhatty, 21
British Standards Institution, 30
Edlbacher, 53
Franzen, Disse, and Eysell, 61
Franzen, Eysell, and Schall, 62, 63
Franzen and Pauli, 64
Friedrich, 67
Gysel, 85

* The numbers which appear after each entry in this table refer to the literature citations in the reference list at the end of the chapter.

20. Other Groups

TABLE 35 (*Continued*)

Alkimide (N-methyl, N-ethyl) (*Cont.*)
Haas, 86, 87
Kuhn and Roth, 136
Sirotenko, 218
Slotta and Haberland, 219
Sudo, Shimoe, and Tsujii, 239

Carbonyl (aldehydes and ketones)
Bennett, May, and Gregory, 14
Berka and Zyka, 19
Bose, 25, 26
Bragg and Hough, 27
Brandstätter, 28
Buděšínský, 33
Buděšínský and Körbl, 37
Duval and Xuong, 51
Falkenhausen, 57
Genevois, 72
Goltz and Glew, 75
Hunter and Potter, 103
Kirsten, 123
Klimova and Zabrodina, 126, 127
Lieb, Schöniger, and Schiviz-Schivizhoffen, 145
Ma, Logun, and Mazzella, 153
Malmberg, Weinstein, Fishel, and Krause, 156
Maute and Owens, 159
Mitchell, 165
Nakamura, 170
Neuberg and Strauss, 172
Petránek and Večeřa, 181
Petrova and Novikova, 182
Prévost and Souchay, 185
Schöniger and Lieb, 205
Schöniger, Lieb, and Gassner, 207, 208
Sobotka and Trutnovsky, 224
Sozzi, 228
Vonesch and Guagnini, 255
Yamamura, 262

Acetals
Mitchell, 166

Hydroxylamine number
Vonesch and Guagnini, 255

Esters and saponification number
Belcher and Phillips, 13

Esters and saponification number (*Cont.*)
Gey and Schön, 73
Goddu, LeBlanc, and Wright, 74
Gorbach, 76, 77
Grützner and Hintermaier, 82
Hack, 88
Hall and Schaefer, 90
Ketchum, 120
Komori, 130
Lee, 140
Leurquin and Delville, 143
Marcali and Rieman, 157
Mitchell, Smith, and Money, 167
Morgan, 169
Smith, Mitchell, and Billmeyer, 221
Van Etten, 251

Methyl groups attached to carbon (C-methyl)
Calderón Martinez, 38
Garbers, Schmid, and Karrer, 71
Kirsten and Stenhagen, 124
Kuhn and Roth, 135
Roth, 189–192
Schöniger, Lieb and El Din Ibrahim, 206
Sudo, Shimoe, and Tsujii, 240
Tashinian, Baker, and Koch, 244
Wiesenberger, 258

Isopropylidene
Kuhn and Roth, 134

Active hydrogen:
  A. General methods for active hydrogen
Arventiev and Figel, 5
Brown and Hafliger, 31
Buděšínský, 34
Hamill, 92
Higuchi, 95
Kainz, Polansky, Schinzel, and Wesseley, 114
McAlpine and Ongley, 160
Perold and Snyman, 178
Roth, 193
Soltys, 225
Wright, 261

TABLE 35 (*Continued*)

Active hydrogen (*Cont.*)

B. Active hydrogen using Grignard reagent

Kainz, Polansky, Schinzel, and Wesseley, 114
Kohler, Stone, and Fuson, 128
McAlpine and Ongley, 160
Perold and Snyman, 178
Roth, 193
Soltys, 225
Souček, 226, 227
Stevens, 232
Zaugg and Lauer, 264

C. Active hydrogen using lithium aluminum hydride

Arjungi, Kulkarni, and Gore, 3
Colson, 48
Higuchi, 95
Hockstein, 97
Höfling, Lieb, and Schöniger, 98
Kainz, Polansky, Schinzel, and Wesseley, 114
Lieb and Schöniger, 144
Schöniger, 204
Souček, 226
Stefanac, 230
Stenmark and Weiss, 231
Subbo Rao, Shah, and Pansare, 238
Ulbrich and Makeš, 250

D. Active hydrogen using aluminum tritide

Chleck, Brousaides, Sullivan and Zeigler, 41

E. Active hydrogen using deuterium or tritium

Eastham and Raaen, 52

F. Active hydrogen using diazomethane

Arndt, 4

Hydroxyl groups and alcohols

Blickenstaff, Schaeffer, and Kathman, 23
Boos, 24
Brunner and Thomas, 32

Hydroxyl groups and alcohols (*Cont.*)

Griffiths and Stock, 81
Johnson, 106
Jurecek, Chladek, Chladkova, Souček, and Srpova, 110
Kabasakalian, Townley, and Yudis, 112
Kepner and Webb, 119
Ma and Burstein, 150
Ma and Moss, 154
Ma and Waldman, 155
Schulek and Burger, 210
Shishikura, 213
Stenmark and Weiss, 231
Stodola, 236
Terent'ev and Kupletskaya, 245

Vinyl ether

Siggia and Edsberg, 215

Peroxide

Ma and Gerstein, 152
Roth and Schuster, 197

α-Epoxy groups

Jungnickel, Peters, Polgar, and Weiss, 109
Ulbrich and Makeš, 250

Bases

Belcher, Berger, and West, 10
Buděšínský and Körbl, 35, 36
Doležil and Bulandr, 50
Gorbach and Kögler, 78
Gutterson and Ma, 84
Kainz and Pohm, 113
Lincoln and Chinnick, 146
Ogawa, 175
Smith, Mitchell, and Billmeyer, 221
Streuli, 237

Amines (secondary and tertiary) and amides

See Chapters 7 and 8
Breyhan, 29
Clark and Morgan, 45
Hillenbrand and Pentz, 96
Jan, Kolsek, and Perpar, 104
Keen and Fritz, 117

20. Other Groups

TABLE 35 (*Continued*)

Amines (secondary and tertiary) and amides (*Cont.*)
Litvenenko and Grekov, 147
Morgan, 168
Roth and Schuster, 196
Terent'ev, Kupletskaya, and Andreeva, 246
Yokoo, 263

Hydrazines, semicarbazides, thiosemicarbazides, etc.
Berka and Zyka, 20
McKennis, Weatherby, and Dellis, 161
Schulek and Burger, 209
Singh, 216
Singh and Sahota, 217
Vulterin and Zyka, 256

Sulfhydryl (mercaptans)
Dal Nogare, 49
Holasek, Lieb, and Merz, 99
Holter and Løvtrup, 100
Jančik, Buben, and Körbl, 105
Karchmer, 115
Kolb and Toennies, 129
Kuhn, Birkofer and Quackenbush, 133
Rosenberg, Perrone, and Kirk, 188
Sahashi and Shibasaki, 198

Organic sulfur compounds in general
Cihalik and Ruzicka, 42
Joshi, 107
Lennartz and Middeldorf, 141

Disulfides, dialkylsulfides
Hubbard, Haines, and Ball, 102
Kies and Weezel, 121
Roth, 194

Isocyanate and isothiocyanate
Karten and Ma, 116
Roth, 194

Dithiocarbamate and thiuramdisulfide
Roth and Beck, 195

Acetylene linkage (triple bond)
Miocque and Gautier, 164
Roth, 191

Gas analysis
See Chapter 18
Aristarkhova, 2
Beckman, McCullough, and Crane, 8

Gas analysis (*Cont.*)
Belcher, 9
Benson, 15, 16
Berg, 17
Blacet, Sellers, and Blaedel, 22
Engineering Unit, 55
Euchen and Knick, 56
Foreman, 60
Furman, 70
Grant, 79, 80
Guldner and Beach, 83
Haden and Luttrop, 89
Hallett, 91
Kenty and Reuter, 118
Kieselbach, 122
Krogh, 131
Kuchler and Weller, 132
Langmuir, 137–139
LeRoy and Steacie, 142
Milton and Waters, 163
Nash, 171
Prescott and Morrison, 184
Salsbury, Cole, and Yoe, 199
Sanderson, 200
Schmit-Jensen, 201
Scholander and Evans, 202
Scholander and Irving, 203
Sebastian and Howard, 211
Sendroy and Granville, 212
Smaller and Hall, 220
Smith and Leighton, 222
Spence, 229
Sutton, 241
Swearingen, Gerbes, and Ellis, 243
Uhrig and Levin, 249
Whiteley, 257
Willits, 260

Apparatus
British Standards Institution, 30
Buděšínský, 34
Franzen and Pauli, 64
Grützner and Hintermaier, 82
Leurquin and Delville, 143
Schöniger, Lieb, and El Din Ibrahim, 206
Sirotenko, 218
Smith, Mitchell, and Billmeyer, 221
Wiesenberger, 258
Zaugg and Lauer, 264

REFERENCES

1. Anger, V., *Mikrochim. Acta,* p. 58 (1960).
2. Aristarkhova, M. V., *Zavodskaya Lab.,* **9**, 1096 (1940).
3. Arjungi, K. N., Kulkarni, R. S., and Gore, T. S., *J. Sci. Ind. Research (India),* **17B**, 459 (1958); *Chem. Abstr.,* **53**, 11095 (1959).
4. Arndt, F. G., in "Organic Analysis" (J. Mitchell, Jr., I. M. Kolthoff, E. S. Proskauer, and A. Weissberger, eds.), Vol. I, p. 197, Interscience, New York, 1953.
5. Arventiev, B., and Figel, S., *Acad. rep. populare Romîne, Filiala Iași, Studii cercetări științ.,* **4**, 225 (1953).
6. Becker, W. W., *Anal. Chem.,* **22**, 185 (1950).
7. Becker, W. W., and Schaefer, W. E., in "Organic Analysis" (J. Mitchell, Jr., I. M. Kolthoff, E. S. Proskauer, and A. Weissberger, eds.), Vol. II, p. 71, Interscience, New York, 1954.
8. Beckman, A. O., McCullough, J. D., and Crane, R. A., *Anal. Chem.,* **20**, 674 (1948).
9. Belcher, R., *Metallurgia,* **35**, 310 (1947).
10. Belcher, R., Berger, J., and West, T. S., *J. Chem. Soc.,* p. 2882 (1959).
11. Belcher, R., Bhatty, M. K., and West, T. S., *J. Chem. Soc.,* p. 2393 (1958).
12. Belcher, R., and Godbert, A. L., "Semi-Micro Quantitative Organic Analysis," 2nd ed., Longmans, Green, London, 1954.
13. Belcher, R., and Phillips, D. F., **BIOS No. 1606** (British Intelligence Objectives Sub-Committee).
14. Bennett, A., May, L. G., and Gregory, R., *J. Lab. Clin. Med.,* **37**, 643 (1951).
15. Benson, S. W., *Ind. Eng. Chem., Anal. Ed.,* **13**, 502 (1941).
16. Benson, S. W., *Ind. Eng. Chem., Anal. Ed.,* **14**, 189 (1942).
17. Berg, W. E., *Science,* **104**, 575 (1946).
18. Berinzaghi, B., *Anales asoc. quím. arg.,* **44**, 120 (1956).
19. Berka, A., and Zyka, J., *Chem. listy,* **50**, 831 (1956).
20. Berka, A., and Zyka, J., *Chem. listy,* **52**, 926 (1958).
21. Bhatty, M. K., *Analyst,* **82**, 458 (1957).
22. Blacet, F. E., Sellers, A. L., and Blaedel, W. J., *Ind. Eng. Chem., Anal. Ed.,* **12**, 356 (1940).
23. Blickenstaff, R. T., Schaeffer, J. R., and Kathman, G. G., *Anal. Chem.,* **26**, 746 (1954).
24. Boos, R. N., *Anal. Chem.,* **20**, 964 (1948).
25. Bose, S., *Anal. Chem.,* **30**, 1526 (1958).
26. Bose, S., *J. Indian Chem. Soc.,* **34**, 739 (1957).
27. Bragg, P. D., and Hough, L., *J. Chem. Soc.,* p. 4347 (1957).
28. Brandstätter, M., *Mikrochemie ver. Mikrochim. Acta,* **32**, 33, 162 (1944).
29. Breyhan, T., *Z. anal. Chem.,* **152**, 412 (1956).
30. British Standards Institution, *Brit. Standards,* **1428**, Pt. C1 (1954).
31. Brown, H. C., and Hafliger, O., *Anal. Chem.,* **26**, 757 (1954).
32. Brunner, H., and Thomas, H. R., *J. Appl. Chem. (London),* **3**, 49 (1953).
33. Buděšínský, B., *Chem. listy,* **52**, 2292 (1958); *Anal. Abstr.,* **6**, No. 3552 (1959).
34. Buděšínský, B., *Collection Czechoslov. Chem. Communs.,* **24**, 2948 (1959).
35. Buděšínský, B., and Körbl, J., *Chem. listy,* **52**, 1513 (1958).
36. Buděšínský, B., and Körbl, J., *Collection Czechoslov. Chem. Communs.,* **25**, 76 (1960).
37. Buděšínský, B., and Körbl, J., *Mikrochim. Acta,* p. 922 (1959).
38. Calderón Martinez, J., *Quím. e ind. (Bilbao),* **3**, 56 (1956).
39. Cheronis, N. D., "Micro and Semimicro Methods," Interscience, New York, 1954.

40. Cheronis, N. D., and Entrikin, J. B., "Semimicro Qualitative Organic Analysis," 2nd ed., Interscience, New York, 1957.
41. Chleck, D. J., Brousaides, F. J., Sullivan, W., and Zeigler, C. A., *Intern. J. Appl. Radiation and Isotopes,* **7**, 182 (1960).
42. Cihalik, J., and Ruzicka, J., *Collection Czechoslov. Chem. Communs.,* **21**, 262 (1956).
43. Clark, E. P., "Semimicro Quantitative Organic Analysis," Academic Press, New York, 1943.
44. Clark, S. J., "Quantitative Methods of Organic Microanalysis," Butterworths, London, 1956.
45. Clark, S. J., and Morgan, D. J., *Mikrochim. Acta,* p. 966 (1956).
46. Clarke, H. T., Johnson, J. R., and Robinson, Sir R., eds., "Chemistry of Penicillin," pp. 113, 289, 583, Princeton Univ. Press, Princeton, New Jersey, 1949.
47. Cohn, M., *in* "Preparation and Measurement of Isotopic Tracers" (D. W. Wilson, A. O. C. Nier, and S. P. Reimann, eds.), p. 51, Edwards, Ann Arbor, Michigan, 1946.
48. Colson, A. F., *Analyst,* **82**, 358 (1957).
49. Dal Nogare, S., *in* "Organic Analysis" (J. Mitchell, Jr., I. M. Kolthoff, E. S. Proskauer, and A. Weissberger, eds.), Vol. I, p. 329, Interscience, New York, 1953.
50. Doležil, M., and Bulandr, J., *Chem. listy,* **51**, 255 (1957).
51. Duval, C., and Xuong, N. D., *Mikrochim. Acta,* p. 747 (1956).
52. Eastham, J. F., and Raaen, V. F., *Anal. Chem.,* **31**, 555 (1959).
53. Edlbacher, S., *Z. physiol. Chem. Hoppe-Seyler's,* **101**, 278 (1918).
54. Ehlers, V. B., and Sease, J. W., *Anal. Chem.,* **31**, 16 (1959).
55. Engineering Unit, Div. Ind. Hyg., Natl. Inst. Health, *Ind. Eng. Chem., Anal. Ed.,* **16**, 346 (1944).
56. Euchen, A., and Knick, H., *Brennstoff-Chem.,* **17**, 241 (1946).
57. Falkenhausen, F. V. v., *Z. anal. Chem.,* **99**, 241 (1935).
58. Fenger-Eriksen, K., Krogh, A., and Ussing, H., *Biochem. J.,* **30**, 1264 (1936).
59. Flaschenträger, B., *Z. physiol. Chem. Hoppe-Seyler's,* **146**, 219 (1925).
60. Foreman, J. K., *Mikrochim. Acta,* p. 1481 (1956).
61. Franzen, F., Disse, W., and Eysell, K., *Mikrochim. Acta,* p. 44 (1953).
62. Franzen, F., Eysell, K., and Schall, H., *Mikrochim. Acta,* p. 708 (1955).
63. Franzen, F., Eysell, K., and Schall, H., *Mikrochim. Acta,* p. 712 (1955).
64. Franzen, F., and Pauli, H., *Mikrochim. Acta,* p. 845 (1955).
65. Friedrich, A., "Die Praxis der quantitativen organischen Mikroanalyse," Deuticke, Vienna and Leipzig, 1933.
66. Friedrich, A., *Mikrochemie,* **7**, 185, 195 (1929).
67. Friedrich, A., *Mikrochemie,* **8**, 94 (1930).
68. Friedrich, A., *Z. physiol. Chem. Hoppe-Seyler's,* **163**, 141 (1927).
69. Fritz, J. S., and Hammond, G. S., "Quantitative Organic Analysis," Wiley, New York, and Chapman & Hall, London, 1957.
70. Furman, N. H., ed., "Scott's Standard Methods of Chemical Analysis," 5th ed., Vol. II, Van Nostrand, New York, 1939.
71. Garbers, C. F., Schmid, H., and Karrer, P., *Helv. Chim. Acta,* **37**, 1336 (1954).
72. Genevois, L., *Chim. anal.,* **29**, 77 (1947).
73. Gey, K. F., and Schön, H., *Z. physiol. Chem. Hoppe-Seyler's,* **305**, 149 (1956).
74. Goddu, R. F., LeBlanc, N. F., and Wright, C. D., *Anal. Chem.,* **27**, 1251 (1955).
75. Goltz, G. E., and Glew, D. N., *Anal. Chem.,* **29**, 816 (1957).
76. Gorbach, G., *Fette u. Seifen,* **52**, 405 (1950).

77. Gorbach, G., *Mikrochemie ver. Mikrochim. Acta,* **31**, 319 (1944).
78. Gorbach, G., and Kögler, H., *Mikrochim. Acta,* p. 573 (1957).
79. Grant, J., "Quantitative Organic Microanalysis, Based on the Methods of Fritz Pregl," 4th ed., Blakiston, Philadelphia, Pennsylvania, 1946.
80. Grant, J., "Quantitative Organic Microanalysis," 5th ed., Blakiston, Philadelphia, Pennsylvania, 1951.
81. Griffiths, V. S., and Stock, D. I., *J. Chem. Soc.,* p. 1633 (1956).
82. Grützner, R., and Hintermaier, A., *Mikrochemie ver. Mikrochim. Acta,* **38**, 66 (1951).
83. Guldner, W. G., and Beach, A. L., *Anal. Chem.,* **26**, 1199 (1954).
84. Gutterson, M., and Ma, T. S., *Mikrochim. Acta,* p. 1 (1960).
85. Gysel, H., *Mikrochim. Acta,* p. 743 (1954).
86. Haas, P., *Mikrochemie,* **7**, 69 (1929).
87. Haas, P., *Mikrochemie,* **8**, 89 (1930).
88. Hack, M. H., *Arch. Biochem. Biophys.,* **58**, 19 (1955).
89. Haden, W. L., and Luttropp, E. S., *Ind. Eng. Chem., Anal. Ed.,* **13**, 571 (1941).
90. Hall, R. T., and Schaefer, W. E., *in* "Organic Analysis" (J. Mitchell, Jr., I. M. Kolthoff, E. S. Proskauer, and A. Weissberger, eds.), Vol. II, p. 19, Interscience, New York, 1954.
91. Hallett, L. T., *Ind. Eng. Chem., Anal. Ed.,* **14**, 956 (1942).
92. Hamill, W. H., *J. Am. Chem. Soc.,* **59**, 1152 (1937).
93. Hibbert, H., and Sudborough, J. J., *J. Chem. Soc.,* **19**, 285 (1903).
94. Hibbert, H., and Sudborough, J. J., *J. Chem. Soc.,* **20**, 165 (1904).
95. Higuchi, T., *in* "Organic Analysis" (J. Mitchell, Jr., I. M. Kolthoff, E. S. Proskauer, and A. Weissberger, eds.), Vol. II, p. 123, Interscience, New York, 1954.
96. Hillenbrand, E. F., Jr., and Pentz, C. A., *in* "Organic Analysis" (J. Mitchell, Jr., I. M. Kolthoff, E. S. Proskauer, and A. Weissberger, eds.), Vol. III, p. 129, Interscience, New York, 1956.
97. Hockstein, F. A., *J. Am. Chem. Soc.,* **71**, 305 (1949).
98. Höfling, E., Lieb, H., and Schöniger, W., *Monatsh. Chem.,* **83**, 60 (1952).
99. Holasek, A., Lieb, H., and Merz, W., *Mikrochim. Acta,* p. 1216 (1956).
100. Holter, H., and Løvtrup, Søren, *Compt. rend. trav. lab. Carlsberg. Sér. chim.,* **27**, 72 (1949).
101. Hörmann, H., Lamberts, J., and Fries, G., *Z. physiol. Chem. Hoppe-Seyler's,* **306**, 42 (1956).
102. Hubbard, R. L., Haines, W. E., and Ball, J. S., *Anal. Chem.,* **30**, 91 (1958).
103. Hunter, I. R., and Potter, E. F., *Anal. Chem.,* **30**, 293 (1958).
104. Jan, J., Kolsek, J., and Perpar, M., *Z. anal. Chem.,* **153**, 4 (1956).
105. Jančik, F., Buben, F., and Körbl, J., *Československ farm.,* **5**, 515 (1956).
106. Johnson, B. L., *Anal. Chem.,* **20**, 777 (1948).
107. Joshi, M. K., *Anal. Chim. Acta,* **14**, 509 (1956).
108. Jucker, A., *Anal. Chim. Acta,* **16**, 210 (1957).
109. Jungnickel, J. L., Peters, E. D., Polgar, A., and Weiss, F. T., *in* "Organic Analysis" (J. Mitchell, Jr., I. M. Kolthoff, E. S. Proskauer, and A. Weissberger, eds.), Vol. I, p. 127, Interscience, New York, 1953.
110. Jurecek, M., Chladek, O., Chladkova, R., Souček, M., and Srpova, B., *Chem. listy,* **51**, 448 (1957).
111. Juvet, R. S., Twickler, M. C., and Afremow, L. C., *Anal. Chim. Acta,* **22**, 87 (1960).
112. Kabasakalian, P., Townley, E. R., and Yudis, M. D., *Anal. Chem.,* **31**, 375 (1959).
113. Kainz, G., and Pohm, M., *Mikrochemie ver. Mikrochim. Acta,* **35**, 189 (1950).

20. Other Groups

114. Kainz, G., Polansky, I., Schinzel, E., and Wesseley, F., *Mikrochim. Acta*, p. 241 (1957).
115. Karchmer, J. H., *Anal. Chem.*, **29**, 425 (1957).
116. Karten, B. S., and Ma, T. S., *Microchem. J.*, **3**, 507 (1959).
117. Keen, R. T., and Fritz, J. S., *Anal. Chem.*, **24**, 564 (1952).
118. Kenty, C., and Reuter, F. W., Jr., *Rev. Sci. Instr.*, **18**, 918 (1947).
119. Kepner, R. E., and Webb, A. D., *Anal. Chem.*, **26**, 925 (1954).
120. Ketchum, D., *Ind. Eng. Chem., Anal. Ed.*, **18**, 273 (1946).
121. Kies, H. L., and Weezel, G. J. van, *Z. anal. Chem.*, **161**, 348 (1958).
122. Kieselbach, R., *Ind. Eng. Chem., Anal. Ed.*, **16**, 764 (1944).
123. Kirsten, W. J., *Microchem. J.*, **2**, 179 (1958).
124. Kirsten, W., and Stenhagen, E., *Acta Chem. Scand.*, **6**, 682 (1952).
125. Kleinzeller, A., and Trim, A. R., *Analyst*, **69**, 241 (1944).
126. Klimova, V. A., and Zabrodina, K. S., *Bull. Acad. Sci. U.S.S.R., Div. Chem. Sci. S.S.R.* (*English translation*), p. 164 (1959).
127. Klimova, V. A., and Zabrodina, K. S., *Izvest. Akad. Nauk S.S.S.R.*, p. 175 (1959); *Anal. Abstr.*, **6**, No. 4444 (1959).
128. Kohler, E. P., Stone, J. F., and Fuson, R. C., *J. Am. Chem. Soc.*, **49**, 3181 (1927).
129. Kolb, J. J., and Toennies, G., *Anal. Chem.*, **24**, 1164 (1952).
130. Komori, S., *Kôgyô Kagaku Zasshi*, **51**, 120 (1948).
131. Krogh, A., *Skand. Arch. Physiol.*, **20**, 279 (1908).
132. Kuchler, L., and Weller, O. E., *Mikrochemie*, **26**, 44 (1939).
133. Kuhn, R., Birkofer, L., and Quackenbush, F. W., *Ber.*, **72**, 407 (1939).
134. Kuhn, R., and Roth, H., *Ber.*, **65**, 1285 (1932).
135. Kuhn, R., and Roth, H., *Ber.*, **66**, 1274 (1933).
136. Kuhn, R., and Roth, H., *Ber.*, **67B**, 1458 (1934).
137. Langmuir, I., *Ind. Eng. Chem.*, **7**, 348 (1915).
138. Langmuir, I., *J. Am. Chem. Soc.*, **34**, 310 (1912).
139. Langmuir, I., *J. Am. Chem. Soc.*, **37**, 1139 (1915).
140. Lee, F. A., *J. Assoc. Offic. Agr. Chemists*, **41**, 899 (1958).
141. Lennartz, T. A., and Middeldorf, R., *Süddeut. Apotheker-Ztg.*, **89**, 593 (1949).
142. LeRoy, D. J., and Steacie, E. W. R., *Ind. Eng. Chem., Anal. Ed.*, **16**, 341 (1944).
143. Leurquin, J., and Delville, J. P., *Experientia*, **6**, 274 (1950).
144. Lieb, H., and Schöniger, W., *Mikrochemie ver. Mikrochim. Acta*, **35**, 400 (1950).
145. Lieb, H., Schöniger, W., and Schiviz-Schivizhoffen, E., *Mikrochemie ver. Mikrochim. Acta*, **35**, 407 (1950).
146. Lincoln, P. A., and Chinnick, C. C. T., *Analyst*, **81**, 100 (1956).
147. Litvenenko, L. M., and Grekov, A. P., *Zhur. Anal. Khim.*, **10**, 164 (1955).
148. Ma, T. S., *Microchem. J.*, **3**, 415 (1959); **4**, 373 (1960).
149. Ma, T. S., "Proceedings of the International Symposium on Microchemistry 1958," p. 151, Pergamon, Oxford, London, New York, and Paris, 1960.
150. Ma, T. S., and Burstein, R., Unpublished work. See R. Burstein, "Microdetermination of Phenols by Bromination," Thesis, Brooklyn College, Brooklyn, New York (1959).
151. Ma, T. S., and Earley, J. V., *Mikrochim. Acta*, p. 129 (1959); *Anal Abstr.*, **6**, No. 3558 (1959).
152. Ma, T. S., and Gerstein, T., Unpublished work. See T. Gerstein, "Microdetermination of Peroxides and Quinones," Thesis, Brooklyn College, Brooklyn, New York (1959).
153. Ma, T. S., Logun, J., and Mazzella, P. P., *Microchem. J.*, **1**, 67 (1957).

154. Ma, T. S., and Moss, H., Unpublished work. See H. Moss, "Microdetermination of Functional Groups Using Periodate," Thesis, Brooklyn College, Brooklyn, New York (1958).
155. Ma, T. S., and Waldman, H., Unpublished work. See H. Waldman, "Microdetermination of Hydroxyl Groups in Organic Compounds," Thesis, Brooklyn College, Brooklyn, New York (1959).
156. Malmberg, E. W., Weinstein, B., Fishel, D. L., and Krause, R. A., *Mikrochim. Acta*, p. 210 (1959).
157. Marcali, K., and Rieman, W., III, *Ind. Eng. Chem., Anal. Ed.*, **18**, 144 (1946).
158. Marrian, P. M., and Marrian, G. F., *Biochem. J.*, **24**, 746 (1930).
159. Maute, R. L., and Owens, M. L., Jr., *Anal. Chem.*, **28**, 1312 (1956).
160. McAlpine, I. M., and Ongley, P. A., *Anal. Chem.*, **27**, 55 (1955).
161. McKennis, H., Jr., Weatherby, H. J., and Dellis, E. P., *Anal. Chem.*, **30**, 499 (1958).
162. Meisenheimer, J., and Schlichenmayer, W., *Ber.*, **61**, 2029 (1928).
163. Milton, R. F., and Waters, W. A., "Methods of Quantitative Microanalysis," Longmans, Green, New York, and Arnold, London, 1949.
164. Miocque, M., and Gautier, J. A., *Bull. soc. chim. France*, p. 467 (1958).
165. Mitchell, J., Jr., *in* "Organic Analysis" (J. Mitchell, Jr., I. M. Kolthoff, E. S. Proskauer, and A. Weissberger, eds.), Vol. I, p. 243, Interscience, New York, 1953.
166. Mitchell, J., Jr., *in* "Organic Analysis" (J. Mitchell, Jr., I. M. Kolthoff, E. S. Proskauer, and A. Weissberger, eds.), Vol. I, p. 309, Interscience, New York, 1953.
167. Mitchell, J., Jr., Smith, D. M., and Money, F. S., *Ind. Eng. Chem., Anal. Ed.*, **16**, 410 (1944).
168. Morgan, D. J., *Mikrochim. Acta*, p. 104 (1958).
169. Morgan, P. W., *Ind. Eng. Chem., Anal. Ed.*, **18**, 500 (1946).
170. Nakamura, N., *Bunseki Kagaku*, **5**, 459 (1956).
171. Nash, L. K., *Ind. Eng. Chem., Anal. Ed.*, **18**, 505 (1946).
172. Neuberg, C., and Strauss, E., *Arch. Biochem.*, p. 211 (1945).
173. Niederl, J. B., and Niederl, V., "Micromethods of Quantitative Organic Elementary Analysis," Wiley, New York, 1938.
174. Niederl, J. B., and Niederl, V., "Micromethods of Quantitative Organic Analysis," 2nd ed., Wiley, New York, 1942.
175. Ogawa, T., *Nippon Kagaku Zasshi*, **76**, 739 (1955).
176. Ogg, C. L., Porter, W. L., and Willits, C. O., *Ind. Eng. Chem., Anal. Ed.*, **17**, 394 (1945).
177. Olleman, E. D., *Anal. Chem.*, **24**, 1425 (1952).
178. Perold, G. W., and Snyman, J. M., *Mikrochim. Acta*, p. 225 (1958).
179. Perpar, M., Tišler, M., Vrbaški, Ž., *Mikrochim. Acta*, p. 64 (1959).
180. Petersen, J. W., Hedberg, K. W., and Christensen, B. E., *Ind. Eng. Chem., Anal. Ed.*, **15**, 225 (1943).
181. Petránek, J., and Večeřa, M., *Chem. listy*, **51**, 1686 (1957); *Chem. Abstr.*, **53**, 980 (1958).
182. Petrova, L. N., and Novikova, E. N., *Zhur. Priklad. Khim.*, **28**, 219 (1955).
183. Pregl, F., "Quantitative Organic Microanalysis" (E. Fyleman, trans., 2nd German ed.), Churchill, London, 1924.
184. Prescott, C. H., Jr., and Morrison, J., *Ind. Eng. Chem., Anal. Ed.*, **11**, 230 (1939).
185. Prévost, C., and Souchay, P., *Chim. anal.*, **37**, 3 (1955).
186. Rachele, J. R., and Melville, D. B., Personal communication.

20. Other Groups

187. Remsen, I., and Orndorff, W. R., "Compounds of Carbon or Organic Chemistry," p. 112, Heath, New York, 1922.
188. Rosenberg, S., Perrone, J. L., and Kirk, P. L., *Anal. Chem.*, **22**, 1186 (1950).
189. Roth, H., "Die quantitative organische Mikroanalyse von Fritz Pregl," 4th ed., Springer, Berlin, 1935.
190. Roth, H., "F. Pregl quantitative organische Mikroanalyse," 5th ed., Springer, Wien, 1947.
191. Roth, H., "Pregl-Roth quantitative organische Mikroanalyse," 7th ed., Springer, Wien, 1958.
192. Roth, H., "Quantitative Organic Microanalysis of Fritz Pregl," 3rd ed., (E. B. Daw, trans., 4th German ed.), Blakiston, Philadelphia, Pennsylvania, 1937.
193. Roth, H., *Mikrochemie,* **11**, 140 (1932).
194. Roth, H., *Mikrochim. Acta,* p. 766 (1958).
195. Roth, H., and Beck, W., *Mikrochim. Acta,* p. 844 (1957).
196. Roth, H., and Schuster, P., *Mikrochim. Acta,* p. 837 (1957).
197. Roth, H., and Schuster, P., *Mikrochim. Acta,* p. 840 (1957).
198. Sahashi, Y., and Shibasaki, H., *Nippon Nôgei-kagaku Kaishi,* **25**, 57 (1951).
199. Salsbury, J. M., Cole, J. W., and Yoe, J. H., *Anal. Chem.*, **19**, 66 (1947).
200. Sanderson, R. T., *Ind. Eng. Chem., Anal. Ed.,* **15**, 76 (1943).
201. Schmit-Jensen, H. O., *Biochem. J.,* **14**, 4 (1920).
202. Scholander, P. F., and Evans, H. J., *J. Biol. Chem.*, **169**, 551 (1947).
203. Scholander, P. F., and Irving, L., *J. Biol. Chem.*, **169**, 561 (1947).
204. Schöniger, W., *Z. anal. Chem.,* **133**, 4 (1951).
205. Schöniger, W., and Lieb, H., *Mikrochemie ver. Mikrochim. Acta,* **38**, 165 (1951).
206. Schöniger, W., Lieb, H., and El Din Ibrahim, M. G., *Mikrochim. Acta,* p. 96 (1954).
207. Schöniger, W., Lieb, H., and Gassner, K., *Mikrochim. Acta,* p. 434 (1953).
208. Schöniger, W., Lieb, H., and Gassner, K., *Z. anal. Chem.,* **134**, 188 (1951).
209. Schulek, E., and Burger, K., *Talanta,* **1**, 344 (1958).
210. Schulek, E., and Burger, K., *Z. anal. Chem.,* **161**, 184 (1958).
211. Sebastian, J. J. S., and Howard, H. C., *Ind. Eng. Chem., Anal. Ed.,* **6**, 172 (1934).
212. Sendroy, J., Jr., and Granville, W. C., *Anal. Chem.*, **19**, 500 (1947).
213. Shishikura, Y., *Seikagaku,* **19**, 145 (1947).
214. Siggia, S., "Quantitative Organic Analysis via Functional Groups," 2nd ed., Wiley, New York, and Chapman & Hall, London, 1954.
215. Siggia, S., and Edsberg, L. R., *Anal. Chem.*, **20**, 762 (1948).
216. Singh, B., *Anal. Chim. Acta,* **17**, 467 (1957).
217. Singh, B., and Sahota, S. S., *Anal. Chim. Acta,* **17**, 285 (1957).
218. Sirotenko, A. A., *Mikrochim. Acta,* p. 1 (1955).
219. Slotta, K. H., and Haberland, G., *Ber.,* **65B**, 127 (1932).
220. Smaller, B., and Hall, J. F., Jr., *Ind. Eng. Chem., Anal. Ed.,* **16**, 64 (1944).
221. Smith, D. M., Mitchell, J., Jr., and Billmeyer, A. M., *Anal. Chem.*, **24**, 1847 (1952).
222. Smith, R. N., and Leighton, P. A., *Ind. Eng. Chem., Anal. Ed.,* **14**, 758 (1942).
223. Sobotka, M., "Proceedings of the International Symposium on Microchemistry 1958," p. 171, Pergamon, Oxford, London, New York, and Paris, 1960.
224. Sobotka, M., and Trutnovsky, H., *Microchem. J.,* **3**, 211 (1959).
225. Soltys, A., *Mikrochemie,* **20**, 107 (1936).
226. Souček, M., *Chem. listy,* **50**, 323 (1956).
227. Souček, M., *Collection Czechoslov. Chem. Communs.*, **23**, 554 (1958).

228. Sozzi, J. A., *Anales farm. y bioquím.*, **14**, 41 (1943).
229. Spence, R., *J. Chem. Soc.*, p. 1300 (1940).
230. Stefanac, Z., *Croat. Chem. Acta*, **28**, 295 (1956).
231. Stenmark, G. A., and Weiss, F. T., *Anal. Chem.*, **28**, 1784 (1956).
232. Stevens, G. D., *Anal. Chem.*, **28**, 1184 (1956).
233. Steyermark, Al, *Anal. Chem.*, **20**, 368 (1948).
234. Steyermark, Al, "Quantitative Organic Microanalysis," Blakiston, Philadelphia, Pennsylvania, 1951.
235. Steyermark, Al, Unpublished work.
236. Stodola, F. H., *Mikrochemie*, **21**, 180 (1936–37).
237. Streuli, C. A., *Anal. Chem.*, **28**, 130 (1956).
238. Subba Rao, D., Shah, G. D., and Pansare, V. S., *Mikrochim. Acta*, p. 81 (1954).
239. Sudo, T., Shimoe, D., and Tsujii, T., *Bunseki Kagaku*, **3**, 403 (1954).
240. Sudo, T., Shimoe, D., and Tsujii, T., *Bunseki Kagaku*, **6**, 494 (1957).
241. Sutton, T. C., *J. Sci. Instr.*, **15**, 133 (1938).
242. Suzuki, S., Muramoto, Y., Ueno, M., and Sugano, T., *Bull. Chem. Soc. Japan*, **30**, 775 (1957).
243. Swearingen, J. S., Gerbes, O., and Ellis, E. W., *Ind. Eng. Chem., Anal. Ed.*, **5**, 369 (1933).
244. Tashinian, V. H., Baker, M. J., and Koch, C. W., *Anal. Chem.*, **28**, 1304 (1956).
245. Terent'ev, A. P., and Kupletskaya, N. B., *Zhur. Obshcheî Khim.*, **26**, 451 (1956).
246. Terent'ev, A. P., Kupletskaya, N. B., and Andreeva, E. V., *Zhur. Obshcheî Khim.*, **26**, 881 (1956).
247. Thomas, Arthur H., Company, Philadelphia, Pennsylvania.
248. Tschugaeff, L., *Ber.*, **35**, 3912 (1902).
249. Uhrig, K., and Levin, H., *Ind. Eng. Chem., Anal. Ed.*, **13**, 90 (1941).
250. Ulbrich, V., and Makeš, J., *Chem. průmsyl*, **8**, 163 (1958); *Anal. Abstr.*, **6**, No. 643 (1959).
251. Van Etten, C. H., *Anal. Chem.*, **23**, 1697 (1951).
252. Veibel, S., "Proceedings of the International Symposium on Microchemistry 1958," p. 159, Pergamon, Oxford, London, New York, and Paris, 1960.
253. Vonesch, E. E., and Guagnini, O. A., *Anales asoc. quím. arg.*, **43**, 62 (1955).
254. Vonesch, E. E., and Guagnini, O. A., *Anales asoc. quím. arg.*, **43**, 101 (1955).
255. Vonesch, E. E., and Guagnini, O. A., *Anales asoc. quím. arg.*, **43**, 185 (1955).
256. Vulterin, J., and Zyka, J., *Chem. listy*, **50**, 364 (1956).
257. Whiteley, A. H., *J. Biol. Chem.*, **174**, 947 (1948).
258. Wiesenberger, E., *Mikrochim. Acta*, p. 127 (1954).
259. Williams, R. J., *J. Am. Chem. Soc.*, **58**, 1819 (1936).
260. Willits, C. O., *Anal. Chem.*, **21**, 132 (1949).
261. Wright, G. F., *in* "Organic Analysis" (J. Mitchell, Jr., I. M. Kolthoff, E. S. Proskauer, and A. Weissberger, eds.), Vol. I, p. 155, Interscience, New York, 1953.
262. Yamamura, S. S., *Dissertation Abstr.*, **17**, 2787 (1957).
263. Yokoo, M., *Chem. & Pharm. Bull. (Tokyo)*, **6**, 64 (1958).
264. Zaugg, H. E., and Lauer, W. M., *Anal. Chem.*, **20**, 1022 (1948).
265. Zerewitinoff, Th., *Ber.*, **40**, 2023 (1907).
266. Zerewitinoff, Th., *Ber.*, **41**, 2233 (1908).
267. Zerewitinoff, Th., *Ber.*, **42**, 4802 (1909).
268. Zerewitinoff, Th., *Ber.*, **43**, 3590 (1910).
269. Zerewitinoff, Th., *Ber.*, **47**, 1659, 2417 (1914).

CHAPTER 21

Microdetermination of Molecular Weight

The microdetermination of the molecular weight of a substance may be accomplished by several different methods,* most of which are small-scale applications of the various macromethods found throughout the literature, such as the ebullioscopic (elevation of boiling point of the solvent), cryoscopic (depression of the freezing or melting point of the solvent), and the vaporimetric (vaporization of the substance in a closed system) methods.[41] To a certain extent, the determination of the neutralization equivalent is one of molecular weight, as explained in Chapter 15. The same holds true of the saponification equivalent[37] (see Chapter 20) and the hydrogenation number (see Chapter 19). In a broad sense, the determination of any single element or group can be interpreted as that of the molecular weight. For example, if a substance is found to contain 10% of nitrogen, its molecular weight must be 140 if only one nitrogen atom is present or 280 if two are present, etc. In other words, it is 140 or some multiple of this figure. Similarly, if a substance contains 10% of acetyl, its molecular weight is 430 or some multiple.

Two methods of determining the molecular weight will be described in detail in this chapter, namely, the Rast[66,67] and the Signer[75] as modified by Clark.[15,16] Both methods have advantages and disadvantages and, in the opinion of the author, neither can be considered to be absolutely reliable. Therefore, it is advisable to use *both* methods and to use *more* than one type of solvent with each method, thereby reducing the possibility of drawing erroneous conclusions from the results. The possibility of decomposition, dissociation, reaction, polymerization, etc., during the determination should not be ignored. The Rast method is more frequently used because of the time factor. However, the results of collaborative study[54] indicated that the Signer method is the better.

THE RAST METHOD FOR THE DETERMINATION OF MOLECULAR WEIGHT

The Rast method,[16,29,30,50,51,66,67,70-73,79] a cryoscopic one, for the determination of molecular weight, is based upon the well-known principle that a solute depresses the melting or freezing point of a solvent proportional to the mol

* Please see references 4, 5, 16, 27, 29, 30, 50, 51, 63, 66, 67, 70–73, 79.

fraction of solute present.[41] However, the solute or substance whose molecular weight is being determined, must not undergo decomposition at the temperature of the melting point of the solvent and also there must be no reaction between the two substances. The solvent selected should melt at some temperature below the melting or boiling point of the sample to be analyzed.

Reagents

Before any of the solvents listed below can be used for a determination with an unknown, they must be used with a test sample such as azobenzene or anthracene going through the regular procedure to obtain their constants (see below).

CAMPHOR, M.P., 176°–180° C.[82]

Very pure camphor is required as a solvent for high-melting samples. It should be resublimed in the laboratory and stored in a wide-mouthed ground glass-stoppered bottle.

EXALTONE (CYCLOPENTADECANONE)[82] M.P., 65.6° C.

This material should be very pure. It is used as a solvent for the substances with melting points *below* that of camphor.

OTHER SOLVENTS—SEE TABLE 36

Camphene, m.p. 49° C.; borneol, m.p. 206° C.; bornylamine, m.p. 164° C.; camphoquinone, m.p. 190° C.; cyclohexanol, m.p. 24.6°–24.8° C.; camphenilone, m.p. 38° C.; and pinene dibromide, m.p. 170° C. may be used following the rule in regard to the relation between melting points of the solute and solvent given above. For additional solvents the reader is referred to the list of twenty-seven compounds compiled by Roth[71,72] that includes comments on the use of each as well as references for their preparation and purification.

Apparatus

MELTING POINT APPARATUS[37,56,82]

The ordinary *mechanically stirred* melting point bath containing Crisco* or silicone fluid is satisfactory for the determination. It should be well illuminated. A set of good grade of Anschütz thermometers is required, one thermometer for each of the ranges in which fall the melting points of the solvents used. A satisfactory reading lens is used for observing the thermometer. Figure 203

---

* Hydrogenated vegetable oil.[64]

shows a new commercially available[82] melting point apparatus particularly suitable for the determination, when the bath assembly modified for molecular weight determinations (Fig. 203b) is used. This bath assembly, which includes an immersion heater, stirrer, armored thermometer, and capillary tube

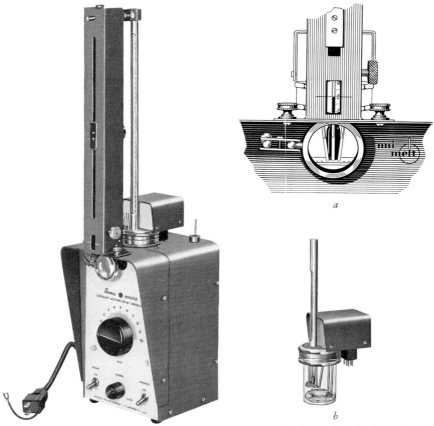

FIG. 203. Thomas-Hoover melting point apparatus, showing enlarged view of periscopic reading device, a, and special bath assembly for molecular weight determinations by Rast method, b.

holder, fits into a well in the top of the control cabinet. A periscopic thermometer reader provides magnified thermometer indication in a window approximately 1.5 inches above the center of the capillary viewing lens so that both the temperature and the capillary contents may be under observation with little shifting of the operator's eyes. This adds to the accuracy of the method. The bath is silicone fluid. It heats rapidly with no lagging, the heat transfer allowing the bath temperature to rise to 350° C. in 6.5 minutes. Rapid cooling is affected by means of a compressed air line, making it possible to cool from 300° C. down to 150° C. in about 11 minutes.

MELTING POINT CAPILLARIES[16,29,30,50,51,66,67,70-73,79]

Pyrex[21] glass tubing is drawn out and sections with tapering walls should be selected having dimensions approximately those shown in Fig. 204, *a*. The

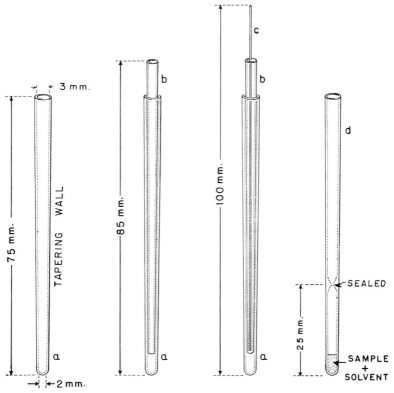

FIG. 204. Capillaries for microdetermination of molecular weight by the Rast method, showing various stages—details of construction, dimensions are approximate.

smaller ends are sealed off as shown. Slightly *longer* and *narrower* sections are also selected which will fit into the closed outer melting tubes, *a*, as shown in Fig. 204, *b*. Two open ends tubes, *b*, are required for each closed end tube, *a*, per sample. Both open ends of the longer tubes, *b*, are fire polished. A glass rod is drawn out and sections selected that are a few centimeters longer than the open tubes, *b*, but of small enough outside diameter to pass through them as shown in Fig. 204, *c*. These small glass rods act as plungers to force the samples from the open end tubes, *b*, into the melting tubes, *a*, as described

21. Molecular Weight

below under Procedure. A number of these units should be prepared—two closed tubes, *a*, three open end tubes, *b*, and two plunger rods being required for a molecular weight determination, not run in duplicate on solid substances. When determining the molecular weight of liquids, the glass rod is not needed but the open end tube should be drawn out into a capillary at the smaller end.

Capillaries of slightly different dimensions are commercially available.[82]

Procedure

In the procedure[16,29,30,50,51,66,67,70-73,79] the closed end capillary, *a*, is weighed on the balance. The solid sample is placed on a small watch glass and the smaller tip of the open end tube, *b*, is tapped several times into the sample so that the material packs tightly in it at this point. The excess sample is wiped from the outside of the open end tube, *b*. The open end tube, *b*, is inserted in the weighed closed end one, *a*, so that the sample end is near the closed bottom. The plunger rod is passed through the inner capillary so that the sample is forced into the bottom of the weighed tube, *a*. Both open end tube, *b*, and the plunger rod are removed together, care being exercised so that no particles of sample are carried along the upper wall of the closed tube. The tube containing the sample is returned to the balance and reweighed to obtain the weight of the sample. (Note: With liquid samples no plunger rod is used and the smaller tip of the open end tube is drawn out into a fine capillary to aid in the insertion of the sample.) A second open end tube, *b*, is used along with a second glass rod for insertion of the solvent on top of the sample in the closed tube, exactly in the same manner as for the introduction of the sample. The amount of solvent depends on the molecular weight of the substance, but in general ten to twenty times the weight of the sample is used. The closed end tube, *a*, is returned to the balance and weighed. The three weighings (empty tube; tube + sample; tube + sample + solvent) give the weights of sample and solvent. (For substances melting above the melting point of camphor, this solvent is used and exaltone is used for the others. If, however, neither of these is a solvent for the sample, which must be determined previously, one of the other solvents listed may be used.) The weighed melting point tube is removed from the balance and sealed off by means of a *needle-point flame* from a blast lamp at a distance of about 25 mm. from the bottom as shown in Fig. 204, *d*. (The tube is allowed to collapse in the flame to form a solid portion which is then given a slight twist to ensure its being sealed.) The upper part of the tube above the seal is kept attached to the bottom for use as a handle. Great care must be exercised in the sealing off process so that neither sample nor solvent is lost by the heating or the determination is ruined.

A second sample capillary is filled as above with approximately the same amount of solvent, but no weighing is necessary since this tube contains no sample and is used only for the purpose of obtaining the melting point of the solvent.

The two sealed tubes (one with solvent alone, the other with solvent plus sample) are attached, *side by side,* to an Anschütz thermometer of proper range in the manner usually prescribed for the determination of melting points[37,56] (material in the sample tube at the same level as the bulb—about one-half way down in the bath). The bath is stirred rapidly and heated to whatever temperature is required to *completely melt the mixture.* It is cooled a few degrees to allow the melt to resolidify and then slowly reheated so that the temperature rises at the rate of about 0.5° C. per minute. The temperature at which the *last crystal disappears* is taken as the *melting point of the mixture.* The temperature is increased at the same rate until the *solvent* has melted. This too is that point at which the last crystal disappears. The bath is cooled a little below the melting point of the mixture and when it has resolidified, the heating is repeated at the same rate to obtain the melting points of the mixture and of the solvent. This process of heating, cooling, reheating, etc., is repeated until successive determinations check to within ± 0.2° C. The repetition is necessary since this is the only means of knowing when the *mixture is homogeneous* (solution is complete). Even though the one tube contains only solvent, the duplication of the melting point of this material is quite in order since it gives practice as well as a check on the value. (If the *freezing point* of the mixture is more clearly defined than the melting point, which is sometimes the case, the former may be noted instead of the latter. In such a case, the temperature is raised to a few degrees above the melting point and the bath cooled at the rate of 0.5° C. per minute. Obviously, the *freezing point* of the solvent must also be recorded rather than the melting point in such cases.) Since the melting (or freezing) points of both the solvent and the mixture are done simultaneously, under identical conditions, there is no need for calibration or stem corrections due to immersion. Only the *depression* of the melting (or freezing) point of the solvent enters into the calculation—not the actual temperatures.

Calculation:

The molecular weight is calculated from the equation:

$$\text{M.W.} = \frac{1000 \times K \times S}{L \times \Delta_t}$$

where $K =$ the molecular melting (or freezing) point depression. (Melting or freezing depression for one mol of solute per 1000 grams of solvent.)

$S =$ weight of sample (solute)

$L =$ weight of solvent

$\Delta_t =$ depression of the melting (or freezing) point of the solvent.

21. Molecular Weight

Table 36 gives the list of some possible solvents together with their melting points, molecular melting (or freezing) point depressions and a few comments. *The values for K should not be accepted as listed in the table, but should be determined by analysis of a standard test sample.* (Since all other members of the equation are known in the test analysis, K, the only unknown, is calculated.) The standard test samples must be of the highest purity. Acetanilide, azobenzene, chloroanthraquinone, anthracene, etc., may be used.

Examples:

a. The following data are obtained during the course of a determination using camphor ($K = 39.1$) as the solvent:

> Wt. sample = 1.004 mg.
> Wt. camphor = 20.807 mg.
> M.P. of camphor = 178.0° C.
> M.P. of mixture = 166.2° C.
> $\therefore \Delta_t = 178.0 - 166.2 = 11.8°$ C.
> $\therefore$ M.W. = $\dfrac{1000 \times 39.1 \times 1.004}{20.807 \times 11.8}$
> = 160

b. The following data are obtained from the same compound as above (used in Example a) but with a different lot of camphor ($K = 38.2$):

> Wt. sample = 2.073 mg.
> Wt. camphor = 38.632 mg.
> M.P. of camphor = 177.5° C.
> M.P. of mixture = 164.6° C.
> $\therefore \Delta_t = 12.9°$ C.
> $\therefore$ M.W. = $\dfrac{1000 \times 38.2 \times 2.073}{38.632 \times 12.9}$
> = 159

c. The following data are obtained from another compound using exaltone ($K = 21.1$) as the solvent:

> Wt. sample = 1.102 mg.
> Wt. exaltone = 12.679 mg.
> M.P. of exaltone = 65.5° C.
> M.P. of mixture = 56.2° C.
> $\therefore \Delta_t = 9.3°$ C.
> $\therefore$ M.W. = $\dfrac{1000 \times 21.1 \times 1.102}{12.679 \times 9.3}$
> = 197

The allowable error for the determination is $\pm 5\%$ of the molecular weight.

TABLE 36*
SOLVENTS

Solvent	M.P.(°C.)	K(°C.)	Comments[16,71,72]
Borneol[50,51,58,71,72]	206	35.8	—
Bornylamine[16,29,30,59,70-73]	164	40.6	Suitable for basic substances
Camphene[29,30,58,70-73]	49	31.0	—
Camphenilone[29,30,62,70-73]	38	64.0	—
Camphoquinone[16,29,30,60,61,70-73]	190	45.7	Used for high-melting substances
Camphor†[16,29,30,50,51,70-73]	176–180	37.0–40.0	Used for high-melting substances
Cyclohexanol[16,85]	24.6–24.8	37.7	Unsuited for heavily halogenated compounds
Exaltone (Cyclopentadecanone)†[16,29,30,70-73]	65.6	21.3	Very good. Used for azo dyes, sterols, carotenoids, quinones
Pinene dibromide[71,72]	170	80.9	—

* For additional information, refer to the table of twenty-seven compounds compiled by Roth.[71,72]

† Commercially available.[82]

THE SIGNER (ISOTHERMAL DISTILLATION) METHOD FOR THE DETERMINATION OF MOLECULAR WEIGHT

The Signer method[15,16,75,79] for the determination of molecular weight is based upon the principle of isothermal distillation. If two solutions, having a common solvent, are placed in an evacuated system and held at a constant temperature, the solvent will distill from the solution of greater vapor pressure to the one of lesser, until equilibrium is established. Then the volumes of the two solutions will be constant and equimolar. If the one solution contains a known substance and the other solution contains an unknown, the molecular weight of the latter can be calculated from the data obtained in such an experiment. The main disadvantage of the method is the time required for equilibrium to be established so that the calculations can be made. After the apparatus has been filled and sealed off, however, a daily reading of the two volumes is all that is required.

21. Molecular Weight

Reagents

SOLVENTS

Any pure volatile solvent with a high vapor pressure such as those listed below may be used, provided that both the unknown sample and the standard are sufficiently soluble in it to give solutions of slightly less than 0.1 molar concentration.

>   Ethyl ether   Acetone
>   Ethyl bromide   Methyl acetate
>   Methylene chloride   Glacial acetic acid
>   Ethyl formate

Obviously, there must be no reaction between the solvent and either (or both) the sample or the standard. Water is a poor solvent for these determinations.

STANDARDS

A variety of standards of highest purity may be used. Colored substances, as for example azobenzene, are preferred, since with them there is little or no chance of confusing the solution of the standard with that of the unknown (or vice versa) and labeling is unnecessary. Sometimes it is desirable to use as a standard a substance of similar structure to that of the unknown.

Apparatus

The apparatus[74] used is that described by Clark,[15,16] a diagram of which is shown in Fig. 205, *A*. It consists of two graduated tubes, each of which is connected to a small bulb. The bulbs in turn are joined by an inverted U-shaped section, at the upper sides of which are two tubes used for introducing the sample, standard, and solvent as well as for evacuation of the system. The capacity of the graduated tubes is 2 ml. each in the apparatus described by Clark. However, the author of this book has found those of 1-ml. capacity each to be quite useful when very small samples are used. With the smaller graduated tubes, the other dimensions are obviously reduced somewhat, with the exception of the sample inlet tubes. If the latter are reduced in diameter, they are too small to accommodate the weighing tubes (Figs. 47–49, Chapter 3).

VACUUM SYSTEM

A suitable vacuum system composed of a good electrically driven pump, traps, and manometer is required for evacuating the apparatus.

BLAST LAMP

A blast lamp of the type used for sealing Carius combustion tubes (see Chapter 10) is required for sealing off the inlet tubes.

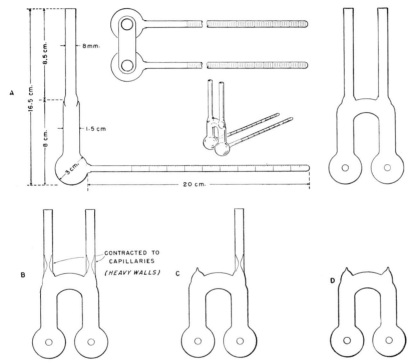

FIG. 205. Signer (isothermal distillation) apparatus for determination of molecular weight. (*A*) Details of construction. (*B, C, D*) Various stages in filling procedure.

Procedure[16,79]

The sample is weighed in a weighing tube (Figs. 47–49, Chapter 3) and introduced into the one bulb. The standard is weighed similarly and introduced into the other bulb. The relative amounts of sample and standard should be in proportion to their molecular weights, if that of the unknown can be estimated. As stated above, the amounts of material should be such that solutions of slightly less than 0.1 molar in concentration will be obtained (if the substance is soluble enough—otherwise less material is used). Care should be exercised so that no particles remain on the walls of the inlet tubes. After introduction of the sample and the standard, *both* inlet tubes are constricted down to capillaries of about 1 mm. I.D., through the use of a blast lamp flame in a manner similar to that employed for sealing the Carius combustion tubes (Chapter 10). The constrictions are made close to the bottom of the inlet tubes near to the upper sides of the inverted U (Fig. 205, B). After cooling, solvent is added through each capillary into the bulbs containing the sample and the standard. When 2-ml. graduated tubes are employed slightly less than 2 ml. of solvent is added to each bulb, and, similarly, when the one-ml. size is used a little less than one ml. is added to each. The contents of the bulbs are chilled somewhat to reduce the danger of fire, and, after warming one of the capillaries gently, to remove adhering solvent, it is sealed off

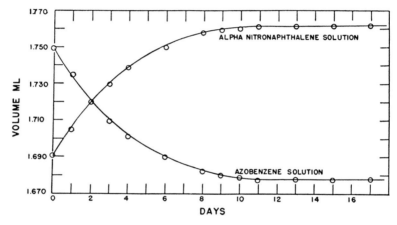

FIG. 206. Curves showing time-volume relationships during determination of molecular weight of α-nitronaphthalene using azobenzene as standard. Solvent, acetone; solutions, 0.0932 molar at equilibrium. Molecular weight of α-nitronaphthalene: Calculated, 173.1; found, 173.7.

with a needle-point flame from a blast lamp (Fig. 205, C). The other inlet tube is attached to the vacuum system and the apparatus pumped out. As the apparatus is evacuated, solvent distills out which is accompanied by a lowering of the temperature so that further external cooling is unnecessary. *After the pressure in the system no longer drops*, the apparatus is tilted so that the solutions flow into the graduated tubes. The contents of each tube is allowed to evaporate at the minimum pressure obtainable until each graduated tube is about one-half to two-thirds full (about 1–1.3 ml. with the 2-ml. tube and 0.5–0.65 ml. with the 1-ml. size). The actual reduction in volumes should occur during the time the solutions are in the bulbs and the unit should be tilted from time to time to note the volumes. The second capillary inlet tube is sealed by carefully applying the flame of the blast lamp at the constricted point and drawing the upper portion away *as soon as it is soft enough to do so* (see Fig. 205, *D*). If the evacuated capillary is heated too soft, there is danger of a hole being sucked in at this point, ruining the experiment. (It is suggested that the beginner practice sealing off an evacuated capillary before performing this part of the determination.)

The volumes in each graduated tube are recorded and the apparatus is set aside under isothermal conditions. If the laboratory is adequately air-conditioned *day and night,* it suffices to place the evacuated tubes in a closed container such as an aluminum pot or pressure cooker. Otherwise, the apparatus must be placed in a constant-temperature cabinet. (If a constant-temperature bath is used, the tubes must be dried quickly upon removal so that there is no chilling effect from evaporation of bath liquid on their surface.)

Each day thereafter the volumes of both graduated tubes are recorded. If desired, a graph may be made plotting the volume of each against the time as shown in Fig. 206. As time goes on, solvent will distill from the solution of greater vapor pressure to the one of lesser, as explained above. Finally, the volumes become constant and equimolar, and the experiment is completed.

Calculation:

The molecular weight of an unknown is calculated from the final data obtained by substitution in the following formula[16,79]:

$$M_1 = \frac{G_1 MV}{GV_1}$$

where

M = M.W. of standard
V = volume of solution of standard
G = wt. of standard
M_1 = M.W. of unknown
V_1 = volume of solution of unknown
G_1 = wt. of unknown

Example:

10.876 mg. of an unknown and 13.013 mg. of azobenzene are used. At equilibrium (after 7 days) the volume of the solution containing the unknown is 1.65 ml. and that containing the azobenzene is 1.05 ml. (Note: M.W. of azobenzene = 182.228.)

$$\therefore M_1 = \frac{10.876 \times 182.228 \times 1.05}{13.013 \times 1.65}$$
$$= 97.$$

The allowable error for the determination is $\pm 5\%$ of the molecular weight.

ADDITIONAL INFORMATION FOR CHAPTER 21

It was stated in the beginning of this chapter that, in the opinion of the author, neither of the two methods described in detail is completely satisfactory. Recently, the use of thermistors is attracting attention and one method has been described using them, which, from reports, appears quite interesting from both the standpoints of speed and accuracy.[26,34,55,83a] As this book goes to press, the unit in the author's laboratory has been in use only a few months, but the indications are that this method and apparatus will prove to be reliable for a variety of compounds.

This method described by Neumayer[48] employs isothermal distillation and requires only 3 minutes for equilibrium. Pure solvent and solution are exposed to the vapor of the solvent. Two thermistors are used, one for the solvent and the other for the solution. The solvent temperature remains constant, being subject to equal rates of evaporation and condensation. The solution, however, increases in temperature due to vapor condensation. The temperature difference is proportional to the mol fraction of the solute. This method is described below.

Apparatus[48,82]

The Wheatstone bridge circuit used is shown in Fig. 207. The various components of the circuit are as follows: B_1, 4.5 volt battery; S_1, single-pole, single-throw switch; R_1 and R_2, 100,000 ohm wire-wound fixed resistors; R_3, 1,000 ohm wire-wound variable resistor; R_4, 20 ohm wire-wound variable resistor; R_5, 1,000 ohm ten-turn, wire-wound micropotentiometer; TM_1 and TM_2, type $GA_{51}P8$ thermistors made by Fenwal Electronics, Inc., Framingham, Massachusetts; M_1, 0–50 microammeter; M_2, Null Detector Model No. 104WIG made by the Minneapolis-Honeywell Regulator Co., Minneapolis, Minnesota.

The solvent-vapor chamber and thermistor assembly are shown in Fig. 208. The various parts are as follows: A, glass solvent-vapor chamber; B, Bakelite cap machined to fit ⚜ 71/25 glass joint; C, 6-mm. stainless steel rod support for thermistor assembly; D, Teflon thermistor support; E, 5-mm. glass tubing; F, Fenwal type $GA_{51}P8$ thermistors; G, cupric oxide-phosphoric acid cement; H, 40-turn platinum wire coils made with 0.3-mm. diameter wire; I, 3-mm. thick absorbent paper lining the total depth and about three-fourths of the circumference of the inner chamber wall and covered on both sides with 1.5-mm. mesh aluminum screen; J, stainless steel crucible; K, solvent; L, constant temperature water bath.

The apparatus is now commercially available.[82] This model has replaced the crucible with a small funnel through which the excess solution may be removed by suction. The medicine droppers are inserted in the cap of the chamber and kept there at all times to prevent temperature variations.

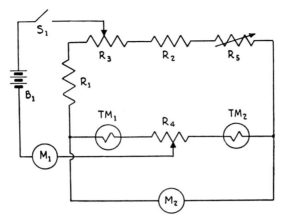

FIG. 207. Neumayer apparatus for determination of molecular weight—Wheatstone bridge circuit.

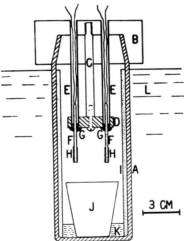

FIG. 208. Neumayer apparatus for determination of molecular weight—solvent-vapor chamber vs. thermistor assembly.

Procedure

The apparatus shown in Fig. 208 is immersed in a water bath thermostated at about 30° C. ± 0.01. Solvent is poured into the vapor chamber so that there is about one cm. of excess solvent in the bottom of the chamber over that required to saturate the absorbent paper lining the inside wall of the chamber. Both platinum wire coils covering the thermistors are rinsed with about 0.1–0.2 ml. of solvent. The solvent can be readily placed on the coils by means of an elongated medicine dropper inserted into the vapor chamber through a hole in the Bakelite cap. The medicine dropper is stored, when not in use, in copper tubing which is sealed on the bottom and immersed in the water bath. The micropotentiometer, R_5, shown in Fig. 207 is set at zero resistance and switch, S_1, is turned on. When temperature equilibrium has been established within the vapor chamber, the bridge is balanced using resistors, R_3 and R_4. The solvent in contact with the sensing thermistor, TM_1, is replaced by rinsing with 0.1–0.2 ml. of solution which has been previously brought to 30° C. by immersion of its container in the water bath. The bridge is balanced using the 1000 ohm micropotentiometer.

Approximately 3 minutes are required for the sensing thermistor to reach its maximum temperature. The maximum resistance of the micropotentiometer is recorded and this value is used to calculate the molecular weight of the solute. The sensing thermistor is rinsed with solvent and the micropotentiometer is returned to zero resistance. Temperature equilibrium will again be established in about 3 minutes, at which time the molecular weight of the next sample may be determined. If desired, the sample may be recovered from the stainless steel crucible within the solvent-vapor chamber.

Calculation:[48]

Thermistor resistance varies logarithmically with temperature but only a maximum of approximately 1% of the total resistance of one of the thermistors

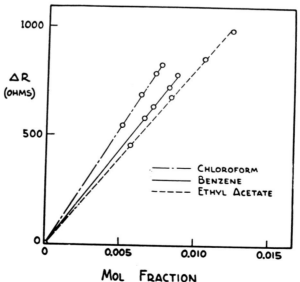

FIG. 209. Relationship of ΔR vs. mol fraction of solute (Neumayer method for determination of molecular weight).

is utilized in these experiments so ΔT is essentially proportional to ΔR. It is also true that $\Delta R = K \times MF$ (see Fig. 209), where ΔT and ΔR are the changes in temperature and resistance, respectively, of the sensing thermistors, K is a proportionality constant and MF is the mol fraction of solute in solution.*[41] Substituting the appropriate terms for MF and rearranging yields the equation:†

$$MW \text{ solute} = \frac{g \text{ solute} \times MW \text{ solvent } (K - \Delta R)}{\Delta R \times g \text{ solvent}}$$

where MW is molecular weight and g is weight in grams. K is determined for *each* solvent by making measurements on known amounts of test compounds.

TABLE 37
ADDITIONAL INFORMATION ON REFERENCES‡ RELATED TO CHAPTER 21

In addition to the above, as well as the procedures presented in detail in the preceding pages of this chapter, the author wishes to call to the attention of the reader the references listed in Table 37. (See statement at top of Table 4 of Chapter 1, regarding completeness of this material.)

Books
Belcher and Godbert, 4, 5
Bonnar, Dimbat, and Stross, 7
Clark, E. P., 16
Clark, S. J., 17
Furman, 27
Grant, 29, 30
Millard, 41
Milton and Waters, 42
Niederl and Niederl, 50, 51
Niederl and Sozzi, 52
Pregl, 63
Roth, 70–73
Steyermark, 79
Sucharda and Bobranski, 80

Collaborative study
Ogg, 54, 55

Review
Wilson, C. L., 84

Ultramicro-, submicro-methods
Brandstätter, 9
Gibson and Currie, 28
Guerrant, 31
Kofler and Brandstätter, 38

Ebullioscopic (boiling point rise)
Belcher and Godbert, 4, 5
Colson, 19, 20
Dimbat and Stross, 24
Furman, 27
Grant, 29, 30
Magee and Wilson, 40
Niederl and Niederl, 50, 51
Pregl, 63
Rieche, 68, 69
Roth, 70–73
Sucharda and Bobranski, 80, 81
Wilson, 84

* Mol fraction $= \dfrac{\text{mols solute}}{\text{mols solute} + \text{mols solvent}}$

† The slope of the line obtained by plotting the resistance against the concentration also may be used in calculating the results.[83a]

‡ The numbers which appear after each entry in this table refer to the literature citations in the reference list at the end of the chapter.

21. Molecular Weight

TABLE 37 (*Continued*)

Cryoscopic (depression of melting or freezing point)

Aluise, 1
Brandstätter, 9
Csokan, 22
Furman, 27
Gibson and Currie, 28
Grant, 29, 30
Iwamoto, 36
Kofler and Brandstätter, 38
Kubota and Yamane, 39
Niederl and Niederl, 50, 51
Pirsch, 57
Rast, 66, 67
Roth, 70–73
Smit, 76
Smit, Ruyter, and Van Wijk, 77
Steyermark, 79
Zscheile and White, 87

Using Kofler micro hot stage

Brandstätter, 9
Gibson and Currie, 28
Kofler and Brandstätter, 38

Osmotic pressure methods

Christiansen and Jensen, 14
Forster and Breitenbach, 25

Effusion method

Nash, 45

Vaporimetric, vapor pressure comparison methods

Blank, 6
Bratton and Lochte, 10
Furman, 27
Gysel and Hamberger, 32
Hawkins and Arthur, 33
Nash, 44
Nernst, 47
Niederl and Niederl, 50, 51
Puddington, 65

Isothermal distillation methods

Barger, 2, 3
Bourdillon, 8
Childs, 13

Isothermal distillation methods (*Cont.*)

Clark, E. P., 15, 16
Francis, 26
Guerrant, 31
Hofstaeder, 34
Hoyer, 35
Morton, Campbell, and Ma, 43
Neumayer, 48
Niederl and Levy, 49
Niederl and Niederl, 50, 51
van Nieuwenburg and van Ligten, 53
Ogg, 54, 55
Roth, 72
Spies, 78
White and Morris, 83
Wilson and Bini, 83a
Wright, 86

Potentiometric, nonaqueous titration

Brockmann and Meyer, 12

Neutralization equivalent

See Chapter 15

Spectrophotometric method

Cunningham, Dawson, and Spring, 23

Apparatus

British Standards Institution, 11
Christiansen and Jensen, 14
Clark, E. P., 16
Clark, S. J., 17
Colson, 18–20
Dimbat and Stross, 24
Hawkins and Arthur, 33
Hoyer, 35
Nash, 45, 46
Niederl and Niederl, 50, 51
van Nieuwenburg and van Ligten, 53
Roth, 70–73
Steyermark, 79
White and Morris, 83

Use of thermistors

Dimbat and Stross, 24
Francis, 26
Hofstaeder, 34
Neumayer, 48
Ogg, 55
Wilson and Bini, 83a

REFERENCES

1. Aluise, V. A., *Ind. Eng. Chem., Anal. Ed.,* **13**, 365 (1941).
2. Barger, G., *Ber.,* **37**, 1754 (1904).
3. Barger, G., *J. Chem. Soc.,* **85**, 286 (1904).
4. Belcher, R., and Godbert, A. L., "Semi-Micro Quantitative Organic Analysis," Longmans, Green, London, New York, and Toronto, 1945.
5. Belcher, R., and Godbert, A. L., "Semi-Micro Quantitative Organic Analysis," 2nd ed., Longmans, Green, London, 1954.
6. Blank, E. W., *Mikrochemie,* **13**, 149 (1933).
7. Bonnar, R. U., Dimbat, M., and Stross, F. H., "Number Average Molecular Weight," Interscience, New York, 1958.
8. Bourdillon, J., *J. Biol. Chem.,* **127**, 617 (1939).
9. Brandstätter, M., *Mikrochemie ver. Mikrochim. Acta,* **36/37**, 291 (1951).
10. Bratton, A. C., and Lochte, H. L., *Ind. Eng. Chem., Anal. Ed.,* **4**, 365 (1932).
11. British Standards Institution, *Brit. Standards,* **1428**, Pt. K1 (1958).
12. Brockmann, H., and Meyer, E., *Ber.,* **86**, 1514 (1953).
13. Childs, C. E., *Anal. Chem.,* **26**, 1963 (1954).
14. Christiansen, J. A., and Jensen, C. E., *Acta Chem. Scand.,* **5**, 849 (1951).
15. Clark, E. P., *Ind. Eng. Chem., Anal. Ed.,* **13**, 820 (1941).
16. Clark, E. P., "Semimicro Quantitative Organic Analysis," Academic Press, New York, 1943.
17. Clark, S. J., "Quantitative Methods of Organic Microanalysis," Butterworths, London, 1956.
18. Colson, A. F., *Analyst,* **77**, 139 (1952).
19. Colson, A. F., *Analyst,* **80**, 690 (1955).
20. Colson, A. F., *Analyst,* **83**, 169 (1958).
21. Corning Glass Works, Corning, New York.
22. Csokan, P., *Magyar Kémi. Folyóirat,* **48**, 56 (1942).
23. Cunningham, K. G., Dawson, W., and Spring, F. S., *J. Chem. Soc.,* p. 2305 (1951).
24. Dimbat, M., and Stross, F. H., *Anal. Chem.,* **29**, 1517 (1957).
25. Forster, E. L., and Breitenbach, J. W., *Mikrochim. Acta,* p. 982 (1956).
26. Francis, H. J., Jr., Personal communication, 1960.
27. Furman, N. H., ed., "Scott's Standard Methods of Chemical Analysis," 5th ed., Vol. II, Van Nostrand, New York, 1939.
28. Gibson, D. T., and Currie, J., *Mikrochim. Acta,* p. 644 (1957).
29. Grant, J., "Quantitative Organic Microanalysis, Based on the Methods of Fritz Pregl," 4th ed., Blakiston, Philadelphia, Pennsylvania, 1946.
30. Grant, J., "Quantitative Organic Microanalysis," 5th ed., Blakiston, Philadelphia, Pennsylvania, 1951.
31. Guerrant, G. O., *Anal. Chem.,* **30**, 143 (1958).
32. Gysel, H., and Hamberger, K., *Mikrochim. Acta,* p. 254 (1957).
33. Hawkins, J. J., and Arthur, P., *Anal. Chem.,* **23**, 533 (1951).
34. Hofstaeder, R., Metropol. Microchem. Soc., New York, January 1960.
35. Hoyer, H., *Mikrochemie ver. Mikrochim. Acta,* **36/37**, 1169 (1951).
36. Iwamoto, K., *Sci. Repts. Tôhoku Imp. Univ. First Ser.,* **17**, 719 (1928).
37. Kamm, O., "Qualitative Organic Analysis," Wiley, New York, 1923.
38. Kofler, L., and Brandstätter, M., *Mikrochemie ver. Mikrochim. Acta,* **33**, 20 (1948).
39. Kubota, B., and Yamane, T., *Bull. Chem. Soc. Japan,* **2**, 209 (1927).
40. Magee, R. J., and Wilson, C. L., *Analyst,* **73**, 597 (1948).

21. Molecular Weight

41. Millard, E. B., "Physical Chemistry for Colleges," 1st ed., McGraw-Hill, New York, 1921.
42. Milton, R. F., and Waters, W. A., "Methods of Quantitative Microanalysis," 2nd ed., Arnold, London, 1955.
43. Morton, J. J., Campbell, A. D., and Ma, T. S., *Analyst,* **78**, 722 (1953).
44. Nash, L. K., *Anal. Chem.,* **19**, 799 (1947).
45. Nash, L. K., *Anal. Chem.,* **20**, 258 (1948).
46. Nash, L. K., *Anal. Chem.,* **23**, 1868 (1951).
47. Nernst, W., *Göttinger Nachrichten,* p. 75 (1903).
48. Neumayer, J. J., *Anal. Chim. Acta,* **20**, 519 (1959).
49. Niederl, J. B., and Levy, A. M., *Science,* **92**, 225 (1940).
50. Niederl, J. B., and Niederl, V., "Micromethods of Quantitative Organic Elementary Analysis," Wiley, New York, 1938.
51. Niederl, J. B., and Niederl, V., "Micromethods of Quantitative Organic Analysis," 2nd ed., Wiley, New York, 1942.
52. Niederl, J. B., and Sozzi, J. A., "Microanálisis Elemental Orgánico," Calle Arcos, Buenos Aires, 1958.
53. Nieuwenburg, C. J. van, and Ligten, J. W. L. van, *Anal. Chim. Acta,* **9**, 66 (1953).
54. Ogg, C. L., *J. Assoc. Offic. Agr. Chemists,* **41**, 294 (1958).
55. Ogg, C. L., Personal communication, 1959; *J. Assoc. Offic. Agr. Chemists,* **43**, 693 (1960).
56. Pharmacopeia of the United States by the Authority of the Pharmacopeial Convention, XIII, XIV, XV, XVI, Mack, Easton, Pennsylvania, 1947, 1950, 1955, and 1960.
57. Pirsch, J., *Angew. Chem.,* **51**, 73 (1938).
58. Pirsch, J., *Ber.,* **65**, 862 (1932).
59. Pirsch, J., *Ber.,* **65**, 1227 (1932).
60. Pirsch, J., *Ber.,* **66**, 349 (1933).
61. Pirsch, J., *Ber.,* **66**, 815 (1933).
62. Pirsch, J., *Ber.,* **66**, 1694 (1933).
63. Pregl, F., "Quantitative Organic Microanalysis," (E. Fyleman, trans. 2nd German ed.), Churchill, London, 1924.
64. Procter & Gamble Company, Cincinnati, Ohio.
65. Puddington, I. E., *Can. J. Research,* **27**, 151 (1949).
66. Rast, K., *Ber.,* **55B**, 1051 (1922).
67. Rast, K., *Ber.,* **55B**, 3727 (1922).
68. Rieche, A., *Ber.,* **59**, 2181 (1926).
69. Rieche, A., *Mikrochemie,* **12**, 129 (1932).
70. Roth, H., "Die quantitative organische Mikroanalyse von Fritz Pregl," 4th ed., Springer, Berlin, 1935.
71. Roth, H., "F. Pregl quantitative organische Mikroanalyse," 5th ed., Springer, Wien, 1947.
72. Roth, H., "Pregl-Roth quantitative organische Mikroanalyse," 7th ed., Springer, Wien, 1958.
73. Roth, H., "Quantative Organic Microanalysis of Fritz Pregl," 3rd ed., (E. B. Daw, trans. 4th German ed.), Blakiston, Philadelphia, Pennsylvania, 1937.
74. Scientific Glass Apparatus Company, Bloomfield, New Jersey.
75. Signer, R., *Ann.,* **478**, 246 (1930).
76. Smit, W. M., *Chem. Weekblad,* **36**, 750 (1939).
77. Smit, W. M., Ruyter, J. H., and Van Wijk, H. F., *Anal. Chim. Acta,* **22**, 8 (1960).

78. Spies, J. R., *J. Am. Chem. Soc.,* **55**, 250 (1933).
79. Steyermark, Al, "Quantitative Organic Microanalysis," Blakiston, Philadelphia, Pennsylvania, 1951.
80. Sucharda, E., and Bobranski, B., "Halbmikromethoden zur automatischen Verbrennung organischer Substanzen und ebullioskopische Molekulargewichtsbestimmung," p. 135, Vieweg, Braunschweig, Germany, 1929.
81. Sucharda, E., and Bobranski, B., *Przemyst Chem.,* **11**, 371 (1927).
82. Thomas, Arthur H., Company, Philadelphia, Pennsylvania.
83. White, L. M., and Morris, R. T., *Anal. Chem.,* **24**, 1063 (1952).
83a. Wilson, A., and Bini, L., Meeting-in-Miniature, North Jersey Section, American Chemical Society, South Orange, New Jersey, 1959; *Anal. Chem.,* in press.
84. Wilson, C. L., *Analyst,* **73**, 585 (1948).
85. Wilson, H. N., and Heron, A. E., *J. Soc. Chem. Ind. (London),* **60**, 168 (1941).
86. Wright, R., *Analyst,* **73**, 387 (1948).
87. Zscheile, F. P., and White, J. W., Jr., *Ind. Eng. Chem., Anal. Ed.,* **12**, 436 (1940).

CHAPTER 22

Microdetermination of Some Physical Constants

   I. MICRODETERMINATION OF MELTING POINT
  II. MICRODETERMINATION OF BOILING POINT
 III. MICRODETERMINATION OF SPECIFIC GRAVITY
 IV. MICRODETERMINATION OF OTHER PHYSICAL CONSTANTS

Although this book deals with quantitative organic microanalysis, it is felt that brief descriptions of the methods of obtaining some of the more common physical constants are in order. Melting point, boiling point, and specific gravity are the constants most commonly determined[46] and consequently methods for these will be described while mere mention of some of the rest will be made.

I. MICRODETERMINATION OF MELTING POINT

The melting point determination performed in the usual manner is in reality a micromethod. It is assumed that the reader has had considerable experience with the above procedure. The author wishes to point out that the above is the most reliable one when the proper corrections are made (calibration of the thermometer and correction for exposed stem[46,72]) and should always be used when possible. A new commercially available apparatus[89] for determining melting points was described in connection with the Rast method for the determination of molecular weight (Fig. 203, Chapter 21). On occasions when material is scarce it is desirable to use another method for which microscopic quantities of substance are required. Even when adequate quantities of material are available, the microscopic method may be used to advantage. For example, with it the presence of more than one compound in a sample is demonstrated easily, since each individual crystal can be observed during the melting process. In such cases, under the microscope each substance is seen to melt sharply at its own melting point, unaffected by the other. Also the method can be used successfully to identify various substances[2,49,51] and the author has used it to advantage on numerous occasions.* The Arthur H. Thomas Company,[89] the

---

* During the period of operation of the author's laboratory, manufacturing problems

manufacturer of the apparatus used, has prepared a pamphlet giving a rather complete bibliography on this subject.

Apparatus[89]

The apparatus assembly required is shown in Fig. 210. It consists of an electrically heated hot stage into which is inserted a thermometer. The sample is placed on a microscopic slide which rests directly upon the heated metal sur-

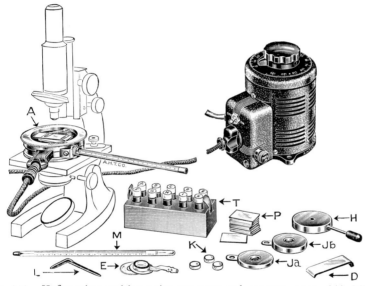

FIG. 210. Kofler micro melting point apparatus and accessory parts. (A) Micro hot stage. (D) Baffle. (L) Tool. (M) Thermometer. (T) Test reagents. (P) Microslides. (E) Vacuum sublimation chamber. (H) Cooling block. (Ja, Jb) Sublimation blocks. (K) Sublimation dishes.

have arisen, particularly in regard to the aging of pharmaceuticals in ampul form used for parenteral injection. Since the amount of active ingredient in each ampul is usually a matter of a few milligrams, products of deterioration usually amount to a matter of micrograms (compare the section on ampul testing by the manometric apparatus, Chapter 18). If precipitates result from the deterioration they are heated routinely with the apparatus described below before proceeding with an analysis. In this way, the author has been able to detect the presence of several substances in a precipitate, weighing a small fraction of a milligram, as well as to identify at least one or two components of it within a matter of minutes. For example, elementary sulfur acts very characteristically—it melts and then sublimes in the form of tiny yellow droplets on the upper microscopic slide (see below). Sometimes the elementary sulfur is present as a contaminant and the melting stage cannot be noticed because of the relatively small quantity but the characteristic sublimate can be seen always.

face. Over the sample is placed a coverglass and the combination covered by a raised baffle (Fig. 211). A glass plate, ground around the outer rim to insure close contact with the metal parts to prevent circulation of air currents, is used as a cover. The assembly is placed on the stage of a microscope so that the

FIG. 211. (*Left*) Kofler micro melting point apparatus, coverglass and baffle removed. (*Right*) Baffle in position on apparatus.

sample can be viewed through the objective lens and eye piece. The hot plate is connected to the electric service through a suitable variable transformer for controlling the temperature. The thermometer is calibrated by determining the melting points of a number of standards whose *corrected* melting points are known. In this manner, the melting points obtained with unknown samples

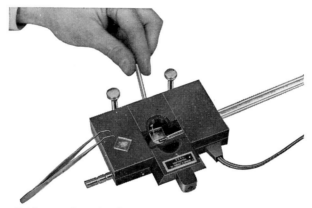

FIG. 212. Cold stage for microdetermination of freezing points for use on stage of microscope similar to Kofler micro melting point apparatus (see Fig. 210).

are the corrected ones. (Note: The hot stage is equipped also with sublimation units—Fig. 210.)

A cold stage for determining freezing points of liquids is also commercially available[89] (Fig. 212), as well as an audiohm thermistor attachment which permits the determinations *without removing the eye from the microscope* (Fig. 213). Other advantages of the thermistor method are faster response, obtained by locating a smaller sensor closer to the field of observation than is possible with a thermometer, and the elimination of stage-calibrated thermometers. Uninterrupted observation of the fusion through all its phases is made possible by measuring the temperature as a function of the thermistor resistance, determined at any instant by nulling the audible bridge signal.

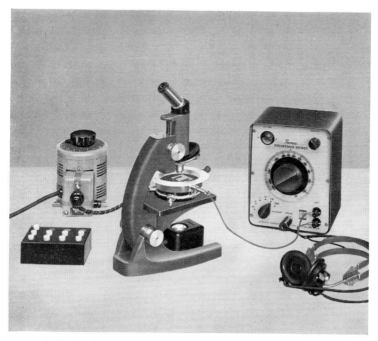

FIG. 213. Audiohm thermistor attachment for Kofler micro melting point apparatus.

Procedure

In the procedure[14,16,37,38,48-52,60,79,80,82,87,89,92,96] the micro hot plate with inserted thermometer is set on the stage of a microscope. A small amount of sample is placed on a microscopic slide (25 × 37 mm.) and covered with a micro coverglass. The combination then is placed directly on the metal micro hot stage, and over the glasses is placed a glass baffle (Fig. 211). Finally, the large circular glass plate with the ground rim is set on top of the metal rim of the hot plate. The source of electricity is attached and the temperature slowly raised. As the melting point is approached, the rate of temperature rise should be reduced to that of 0.5° C. per minute. The micro melting point is that temperature at which the smaller crystals have changed into droplets and the outlines of the larger crystals start to round off. In this manner the micro melting point obtained is that at which the solid and liquid phases are in equilibrium. If the substance is stable at its melting point, cooling about 0.5° C. causes the crystals to start growing.

II. MICRODETERMINATION OF BOILING POINT

When enough material is available, miniature distillation flasks and columns of the same designs as ordinarily employed in the organic laboratory, may be used to obtain the boiling range of the liquid in question. However, when only fractions of a milliliter are available the boiling point is obtained readily by use of the following procedure. The method is of limited value since it is adaptable only to pure compounds.[46]

Apparatus[46,87]

MELTING POINT BATH

An ordinary melting point bath, mechanically stirred, of the type recommended for the Rast molecular weight determination is required.

THERMOMETERS

Calibrated Anschütz thermometers of the desired ranges are required. An auxiliary thermometer, 0–250° C. range, is used when corrections for the exposed stem are made.

BOILING POINT CAPILLARIES

Glass tubing is drawn out so as to obtain a section of about 2–3 mm. I.D. and about 150 mm. in length. A needle-point flame from a blast lamp is used to seal off a small length about 40–50 mm. from the one end. The smaller end is then cut off with a file so that the sealed off portion is about 6–8 mm. in length from the open end up to the seal (see Fig. 214). (It is preferable to make the capillary boiling tube in this manner rather than to attempt to seal off the small section at a distance of about 6–8 mm. from the open end.) This small tube acts as a condenser.

SMALL TEST TUBE

A small Pyrex[21] test tube (approximately 7 mm. $\times$ 50 mm.) is used as the distilling vessel. If this size is too large for the amount of sample, a smaller tube may be made by sealing off the proper size of glass tubing.

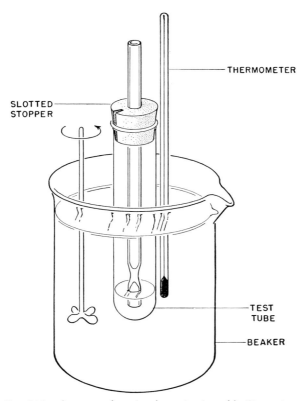

FIG. 214. Apparatus for microdetermination of boiling points.

Procedure[46,87]

The sample is placed in the small test tube and the capillary or condenser tube is put into place with the aid of a *slotted* stopper. (The test tube must not be stoppered shut on inserting the condenser. It must be open to the atmosphere.) The condenser tube should be immersed to a depth of about 2–3 mm. in the liquid as shown in the Fig. 214. The test tube and condenser capillary are mounted in the heating bath as shown. The depth of the liquid (Crisco* or silicone fluid) in the bath should be great enough (about 7 cm.) so that it extends about 3 cm. below the bottom of the test tube and the same distance above the glass seal in the condenser. The Anschütz thermometer is attached to the test tube with the center of the mercury-filled bulb at the same level as that of the liquid under test. The bath is vigorously stirred

* Hydrogenated vegetable oil.[74]

(mechanically) and heated. As the temperature rises, the air trapped in the condenser tube expands and escapes from the lower open end. The temperature is raised until it is *above* the boiling point of the liquid under test as shown by the *cessation* of escaping air. The temperature of the bath is lowered slowly (0.5° C. per minute), during which vapor bubbles cease to emerge from the lower end, and then the liquid begins to suck back into the condenser tube. The boiling point is that temperature at which the level of the liquid *in* the condenser capillary is the same as that *outside* of it. At this temperature the liquid is in equilibrium with the vapor. The apparatus should be calibrated using several pure substances of known boiling points covering the desired range, after which the necessary corrections should be made when working with unknowns. It is also advisable to use corrections for the exposed stem accomplished in the usual manner. For this the bulb of an auxiliary thermometer is placed against the Anschütz, midway between the surface of the liquid bath and the mark representing the boiling point. The average stem temperature is obtained thusly and used in the following formula[46,72]:

$$\text{Correction for exposed stem} = +N(t - t')0.000154$$

where

$N =$ the number of degrees on the stem of the thermometer between the surface of the bath liquid and the temperature read

$t =$ the temperature read on the Anschütz thermometer

$t' =$ the average temperature of the exposed mercury column (as read on the auxiliary thermometer)

$0.000154 =$ apparent coefficient of expansion of mercury in glass

The correction obtained by calculation is added to the recorded temperature in addition to whatever corrections ($+$ or $-$) are necessary from the calibration of the thermometer.

III. MICRODETERMINATION OF SPECIFIC GRAVITY USING THE GRAVITOMETER

The specific gravity of liquids may be determined quickly, making use of the principle of hydrostatic balance.[23,32] When two U-tube manometers containing liquids of different densities are connected to a common source of reduced pressure, the heights of the liquids in the connecting manometers are inversely proportional to their densities (see Fig. 215, *left*) as expressed by the following equation:

$$h_1 d_1 = h_2 d_2$$

or

$$\frac{h_1}{h_2} = \frac{d_2}{d_1}$$

where

h_1 = height of liquid No. 1
d_1 = density of liquid No. 1
h_2 = height of liquid No. 2
d_2 = density of liquid No. 2

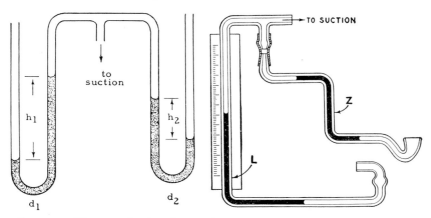

FIG. 215. Diagrams showing details of Fisher-Davidson gravitometer. (*Left*) Principle. (*Right*) Diagram of apparatus.

The apparatus shown in the Figs. 215, *right*, and 216 has been designed so that a direct reading of specific gravity is obtained when a reference liquid is used along with the unknown. The height of the liquid No. 2, the un-

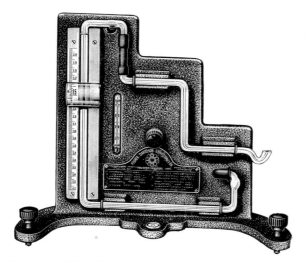

FIG. 216. Fisher-Davidson gravitometer, commercial model.

known, is held constant at a height equal to S in the Z tube. When the reference liquid No. 1 is ethylbenzene ($d_4^{20} = 0.867$), the above equation becomes

$$h_1 \times 0.867 = S d_2$$

or
$$h_1 = \frac{S d_2}{0.867}$$

Now if S is equal to 0.867 unit of the scale attached to the vertical portion of the tube, L, the equation becomes

$$h_1 = \frac{0.867 d_2}{0.867}$$

or
$$h_1 = d_2$$

The tube, Z, containing the unknown is a simple manostat which, within limits, insures a fixed pressure difference between the atmosphere and the tube connecting the two tubes, L and Z. The value of this fixed pressure varies directly with the density of the liquid contained in the tube, Z, and is measured by means of the L-tube manometer, which is fitted with a scale graduated in terms of d_4^{20}. Consequently, reading the scale attached to the tube, L, gives a direct reading of the specific gravity of the liquid at 20° C. compared with water at 4° C. (d_4^{20}) over the range 0.600–2.000, when ethylbenzene is the reference liquid. If some other reference is placed in the tube, L, a factor must be applied. For example, for specific gravities above 2.000, carbon tetrachloride is used, and then the scale readings must be multiplied by the ratio

$$\frac{\text{sp. gr. of } CCl_4}{\text{sp. gr. of } C_6H_5C_2H_5} = \frac{1.595}{0.867}$$

$$= 1.84$$

For most organic liquids the readings obtained correspond closely for d_4^{20} even if made at temperatures several degrees above or below 20° C. This is true since the measurement of one organic liquid (the unknown) is made relative to another one (the reference), and most organic liquids have temperature coefficients of density which lie within a narrow range.[9,32] On the average, the values of specific gravity observed at 20 ± 5° C. will not differ by more than 0.1% from those observed at 20° C.*

* Where the coefficient of expansion of the liquid under test is known, the correction, calculated from the following formula, is *added* to the observed value:

$$\text{Correction} = \frac{(t - 20°)(C_x - 0.101)(d_{\text{Obs.}})}{100}$$

where t = temperature at which the observation is made

C_x = coefficient of expansion of the liquid

0.101 = percentage change in the density of ethylbenzene in the vicinity of 20° C.

$d_{\text{Obs.}}$ = observed density reading.

Reagents

ETHYLBENZENE[32]

Pure ethylbenzene ($d_4^{20} = 0.867$) is required as the reference liquid for measurements on unknowns whose specific gravities are in the range of 0.600–2.000.

CARBON TETRACHLORIDE[32]

Pure carbon tetrachloride ($d_4^{20} = 1.595$) is required as the reference liquid, in place of ethylbenzene, for measurements on unknowns whose specific gravities are above 2.000. When carbon tetrachloride is used, the scale readings must be multiplied by

$$\frac{1.595}{0.867} \text{ or } 1.84$$

Apparatus

The apparatus[22,32,87] used is the Fisher-Davidson gravitometer and is shown in Figs. 215, *right*, and 216. It consists of the tube, L, which contains the reference liquid, No. 1, and a tube, Z, for the unknown. The tubes are connected to a common source of reduced pressure produced by the movement of a milled knob counterclockwise along a section of rubber tubing.

For most organic liquids the standard tube, Z, which is approximately 1.9 mm. I.D., requiring about 0.3 ml. of sample, is used. For samples of low viscosity, the bore of the tube may be even less. For liquids of high viscosity, tubes of approximately 2.6–3.0 mm. I.D. or even 4 mm. may be used. The former requires about 1.0 ml. and the latter about 2.0 ml. of sample, respectively. The larger bores are necessary because equilibrium is obtained more slowly with this type of material.

Procedure[23,32,87]

The reference liquid (ethylbenzene for the range 0.600–2.000, or carbon tetrachloride for the range above 2.000) is added to the tube, L, through the lower end until the liquid level stands between the two marks of the receptacle on the right arm. (If air pockets are present, the milled knob on the suction device is turned counterclockwise as far as possible. The rubber connector between the two tubes, L and Z, is pinched and the milled knob turned slowly in a clockwise direction, forcing the liquid back into the receptacle where the air will be expelled. Then the milled knob is slowly turned counterclockwise to return the liquid free from air pockets.)

The milled knob of the suction device is turned clockwise as far as possible. The tube, Z, is connected to the tube, L, by means of the rubber connector and then clamped into place on the support. The instrument is made level by means of the leveling screws provided for this purpose. Sufficient unknown sample is added to fill completely the cup of the tube, Z. Then the milled knob is turned slowly counterclockwise drawing the liquid up into the tube, Z, so that the menisci stand approximately in the middle of the horizontal arms of the tube, Z. (If the menisci do not reach the above points, the liquid is returned to the cup by reversing the knob and making the necessary adjustment. If air pockets are present in the capillary of the tube, Z, the receptacle of the tube, L, is stoppered and the milled knob turned so as to return the liquid in the tube, Z, to the cup momentarily.) The slide on the scale is moved until the index coincides with the meniscus of the reference liquid in the tube, L, and the position of the index on the scale is read, estimating the third decimal place. The result is the specific gravity of the liquid at 20° C. compared with water at 4° C. (d_4^{20}). It is good practice to move the liquid in the tube, Z, forward and backward, slightly, by means of the milled knob, and then check the reading. This insures that equilibrium has been reached.

A summary of the operations is given by the following:

1. The level of the standard in the right arm of the tube, L, should be between the two etched lines
2. Turn the milled knob to the extreme right (clockwise)
3. Connect the tube, Z, to tube, L
4. Level instrument
5. Add sufficient liquid of unknown specific gravity to fill completely the cup of the tube, Z
6. Turn the milled knob to the left (counterclockwise) until the liquid is drawn up into the tube, Z, so that the menisci stand approximately in the middle of the horizontal arms of the tube, Z
7. Read the level of the standard in the tube, L, to the third decimal.

USING MICRO WEIGHING PIPETTES, DENSITY-TYPE (PYCNOMETERS)

The specific gravity may be determined with the aid of weighing pipettes or pycnometers[1,69,88] of the types shown in Fig. 217. These are used in the same manner as are regular pycnometers with which the reader should be familiar, being weighed empty and then full. Knowing the weight, temperature, and volume, the specific gravity is obtained.

Three sizes are specified, designed for the determination of density on small amounts of volatile, viscous, or hygroscopic liquids. Most of the substance can easily be recovered after the determination. The decigram size has a capacity of 100 microliters and is especially suitable for highly viscous liquids. The centigram size has a capacity of 40–80 microliters, and the milligram size a capacity of 10–30 microliters. The centigram and milligram sizes are graduated in one mm. divisions and can be used even when the total amount

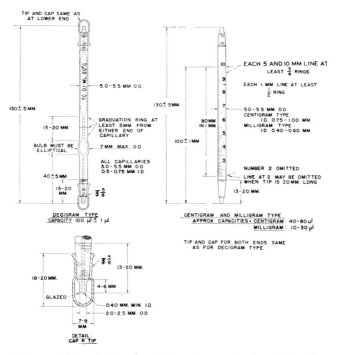

FIG. 217. Micro weighing pipettes, density-type (pycnometers)—details of construction.

of sample available is less than the maximum capacity of the micropipette. The sample may occupy any portion of the graduated stem, thereby eliminating the necessity of making a precise adjustment to a fixed calibration mark.

The pipettes are calibrated according to the directions given in Chapter 5 or by the methods proposed by Alber.[1] Water, bromoform, or mercury are used for the decigram size (with mercury, the operation is more difficult). With the centigram and milligram sizes, measurement of the bore by means of a microscope followed by the use of distilled water is sufficient. The reader is referred to the original literature.[1,88]

IV. MICRODETERMINATION OF OTHER PHYSICAL CONSTANTS

The determination of other constants such as the refractive index, optical rotation, etc., can be determined also on a microscale and will be briefly discussed. For details of the methods the reader is referred to the literature cited.

REFRACTIVE INDEX

The refractive index as determined by the ordinary refractometers found in the laboratory may be considered to be a microdetermination because one drop of liquid is required. However, a truly microrefractometer[2,3,36,68,84,87,91] is commercially available[89] which requires 6–8 cu. mm. of sample, 5–6 of which can

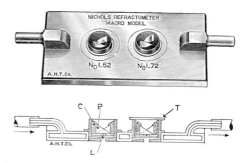

FIG. 218. (*Top*) Nichols refractometer. (*Bottom*) Cross-section of Nichols refractometer, indicating water circulation within jacket. (C) Cement. (P) Prism. (L) Engraved line. (T) Coverglass.

be recovered, depending upon the physical properties. The instrument is shown in Fig. 218. Jelley[44] also designed an instrument which is rapid and simple and uses 0.1 cu. mm. of liquid.

The refractive index of a *solid* may be determined by a comparison under the microscope with standard samples prepared from a variety of powdered glass[30,76].

OPTICAL ROTATION

The optical rotation of an optically active compound may be done on a microscale by using a micropolarimeter tube and an ordinary polarimeter.[24,27,28,38,79–81]

TABLE 38
ADDITIONAL INFORMATION ON REFERENCES* RELATED TO CHAPTER 22

In addition to the procedures presented in the preceding pages of this chapter, the author wishes to call to the attention of the reader the references listed in Table 38. (See statement at top of Table 4 of Chapter 1, regarding completeness of this material.)

Books
Belcher and Godbert, 4, 5
Chamot and Mason, 14
Cheronis, 15
Cheronis and Entrikin, 16, 17
Clark, E. P., 18
Clark, S. J., 19
Emich and Schneider, 25
Gibb, 36
Grant, 37, 38
Kamm, 46
Kofler and Kofler, 49, 50
Kofler and Kofler with Brandstätter, 51
Kofler, Kofler, and Mayrhofer, 52
Pharmacopeia, 72
Roth, 78–81
Schneider, 82
Shriner and Fuson, 84
Shriner, Fuson, and Curtin, 85
Steyermark, 87
Weissberger, 91, 92

Melting point (freezing point)
Brandenburg and Frobose, 10
Brandstätter and Thaler, 12
Chakravarti and Chaudhuri, 13
Fischer, R., 29
Fischer and Reichel, 31
Hewitt, 39
Hilbck, 40
Hippenmeyer, 41, 42
Hozumi and Morita, 43
Kofler and Kofler, 53
Kofler and Sitte, 54
Kofler, 55
Maffei and Wasicki, 57
Mastrangelo and Aston, 58
Matthews, 59
McCrone, 60, 61
McCrone and Massenberg, 62
McCullough and Waddington, 63

Melting point (freezing point) (Cont.)
Morton and Mahoney, 64
Müller and Zenchelsky, 65
Opfer-Schaum, 70
Owen and Reid, 71
Pinkus and Waldrep, 73
Sekera and Pokorny, 83
Somereyns, 86
Tschamler, 90
Wiegand, 94
Yanagimoto, 95

Boiling point
Böhme and Böhm, 7
Emich and Schneider, 25
Garcia, 35
Morton and Mahoney, 64
Owen and Reid, 71
Wiberley, Siegfriedt, and Benedetti-Pichler, 93
Wiegand, 94

Specific gravity (density)
Alber, 1
Belcher and Godbert, 4, 5
Blank and Willard, 6
Clemo and McQuillen, 20
Fenger-Eriksen, Krogh, and Ussing, 26
Furter, 34
Nettesheim, 67
Rosebury and Van Heyningen, 77

Optical rotation
Donau, 24
Fischer, E., 27, 28
Kacser and Ubbelohde, 45

Vapor pressure
Bonhorst, Althouse, and Triebold, 8
Nash, 66

* The numbers which appear after each entry in this table refer to the literature citations in the reference list at the end of the chapter.

TABLE 38 (*Continued*)

Refraction, molecular refraction, refractive index	Refraction, molecular refraction, refractive index (*Cont.*)
Alber and Bryant, 3	Reimers, 76
Brandstätter-Kuhnert and Martinek, 11	Roth, 78–81
Frediani, 33	
Kirk and Gibson, 47	**Viscosity**
Reihlen, 75	Labout and van Oort, 56

REFERENCES

1. Alber, H. K., *Ind. Eng. Chem., Anal. Ed.*, **12**, 764 (1940).
2. Alber, H. K., *Mikrochemie ver. Mikrochim. Acta*, **29**, 294 (1941).
3. Alber, H. K., and Bryant, J. T., *Ind. Eng. Chem., Anal. Ed.*, **12**, 305 (1940).
4. Belcher, R., and Godbert, A. L., "Semi-Micro Quantitative Organic Analysis," Longmans, Green, London, New York, and Toronto, 1945.
5. Belcher, R., and Godbert, A. L., "Semi-Micro Quantitative Organic Analysis," 2nd ed., Longmans, Green, London, 1954.
6. Blank, E. W., and Willard, M. L., *J. Chem. Educ.*, **10**, 109 (1933).
7. Böhme, H., and Böhm, R. H., *Mikrochim Acta*, p. 270 (1959).
8. Bonhorst, C. W., Althouse, P. M., and Triebold, H. O., *J. Am. Oil Chemists' Soc.*, **26**, 375 (1949).
9. Bosart, L. W., *Perfumery Essent. Oil Record*, **30**, 145 (1939).
10. Brandenburg, V. W., and Frobose, H., *Pharmazie*, **9**, 1009 (1954).
11. Brandstätter-Kuhnert, M., and Martinek, A., *Mikrochim. Acta*, p. 803 (1958).
12. Brandstätter, M., and Thaler, H., *Mikrochemie ver. Mikrochim. Acta*, **38**, 358 (1951).
13. Chakravarti, R. N., and Chaudhuri, K. N., *Bull. Calcutta School Trop. Med.*, **4**, 16 (1956).
14. Chamot, E. M., and Mason, C. W., "Handbook of Chemical Microscopy," 2nd ed., Vol. I, p. 207, Wiley, New York, and Chapman & Hall, London, 1938.
15. Cheronis, N. D., "Semimicro and Macro Organic Chemistry," Crowell, New York, 1942.
16. Cheronis, N. D., and Entrikin, J. B., "Semimicro Qualitative Organic Analysis," p. 36, Crowell, New York, 1947.
17. Cheronis, N. D., and Entrikin, J. B., "Semimicro Qualitative Organic Analysis," 2nd ed., Interscience, New York, 1957.
18. Clark, E. P., "Semimicro Quantitative Organic Analysis," Academic Press, New York, 1943.
19. Clark, S. J., "Quantitative Methods of Organic Microanalysis," Butterworths, London, 1956.
20. Clemo, G. R., and McQuillen, A., *J. Chem. Soc.*, p. 1220 (1935).
21. Corning Glass Works, Corning, New York.
22. Davidson, D., Popowsky, M., and Rosenblatt, P., U. S. Patent 2,328,787 (1943).
23. Davidson, D., and Popowsky, M., 102nd Meeting of the American Chemical Society, Division of Analytical and Micro Chemistry, p. L17, Atlantic City, New Jersey, September 1941.
24. Donau, J., *Monatsh. Chem.*, **29**, 333 (1908).
25. Emich, F., and Schneider, F., "Microchemical Manual," Wiley, New York, 1934.

26. Fenger-Eriksen, K., Krogh, A., and Ussing, H., *Biochem. J.,* **30**, 1264 (1936).
27. Fischer, E., *Ber.,* **44**, 129 (1911).
28. Fischer, E., *Ber.,* **54**, 1979 (1921).
29. Fischer, R., *Mikrochemie ver. Mikrochim. Acta,* **36/37**, 296 (1951).
30. Fischer, R., and Kocher, C., *Mikrochemie ver. Mikrochim. Acta,* **32**, 173 (1944).
31. Fischer, R., and Reichel, Z., *Mikrochemie ver. Mikrochim. Acta,* **31**, 102 (1944).
32. Fisher Scientific Company, Tech. Serv. Dept. Pamphlet on the Fisher-Davidson Gravitometer No. 11-509, New York, and Pittsburgh, Pennsylvania.
33. Frediani, H. A., *Ind. Eng. Chem., Anal. Ed.,* **14**, 439 (1942).
34. Furter, M. F., *Helv. Chim. Acta,* **21**, 1674 (1938).
35. Garcia, C. R., *Ind. Eng. Chem., Anal. Ed.,* **15**, 648 (1943).
36. Gibb, T. R. P., Jr., "Optical Methods of Chemical Analysis," pp. 208, 209, 336, McGraw-Hill, New York, 1942.
37. Grant, J., "Quantitative Organic Microanalysis," Based on the Methods of Fritz Pregl," 4th ed., Blakiston, Philadelphia, Pennsylvania, 1946.
38. Grant, J., "Quantitative Organic Microanalysis," 5th ed., Blakiston, Philadelphia, Pennsylvania, 1951.
39. Hewitt, E. J., *Chem. & Ind. (London),* p. 42 (1947).
40. Hilbck, H., *Mikrochemie ver. Mikrochim. Acta,* **36/37**, 307 (1951).
41. Hippenmeyer, F., *Mikrochemie ver. Mikrochim. Acta,* **39**, 409 (1952).
42. Hippenmeyer, F., *Pharm. Acta Helv.,* **25**, 161 (1950).
43. Hozumi, K., and Morita, N., *Yakugaku Zasshi,* **79**, 1028 (1959).
44. Jelley, E. E., *J. Roy. Microscop. Soc.,* **54**, 234 (1934).
45. Kacser, H., and Ubbelohde, A. R., *J. Soc. Chem. Ind. (London),* **68**, 135 (1949).
46. Kamm, O., "Qualitative Organic Analysis," pp. 116, 119, Wiley, New York, 1923.
47. Kirk, P. L., and Gibson, C. S., *Ind. Eng. Chem., Anal. Ed.,* **11**, 403 (1939).
48. Kofler, L., *Mikrochemie,* **15**, 242 (1934).
49. Kofler, L., and Kofler, A., "Mikro-Methoden zur Kennzeichnung organischer Stoffe und Stoffgemische," Wagner, Innsbruck, 1948.
50. Kofler, L., and Kofler, A., "Mikroskopische Methoden," *in* "Handbuch der mikrochemischen Methoden" (F. Hecht, and M. K. Zacherl, eds.), Band I, Tl. 1, Springer, Wien, 1954.
51. Kofler, L., and Kofler, A., in cooperation with Brandstätter, M., "Thermo-Mikro-Methoden zur Kennzeichnung organischer Stoffe und Stoffgemische," 3rd ed., Verlag Chemie, Weinheim/Bergstrasse, 1954
52. Kofler, L., Kofler, A., and Mayrhofer, A., "Mikroskopische Methoden in der Mikrochemie," Haim, Vienna, 1936.
53. Kofler, L., and Kofler, W., *Mikrochemie ver. Mikrochim. Acta,* **34**, 374 (1949).
54. Kofler, L., and Sitte, H., *Monatsh. Chem.,* **81**, 619 (1950).
55. Kofler, W., *Mikrochemie ver. Mikrochim. Acta,* **39**, 84 (1952).
56. Labout, J. W. A., and Oort, W. P. van, *Anal. Chem.,* **28**, 1147 (1956).
57. Maffei, F. J., and Wasicki, R., *Anais assoc. quím. Brasil,* **7**, 111 (1948).
58. Mastrangelo, S. V. R., and Aston, J. G., *Anal. Chem.,* **26**, 764 (1954).
59. Matthews, F. W., *Anal. Chem.,* **20**, 1112 (1948).
60. McCrone, W. C., *Anal. Chem.,* **21**, 436 (1949).
61. McCrone, W. C., *Mikrochemie ver. Mikrochim. Acta,* **38**, 476 (1951).
62. McCrone, W. C., and Massenberg, S., *Anal. Chem.,* **28**, 1038 (1956).
63. McCullough, J. P., and Waddington, G., *Anal. Chim. Acta,* **17**, 80 (1957).
64. Morton, A. A., and Mahoney, J. F., *Ind. Eng. Chem., Anal. Ed.,* **13**, 498 (1941).
65. Müller, R. H., and Zenchelsky, S. T., *Anal. Chem.,* **24**, 844 (1952).

66. Nash, L. K., *Anal. Chem.*, **21**, 1405 (1949).
67. Nettesheim, G., *Erdöl u. Kohle*, **10**, 73 (1957).
68. Nichols, L., *Natl. Paint Bull.*, **1**, 12, 14 (1937).
69. Niederl, J. B., and Niederl, V., "Micromethods of Quantitative Organic Analysis," 2nd ed., Wiley, New York, 1942.
70. Opfer-Schaum, R., *Süddeut. Apotheker-Ztg.*, **89**, 269 (1949).
71. Owen, W. S., and Reid, W. M., *Mikrochim. Acta*, p. 1373 (1956).
72. Pharmacopeia of the United States by the Authority of the Pharmacopeial Convention, XIII, XIV, XV, and XVI, Mack, Easton, Pennsylvania, 1947, 1950, 1955, and 1960.
73. Pinkus, A. G., and Waldrep, P. G., *Mikrochim. Acta*, p. 772 (1959).
74. Procter & Gamble Company, Cincinnati, Ohio.
75. Reihlen, H., *Mikrochemie*, **23**, 285 (1938).
76. Reimers, F., *Dansk. Tidskr. Farm.*, **15**, 81 (1941).
77. Rosebury, F., and Van Heyningen, W. E., *Ind. Eng. Chem., Anal. Ed.*, **14**, 363 (1942).
78. Roth, H., "Die quantitative organische Mikroanalyse von Fritz Pregl," 4th ed., Springer, Berlin, 1935.
79. Roth, H., "F. Pregl quantitative organische Mikroanalyse," 5th ed., Springer, Wien, 1947.
80. Roth, H., "Pregl-Roth quantitative organische Mikroanalyse," 7th ed., Springer, Wien, 1958.
81. Roth, H., "Quantitative Organic Microanalysis of Fritz Pregl," 3rd ed., (E. B. Daw, trans. 4th German ed.), Blakiston, Philadelphia, Pennsylvania, 1937.
82. Schneider, F., "Organic Qualitative Microanalysis," pp. 89–92, Wiley, New York, 1946.
83. Sekera, A., and Pokorny, J., *Mikrochim. Acta*, p. 103 (1957).
84. Shriner, R. L., and Fuson, R. C., "The Systematic Identification of Organic Compounds," pp. 40–45, Wiley, New York, 1948.
85. Shriner, R. L., Fuson, R. C., and Curtin, D. Y., "The Systematic Identification of Organic Compounds," Wiley, New York, and Chapman & Hall, London, 1956.
86. Somereyns, G., *Mikrochim. Acta*, p. 332 (1953).
87. Steyermark, Al, "Quantitative Organic Microanalysis," Blakiston, Philadelphia, Pennsylvania, 1951.
88. Steyermark, Al, Alber, H. K., Aluise, V. A., Huffman, E. W. D., Jolley, E. L., Kuck, J. A., Moran, J. J., Ogg, C. L., and Pietri, C. E., *Anal. Chem.*, **32**, 1045 (1960).
89. Thomas, Arthur H., Company, Philadelphia, Pennsylvania.
90. Tschamler, H., *Mikrochemie ver. Mikrochim. Acta*, **35**, 353 (1950).
91. Weissberger, A., "Physical Methods of Organic Chemistry," Vol. I, pp. 499, 733–734, Interscience, New York, 1945.
92. Weissberger, A., "Physical Methods of Organic Chemistry," 2nd ed., Pt. I, pp. 78–79, 883–885, Interscience, New York, 1949.
93. Wiberley, J. S., Siegfriedt, R. K., and Benedetti-Pichler, A. A., *Mikrochemie ver. Mikrochim. Acta*, **38**, 471 (1951).
94. Wiegand, C., *Angew. Chem.*, **67**, 77 (1955).
95. Yanagimoto Co. Ltd., Bull. on Micro Melting Point Apparatus, Catalogue No. A-111, Japan.
96. Zscheile, F. P., and White, J. W., Jr., *Ind. Eng. Chem., Anal. Ed.*, **12**, 436 (1940).

CHAPTER 23

Calculations

At the end of each method, described in the preceding chapters, factors and formulas are given for calculating the results, and specific examples are calculated. In this chapter some general formulas are given and the factors are collected into tables, one for the gravimetric (Table 40) and one for the volumetric (Table 41). There is included a table of multiples of atomic and molecular weights (Table 42) which is useful for calculating percentages in empirical formulas.* There are also included discussions on the calculation of factors and on the calculation of percentages from the empirical formula and vice versa. The chapter is concluded with two sets of logarithmic tables. The first of these, Table 43, the Nitrogen Reduction Table,[9,10] gives the logarithms of the weight of one ml. of nitrogen at the various temperatures and pressures encountered in the laboratory and is used for calculating the results of the Dumas nitrogen determination. The second, Table 44, is a regular table of five-place logarithms.

GENERAL FORMULAS

Gravimetric:

$$\frac{\text{Wt. of precipitate} \times \text{factor} \times 100}{\text{Wt. sample}} = \% \text{ Element}$$

Volumetric:

$$\frac{\text{Ml. of standard solution} \times \text{factor} \times 100}{\text{Wt. sample}} = \% \text{ Element}$$

Gasometric:

$$\frac{\text{Corrected ml. of gas} \times \text{factor} \times 100}{\text{Wt. sample}} = \% \text{ Element}$$

Manometric:

$$\frac{\text{Corrected pressure of gas} \times \text{factor} \times 100}{\text{Wt. sample}} = \% \text{ Element}$$

* The atomic weights proposed by the Commission on Atomic Weights of the I.U.P.A.C., 1957,[6] have been used throughout. Tables showing the percentage composition of many organic compounds have been published by Gysel[3] and by Kuffner and Gruber.[7] Thompson[12] described a calculator for selection of sample weights in the microdetermination of carbon and hydrogen.

23. Calculations

MOLECULAR WEIGHT:

(a) *Rast:*

$$\text{M.W.} = \frac{1000 \times K \times S}{L \times \Delta_t}$$

where K = the molecular melting (or freezing) point depression (melting or freezing point depression for one mol of solute per 1000 grams of solvent)

S = wt. of sample (solute)
L = wt. of solvent
Δ_t = depression of the melting point of the solvent

(b) *Signer:*

$$M_1 = \frac{G_1 M V}{G V_1}$$

where M = M.W. of standard
V = vol. of solution of standard
G = wt. of standard
M_1 = M.W. of unknown
V_1 = vol. of solution of unknown
G_1 = wt. of unknown

(c) *Neutralization Equivalent:*

$$\text{Neut. Equiv.} = \frac{\text{Wt. of sample in mg.} \times 100}{\text{Ml. of 0.01}N \text{ alkali used}}$$

(d) *Hydrogen Number:*

$$\text{H No.} = \frac{\text{Wt. of sample in mg.} \times 22.415}{\text{Vol. of } H_2 \text{ at } 0°\text{ C., 760 mm.}}$$

Calculation of Factors[11]

Factors, such as given in Tables 40 and 41, usually are found in the various books on analytical chemistry[2,13] and in the handbooks.[4,5,8] However, the author believes that an understanding of their methods of calculation is in order since it allows the analyst to calculate factors if the particular ones desired are not listed. The following examples will serve the purpose.

GRAVIMETRIC:

Examples:

  a. From the equations given in Chapter 20, it is seen that one methyl group attached to nitrogen is equivalent to one molecule of silver iodide

$$CH_3 \rightleftharpoons AgI$$
$$\text{M.W. } 15.035 \qquad \text{M.W. } 234.790$$

From the proportion

$$\text{M.W. of } CH_3 : \text{M.W. of } AgI :: x : 1$$

where

$$x = \text{mg. of } CH_3$$

we get

$$\therefore 15.035 : 234.790 :: x : 1$$

or $\quad x = \dfrac{15.035 \times 1}{234.790} = 0.06404$ mg. of CH_3 equivalent to 1 mg. of AgI

  b. Likewise, from Chapter 20 it is seen that one ethyl group attached to nitrogen is equivalent to one molecule of silver iodide

$$C_2H_5 \rightleftharpoons AgI$$
$$\text{M.W. } 29.062 \qquad \text{M.W. } 234.790$$

$$\therefore \text{M.W. of } C_2H_5 : \text{M.W. of } AgI :: x : 1$$

where $\quad x = \text{mg. of } C_2H_5$

$$\therefore 29.062 : 234.790 :: x : 1$$

or $\quad x = \dfrac{29.062 \times 1}{234.790} = 0.1238$ mg. of C_2H_5 equivalent to 1 mg. of AgI

VOLUMETRIC:

Examples:

  a. From the equations in Chapter 20, it is seen that one methyl group attached to nitrogen is equivalent to six molecules of thiosulfate.

23. Calculations

$$\underset{\text{M.W. 15.035}}{CH_3} \rightleftharpoons \underset{\text{M.W. 948.684}}{6\ Na_2S_2O_3}$$

Now in 1 ml. of 0.01N thiosulfate there are 1.58114 mg. of $Na_2S_2O_3$

$\therefore$ M.W. of CH_3 : M.W. of 6 $Na_2S_2O_3$ :: x : 1.58114

where $x =$ mg. of CH_3

$\therefore$ 15.035 : 948.684 : x : 1.58114

or $\quad x = \dfrac{15.035 \times 1.58114}{948.684} = 0.02506$ mg. of CH_3 equivalent to 1 ml. of 0.01N $Na_2S_2O_3$

b. Similarly, from the equations of Chapter 20, it is seen that one ethyl group attached to nitrogen is equivalent to six molecules of thiosulfate.

$$\underset{\text{M.W. 29.062}}{C_2H_5} \rightleftharpoons \underset{\text{M.W. 948.684}}{6Na_2S_2O_3}$$

$\therefore$ 29.062 : 948.684 :: x : 1.58114
(see *Example a* above)

or $\quad x = \dfrac{29.062 \times 1.58114}{948.684} = 0.04844$ mg. of C_2H_5 equivalent to 1 ml. of 0.01N $Na_2S_2O_3$

c. From Chapter 14 it is seen that five atoms of oxygen are equivalent to twelve molecules of thiosulfate.

$$\underset{\text{M.W. 80}}{5[O]} \rightleftharpoons \underset{\text{M.W. 1897.368}}{12\ Na_2S_2O_3}$$

$\therefore$ 80 : 1897.368 :: x : 1.58114 (see *Example a* above)

or $\quad x = \dfrac{80 \times 1.58114}{1897.368} = 0.06667$ mg. of oxygen equivalent to 1 ml. of 0.01N $Na_2S_2O_3$

Calculation of Percentages from the Empirical Formula and Vice Versa

For many of the samples analyzed the empirical formulas are known and the theoretical values of each element may be calculated. This is accomplished by use of the following formula:

Theoretical per cent of the element =

$$\dfrac{\text{At. wt. of element} \times \text{no. of these elements present} \times 100}{\text{M.W. of compound}}$$

Likewise, for groups, the following is used:

Theoretical per cent of the group =

$$\dfrac{\text{Sum of at. wts. of elements in group} \times \text{no. of these groups present} \times 100}{\text{M.W. of compound}}$$

Percentages

Example:

The percentages of carbon, hydrogen, and carboxyl in benzoic acid, C_6H_5COOH, are obtained by

$$\frac{12.011 \times 7 \times 100}{122.125} = 68.85\% \text{ C}$$

$$\frac{1.008 \times 6 \times 100}{122.125} = 4.95\% \text{ H}$$

and

$$\frac{45.019 \times 1 \times 100}{122.125} = 36.86\% \text{ COOH}$$

On the other hand, for many of the samples the empirical formulas are not known and must be calculated from the found values. The method of calculation given here[11] is that discussed in detail by Clark.[1] For an example, suppose the elementary analyses of a compound showed it to have the following percentages of carbon, hydrogen, and oxygen:

$$C = 58.72\%$$
$$H = 5.92\%$$
$$O = 35.36\%$$

Therefore, in 100 grams of substance there are

58.72 grams C
5.92 grams H
35.36 grams O

The atomic weights of the three elements are, respectively, 12.011, 1.008, and 16.000 and if these are called grams, 100 grams of the above compound contains

$$\frac{58.72}{12.011} = 4.89 \text{ units of C}$$

$$\frac{5.92}{1.008} = 5.87 \text{ units of H}$$

$$\frac{35.36}{16.000} = 2.21 \text{ units of O}$$

Therefore, the compound contains carbon, hydrogen, and oxygen in the proportion of 4.89 to 5.87 to 2.21 units respectively. By dividing these three values by that of the smallest, namely, 2.21, the proportion becomes 2.21 units of carbon to 2.66 units of hydrogen to 1 unit of oxygen. Whole numbers are required and in this case it is not possible to see on inspection what factor is necessary to use in order to have all three come out whole numbers. Therefore, Table 39 is set up in which all of the values in the proportion 2.21 units of C:2.66 units of H:1 unit of O are multiplied by 1, 2, 3, 4, etc.

23. Calculations

TABLE 39
Type of Series Used in Calculation of Empirical Formulae

Series	C	H	O
(a)	2.21	2.66	1
(b)	4.42	5.32	2
(c)	6.63	7.98	3
(d)	8.84	10.64	4
(e)	11.05	13.30	5
(f)	13.26	15.96	6
(g)	15.47	18.62	7
(h)	17.68	21.28	8
(i)	19.89	23.94	9
(j)	22.10	26.60	10
(k)	24.31	29.26	11
(l)	26.52	31.92	12
(m)	28.73	34.58	13

Before any attempt is made to select a series in which all of the numbers are closest to being whole, the rule concerning the composition of molecules with elements having odd valences must be considered. The number of elements in a molecule having odd valences must be equal to an even number. This must be true since carbon is tetravalent. This rule eliminates immediately each of the series in which the hydrogen is closest to odd numbers and only those closest to the even ones should be considered, namely, (c), (f), (i), and (l). Of this group (i) has both of the values for C and H closest to whole numbers. Consequently, the empirical formula is $C_{20}H_{24}O_9$ which has the theoretical values of carbon, hydrogen, and oxygen equal to the following

$$C = 58.82\%, \quad H = 5.92\%, \quad O = 35.26\%$$

which are in excellent agreement with the found figures.

Other data such as the values obtained from carboxyl (neutralization equivalent), methoxyl, acetyl, molecular weight, etc., are helpful in selecting the proper empirical formula.

TABLE 40
Table of Gravimetric Factors

Chapter	Sought	Weighed	Factor
6	Na	Na_2SO_4	0.3237
6	K	K_2SO_4	0.4487
6	Ba	$BaSO_4$	0.5885
6	Ca	$CaSO_4$	0.2944
6	Cu	CuO	0.7988
6	Fe	Fe_2O_3	0.6994
6	Pt	Pt	1.000
6	Au	Au	1.000
6	Ag	Ag	1.000*
9	H	H_2O	0.1119
9	C	CO_2	0.2729
10	S	$BaSO_4$	0.1374
11	Cl	AgCl	0.2474
11	Br	AgBr	0.4255
11	I	AgI	0.5405
12	P	$(NH_4)_3PO_4 \cdot 14MoO_3$	0.014524
13	As	$Mg_2As_2O_7$	0.4826
14	O	CO_2	0.3635
16	OCH_3	AgI	0.1322
16	OC_2H_5	AgI	0.1919
20	CH_3	AgI	0.06404
20	C_2H_5	AgI	0.1238

* Also see footnote on page 133.

TABLE 41
Table of Volumetric Factors

Chapter	Sought	Standard solution	1 ml. Equivalent (mg.)
8	N	0.01N HCl	0.14008
10	S	0.01N $BaCl_2$	0.1603
10 and 15	S	0.01N NaOH	0.1603
13	As	0.01N $Na_2S_2O_3$	0.3748
14	O	0.01N $Na_2S_2O_3$	0.06667
15	COOH	0.01N NaOH	0.4502
15	Cl	0.01N NaOH	0.3546
15	Br	0.01N NaOH	0.7992
16	OCH_3	0.01N $Na_2S_2O_3$	0.05173
16	OC_2H_5	0.01N $Na_2S_2O_3$	0.07510
17	CH_3CO	0.01N $Na_2S_2O_3$	0.4305
17	HCO	0.01N $Na_2S_2O_3$	0.2902
20	CH_3	0.01N $Na_2S_2O_3$	0.02506
20	C_2H_5	0.01N $Na_2S_2O_3$	0.04844

23. Calculations

TABLE 42
SOME ATOMIC AND MOLECULAR WEIGHTS AND SOME MULTIPLES OF THEM

Hydrogen					Oxygen			
H	1.008	26	26.208	O	16.00	7	112.00
2	2.016	27	27.216	2	32.00	8	128.00
3	3.024	28	28.224	3	48.00	9	144.00
4	4.032	29	29.232	4	64.00	10	160.00
5	5.040	30	30.240	5	80.00	11	176.00
6	6.048	31	31.248	6	96.00	12	192.00
7	7.056	32	32.256		Nitrogen		
8	8.064	33	33.264				
9	9.072	34	34.272	N	14.008	4	56.032
10	10.080	35	35.280	2	28.016	5	70.040
11	11.088	36	36.288	3	42.024	6	84.048
12	12.096	37	37.296		Sulfur		
13	13.104	38	38.304				
14	14.112	39	39.312	S	32.066	3	96.198
15	15.120	40	40.320	2	64.132	4	128.264
16	16.128	41	41.328		Chlorine		
17	17.136	42	42.336	Cl	35.457	4	141.828
18	18.144	43	43.344	2	70.914	5	177.285
19	19.152	44	44.352	3	106.371	6	212.742
20	20.160	45	45.360		Bromine		
21	21.168	46	46.368				
22	22.176	47	47.376	Br	79.916	4	319.664
23	23.184	48	48.384	2	159.832	5	399.580
24	24.192	49	49.392	3	239.748	6	479.496
25	25.200	50	50.400		Fluorine		
Carbon							
				F	19.000	3	57.000
C	12.011	21	252.231	2	38.000	4	76.000
2	24.022	22	264.242		Iodine		
3	36.033	23	276.253				
4	48.044	24	288.264	I	126.91	4	507.64
5	60.055	25	300.275	2	253.82	5	634.55
6	72.066	26	312.286	3	380.73	6	761.46
7	84.077	27	324.297		Sodium		
8	96.088	28	336.308				
9	108.099	29	348.319	Na	22.991	3	68.973
10	120.110	30	360.330	2	45.982	4	91.964
11	132.121	31	372.341		Potassium		
12	144.132	32	384.352	K	39.100	3	117.300
13	156.143	33	396.363	2	78.200	4	156.400
14	168.154	34	408.374		Calcium		
15	180.165	35	420.385				
16	192.176	36	432.396	Ca	40.08	3	120.24
17	204.187	37	444.407	2	80.16		
18	216.198	38	456.418		Barium		
19	228.209	39	468.429	Ba	137.36	3	412.08
20	240.220	40	480.440	2	274.72		

TABLE 42 (*Continued*)

Phosphorus				Acetyl			
P	30.975	3	92.925	CH$_3$CO	43.046	4	172.184
2	61.950	4	123.900	2	86.092	5	215.230
				3	129.138	6	258.276
Arsenic				Formyl			
As	74.91	3	224.73	HCO	29.019	3	87.057
2	149.82			2	58.038		
Silver				Methoxyl			
Ag	107.880	3	323.640	OCH$_3$	31.035	4	124.140
2	215.760			2	62.070	5	155.175
				3	93.105	6	186.210
Carboxyl				Ethoxyl			
COOH	45.019	4	180.076	OC$_2$H$_5$	45.062	4	180.248
2	90.038	5	225.095	2	90.124	5	225.310
3	135.057	6	270.114	3	135.186	6	270.372

23. Calculations

TABLE 43
Nitrogen Reduction Table*

At standard temperature and pressure (0° C., 760 mm.), 1 ml. of nitrogen weighs 1.2505 mg. The following table gives the logarithm of the weight of 1 ml. of nitrogen at a temperature of $t°$ C. and a pressure of p mm. (Taken from Niederl and Niederl.[9,10])

$t°$	$p=690$	691	692	693	694	$t°$	$p=695$	696	697	698	699
10	03 946	04 009	04 072	04 135	04 197	10	04 260	04 322	04 385	04 447	04 509
11	03 793	03 856	03 919	03 982	04 044	11	04 107	04 169	04 232	04 294	04 356
12	03 640	03 703	03 766	03 829	03 891	12	03 954	04 016	04 079	04 141	04 203
13	03 488	03 551	03 614	03 677	03 739	13	03 802	03 864	03 927	03 989	04 051
14	03 336	03 399	03 462	03 525	03 587	14	03 650	03 712	03 775	03 837	03 899
15	03 184	03 247	03 310	03 373	03 435	15	03 498	03 560	03 623	03 685	03 747
16	03 033	03 096	03 159	03 222	03 284	16	03 347	03 409	03 472	03 534	03 596
17	02 883	02 946	03 009	03 072	03 134	17	03 197	03 259	03 322	03 384	03 446
18	02 733	02 796	02 859	02 922	02 984	18	03 047	03 109	03 172	03 234	03 296
19	02 584	02 647	02 710	02 773	02 835	19	02 898	02 960	03 023	03 085	03 147
20	02 435	02 498	02 561	02 624	02 686	20	02 749	02 811	02 874	02 936	02 998
21	02 287	02 350	02 413	02 476	02 538	21	02 601	02 663	02 726	02 788	02 850
22	02 139	02 202	02 265	02 328	02 390	22	02 453	02 515	02 578	02 640	02 702
23	01 992	02 055	02 118	02 181	02 243	23	02 306	02 368	02 431	02 493	02 555
24	01 846	01 909	01 972	02 035	02 097	24	02 160	02 222	02 285	02 347	02 409
25	01 700	01 763	01 826	01 889	01 951	25	02 014	02 076	02 139	02 201	02 263
26	01 555	01 618	01 681	01 744	01 806	26	01 869	01 931	01 994	02 056	02 118
27	01 410	01 473	01 536	01 599	01 661	27	01 724	01 786	01 849	01 911	01 973
28	01 266	01 329	01 392	01 455	01 517	28	01 580	01 642	01 705	01 767	01 829
29	01 122	01 185	01 248	01 311	01 373	29	01 436	01 498	01 561	01 623	01 683
30	00 979	01 042	01 105	01 168	01 230	30	01 293	01 355	01 418	01 480	01 540
31	00 836	00 899	00 962	01 025	01 087	31	01 150	01 212	01 275	01 337	01 397
32	00 694	00 757	00 820	00 883	00 945	32	01 008	01 070	01 133	01 195	01 255
33	00 552	00 615	00 678	00 741	00 803	33	00 866	00 928	00 991	01 053	01 113
34	00 411	00 474	00 537	00 600	00 662	34	00 725	00 787	00 850	00 912	00 972
35	00 270	00 333	00 396	00 459	00 521	35	00 584	00 646	00 709	00 771	00 831
36	00 130	00 193	00 256	00 319	00 381	36	00 444	00 506	00 569	00 631	00 691

* The logarithms up to and including 24° C. were taken from Küster-Thiel, "Logarithmische Rechentafeln für Chemiker," de Gruyter, Berlin, and Leipzig, 1935, and the logarithms from 25° up to and including 36° C. were interpolated.

NITROGEN REDUCTION TABLE (*Continued*)

At standard temperature and pressure (0° C., 760 mm.), 1 ml. of nitrogen weighs 1.2505 mg. The following table gives the logarithm of the weight of 1 ml. of nitrogen at a temperature of $t°$ C. and a pressure of p mm. (Taken from Niederl and Niederl.[9,10])

$t°$	$p=700$	701	702	703	704	$t°$	$p=705$	706	707	708	709
10	04 571	04 633	04 695	04 757	04 819	10	04 880	04 942	05 003	05 065	05 126
11	04 418	04 480	04 542	04 604	04 666	11	04 727	04 789	04 850	04 912	04 973
12	04 265	04 327	04 389	04 451	04 513	12	04 574	04 636	04 697	04 759	04 820
13	04 113	04 175	04 237	04 299	04 361	13	04 422	04 484	04 545	04 607	04 668
14	03 961	04 023	04 085	04 147	04 209	14	04 270	04 332	04 393	04 455	04 516
15	03 809	03 871	03 933	03 995	04 057	15	04 118	04 180	04 241	04 303	04 364
16	03 658	03 720	03 782	03 844	03 906	16	03 967	04 029	04 090	04 152	04 213
17	03 508	03 570	03 632	03 694	03 756	17	03 817	03 879	03 940	04 002	04 063
18	03 358	03 420	03 482	03 544	03 606	18	03 667	03 729	03 790	03 852	03 913
19	03 209	03 271	03 333	03 395	03 457	19	03 518	03 580	03 641	03 703	03 764
20	03 060	03 122	03 184	03 246	03 308	20	03 369	03 431	03 492	03 554	03 615
21	02 912	02 974	03 036	03 098	03 160	21	03 221	03 283	03 344	03 406	03 467
22	02 764	02 826	02 888	02 950	03 012	22	03 073	03 135	03 196	03 258	03 319
23	02 617	02 679	02 741	02 803	02 865	23	02 926	02 988	03 049	03 111	03 172
24	02 471	02 533	02 595	02 657	02 719	24	02 780	02 842	02 903	02 965	03 026
25	02 325	02 387	02 449	02 511	02 573	25	02 634	02 696	02 757	02 819	02 880
26	02 180	02 242	02 304	02 366	02 428	26	02 489	02 551	02 612	02 674	02 735
27	02 035	02 097	02 159	02 221	02 283	27	02 344	02 406	02 467	02 529	02 590
28	01 891	01 953	02 015	02 077	02 139	28	02 200	02 262	02 323	02 385	02 446
29	01 747	01 809	01 871	01 933	01 995	29	02 056	02 118	02 179	02 241	02 302
30	01 604	01 666	01 728	01 790	01 852	30	01 913	01 975	02 036	02 098	02 159
31	01 461	01 523	01 585	01 647	01 709	31	01 770	01 832	01 893	01 955	02 016
32	01 319	01 381	01 443	01 505	01 567	32	01 628	01 690	01 751	01 813	01 874
33	01 177	01 239	01 301	01 363	01 425	33	01 486	01 548	01 609	01 671	01 732
34	01 036	01 098	01 160	01 222	01 284	34	01 345	01 407	01 468	01 530	01 591
35	00 895	00 957	01 019	01 081	01 143	35	01 205	01 267	01 327	01 389	01 450
36	00 755	00 817	00 879	00 941	01 003	36	01 065	01 127	01 187	01 249	01 310

23. Calculations

Nitrogen Reduction Table (*Continued*)

At standard temperature and pressure (0° C., 760 mm.), 1 ml. of nitrogen weighs 1.2505 mg. The following table gives the logarithm of the weight of 1 ml. of nitrogen at a temperature of $t°$ C. and a pressure of p mm. (Taken from Niederl and Niederl.[9,10])

$t°$	$p=710$	711	712	713	714	$t°$	$p=715$	716	717	718	719
10	05 189	05 250	05 311	05 372	05 433	10	05 494	05 554	05 615	05 675	05 736
11	05 035	05 096	05 157	05 218	05 279	11	05 340	05 400	05 461	05 521	05 582
12	04 882	04 943	05 004	05 065	05 126	12	05 187	05 247	05 308	05 368	05 429
13	04 730	04 791	04 852	04 913	04 974	13	05 035	05 095	05 156	05 216	05 277
14	04 578	04 639	04 700	04 761	04 822	14	04 883	04 943	05 004	05 064	05 125
15	04 427	04 488	04 549	04 610	04 671	15	04 732	04 792	04 853	04 913	04 974
16	04 276	04 337	04 398	04 459	04 520	16	04 581	04 641	04 702	04 762	04 823
17	04 126	04 187	04 248	04 309	04 370	17	04 431	04 491	04 552	04 612	04 673
18	03 976	04 037	04 098	04 159	04 220	18	04 281	04 341	04 402	04 462	04 523
19	03 827	03 888	03 949	04 010	04 071	19	04 132	04 192	04 253	04 313	04 374
20	03 678	03 739	03 800	03 861	03 922	20	03 983	04 043	04 104	04 164	04 225
21	03 530	03 591	03 652	03 713	03 774	21	03 835	03 895	03 956	04 016	04 077
22	03 382	03 443	03 504	03 565	03 626	22	03 687	03 747	03 808	03 868	03 929
23	03 235	03 296	03 357	03 418	03 479	23	03 540	03 600	03 661	03 721	03 782
24	03 088	03 149	03 210	03 271	03 332	24	03 393	03 453	03 514	03 574	03 635
25	02 942	03 003	03 064	03 125	03 186	25	03 247	03 307	03 368	03 428	03 489
26	02 797	02 858	02 919	02 980	03 041	26	03 102	03 162	03 223	03 283	03 344
27	02 652	02 713	02 774	02 835	02 896	27	02 957	03 017	03 078	03 138	03 199
28	02 508	02 569	02 630	02 691	02 752	28	02 813	02 873	02 934	02 994	03 055
29	02 364	02 425	02 486	02 547	02 608	29	02 669	02 729	02 790	02 850	02 911
30	02 221	02 282	02 343	02 404	02 465	30	02 526	02 586	02 647	02 707	02 768
31	02 078	02 139	02 203	02 261	02 322	31	02 383	02 443	02 504	02 564	02 625
32	01 936	01 997	02 061	02 119	02 180	32	02 241	02 301	02 362	02 422	02 483
33	01 794	01 855	01 919	01 977	02 038	33	02 099	02 159	02 220	02 280	02 341
34	01 653	01 714	01 778	01 836	01 897	34	01 958	02 018	02 079	02 139	02 200
35	01 512	01 573	01 637	01 695	01 756	35	01 817	01 877	01 938	01 998	02 059
36	01 372	01 433	01 497	01 555	01 616	36	01 677	01 737	01 798	01 858	01 919

Nitrogen Reduction Table (*Continued*)

At standard temperature and pressure (0° C., 760 mm.), 1 ml. of nitrogen weighs 1.2505 mg. The following table gives the logarithm of the weight of 1 ml. of nitrogen at a temperature of $t°$ C. and a pressure of p mm. (Taken from Niederl and Niederl.[9,10])

$t°$	$p=720$	721	722	723	724	$t°$	$p=725$	726	727	728	729
10	05 796	05 857	05 917	05 977	06 037	10	06 097	06 157	06 216	06 276	06 336
11	05 642	05 703	05 763	05 823	05 883	11	05 943	06 003	06 062	06 122	06 182
12	05 489	05 550	05 610	05 670	05 730	12	05 790	05 850	05 909	05 969	06 029
13	05 337	05 398	05 458	05 518	05 578	13	05 638	05 698	05 757	05 817	05 877
14	05 185	05 246	05 306	05 366	05 426	14	05 486	05 546	05 605	05 665	05 725
15	05 034	05 095	05 155	05 215	05 275	15	05 335	05 395	05 454	05 514	05 574
16	04 883	04 944	05 004	05 064	05 124	16	05 184	05 244	05 303	05 363	05 423
17	04 733	04 794	04 854	04 914	04 974	17	05 034	05 094	05 153	05 213	05 273
18	04 583	04 644	04 704	04 764	04 824	18	04 884	04 944	05 003	05 063	05 123
19	04 434	04 495	04 555	04 615	04 675	19	04 735	04 795	04 854	04 914	04 974
20	04 285	04 346	04 406	04 466	04 526	20	04 586	04 646	04 705	04 765	04 825
21	04 137	04 198	04 258	04 318	04 378	21	04 438	04 498	04 557	04 617	04 677
22	03 989	04 050	04 110	04 170	04 230	22	04 290	04 350	04 409	04 469	04 529
23	03 842	03 903	03 963	04 023	04 083	23	04 143	04 203	04 262	04 322	04 382
24	03 695	03 756	03 816	03 876	03 936	24	03 996	04 056	04 115	04 175	04 235
25	03 549	03 610	03 670	03 730	03 790	25	03 850	03 910	03 969	04 029	04 089
26	03 404	03 465	03 525	03 585	03 645	26	03 705	03 765	03 824	03 884	03 944
27	03 259	03 320	03 380	03 440	03 500	27	03 560	03 620	03 679	03 739	03 799
28	03 115	03 176	03 236	03 296	03 356	28	03 416	03 476	03 535	03 595	03 655
29	02 971	03 032	03 092	03 152	03 212	29	03 272	03 332	03 391	03 451	03 511
30	02 828	02 889	02 949	03 009	03 069	30	03 129	03 189	03 248	03 308	03 368
31	02 685	02 746	02 806	02 866	02 926	31	02 986	03 046	03 105	03 165	03 225
32	02 543	02 604	02 664	02 724	02 784	32	02 844	02 904	02 963	03 023	03 083
33	02 401	02 462	02 522	02 582	02 642	33	02 702	02 762	02 821	02 881	02 941
34	02 260	02 321	02 381	02 441	02 501	34	02 561	02 621	02 680	02 740	02 800
35	02 119	02 180	02 240	02 300	02 360	35	02 420	02 480	02 539	02 599	02 659
36	01 979	02 040	02 100	02 160	02 220	36	02 280	02 340	02 399	02 459	02 519

23. Calculations

Nitrogen Reduction Table (*Continued*)

At standard temperature and pressure (0° C., 760 mm.), 1 ml. of nitrogen weighs 1.2505 mg. The following table gives the logarithm of the weight of 1 ml. of nitrogen at a temperature of $t°$ C. and a pressure of p mm. (Taken from Niederl and Niederl.[9,10])

$t°$	$p=730$	731	732	733	734	$t°$	$p=735$	736	737	738	739
10	06 395	06 455	06 514	06 573	06 633	10	06 692	06 751	06 810	06 869	06 927
11	06 241	06 301	06 360	06 419	06 479	11	06 538	06 597	06 656	06 715	06 773
12	06 088	06 148	06 207	06 266	06 326	12	06 385	06 444	06 503	06 562	06 620
13	05 936	05 996	06 055	06 114	06 174	13	06 233	06 292	06 351	06 410	06 468
14	05 784	05 844	05 903	05 962	06 022	14	06 081	06 140	06 199	06 258	06 316
15	05 633	05 693	05 752	05 811	05 871	15	05 930	05 989	06 048	06 107	06 165
16	05 482	05 542	05 601	05 660	05 720	16	05 779	05 838	05 897	05 956	06 014
17	05 332	05 392	05 451	05 510	05 570	17	05 629	05 688	05 747	05 806	05 864
18	05 182	05 242	05 301	05 360	05 420	18	05 479	05 538	05 597	05 656	05 714
19	05 033	05 093	05 152	05 211	05 271	19	05 330	05 389	05 448	05 507	05 565
20	04 884	04 944	05 003	05 062	05 122	20	05 181	05 240	05 299	05 358	05 416
21	04 736	04 796	04 855	04 914	04 974	21	05 033	05 092	05 151	05 210	05 268
22	04 588	04 648	04 707	04 766	04 826	22	04 885	04 944	05 003	05 062	05 120
23	04 441	04 501	04 560	04 619	04 679	23	04 738	04 797	04 856	04 915	04 973
24	04 294	04 354	04 413	04 472	04 532	24	04 591	04 650	04 709	04 768	04 826
25	04 148	04 208	04 267	04 326	04 386	25	04 445	04 504	04 563	04 622	04 680
26	04 003	04 063	04 122	04 181	04 241	26	04 300	04 359	04 418	04 477	04 535
27	03 858	03 918	03 977	04 036	04 096	27	04 155	04 214	04 273	04 332	04 390
28	03 714	03 774	03 833	03 892	03 952	28	04 011	04 070	04 129	04 188	04 246
29	03 570	03 630	03 689	03 748	03 808	29	03 867	03 926	03 985	04 044	04 102
30	03 427	03 487	03 546	03 605	03 665	30	03 724	03 783	03 842	03 901	03 959
31	03 284	03 344	03 403	03 462	03 522	31	03 581	03 640	03 699	03 758	03 816
32	03 142	03 202	03 261	03 320	03 380	32	03 439	03 498	03 557	03 616	03 674
33	03 000	03 060	03 119	03 178	03 238	33	03 297	03 356	03 415	03 474	03 532
34	02 859	02 919	02 978	03 037	03 097	34	03 156	03 215	03 274	03 333	03 391
35	02 718	02 778	02 837	02 896	02 956	35	03 015	03 074	03 133	03 192	03 250
36	02 578	02 638	02 697	02 756	02 816	36	02 875	02 934	02 993	03 052	03 110

Nitrogen Reduction Table (*Continued*)

At standard temperature and pressure (0° C., 760 mm.), 1 ml. of nitrogen weighs 1.2505 mg. The following table gives the logarithm of the weight of 1 ml. of nitrogen at a temperature of $t°$ C. and a pressure of p mm. (Taken from Niederl and Niederl.[9,10])

$t°$	$p=740$	741	742	743	744	$t°$	$p=745$	746	747	748	749
10	06 986	07 045	07 103	07 162	07 220	10	07 279	07 337	07 395	07 453	07 511
11	06 832	06 891	06 949	07 008	07 066	11	07 125	07 183	07 241	07 299	07 357
12	06 679	06 738	06 796	06 855	06 913	12	06 972	07 030	07 088	07 146	07 204
13	06 527	06 586	06 644	06 703	06 761	13	06 820	06 878	06 936	06 994	07 052
14	06 375	06 434	06 492	06 551	06 609	14	06 668	06 726	06 784	06 842	06 900
15	06 224	06 283	06 341	06 400	06 458	15	06 517	06 575	06 633	06 691	06 749
16	06 073	06 132	06 190	06 249	06 307	16	06 366	06 424	06 482	06 540	06 598
17	05 923	05 982	06 040	06 099	06 157	17	06 216	06 274	06 332	06 390	06 448
18	05 773	05 832	05 890	05 949	06 007	18	06 066	06 124	06 182	06 240	06 298
19	05 624	05 683	05 741	05 800	05 858	19	05 917	05 975	06 033	06 091	06 149
20	05 475	05 534	05 592	05 651	05 709	20	05 768	05 826	05 884	05 942	06 000
21	05 327	05 386	05 444	05 503	05 561	21	05 620	05 678	05 736	05 794	05 852
22	05 179	05 238	05 296	05 355	05 413	22	05 472	05 530	05 588	05 646	05 704
23	05 032	05 091	05 149	05 208	05 266	23	05 325	05 383	05 441	05 499	05 557
24	04 885	04 944	05 002	05 061	05 119	24	05 178	05 236	05 294	05 352	05 410
25	04 739	04 798	04 856	04 915	04 973	25	05 032	05 090	05 148	05 206	05 264
26	04 594	04 653	04 711	04 770	04 828	26	04 887	04 945	05 003	05 061	05 119
27	04 449	04 508	04 566	04 625	04 683	27	04 742	04 800	04 858	04 916	04 974
28	04 305	04 364	04 422	04 481	04 539	28	04 598	04 656	04 714	04 772	04 830
29	04 161	04 220	04 278	04 337	04 395	29	04 454	04 512	04 570	04 628	04 686
30	04 018	04 078	04 135	04 194	04 252	30	04 311	04 369	04 428	04 485	04 543
31	03 875	03 935	03 992	04 051	04 109	31	04 168	04 226	04 285	04 342	04 400
32	03 733	03 793	03 850	03 909	03 967	32	04 026	04 084	04 143	04 200	04 258
33	03 591	03 651	03 708	03 767	03 825	33	03 884	03 942	04 001	04 058	04 116
34	03 450	03 510	03 567	03 626	03 684	34	03 743	03 801	03 860	03 917	03 975
35	03 309	03 369	03 426	03 485	03 543	35	03 602	03 660	03 719	03 776	03 834
36	03 169	03 229	03 286	03 345	03 403	36	03 462	03 520	03 579	03 636	03 694

23. Calculations

NITROGEN REDUCTION TABLE (*Continued*)

At standard temperature and pressure (0° C., 760 mm.), 1 ml. of nitrogen weighs 1.2505 mg. The following table gives the logarithm of the weight of 1 ml. of nitrogen at a temperature of $t°$ C. and a pressure of p mm. (Taken from Niederl and Niederl.[9,10])

$t°$	$p=750$	751	752	753	754	$t°$	$p=755$	756	757	758	759
10	07 569	07 627	07 685	07 742	07 800	10	07 858	07 915	07 973	08 030	08 087
11	07 415	07 473	07 531	07 588	07 646	11	07 704	07 761	07 819	07 876	07 933
12	07 262	07 320	07 378	07 435	07 493	12	07 551	07 608	07 666	07 723	07 780
13	07 110	07 168	07 226	07 283	07 341	13	07 399	07 456	07 514	07 571	07 628
14	06 958	07 016	07 074	07 131	07 189	14	07 247	07 304	07 362	07 419	07 476
15	06 807	06 865	06 923	06 980	07 038	15	07 096	07 153	07 211	07 268	07 325
16	06 656	06 714	06 772	06 829	06 887	16	06 945	07 002	07 060	07 117	07 174
17	06 506	06 564	06 622	06 679	06 737	17	06 795	06 852	06 910	06 967	07 024
18	06 356	06 414	06 472	06 529	06 587	18	06 645	06 702	06 760	06 817	06 874
19	06 207	06 265	06 323	06 380	06 438	19	06 496	06 553	06 611	06 668	06 725
20	06 058	06 116	06 174	06 231	06 289	20	06 347	06 404	06 462	06 519	06 576
21	05 910	05 968	06 026	06 083	06 141	21	06 199	06 256	06 314	06 371	06 428
22	05 762	05 820	05 878	05 935	05 993	22	06 051	06 108	06 166	06 223	06 280
23	05 615	05 673	05 731	05 788	05 846	23	05 904	05 961	06 019	06 076	06 133
24	05 468	05 526	05 584	05 641	05 699	24	05 757	05 814	05 872	05 929	05 986
25	05 322	05 380	05 438	05 495	05 553	25	05 611	05 668	05 726	05 783	05 840
26	05 177	05 235	05 293	05 350	05 408	26	05 466	05 523	05 581	05 638	05 695
27	05 032	05 090	05 148	05 205	05 263	27	05 321	05 378	05 436	05 493	05 550
28	04 888	04 946	05 004	05 061	05 119	28	05 177	05 234	05 292	05 349	05 406
29	04 744	04 802	04 860	04 917	04 975	29	05 033	05 090	05 148	05 205	05 262
30	04 601	04 659	04 717	04 774	04 832	30	04 890	04 947	05 005	05 062	05 119
31	04 458	04 516	04 574	04 631	04 689	31	04 747	04 804	04 862	04 919	04 976
32	04 316	04 374	04 432	04 489	04 547	32	04 605	04 662	04 720	04 777	04 834
33	04 174	04 232	04 290	04 347	04 405	33	04 463	04 520	04 578	04 635	04 692
34	04 033	04 091	04 149	04 206	04 264	34	04 322	04 379	04 437	04 494	04 551
35	03 892	03 950	04 008	04 065	04 123	35	04 181	04 238	04 296	04 353	04 410
36	03 752	03 810	03 868	03 925	03 983	36	04 041	04 098	04 156	04 213	04 270

Nitrogen Reduction Table (*Continued*)

At standard temperature and pressure (0° C., 760 mm.), 1 ml. of nitrogen weighs 1.2505 mg. The following table gives the logarithm of the weight of 1 ml. of nitrogen at a temperature of $t°$ C. and a pressure of p mm. (Taken from Niederl and Niederl.[9,10])

$t°$	$p=760$	761	762	763	764	$t°$	$p=765$	766	767	768	769
10	08 144	08 201	08 258	08 315	08 372	10	08 429	08 486	08 543	08 599	08 656
11	07 990	08 047	08 104	08 161	08 218	11	08 275	08 332	08 389	08 445	08 502
12	07 837	07 894	07 951	08 008	08 065	12	08 122	08 179	08 236	08 292	08 349
13	07 685	07 742	07 799	07 856	07 913	13	07 970	08 027	08 084	08 140	08 197
14	07 533	07 590	07 647	07 704	07 761	14	07 818	07 875	07 932	07 988	08 045
15	07 382	07 439	07 496	07 553	07 610	15	07 667	07 724	07 781	07 837	07 894
16	07 231	07 288	07 345	07 402	07 459	16	07 516	07 573	07 630	07 686	07 743
17	07 081	07 138	07 195	07 252	07 309	17	07 366	07 423	07 480	07 536	07 593
18	06 931	06 988	07 045	07 102	07 159	18	07 216	07 273	07 330	07 386	07 443
19	06 782	06 839	06 896	06 953	07 010	19	07 067	07 124	07 181	07 237	07 294
20	06 633	06 690	06 747	06 804	06 861	20	06 918	06 975	07 032	07 088	07 145
21	06 485	06 542	06 599	06 656	06 713	21	06 770	06 827	06 884	06 940	06 997
22	06 337	06 394	06 451	06 508	06 565	22	06 622	06 679	06 736	06 792	06 849
23	06 190	06 247	06 304	06 361	06 418	23	06 475	06 532	06 589	06 645	06 702
24	06 043	06 100	06 157	06 214	06 271	24	06 328	06 385	06 442	06 498	06 555
25	05 897	05 954	06 011	06 068	06 125	25	06 182	06 239	06 296	06 352	06 409
26	05 752	05 809	05 866	05 923	05 980	26	06 037	06 094	06 151	06 207	06 264
27	05 607	05 664	05 721	05 778	05 835	27	05 892	05 949	06 006	06 062	06 119
28	05 463	05 520	05 577	05 634	05 691	28	05 748	05 805	05 862	05 918	05 975
29	05 319	05 376	05 433	05 490	05 547	29	05 604	05 661	05 718	05 774	05 831
30	05 176	05 233	05 290	05 347	05 404	30	05 461	05 518	05 575	05 631	05 688
31	05 033	05 090	05 147	05 204	05 261	31	05 318	05 375	05 432	05 488	05 545
32	04 891	04 948	05 005	05 062	05 119	32	05 176	05 233	05 290	05 346	05 403
33	04 749	04 806	04 863	04 920	04 977	33	05 034	05 091	05 148	05 204	05 261
34	04 608	04 665	04 722	04 779	04 836	34	04 893	04 950	05 007	05 063	05 120
35	04 467	04 524	04 581	04 638	04 695	35	04 752	04 809	04 866	04 922	04 979
36	04 327	04 384	04 441	04 498	04 555	36	04 612	04 669	04 726	04 782	04 839

23. Calculations

Nitrogen Reduction Table (*Continued*)

At standard temperature and pressure (0° C., 760 mm.), 1 ml. of nitrogen weighs 1.2505 mg. The following table gives the logarithm of the weight of 1 ml. of nitrogen at a temperature of $t°$ C. and a pressure of p mm. (Taken from Niederl and Niederl.[9,10])

$t°$	$p=770$	771	772	773	774	$t°$	$p=775$	776	777	778	779
10	08 712	08 768	08 825	08 881	08 937	10	08 993	09 049	09 105	09 161	09 217
11	08 558	08 614	08 671	08 727	08 783	11	08 839	08 895	08 951	09 007	09 063
12	08 405	08 461	08 518	08 574	08 630	12	08 686	08 742	08 798	08 854	08 910
13	08 253	08 309	08 366	08 422	08 478	13	08 534	08 590	08 646	08 702	08 758
14	08 101	08 157	08 214	08 270	08 326	14	08 382	08 438	08 494	08 550	08 606
15	07 949	08 005	08 062	08 118	08 174	15	08 230	08 286	08 342	08 398	08 454
16	07 798	07 854	07 911	07 967	08 023	16	08 079	08 135	08 191	08 247	08 303
17	07 648	07 704	07 761	07 817	07 873	17	07 929	07 985	08 041	08 097	08 153
18	07 498	07 554	07 611	07 667	07 723	18	07 779	07 835	07 891	07 947	08 003
19	07 349	07 405	07 462	07 518	07 574	19	07 630	07 686	07 742	07 798	07 854
20	07 200	07 256	07 313	07 369	07 425	20	07 481	07 537	07 593	07 649	07 705
21	07 052	07 108	07 165	07 221	07 277	21	07 333	07 389	07 445	07 501	07 557
22	06 904	06 960	07 017	07 073	07 129	22	07 185	07 241	07 297	07 353	07 409
23	06 757	06 813	06 870	06 926	06 982	23	07 038	07 094	07 150	07 206	07 262
24	06 611	06 667	06 724	06 780	06 836	24	06 892	06 948	07 004	07 060	07 116
25	06 464	06 523	06 578	06 634	06 690	25	06 746	06 802	06 858	06 914	06 970
26	06 320	06 378	06 433	06 489	06 545	26	06 601	06 657	06 713	06 769	06 825
27	06 175	06 233	06 288	06 344	06 400	27	06 456	06 512	06 568	06 624	06 680
28	06 031	06 089	06 144	06 200	06 256	28	06 312	06 368	06 424	06 480	06 536
29	05 887	05 945	06 000	06 056	06 112	29	06 168	06 224	06 280	06 336	06 392
30	05 744	05 802	05 857	05 913	05 969	30	06 025	06 081	06 137	06 193	06 249
31	05 601	05 659	05 714	05 770	05 826	31	05 882	05 938	05 994	06 050	06 106
32	05 459	05 517	05 573	05 628	05 684	32	05 740	05 796	05 852	05 908	05 964
33	05 317	05 375	05 431	05 486	05 542	33	05 598	05 654	05 710	05 756	05 822
34	05 176	05 234	05 290	05 345	05 401	34	05 457	05 513	05 569	05 615	05 681
35	05 035	05 093	05 149	05 204	05 260	35	05 316	05 372	05 428	05 474	05 540
36	04 895	04 953	05 009	05 064	05 120	36	05 176	05 232	05 288	05 334	05 400

NITROGEN REDUCTION TABLE (*Continued*)

At standard temperature and pressure (0° C., 760 mm.), 1 ml. of nitrogen weighs 1.2505 mg. The following table gives the logarithm of the weight of 1 ml. of nitrogen at a temperature of $t°$ C. and a pressure of p mm. (Taken from Niederl and Niederl.[9,10])

$t°$	$p=780$	781	782	783	784	$t°$	$p=785$	786	787	788	789
10	09 272	09 328	09 384	09 439	09 495	10	09 550	09 605	09 660	09 716	09 771
11	09 118	09 174	09 230	09 285	09 341	11	09 396	09 451	09 506	09 562	09 617
12	08 965	09 021	09 077	09 132	09 188	12	09 243	09 298	09 353	09 409	09 464
13	08 813	08 869	08 925	08 980	09 036	13	09 091	09 146	09 201	09 257	09 312
14	08 661	08 717	08 773	08 828	08 884	14	08 939	08 994	09 049	09 105	09 162
15	08 509	08 565	08 621	08 676	08 732	15	08 787	08 842	08 897	08 953	09 008
16	08 358	08 414	08 470	08 525	08 581	16	08 636	08 691	08 746	08 802	08 857
17	08 208	08 264	08 320	08 375	08 431	17	08 486	08 541	08 596	08 652	08 707
18	08 058	08 114	08 170	08 225	08 281	18	08 336	08 391	08 446	08 502	08 557
19	07 909	07 965	08 021	08 076	08 132	19	08 187	08 242	08 297	08 353	08 408
20	07 760	07 816	07 872	07 927	07 983	20	08 038	08 093	08 148	08 204	08 259
21	07 612	07 668	07 724	07 779	07 835	21	07 890	07 945	08 000	08 056	08 111
22	07 464	07 520	07 576	07 631	07 687	22	07 742	07 797	07 852	07 908	07 963
23	07 317	07 373	07 429	07 484	07 540	23	07 595	07 650	07 705	07 761	07 816
24	07 171	07 227	07 283	07 338	07 394	24	07 449	07 504	07 559	07 615	07 670
25	07 025	07 081	07 137	07 192	07 248	25	07 303	07 358	07 413	07 469	07 524
26	06 880	06 936	06 992	07 047	07 103	26	07 158	07 213	07 268	07 324	07 379
27	06 735	06 771	06 847	06 902	06 958	27	07 013	07 068	07 123	07 179	07 234
28	06 591	06 627	06 703	06 758	06 814	28	06 869	06 924	06 979	07 035	07 090
29	06 447	06 483	06 559	06 614	06 670	29	06 725	06 780	06 835	06 891	06 946
30	06 304	06 340	06 416	06 471	06 527	30	06 582	06 637	06 692	06 748	06 803
31	06 161	06 197	06 273	06 328	06 387	31	06 439	06 494	06 549	06 605	06 660
32	06 019	06 055	06 131	06 186	06 245	32	06 297	06 352	06 407	06 463	06 518
33	05 877	05 913	05 989	06 044	06 103	33	06 155	06 210	06 265	06 321	06 376
34	05 736	05 772	05 848	05 903	05 962	34	06 014	06 069	06 124	06 180	06 235
35	05 595	05 631	05 707	05 762	05 821	35	05 873	05 928	05 983	06 039	06 094
36	05 455	05 491	05 567	05 622	05 681	36	05 733	05 788	05 843	05 899	05 954

TABLE 44
COMMON LOGARITHMS OF THE NATURAL NUMBERS

No.	0	1	2	3	4	5	6	7	8	9
100	00 000	00 043	00 087	00 130	00 173	00 217	00 260	00 303	00 346	00 389
101	00 432	00 475	00 518	00 561	00 604	00 647	00 689	00 732	00 775	00 817
102	00 860	00 903	00 945	00 988	01 030	01 072	01 115	01 157	01 199	01 242
103	01 284	01 326	01 368	01 410	01 452	01 494	01 536	01 578	01 620	01 662
104	01 703	01 745	01 787	01 828	01 870	01 912	01 953	01 995	02 036	02 078
105	02 119	02 160	02 202	02 243	02 284	02 325	02 366	02 407	02 449	02 490
106	02 531	02 572	02 612	02 653	02 694	02 735	02 776	02 816	02 857	02 898
107	02 938	02 979	03 019	03 060	03 100	03 141	03 181	03 222	03 262	03 302
108	03 342	03 383	03 423	03 463	03 503	03 543	03 583	03 623	03 663	03 703
109	03 743	03 782	03 822	03 862	03 902	03 941	03 981	04 021	04 060	04 100
110	04 139	04 179	04 218	04 258	04 297	04 336	04 376	04 415	04 454	04 493
111	04 532	04 571	04 610	04 650	04 689	04 727	04 766	04 805	04 844	04 883
112	04 922	04 961	04 999	05 038	05 077	05 115	05 154	05 192	05 231	05 269
113	05 308	05 346	05 385	05 423	05 461	05 500	05 538	05 576	05 614	05 652
114	05 690	05 729	05 767	05 805	05 843	05 881	05 918	05 956	05 994	06 032
115	06 070	06 108	06 145	06 183	06 221	06 258	06 296	06 333	06 371	06 408
116	06 446	06 483	06 521	06 558	06 595	06 633	06 670	06 707	06 744	06 781
117	06 819	06 856	06 893	06 930	06 967	07 004	07 041	07 078	07 115	07 151
118	07 188	07 225	07 262	07 298	07 335	07 372	07 408	07 445	07 482	07 518
119	07 555	07 591	07 628	07 664	07 700	07 737	07 773	07 809	07 846	07 882
120	07 918	07 954	07 990	08 027	08 063	08 099	08 135	08 171	08 207	08 243
121	08 279	08 314	08 350	08 386	08 422	08 458	08 493	08 529	08 565	08 600
122	08 636	08 672	08 707	08 743	08 778	08 814	08 849	08 884	08 920	08 955
123	08 991	09 026	09 061	09 096	09 132	09 167	09 202	09 237	09 272	09 307
124	09 342	09 377	09 412	09 447	09 482	09 517	09 552	09 587	09 621	09 656
125	09 691	09 726	09 760	09 795	09 830	09 864	09 899	09 934	09 968	10 003
126	10 037	10 072	10 106	10 140	10 175	10 209	10 243	10 278	10 312	10 346
127	10 380	10 415	10 449	10 483	10 517	10 551	10 585	10 619	10 653	10 687
128	10 721	10 755	10 789	10 823	10 857	10 890	10 924	10 958	10 992	11 025
129	11 059	11 093	11 126	11 160	11 193	11 227	11 261	11 294	11 327	11 361
130	11 394	11 428	11 461	11 494	11 528	11 561	11 594	11 628	11 661	11 694
131	11 727	11 760	11 793	11 826	11 860	11 893	11 926	11 959	11 992	12 024
132	12 057	12 090	12 123	12 156	12 189	12 222	12 254	12 287	12 320	12 352
133	12 385	12 418	12 450	12 483	12 516	12 548	12 581	12 613	12 646	12 678
134	12 710	12 743	12 775	12 808	12 840	12 872	12 905	12 937	12 969	13 001
135	13 033	13 066	13 098	13 130	13 162	13 194	13 226	13 258	13 290	13 322
136	13 354	13 386	13 418	13 450	13 481	13 513	13 545	13 577	13 609	13 640
137	13 672	13 704	13 735	13 767	13 799	13 830	13 862	13 893	13 925	13 956
138	13 988	14 019	14 051	14 082	14 114	14 145	14 176	14 208	14 239	14 270
139	14 301	14 333	14 364	14 395	14 426	14 457	14 489	14 520	14 551	14 582
140	14 613	14 644	14 675	14 706	14 737	14 768	14 799	14 829	14 860	14 891
141	14 922	14 953	14 983	15 014	15 045	15 076	15 106	15 137	15 168	15 198
142	15 229	15 259	15 290	15 320	15 351	15 381	15 412	15 442	15 473	15 503
143	15 534	15 564	15 594	15 625	15 655	15 685	15 715	15 746	15 776	15 806
144	15 836	15 866	15 897	15 927	15 957	15 987	16 017	16 047	16 077	16 107
145	16 137	16 167	16 197	16 227	16 256	16 286	16 316	16 346	16 376	16 406
146	16 435	16 465	16 495	16 524	16 554	16 584	16 613	16 643	16 673	16 702
147	16 732	16 761	16 791	16 820	16 850	16 879	16 909	16 938	16 967	16 997
148	17 026	17 056	17 085	17 114	17 143	17 173	17 202	17 231	17 260	17 289
149	17 319	17 348	17 377	17 406	17 435	17 464	17 493	17 522	17 551	17 580
150	17 609	17 638	17 667	17 696	17 725	17 754	17 782	17 811	17 840	17 869

Common Logarithms of the Natural Numbers (*Continued*)

No.	0	1	2	3	4	5	6	7	8	9
150	17 609	17 638	17 667	17 696	17 725	17 754	17 782	17 811	17 840	17 869
151	17 898	17 926	17 955	17 984	18 013	18 041	18 070	18 099	18 127	18 156
152	18 184	18 213	18 241	18 270	18 298	18 327	18 355	18 384	18 412	18 441
153	18 469	18 498	18 526	18 554	18 583	18 611	18 639	18 667	18 696	18 724
154	18 752	18 780	18 808	18 837	18 865	18 893	18 921	18 949	18 977	19 005
155	19 033	19 061	19 089	19 117	19 145	19 173	19 201	19 229	19 257	19 285
156	19 312	19 340	19 368	19 396	19 424	19 451	19 479	19 507	19 535	19 562
157	19 590	19 618	19 645	19 673	19 700	19 728	19 756	19 783	19 811	19 838
158	19 866	19 893	19 921	19 948	19 976	20 003	20 030	20 058	20 085	20 112
159	20 140	20 167	20 194	20 222	20 249	20 276	20 303	20 330	20 358	20 385
160	20 412	20 439	20 466	20 493	20 520	20 548	20 575	20 602	20 629	20 656
161	20 683	20 710	20 737	20 763	20 790	20 817	20 844	20 871	20 898	20 925
162	20 952	20 978	21 005	21 032	21 059	21 085	21 112	21 139	21 165	21 192
163	21 219	21 245	21 272	21 299	21 325	21 352	21 378	21 405	21 431	21 458
164	21 484	21 511	21 537	21 564	21 590	21 617	21 643	21 669	21 696	21 722
165	21 748	21 775	21 801	21 827	21 854	21 880	21 906	21 932	21 958	21 985
166	22 011	22 037	22 063	22 089	22 115	22 141	22 167	22 194	22 220	22 246
167	22 272	22 298	22 324	22 350	22 376	22 401	22 427	22 453	22 479	22 505
168	22 531	22 557	22 583	22 608	22 634	22 660	22 686	22 712	22 737	22 763
169	22 789	22 814	22 840	22 866	22 891	22 917	22 943	22 968	22 994	23 019
170	23 045	23 070	23 096	23 121	23 147	23 172	23 198	23 223	23 249	23 274
171	23 300	23 325	23 350	23 376	23 401	23 426	23 452	23 477	23 502	23 528
172	23 553	23 578	23 603	23 629	23 654	23 679	23 704	23 729	23 754	23 779
173	23 805	23 830	23 855	23 880	23 905	23 930	23 955	23 980	24 005	24 030
174	24 055	24 080	24 105	24 130	24 155	24 180	24 204	24 229	24 254	24 279
175	24 304	24 329	24 353	24 378	24 403	24 428	24 452	24 477	24 502	24 527
176	24 551	24 576	24 601	24 625	24 650	24 674	24 699	24 724	24 748	24 773
177	24 797	24 822	24 846	24 871	24 895	24 920	24 944	24 969	24 993	25 018
178	25 042	25 066	25 091	25 115	25 139	25 164	25 188	25 212	25 237	25 261
179	25 285	25 310	25 334	25 358	25 382	25 406	25 431	25 455	25 479	25 503
180	25 527	25 551	25 575	25 600	25 624	25 648	25 672	25 696	25 720	25 744
181	25 768	25 792	25 816	25 840	25 864	25 888	25 912	25 935	25 959	25 983
182	26 007	26 031	26 055	26 079	26 102	26 126	26 150	26 174	26 198	26 221
183	26 245	26 269	26 293	26 316	26 340	26 364	26 387	26 411	26 435	26 458
184	26 482	26 505	26 529	26 553	26 576	26 600	26 623	26 647	26 670	26 694
185	26 717	26 741	26 764	26 788	26 811	26 834	26 858	26 881	26 905	26 928
186	26 951	26 975	26 998	27 021	27 045	27 068	27 091	27 114	27 138	27 161
187	27 184	27 207	27 231	27 254	27 277	27 300	27 323	27 346	27 370	27 393
188	27 416	27 439	27 462	27 485	27 508	27 531	27 554	27 577	27 600	27 623
189	27 646	27 669	27 692	27 715	27 738	27 761	27 784	27 807	27 830	27 852
190	27 875	27 898	27 921	27 944	27 967	27 989	28 012	28 035	28 058	28 081
191	28 103	28 126	28 149	28 171	28 194	28 217	28 240	28 262	28 285	28 307
192	28 330	28 353	28 375	28 398	28 421	28 443	28 466	28 488	28 511	28 533
193	28 556	28 578	28 601	28 623	28 646	28 668	28 691	28 713	28 735	28 758
194	28 780	28 803	28 825	28 847	28 870	28 892	28 914	28 937	28 959	28 981
195	29 003	29 026	29 048	29 070	29 092	29 115	29 137	29 159	29 181	29 203
196	29 226	29 248	29 270	29 292	29 314	29 336	29 358	29 380	29 403	29 425
197	29 447	29 469	29 491	29 513	29 535	29 557	29 579	29 601	29 623	29 645
198	29 667	29 688	29 710	29 732	29 754	29 776	29 798	29 820	29 842	29 863
199	29 885	29 907	29 929	29 951	29 973	29 994	30 016	30 038	30 060	30 081
200	30 103	30 125	30 146	30 168	30 190	30 211	30 233	30 255	30 276	30 298

COMMON LOGARITHMS OF THE NATURAL NUMBERS (*Continued*)

No.	0	1	2	3	4	5	6	7	8	9
200	30 103	30 125	30 146	30 168	30 190	30 211	30 233	30 255	30 276	30 298
201	30 320	30 341	30 363	30 384	30 406	30 428	30 449	30 471	30 492	30 514
202	30 535	30 557	30 578	30 600	30 621	30 643	30 664	30 685	30 707	30 728
203	30 750	30 771	30 792	30 814	30 835	30 856	30 878	30 899	30 920	30 942
204	30 963	30 984	31 006	31 027	31 048	31 069	31 091	31 112	31 133	31 154
205	31 175	31 197	31 218	31 239	31 260	31 281	31 302	31 323	31 345	31 366
206	31 387	31 408	31 429	31 450	31 471	31 492	31 513	31 534	31 555	31 576
207	31 597	31 618	31 639	31 660	31 681	31 702	31 723	31 744	31 765	31 785
208	31 806	31 827	31 848	31 869	31 890	31 911	31 931	31 952	31 973	31 994
209	32 015	32 035	32 056	32 077	32 098	32 118	32 139	32 160	32 181	32 201
210	32 222	32 243	32 263	32 284	32 305	32 325	32 346	32 366	32 387	32 408
211	32 428	32 449	32 469	32 490	32 510	32 531	32 552	32 572	32 593	32 613
212	32 634	32 654	32 675	32 695	32 715	32 736	32 756	32 777	32 797	32 818
213	32 838	32 858	32 879	32 899	32 919	32 940	32 960	32 980	33 001	33 021
214	33 041	33 062	33 082	33 102	33 122	33 143	33 163	33 183	33 203	33 224
215	33 244	33 264	33 284	33 304	33 325	33 345	33 365	33 385	33 405	33 425
216	33 445	33 465	33 486	33 506	33 526	33 546	33 566	33 586	33 606	33 626
217	33 646	33 666	33 686	33 706	33 726	33 746	33 766	33 786	33 806	33 826
218	33 846	33 866	33 885	33 905	33 925	33 945	33 965	33 985	34 005	34 025
219	34 044	34 064	34 084	34 104	34 124	34 143	34 163	34 183	34 203	34 223
220	34 242	34 262	34 282	34 301	34 321	34 341	34 361	34 380	34 400	34 420
221	34 439	34 459	34 479	34 498	34 518	34 537	34 557	34 577	34 596	34 616
222	34 635	34 655	34 674	34 694	34 713	34 733	34 753	34 772	34 792	34 811
223	34 830	34 850	34 869	34 889	34 908	34 928	34 947	34 967	34 986	35 005
224	35 025	35 044	35 064	35 083	35 102	35 122	35 141	35 160	35 180	35 199
225	35 218	35 238	35 257	35 276	35 295	35 315	35 334	35 353	35 372	35 392
226	35 411	35 430	35 449	35 468	35 488	35 507	35 526	35 545	35 564	35 583
227	35 603	35 622	35 641	35 660	35 679	35 698	35 717	35 736	35 755	35 774
228	35 793	35 813	35 832	35 851	35 870	35 889	35 908	35 927	35 946	35 965
229	35 984	36 003	36 021	36 040	36 059	36 078	36 097	36 116	36 135	36 154
230	36 173	36 192	36 211	36 229	36 248	36 267	36 286	36 305	36 324	36 342
231	36 361	36 380	36 399	36 418	36 436	36 455	36 474	36 493	36 511	36 530
232	36 549	36 568	36 586	36 605	36 624	36 642	36 661	36 680	36 698	36 717
233	36 736	36 754	36 773	36 791	36 810	36 829	36 847	36 866	36 884	36 903
234	36 922	36 940	36 959	36 977	36 996	37 014	37 033	37 051	37 070	37 088
235	37 107	37 125	37 144	37 162	37 181	37 199	37 218	37 236	37 254	37 273
236	37 291	37 310	37 328	37 346	37 365	37 383	37 401	37 420	37 438	37 457
237	37 475	37 493	37 511	37 530	37 548	37 566	37 585	37 603	37 621	37 639
238	37 658	37 676	37 694	37 712	37 731	37 749	37 767	37 785	37 803	37 822
239	37 840	37 858	37 876	37 894	37 912	37 931	37 949	37 967	37 985	38 003
240	38 021	38 039	38 057	38 075	38 093	38 112	38 130	38 148	38 166	38 184
241	38 202	38 220	38 238	38 256	38 274	38 292	38 310	38 328	38 346	38 364
242	38 382	38 399	38 417	38 435	38 453	38 471	38 489	38 507	38 525	38 543
243	38 561	38 578	38 596	38 614	38 632	38 650	38 668	38 686	38 703	38 721
244	38 739	38 757	38 775	38 792	38 810	38 828	38 846	38 863	38 881	38 899
245	38 917	38 934	38 952	38 970	38 987	39 005	39 023	39 041	39 058	39 076
246	39 094	39 111	39 129	39 146	39 164	39 182	39 199	39 217	39 235	39 252
247	39 270	39 287	39 305	39 322	39 340	39 358	39 375	39 393	39 410	39 428
248	39 445	39 463	39 480	39 498	39 515	39 533	39 550	39 568	39 585	39 602
249	39 620	39 637	39 655	39 672	39 690	39 707	39 724	39 742	39 759	39 777
250	39 794	39 811	39 829	39 846	39 863	39 881	39 898	39 915	39 933	39 950

COMMON LOGARITHMS OF THE NATURAL NUMBERS (*Continued*)

No.	0	1	2	3	4	5	6	7	8	9
250	39 794	39 811	39 829	39 846	39 863	39 881	39 898	39 915	39 933	39 950
251	39 967	39 985	40 002	40 019	40 037	40 054	40 071	40 088	40 106	40 123
252	40 140	40 157	40 175	40 192	40 209	40 226	40 243	40 261	40 278	40 295
253	40 312	40 329	40 346	40 364	40 381	40 398	40 415	40 432	40 449	40 466
254	40 483	40 500	40 518	40 535	40 552	40 569	40 586	40 603	40 620	40 637
255	40 654	40 671	40 688	40 705	40 722	40 739	40 756	40 773	40 790	40 807
256	40 824	40 841	40 858	40 875	40 892	40 909	40 926	40 943	40 960	40 976
257	40 993	41 010	41 027	41 044	41 061	41 078	41 095	41 111	41 128	41 145
258	41 162	41 179	41 196	41 212	41 229	41 246	41 263	41 280	41 296	41 313
259	41 330	41 347	41 363	41 380	41 397	41 414	41 430	41 447	41 464	41 481
260	41 497	41 514	41 531	41 547	41 564	41 581	41 597	41 614	41 631	41 647
261	41 664	41 681	41 697	41 714	41 731	41 747	41 764	41 780	41 797	41 814
262	41 830	41 847	41 863	41 880	41 896	41 913	41 929	41 946	41 963	41 979
263	41 996	42 012	42 029	42 045	42 062	42 078	42 095	42 111	42 127	42 144
264	42 160	42 177	42 193	42 210	42 226	42 243	42 259	42 275	42 292	42 308
265	42 325	42 341	42 357	42 374	42 390	42 406	42 423	42 439	42 455	42 472
266	42 488	42 504	42 521	42 537	42 553	42 570	42 586	42 602	42 619	42 635
267	42 651	42 667	42 684	42 700	42 716	42 732	42 749	42 765	42 781	42 797
268	42 813	42 830	42 846	42 862	42 878	42 894	42 911	42 927	42 943	42 959
269	42 975	42 991	43 008	43 024	43 040	43 056	43 072	43 088	43 104	43 120
270	43 136	43 152	43 169	43 185	43 201	43 217	43 233	43 249	43 265	43 281
271	43 297	43 313	43 329	43 345	43 361	43 377	43 393	43 409	43 425	43 441
272	43 457	43 473	43 489	43 505	43 521	43 537	43 553	43 569	43 584	43 600
273	43 616	43 632	43 648	43 664	43 680	43 696	43 712	43 727	43 743	43 759
274	43 775	43 791	43 807	43 823	43 838	43 854	43 870	43 886	43 902	43 917
275	43 933	43 949	43 965	43 981	43 996	44 012	44 028	44 044	44 059	44 075
276	44 091	44 107	44 122	44 138	44 154	44 170	44 185	44 201	44 217	44 232
277	44 248	44 264	44 279	44 295	44 311	44 326	44 342	44 358	44 373	44 389
278	44 404	44 420	44 436	44 451	44 467	44 483	44 498	44 514	44 529	44 545
279	44 560	44 576	44 592	44 607	44 623	44 638	44 654	44 669	44 685	44 700
280	44 716	44 731	44 747	44 762	44 778	44 793	44 809	44 824	44 840	44 855
281	44 871	44 886	44 902	44 917	44 932	44 948	44 963	44 979	44 994	45 010
282	45 025	45 040	45 056	45 071	45 086	45 102	45 117	45 133	45 148	45 163
283	45 179	45 194	45 209	45 225	45 240	45 255	45 271	45 286	45 301	45 317
284	45 332	45 347	45 362	45 378	45 393	45 408	45 423	45 439	45 454	45 469
285	45 484	45 500	45 515	45 530	45 545	45 561	45 576	45 591	45 606	45 621
286	45 637	45 652	45 667	45 682	45 697	45 712	45 728	45 743	45 758	45 773
287	45 788	45 803	45 818	45 834	45 849	45 864	45 879	45 894	45 909	45 924
288	45 939	45 954	45 969	45 984	46 000	46 015	46 030	46 045	46 060	46 075
289	46 090	46 105	46 120	46 135	46 150	46 165	46 180	46 195	46 210	46 225
290	46 240	46 255	46 270	46 285	46 300	46 315	46 330	46 345	46 359	46 374
291	46 389	46 404	46 419	46 434	46 449	46 464	46 479	46 494	46 509	46 523
292	46 538	46 553	46 568	46 583	46 598	46 613	46 627	46 642	46 657	46 672
293	46 687	46 702	46 716	46 731	46 746	46 761	46 776	46 790	46 805	46 820
294	46 835	46 850	46 864	46 879	46 894	46 909	46 923	46 938	46 953	46 967
295	46 982	46 997	47 012	47 026	47 041	47 056	47 070	47 085	47 100	47 114
296	47 129	47 144	47 159	47 173	47 188	47 202	47 217	47 232	47 246	47 261
297	47 276	47 290	47 305	47 319	47 334	47 349	47 363	47 378	47 392	47 407
298	47 422	47 436	47 451	47 465	47 480	47 494	47 509	47 524	47 538	47 553
299	47 567	47 582	47 596	47 611	47 625	47 640	47 654	47 669	47 683	47 698
300	47 712	47 727	47 741	47 756	47 770	47 784	47 799	47 813	47 828	47 842

23. Calculations

Common Logarithms of the Natural Numbers (*Continued*)

No.	0	1	2	3	4	5	6	7	8	9
300	47 712	47 727	47 741	47 756	47 770	47 784	47 799	47 813	47 828	47 842
301	47 857	47 871	47 885	47 900	47 914	47 929	47 943	47 958	47 972	47 986
302	48 001	48 015	48 029	48 044	48 058	48 073	48 087	48 101	48 116	48 130
303	48 144	48 159	48 173	48 187	48 202	48 216	48 230	48 244	48 259	48 273
304	48 287	48 302	48 316	48 330	48 344	48 359	48 373	48 387	48 401	48 416
305	48 430	48 444	48 458	48 473	48 487	48 501	48 515	48 530	48 544	48 558
306	48 572	48 586	48 601	48 615	48 629	48 643	48 657	48 671	48 686	48 700
307	48 714	48 728	48 742	48 756	48 770	48 785	48 799	48 813	48 827	48 841
308	48 855	48 869	48 883	48 897	48 911	48 926	48 940	48 954	48 968	48 982
309	48 996	49 010	49 024	49 038	49 052	49 066	49 080	49 094	49 108	49 122
310	49 136	49 150	49 164	49 178	49 192	49 206	49 220	49 234	49 248	49 262
311	49 276	49 290	49 304	49 318	49 332	49 346	49 360	49 374	49 388	49 402
312	49 415	49 429	49 443	49 457	49 471	49 485	49 499	49 513	49 527	49 541
313	49 554	49 568	49 582	49 596	49 610	49 624	49 638	49 651	49 665	49 679
314	49 693	49 707	49 721	49 734	49 748	49 762	49 776	49 790	49 803	49 817
315	49 831	49 845	49 859	49 872	49 886	49 900	49 914	49 927	49 941	49 955
316	49 969	49 982	49 996	50 010	50 024	50 037	50 051	50 065	50 079	50 092
317	50 106	50 120	50 133	50 147	50 161	50 174	50 188	50 202	50 215	50 229
318	50 243	50 256	50 270	50 284	50 297	50 311	50 325	50 338	50 352	50 365
319	50 379	50 393	50 406	50 420	50 433	50 447	50 461	50 474	50 488	50 501
320	50 515	50 529	50 542	50 556	50 569	50 583	50 596	50 610	50 623	50 637
321	50 651	50 664	50 678	50 691	50 705	50 718	50 732	50 745	50 759	50 772
322	50 786	50 799	50 813	50 826	50 840	50 853	50 866	50 880	50 893	50 907
323	50 920	50 934	50 947	50 961	50 974	50 987	51 001	51 014	51 028	51 041
324	51 055	51 068	51 081	51 095	51 108	51 121	51 135	51 148	51 162	51 175
325	51 188	51 202	51 215	51 228	51 242	51 255	51 268	51 282	51 295	51 308
326	51 322	51 335	51 348	51 362	51 375	51 388	51 402	51 415	51 428	51 441
327	51 455	51 468	51 481	51 495	51 508	51 521	51 534	51 548	51 561	51 574
328	51 587	51 601	51 614	51 627	51 640	51 654	51 667	51 680	51 693	51 706
329	51 720	51 733	51 746	51 759	51 772	51 786	51 799	51 812	51 825	51 838
330	51 851	51 865	51 878	51 891	51 904	51 917	51 930	51 943	51 957	51 970
331	51 983	51 996	52 009	52 022	52 035	52 048	52 061	52 075	52 088	52 101
332	52 114	52 127	52 140	52 153	52 166	52 179	52 192	52 205	52 218	52 231
333	52 244	52 257	52 270	52 284	52 297	52 310	52 323	52 336	52 349	52 362
334	52 375	52 388	52 401	52 414	52 427	52 440	52 453	52 466	52 479	52 492
335	52 504	52 517	52 530	52 543	52 556	52 569	52 582	52 595	52 608	52 621
336	52 634	52 647	52 660	52 673	52 686	52 699	52 711	52 724	52 737	52 750
337	52 763	52 776	52 789	52 802	52 815	52 827	52 840	52 853	52 866	52 879
338	52 892	52 905	52 917	52 930	52 943	52 956	52 969	52 982	52 994	53 007
339	53 020	53 033	53 046	53 058	53 071	53 084	53 097	53 110	53 122	53 135
340	53 148	53 161	53 173	53 186	53 199	53 212	53 224	53 237	53 250	53 263
341	53 275	53 288	53 301	53 314	53 326	53 339	53 352	53 364	53 377	53 390
342	53 403	53 415	53 428	53 441	53 453	53 466	53 479	53 491	53 504	53 517
343	53 529	53 542	53 555	53 567	53 580	53 593	53 605	53 618	53 631	53 643
344	53 656	53 668	53 681	53 694	53 706	53 719	53 732	53 744	53 757	53 769
345	53 782	53 794	53 807	53 820	53 832	53 845	53 857	53 870	53 882	53 895
346	53 908	53 920	53 933	53 945	53 958	53 970	53 983	53 995	54 008	54 020
347	54 033	54 045	54 058	54 070	54 083	54 095	54 108	54 120	54 133	54 145
348	54 158	54 170	54 183	54 195	54 208	54 220	54 233	54 245	54 258	54 270
349	54 283	54 295	54 307	54 320	54 332	54 345	54 357	54 370	54 382	54 394
350	54 407	54 419	54 432	54 444	54 456	54 469	54 481	54 494	54 506	54 518

COMMON LOGARITHMS OF THE NATURAL NUMBERS (*Continued*)

No.	0	1	2	3	4	5	6	7	8	9
350	54 407	54 419	54 432	54 444	54 456	54 469	54 481	54 494	54 506	54 518
351	54 531	54 543	54 555	54 568	54 580	54 593	54 605	54 617	54 630	54 642
352	54 654	54 667	54 679	54 691	54 704	54 716	54 728	54 741	54 753	54 765
353	54 777	54 790	54 802	54 814	54 827	54 839	54 851	54 864	54 876	54 888
354	54 900	54 913	54 925	54 937	54 949	54 962	54 974	54 986	54 998	55 011
355	55 023	55 035	55 047	55 060	55 072	55 084	55 096	55 108	55 121	55 133
356	55 145	55 157	55 169	55 182	55 194	55 206	55 218	55 230	55 242	55 255
357	55 267	55 279	55 291	55 303	55 315	55 328	55 340	55 352	55 364	55 376
358	55 388	55 400	55 413	55 425	55 437	55 449	55 461	55 473	55 485	55 497
359	55 509	55 522	55 534	55 546	55 558	55 570	55 582	55 594	55 606	55 618
360	55 630	55 642	55 654	55 666	55 678	55 691	55 703	55 715	55 727	55 739
361	55 751	55 763	55 775	55 787	55 799	55 811	55 823	55 835	55 847	55 859
362	55 871	55 883	55 895	55 907	55 919	55 931	55 943	55 955	55 967	55 979
363	55 991	56 003	56 015	56 027	56 038	56 050	56 062	56 074	56 086	56 098
364	56 110	56 122	56 134	56 146	56 158	56 170	56 182	56 194	56 205	56 217
365	56 229	56 241	56 253	56 265	56 277	56 289	56 301	56 312	56 324	56 336
366	56 348	56 360	56 372	56 384	56 396	56 407	56 419	56 431	56 443	56 455
367	56 467	56 478	56 490	56 502	56 514	56 526	56 538	56 549	56 561	56 573
368	56 585	56 597	56 608	56 620	56 632	56 644	56 656	56 667	56 679	56 691
369	56 703	56 714	56 726	56 738	56 750	56 761	56 773	56 785	56 797	56 808
370	56 820	56 832	56 844	56 855	56 867	56 879	56 891	56 902	56 914	56 926
371	56 937	56 949	56 961	56 972	56 984	56 996	57 008	57 019	57 031	57 043
372	57 054	57 066	57 078	57 089	57 101	57 113	57 124	57 136	57 148	57 159
373	57 171	57 183	57 194	57 206	57 217	57 229	57 241	57 252	57 264	57 276
374	57 287	57 299	57 310	57 322	57 334	57 345	57 357	57 368	57 380	57 392
375	57 403	57 415	57 426	57 438	57 449	57 461	57 473	57 484	57 496	57 507
376	57 519	57 530	57 542	57 553	57 565	57 576	57 588	57 600	57 611	57 623
377	57 634	57 646	57 657	57 669	57 680	57 692	57 703	57 715	57 726	57 738
378	57 749	57 761	57 772	57 784	57 795	57 807	57 818	57 830	57 841	57 852
379	57 864	57 875	57 887	57 898	57 910	57 921	57 933	57 944	57 955	57 967
380	57 978	57 990	58 001	58 013	58 024	58 035	58 047	58 058	58 070	58 081
381	58 092	58 104	58 115	58 127	58 138	58 149	58 161	58 172	58 184	58 195
382	58 206	58 218	58 229	58 240	58 252	58 263	58 274	58 286	58 297	58 309
383	58 320	58 331	58 343	58 354	58 365	58 377	58 388	58 399	58 410	58 422
384	58 433	58 444	58 456	58 467	58 478	58 490	58 501	58 512	58 524	58 535
385	58 546	58 557	58 569	58 580	58 591	58 602	58 614	58 625	58 636	58 647
386	58 659	58 670	58 681	58 692	58 704	58 715	58 726	58 737	58 749	58 760
387	58 771	58 782	58 794	58 805	58 816	58 827	58 838	58 850	58 861	58 872
388	58 883	58 894	58 906	58 917	58 928	58 939	58 950	58 961	58 973	58 984
389	58 995	59 006	59 017	59 028	59 040	59 051	59 062	59 073	59 084	59 095
390	59 106	59 118	59 129	59 140	59 151	59 162	59 173	59 184	59 195	59 207
391	59 218	59 229	59 240	59 251	59 262	59 273	59 284	59 295	59 306	59 318
392	59 329	59 340	59 351	59 362	59 373	59 384	59 395	59 406	59 417	59 428
393	59 439	59 450	59 461	59 472	59 483	59 494	59 506	59 517	59 528	59 539
394	59 550	59 561	59 572	59 583	59 594	59 605	59 616	59 627	59 638	59 649
395	59 660	59 671	59 682	59 693	59 704	59 715	59 726	59 737	59 748	59 759
396	59 770	59 780	59 791	59 802	59 813	59 824	59 835	59 846	59 857	59 868
397	59 879	59 890	59 901	59 912	59 923	59 934	59 945	59 956	59 966	59 977
398	59 988	59 999	60 010	60 021	60 032	60 043	60 054	60 065	60 076	60 086
399	60 097	60 108	60 119	60 130	60 141	60 152	60 163	60 173	60 184	60 195
400	60 206	60 217	60 228	60 239	60 249	60 260	60 271	60 282	60 293	60 304

23. Calculations

COMMON LOGARITHMS OF THE NATURAL NUMBERS (*Continued*)

No.	0	1	2	3	4	5	6	7	8	9
400	60 206	60 217	60 228	60 239	60 249	60 260	60 271	60 282	60 293	60 304
401	60 314	60 325	60 336	60 347	60 358	60 369	60 379	60 390	60 401	60 412
402	60 423	60 433	60 444	60 455	60 466	60 477	60 487	60 498	60 509	60 520
403	60 531	60 541	60 552	60 563	60 574	60 584	60 595	60 606	60 617	60 627
404	60 638	60 649	60 660	60 670	60 681	60 692	60 703	60 713	60 724	60 735
405	60 746	60 756	60 767	60 778	60 788	60 799	60 810	60 821	60 831	60 842
406	60 853	60 863	60 874	60 885	60 895	60 906	60 917	60 927	60 938	60 949
407	60 959	60 970	60 981	60 991	61 002	61 013	61 023	61 034	61 045	61 055
408	61 066	61 077	61 087	61 098	61 109	61 119	61 130	61 140	61 151	61 162
409	61 172	61 183	61 194	61 204	61 215	61 225	61 236	61 247	61 257	61 268
410	61 278	61 289	61 300	61 310	61 321	61 331	61 342	61 352	61 363	61 374
411	61 384	61 395	61 405	61 416	61 426	61 437	61 448	61 458	61 469	61 479
412	61 490	61 500	61 511	61 521	61 532	61 542	61 553	61 563	61 574	61 584
413	61 595	61 606	61 616	61 627	61 637	61 648	61 658	61 669	61 679	61 690
414	61 700	61 711	61 721	61 731	61 742	61 752	61 763	61 773	61 784	61 794
415	61 805	61 815	61 826	61 836	61 847	61 857	61 868	61 878	61 888	61 899
416	61 909	61 920	61 930	61 941	61 951	61 962	61 972	61 982	61 993	62 003
417	62 014	62 024	62 034	62 045	62 055	62 066	62 076	62 086	62 097	62 107
418	62 118	62 128	62 138	62 149	62 159	62 170	62 180	62 190	62 201	62 211
419	62 221	62 232	62 242	62 252	62 263	62 273	62 284	62 294	62 304	62 315
420	62 325	62 335	62 346	62 356	62 366	62 377	62 387	62 397	62 408	62 418
421	62 428	62 439	62 449	62 459	62 469	62 480	62 490	62 500	62 511	62 521
422	62 531	62 542	62 552	62 562	62 572	62 583	62 593	62 603	62 613	62 624
423	62 634	62 644	62 655	62 665	62 675	62 685	62 696	62 706	62 716	62 726
424	62 737	62 747	62 757	62 767	62 778	62 788	62 798	62 808	62 818	62 829
425	62 839	62 849	62 859	62 870	62 880	62 890	62 900	62 910	62 921	62 931
426	62 941	62 951	62 961	62 972	62 982	62 992	63 002	63 012	63 022	63 033
427	63 043	63 053	63 063	63 073	63 083	63 094	63 104	63 114	63 124	63 134
428	63 144	63 155	63 165	63 175	63 185	63 195	63 205	63 215	63 225	63 236
429	63 246	63 256	63 266	63 276	63 286	63 296	63 306	63 317	63 327	63 337
430	63 347	63 357	63 367	63 377	63 387	63 397	63 407	63 417	63 428	63 438
431	63 448	63 458	63 468	63 478	63 488	63 498	63 508	63 518	63 528	63 538
432	63 548	63 558	63 568	63 579	63 589	63 599	63 609	63 619	63 629	63 639
433	63 649	63 659	63 669	63 679	63 689	63 699	63 709	63 719	63 729	63 739
434	63 749	63 759	63 769	63 779	63 789	63 799	63 809	63 819	63 829	63 839
435	63 849	63 859	63 869	63 879	63 889	63 899	63 909	63 919	63 929	63 939
436	63 949	63 959	63 969	63 979	63 988	63 998	64 008	64 018	64 028	64 038
437	64 048	64 058	64 068	64 078	64 088	64 098	64 108	64 118	64 128	64 137
438	64 147	64 157	64 167	64 177	64 187	64 197	64 207	64 217	64 227	64 237
439	64 246	64 256	64 266	64 276	64 286	64 296	64 306	64 316	64 326	64 335
440	64 345	64 355	64 365	64 375	64 385	64 395	64 404	64 414	64 424	64 434
441	64 444	64 454	64 464	64 473	64 483	64 493	64 503	64 513	64 523	64 532
442	64 542	64 552	64 562	64 572	64 582	64 591	64 601	64 611	64 621	64 631
443	64 640	64 650	64 660	64 670	64 680	64 689	64 699	64 709	64 719	64 729
444	64 738	64 748	64 758	64 768	64 777	64 787	64 797	64 807	64 816	64 826
445	64 836	64 846	64 856	64 865	64 875	64 885	64 895	64 904	64 914	64 924
446	64 933	64 943	64 953	64 963	64 972	64 982	64 992	65 002	65 011	65 021
447	65 031	65 040	65 050	65 060	65 070	65 079	65 089	65 099	65 108	65 118
448	65 128	65 137	65 147	65 157	65 167	65 176	65 186	65 196	65 205	65 215
449	65 225	65 234	65 244	65 254	65 263	65 273	65 283	65 292	65 302	65 312
450	65 321	65 331	65 341	65 350	65 360	65 369	65 379	65 389	65 398	65 408

COMMON LOGARITHMS OF THE NATURAL NUMBERS (*Continued*)

No.	0	1	2	3	4	5	6	7	8	9
450	65 321	65 331	65 341	65 350	65 360	65 369	65 379	65 389	65 398	65 408
451	65 418	65 427	65 437	65 447	65 456	65 466	65 475	65 485	65 495	65 504
452	65 514	65 523	65 533	65 543	65 552	65 562	65 571	65 581	65 591	65 600
453	65 610	65 619	65 629	65 639	65 648	65 658	65 667	65 677	65 686	65 696
454	65 706	65 715	65 725	65 734	65 744	65 753	65 763	65 772	65 782	65 792
455	65 801	65 811	65 820	65 830	65 839	65 849	65 858	65 868	65 877	65 887
456	65 896	65 906	65 916	65 925	65 935	65 944	65 954	65 963	65 973	65 982
457	65 992	66 001	66 011	66 020	66 030	66 039	66 049	66 058	66 068	66 077
458	66 087	66 096	66 106	66 115	66 124	66 134	66 143	66 153	66 162	66 172
459	66 181	66 191	66 200	66 210	66 219	66 229	66 238	66 247	66 257	66 266
460	66 276	66 285	66 295	66 304	66 314	66 323	66 332	66 342	66 351	66 361
461	66 370	66 380	66 389	66 398	66 408	66 417	66 427	66 436	66 445	66 455
462	66 464	66 474	66 483	66 492	66 502	66 511	66 521	66 530	66 539	66 549
463	66 558	66 567	66 577	66 586	66 596	66 605	66 614	66 624	66 633	66 642
464	66 652	66 661	66 671	66 680	66 689	66 699	66 708	66 717	66 727	66 736
465	66 745	66 755	66 764	66 773	66 783	66 792	66 801	66 811	66 820	66 829
466	66 839	66 848	66 857	66 867	66 876	66 885	66 894	66 904	66 913	66 922
467	66 932	66 941	66 950	66 960	66 969	66 978	66 987	66 997	67 006	67 015
468	67 025	67 034	67 043	67 052	67 062	67 071	67 080	67 089	67 099	67 108
469	67 117	67 127	67 136	67 145	67 154	67 164	67 173	67 182	67 191	67 201
470	67 210	67 219	67 228	67 237	67 247	67 256	67 265	67 274	67 284	67 293
471	67 302	67 311	67 321	67 330	67 339	67 348	67 357	67 367	67 376	67 385
472	67 394	67 403	67 413	67 422	67 431	67 440	67 449	67 459	67 468	67 477
473	67 486	67 495	67 504	67 514	67 523	67 532	67 541	67 550	67 560	67 569
474	67 578	67 587	67 596	67 605	67 614	67 624	67 633	67 642	67 651	67 660
475	67 669	67 679	67 688	67 697	67 706	67 715	67 724	67 733	67 742	67 752
476	67 761	67 770	67 779	67 788	67 797	67 806	67 815	67 825	67 834	67 843
477	67 852	67 861	67 870	67 879	67 888	67 897	67 906	67 916	67 925	67 934
478	67 943	67 952	67 961	67 970	67 979	67 988	67 997	68 006	68 015	68 024
479	68 034	68 043	68 052	68 061	68 070	68 079	68 088	68 097	68 106	68 115
480	68 124	68 133	68 142	68 151	68 160	68 169	68 178	68 187	68 196	68 205
481	68 215	68 224	68 233	68 242	68 251	68 260	68 269	68 278	68 287	68 296
482	68 305	68 314	68 323	68 332	68 341	68 350	68 359	68 368	68 377	68 386
483	68 395	68 404	68 413	68 422	68 431	68 440	68 449	68 458	68 467	68 476
484	68 485	68 494	68 502	68 511	68 520	68 529	68 538	68 547	68 556	68 565
485	68 574	68 583	68 592	68 601	68 610	68 619	68 628	68 637	68 646	68 655
486	68 664	68 673	68 681	68 690	68 699	68 708	68 717	68 726	68 735	68 744
487	68 753	68 762	68 771	68 780	68 789	68 797	68 806	68 815	68 824	68 833
488	68 842	68 851	68 860	68 869	68 878	68 886	68 895	68 904	68 913	68 922
489	68 931	68 940	68 949	68 958	68 966	68 975	68 984	68 993	69 002	69 011
490	69 020	69 028	69 037	69 046	69 055	69 064	69 073	69 082	69 090	69 099
491	69 108	69 117	69 126	69 135	69 144	69 152	69 161	69 170	69 179	69 188
492	69 197	69 205	69 214	69 223	69 232	69 241	69 249	69 258	69 267	69 276
493	69 285	69 294	69 302	69 311	69 320	69 329	69 338	69 346	69 355	69 364
494	69 373	69 381	69 390	69 399	69 408	69 417	69 425	69 434	69 443	69 452
495	69 461	69 469	69 478	69 487	69 496	69 504	69 513	69 522	69 531	69 539
496	69 548	69 557	69 566	69 574	69 583	69 592	69 601	69 609	69 618	69 627
497	69 636	69 644	69 653	69 662	69 671	69 679	69 688	69 697	69 705	69 714
498	69 723	69 732	69 740	69 749	69 758	69 767	69 775	69 784	69 793	69 801
499	69 810	69 819	69 827	69 836	69 845	69 854	69 862	69 871	69 880	69 888
500	69 897	69 906	69 914	69 923	69 932	69 940	69 949	69 958	69 966	69 **975**

23. Calculations

COMMON LOGARITHMS OF THE NATURAL NUMBERS (*Continued*)

No.	0	1	2	3	4	5	6	7	8	9
500	69 897	69 906	69 914	69 923	69 932	69 940	69 949	69 958	69 966	69 975
501	69 984	69 992	70 001	70 010	70 018	70 027	70 036	70 044	70 053	70 062
502	70 070	70 079	70 088	70 096	70 105	70 114	70 122	70 131	70 140	70 148
503	70 157	70 165	70 174	70 183	70 191	70 200	70 209	70 217	70 226	70 234
504	70 243	70 252	70 260	70 269	70 278	70 286	70 295	70 303	70 312	70 321
505	70 329	70 338	70 346	70 355	70 364	70 372	70 381	70 389	70 398	70 406
506	70 415	70 424	70 432	70 441	70 449	70 458	70 467	70 475	70 484	70 492
507	70 501	70 509	70 518	70 526	70 535	70 544	70 552	70 561	70 569	70 578
508	70 586	70 595	70 603	70 612	70 621	70 629	70 638	70 646	70 655	70 663
509	70 672	70 680	70 689	70 697	70 706	70 714	70 723	70 731	70 740	70 749
510	70 757	70 766	70 774	70 783	70 791	70 800	70 808	70 817	70 825	70 834
511	70 842	70 851	70 859	70 868	70 876	70 885	70 893	70 902	70 910	70 919
512	70 927	70 935	70 944	70 952	70 961	70 969	70 978	70 986	70 995	71 003
513	71 012	71 020	71 029	71 037	71 046	71 054	71 063	71 071	71 079	71 088
514	71 096	71 105	71 113	71 122	71 130	71 139	71 147	71 155	71 164	71 172
515	71 181	71 189	71 198	71 206	71 214	71 223	71 231	71 240	71 248	71 257
516	71 265	71 273	71 282	71 290	71 299	71 307	71 315	71 324	71 332	71 341
517	71 349	71 357	71 366	71 374	71 383	71 391	71 399	71 408	71 416	71 425
518	71 433	71 441	71 450	71 458	71 466	71 475	71 483	71 492	71 500	71 508
519	71 517	71 525	71 533	71 542	71 550	71 559	71 567	71 575	71 584	71 592
520	71 600	71 609	71 617	71 625	71 634	71 642	71 650	71 659	71 667	71 675
521	71 684	71 692	71 700	71 709	71 717	71 725	71 734	71 742	71 750	71 759
522	71 767	71 775	71 784	71 792	71 800	71 809	71 817	71 825	71 834	71 842
523	71 850	71 858	71 867	71 875	71 883	71 892	71 900	71 908	71 917	71 925
524	71 933	71 941	71 950	71 958	71 966	71 975	71 983	71 991	71 999	72 008
525	72 016	72 024	72 032	72 041	72 049	72 057	72 066	72 074	72 082	72 090
526	72 099	72 107	72 115	72 123	72 132	72 140	72 148	72 156	72 165	72 173
527	72 181	72 189	72 198	72 206	72 214	72 222	72 230	72 239	72 247	72 255
528	72 263	72 272	72 280	72 288	72 296	72 304	72 313	72 321	72 329	72 337
529	72 346	72 354	72 362	72 370	72 378	72 387	72 395	72 403	72 411	72 419
530	72 428	72 436	72 444	72 452	72 460	72 469	72 477	72 485	72 493	72 501
531	72 509	72 518	72 526	72 534	72 542	72 550	72 558	72 567	72 575	72 583
532	72 591	72 599	72 607	72 616	72 624	72 632	72 640	72 648	72 656	72 665
533	72 673	72 681	72 689	72 697	72 705	72 713	72 722	72 730	72 738	72 746
534	72 754	72 762	72 770	72 779	72 787	72 795	72 803	72 811	72 819	72 827
535	72 835	72 843	72 852	72 860	72 868	72 876	72 884	72 892	72 900	72 908
536	72 916	72 925	72 933	72 941	72 949	72 957	72 965	72 973	72 981	72 989
537	72 997	73 006	73 014	73 022	73 030	73 038	73 046	73 054	73 062	73 070
538	73 078	73 086	73 094	73 102	73 111	73 119	73 127	73 135	73 143	73 151
539	73 159	73 167	73 175	73 183	73 191	73 199	73 207	73 215	73 223	73 231
540	73 239	73 247	73 255	73 263	73 272	73 280	73 288	73 296	73 304	73 312
541	73 320	73 328	73 336	73 344	73 352	73 360	73 368	73 376	73 384	73 392
542	73 400	73 408	73 416	73 424	73 432	73 440	73 448	73 456	73 464	73 472
543	73 480	73 488	73 496	73 504	73 512	73 520	73 528	73 536	73 544	73 552
544	73 560	73 568	73 576	73 584	73 592	73 600	73 608	73 616	73 624	73 632
545	73 640	73 648	73 656	73 664	73 672	73 679	73 687	73 695	73 703	73 711
546	73 719	73 727	73 735	73 743	73 751	73 759	73 767	73 775	73 783	73 791
547	73 799	73 807	73 815	73 823	73 830	73 838	73 846	73 854	73 862	73 870
548	73 878	73 886	73 894	73 902	73 910	73 918	73 926	73 933	73 941	73 949
549	73 957	73 965	73 973	73 981	73 989	73 997	74 005	74 013	74 020	74 028
550	74 036	74 044	74 052	74 060	74 068	74 076	74 084	74 092	74 099	74 107

Common Logarithms of the Natural Numbers (*Continued*)

No.	0	1	2	3	4	5	6	7	8	9
550	74 036	74 044	74 052	74 060	74 068	74 076	74 084	74 092	74 099	74 107
551	74 115	74 123	74 131	74 139	74 147	74 155	74 162	74 170	74 178	74 186
552	74 194	74 202	74 210	74 218	74 225	74 233	74 241	74 249	74 257	74 265
553	74 273	74 280	74 288	74 296	74 304	74 312	74 320	74 327	74 335	74 343
554	74 351	74 359	74 367	74 374	74 382	74 390	74 398	74 406	74 414	74 421
555	74 429	74 437	74 445	74 453	74 461	74 468	74 476	74 484	74 492	74 500
556	74 507	74 515	74 523	74 531	74 539	74 547	74 554	74 562	74 570	74 578
557	74 586	74 593	74 601	74 609	74 617	74 624	74 632	74 640	74 648	74 656
558	74 663	74 671	74 679	74 687	74 695	74 702	74 710	74 718	74 726	74 733
559	74 741	74 749	74 757	74 764	74 772	74 780	74 788	74 796	74 803	74 811
560	74 819	74 827	74 834	74 842	74 850	74 858	74 865	74 873	74 881	74 889
561	74 896	74 904	74 912	74 920	74 927	74 935	74 943	74 950	74 958	74 966
562	74 974	74 981	74 989	74 997	75 005	75 012	75 020	75 028	75 035	75 043
563	75 051	75 059	75 066	75 074	75 082	75 089	75 097	75 105	75 113	75 120
564	75 128	75 136	75 143	75 151	75 159	75 166	75 174	75 182	75 189	75 197
565	75 205	75 213	75 220	75 228	75 236	75 243	75 251	75 259	75 266	75 274
566	75 282	75 289	75 297	75 305	75 312	75 320	75 328	75 335	75 343	75 351
567	75 358	75 366	75 374	75 381	75 389	75 397	75 404	75 412	75 420	75 427
568	75 435	75 442	75 450	75 458	75 465	75 473	75 481	75 488	75 496	75 504
569	75 511	75 519	75 526	75 534	75 542	75 549	75 557	75 565	75 572	75 580
570	75 587	75 595	75 603	75 610	75 618	75 626	75 633	75 641	75 648	75 656
571	75 664	75 671	75 679	75 686	75 694	75 702	75 709	75 717	75 724	75 732
572	75 740	75 747	75 755	75 762	75 770	75 778	75 785	75 793	75 800	75 808
573	75 815	75 823	75 831	75 838	75 846	75 853	75 861	75 868	75 876	75 884
574	75 891	75 899	75 906	75 914	75 921	75 929	75 937	75 944	75 952	75 959
575	75 967	75 974	75 982	75 989	75 997	76 005	76 012	76 020	76 027	76 035
576	76 042	76 050	76 057	76 065	76 072	76 080	76 087	76 095	76 103	76 110
577	76 118	76 125	76 133	76 140	76 148	76 155	76 163	76 170	76 178	76 185
578	76 193	76 200	76 208	76 215	76 223	76 230	76 238	76 245	76 253	76 260
579	76 268	76 275	76 283	76 290	76 298	76 305	76 313	76 320	76 328	76 335
580	76 343	76 350	76 358	76 365	76 373	76 380	76 388	76 395	76 403	76 410
581	76 418	76 425	76 433	76 440	76 448	76 455	76 462	76 470	76 477	76 485
582	76 492	76 500	76 507	76 515	76 522	76 530	76 537	76 545	76 552	76 559
583	76 567	76 574	76 582	76 589	76 597	76 604	76 612	76 619	76 626	76 634
584	76 641	76 649	76 656	76 664	76 671	76 678	76 686	76 693	76 701	76 708
585	76 716	76 723	76 730	76 738	76 745	76 753	76 760	76 768	76 775	76 782
586	76 790	76 797	76 805	76 812	76 819	76 827	76 834	76 842	76 849	76 856
587	76 864	76 871	76 879	76 886	76 893	76 901	76 908	76 916	76 923	76 930
588	76 938	76 945	76 953	76 960	76 967	76 975	76 982	76 989	76 997	77 004
589	77 012	77 019	77 026	77 034	77 041	77 048	77 056	77 063	77 070	77 078
590	77 085	77 093	77 100	77 107	77 115	77 122	77 129	77 137	77 144	77 151
591	77 159	77 166	77 173	77 181	77 188	77 195	77 203	77 210	77 217	77 225
592	77 232	77 240	77 247	77 254	77 262	77 269	77 276	77 283	77 291	77 298
593	77 305	77 313	77 320	77 327	77 335	77 342	77 349	77 357	77 364	77 371
594	77 379	77 386	77 393	77 401	77 408	77 415	77 422	77 430	77 437	77 444
595	77 452	77 459	77 466	77 474	77 481	77 488	77 495	77 503	77 510	77 517
596	77 525	77 532	77 539	77 546	77 554	77 561	77 568	77 576	77 583	77 590
597	77 597	77 605	77 612	77 619	77 627	77 634	77 641	77 648	77 656	77 663
598	77 670	77 677	77 685	77 692	77 699	77 706	77 714	77 721	77 728	77 735
599	77 743	77 750	77 757	77 764	77 772	77 779	77 786	77 793	77 801	77 808
600	77 815	77 822	77 830	77 837	77 844	77 851	77 859	77 866	77 873	77 880

23. Calculations

COMMON LOGARITHMS OF THE NATURAL NUMBERS (*Continued*)

No.	0	1	2	3	4	5	6	7	8	9
600	77 815	77 822	77 830	77 837	77 844	77 851	77 859	77 866	77 873	77 880
601	77 887	77 895	77 902	77 909	77 916	77 924	77 931	77 938	77 945	77 952
602	77 960	77 967	77 974	77 981	77 988	77 996	78 003	78 010	78 017	78 025
603	78 032	78 039	78 046	78 053	78 061	78 068	78 075	78 082	78 089	78 097
604	78 104	78 111	78 118	78 125	78 132	78 140	78 147	78 154	78 161	78 168
605	78 176	78 183	78 190	78 197	78 204	78 211	78 219	78 226	78 233	78 240
606	78 247	78 254	78 262	78 269	78 276	78 283	78 290	78 297	78 305	78 312
607	78 319	78 326	78 333	78 340	78 347	78 355	78 362	78 369	78 376	78 383
608	78 390	78 398	78 405	78 412	78 419	78 426	78 433	78 440	78 447	78 455
609	78 462	78 469	78 476	78 483	78 490	78 497	78 504	78 512	78 519	78 526
610	78 533	78 540	78 547	78 554	78 561	78 569	78 576	78 583	78 590	78 597
611	78 604	78 611	78 618	78 625	78 633	78 640	78 647	78 654	78 661	78 668
612	78 675	78 682	78 689	78 696	78 704	78 711	78 718	78 725	78 732	78 739
613	78 746	78 753	78 760	78 767	78 774	78 781	78 789	78 796	78 803	78 810
614	78 817	78 824	78 831	78 838	78 845	78 852	78 859	78 866	78 873	78 880
615	78 888	78 895	78 902	78 909	78 916	78 923	78 930	78 937	78 944	78 951
616	78 958	78 965	78 972	78 979	78 986	78 993	79 000	79 007	79 014	79 021
617	79 029	79 036	79 043	79 050	79 057	79 064	79 071	79 078	79 085	79 092
618	79 099	79 106	79 113	79 120	79 127	79 134	79 141	79 148	79 155	79 162
619	79 169	79 176	79 183	79 190	79 197	79 204	79 211	79 218	79 225	79 232
620	79 239	79 246	79 253	79 260	79 267	79 274	79 281	79 288	79 295	79 302
621	79 309	79 316	79 323	79 330	79 337	79 344	79 351	79 358	79 365	79 372
622	79 379	79 386	79 393	79 400	79 407	79 414	79 421	79 428	79 435	79 442
623	79 449	79 456	79 463	79 470	79 477	79 484	79 491	79 498	79 505	79 511
624	79 518	79 525	79 532	79 539	79 546	79 553	79 560	79 567	79 574	79 581
625	79 588	79 595	79 602	79 609	79 616	79 623	79 630	79 637	79 644	79 650
626	79 657	79 664	79 671	79 678	79 685	79 692	79 699	79 706	79 713	79 720
627	79 727	79 734	79 741	79 748	79 754	79 761	79 768	79 775	79 782	79 789
628	79 796	79 803	79 810	79 817	79 824	79 831	79 837	79 844	79 851	79 858
629	79 865	79 872	79 879	79 886	79 893	79 900	79 906	79 913	79 920	79 927
630	79 934	79 941	79 948	79 955	79 962	79 969	79 975	79 982	79 989	79 996
631	80 003	80 010	80 017	80 024	80 030	80 037	80 044	80 051	80 058	80 065
632	80 072	80 079	80 085	80 092	80 099	80 106	80 113	80 120	80 127	80 134
633	80 140	80 147	80 154	80 161	80 168	80 175	80 182	80 188	80 195	80 202
634	80 209	80 216	80 223	80 229	80 236	80 243	80 250	80 257	80 264	80 271
635	80 277	80 284	80 291	80 298	80 305	80 312	80 318	80 325	80 332	80 339
636	80 346	80 353	80 359	80 366	80 373	80 380	80 387	80 393	80 400	80 407
637	80 414	80 421	80 428	80 434	80 441	80 448	80 455	80 462	80 468	80 475
638	80 482	80 489	80 496	80 502	80 509	80 516	80 523	80 530	80 536	80 543
639	80 550	80 557	80 564	80 570	80 577	80 584	80 591	80 598	80 604	80 611
640	80 618	80 625	80 632	80 638	80 645	80 652	80 659	80 665	80 672	80 679
641	80 686	80 693	80 699	80 706	80 713	80 720	80 726	80 733	80 740	80 747
642	80 754	80 760	80 767	80 774	80 781	80 787	80 794	80 801	80 808	80 814
643	80 821	80 828	80 835	80 841	80 848	80 855	80 862	80 868	80 875	80 882
644	80 889	80 895	80 902	80 909	80 916	80 922	80 929	80 936	80 943	80 949
645	80 956	80 963	80 969	80 976	80 983	80 990	80 996	81 003	81 010	81 017
646	81 023	81 030	81 037	81 043	81 050	81 057	81 064	81 070	81 077	81 084
647	81 090	81 097	81 104	81 111	81 117	81 124	81 131	81 137	81 144	81 151
648	81 158	81 164	81 171	81 178	81 184	81 191	81 198	81 204	81 211	81 218
649	81 224	81 231	81 238	81 245	81 251	81 258	81 265	81 271	81 278	81 285
650	81 291	81 298	81 305	81 311	81 318	81 325	81 331	81 338	81 345	81 351

Common Logarithms of the Natural Numbers (*Continued*)

No.	0	1	2	3	4	5	6	7	8	9
650	81 291	81 298	81 305	81 311	81 318	81 325	81 331	81 338	81 345	81 351
651	81 358	81 365	81 371	81 378	81 385	81 391	81 398	81 405	81 411	81 418
652	81 425	81 431	81 438	81 445	81 451	81 458	81 465	81 471	81 478	81 485
653	81 491	81 498	81 505	81 511	81 518	81 525	81 531	81 538	81 544	81 551
654	81 558	81 564	81 571	81 578	81 584	81 591	81 598	81 604	81 611	81 617
655	81 624	81 631	81 637	81 644	81 651	81 657	81 664	81 671	81 677	81 684
656	81 690	81 697	81 704	81 710	81 717	81 723	81 730	81 737	81 743	81 750
657	81 757	81 763	81 770	81 776	81 783	81 790	81 796	81 803	81 809	81 816
658	81 823	81 829	81 836	81 842	81 849	81 856	81 862	81 869	81 875	81 882
659	81 889	81 895	81 902	81 908	81 915	81 921	81 928	81 935	81 941	81 948
660	81 954	81 961	81 968	81 974	81 981	81 987	81 994	82 000	82 007	82 014
661	82 020	82 027	82 033	82 040	82 046	82 053	82 060	82 066	82 073	82 079
662	82 086	82 092	82 099	82 105	82 112	82 119	82 125	82 132	82 138	82 145
663	82 151	82 158	82 164	82 171	82 178	82 184	82 191	82 197	82 204	82 210
664	82 217	82 223	82 230	82 236	82 243	82 249	82 256	82 263	82 269	82 276
665	82 282	82 289	82 295	82 302	82 308	82 315	82 321	82 328	82 334	82 341
666	82 347	82 354	82 360	82 367	82 373	82 380	82 387	82 393	82 400	82 406
667	82 413	82 419	82 426	82 432	82 439	82 445	82 452	82 458	82 465	82 471
668	82 478	82 484	82 491	82 497	82 504	82 510	82 517	82 523	82 530	82 536
669	82 543	82 549	82 556	82 562	82 569	82 575	82 582	82 588	82 595	82 601
670	82 607	82 614	82 620	82 627	82 633	82 640	82 646	82 653	82 659	82 666
671	82 672	82 679	82 685	82 692	82 698	82 705	82 711	82 718	82 724	82 730
672	82 737	82 743	82 750	82 756	82 763	82 769	82 776	82 782	82 789	82 795
673	82 802	82 808	82 814	82 821	82 827	82 834	82 840	82 847	82 853	82 860
674	82 866	82 872	82 879	82 885	82 892	82 898	82 905	82 911	82 918	82 924
675	82 930	82 937	82 943	82 950	82 956	82 963	82 969	82 975	82 982	82 988
676	82 995	83 001	83 008	83 014	83 020	83 027	83 033	83 040	83 046	83 052
677	83 059	83 065	83 072	83 078	83 085	83 091	83 097	83 104	83 110	83 117
678	83 123	83 129	83 136	83 142	83 149	83 155	83 161	83 168	83 174	83 181
679	83 187	83 193	83 200	83 206	83 213	83 219	83 225	83 232	83 238	83 245
680	83 251	83 257	83 264	83 270	83 276	83 283	83 289	83 296	83 302	83 308
681	83 315	83 321	83 327	83 334	83 340	83 347	83 353	83 359	83 366	83 372
682	83 378	83 385	83 391	83 398	83 404	83 410	83 417	83 423	83 429	83 436
683	83 442	83 448	83 455	83 461	83 467	83 474	83 480	83 487	83 493	83 499
684	83 506	83 512	83 518	83 525	83 531	83 537	83 544	83 550	83 556	83 563
685	83 569	83 575	83 582	83 588	83 594	83 601	83 607	83 613	83 620	83 626
686	83 632	83 639	83 645	83 651	83 658	83 664	83 670	83 677	83 683	83 689
687	83 696	83 702	83 708	83 715	83 721	83 727	83 734	83 740	83 746	83 753
688	83 759	83 765	83 771	83 778	83 784	83 790	83 797	83 803	83 809	83 816
689	83 822	83 828	83 835	83 841	83 847	83 853	83 860	83 866	83 872	83 879
690	83 885	83 891	83 897	83 904	83 910	83 916	83 923	83 929	83 935	83 942
691	83 948	83 954	83 960	83 967	83 973	83 979	83 985	83 992	83 998	84 004
692	84 011	84 017	84 023	84 029	84 036	84 042	84 048	84 055	84 061	84 067
693	84 073	84 080	84 086	84 092	84 098	84 105	84 111	84 117	84 123	84 130
694	84 136	84 142	84 148	84 155	84 161	84 167	84 173	84 180	84 186	84 192
695	84 198	84 205	84 211	84 217	84 223	84 230	84 236	84 242	84 248	84 255
696	84 261	84 267	84 273	84 280	84 286	84 292	84 298	84 305	84 311	84 317
697	84 323	84 330	84 336	84 342	84 348	84 354	84 361	84 367	84 373	84 379
698	84 386	84 392	84 398	84 404	84 410	84 417	84 423	84 429	84 435	84 442
699	84 448	84 454	84 460	84 466	84 473	84 479	84 485	84 491	84 497	84 504
700	84 510	84 516	84 522	84 528	84 535	84 541	84 547	84 553	84 559	84 566

23. Calculations

Common Logarithms of the Natural Numbers (*Continued*)

No.	0	1	2	3	4	5	6	7	8	9
700	84 510	84 516	84 522	84 528	84 535	84 541	84 547	84 553	84 559	84 566
701	84 572	84 578	84 584	84 590	84 597	84 603	84 609	84 615	84 621	84 628
702	84 634	84 640	84 646	84 652	84 658	84 665	84 671	84 677	84 683	84 689
703	84 696	84 702	84 708	84 714	84 720	84 726	84 733	84 739	84 745	84 751
704	84 757	84 763	84 770	84 776	84 782	84 788	84 794	84 800	84 807	84 813
705	84 819	84 825	84 831	84 837	84 844	84 850	84 856	84 862	84 868	84 874
706	84 880	84 887	84 893	84 899	84 905	84 911	84 917	84 924	84 930	84 936
707	84 942	84 948	84 954	84 960	84 967	84 973	84 979	84 985	84 991	84 997
708	85 003	85 009	85 016	85 022	85 028	85 034	85 040	85 046	85 052	85 058
709	85 065	85 071	85 077	85 083	85 089	85 095	85 101	85 107	85 114	85 120
710	85 126	85 132	85 138	85 144	85 150	85 156	85 163	85 169	85 175	85 181
711	85 187	85 193	85 199	85 205	85 211	85 217	85 224	85 230	85 236	85 242
712	85 248	85 254	85 260	85 266	85 272	85 278	85 285	85 291	85 297	85 303
713	85 309	85 315	85 321	85 327	85 333	85 339	85 345	85 352	85 358	85 364
714	85 370	85 376	85 382	85 388	85 394	85 400	85 406	85 412	85 418	85 425
715	85 431	85 437	85 443	85 449	85 455	85 461	85 467	85 473	85 479	85 485
716	85 491	85 497	85 503	85 509	85 516	85 522	85 528	85 534	85 540	85 546
717	85 552	85 558	85 564	85 570	85 576	85 582	85 588	85 594	85 600	85 606
718	85 612	85 618	85 625	85 631	85 637	85 643	85 649	85 655	85 661	85 667
719	85 673	85 679	85 685	85 691	85 697	85 703	85 709	85 715	85 721	85 727
720	85 733	85 739	85 745	85 751	85 757	85 763	85 769	85 775	85 781	85 788
721	85 794	85 800	85 806	85 812	85 818	85 824	85 830	85 836	85 842	85 848
722	85 854	85 860	85 866	85 872	85 878	85 884	85 890	85 896	85 902	85 908
723	85 914	85 920	85 926	85 932	85 938	85 944	85 950	85 956	85 962	85 968
724	85 974	85 980	85 986	85 992	85 998	86 004	86 010	86 016	86 022	86 028
725	86 034	86 040	86 046	86 052	86 058	86 064	86 070	86 076	86 082	86 088
726	86 094	86 100	86 106	86 112	86 118	86 124	86 130	86 136	86 141	86 147
727	86 153	86 159	86 165	86 171	86 177	86 183	86 189	86 195	86 201	86 207
728	86 213	86 219	86 225	86 231	86 237	86 243	86 249	86 255	86 261	86 267
729	86 273	86 279	86 285	86 291	86 297	86 303	86 308	86 314	86 320	86 326
730	86 332	86 338	86 344	86 350	86 356	86 362	86 368	86 374	86 380	86 386
731	86 392	86 398	86 404	86 410	86 415	86 421	86 427	86 433	86 439	86 445
732	86 451	86 457	86 463	86 469	86 475	86 481	86 487	86 493	86 499	86 504
733	86 510	86 516	86 522	86 528	86 534	86 540	86 546	86 552	86 558	86 564
734	86 570	86 576	86 581	86 587	86 593	86 599	86 605	86 611	86 617	86 623
735	86 629	86 635	86 641	86 646	86 652	86 658	86 664	86 670	86 676	86 682
736	86 688	86 694	86 700	86 705	86 711	86 717	86 723	86 729	86 735	86 741
737	86 747	86 753	86 759	86 764	86 770	86 776	86 782	86 788	86 794	86 800
738	86 806	86 812	86 817	86 823	86 829	86 835	86 841	86 847	86 853	85 859
739	86 864	86 870	86 876	86 882	86 888	86 894	86 900	86 906	86 911	86 917
740	86 923	86 929	86 935	86 941	86 947	86 953	86 958	86 964	86 970	86 976
741	86 982	86 988	86 994	86 999	87 005	87 011	87 017	87 023	87 029	87 035
742	87 040	87 046	87 052	87 058	87 064	87 070	87 075	87 081	87 087	87 093
743	87 099	87 105	87 111	87 116	87 122	87 128	87 134	87 140	87 146	87 151
744	87 157	87 163	87 169	87 175	87 181	87 186	87 192	87 198	87 204	87 210
745	87 216	87 221	87 227	87 233	87 239	87 245	87 251	87 256	87 262	87 268
746	87 274	87 280	87 286	87 291	87 297	87 303	87 309	87 315	87 320	87 326
747	87 332	87 338	87 344	87 349	87 355	87 361	87 367	87 373	87 379	87 384
748	87 390	87 396	87 402	87 408	87 413	87 419	87 425	87 431	87 437	87 442
749	87 448	87 454	87 460	87 466	87 471	87 477	87 483	87 489	87 495	87 500
750	87 506	87 512	87 518	87 523	87 529	87 535	87 541	87 547	87 552	87 558

COMMON LOGARITHMS OF THE NATURAL NUMBERS (*Continued*)

No.	0	1	2	3	4	5	6	7	8	9
750	87 506	87 512	87 518	87 523	87 529	87 535	87 541	87 547	87 552	87 558
751	87 564	87 570	87 576	87 581	87 587	87 593	87 599	87 604	87 610	87 616
752	87 622	87 628	87 633	87 639	87 645	87 651	87 656	87 662	87 668	87 674
753	87 679	87 685	87 691	87 697	87 703	87 708	87 714	87 720	87 726	87 731
754	87 737	87 743	87 749	87 754	87 760	87 766	87 772	87 777	87 783	87 789
755	87 795	87 800	87 806	87 812	87 818	87 823	87 829	87 835	87 841	87 846
756	87 852	87 858	87 864	87 869	87 875	87 881	87 887	87 892	87 898	87 904
757	87 910	87 915	87 921	87 927	87 933	87 938	87 944	87 950	87 955	87 961
758	87 967	87 973	87 978	87 984	87 990	87 996	88 001	88 007	88 013	88 018
759	88 024	88 030	88 036	88 041	88 047	88 053	88 058	88 064	88 070	88 076
760	88 081	88 087	88 093	88 098	88 104	88 110	88 116	88 121	88 127	88 133
761	88 138	88 144	88 150	88 156	88 161	88 167	88 173	88 178	88 184	88 190
762	88 195	88 201	88 207	88 213	88 218	88 224	88 230	88 235	88 241	88 247
763	88 252	88 258	88 264	88 270	88 275	88 281	88 287	88 292	88 298	88 304
764	88 309	88 315	88 321	88 326	88 332	88 338	88 343	88 349	88 355	88 360
765	88 366	88 372	88 377	88 383	88 389	88 395	88 400	88 406	88 412	88 417
766	88 423	88 429	88 434	88 440	88 446	88 451	88 457	88 463	88 468	88 474
767	88 480	88 485	88 491	88 497	88 502	88 508	88 513	88 519	88 525	88 530
768	88 536	88 542	88 547	88 553	88 559	88 564	88 570	88 576	88 581	88 587
769	88 593	88 598	88 604	88 610	88 615	88 621	88 627	88 632	88 638	88 643
770	88 649	88 655	88 660	88 666	88 672	88 677	88 683	88 689	88 694	88 700
771	88 705	88 711	88 717	88 722	88 728	88 734	88 739	88 745	88 750	88 756
772	88 762	88 767	88 773	88 779	88 784	88 790	88 795	88 801	88 807	88 812
773	88 818	88 824	88 829	88 835	88 840	88 846	88 852	88 857	88 863	88 868
774	88 874	88 880	88 885	88 891	88 897	88 902	88 908	88 913	88 919	88 925
775	88 930	88 936	88 941	88 947	88 953	88 958	88 964	88 969	88 975	88 981
776	88 986	88 992	88 997	89 003	89 009	89 014	89 020	89 025	89 031	89 037
777	89 042	89 048	89 053	89 059	89 064	89 070	89 076	89 081	89 087	89 092
778	89 098	89 104	89 109	89 115	89 120	89 126	89 131	89 137	89 143	89 148
779	89 154	89 159	89 165	89 170	89 176	89 182	89 187	89 193	89 198	89 204
780	89 209	89 215	89 221	89 226	89 232	89 237	89 243	89 248	89 254	89 260
781	89 265	89 271	89 276	89 282	89 287	89 293	89 298	89 304	89 310	89 315
782	89 321	89 326	89 332	89 337	89 343	89 348	89 354	89 360	89 365	89 371
783	89 376	89 382	89 387	89 393	89 398	89 404	89 409	89 415	89 421	89 426
784	89 432	89 437	89 443	89 448	89 454	89 459	89 465	89 470	89 476	89 481
785	89 487	89 492	89 498	89 504	89 509	89 515	89 520	89 526	89 531	89 537
786	89 542	89 548	89 553	89 559	89 564	89 570	89 575	89 581	89 586	89 592
787	89 597	89 603	89 609	89 614	89 620	89 625	89 631	89 636	89 642	89 647
788	89 653	89 658	89 664	89 669	89 675	89 680	89 686	89 691	89 697	89 702
789	89 708	89 713	89 719	89 724	89 730	89 735	89 741	89 746	89 752	89 757
790	89 763	89 768	89 774	89 779	89 785	89 790	89 796	89 801	89 807	89 812
791	89 818	89 823	89 829	89 834	89 840	89 845	89 851	89 856	89 862	89 867
792	89 873	89 878	89 883	89 889	89 894	89 900	89 905	89 911	89 916	89 922
793	89 927	89 933	89 938	89 944	89 949	89 955	89 960	89 966	89 971	89 977
794	89 982	89 988	89 993	89 998	90 004	90 009	90 015	90 020	90 026	90 031
795	90 037	90 042	90 048	90 053	90 059	90 064	90 069	90 075	90 080	90 086
796	90 091	90 097	90 102	90 108	90 113	90 119	90 124	90 129	90 135	90 140
797	90 146	90 151	90 157	90 162	90 168	90 173	90 179	90 184	90 189	90 195
798	90 200	90 206	90 211	90 217	90 222	90 227	90 233	90 238	90 244	90 249
799	90 255	90 260	90 266	90 271	90 276	90 282	90 287	90 293	90 298	90 304
800	90 309	90 314	90 320	90 325	90 331	90 336	90 342	90 347	90 352	90 358

23. Calculations

COMMON LOGARITHMS OF THE NATURAL NUMBERS (*Continued*)

No.	0	1	2	3	4	5	6	7	8	9
800	90 309	90 314	90 320	90 325	90 331	90 336	90 342	90 347	90 352	90 358
801	90 363	90 369	90 374	90 380	90 385	90 390	90 396	90 401	90 407	90 412
802	90 417	90 423	90 428	90 434	90 439	90 445	90 450	90 455	90 461	90 466
803	90 472	90 477	90 482	90 488	90 493	90 499	90 504	90 509	90 515	90 520
804	90 526	90 531	90 536	90 542	90 547	90 553	90 558	90 563	90 569	90 574
805	90 580	90 585	90 590	90 596	90 601	90 607	90 612	90 617	90 623	90 628
806	90 634	90 639	90 644	90 650	90 655	90 660	90 666	90 671	90 677	90 682
807	90 687	90 693	90 698	90 703	90 709	90 714	90 720	90 725	90 730	90 736
808	90 741	90 747	90 752	90 757	90 763	90 768	90 773	90 779	90 784	90 789
809	90 795	90 800	90 806	90 811	90 816	90 822	90 827	90 832	90 838	90 843
810	90 849	90 854	90 859	90 865	90 870	90 875	90 881	90 886	90 891	90 897
811	90 902	90 907	90 913	90 918	90 924	90 929	90 934	90 940	90 945	90 950
812	90 956	90 961	90 966	90 972	90 977	90 982	90 988	90 993	90 998	91 004
813	91 009	91 014	91 020	91 025	91 030	91 036	91 041	91 046	91 052	91 057
814	91 062	91 068	91 073	91 078	91 084	91 089	91 094	91 100	91 105	91 110
815	91 116	91 121	91 126	91 132	91 137	91 142	91 148	91 153	91 158	91 164
816	91 169	91 174	91 180	91 185	91 190	91 196	91 201	91 206	91 212	91 217
817	91 222	91 228	91 233	91 238	91 243	91 249	91 254	91 259	91 265	91 270
818	91 275	91 281	91 286	91 291	91 297	91 302	91 307	91 312	91 318	91 323
819	91 328	91 334	91 339	91 344	91 350	91 355	91 360	91 365	91 371	91 376
820	91 381	91 387	91 392	91 397	91 403	91 408	91 413	91 418	91 424	91 429
821	91 434	91 440	91 445	91 450	91 455	91 461	91 466	91 471	91 477	91 482
822	91 487	91 492	91 498	91 503	91 508	91 514	91 519	91 524	91 529	91 535
823	91 540	91 545	91 551	91 556	91 561	91 566	91 572	91 577	91 582	91 587
824	91 593	91 598	91 603	91 609	91 614	91 619	91 624	91 630	91 635	91 640
825	91 645	91 651	91 656	91 661	91 666	91 672	91 677	91 682	91 687	91 693
826	91 698	91 703	91 709	91 714	91 719	91 724	91 730	91 735	91 740	91 745
827	91 751	91 756	91 761	91 766	91 772	91 777	91 782	91 787	91 793	91 798
828	91 803	91 808	91 814	91 819	91 824	91 829	91 834	91 840	91 845	91 850
829	91 855	91 861	91 866	91 871	91 876	91 882	91 887	91 892	91 897	91 903
830	91 908	91 913	91 918	91 924	91 929	91 934	91 939	91 944	91 950	91 955
831	91 960	91 965	91 971	91 976	91 981	91 986	91 991	91 997	92 002	92 007
832	92 012	92 018	92 023	92 028	92 033	92 038	92 044	92 049	92 054	92 059
833	92 065	92 070	92 075	92 080	92 085	92 091	92 096	92 101	92 106	92 111
834	92 117	92 122	92 127	92 132	92 137	92 143	92 148	92 153	92 158	92 163
835	92 169	92 174	92 179	92 184	92 189	92 195	92 200	92 205	92 210	92 215
836	92 221	92 226	92 231	92 236	92 241	92 247	92 252	92 257	92 262	92 267
837	92 273	92 278	92 283	92 288	92 293	92 298	92 304	92 309	92 314	92 319
838	92 324	92 330	92 335	92 340	92 345	92 350	92 355	92 361	92 366	92 371
839	92 376	92 381	93 387	92 392	92 397	92 402	92 407	92 412	92 418	92 423
840	92 428	92 433	92 438	92 443	92 449	92 454	92 459	92 464	92 469	92 474
841	92 480	92 485	92 490	92 495	92 500	92 505	92 511	92 516	92 521	92 526
842	92 531	92 536	92 542	92 547	92 552	92 557	92 562	92 567	92 572	92 578
843	92 583	92 588	92 593	92 598	92 603	92 609	92 614	92 619	92 624	92 629
844	92 634	92 639	92 645	92 650	92 655	92 660	92 665	92 670	92 675	92 681
845	92 686	92 691	92 696	92 701	92 706	92 711	92 716	92 722	92 727	92 732
846	92 737	92 742	92 747	92 752	92 758	92 763	92 768	92 773	92 778	92 783
847	92 788	92 793	92 799	92 804	92 809	92 814	92 819	92 824	92 829	92 834
848	92 840	92 845	92 850	92 855	92 860	92 865	92 870	92 875	92 881	92 886
849	92 891	92 896	92 901	92 906	92 911	92 916	92 921	92 927	92 932	92 937
850	92 942	92 947	92 952	92 957	92 962	92 967	92 973	92 978	92 983	92 988

COMMON LOGARITHMS OF THE NATURAL NUMBERS (*Continued*)

No.	0	1	2	3	4	5	6	7	8	9
850	92 942	92 947	92 952	92 957	92 962	92 967	92 973	92 978	92 983	92 988
851	92 993	92 998	93 003	93 008	93 013	93 018	93 024	93 029	93 034	93 039
852	93 044	93 049	93 054	93 059	93 064	93 069	93 075	93 080	93 085	93 090
853	93 095	93 100	93 105	93 110	93 115	93 120	93 125	93 131	93 136	93 141
854	93 146	93 151	93 156	93 161	93 166	93 171	93 176	93 181	93 186	93 192
855	93 197	93 202	93 207	93 212	93 217	93 222	93 227	93 232	93 237	93 242
856	93 247	93 252	93 258	93 263	93 268	93 273	93 278	93 283	93 288	93 293
857	93 298	93 303	93 308	93 313	93 318	93 323	93 328	93 334	93 339	93 344
858	93 349	93 354	93 359	93 364	93 369	93 374	93 379	93 384	93 389	93 394
859	93 399	93 404	93 409	93 414	93 420	93 425	93 430	93 435	93 440	93 445
860	93 450	93 455	93 460	93 465	93 470	93 475	93 480	93 485	93 490	93 495
861	93 500	93 505	93 510	93 515	93 520	93 526	93 531	93 536	93 541	93 546
862	93 551	93 556	93 561	93 566	93 571	93 576	93 581	93 586	93 591	93 596
863	93 601	93 606	93 611	93 616	93 621	93 626	93 631	93 636	93 641	93 646
864	93 651	93 656	93 661	93 666	93 671	93 676	93 682	93 687	93 692	93 697
865	93 702	93 707	93 712	93 717	93 722	93 727	93 732	93 737	93 742	93 747
866	93 752	93 757	93 762	93 767	93 772	93 777	93 782	93 787	93 792	93 797
867	93 802	93 807	93 812	93 817	93 822	93 827	93 832	93 837	93 842	93 847
868	93 852	93 857	93 862	93 867	93 872	93 877	93 882	93 887	93 892	93 897
869	93 902	93 907	93 912	93 917	93 922	93 927	93 932	93 937	93 942	93 947
870	93 952	93 957	93 962	93 967	93 972	93 977	93 982	93 987	93 992	93 997
871	94 002	94 007	94 012	94 017	94 022	94 027	94 032	94 037	94 042	94 047
872	94 052	94 057	94 062	94 067	94 072	94 077	94 082	94 086	94 091	94 096
873	94 101	94 106	94 111	94 116	94 121	94 126	94 131	94 136	94 141	94 146
874	94 151	94 156	94 161	94 166	94 171	94 176	94 181	94 186	94 191	94 196
875	94 201	94 206	94 211	94 216	94 221	94 226	94 231	94 236	94 240	94 245
876	94 250	94 255	94 260	94 265	94 270	94 275	94 280	94 285	94 290	94 295
877	94 300	94 305	94 310	94 315	94 320	94 325	94 330	94 335	94 340	94 345
878	94 349	94 354	94 359	94 364	94 369	94 374	94 379	94 384	94 389	94 394
879	94 399	94 404	94 409	94 414	94 419	94 424	94 429	94 433	94 438	94 443
880	94 448	94 453	94 458	94 463	94 468	94 473	94 478	94 483	94 488	94 493
881	94 498	94 503	94 507	94 512	94 517	94 522	94 527	94 532	94 537	94 542
882	94 547	94 552	94 557	94 562	94 567	94 571	94 576	94 581	94 586	94 591
883	94 596	94 601	94 606	94 611	94 616	94 621	94 626	94 630	94 635	94 640
884	94 645	94 650	94 655	94 660	94 665	94 670	94 675	94 680	94 685	94 689
885	94 694	94 699	94 704	94 709	94 714	94 719	94 724	94 729	94 734	94 738
886	94 743	94 748	94 753	94 758	94 763	94 768	94 773	94 778	94 783	94 787
887	94 792	94 797	94 802	94 807	94 812	94 817	94 822	94 827	94 832	94 836
888	94 841	94 846	94 851	94 856	94 861	94 866	94 871	94 876	94 880	94 885
889	94 890	94 895	94 900	94 905	94 910	94 915	94 919	94 924	94 929	94 934
890	94 939	94 944	94 949	94 954	94 959	94 963	94 968	94 973	94 978	94 983
891	94 988	94 993	94 998	95 002	95 007	95 012	95 017	95 022	95 027	95 032
892	95 036	95 041	95 046	95 051	95 056	95 061	95 066	95 071	95 075	95 080
893	95 085	95 090	95 095	95 100	95 105	95 109	95 114	95 119	95 124	95 129
894	95 134	95 139	95 143	95 148	95 153	95 158	95 163	95 168	95 173	95 177
895	95 182	95 187	95 192	95 197	95 202	95 207	95 211	95 216	95 221	95 226
896	95 231	95 236	95 240	95 245	95 250	95 255	95 260	95 265	95 270	95 274
897	95 279	95 284	95 289	95 294	95 299	95 303	95 308	95 313	95 318	95 323
898	95 328	95 332	95 337	95 342	95 347	95 352	95 357	95 361	95 366	95 371
899	95 376	95 381	95 386	95 390	95 395	95 400	95 405	95 410	95 415	95 419
900	95 424	95 429	95 434	95 439	95 444	95 448	95 453	95 458	95 463	95 468

23. Calculations

COMMON LOGARITHMS OF THE NATURAL NUMBERS (*Continued*)

No.	0	1	2	3	4	5	6	7	8	9
900	95 424	95 429	95 434	95 439	95 444	95 448	95 453	95 458	95 463	95 468
901	95 472	95 477	95 482	95 487	95 492	95 497	95 501	95 506	95 511	95 516
902	95 521	95 525	95 530	95 535	95 540	95 545	95 550	95 554	95 559	95 564
903	95 569	95 574	95 578	95 583	95 588	95 593	95 598	95 602	95 607	95 612
904	95 617	95 622	95 626	95 631	95 636	95 641	95 646	95 650	95 655	95 660
905	95 665	95 670	95 674	95 679	95 684	95 689	95 694	95 698	95 703	95 708
906	95 713	95 718	95 722	95 727	95 732	95 737	95 742	95 746	95 751	95 756
907	95 761	95 766	95 770	95 775	95 780	95 785	95 789	95 794	95 799	95 804
908	95 809	95 813	95 818	95 823	95 828	95 832	95 837	95 842	95 847	95 852
909	95 856	95 861	95 866	95 871	95 875	95 880	95 885	95 890	95 895	95 899
910	95 904	95 909	95 914	95 918	95 923	95 928	95 933	95 938	95 942	95 947
911	95 952	95 957	95 961	95 966	95 971	95 976	95 980	95 985	95 990	95 995
912	95 999	96 004	96 009	96 014	96 019	96 023	96 028	96 033	96 038	96 042
913	96 047	96 052	96 057	96 061	96 066	96 071	96 076	96 080	96 085	96 090
914	96 095	96 099	96 104	96 109	96 114	96 118	96 123	96 128	96 133	96 137
915	96 142	96 147	96 152	96 156	96 161	96 166	96 171	96 175	96 180	96 185
916	96 190	96 194	96 199	96 204	96 209	96 213	96 218	96 223	96 227	96 232
917	96 237	96 242	96 246	96 251	96 256	96 261	96 265	96 270	96 275	96 280
918	96 284	96 289	96 294	96 298	96 303	96 308	96 313	96 317	96 322	96 327
919	96 332	96 336	96 341	96 346	96 350	96 355	96 360	96 365	96 369	96 374
920	96 379	96 384	96 388	96 393	96 398	96 402	96 407	96 412	96 417	96 421
921	96 426	96 431	96 435	96 440	96 445	96 450	96 454	96 459	96 464	96 468
922	96 473	96 478	96 483	96 487	96 492	96 497	96 501	96 506	96 511	96 515
923	96 520	96 525	96 530	96 534	96 539	96 544	96 548	96 553	96 558	96 562
924	96 567	96 572	96 577	96 581	96 586	96 591	96 595	96 600	96 605	96 609
925	96 614	96 619	96 624	96 628	96 633	96 638	96 642	96 647	96 652	96 656
926	96 661	96 666	96 670	96 675	96 680	96 685	96 689	96 694	96 699	96 703
927	96 708	96 713	96 717	96 722	96 727	96 731	96 736	96 741	96 745	96 750
928	96 755	96 759	96 764	96 769	96 774	96 778	96 783	96 788	96 792	96 797
929	96 802	96 806	96 811	96 816	96 820	96 825	96 830	96 834	96 839	96 844
930	96 848	96 853	96 858	96 862	96 867	96 872	96 876	96 881	96 886	96 890
931	96 895	96 900	96 904	96 909	96 914	96 918	96 923	96 928	96 932	96 937
932	96 942	96 946	96 951	96 956	96 960	96 965	96 970	96 974	96 979	96 984
933	96 988	96 993	96 997	97 002	97 007	97 011	97 016	97 021	97 025	97 030
934	97 035	97 039	97 044	97 049	97 053	97 058	97 063	97 067	97 072	97 077
935	97 081	97 086	97 090	97 095	97 100	97 104	97 109	97 114	97 118	97 123
936	97 128	97 132	97 137	97 142	97 146	97 151	97 155	97 160	97 165	97 169
937	97 174	97 179	97 183	97 188	97 192	97 197	97 202	97 206	97 211	97 216
938	97 220	97 225	97 230	97 234	97 239	97 243	97 248	97 253	97 257	97 262
939	97 267	97 271	97 276	97 280	97 285	97 290	97 294	97 299	97 304	97 308
940	97 313	97 317	97 322	97 327	97 331	97 336	97 340	97 345	97 350	97 354
941	97 359	97 364	97 368	97 373	97 377	97 382	97 387	97 391	97 396	97 400
942	97 405	97 410	97 414	97 419	97 424	97 428	97 433	97 437	97 442	97 447
943	97 451	97 456	97 460	97 465	97 470	97 474	97 479	97 483	97 488	97 493
944	97 497	97 502	97 506	97 511	97 516	97 520	97 525	97 529	97 534	97 539
945	97 543	97 548	97 552	97 557	97 562	97 566	97 571	97 575	97 580	97 585
946	97 589	97 594	97 598	97 603	97 607	97 612	97 617	97 621	97 626	97 630
947	97 635	97 640	97 644	97 649	97 653	97 658	97 663	97 667	97 672	97 676
948	97 681	97 685	97 690	97 695	97 699	97 704	97 708	97 713	97 717	97 722
949	97 727	97 731	97 736	97 740	97 745	97 749	97 754	97 759	97 763	97 768
950	97 772	97 777	97 782	97 786	97 791	97 795	97 800	97 804	97 809	97 813

Common Logarithms of the Natural Numbers (*Continued*)

No.	0	1	2	3	4	5	6	7	8	9
950	97 772	97 777	97 782	97 786	97 791	97 795	97 800	97 804	97 809	97 813
951	97 818	97 823	97 827	97 832	97 836	97 841	97 845	97 850	97 855	97 859
952	97 864	97 868	97 873	97 877	97 882	97 886	97 891	97 896	97 900	97 905
953	97 909	97 914	97 918	97 923	97 928	97 932	97 937	97 941	97 946	97 950
954	97 955	97 959	97 964	97 968	97 973	97 978	97 982	97 987	97 991	97 996
955	98 000	98 005	98 009	98 014	98 019	98 023	98 028	98 032	98 037	98 041
956	98 046	98 050	98 055	98 059	98 064	98 068	98 073	98 078	98 082	98 087
957	98 091	98 096	98 100	98 105	98 109	98 114	98 118	98 123	98 127	98 132
958	98 137	98 141	98 146	98 150	98 155	98 159	98 164	98 168	98 173	98 177
959	98 182	98 186	98 191	98 195	98 200	98 204	98 209	98 214	98 218	98 223
960	98 227	98 232	98 236	98 241	98 245	98 250	98 254	98 259	98 263	98 268
961	98 272	98 277	98 281	98 286	98 290	98 295	98 299	98 304	98 308	98 313
962	98 318	98 322	98 327	98 331	98 336	98 340	98 345	98 349	98 354	98 358
963	98 363	98 367	98 372	98 376	98 381	98 385	98 390	98 394	98 399	98 403
964	98 408	98 412	98 417	98 421	98 426	98 430	98 435	98 439	98 444	98 448
965	98 453	98 457	98 462	98 466	98 471	98 475	98 480	98 484	98 489	98 493
966	98 498	98 502	98 507	98 511	98 516	98 520	98 525	98 529	98 534	98 538
967	98 543	98 547	98 552	98 556	98 561	98 565	98 570	98 574	98 579	98 583
968	98 588	98 592	98 597	98 601	98 605	98 610	98 614	98 619	98 623	98 628
969	98 632	98 637	98 641	98 646	98 650	98 655	98 659	98 664	98 668	98 673
970	98 677	98 682	98 686	98 691	98 695	98 700	98 704	98 709	98 713	98 717
971	98 722	98 726	98 731	98 735	98 740	98 744	98 749	98 753	98 758	98 762
972	98 767	98 771	98 776	98 780	98 784	98 789	98 793	98 798	98 802	98 807
973	98 811	98 816	98 820	98 825	98 829	98 834	98 838	98 843	98 847	98 851
974	98 856	98 860	98 865	98 869	98 874	98 878	98 883	98 887	98 892	98 896
975	98 900	98 905	98 909	98 914	98 918	98 923	98 927	98 932	98 936	98 941
976	98 945	98 949	98 954	98 958	98 963	98 967	98 972	98 976	98 981	98 985
977	98 989	98 994	98 998	99 003	99 007	99 012	99 016	99 021	99 025	99 029
978	99 034	99 038	99 043	99 047	99 052	99 056	99 061	99 065	99 069	99 074
979	99 078	99 083	99 087	99 092	99 096	99 100	99 105	99 109	99 114	99 118
980	99 123	99 127	99 131	99 136	99 140	99 145	99 149	99 154	99 158	99 162
981	99 167	99 171	99 176	99 180	99 185	99 189	99 193	99 198	99 202	99 207
982	99 211	99 216	99 220	99 224	99 229	99 233	99 238	99 242	99 247	99 251
983	99 255	99 260	99 264	99 269	99 273	99 277	99 282	99 286	99 291	99 295
984	99 300	99 304	99 308	99 313	99 317	99 322	99 326	99 330	99 335	99 339
985	99 344	99 348	99 352	99 357	99 361	99 366	99 370	99 374	99 379	99 383
986	99 388	99 392	99 396	99 401	99 405	99 410	99 414	99 419	99 423	99 427
987	99 432	99 436	99 441	99 445	99 449	99 454	99 458	99 463	99 467	99 471
988	99 476	99 480	99 484	99 489	99 493	99 498	99 502	99 506	99 511	99 515
989	99 520	99 524	99 528	99 533	99 537	99 542	99 546	99 550	99 555	99 559
990	99 564	99 568	99 572	99 577	99 581	99 585	99 590	99 594	99 599	99 603
991	99 607	99 612	99 616	99 621	99 625	99 629	99 634	99 638	99 642	99 647
992	99 651	99 656	99 660	99 664	99 669	99 673	99 677	99 682	99 686	99 691
993	99 695	99 699	99 704	99 708	99 712	99 717	99 721	99 726	99 730	99 734
994	99 739	99 743	99 747	99 752	99 756	99 760	99 765	99 769	99 774	99 778
995	99 782	99 787	99 791	99 795	99 800	99 804	99 808	99 813	99 817	99 822
996	99 826	99 830	99 835	99 839	99 843	99 848	99 852	99 856	99 861	99 865
997	99 870	99 874	99 878	99 883	99 887	99 891	99 896	99 900	99 904	99 909
998	99 913	99 917	99 922	99 926	99 930	99 935	99 939	99 944	99 948	99 952
999	99 957	99 961	99 965	99 970	99 974	99 978	99 983	99 987	99 991	99 996
1000	00 000	00 004	00 009	00 013	00 017	00 022	00 026	00 030	00 035	00 039

REFERENCES

1. Clark, E. P., "Semimicro Quantitative Organic Analysis," Academic Press, New York, 1943.
2. Furman, N. H., ed., "Scott's Standard Methods of Chemical Analysis," 5th ed., Vols. I, and II, Van Nostrand, New York, 1939.
3. Gysel, H., "Tables of Percentage Composition of Organic Compounds," Birkhäuser, Basel, 1951.
4. Hodgman, C. D., "Handbook of Chemistry and Physics," 28 ed., Chemical Rubber, Cleveland, Ohio, 1944.
5. Hodgman, C. D., and Lange, N. A., "Handbook of Chemistry and Physics," 9th ed., Chemical Rubber, Cleveland, Ohio, 1922.
6. International Union of Pure and Applied Chemistry, Commission on Atomic Weights, "Table of Atomic Weights, 1957," Basel.
7. Kuffner, F., and Gruber, W., *Mikrochemie ver. Mikrochim. Acta,* **36/37**, 1177 (1951).
8. Lange, N. A., "Handbook of Chemistry," Handbook Publishers, Sandusky, Ohio, 1940.
9. Niederl, J. B., and Niederl, V., "Micromethods of Quantitative Organic Elementary Analysis," Wiley, New York, 1938.
10. Niederl, J. B., and Niederl, V., "Micromethods of Quantitative Organic Analysis," 2nd ed., Wiley, New York, 1942.
11. Steyermark, Al, "Quantitative Organic Microanalysis," Blakiston, Philadelphia, Pennsylvania, 1951.
12. Thompson, R. C., *Anal. Chem.,* **25**, 535 (1953).
13. Treadwell, F. P., and Hall, W. T., "Analytical Chemistry," 6th ed., Vol. II, "Quantitative Analysis," Wiley, New York, and Chapman & Hall, London, 1924.

AUTHOR INDEX

Numbers in parentheses are reference numbers and are included to assist in locating references in which authors' names are not mentioned in the text. Numbers in italics refer to pages on which the references are listed.

Abdel-Wahhab, S. M., *99*
Abdine, H., *128*
Abrahamczik, E., *23, 265,* 277 (1), 278 (1), *309, 342*
Abramson, E., *181, 265*
Acree, F., Jr., 4 (2), *23, 215*
Adams, C. I., *215*
Adams, J. E., *265*
Adams, R., 496 (1, 9, 47), *505, 506*
Adibek-Melikyan, A. I., *23*
Adkins, H., 495 (2), *505*
Adler, S., 151 (66), *183*
Afremow, L. C., *523*
Agazzi, E. J., *185, 274, 309, 313, 342,* 377 (11), 398 (11), 399 (11), 400 (11), *407*
Agiza, A. H., *265, 420*
Aguayo, N., *353*
Ahmed, M. N., *309*
Ainsworth, A. W., 30 (1), *47*
Akaza (*nee* Kishi), I., *311*
Alber, H. K., 2 (177), 3 (103, 148, 149, 150, 175, 176, 177, 178, 179, 180, 181, 182, 183, 184), 10, *23, 26, 27, 28,* 30 (5), 31 (5), *47, 48,* 58 (63, 64), 59 (63, 64), 60 (63), 62 (63), 63 (63), 64 (63), 65 (63), 66 (4, 63), 67 (63), 68 (2, 63, 64), 69 (64), 70 (64), 73 (63, 64), 74 (63), 75 (63, 64), *80, 81, 82,* 83 (118, 119, 120), 88 (121), 91 (1, 121), *98, 101,* 113 (195, 196, 197), 116 (2, 195, 196, 197), *127, 132,* 134 (1, 224, 225), *144, 149,* 160 (189, 190), 161 (167, 168, 190), 171 (190), 172 (190), *186, 187,* 192 (194), 193 (194), 195 (194), 196 (194), 199 (194), 200 (194), 211 (194), *220,* 225 (351), 228 (351), 230 (351), 238 (351), 240 (351), 255 (351), 257 (351), 258 (351), *273, 274,* 278 (211), 280 (211), 281 (211), 287 (212), 297 (212), 298 (212), 302 (212), *313, 314,* 317 (380, 381), 318 (381), 319 (381), 322 (381), 324 (381), *350, 352,* 356 (91), 360 (90), *366,* 368 (58), 369 (58), 370 (59), 371 (59), *376,* 389 (4), 393 (90), 401 (90), 402(4), *407, 409,* 423 (101), 426 (101), 429 (101), 431 (101), 437 (101), *443,* 447 (45, 46), *453,* 548 (2), 558 (1, 88), 559 (1, 88), 560 (2, 3), *562, 564*
Albrink, M. J., *127, 416*
Alexander, A. P., *440*
Alford, W. C., *181*
Alicino, J. F., *23,* 104 (4), *127,* 276 (5, 6, 7), 277 (5), 278 (5), *309,* 316 (5), 323 (4), 325 (4), *342, 416,* 444 (2), *451*
Alimarin, I. P., *23, 127*
Allam, F., *265*
Allan, J. C., *127*
Allen, H., Jr., *144, 265*
Allen, K. A., *127*
Allen, S. A., *132*
Almassy, G., *144*
Almström, G. K., *265*
Alperowicz, I., *347*
Althouse, P. M., *562*
Altieri, P. L., *48, 81, 184, 270*
Aluise, V. A., 2 (177), 3 (175-184), *28,* 58 (63, 64), 59 (63, 64), 60 (63), 62 (63), 63 (63), 64 (63), 65 (63), 66 (63), 67 (63), 68 (63, 64), 69 (64), 70 (64), 73 (63, 64), 74 (63), 75 (63, 64), *81, 82,* 83 (118, 119, 120), 88 (121), 91 (121), *101, 103,* 113 (195, 196, 197), 116 (195, 196, 197), *127, 132,* 134 (224, 225), *149,* 160 (189, 190), 161 (190), 171 (190), 172 (190), 174, *181, 186, 187,* 192 (194), 193 (194), 195 (194), 196 (194), 199 (194), 200 (194), 211 (194), *220,* 225 (351), 228 (351), 230 (351) 235 (8), 236 (8), 238 (351), 240 (351), 255

Author Index

(351), 257 (351), 258 (351), 265, 274, 278 (211), 280 (211), 281 (211), 287 (212), 297 (212), 298 (212), 302 (212), 314, 317 (380, 381), 318 (381), 319 (381), 322 (381), 324 (381), 352, 356 (91), 360 (90), 366, 368 (58), 369 (58), 370 (59), 371 (59), 376, 377, 379 (3, 5), 383 (2), 389 (3, 4), 391 (3), 392, 393 (90), 397 (5), 398, 401 (90), 402 (5), 407, 409, 423 (101), 426 (101), 429 (101), 431 (101), 437 (101), 443, 447 (45, 46), 453, 545, 558 (88), 559 (88), 564

Alvarez Querol, M. C., 23
Alving, R. E., 220
Amako, S., 183
Ambrosino, C., 23, 47, 80
Amdur, E., 83 (65), 99, 375
Amin, A. M., 128, 144, 145
Anbar, M., 407
Andersch, M. A., 144
Anderson, C. W., 144
Anderson, D. M. W., 440
Anderson, H. H., 127
Anderson, L., 220, 309
Andersson, O., 309
Andreeva, E. V., 527
Anger, V., 521
Ani, M., 23
Anisimova, G. F., 184, 269
Annino, R., 181, 265
Apelgot, S., 265
Aquist, S. E. G., 492
Archer, E. E., 265, 342
Ariel, M., 145
Arison, B. H., 274
Aristarkhova, M. V., 521
Arjungi, K. N., 521
Arlt, H. G., Jr., 440
Armstrong, W. D., 493
Arndt, F. G., 521
Arnold, E. A., 144
Arnold, M., 270
Arthur, P., 181, 265, 545
Arventiev, B., 521
Asami, T., 215
Asbury, H., 47
Ashby, R. O., 144
Ashworth, M. R. F., 507
Ašperger, S., 144

Aston, J. G., 563
Austin, J. H., 454 (3), 488 (3), 492
Awad, W. I., 419, 493
Ayers, C. W., 265, 416

Babcock, M. J., 98
Bachofer, M. D., 364
Backeberg, O. G., 265
Badley, J. H., 377 (11), 398 (11), 399 (11), 400 (11), 407
Bähr, G., 144, 374
Baernstein, H. D., 440
Baertschi, P., 265
Bagreeva, M. R., 150
Baier, R., 364
Bailey, A. J., 440
Bakay, V., 314
Baker, B. B., 342
Baker, M. J., 492, 527
Baker, P. R. W., 215, 416
Balaz, F., 440
Baldeschwieler, E. L., 183
Baldwin, R. R., 47
Balis, E. W., 265
Ball, F. L., 350
Ball, J. S., 523
Ballard, A. E., 150, 270
Ballard, D. G. H., 127
Ballczo, H., 342
Ballentine, R., 196 (10), 215
Ballinger, D. G., 348
Bamford, C. H., 127
Banerjee, N. G., 272
Bang, I., 374
Banks, R. E., 181, 343
Banks, W. F., Jr., 312
Barakat, M. Z., 343
Barber, H. H., 144
Barcroft, J., 492, 505
Barger, G., 1, 23, 545
Barham, H. N., 266
Barnard, A. J., 128
Barnum, D. W., 344
Barraclough, K. C., 91 (7), 98
Barrett, D. G., 348
Barrett, F. R., 144
Bartels, U., 346
Barth, L. G., 127
Barusch, M. R., 506

Bass, E. A., 103 (199), 104 (198), 105 (198), 106 (199), 110 (198), 112 (199), 113 (199), *132,* 189 (195), 190 (195), 191 (195), 192 (195), 210 (195), *220,* 276 (213, 214), 277 (213, 214), 278 213, 214), 282 (214), 283 (214), 291 (213), 292 (213), *314,* 326 (385), 327 (385), 331 (385), 332 (385), *352*
Bastin, E. L., *98, 185, 271*
Batalin, A. Kh., *23*
Batdorf, D. K., *348*
Bather, J. M., *309, 343*
Batt, W. G., *98, 181, 309, 343, 364*
Battles, W., *309*
Battley, E. H., *492*
Baudet, P., *215, 416*
Baum, H., 276 (160), 277 (160), 290 (160), *312,* 316 (292), *349*
Baumann, M. L., *421*
Bawden, A. T., *145*
Baxley, W. H., 316 (252), *348*
Baxter, G. P., *265*
Bayer, E., *451*
Bazen, J., *128*
Beach, A. L., *523*
Beadle, L. C., *219*
Beamish, F. E., 334 (23), *343, 374*
Bear, F. E., 188 (153), *219*
Beatty, C., III, 188 (12), *215*
Beauchene, R. E., *144, 182*
Beaucourt, J. H., *23, 343*
Beaucourt, K., 257 (41), *266*
Beazley, C. W., *182,* 276 (15), *309,* 316 (26), *343*
Bechet, J., *309*
Beck, W., *526*
Becker, W. W., *23,* 103, *127,* 377 (5), 379 (5), 383 (5), 392 (5), 397 (5), 398, 402 (5), *407, 521*
Beckman, A. O., *521*
Bedekian, A., *187*
Beet, A. E., *215*
Behnke, U., *420*
Belcher, R., 2 (24, 25), *23,* 30 (10, 11), *47,* 50 (7, 8), 51 (7, 8), 54 (7, 8), 57 (7, 8), 70 (7, 8), 71 (7, 8), 75 (7, 8), *80, 98, 127,* 133 (17, 18), 134 (17, 18), 137 (17, 18), *144,* 151 (11, 12), 174 (11, 12), 177 (11, 12), *182, 215,* 230 (27), 244 (24, 25), 247 (24, 25), 259, *265,*
309, 334 (32, 39, 45), 343, *364,* 367 5, 6), *374, 407, 416,* 422 (12, 13), *440, 451, 521,* 528 (4, 5), *545, 562*
Bell, R. D., 355 (9), *364*
Belluco, U., *364, 374*
Benedetti-Pichler, A. A., 1 (28), 2 (105), 3 (103, 140), *23, 26, 27,* 30 (12, 76), 31 (67, 76), *47, 48,* 50 (11), 53 (50), 54 (50), 56 (50), 75 (9), *80, 81, 98, 127, 129, 130, 144, 147, 149, 312, 441, 564*
Bennett, A., *265, 521*
Bennett, C. E., *343*
Bennewitz, R., *343, 364*
Benson, S. W., *521*
Bentley, G. T., *129, 147*
Berck, B., *215*
Bereznitskaya, E. G., *147, 269, 311, 347*
Berg, L. G., *182, 265*
Berg, W. E., *521*
Bergamini, C., *309*
Berger, J., *127, 416, 521*
Berger, L., *131*
Berggren, A., *129*
Bergmann, G., *416*
Bergold, G., *416*
Bering, P., *98*
Berinzaghi, B., *521*
Berka, A., *521*
Berneking, A. D., *144, 182*
Bernhart, D. N., *364*
Bernhauer, K., *98*
Bernstorf, E., *216*
Beroza, M., *416*
Berret, R., *265, 407*
Bertescu, L., *272*
Bertrand, P., *309*
Bessey, O. A., *419*
Bethge, P. O., *309, 440*
Bettzieche, F., *416*
Bevington, J. C., *23*
Bhasin, R. L., *215, 309*
Bhatty, M. K., *182, 215, 440, 492, 521*
Biava, F. P., *314,* 334 (382), *352,* 383 (91), 389 (91), *409*
Bickford, C. F., *98*
Bieling, H., *144, 374*
Bien, G. S., *127*
Billitzer, A. W., *440*
Billmeyer, A. M., *420, 526*
Bini, L., 540 (83a), 543 (83a), *547*

Author Index

Birket-Smith, E., *127*
Birkofer, L., *442, 524*
Blabolil, K., *259* (200), *270*
Blacet, F. E., *521*
Black, S., *417*
Blackburn, S., *417*
Bladh, E., *309*
Blaedel, W. J., *127, 343, 521*
Blake, G. G., *127*
Blanc, P., *309*
Blank, E. W., *545, 562*
Blickenstaff, R. T., *521*
Blinn, R. C., *343*
Block, R. J., *417*
Blom, J., *215*
Blount, B. K., *98*
Blümel, F., *277* (1), *278* (1), *309*
Blumendal, H. B., *148*
Bobranski, B., *266, 547*
Bobtelsky, M., *127, 144, 417*
Bode, E., *343*
Bode, F., *215*
Bodenheimer, W., *266*
Bodnar, J., *374*
Böck, F., *257* (41), *266*
Böe, J., *316* (244), *348*
Böhm, R. H., *562*
Böhme, H., *562*
Boëtius, M., *23, 144, 266, 309, 343*
Bogan, E. J., *309*
Bogin, D., *145*
Boguth, W., *127*
Boikina, B. S., *350*
Boissonnas, R. A., *215*
Boivin, A., *272*
Bolling, D., *417*
Bondarevskaya, E. A., *347, 408, 443*
Bone, A. D., *421*
Bonhorst, C. W., *562*
Bonnar, R. U., *344, 545*
Bonstein, T. E., *348*
Bonting, S. L., *417*
Boos, R. N., *10, 24, 266, 309, 364, 492, 521*
Borgstrom, S., *146*
Borner, K., *130*
Borsook, H., *492*
Bosart, L. W., *556* (9), *562*
Bosch, F. de A., *343*
Bose, S., *521*
Bothner-By, A. A., *266*

Boulet, M. A., *147*
Bourdillon, J., *545*
Bournique, R. A., *147, 440*
Bovee, H. H., *309*
Bowman, R. L., *128, 332, 344*
Boyer, R. Q., *31* (44), *48*
Boyle, A. J., *315*
Bradbury, J. H., *127*
Bradbury, R. B., *451*
Bradley, R. S., *47, 80*
Bradstreet, R. B., *188* (31), *215, 216*
Braganca, B. M., *492*
Bragg, P. D., *521*
Brame, E. G. Jr., *343*
Brancone, L. M., *182, 505*
Brand, E., *417*
Brand, R. W., *103* (149), *130, 188* (144), *190* (144), *191* (144), *192* (144), *218*
Brandenburg, V. W., *562*
Brandstätter-Kuhnert, M. (*also* Brandstätter, M.), *521, 545, 562*
Brandt, C. S., *128*
Brandt, K., *343*
Brandt, W. W., *266, 343*
Brant, J. H., *216*
Brantner, H., *144*
Bratton, A. C., *545*
Braukman, B., *407*
Braun, J. C., *185*
Braverman, M. M., *311*
Bray, R. H., *364*
Brechbühler, T., *351*
Brecher, C., *103* (36), *127, 422, 443*
Breitenbach, J. W., *545*
Bretschneider, G., *310*
Breusch, F. L., *417*
Breyhan, T., *216, 521*
Bricker, C. E., *374*
Brignoni, J. M., *505*
Brite, D. W., *364*
Broad, W. C., *128*
Brochet, A., *265*
Brockmann, H., *545*
Broderson, H. H., *348*
Broeker, A. G., *128*
Broekhuysen, J., *309*
Bromund, W. H., *47, 80*
Bronk, L. B., *265*
Brooks, F. R., *185, 274, 309, 342, 377* (11), *398* (11), *399* (11), *400* (11), *407*

Author Index

Brooks, S. C., 128
Brousaides, F. J., *522*
Brown, D. R., *349*
Brown, F., *182, 344*
Brown, F. M., *49*
Brown, G. H., *348*
Brown, H. C., *521*
Brown, L. E., *47*
Brown, S., *421*
Brüel, D., *216*
Brunisholz, G., *344*
Brunner, H., *521*
Bruno, M., *364, 374*
Bruns, B. P., *99*
Bryant, F., *417*
Bryant, J. T., 560 (3), *562*
Bryant, W. M. D., 92 (111), 94 (111), *101*
Buben, F., *523*
Buchanan, B. B., *182*
Buchka, M., *219*
Buck, J. B., *127*
Buděšinský, B., *452, 505, 521*
Bühler, F., *268, 311, 353*
Bühn, T., *442*
Buell, B. E., *144*
Bürger, K., 316 (61), *344, 407, 440*
Buffa, A., *417*
Bulandr, J., *522*
Bullock, A. B., *98*
Bullock, B., *145, 344*
Bulušek, J., *353*
Bumsted, H. S., *344*
Bunzell, H. H., *98*
Burger, K., *219, 526*
Burr, J. G., Jr., *492*
Burstein, R., *524*
Burton, F., 83 (20), *98*
Burton, J. D., *364*
Bussmann, G., *182, 183, 309*
Butler, E. J., *145*
Butler, J. P., *417*
Bystrov, S. P., *374*

Cadman, W. J., *80*
Cady, G. H., *351*
Cahn, L., *80*
Cahnmann, H. J., *146*
Caldas, E. F., 334 (32), *343*
Calderón Martinez, J., *521*
Caldwell, J. R., *344*

Caldwell, M. J., *266*
Caley, E. R., *145*
Calkins, V. P., *417*
Callan, T. P., *309*
Campanile, V. A., *274,* 377, 398 (11), 399 (11), 400, *407*
Campbell, A. D., *440, 546*
Canal, F., *266*
Canales, A. M., 377, 379 (12), 391 (12), 392 (12), 398 (12), 399 (12), 400 (12), *407*
Cannon, J. H., *98, 344*
Cannon, W. A., *182*
Caramanos, S., *219*
Carballido Ramallo, O., *217*
Carius, L., 276 (36-40), 287 (37-40), *309,* 316, 317 (64-67), *344, 367, 374*
Carles, J., *364*
Carlson, O. T., *440*
Carlsson, M. E., *365*
Carmichael, H., 3 (103), *26, 48, 81*
Carney, A. S., 276 (136), *312*
Carothers, W. H., 496 (9), *505*
Carruthers, C., *145*
Carson, W. N., Jr., *417, 492*
Carstens, C., *311*
Cassil, C. C., *374*
Cavallini, D., *417*
Cepciansky, I., *216*
Cervinka, V., *418*
Cevolani, F., *344*
Chain, E. B., *350*
Chakravarti, R. N., *562*
Chalmers, A., *309*
Chalmers, R. A., *364*
Chamot, E. M., 551 (14), *562*
Chaney, A., *452*
Chaney, A. L., *374*
Chaphekar, M. R., *505*
Chapman, G. W., *145*
Chapman, N. B., *344*
Chapon, L., *492*
Charest, M. P., *492*
Charlton, F. E., *182, 266*
Chatagnon, C., *145*
Chatagnon, P., *145*
Chateau, H., *344*
Chaudhuri, K. N., *562*
Cheek, D. B., *344*
Chen, P. S., *364*

Chen, S. L., *266*
Chen, W. K., *310*
Cheng, K. L., *130*
Cheng, W., *344*
Cherbuliez, E., *215, 416*
Cherkin, A., *417*
Cheronis, N. D., 2 (37, 38), *24, 98, 521, 522,* 551 (16), *562*
Chesbro, R. M., *351*
Chettleburgh, V. J., *440*
Chiang, Fan-Tih, *344*
Childs, C. E., *182, 545*
Chinnick, C. C. T., *524*
Chinoy, J. J., *127, 440*
Chirnoaga, E., *311,* 316 (166), *346*
Chiu, J., *269*
Chladek, O., *523*
Chladkova, R., *523*
Chleck, D. J., *522*
Chopard-dit-Jean, L. H., *131*
Christensen, B. E., *266, 440, 441, 452,* 512 (180), 513 (180), *525*
Christensen, B. G., *344*
Christiansen, J. A., *545*
Christman, D. R., *266*
Chromcova, L., *216*
Chudakova, I. K., *315, 353*
Chumachenko, M. N., *25, 147, 311, 346, 347*
Churáček, J., *418*
Churchill, H. V., *350*
Churmanteev, S. V., *275, 353*
Ciaranfi, E., *417*
Cieleszky, V., *374*
Číhalík, J., *344, 522*
Cimerman, C., *145, 150, 417*
Ciucani, M., *417*
Ciusa, W., *98*
Claborn, H. V., *417*
Claff, C. L., *127, 128*
Clark, E. P., 2 (43), 4 (41, 42), *24,* 31 (18), *47,* 57 (17), *80,* 88 (26), *98,* 102 (42), *128,* 137 (59), *145,* 177 (25), *182,* 188 (40, 41), 189 (41), 192 (41), *216,* 222 (54), 247 (54), *266,* 276 (42), 287 (42), *309,* 316 (78), *344, 364,* 410 (32), *412, 417,* 422 (28), *441, 452, 522,* 528, 531 (16), 532 (16), 535 (15, 16), 536, 537 (16), 539 (16), *545, 562,* 569, *602*
Clark, H. A., *147*
Clark, H. S., *182, 266, 344*

Clark, J. W., *47*
Clark, R. H., 276 (174), *313,* 316 (306), *350*
Clark, R. O., 235 (59), 236 (59), *266*
Clark, S. J., 2 (44), 5 (44), *24,* 30 (20), *47,* 50 (18), *80,* 88 (27), *98, 128, 145,* 151 (27), *182, 216, 266,* 276 (43), *310,* 334 (32), *343, 344, 364, 375, 407, 417, 441, 452, 522, 545, 562*
Clark, W. M., 123 (44), *128,* 410 (34), *417*
Clarke, B. L., *98*
Clarke, H. T., 514, *522*
Clarke, R., *452*
Clarke, R. G., *182*
Clegg, D. L., 174 (138), *185*
Clemo, G. R., *562*
Clifford, P. A., *344*
Close, R. A., *127*
Coe, M. R., 103 (225), *132,* 188 (214), 190 (214), 191 (214), 192 (214), *220*
Coe, R. R., *346, 349*
Cogbill, E. C., *344*
Cogbill, E. G., *364*
Cohen, J., 103 (190), *131*
Cohen, L. E., *364*
Cohn, M., *522*
Colaitis, D., *266*
Colarusso, R. J., *132*
Colas, R., *417*
Cole, J. O., 188 (43), 190 (43), *216*
Cole, J. W., *526*
Cole, M. S., *219*
Collier, P. R., *216*
Collins, H. L., *374*
Colson, A. F., *182, 266,* 276 (44), *310, 407, 441, 505, 522, 545*
Conn, J. B., *364*
Connelly, J. A., *272*
Conolly J. M., *98*
Conway, H. S., *377,* 379 (15, 65), 389 (4), 391 (15, 65), 392 (15, 65), 399 (65), 402 (4), *407, 408*
Cooke, W. D., *130*
Coombs, H. I., 134 (61), *145*
Cooper, F. J., 104 (150), 105 (150), 110 (150), *130,* 276 (169), 277 (169), 278 (169), 282, *313,* 495 (32), 496 (32), 497 (32), 499 (32), *506*
Cooper, S. S., *128*
Corbaz, A., *150*

Corbitt, H. B., *419*
Corliss, J. M., *182*
Corner, M., 56 (20), *80, 145, 344, 364, 375*
Cornwell, R. T. K., *266*
Corwin, A. H., 3 (103, 140), *26, 27,* 30 (21, 76), 31 (76), *47, 48, 49,* 53 (50), 54 (50), 56 (50), *81, 266*
Cotlove, E., *128,* 332, *344*
Couchoud, M., *418*
Cousin, B., *270*
Cowell, W. H., *218, 266*
Craig, H. C., *352*
Craig, L. C., *98*
Craig, R., 31 (44), *48*
Cramer, F. B., *452*
Crane, R. A., *521*
Crawford, A., *375*
Crespi, V., *344*
Crickenberger, A., 316 (5), *342*
Cropper, F. R., *182, 266*
Cross, C. K., *266*
Crowder, M. M., *130*
Crowell, W. R., *147*
Cruikshank, S. S., *186, 219, 313, 351,* 354 (81), *366, 408*
Csokan, P., *545*
Cundiff, R. D., *420*
Cunningham, B., *128, 494*
Cunningham, K. G., *545*
Currey, R. P., *344*
Currie, J., *545*
Curtin, D. Y., *564*
Custer, J. J., *344*
Cuthbertson, F., *181, 343*
Czanderna, A. W., *47*
Czech, F. W., *364*
Czepiel, T. P., *417*
Czonka, F. A., *93, 98*

Dahle, D., *344*
Dahlenborg, H., *343*
Dahmen, E. A. M. F., *421*
Dahn, H., *407*
Daimler, B. H., *128*
Dalen, E. van, *274*
Dal Nogare, S., *522*
Dalrymple, R. S., *216*
Damiens, R., *266*
Damon, G. F., *347*
Dando, B., *182*
Danford, V., *272*
Danielson, M., *99*
Daudel, P., *344*
Daugherty, M., *348*
Davenport, J. E., *267*
Davidson, D., 554 (23), 557 (22, 23), *562*
Davidson, J., *145*
Davis, O. L., *220*
Dawson, W., *545*
Day, H. G., *216*
Dean, R. B., *344*
Debal, E., *344*
Debbrecht, F. J., *343*
DeFord, D. D., *268, 506*
De Francesco, F., *145*
Degens, P. N., Jr., *407*
Deibner, L., *145*
Deinum, H. W., *407*
De Jong, R. N., *220*
Delaby, R., *266*
Delande, N., *419*
Dell, J. C., 276 (213), 277 (213), 278 (213), 291 (213), 292 (213), *314*
Dellis, E. P., *525*
Deltour, G., *349*
Delville, J. P., *218, 524*
Demmel, H., *148*
Demoen, P., *314*
Dermelj, M., *216*
Dern, R. J., *128*
Desassis, A., *344*
Deschamps, P., *186*
Deturk, E. E., *220*
Deveraux, H., 292 (146), *312*
Dezsö, I., *145*
DiBacco, G., *364, 375*
Dickinson, W. E., 189 (49, 50), 191 (50), *216*
Dickman, S. R., *364*
Diehl, H., *145*
Diemair, W., *364*
Dieterle, H., *266*
Dillon, R. T., *421, 494*
Dimbat, M., *545*
Dinnerstein, R. A., *407*
Di Pietro, C., *182*
Dippel, W. A., *364*
Dirscherl, A., *182, 310*
Disse, W., *441, 522*
Ditrych, Z., *441*

Dittrich, S., *216*
Dixon, J. P., *216, 310, 407*
Doadrio, A., *25, 312*
Doan, D. J., *273*
Doering, H., 316 (97), *345*
Doherty, D. G., *216, 492*
Doisy, E. A., 355 (9), *364*
Doležil, M., *310, 522*
Dombrowski, A., *266*
Donahoo, W. P., *181, 265*
Donau, J., 1 (51), *24*, 31 (23), *147*, 276 (53), 287 (53), *310*, 316 (108), *345*, 560 (24), *562*
Donia, R. A., *442*
Doppler, G., *342*
Dorf, H., *128*
Dorfman, L., *273*
Dory, I., *24*
Dostrovsky, I., *407*
Downing, R., 163 (39), *182*
Doyle, W. L., *216*
Drekopf, K., *407*
Druzhinin, I. G., *145*
Dubbs, C. A., *98*
Dubinina, I. F., *184*
Dubnoff, J. W., *492*
Dubouloz, P., *345*
Dubowski, K. M., *98*
Dubravcic, M., *350*
Dubsky, J. V., *182*
Düsing, W., *128*
Duffield, J., *409*
Duncan, G. D., *345*
Duncan, J. L., *440*
Dundy, M., *377, 407*
Dunicz, B. L., *310*
Dunn, M. S., *417*
Dunne, T., *349*
Duschinsky, R. C., *420*
Duswalt, A. A., *266*
Duswalt, A. A., Jr., *343*
Duval, C., *145, 522*
Duxbury, McD., *345*
Dvorszky, M., *146, 365*
Dyer, D. C., *417*

Earle, A., *420*
Earley, J. V., *524*
Easterbrook, W. C., *441*
Eastham, J. F., *522*

Eastroe, J. E., *417*
Eccleston, B. H., *310*
Eckhard, S., *270, 311*
Eder, K., *81, 182, 216, 267*
Edlbacher, S., *522*
Edsberg, L. R., *526*
Edwards, A. H., *216*
Edwards, F. C., *147*
Edwards, G. A., *131*
Edwards, J. D., *418*
Edwards, J. W., *26*
Egami, F., *315*
Eger, C., *345, 364*
Egorova, N. F., *267, 271, 312*
Ehlers, V. B., *522*
Ehrenberger, F., *345, 407*
Ehret, W. F., 74 (67), *82*, 259 (369), *274*
Ehrlich-Rogozinsky, S., *347, 442*
Eichhorn, F., *493*
Eisenstadter, J., *144*
Eitingon, M., *130*
El-Badry, H. M., *24*
El Din Ibrahim, M. G., *452, 526*
Elek, A., 102 (53), *103, 128, 216, 267*, 276 (51, 52), *310*, 316 (103, 104), 323 (103, 104), 325 (103, 104), *345, 417*, 422 (33), 423 (33), 425 (33), 433 (33), 434 (33), *435, 441*, 444, 445 (19), 447 (19), 449 (19), *452*
Ellenbogen, E., *417*
Ellenburg, J. Y., *99*
Ellis, C., 495 (12), *505*
Ellis, E. W., *527*
Ellis, G. H., *128, 345*
Ellis, K. W., *145*
Elving, P. J., *267*, 316 (106), *345, 350, 377, 407, 417*
El-Wahab, M. R., *343*
Emanuel, C. F., *99*
Embden, G., *365*
Eméleus, H. J., *145*
Emi, K., *345*
Emich, F., 1, *24*, 31 (25, 26), *47, 99, 276* (53), 287 (53), *310*, 316 (108), *345, 562*
Engelbrecht, R. M., *505*
Entrikin, J. B., *98, 522*, 551 (16), *562*
Epstein, J., *417*
Erdey, L., *345*
Erdos, José, *99, 310, 505*
Ermolenko, N. V., *352*

Esafox, V. I., *216*
Estevan, J., *345*
Etcheverry, D. H., *272*
Étienne, A., *182, 267, 310, 345*
Euchen, A., *522*
Evans, H. J., *526*
Evans, R. N., *267*
Exley, D., *216*
Eysell, K., *441, 522*

Falkenhausen, F. V. v., *522*
Fanto, R., *443*
Farley, L. L., *349*
Farlow, N. H., *345*
Farrington, P. S., *131, 492*
Faulconer, W. B. M., *217*
Faulkner, M. B., 4 (185), *28,* 316 (383), 323 (383), *352*
Fawcett, J. K., *216*
Federico, K., *417*
Fedoseev, P. N., *183, 187, 267, 310, 345*
Feeney, R. E., *420*
Fehér, M., *219*
Feigl, F., *375*
Feld, U., *421*
Feldman, C., *99*
Fellenberg, Th. v., 316 (118), *345*
Fels, G., *216*
Fenger-Eriksen, K., 515, *522, 563*
Fennell, T. R. F. W., *145, 216,* 354, *365*
Ferguson, B., *99*
Fernandez, J. B., *185*
Fertel'meister, Ya. N., *48*
Fertig, J. W., *146*
Feuer, I., *47, 128*
Fichera, G., *345*
Fieser, L. F., *267*
Figala, N., *271*
Figel, S., *521*
Fildes, J. E., *265, 343, 345, 440*
Filipovič, L., *441*
Fill, M. A., *28, 49, 132, 187*
Fine, L., *345*
Fischer, E., 560 (27, 28), *563*
Fischer, F. O., *267*
Fischer, Karl, 92, 94, *99*
Fischer, R., *99,* 560 (30), *563*
Fischer, R. B., *310*
Fischer, V., *366*
Fish, V. B., *216*

Fishel, D. L., *525*
Fiske, C. H., *310, 365*
Fitzgerald, D. M., *315*
Flaschenträger, B., *99, 267,* 514 (59), *522*
Flaschka, H., *128, 145, 345, 505*
Fleischer, K. D., *149, 365, 376*
Flon, M., *344*
Floramo, N. A., *420*
Folch, J., 67 (66), *82,* 111 (219), *132,* 222 (368), *274, 312, 421,* 454 (92), 455 (92), 458 (92), 465 (92), 473 (92), 476 (92), *494*
Folin, O., 116 (65), *128*
Fondarai, J., *345*
Fong, J., *366*
Fonnesu, A., *417*
Fontaine, T. D., *365*
Foreman, J. K., *267, 407, 522*
Forsblad, I., *267*
Forsee, W. T., *272*
Forster, E. L., *545*
Fort, R., 8 (54), *24, 407*
Forziati, F. H., *417*
Foster, F. J., 155 (169), 161 (169), 174, *186*
Foster, H., *345*
Foster, R. H. K., *128*
Fóti, G., *219, 314*
Foulk, C. W., *145*
Foulke, D. G., *420*
Fowler, R. M., 3 (103), *26, 48, 81*
Francis, H. J., Jr., *345,* 540 (26), *545*
Francotte, C., *419*
Franzen, F., *441, 522*
Frauenfelder, L. J., *348*
Frazer, J. W., *183, 267*
Fredericks, E. M., *309, 342*
Frediani, H. A., *128, 130, 348, 563*
Freeland, M. Q., *310*
Freier, H. E., *267, 345*
Frenkel, S., *145*
Freri, M., *310*
Frey, H. M., *267*
Friedman, A. F., *218*
Friedman, L., *440*
Friedmann, T. E., *267*
Friedrich, A., 62, *80,* 189 (65), *216,* 255 (97), *258, 266, 310,* 316 (127), *345, 417,* 437 (37, 38, 39), *441,* 509 (65, 66, 68), 511 (65, 66, 68), *522*
Fries, G., *523*

Fritz, J. S., *24, 128, 145, 183, 310, 417, 493, 505, 522, 524*
Frobose, H., *562*
Frodyma, M. M., *272*
Fromageot, C., *417*
Frontali, N., *417*
Fuchs, F., *417*
Fuchs, H. J., *220*
Fülöp, T., *145*
Fürst, H., *350*
Fujimoto, R., *267, 310, 345*
Fujiwara, H., *267*
Fukker, F. K., *146*
Fukuda, M., *183, 441*
Fulmor, W., *182*
Fulton, J. W., *408*
Funasaka, W., *345*
Funnell, H. S., *375*
Furman, N. H., 12 (56), *24,* 30 (29), *47,* 50 (22), 51 (22), 53 (22), 54 (22), 73 (22), 76 (22), *80,* 88 (44), *99,* 102 (72), 111 (72), *128,* 133 (82), *145,* 151 (48), 152 (48), 161 (48), 163, 177 (48), *183,* 188 (66), 189 (66), 191 (66), *216,* 244 (102), *267,* 276 (67), *310,* 316 (130), *345,* 354, 359 (36), *365,* 367 (22), 370 (22), 372 (22), *375,* 411 (50), *417, 441, 522,* 528 (27), *545,* 567 (2), *602*
Furter, M. F., 30 (30), *47, 183,* 188 (67), *216,* 222 (103), 225 (106), 230 (104), 235 (104), 236 (104), 247 (104), 253 (104), *267, 418,* 422 (44, 45), 433 (44, 45), 435, *441, 563*
Furuhashi, K., *274*
Fuson, R. C., *524,* 560 (84), *564*

Gaberman, V., *216*
Gabourel, J. D., *492*
Gage, D. G., *24*
Gagnon, A., *492*
Gainer, H., *420*
Gambrill, C. M., *350*
Garbers, C. F., *522*
Garch, J., *183, 267, 407*
Garcia, C. R., *563*
Gardner, D. E., *351*
Gardner, T. S., *452*
Gardon, J. L., *216*
Gargiulo, M., *346*

Garner, M. W., 4 (186), *28,* 316 (384), 323 (384), *352*
Garschagen, H., *128*
Gassner, K., *526*
Gates, O. R., *365*
Gatterer, A., *345*
Gautier, J. A., *145, 146, 525*
Gedalia, I., *347*
Geilmann, W., 108 (74), *128, 310*
Gellhorn, A., *146*
Gel'man, N. E., *183, 184, 267, 270, 310, 311, 345, 346, 347*
Genevois, L., *183, 522*
Gerbaulet, K., *346*
Gerbes, O., *527*
Gerhard, E. R., *310*
Gerritsma, K. W., *421*
Gerstein, T., *524*
Gettler, A. O., *99, 441*
Gey, K. F., *418, 522*
Gheorghiu, C., *365*
Gibb, T. R. P., Jr., 560 (36), *563*
Gibbons, D., *23, 144*
Gibson, C. S., *563*
Gibson, D. T., *545*
Gibson, N. A., *145*
Gierlinger, W., *311*
Gilbert, A. B., *146*
Gildenberg, L., *310*
Gillam, W. S., *146*
Gillis, J., *24*
Gilmont, R., *128*
Giraut-Erler, L., *506*
Giri, K. V., *418*
Glagoleva-Malikova, E. M., *418*
Glazova, K. I., *270, 347*
Glazunova, Z. I., *147*
Glew, D. N., *522*
Gloss, K., 316 (244), *348*
Godbert, A. L., 2 (24, 25), *23,* 30 (10, 11), *47,* 50 (7, 8), 51 (7, 8), 54 (7, 8), 57 (7, 8), 70 (7, 8), 71 (7, 8), 75 (7, 8), *81, 98, 127,* 133 (17, 18), 134 (17, 18), 137 (17, 18), *144,* 151 (11, 12), 174 (11, 12), 177 (11, 12), *182, 215,* 244 (24, 25), 247 (24, 25), *265, 309, 343, 364,* 367 (5, 6), *374, 407, 416,* 422 (12, 13), *440, 451, 521,* 528 (4, 5), *545, 562*
Goddu, R. F., *128, 522*

Godfrey, P. R., 104 (78), *128, 282* (72), *310*
Goia, I., *346*
Goiffon, R., *418*
Goldbach, H. J., *419*
Goldbaum, L., *99*
Goldberger, H., *80*
Goldspink, A. A., *346*
Goldstein, M., *266*
Goltz, G. E., *522*
Gómez Vigide, R. F., *216, 217*
Gonick, H., 158 (56), *183*
González Carreró, J., *217*
Gora, E. K., *492*
Gorbach, G., *24, 47, 99, 128, 129, 418, 522, 523*
Gorbach, S., *407*
Gordon, H. T., *346, 418*
Gore, T. S., *505, 521*
Gorsuch, T. T., *310*
Gottfried, S. P., 188 (189), 190 (189), *220*
Gottschalk, G., *365*
Gould, C. W., Jr., *99*
Gould, E. S., *146*
Goulden, F., *267*
Goulden, R., *265, 343*
Gouverneur, P., *310, 346, 407*
Govaert, F., *183*
Gränacher, C. H., *267*
Graham, I., *47*
Grail, G. F., *420*
Gran, G., *441*
Granatelli, L., *310*
Grangaud, R., *346*
Grant, C. L., *346*
Grant, G. A., *407*
Grant, J., 2 (61, 62), 5 (61, 62), *24,* 30 (34, 35), *47,* 50 (24, 25), 51 (24, 25), 60 (25), 71 (24, 25), 73 (24, 25), 76 (24, 25), *80,* 88 (49, 50), *99,* 102 (83, 84), 103 (84), 104 (83, 84), 116 (84), *129,* 133 (90, 91), 134 (90, 91), 137 (90, 91), *146,* 151 (58, 59), 152 (58, 59), 161 (58, 59), 163, 166 (58, 59), 168 (58, 59), 171 (58, 59), 174 (58, 59), 175 (58, 59), 177 (58, 59), *183,* 188 (73, 74), 189 (73, 74), 190 (73, 74), 196 (73, 74), 199 (73, 74), *217,* 222 (112, 113), 224 (112), 228 (112, 113), 230 (112, 113), 233 (112), 238 (112, 113), 241 (112, 113), 243 (112, 113), 244 (112, 113), 247 (112, 113), 248 (112, 113), 250 (112, 113), 252 (112, 113), 255 (112, 113), 257 (112, 113), *268,* 276 (76, 77), 295 (76, 77), 296 (76, 77), 297 (76, 77), *310,* 316 (139, 140), 317 (139, 140), 324 (139, 140), 325 (139, 140), *346,* 356 (41, 42), 359 (42), *365,* 367 (23, 24), 370 (23, 24), 372 (23), *375, 407,* 412, *418,* 422 (50, 51), 423 (50, 51), 424 (50, 51), 433 (50, 51), 434 (50, 51), 437 (50, 51), *441, 452, 492, 506,* 508 (79, 80), 509 (79, 80), 510 (79, 80), *523,* 528 (29, 30), 531 (29, 30), 532 (29, 30), 535 (29, 30), *545,* 551 (37, 38), 560 (38), *563*
Grant, W. M., *418*
Granville, W. C., *526*
Grassmann, W., *418*
Grassner, F., *310*
Graue, G., *310, 419*
Graus, B., *144, 417*
Green, I. J., *351*
Green, L. G., *217*
Gregg, J. R., 196 (10), *215*
Gregory, R., *521*
Grekov, A. P., *524*
Griffel, M., *312,* 316 (219), *348*
Griffiths, V. S., *523*
Grim, E. C., *147, 312*
Grimberg, J., *506*
Grimes, M. D., *182*
Grodsky, J., *346*
Grogan, C. H., *146*
Gromakova, L. M., *182, 265*
Gross, W. G., *346*
Grosse, A. V., *407*
Grote, W., 276 (80), *310,* 316 (143, 144), *346*
Grube, G., *407*
Gruber, K., *442*
Gruber, W., *453,* 565, *602*
Grünberger, D., *268*
Grützner, R., *407, 523*
Grunbaum, B. J., *187*
Grunbaum, B. W., *129, 184, 217, 418, 506*
Grutsch, J. F., *506*
Guagnini, O. A., *527*
Guba, F., *314*
Gubanova, A. V., *275, 353*
Gubser, H., *418*

Guerrant, G. O., *545*
Guerrero, A. H., *313*
Guldner, W. G., *523*
Gullberg, J. E., 31 (44), *48*
Gusev, S. I., *146*
Guss, H. C., *183*
Gustin, G. M., 174, 175, *183*
Gutbier, B., *309, 343*
Gutmann, H. R., *272*
Gutterson, M., *523*
Gutzeit, M., *375*
Guzman, G. M., *268*
Gwirtsman, J., 103 (127), 105 (127, 137), 112 (127, 137), 113 (127, 137), *130,* 328 (268), 332 (248), 334 (248), *346, 348, 349, 353*
Gysel, H., *24, 25, 80, 183, 346, 441, 523, 545,* 565, *602*

Haack, A., *219*
Haagen-Smit, A. J., 1 (65), *25, 506*
Haar, K. ter, *128*
Haas, P., *523*
Haberland, G., *442, 526*
Habib-Labib, G., *99*
Hack, M. H., *523*
Haden, W. L., *523*
Hadorn, H., *352*
Haendler, H. M., *99, 346*
Hafliger, O., *521*
Hagen, G., *313*
Haight, G. P., *146, 375*
Hainberger, L., *269, 342*
Haines, R. L., *407*
Haines, W. E., *523*
Haldane, J. S., *492, 505*
Hale, A. H., *265*
Hale, C. H., *217*
Hale, M. N., *217*
Hales, J. L., *48*
Hall, J. F., Jr., *526*
Hall, J. I., *346*
Hall, R. T., 103, *127,* 377 (5), 379 (5), 383 (5), 392 (5), 397 (5), 398, 402 (5), *407, 452, 523*
Hall, W. T., 102 (215), *132,* 133 (235), *149,* 316 (399), *353,* 354 (95), *366,* 367 (61), 370 (61), *376,* 567 (13), *602*
Hallett, L. T., 3 (148, 149, 150), *25, 27,* 30 (37), *48,* 53 (28), 56 (28), 76 (28), *80,*
84 (52), *99,* 104 (87), *129,* 151 (65), 155 (65), 161 (65, 167, 168), *183, 186, 217,* 235 (115, 116), 236 (116), *268, 269, 273,* 276 (81, 82), 277 (82), 278 (82), 296 (82, 83), 304 (82, 83), *310, 311, 313,* 316 (147, 148), *346, 350, 418, 523*
Halpern, A., *349*
Hamberger, K., *545*
Hambly, A. N., *127*
Hamill, W. H., *268,* 516, *523*
Hamilton, J. B., *441*
Hamilton, P., *421, 494*
Hammond, C. W., *418*
Hammond, G. S., *24, 145, 183, 417, 505, 522*
Hammond, W. A., *268*
Hanafusa, H., *218*
Hankes, L. H., 430 (53), 435 (53), *441, 492*
Hansen, A. E., *421*
Hapern, M., *144*
Hara, R., *129, 418*
Harada, S., *443*
Harand, J., 10, *23, 144*
Hardin, L. J., *366*
Harington, C. R., 454 (24), 488 (24), *492*
Harlay, V., *346*
Harlow, G. A., *418*
Harman, J. W., *129*
Harms, J., *346*
Harper, P. V., Jr., *268, 492*
Harris, C. C., 389 (4), 402 (4), *407*
Harris, T. H., 151 (181), 163 (181), *186*
Harris, W. W., *350*
Harrison, A. P., 190 (168), *219*
Harrison, G. E., *146*
Harrison, H. C., *146*
Harrison, H. E., *146*
Harte, R. H., 102 (53), 103 (53), *128, 217,* 276 (51), *310,* 316 (103), 323 (103), 324 (103), *345, 417,* 444, 445 (19), 447 (19), 449 (19), *452*
Hartley, O., 188 (80), *217*
Harvey, H. W., *217, 365*
Hasegewa, R., *346*
Haselbach, C. H., *215*
Hashmi, H. H., *346*
Haslam, J., *25, 346*
Hasselmann, M., *346*
Hassid, W. Z., *271*
Hastings, A. B., *131*

Haszeldine, R. N., *145*
Hauschildt, J. D., *148*
Hawkins, J. A., 488 (93), *494*
Hawkins, J. J., *545*
Hawley, R. L., *344*
Hawthorne, M. F., *149*
Hayami, T., *345*
Hayashi, N., *187*
Hayatsu, R., *146, 217*
Hayazu, R., *217, 311, 313, 366*
Hayman, D. F., 12 (70), *25,* 58 (31), 60, 63, 74 (29), *80,* 151 (66), *183,* 259 (120), *268*
Hazenberg, W. M., *268*
Head, E., *268*
Heald, P. J., *493*
Heap, R., *344*
Hecht, F., *25, 144, 146*
Heckly, R. J., *99*
Hedberg, K. W., 512 (180), 513 (180), *525*
Hegedüs, A. J., *146, 148, 365*
Heilbronner, E., *131*
Hein, F., *183*
Heine, V. E., *311, 346*
Heinemann, H., *268*
Heinrich, B. J., *182*
Heller, K., *311, 365, 375*
Hempel, W., *311*
Henlein, R., *351*
Hennart, C., *346*
Henne, A. L., *273, 346*
Hepner, L. S., *218*
Herb, S. F., *418*
Herbain, M., *183*
Herczeg, C., *344*
Herd, R. L., *146*
Hermance, H. W., *98*
Hernler, F., *146, 268, 271,* 316 (156), *346*
Heron, A. E., *183, 268,* 424 (54), *438, 441,* 535 (85), *547*
Herrington, E. F. G., *418*
Herrmann, J., *182, 266, 310, 345*
Herrmann, R., *146*
Hershberg, E. B., 160 (70), 172 (71), *183*
Hershenson, H. M., *493*
Hervier, B., *344*
Heslinga, J., 276 (151), *312, 408*
Hewitt, E. J., *563*
Heyde, W., *418*
Heyrovský, A., *146*

Hibbard, P. L., *146*
Hibbert, H., 514 (93, 94), *523*
Hickey, F. C., *492*
Hickman, K. C. D., *99*
Higashi, S., *146*
Higuchi, T., *523*
Hilbck, H., *563*
Hill, D. W., 276 (52), *310,* 316 (104), 323 (104), 324 (104), *345*
Hill, R. T., *216*
Hill, U. T., *418*
Hillenbrand, E. F., Jr., *493, 523*
Hiller, A., 488 (94), *494*
Hinkel, R. D., *407*
Hintermaier, A., *407, 523*
Hippenmeyer, F., *99, 563*
Hirai, M., *146, 217*
Hirayama, H., *80*
Hirooka, S., *268*
Hitchcock, A. E., *349, 353*
Hoagland, C. L., *146,* 276 (88), *311, 365,* 489 (30, 31), 490 (30,31), *493*
Hochenegger, M., *505*
Hochheiser, S., *311*
Hockenhull, D. J. D., *268, 493*
Hockstein, F. A., *523*
Hodecker, J. H., *149, 365, 376*
Hodgman, C. D., 93 (56), *99,* 111 (90), *129,* 133 (107, 108), 137 (107, 108), *146,* 168 (72), 169 (72), *183,* 224 (128), *268,* 370 (29), *375,* 410 (74), *418,* 567 (4, 5), *602*
Hodgson, W. G., *348*
Hodsman, G. F., *48, 80*
Höfling, E., *523*
Hoekstra, H. R., *407*
Höltje, R., 108 (74), *128, 146*
Hörmann, H., *523*
Hoesli, H., 236 (129), 238 (129), *268,* 368 (30), *375*
Hoffman, D. O., *441*
Hofstaeder, R., 540 (34), *545*
Hohenberg, E., *148*
Holasek, A., *523*
Holeton, R. E., *311*
Holley, C., *268*
Holmes, B., *146*
Holowchak, J., *183, 377, 407*
Holt, B. D., *268, 493*
Holter, H., 122 (120), *130, 215, 311, 523*

Honig, J. M., *47*
Honma, M., *183*
Hopkinson, F. J., *150*
Hopton, J. W., *418*
Horáček, J., *268, 311*
Horeischy, K., *268, 311*
Horiuchi, N., *346*
Horning, E. C., *268*
Horning, M. G., *268*
Horton, C. A., *353*
Hough, L., *521*
Houghton, A. A., *441*
Hoverath, A., *408*
How, A. E., *375*
Howard, H. C., *25, 526*
Howton, D. R., *506*
Hoyer, H., *545*
Hoyme, H., *364*
Hozumi, K., *183, 268, 563*
Hu, C.-C., *217*
Huang, R. L., *441*
Hubbard, D. M., *146, 346*
Hubbard, R. L., *523*
Huber, H., *418, 493*
Huber, W., *407*
Hudson, C. S., *452*
Hudy, J. A., *311*
Hülsen, W., *346*
Huffman, E. W. D., 2 (177), 3 (103, 140, 175-184), *26, 27, 28,* 30 (76), 31 (76), *48, 49,* 53 (50), 54 (50), 56 (50), 58 (63, 64), 59 (63, 64), 60 (63), 62 (63), 63 (63), 64 (63), 65 (63), 66 (63), 67 (63), 68 (63, 64), 69 (64), 70 (64), 73 (63, 64), 74 (63), 75 (63, 64), *81, 82,* 83 (118, 119, 120), 88 (121), 91 (121), *101,* 113 (195, 196, 197), 116 (195, 196, 197), *132,* 134 (224, 225), *149,* 160 (189, 190), 161 (190), 171 (190), 172 (190), *186, 187,* 192 (194), 193 (194), 195 (194), 196 (194), 199 (194), 200 (194), 211 (194), *220,* 222 (146), 225 (351), 228 (351), 230 (351), 234 (146), 235 (146), 238 (351), 240 (351), 255 (351), 257 (351), 258 (351), *268, 274,* 276 (94), 278 (211), 280 (211), 281 (211), 287 (212), 297 (212), 298 (212), 302 (212), *311, 314,* 317 (380, 381), 318 (381), 319 (381), 322 (381), 324 (381), *352,* 356 (91), 360 (90), *366,* 368 (58), 369 (58), 370 (59), 371 (59), *376,* 393 (90), 401 (90), *409,* 423 (101), 426 (101), 429 (101), 431 (101), 437 (101), *443,* 447 (45, 46), *453,* 558 (88), 559 (88), *564*
Hughes, J. C., 162 (81), 168 (81), *184*
Hull, D. E., *48, 80*
Hullin, R. P., *418*
Hume, D. N., *128, 493*
Hunter, G., *146, 346*
Hunter, G. J., *348*
Hunter, H., 56 (20), *80*
Hunter, I. R., *523*
Hunter, W. H., *418*
Huntress, E. H., 93 (57), *99*
Hurka, W., *418, 493*
Hussey, A. S., *217, 268, 493*
Hutton, G. H., *274*

Ide, W. S., *184*
Ieki, T., *80, 271, 272, 312, 452*
Ignatenko, L. S., *183, 267*
Ikeda, S., *146*
Imaeda, K., *183, 268, 408*
Inada, M., *184*
Ingalls, E. D., 14 (187), 15 (187), 19 (187)
Ingber, N. M., *312*
Inglis, A. S., *311, 346, 441, 452*
Ingold, W., *129, 418*
Ingram, G., *25, 48, 184,* 230 (27), *259, 265, 268, 269, 309, 311, 343, 346*
Intonti, R., *346*
Ionescu, M., *346*
Irimescu, I., *269, 311,* 316 (166), *346*
Iritani, N., *311*
Irlin, A. L., *99*
Irving, L., *131, 526*
Ishibashi, M., *146*
Ishihara, K., *345*
Ishii, M., *185*
Isobe, I., *408*
Israelstam, S. S., *265, 269*
Itai, T., *346*
Ivanov, C. P., *314*
Ivasheva, N. P., *187*
Ivashova, N. P., *183, 310, 345*
Iwamoto, K., *545*
Iwasaki, I., *346*
Iwasaki, K., *184*

Jackson, R. P., *267*
Jacobi, M., *147*
Jacobs, M. B., *217, 219, 311, 375*
Jacobs, R. B., *269*
Jadot, J., *408*
Jakl, F., *346*
James, A. T., *418*
Jan, J., *523*
Jančik, F., *523*
Jander, G., *346*
Jelley, E. E., 560, *563*
Jenden, D. J., *217*
Jenik, J., *146, 346, 365, 375*
Jensen, C. E., *545*
Jensen, R., *146*
Jerie, H., *129, 418*
Jett, L. M., *273*
Johanson, R., *217*
Johansson, A., *99, 269*
Johansson, G., *418*
John, H. J., 188 (215), *220*
Johncock, P., *346*
Johns, I. B., *269, 506*
Johnson, B. H., *347*
Johnson, B. L., *523*
Johnson, C. A., *347*
Johnson, C. M., *311*
Johnson, J. R., 514, *522*
Johnson, M. J., *419*
Johnson, W. B., *99*
Johnson, W. H., *132*
Johnston, C. C., 276 (213), 277 (213), 278 (213), 291 (213), 292 (213), *314*
Johnston, C. K., *182*
Johnstone, H. F., *310*
Jolley, E. L., 2 (177), 3 (175-180), *28,* 58 (63), 59 (63), 60 (63), 62 (63), 63 (63), 64 (63), 65 (63), 66 (63), 67 (63), 68 (63), 73 (63), 74 (63), 75 (63), *81,* 83 (118, 119, 120), 88 (121), 91 (121), *101,* 113 (195, 196, 197), 116 (195, 196, 197), *132,* 133 (224), *149,* 278 (211), 280 (211), 281 (211), *314,* 317 (380), *352,* 360 (90), *366,* 368 (58), 369 (58), *376,* 423 (101), 426 (101), 429 (101), 431 (101), 437 (101), *443,* 558 (88), 559 (88), *564*
Jones, A. R., *81*
Jones, A. S., *311*
Jones, S. L., *266, 492*
Jones, W. H., *217,* 389 (4), 402 (4), *407, 408*
Joshi, M. K., *523*
Jucker, A., *523*
Judah, J. D., *347*
Juliard, A. L., *347*
Jung, G. F., *352*
Jungnickel, J. L., *100, 506, 523*
Jungreis, E., *144, 347*
Juránek, J., *269*
Jurány, H., *347*
Jureček, M., *146,* 316 (178), *346, 347, 365, 375, 418, 523*
Jurinka, A., *99*
Juvet, R. S., *269, 523*

Kabasakalian, P., *523*
Kacser, H., *563*
Kadowaki, H., *146*
Kagan, B. M., *418*
Kahane, E., *25, 146*
Kahovec, L., *441*
Kaimowitz, I., *312*
Kainz, G., *184, 217, 269, 311, 347, 408, 418, 452, 493, 523, 524*
Kaiser, E., *418*
Kakabadse, G. J., *130*
Kamada, H., *131*
Kamm, O., 93 (61), *99,* 410, *418,* 528 (37), 529 (37), 533 (37), *545,* 548 (46), 552 (46), 553 (46), 554 (46), *563*
Kan, M., *146, 267*
Kao, S. S., *184*
Karchmer, J. H., *524*
Karrer, P., *522*
Karrman, K. J., *146, 309, 418*
Karten, B. S., *524*
Kasagi, M., *269*
Kasler, F., *418, 493*
Kassner, E. E., 108 (93), *129*
Kassner, J. L., 108 (93), *129*
Kathman, G. G., *521*
Katz, J. J., *407, 409*
Katz, M., *407*
Kaufmann, O., *342*
Kaup, R. R., 103 (199), 106 (199), 112 (199), 113 (199), *132,* 189 (195), 190 (195), 191 (195), 192 (195), 210 (195), 220, 326 (385), 327 (385), 331 (385), 332 (385), *352,* 485 (81), 488 (81), *494*

Kavai, M. Z., 144
Kawane, M., 345, 409
Kawano, T., 269
Kay, H., 269
Kaye, I. A., 103 (94), 129, 217
Keen, R. T., 493, 524
Keister, M. L., 127
Kemmerer, G., 217, 269
Kent, P. W., 311
Kenty, C., 524
Kepner, R. E., 524
Ketchum, D. F., 270, 524
Kholodkovskaya, K. B., 312
Kiba, T., 311
Kies, H. L., 524
Kieselbach, R., 184, 524
Kilpatrick, M. D., 219, 316 (420), 353
Kimball, R. H., 311, 347
Kimbel, K. H., 146
Kimoto, K., 132
King, A., 441
King, E. J., 365
King, G. B., 216
King, J. G., 408
King, R. W., 217
Kingsley, A., 103 (109), 129, 217
Kingsley, G. R., 146, 375
Kinnunen, J., 129, 146
Kinoshita, S., 183
Kinsey, D., 103 (109), 129, 217
Kinsey, R. H., 419
Kinsey, V. E., 129
Kinsman, S., 441
Kirby, H., 269
Kirk, E., 365, 366, 488 (95), 490 (46, 99), 493, 494
Kirk, P. L., 2 (77), 3 (103), 25, 26, 31 (44), 48, 81, 99, 123, 128, 129, 130, 132, 145, 146, 147, 188 (95, 98), 217, 269, 344, 365, 366, 418, 494, 506, 526, 563
Kirkland, J. J., 344
Kirner, W. R., 10, 25, 30 (45), 48, 149, 222 (174, 175), 269, 273, 408
Kirpal, A., 442
Kirshenbaum, A. D., 407
Kirsten, W., 151 (93), 155 (93), 161 (93), 175, 184, 217, 259, 269, 311, 365, 377, 408, 418, 524

Kirsten, W. J., 25, 129, 184, 217, 259, 269, 311, 316 (193, 197), 347, 365, 442, 524
Kishi, I. (also Akaza, I.), 311
Kislitsin, P. S., 145
Kissa, E., 25, 269
Klatzkin, C., 418
Klein, B., 147, 311
Klein, E., 347
Klein, F., 407
Klein, G., 84 (63), 99
Kleinzeller, A., 514 (125), 524
Klimenko, V. G., 311
Klimova, V. A., 147, 184, 269, 270, 311, 347, 365, 524
Klingmüller, V., 217
Klipp, R. W., 407
Knick, H., 522
Knight, C. A., 99
Knights, E. M., Jr., 129
Knižáková, E., 270
Knoll, A., 424 (64), 442
Knop, J., 147
Ko, R., 417
Kobayashi, E., 366
Koch, C. W., 184, 217, 527
Koch, F. C., 104 (106), 129, 411 (90, 91), 418
Koch, M. J., 492
Koch, P., 492
Koch, W., 270, 311
Kocher, C., 560 (30), 563
Koegel, R., 133 (176), 134 (176), 136, 137, 148
Kögler, H., 523
Koepsell, H. J., 418
Körbl, J., 259, 268, 270, 273, 311, 521, 523
Kofler, A., 548 (49, 51), 551 (49-52), 563
Kofler, L., 545, 548 (49, 51), 551 (48-52), 563
Kofler, W., 563
Kohler, E. P., 524
Kohn, M., 25
Kojima, R., 347
Kojima, T., 345
Kokotailo, G. T., 347
Kolb, J. J., 524
Kolka, A. G., 442
Kolsek, J., 523
Kolthoff, I. M., 26, 83 (65), 99, 144, 347, 375, 418

Komárek, K., *48*
Komers, R., 259 (201), *270*
Kometiani, P. A., *419*
Komori, S., *506, 524*
Kondo, A., *147,* 311, *347*
Kondo, H., *271, 312*
Konig, O., *147*
Konigsberg, M., *453*
Konikov, A. P., *493*
Kono, T., *25, 184,* 311, *408*
Konovalov, A., *184, 217, 347*
Korach, S., *146*
Koren, H., *311*
Korenman, I. M., *48, 129, 147*
Koretsky, H., *349*
Koros, E., *129*
Korshun, M. O., *25, 147, 183, 184, 267, 269, 270,* 311, *312, 345, 346, 347, 365,* 377, *408*
Kotionis, A. Z., *421*
Kotlyarov, I. I., *217*
Kottmeyer, G., *419*
Kováks, E., *131*
Koyama, C., *452*
Kozawa, A., *147*
Krahl, M. E., *146*
Krainick, H. G., *270, 353, 443*
Kramer, H. P., *348*
Kratzl, K., *442*
Krause, R. A., *525*
Kreisky, F., *147*
Krekeler, H., 276 (80), *310,* 316 (143, 144, 216), *346, 348*
Kreshkov, A. P., *147*
Kress, K. E., *312*
Kröcker, F., *147, 184*
Krogh, A., 515, *522, 524, 563*
Kubelkova, O., *147*
Kubo, M., *313*
Kubota, B., *545*
Kubota, H., *348*
Kuchler, L., *524*
Kuck, J. A., 2 (177), 3 (140, 148, 149, 150, 175-184), *25,* 27, 28, 30 (49, 76), 31 (76), 34 (49), *48, 49,* 53 (50), 54 (50), 56 (50), 58 (63, 64), 59 (63, 64), 60 (63), 62 (63), 63 (63), 64 (63), 65 (63), 66 (63), 67 (63), 68 (63, 64), 69 (64), 70 (64), 73 (63, 64), 74 (63), 75 (63, 64), *81, 82,* 83 (118, 119, 120),

88 (121), 91 (121), *101,* 103 (109), 113 (195, 196, 197), 116 (195, 196, 197), *129, 132,* 134 (224, 225), *147, 149,* 160 (189, 190), 161 (167, 168, 190), 171 (190), 172 (190), *184, 186, 187,* 192 (194), 193 (194), 195 (194), 196 (194), 199 (194), 200 (194), 211 (194), *217, 220, 225* (351), 228 (351), 230 (351), 235 (212), 238 (351), 240 (212, 351), 255 (351), 257 (351), 258 (351), *270, 273, 274,* 276 (160), 277 (160), 278 (211), 280 (211), 281 (211), 287 (212), 290 (160), 297 (212), 298 (212), 302 (212), *312, 313, 314,* 316 (219, 292), 317 (380, 381), 318 (381), 319 (381), 322 (381), 324 (381), *348, 349, 350, 352,* 356 (91), 360 (90), *366,* 368 (58), 369 (58), 370 (59), 371 (59), *376,* 393 (90), 401 (90), *408, 409,* 423 (101), 426 (101), 429 (101), 431 (101), 437 (101), *442, 443,* 447 (45, 46), *453,* 558 (88), 559 (88), *564*
Küster, W., *442*
Kuffner, F., 565, *602*
Kugel, V. H., *220,* 488 (96), *494*
Kuhn, R., *147,* 348, *442, 452, 506, 524*
Kuipers, J. W., 104 (87), *129,* 276 (82), 277 (82), 278 (82), 296 (82, 83), 304 (82, 83), *310,* 311
Kul'berg, L. M., *419*
Kulkarni, R. S., *521*
Kumov, V. I., *146*
Kumpan, P., *270*
Kunori, M., *348*
Kunz, A., *452*
Kupletskaya, N. B., *527*
Kurihara, B., *270*
Kurtz, L. T., *99*
Kusaka, Y., *348*

Labout, J. W. A., *563*
Lachele, C. E., *375*
Lachiver, F., *348*
Lacourt, A., 62 (21), *80, 129, 147, 270, 348, 408, 419*
Lada, Z., *217*
La Force, J. R., *270*
Lake, G. R., *217*
Lambert, J. L., *353*
Lambert, R. H., *147*
Lamberts, J., *523*

Author Index

Lamm, G. G., *147*
Lamo, M. A. de, *25, 312*
Lancaster, J. E., *348*
Lane, E. S., *129*
Lane, M., *348*
Lang, C. A., *217*
Lang, K., *419*
Lang, R., *147*
Lang, R. E., *189* (116), *218*
Lange, N. A., *93* (56), *99, 111* (90), *129, 133* (108), *137* (108), *146,* 370 (29), *375,* 410 (74), *418,* 567 (5, 8), *602*
Langejan, M., *442*
Langer, A., *270*
Langer, S. H., *129*
Langmuir, I., *524*
Laniece, M., *351*
Lanik, A., *342*
Lapin, L. N., *147, 348*
Lappin, G. R., *99*
Lardy, H. A., *419*
Larina, N. I., *346*
Larsen, R. P., *312*
Lascalzo, A. G., *129*
Lashof, T. W., 3 (103), *26, 48, 81*
Latimer, R. A., *184*
Lauer, K. J. H., *266*
Lauer, W. M., *507, 527*
Laurent, J., *181*
Lauro, M. F., *188* (109), *218*
Laustriat, G., *346*
Lavrovskaya, E. V., *147*
Lawson, G. J., *309*
Lazarow, A., *129*
LeBlanc, N. F., *522*
Lee, F. A., *524*
Lee, T. S., *270, 408*
Lees, M. B., *312*
Lefferts, D. T., *348*
Léger, J., *310*
Leighton, P. A., *526*
Lein, A., *348*
Leipert, T., 316 (230), *348*
Leisey, F. A., *506*
Leithe, W., *184, 348*
Lemp, J. F., *348*
Lennartz, T. A., *524*
Lenz, W., 31 (52), *48*
Leonard, G. W., Jr., *147*
Leopold, B., *216*

Leroux, J., *147, 348*
LeRoy, D. J., *524*
Lesbre, M., *266*
Lescher, V. L., *270*
Lesesne, S. D., *99*
Lestra, H., *218*
Letham, D. S., *311*
Lethco, E., *146*
Leurquin, J., *218, 524*
Levene, P. A., 475 (48), *493*
Levin, B., *218*
Levin, H., *100, 527*
Levina, S. Ya., *275*
Levine, J., *452*
Levvy, G. A., *129, 375*
Levy, A. M., *546*
Levy, G. B., 92 (69), 94 (69), 95, *99*
Lévy, R., *25, 130, 184, 270, 312, 344, 348, 365, 375*
Lewis, J. B., *408*
Lewis, L. G., *269*
Lewis, R. P., *147*
Lewis, W. B., *343*
Ley, J. de, *493*
Liandier, L., *309*
Lieb, H., 1 (96), *25, 99, 270, 352, 452, 523, 524, 526*
Liebhafsky, H. A., *265*
Lieff, M., *442*
Ligett, W. B., 316 (106), *345, 377, 407*
Linch, A. L., *311*
Lincoln, P. A., *524*
Lincoln, R. M., 276 (136), *312*
Lindenbaum, A., *493*
Linderstrøm-Lang, K., 122 (120), *130*
Lindner, J., *100, 130, 270, 271*
Lindner, R., *147, 365*
Lindstrom, O., *147*
Lingane, J. J., *348*
Linhardt, K., *419*
Lipke, J., *345, 364*
Lipscomb, W. N., *147*
Littman, B., 104 (198), 105 (198), 110 (198), *132,* 276 (214), 277 (214), 278 (214), 282 (214), 283 (214), *314*
Litvenenko, L. M., *524*
Liu, S. H., 488 (72), *494*
Llacer, A. J., *130, 147, 218*
Lochte, H. L., *99, 545*
Löffler, J. E., *419*

Loeschauer, E. E., 4 (188), *28, 443, 453*
Loewenstein, E., 30 (49), 34 (49), *48*
Logun, J., *524*
Lohr, L. J., *348*
Long, M. D. C., *219, 220*
Lonsdale, M., *269*
Lopez, J. A., *419*
López Capont, F., *219*
Lorenz, N., *365*
Lortz, H. J., *442*
Loscalzo, A. G., *81*
Loseva, K. T., *351*
Lothe, J. J., *348*
Lothringer, R. L., *442*
Lott, P. F., *130*
Løvtrup, Søren, *311, 523*
Lowry, O. H., *419*
Lubochinsky, B., *218*
Ludemann, W. D., *366*
Lundbak, A., *130*
Lunde, G., *271,* 316 (244), *348*
Luskina, B. M., *220, 274, 352*
Luttropp, E. S., *523*
Lykken, L., *26, 130,* 158 (56), *183, 274*
Lysyj, I., *271, 312, 348*

Ma, T. S., *25, 26, 81, 100, 101,* 103 (127, 128), 106 (127), 112 (127), 113 (127), *130, 184,* 189 (116), 190 (117), 191 (117), *218, 271, 312,* 326 (246, 247), *328,* 332 (248), 334 (247, 248), *348, 349, 365, 419, 442, 523, 524, 525, 546*
Maag, W., *442*
Macdonald, A. M. G., *144, 182, 265, 271, 309,* 334 (32, 39), *343, 348, 364*
MacFadyen, D. A., *421, 494*
Macheboeuf, M., *344*
Machemer, P. E., 190 (118), *218*
MacIntire, W. H., *366*
MacNevin, W. M., *147,* 253 (252), *258, 271,* 316 (252), *348*
MacNulty, B. J., *348*
Macurdy, L. B., 3 (103), *26, 48, 81*
Mader, W. J., *130, 348, 420*
Maffei, F. J., *563*
Maffett, P. A., *147, 348*
Magee, R. J., *545*
Magnuson, H. J., *374*
Mahoney, J. F., *100,* 276 (142), *312, 563*
Mair, R. D., *311*

Makarova, V. P., *147*
Makens, R. F., *442*
Makeš, J., *527*
Makineni, S., *349*
Makovetskii, P. S., *312*
Malissa, H., 2 (105), *26, 100, 130, 270, 271, 311, 312*
Malmberg, E. W., *525*
Malmstadt, H. V., *127, 349*
Maltagliati, M., *309*
Manci, C., *409*
Mandl, J., *149*
Mangeney, G., *184, 271*
Mangravite, R., *348*
Mann, W., *407*
Mannelli, G., *419*
Manohin, B., *130*
Marberg, C. M., *100*
Marcali, K., *218, 525*
Maresh, C., *274*
Margolis, E. I., *271, 312*
Margosis, M., *130, 419*
Marier, J. R., *147*
Marino, S. P., 196 (120), *218*
Markert, F., *453*
Marks, C., *442*
Marks, M. E., *408*
Markunas, P. C., *420*
Marrian, G. F., 514 (158), *525*
Marrian, P. M., 514 (158), *525*
Mars, S., *265*
Marsh, G. E., *349*
Marsh, O., *130*
Martin, A. J., 292 (146), *312*
Martin, A. J. P., *418, 419*
Martin, F., *48, 147, 349, 443*
Martin, G., *147*
Martin, G. E., *344*
Martinek, A., *562*
Maruyama, M., *349*
Marzadro, M., *218*
Mason, A. C., *130*
Mason, C. W., 551 (14), *562*
Massenberg, S., *563*
Massie, W. H. S., *312*
Massingham, W. E., *352*
Massoni, C. J., *148*
Mastrangelo, S. V. R., *563*
Matchett, J. R., *452*
Mathers, A. P., *442*

Mathieu, P., *184*
Mattenheimer, H., *130*
Matthews, F. W., *563*
Matuszak, M. P., *349*
Maurer, J. E., *217, 493*
Maurer, W., *346*
Maurmeyer, R. K., *100, 130, 349, 419*
Maute, R. L., *419, 525*
Mavrodineanu, R., 106 (137), 112 (137), 113 (137), *130,* 328 (268), *346, 349*
May, L. G., *521*
Mayer, A. M., 188 (189), 190 (189), *220*
Maylott, A. O., *408*
Mayrhofer, A., 551 (52), *563*
Mázor, L., *271, 345, 349,* 377 (60), *408, 452*
Mazzella, P. P., *524*
McAlpine, I. M., *525*
McArdle, B., *419*
McBeth, C. H., *218*
McCaldin, D. J., *419, 493*
McCann, D. S., *315*
McChesney, E. W., *312*
McClendon, J. F., *419*
McComb, E. A., *452*
McCorkindale, W., *349*
McCoy, J. S., 276 (160), 277 (160), 290 (160), *312,* 316 (292), *349*
McCoy, R. N., *271*
McCready, R. M., *271, 452*
McCrone, W. C., 551 (60), *563*
McCulloch, F. W., *383, 409*
McCullough, J. D., *521*
McCullough, J. P., *563*
McCurdy, W. H., Jr., *131*
McDonald, E., *100*
McDonald, R. R., *129*
McElroy, W. R., *267*
McFarren, E. F., *419*
McGee, B. E., *28,* 189 (195), 190 (195), 191 (195), 192 (195), 210 (195), *220*
McGregor, A. J., *147*
McHard, J. A., *147, 442*
McHargue, J. S., *346*
McIlwraith, C. G., *26*
McKenna, F. E., *274,* 316 (394), *349, 352*
McKennis, H., Jr., *525*
McKenzie, H. A., *218*
McKinley, J. D., Jr., 189 (116), *218, 365*
McKinney, R. W., *419*

McNabb, W. M., *81, 149,* 190 (118), *218, 376*
McNally, M. J., 383 (91), 389 (91), *409*
McQuillen, A., *562*
Mead, J. F., *506*
Megregian, S., *130, 349*
Mehlig, J. P., *147*
Mehta, R. K. A., *271*
Meisel, T., *345, 452*
Meisenheimer, J., 514 (162), *525*
Meixner, A., *147, 184*
Mellon, M. G., *344*
Mellor, J. W., 133 (165), *148,* 223 (262), 291, 295 (149), *271, 312,* 316 (275), *321, 349,* 424 (76), 436 (76), *442,* 475 (53), *493*
Melville, D. B., 514, *525*
Menkyna, M., *349*
Menschfreund, D., *349*
Merikanto, B., *146*
Merkulova, E. N., *269*
Merlin, E., *346*
Merritt, C., Jr., *182*
Merz, W., *342, 349, 365, 375, 523*
Messmer, A., *24*
Meulen, H. ter, 276 (150, 151), *312, 408*
Meyer, E., *545*
Meyer, F., *100*
Meyer, H., 93 (79), *100*
Meyer, R., *270, 408*
Meyers, E. E., *182*
Meyrowitz, R., *148*
Michel, O., *349*
Michell, J. H., 276 (142), *312*
Michod, J., *344*
Middeldorf, R., *524*
Midgley, T., Jr., *273*
Mikl, O., *312*
Miksch, R., *326, 349*
Miles, S. H., *315*
Mileur, R., *267*
Millard, E. B., 528 (41), 529 (41), 543 (41), *546*
Miller, C. W., *366*
Miller, D. M., *184*
Miller, H. S., *218*
Miller, J. I., *366*
Miller, J. W., *506*
Miller, S. P., *218*
Miller, V. L., *148*

Mills, J. A., *419*
Milner, G. W. C., *26*
Milner, O. I., *130, 218, 312*
Milner, R. T., 88 (80), *100,* 151 (120), 163 (120), *184*
Milton, R. F., *26,* 50 (41, 42), 51 (41, 42), 73 (41, 42), 74 (41, 42), *81,* 88 (81, 82), *100, 130, 148,* 151 (121, 122), 152 (121, 122), 158 (121, 122), 161 (121, 122), 174 (121, 122), *184, 185,* 188 (130, 131), *218, 271,* 276 (155, 156), *312, 349, 365,* 367 (43, 44), *375, 408, 419, 442, 452, 493, 506, 525, 546*
Miocque, M., *525*
Miroshina, V. P., *275*
Mitchell, H. L., *144, 182*
Mitchell, J., Jr., *26,* 92 (84, 111), 94 (84, 111), *100, 101, 272, 407, 419, 420, 421, 525, 526*
Mitsui, T., *48, 81, 185, 271, 349, 409*
Mittlemann, R., *419*
Mitulski, J. D., *182*
Miura, H., *148*
Miwa, M., *366*
Miyahara, F., *149*
Miyahara, K., *80, 185*
Mizukami, S., *185, 271, 272, 312, 452*
Mladenović, S., *149, 376*
Möhlau, E., *218*
Möhle, W., *493*
Moelants, L., *26*
Moelants, L. J., *26, 272, 408*
Möller, E. F., *506*
Moga, V., *272*
Moll, H., *407*
Momose, T., *272*
Monand, P., *312, 349*
Money, F. S., *525*
Monge-Hedde, M. F., *345*
Monkman, J. L., *147, 348*
Montague, B. A., *419*
MonteBovi, A. J., *349*
Mooi, J., *409*
Moore, L. A., 316 (398), *352*
Moore, R. W., *314*
Moore, V. A., *182*
Moore, W. A., *348*
Moran, J. J., 2 (177), 3 (175-184), *28,* 58 (63, 64), 59 (63, 64), 60 (63) 62 (63), 63 (63), 64 (63), 65 (63), 66 (63), 67 (63), 68 (63, 64), 69 (64), 70 (64), 73 (63, 64), 74 (63), 75 (63, 64), *81, 82,* 83 (118, 119, 120), 88 (121), 91 (121), *101,* 113 (195, 196, 197), 116 (195, 196, 197), *132,* 134 (224, 225), *149,* 160 (189, 190), 161 (190), 171 (190), 172 (190), *186, 187,* 192 (194), 193 (194), 195 (194), 196 (194), 199 (194), 200 (194), 211 (194), *220,* 225 (351), 228 (351), 230 (351), 238 (351), 240 (351), 255 (351), 257 (351), 258 (351), *274,* 278 (211), 280 (211), 281 (211), 287 (212), 297 (212), 298 (212), 302 (212), *314,* 317 (380, 381), 318 (381), 319 (381), 322 (381), 324 (381), *352,* 356 (91), 360 (90), *366,* 368 (58), 369 (58), 370 (59), 371 (59), *376,* 393 (90), 401 (90), *409,* 423 (101), 426 (101), 429 (101), 431 (101), 437 (101), *443,* 447 (45, 46), *453,* 558 (88), 559 (88), *564*
Morgan, D. J., *522, 525*
Morgan, G. T., *272*
Morgan, P. W., *442, 525*
Morgulis, S. M., *218*
Morikofer, A., *131*
Morita, N., *272, 563*
Moriya, K., *48*
Moroni, E., *98*
Morris, R. T., *547*
Morrison, G. H., *417*
Morrison, G. R., *220*
Morrison, J., *525*
Morrison, J. D., *342*
Morsingh, F., *441*
Morton, A. A., *100, 563*
Morton, E. S., *187*
Morton, J. J., *546*
Moser, L., 326, *349*
Moss, H., *525*
Mott, R. A., *218*
Moubasher, R., *419, 493*
Moureau, H., *344*
Moyer, H. V., *344*
Müller, E., *272*
Müller, G., *48*
Müller, R. H., 45 (61), *48, 272, 563*
Mukai, Y., *272*
Mulay, V. V., *186*

Mulligan, G. C., 162 (133), 168 (133), *185*
Mulliken, S. P., 93 (57), *99*
Muramatsu, H., *347*
Muramoto, Y., *527*
Murati, I., *144*
Murtaugh, J. J., 92 (69), 94 (69), 95, *99*
Murty, G. V. L. N., *349*
Musgrave, W. K. R., *181, 182, 343, 344, 346*

Nagase, S., *347*
Nagel, R. H., 31 (67), *48*
Nagler, J., *375*
Nair, J. H., III, *419*
Nakai, H., *185*
Nakajima, M., *146*
Nakamura, G. R., *365*
Nakamura, N., *525*
Nakashima, T., *346*
Nakayama, M., *269*
Nara, A., 453
Narayana Rao, D., *132, 149, 353*
Narita, K., *185*
Nash, L. K., *525, 546, 564*
Natelson, S., *26, 130, 148, 344, 349, 365*
Naughton, J. J., *272*
Naylor, H., *349*
Neal, W. B., Jr., *268*
Neill, J. M., *421,* 454 (97), 464 (97), 488 (97), *494*
Nekhorocheff, J., *419*
Nemes, T., *273*
Németh, S., *349*
Nernst, W., 1, *26,* 31 (63, 64), *48, 546*
Nesh, F., *349*
Nessonova, G. D., *442*
Nettesheim, G., *564*
Neuberg, C., *525*
Neudorffer, J., *312*
Neumann, F., *442*
Neumayer, J. J., 540, 542 (48), *546*
Neustadt, M. H., *100*
Neville, O. K., *493*
Nichols, L., 560 (68), *564*
Nichols, M. L., 152 (137), 162 (137), 168 (137), 174 (137), 175 (137), *185, 349*
Nicksic, S. W., *349*
Nicloux, M., *272*

Niederl, J. B., 2 (119, 120), 5 (119, 120), 10, 12 (119, 120), 13 (119, 120), *26,* 30 (65, 66, 68), 31 (66, 67), 34 (65, 66), *48,* 50 (44, 45, 46), 51 (44, 45), 53 (44, 45), 54 (44, 45), 56 (44, 45), 57 (44, 45), 68 (44, 45), 70 (44, 45), 71 (44, 45), 73 (44, 45), 74 (44, 45), 75 (44, 45), 76 (44, 45), *81,* 88 (88, 89), *100,* 102 (144, 145, 147), 103 (144, 145), 104 (144, 145), 111 (145), 123 (145, 146, 147), *130,* 133 (171, 172, 173), 134 (171, 172, 173), 137 (171, 172, 173), *148,* 151 (139, 140, 141), 152 (139, 140, 141), 161 (139, 140, 141), 163 (139, 140), 166 (139, 140), 168 (139), 169 (139, 140), 171 (139, 140), 172 (139, 140), 174 (138, 139), 175 (139, 140), 177, *185, 187,* 188 (137, 138), 190 (137, 138), 196 (137, 138), 199 (137, 138), *218,* 222 (284, 285, 288), 224 (284, 285), 228 (284, 285), 230 (284, 285), 233 (284, 285), 235 (285), 238 (284, 285), 241 (284, 285), 243 (284, 285), 244 (284, 285), 247 (284, 285), 248 (284, 285), 250 (284, 285), 252 (284, 285), 255 (284, 285), 257 (284, 285), 259 (285), *272, 273,* 276 (160, 161, 162), 277 (160, 162), 283 (162), 287 (161, 162), 290 (160, 162), 295 (161, 162), 296 (161, 162), 297 (161, 162), 303 (161, 162), *312, 313,* 316 (292, 293, 294), 317 (293, 294), 323 (293, 294), 324 (293, 294), 325 (293, 294), 326 (294), *349, 350,* 354 (66, 67), 355 (66, 67), 356 (66, 67), 357 (66, 67), 359 (66, 67), *365, 366,* 367 (45, 46), 368 (45), 369 (45), 370 (45, 46), 372 (45-46), *375,* 410 (120, 121), 412, *419, 422* (82, 83), 423 (82, 83), 424 (82), 433 (82, 83), 434 (82, 83), 437 (82), *441, 442,* 445 (33, 34), 447 (33, 34), *452,* 508 (173, 174), 509 (173, 174), 510 (173, 174), 511 (173), 514 (173, 174), *525,* 528 (50, 51), 531 (50, 51), 532 (50, 51), 535 (50, 51), *546,* 558 (69), *564,* 565 (9, 10), 574, 575, 576, 577, 578, 579, 580, 581, 582, 583, *602*

Niederl, V., 2 (119, 120), 5 (119, 120), 10, 12 (119, 120), 13 (119, 120), *26,* 30 (65, 66), 31 (66, 67), 34 (65, 66), *48,* 50 (44, 45), 51 (44, 45), 53 (44, 45),

54 (44, 45), 56 (44, 45), 57 (44, 45), 68 (44, 45), 70 (44, 45), 71 (44, 45), 73 (44, 45), 74 (44, 45), 75 (44, 45), 76 (44, 45), *81,* 88 (88, 89), *100,* 102 (144, 145), 103 (144, 145), 104 (144, 145), 111 (145), 123 (145, 146), *130,* 133 (171, 172), 134 (171, 172), 137 (171, 172), *148,* 151 (139, 140), 152 (139, 140), 161 (139, 140), 163 (139, 140), 166 (139, 140), 168 (139), 169 (139, 140), 171 (139, 140), 172 (139, 140), 174 (139), 175 (139, 140), 177, *185,* 188 (137, 138), 190 (137, 138), 196 (137, 138), 199 (137, 138), *218,* 222 (284, 285), 224 (284, 285), 228 (284, 285), 230 (284, 285), 233 (284, 285), 235 (285), 238 (284, 285), 241 (284, 285), 243 (284, 285), 244 (284, 285), 247 (284, 285), 248 (284, 285), 250 (284, 285), 252 (284, 285), 255 (284, 285), 257 (284, 285), 259 (285), *272,* 276 (161, 162), 277 (162), 283 (162), 287 (161, 162), 290 (162), 295 (161, 162), 296 (161, 162), 297 (161, 162), 303 (161, 162), *312, 313,* 316 (293, 294), 317 (293, 294), 323 (293, 294), 324 (293, 294), 325 (293, 294), 326 (294), *349, 350,* 354 (66, 67), 355 (66, 67), 356 (66, 67), 357 (66, 67), 359 (66, 67), *365,* 367 (45, 46), 368 (45), 369 (45), 370 (45, 46), 372 (45-46), *375,* 410 (120, 121), 412, *419,* 422 (82, 83), 423 (82, 83), 424 (82), 433 (82, 83), 434 (82, 83), 437 (82), *442,* 445 (33, 34), 447 (33, 34), *452,* 508 (173, 174), 509 (173, 174), 510 (173, 174), 511 (173), 514 (173, 174), *525,* 528 (50, 51), 531 (50, 51), 532 (50, 51), 535 (50, 51), *546,* 558 (69), *564,* 565 (9, 10), 574, 575, 576, 577, 578, 579, 580, 581, 582, 583, *602*
Nielsen, H. M., *350*
Niemann, C., *131, 272, 492*
Nieuwenburg, C. J. van, *24, 148, 546*
Nieuwland, J. A., 316 (402), *353*
Nijkamp, H. J., 419
Nikelly, J. B., *130*
Nikolayeva, N. A., *352*
Nilsson, K., *129*
Nippoldt, B. W., 164 (142), *185, 267, 345*
Nishimura, A., *185*
Nishita, H., *311*
Nivens, R., 383 (91), 389 (91), *409*
Noble, E. D., *218*
Noble, R. L., *418*
Nógrády, G., *218*
Noller, C. R., *441, 506*
Nonowa, D. C., *148*
Norimasa, K., *313*
Northrop, J. H., *350*
Norton, A. R., 133 (176), 134 (176), 136, 137, *148,* 155 (169), 161 (169), 174, *186,* 236 (322), 253 (322), *273*
Novikova, E. N., *525*
Nozaki, T., *148*
Nunemakes, R. B., *272*
Nutman, M., *493*
Nutten, A. J., *23, 265, 309, 343, 350, 440*
Nystrom, R. F., *348*

Oberhauser, B. F., *272*
Oberholzer, V. G., *218*
Obtemperanskaya, S. I., *352*
O'Connor, D. J., *417*
Oda, N., *313*
Oelsen, W., *419*
Oemler, A. N., *272*
Oerin, S., *272*
Oetzel, M., *421, 494*
Ogawa, T., *525*
Ogg, C. L., 2 (177), 3 (175, 176, 177, 178, 180), 4 (124-129, 205, 206, 207), *26, 28, 29,* 58 (63), 59 (63), 60 (63), 62 (63, 71), 63 (63), 64 (63), 65 (63), 66 (63), 67 (63), 68 (63), 74 (63), 75 (63), *81, 82,* 83 (118, 119, 120), 88 (121, 127), 91 (121, 127), *101,* 103 (149, 225, 226), 104 (150), 105 (150), 110 (150), 113 (195, 196, 197), 116 (195, 196, 197), *130, 132,* 134 (224), *149,* 154 (213), *185, 187,* 188 (142-146, 214, 216-221), 189 (143, 218), 190 (143, 144, 214, 216, 218, 221), 191 (144, 214, 216, 218, 221), 192 (143, 144, 214, 216, 218), 205 (143, 221), 207 (221), *216, 218, 220,* 221 (383, 384), 230 (385), 235 (383, 384, 385), 236 (383, 384, 385), 253 (383, 384, 385), *272, 274, 275,* 276 (165-169), 277 (169), 278 (169), 282, 295 (165-168, 247), 302 (247), 304 (165-168), *313, 315,* 316 (299), 323 (299), *350,* 354 (69,

70), *366,* 378 (102), 390 (102), 393 (102), *409,* 423 (101), 426 (101), 429 (101), 431 (101), 437 (101), *443, 492,* 495 (32), 496 (32), 497 (32), 499 (32), *506, 507,* 512 (176), 514, *525,* 540 (55), *546,* 558 (88), 559 (88), *564*
Ohashi, S., *185, 272*
Oita, I. J., *185,* 377, 379 (65), 391 (65), 392 (65), 399 (65), *408*
Okáč, A., *272*
Okada, Y., *218*
Okamoto, J., *146*
Oldham, G., *98, 100*
Oliver, F. H., 377, 379 (66), 391 (66), 392 (66), 399 (66), *408*
Oliver, W. T., *375*
Olivier, S. C., *366*
Olleman, E. D., *26, 525*
Olsen, J. S., *349*
Olson, E. C., *350*
Olson, P. B., *267, 345*
Omoto, J. H., *216*
Ongley, P. A., *525*
Onishi, H., *148*
Oort, W. P. van, *563*
Opfer-Schaum, R., *564*
Orekhovich, V. N., *419*
Orestova, V. A., *352*
Orndorff, W. R., 514 (187), *526*
Ory, H. A., *313*
Osborn, R. A., 188 (148), 190 (148), *218*
Ose, S., *267, 310, 345*
Otter, I. K. H., *26, 185, 350*
Otting, W., *408*
Ottolenghi, L., *420*
Ottosson, R., *313*
Overell, B. T., *419*
Owen, W. S., *564*
Owens, M. L., Jr., *419, 525*
Ozawa, T., *346*

Padowetz, W., *182, 313*
Pääbo, K., *350*
Page, I. H., *366,* 490 (99), *494*
Pagel, H. A., *185*
Palmer, J. G., *375*
Pályi, E., *314*
Panicker, A. R., *272*
Pansare, V. S., *186, 527*
Pantages, P., *129*

Pápay, M., *345*
Park, T. O., *82*
Parker, J. A., *220*
Parkin, B. A., *185*
Parks, C. R., 188 (43), 190 (43), *216*
Parks, T. D., *26, 130, 185, 309, 342, 351,* 377, 379 (12), 391 (12), 392 (12), 398 (12), 399 (12), 400 (12), *407*
Parnas, J. K., *219*
Parnas, Ya. O., *219*
Parr, S. W., 276 (173), *313,* 316 (303) *350, 366*
Parrette, R. L., *506*
Parry, E. P., *148*
Parshikov, Y. I., *374*
Parsons, J. S., *352*
Pascher, G., *420*
Patek, V., *346*
Patel, H. R., *148*
Patterson, W. I., *417, 420*
Pauli, H., *522*
Paulson, R. A., 30 (12), *47*
Pavelka, F., *100*
Pavlov, D. N., *314*
Peacock, W. C., *349*
Peacocke, T. A. H., *130*
Pearson, R. M., *315*
Pećar, M., *130*
Pech, J., *312*
Pecherer, B., *350*
Peel, E. W., 276 (174), *313,* 316 (306), *350*
Pella, E., *26, 272*
Pellerin, F., *145*
Penniall, R., *148*
Pentz, C. A., *493, 523*
Pepkowitz, L. P., 188 (153), *219, 272, 313*
Peregud, E. A., *350*
Pereira, A., *419*
Perlman, D., *419*
Perold, G. W., 93 (93), *100, 525*
Perpar, M., *523, 525*
Perrin, C. H., *219*
Perrine, T. D., *185*
Perrone, J. L., *526*
Perthel, R., Jr., *219*
Peter, F., *493*
Peters, E. D., 158 (56), *100, 183, 274, 313,* 377 (11), 398 (11), 399 (11), 400 (11), *407, 523*
Peters, J. H., *272*

Peters, J. P., 488 (61), *493*
Petersen, J. W., 512 (180), 513 (180), *525*
Petránek, J., *525*
Petras, D. A., 103 (199), 106 (199), 112 (199), 113 (199), *132,* 326 (385), 327 (385), 331 (385), 332 (385), *352*
Petrikova, M. N., *23*
Petrova, L. N., *525*
Pettersson, H., 31 (70), *48*
Pfeil, E., *419*
Pfenningberger, R., 316 (156), *346*
Pflaum, D. J., *148*
Pfleger, K., *419*
Pfundt, O., *48*
Phillips, D. F., *265, 521*
Phillips, W. M., *506*
Pickhardt, W. P., *272*
Piehl, F. J., *269*
Pierotti, R. A., *409*
Pietri, C. E., 3 (178), *28,* 83 (120), *101,* 113 (197), 116 (197), *132,* 558 (88), 559 (88), *564*
Pietsch, H., *408*
Pifer, C. W., *131, 132, 420*
Pignard, P., *146*
Pilz, W., *314, 350*
Pinkus, A. G., *564*
Pirsch, J., 535 (58-62), *546*
Pirt, S. J., *350*
Pister, L., *416*
Pitiot, J., *186*
Pizarro, A. V., *421*
Plazin, J., 67 (66), *82,* 111 (219), *132,* 222 (368), *274, 421,* 454 (92), 455 (92), 458 (92), 465 (92), 473 (92), 476 (92), *494*
Ploompuu, J., *129*
Plyler, E. K., *417*
Podbielniak, W. J., *100*
Pöhm, M., *80, 523*
Pogosyants, E. K., *442*
Pohl, E. A., *148*
Pohloudek-Fabini, R., *420*
Poirier, P., *265, 407*
Pokorny, J., *564*
Polansky, I., *524*
Polgár, A., *506, 523*
Pollard, C. B., *272*
Polley, D., *148*
Polley, J. R., *185*

Pomatti, R., 175, *185*
Popescu, B., *269*
Popowsky, M., 554 (23), 557 (22, 23), *562*
Porter, W. L., 512 (176), 514, *525*
Portner, P. E., *219*
Poth, E. J., *185*
Potter, E. F., *523*
Power, F. W., 152 (157), 162 (157), *186,* 222 (305), *272*
Praeger, K., *350*
Prater, A. N., *272, 506*
Pratt, E. L., *419*
Pray, A. R., *144*
Pregl, F., 1, 2 (136), 5 (136), 6 (136), 27, 30 (73), *48,* 50 (48), 51 (48), 68 (48), 71 (48), 73 (48), 76 (48), *81, 100,* 102 (161), *131,* 133 (184), 134 (184), 137 (184), *148,* 151 (158), 152 (158), 161 (158), 166 (158), 167 (158), 168 (158), 171 (158), 175 (158), 177 (158), *186,* 188 (157), 190 (157), 196 (157), 199 (157), *219,* 222 (308), 224 (308), 228 (308), 230 (308), 233 (308), 238 (308), 241 (308), 243 (308), 244 (308), 247 (308), 248 (308), 250 (308), 252 (308), 255 (308), 258, *272,* 276 (180), 295 (180), 296 (180), 303 (180), *313,* 316, 317 (313), 318 (313), 323 (313), 324 (313), 325 (313), *350,* 354 (74), 356 (74), 357 (74), 359 (74), 360 (74), *366, 375,* 411 (132), *419,* 422 (84), 423 (84), 433 (84), 434 (84), *442,* 508 (183), 509 (183), 510 (183), *525,* 528 (63), *546*
Preisler, P. W., *131*
Prescott, C. H., Jr., *525*
Prévost, C., *27, 525*
Pribil, R., 259 (202), *270, 311*
Price, J. W., *148*
Prince, A. L., 188 (153), *219*
Pringsheim, H., *350*
Pro, M. J., *442*
Proskauer, E. S., *26*
Proud, E. R., *272*
Pucher, G. W., *419, 420*
Puddington, I. E., *546*
Pullman, T. N., *128*
Pungor, E., *148, 314, 351*
Purves, C. B., *452*

Quackenbush, F. W., *442, 524*
Quastel, J. H., *492*

Raaen, V. F., *522*
Raben, M. S., *350*
Rabinowitz, J. L., *272*
Rachele, J. R., 514, *525*
Radhakrishnan, A. N., *418*
Radimer, K. J., *351*
Radmacher, W., *408*
Radulescu, E., *365*
Rafailoff, R., *144*
Ramakrishna, V., *349*
Ramanthan, A. N., *132, 149, 353*
Ramberg, L., 30 (74), 34 (74), *49*
Ramirez, A., *353*
Ramsey, L. L., *420*
Rao, D. S., *350*
Rappaport, D., *148*
Rappaport, F., *148, 219, 493*
Rast, K., 528, 531 (66, 67), 532 (66, 67), *546*
Rauen, H. M., *219*
Rauscher, W. H., *148, 186, 350*
Rautanen, N., *494*
Raymond, R., *407*
Raymond, W. H. A., *146*
Reed, R. H., *182,* 424 (54), 438 (54), *441*
Rees, O. W., *182, 266*
Reese, M., *352*
Rehberg, P. B., 122 (163), *131*
Reichel, Z., *563*
Reichl, E. H., *419*
Reichold, E., *419*
Reid, W. M., *564*
Reif, W., *149*
Reihlen, H., *272, 564*
Reilley, C. N., *131*
Reimers, F., 560 (76), *564*
Reissner, R., *146*
Reith, H., *309, 343*
Rejhova, H., *441*
Remport-Horvath, Z., *129*
Remsen, I., 514 (187), *526*
Remy, J., *186*
Renard, M., *186, 273, 408*
Renoll, M. W., *273*
Resch, A., *269, 347*
Reuter, F. W., Jr., *524*
Reuther, K. H., *451*

Revukas, A. J., *267*
Reynolds, B., 316 (5), *342*
Reynolds, C. A., *419*
Reznik, B. A., *313*
Rhodes, D. N., *366*
Rhodes, E. J. H., *186, 219, 313, 351,* 354 (81), *366, 408*
Ribas, I., *219*
Ricciuti, C., *100, 272*
Richards, F. M., *49*
Richardson, G., *100*
Richter, K. M., *131*
Rickard, R. R., *350*
Riddick, J. A., *131*
Rieche, A., *546*
Rieman, Wm. III, *131, 218, 313, 525*
Riemenschneider, R. W., *418*
Riesenfeld, E. H., 1, *26,* 31 (64), *48, 273*
Rietz, E. G., *185*
Rigakos, D. R., *442*
Riley, J. D., *364*
Ritzer, J. E., *101*
Rivera, R., *272*
Roark, J. N., *187*
Roberts, F. M., *100*
Roberts, M., *144*
Roberts, M. W., 354, *365*
Robertson, G. I., *273*
Robinson, J. W., *144*
Robinson, R. J., *148, 273, 309*
Robinson, Sir R., 514, *522*
Robson, A., *417*
Rodden, C. J., 3 (140), 27, 30 (76), 31 (76), *49,* 53 (50), 54 (50), 56 (50), 74 (49), *81,* 134 (192), *148,* 259 (316), *273, 313, 350*
Rodder, J., *49*
Rogers, G. R., *268, 492*
Rogers, L. B., 27
Rogers, R. N., *350*
Rogina, B., *350*
Rolfson, R. B., *130*
Romyn, H. M., *313*
Ronzio, A. R., 151 (162), *186*
Rose, A., *100*
Rose, E., *100*
Rosebury, F., *564*
Rosenberg, S., *526*
Rosenblatt, M., 92 (69), 94 (69), *95, 99*
Rosenblatt, P., 557 (22), *562*

Author Index

Rosenfeld, B., *27*
Rosenqvist, T., *310*
Rosenthaler, L., *100, 313*
Ross, L. E., *312*
Ross, L. R., 188 (215), *220*
Rossmann, E., *506*
Rostokin, A. P., *48, 129*
Roswell, C. A., 83 (101), *100*
Roth, F. J., *148*
Roth, H., 2 (144-147), 5 (144-147), 12 (144-147), 13 (144-147), *27*, 30 (78-81), *49*, 50 (51-54), 51 (51-54), 53 (54), 60 (52), 68 (51-54), 70 (51-54), 71 (51-54), 73 (51-54), 74 (54), 75 (51-54), 76 (51-54), *81*, 88 (102-105), *100*, 102 (168-171), 104 (168-171), *131*, 133 (195-198), 134 (195-198), 137 (195-198), *148*, 151 (163-166), 152 (163-166), 155 (165), 161 (163-166), 163, 166 (163-166), 168 (163-166), 171 (163, 166), 174 (163-166), 175 (163-166), 177 (163-166), *186*, 188 (163-166), 189 (164, 165), 196 (163-166), 199 (163-166), *219, 222* (317-320), 224 (317-320), 228 (317-320), 230 (317-320), 233 (317-320), 235 (318, 319), 238 (317-320), 241 (317-320), 243 (317-320), 244 (317-320), 247 (317-320), 248 (317-320), 250 (317-320), 252 (317-320), 255 (317-320), 257 (317-320), *272, 273*, 276 (186-189), 287 (186-189), 295 (186-189), 296 (186-189), 297 (186-189), 298 (187, 188), *313*, 316 (322-325), 317 (322-325), 324 (322-325), 325 (322-325), 326 (325), *350*, 356 (77-80), 359 (77-80), 361 (79-80), *366*, 367 (50-53), 368 (50-53), 369 (50-53), 370 (50-53), 372 (50-53), *375, 376*, 377, *408*, 412, 420, 422 (86-89), 423 (86-89), 424 (86-89), 433 (86-89), 434 (86-89), 437 (86, 87, 89), *442, 452, 493, 494, 506*, 508 (189-192), 509 (189-192), 510 (189-192), 514 (190-193), 516 (191), *524, 526*, 528 (70-73), 529, 531 (70-73), 532 (70-73), 535 (70, 73), *546*, 551 (79, 80), 560 (79-81), *564*
Roth, W. A., 503 (42), *506*
Rothwell, R., *182*
Rottenberg, M., *350*
Rounds, G. C., *313*
Roux, G., *218*
Rowen, J. W., *417*

Rowley, R. J., *131, 350*
Royer, G. L., 3 (148, 149, 150), *27*, 104 (210), *132*, 133 (176), 134 (176), 136, 137, *148*, 155 (169), 161 (167, 168, 169), 174, *186*, 236 (322), 253 (322), *273, 276* (220), 277 (220), 278 (220), *313, 314, 350, 352*
Rubia Pacheco, J. de la, *343*
Rudloff, E. von, *442*
Ruf, E., *273*
Ruff, O., *350*
Rush, C. A., *186, 219*, 234 (324), 235 (324), *273, 313, 351*, 354 (81), *366, 377, 408*
Russell, W. W., *408*
Rutgers, J. J., *148*
Ruttloff, H., *421*
Ruyter, J. H., *546*
Ruzicka, J., *522*
Ruziczka, W., *506*
Ryan, H., *420*
Rzymowska, C. J., *145*

Saccardi, P., *442*
Sachs, G., *148*
Safford, H. W., *314, 351*
Safranski, L. W., *272, 421*
Sahashi, Y., *526*
Sahota, S. S., *526*
Sai, T., *441*
Sakaguchi, T., *351*
Sakamoto, S., *273, 313, 366*
Sakuraba, S., *148*
Sakurai, H., *132*
Salach, J. I., *149*
Salsbury, J. M., *526*
Saltzman, B. E., *149*
Salvioni, E., 31 (82), *49*
Samachson, J., *351*
Samsel, E. P., *442*
Samuel, D., *407*
Sandell, E. B., *100*
Sanderson, R. T., *526*
Sandkuhle, J., *494*
Sanford, C. R., *99*
Sanford, W. W., *349*
Šantavý, F., *352*
Sanz, M. C., *351*
Sarakhov, A. I., *49, 81*
Sarkisyan, R. S., *23*

Sasaki, K., *147*
Sasaki, Y., *421*
Saschek, W., *313, 372* (54), *376*
Sass, S., *366*
Sassaman, W. A., *182*
Šatava, V., *273*
Sato, H., *349*
Sato, K., *408*
Sato, Y., *440*
Saunders, B. C., *344, 351*
Saunders, J. A., *131*
Savacool, R. V., *506*
Savchenko, A. Ya., *351*
Savell, W. L., *186*
Sax, K. J., *273*
Scalamandre, A. A., *313*
Scales, F. M., 190 (168), *219*
Scandrett, F. J., *219*
Schadendorff, E., *273*
Schaefer, W. E., *521, 523*
Schaeffer, J. R., *521*
Schaffer, F. L., *217, 219, 366, 418*
Schaffert, R. R., *146, 375*
Schall, E. D., *351*
Schall, H., *522*
Scheel, K., 503 (42), *506*
Scheibl, F., *420*
Schenck, W. J., *186*
Scheurer, P. G., *219*
Schinzel, E., *524*
Schiviz-Schivizhoffen, E., *524*
Schlichenmayer, W., 514 (162), *525*
Schloemer, A., *351*
Schmall, M., *131, 132, 420*
Schmid, H., *522*
Schmidt, G., *365*
Schmidt-Nielson, K., *420*
Schmit-Jensen, H. O., *526*
Schmitt, R. B., *273*
Schneider, F., 1 (52), *24*, 83 (107), *99, 101, 131, 144, 420,* 551 (82), *562, 564*
Schnerb, I., *347*
Schniffner, H., *342*
Schöberl, A., 316 (339), *351*
Schöller, F., *269, 347, 493*
Schön, H., *522*
Schöniger, W., *27, 99, 131, 186, 219, 273, 276* (196-200), *291* (196, 198, 199, 200), *292* (198, 199, 200), *294* (198, 199, 200),
313, 314, 316 (341-344), *351, 408, 452, 506, 523, 524, 526*
Schoklitsch, K., *149*
Scholander, P. F., *131, 526*
Scholtis, K., *149*
Schouten, A., *407*
Schreiner, H., *131*
Schrenk, W. G., *144, 182*
Schretzmann, H., *147, 348*
Schreuders, M. A., *407*
Schröpl, E., *421*
Schubert, J., *493*
Schuele, W. J., *81*
Schütze, M., *377, 408, 409*
Schuhecker, K., *366*
Schuhknecht, W., *149*
Schulek, E., *149, 219, 314, 351, 376, 526*
Schulitz, P. H., *149*
Schultz, H. P., *219*
Schultz, Y. O., *149*
Schulze, H. O., *131*
Schumacher, E., *27*
Schumb, W. C., *351*
Schuster, P., *526*
Schwab, G. M., *219*
Schwappach, A., *443*
Schwartz, N., *348*
Schwarz, K., *131*
Schwarzkopf, O., 222 (338), 224 (338), 230 (338), 235 (338), *273, 351*
Scott, F. W., *149*
Seaman, W., *506*
Sease, J. W., *522*
Sebastian, J. J. S., *526*
Secor, G. E., *219, 220,* 316 (421), *353*
Segal, F., *416*
Seiferle, E. J., *506*
Sekera, A., *564*
Seligson, D., *219, 420*
Seligson, H., *219*
Sellers, A. L., *521*
Sellers, D. E., *147*
Selzer, G., *145*
Selzer, M., *145, 417*
Senda, J., *220*
Sendroy, J., *131*
Sendroy, J., Jr., 488 (72), *494, 526*
Senkowski, B., *351*
Seno, S., *349*
Sensabaugh, A. J., *420*

Serra, J., *345*
Serra, J. A., *419*
Servais, P. C., *147*
Servigne, Y., *351*
Sevag, M. G., *420*
Sezerat, A., *351*
Sfortunato, T., *420*
Shaefer, W. E., *452*
Shafer, E. G. E., *351*
Shaffer, P. A., Jr., *131*
Shah, G. D., *186, 350, 527*
Shah, R. A., *309, 343*
Shakrokh, B. K., *351*
Shapiro, B., *420*
Sharpe, E. S., *418*
Shaw, B. W., *442*
Shaw, J., *219*
Shead, A. C., *101*
Sheehan, F., 103 (109), *129, 217*
Sheers, E. H., *219*
Sheft, I., *409*
Shelberg, E. F., *186*
Shemyatenkova, V. T., *147*
Shepard, D. L., *219*
Sher, I. H., *131, 219*
Sheridan, J., *23*
Sherman, C. C., *420*
Sherman, M. S., 88 (80), *100,* 151 (120), 163 (120), *184*
Sherman, R. E., *149*
Sheveleva, N. S., *267, 270, 347*
Sheyanova, F. R., *147*
Shibasaki, H., *526*
Shimoe, D., *149, 314, 420, 453, 494, 527*
Shimosawa, T., *420*
Shinn, L. A., 316 (398), *352*
Shirley, E. L., *313*
Shishikura, Y., *526*
Shive, W. J., *219*
Shorr, E., *366*
Shrader, S. A., *101, 272*
Shrewsbury, C. L., 104 (78), *128,* 282 (72), *310*
Shriner, R. L., 496 (1, 47), *505, 506,* 560 (84), *564*
Shu, P., *274*
Shukis, A., Jr., *149*
Shupe, L. M., *98*
Sickels, J. P., *219*
Siegel, H., *98*

Siegfriedt, R. K., *314, 564*
Sievers, D. C., *216*
Siggia, S., *420, 442, 452, 506, 526*
Signer, R., 528 (75), 535 (75), *546*
Silbert, F. C., *273*
Silker, R. E., *144, 182*
Silver, S. D., *417*
Silverstein, R. M., *219*
Silvert, F. C., *149*
Silvey, G. A., *351*
Simon, V., *351*
Simon, W., *131, 420*
Simons, J. H., *351*
Simonson, T. R., *184*
Singer, E., *81*
Singh, B., *526*
Siniramed, C., *409*
Sirotenko, A. A., *101, 273, 314, 526*
Sisco, R. C., *494*
Sisido, K., 316 (358), *351*
Sisti, A. J., *366*
Sitte, H., *563*
Sjöquist, J., *420*
Skidmore, D. W., *219*
Skinner, C. G., *81*
Skrube, H., *352*
Slavik, K., *219*
Slotta, K. H., *442, 526*
Slovik, N., *351*
Sloviter, H. A., *149, 376*
Small, L. A., *348*
Smaller, B., *526*
Smetana, R., *219*
Smirk, F., *351*
Smit, W. M., *546*
Smith, A. A., *150*
Smith, A. M., *420*
Smith, C. L., *183*
Smith, D. E., *182*
Smith, D. M., 92 (84, 111), 94 (84, 111), *100, 101, 407, 420, 525, 526*
Smith, F., *219*
Smith, F. A., *351*
Smith, G. F., *101, 145, 149, 220, 273*
Smith, G. McPhail, 50 (59), 51 (59), 57 (59), *81,* 133 (218), *149,* 316 (401), *351, 353*
Smith, J., *314*
Smith, O. D., *351*
Smith, R. J., *420*

Smith, R. N., *409, 526*
Smith, W. H., 9 (157), *27, 273,* 383, *389* (4), 402 (4), *407, 409*
Smits, R., *506*
Snedecor, G. W., 56 (60), *81*
Snellman, O., *313*
Šnobl, D., *28, 187, 274, 314, 353*
Snyman, J. M., *525*
Sobel, A. E., 103 (190), *131,* 188 (189), 190 (189), *220, 351*
Sober, H. A., *417*
Sobko, M. Ya., *345*
Sobotka, H., *216*
Sobotka, M., *27, 526*
Socolar, S. J., *149*
Soeda, Y., *314*
Soep, H., *314, 351*
Soibel'man, B. I., *314*
Sokolova, N. V., *273, 351, 352*
Solomon, A. K., *149*
Soltys, A., 1 (159), *27, 101,* 514 (225), *526*
Soltzberg, S., *453*
Somereyns, G., *564*
Sômiya, N., *131*
Sommereyns, G., *419*
Sorensen, J. H., *268*
Souček, M., *523, 526*
Souchay, P., *27, 525*
Southworth, B. C., *149, 365, 376*
Southworth, L., 160 (70), *183*
Sozzi, J. A., *26,* 30 (68), *48,* 50 (46), *81, 100,* 102 (147), 123 (147), *130,* 133 (173), 134 (173), 137 (173), *147, 148,* 151 (141), 152 (141), 161 (141), *185, 218,* 222 (288), *272, 313, 350, 527, 546*
Späth, E., *453*
Spaulding, G. H., *215*
Spence, R., *527*
Spettelm, E. C., *146*
Spěvák, A., *315, 353, 443*
Spiedel, H., *407*
Spier, H. W., *420*
Spies, J. R., 151 (181), 163 (181), *186, 547*
Spikes, W. F., 3 (150), *27, 98,* 161 (168), *186, 273*
Spingler, H., *453*
Spitzy, H.; *352*
Spooner, C. E., 259, *265, 309, 343, 409*
Sporek, K. F., *131*

Sprague, R. S., *310*
Sprecher, J. C., *219*
Spring, F. S., *545*
Squirrell, D. C. M., *25*
Srpova, B., *523*
Staats, F. C., 103, 127, 377 (5), 379 (5), 383 (5), 392 (5) 397 (5), 398, 402 (5), *407*
Stacey, M., *352*
Staemmler, H. J., *352*
Stagg, H. E., *27,* 424 (54), 438 (54), *441*
Stanek, V., *273*
State, H. M., 354, *365*
Steacie, E. W. R., *524*
Steele, R., *420*
Steers, E., *420*
Štefanac, Z., *441, 527*
Stegemann, H., *220, 352*
Stehr, E., 155 (182), *186, 274,* 377, *407*
Steinitz, K., *131, 352*
Stelt, C. van der, *420*
Stempel, B., *352*
Stenhagen, E., *524*
Stenmark, G. A., *527*
Stephen, W. I., *23, 27, 309, 442*
Sterges, A. J., *366*
Stern, A., *366*
Sternberg, H., *267, 274*
Sternglanz, P. D., *186*
Stetten, D., Jr., *220, 420*
Stevens, G. D., *527*
Steyermark, Al, 2 (174, 177), 3 (163, 175-184), 4 (165-170a, 185, 186, 188), 5 (174), 10, 12 (164, 174), 13 (164, 174), 14 (187), 15 (187), 19 (164, 174, 187), 20 (164, 174), *27,* 28, 30 (87, 88), *49,* 50 (62), 51 (62), 53 (61, 62), 54 (62), 57 (62), 58 (62, 63, 64), 59 (62, 63, 64), 60 (62, 63), 62 (62, 63), 63 (62, 63), 64 (63), 65 (62, 63), 66 (62, 63), 67 (62, 63), 68 (62, 63, 64), 69 (62, 64), 70 (62, 64), 71 (62), 73 (62, 63, 64), 74 (61, 62, 63), 75 (62, 63, 64), 76 (62), *81, 82,* 83 (115, 116, 118, 119, 120), 84 (115), 88 (115, 121), 91 (121), 92 (115, 117), 93 (115, 117), *101,* 102 (194), 103 (194, 199), 104 (194, 198), 105 (194, 198), 106 (199), 108 (194), 110 (194, 198), 111 (194), 112 (199), 113 (195, 196, 197, 199), 116 (195, 196, 197), *132,*

133 (223), 134 (223, 224, 225), 136 (222) *149,* 151 (186, 188), 152 (186, 188), 155 (185, 186), 160 (189, 190), 161 (186, 190), 162 (186), 163 (186), 166 (186), 168 (186), 171 (190), 172 (186, 187, 190), *186, 187,* 188 (193), 189 (192, 193, 195), 190 (193, 195), 191 (193, 195), 192 (193, 194, 195), 193 (193, 194), 195 (193, 194), 196 (193, 194), 199 (193, 194), 200 (193, 194), 202 (193), 205 (192, 193), 210 (195), 211 (193, 194), *220,* 222 (348, 350), 223 (348, 350), 224 (348, 350), 225 (106, 350, 351), 228 (350, 351), 230 (348, 349, 350, 351), 233 (350), 234 (348, 350), 235 (348, 349, 350), 236 (349, 350), 238 (350, 351), 240 (350, 351), 241 (350), 243 (350), 244 (350), 247 (349, 350), 248 (350), 250 (350), 251 (348), 252 (350), 255 (348, 350, 351), 257 (350, 351), 258 (349, 350, 351), *274,* 276 (209, 213, 214), 277 (209, 213, 214), 278 (209, 211, 213, 214), 280 (211), 281 (211), 282 (209, 214), 283 (209, 214), 287 (209, 212), 291 (213), 292 (213), 295 (209), 296 (209), 297 (209, 212), 298 (209, 210, 212), 302 (209, 212), *314,* 316 (375, 376, 379, 383, 384), 317 (379, 380, 381), 318 (379, 381), 319 (374, 379, 381), 322 (379, 381), 323 (379, 383, 384), 324 (379, 381), 325 (379), 326 (378, 385), 327 (377a, 385), 331 (385), 332 (385), 334 (382), *352,* 354 (89), 355 (89), 356 (89, 91), 357 (88, 89), 359 (89), 360 (90), *366,* 367 (57), 368 (57, 58), 369 (57, 58), 370 (57, 59), 371 (57, 59), 372 (57), *376,* 377 (86, 87), 378 (88, 89), 379 (89), 383 (88, 89, 91), 389 (91), 390 (89), 393 (88, 89, 90), 394 (88, 89), 396 (88), 397 (88), 399 (89), 401 (88, 89, 90), 402 (88, 89), *409,* 410 (162, 163), 411 (163), 412 (162, 163), *420,* 422 (97-100), 423 (97-101), 424 (97, 99, 100), 425 (97, 100), 426 (99, 101), 429 (101), 431 (99, 101), 432 (97, 100), 433 (97-100), 434 (97, 100), 435 (97, 100), 436 (97, 100), 437 (99, 100, 101), 438 (97, 99, 100), *442, 443,* 444 (44), 445 (44), 447 (44, 45, 46), *453,* 458 (80),

465 (80), 476 (80), 477 (80), 485 (78, 80, 81), 486 (78, 80), 488 (81), *494, 506,* 508 (234), 509 (233, 234), 510 (233, 234), 511 (234), 512 (234), 513 (234), *527,* 528 (79), 531 (79), 532 (79), 535 (79), 537 (79), 539 (79), *547,* 551 (87), 552 (87), 553 (87), 557 (87), 558 (88), 560 (87), *564,* 569 (11), *602*

Stillson, G. H., 235 (59), 236 (59), *266*
Stock, D. I., *523*
Stock, J. T., *28, 49, 132, 187*
Stodola, F. H., 512 (236), *527*
Stöckl, W., *353*
Stöhr, R., *420*
Stoffyn, P., *129*
Stone, E. W., *220*
Stone, J. F., *524*
Stragand, G. L., *314, 351*
Strahm, R. D., *149, 352*
Stratmann, H., *314*
Straub, J., *149*
Strauch, L., *216*
Strauss, E., *525*
Straw, H. T., *315*
Strebel, W., *25*
Strebinger, R., *149*
Streiff, H. J., 2 (174, 177), *27*
Streuli, C. A., *527*
Stroganova, A. M., *146*
Strong, F. M., *420*
Stross, F. H., *273, 545*
Stubblefield, F. M., *220*
Stuber, J. E., *266*
Sturdy, G. E., *82*
Sturua, G. G., *419*
Subba Rao, D., *527*
Subbarow, Y., *364*
Sucharda, E., *547*
Sucher, R., *492*
Sudborough, J. J., 514 (93, 94), *523*
Sudo, T., *149, 314, 420, 453, 527*
Sugano, T., *527*
Sugiyama, N., *274*
Sullivan, P., *24*
Sullivan, W., *522*
Sundberg, O. E., 104 (210), *132,* 236 (322), 253 (322), *273,* 274, 276 (220), 277 (220), 278 (220), *314, 352*
Surak, J. G., *348*

Susano, C. D., *364*
Suter, H., *352*
Sutton, T. C., *527*
Suzuki, M., *408*
Suzuki, S., *527*
Swearingen, J. S., *527*
Sweeny, R. F., *26, 271*
Sweetser, P. B., *352, 374*
Swift, E. H., *492*
Swift, H., *187*
Swigert, G. F., 103 (109), *129, 217*
Swim, L. E., *147*
Syavtsillo, S. V., *147, 443*
Sykes, A., *23, 144, 149*
Sykes, W. Y., *275*
Syme, A. C., *314, 349*
Šynek, L., *187, 274, 315*
Szalkowski, C. R., *420*
Sze, L. C., *127*
Szekeres, L., *314*
Szep, Ö., *374*

Tänzer, I., *364*
Tait, P. C., *274*
Takagi, T., *187*
Takahashi, M., *409*
Takahashi, T., *132*
Takayama, Y., *185*
Takeda, A., *220*
Takenaka, R., *313*
Taki, S., *185, 311*
Takino, Y., *443*
Takiura, K., *443*
Tallman, R. C., *149*
Tamiya, N., *314*
Tanaka, J., *185*
Tanaka, M., *147, 268*
Tanaka, Y., *311, 314*
Tani, H., *453*
Tannenbaum, S., *144, 265*
Tashinian, V. H., *184, 527*
Tatlow, J. C., 334 (45), *343, 352*
Taufel, K., *420, 421*
Taussky, H. H., *366, 421*
Taylor, A. E., *366*
Taylor, D. B., *217*
Taylor, T. G., *421*
Taylor, W. H., *220, 273*
Teeri, A. E., *149*
Tělupilová-Krestýnová, O., *352*

Ten Eyck Schenck, R., *101*
Terent'ev, A. P., *28, 187, 220, 274, 352, 527*
Terent'eva, E. A., *269, 270, 365*
Terent'eva, Ev. A., *28, 312*
TeSelle, L. P., *274*
Teston, R. O'D., *274,* 316 (394), *352*
Tettweiler, K., *314*
Thaler, H., *562*
Thege, I. K., *148*
Thiele, K. H., *144, 374*
Thomas, H. R., *521*
Thomas, J. F., *352*
Thomas, J. W., 316 (398), *343, 352*
Thomis, G. N., *421*
Thompson, J. F., *220*
Thompson, J. H., *98, 265*
Thompson, R. C., *186, 274,* 565, *602*
Thompson, W. R., *132*
Thomson, D. A., *364*
Thomson, M. L., *28, 314*
Thorn, J. A., *274*
Throckmorton, W. H., *274*
Thürkauf, M., *265*
Thurnwald, H., *149*
Timmermans, A. M., *129, 270, 348*
Tišler, M., *525*
Toennies, G., *309, 314, 524*
Tompkins, E. R., *132*
Tompkins, P. C., *147*
Toribara, T. Y., *149, 364*
Tourtellotte, W. W., *220*
Tous, J. G., *421*
Towne, A. K., *48, 81, 270*
Townley, E. R., *523*
Toya, H., *185*
Tracey, M. V., *421*
Trantham, H. V., *128, 332, 344*
Trautz, O., 172 (199, 200), *187*
Treadwell, F. P., 102 (215), *132,* 133 (235), *149,* 316 (399), *353,* 354 (95), *366,* 367 (61), 370 (61), *376,* 567 (13), *602*
Trenner, N. R., *266, 274, 492*
Triebold, H. O., *562*
Trifonov, A., *314*
Trim, A. R., 514 (125), *524*
Trutnovsky, H., *187, 526*
Tsao, M. U., *421*
Tsatsas, G., *266*
Tschamler, H., *564*

Tschugaeff, L., 514 (248), *527*
Tsuji, T., *314*
Tsujii, T., *420, 453, 527*
Tsurumi, S., *421*
Tsuzuki, Y., *366*
Tuckerman, M. M., *365, 376*
Tuemmler, F. D., *274*
Tufts, L. E., *311, 347*
Tulus, R., *417*
Tunnicliff, D. D., 158 (56), *183, 274*
Tunnicliffe, M. E., *149, 366*
Turner, A. R., *48*
Turnstall, R. B., *272*
Tustanovskii, A. A., *419*
Tuttle, C., *49*
Tutundžić, P. S., *149, 376*
Twickler, M. C., *523*

Ubbelohde, A. R., *563*
Ueda, Y., *272*
Ueno, M., *527*
Uhrig, K., *527*
Ulbrich, V., *441, 527*
Ullyot, G. E., *506*
Umberger, C. J., *99*
Umland, F., *149*
Underwood, H. G., *132*
Unger, E. H., *507*
Unterzaucher, J., *187, 259, 274, 377, 409*
Upson, U. L., *132*
Urrutia, H., *353*
Usiekniewicz, K., *217*
Ussing, H., 515, *522, 563*
Utsui, Y., *267, 310, 345*
Utsumi, S., *346*

Vagnina, L. L., *28*
Vaidyanathan, C. S., *418*
Valdener, G., *183, 267, 407*
Vallentyne, J. R., *220*
Van Atta, R. E., *417*
Vance, J. E., *274*
Van de Kamer, J. H., *421*
Vandenheuvel, F. A., *507*
Van Dijk, H., *310, 346*
Van Etten, C. H., *149, 527*
Vango, S. P., *187*
Van Heyningen, W. E., *564*
Vaníčková, E., *505*
Van Meurs, N., *421*

Van Slyke, D. D., 67 (66), *82*, 111 (219), *132, 220, 222, 274, 366, 421*, 454 (24, 92, 97), 455 (92), 458 (92), 464 (97), 465 (92), 473 (92), 474 (85, 88), 475 (48, 85, 86), 476 (88, 92), 485 (88), 488 (24, 61, 88, 93, 94, 95, 96, 97), 489 (89), 490 (99), *492, 493, 494*
Van Straten, F. W., 74 (67), *82*, 259 (369), *274*
Van Wijk, H. F., *546*
Van Winkle, W. A., 316 (401), *353*
Varner, J. E., 253 (252), *258, 271*
Vásquez-Gesto, D., *219*
Vastagh, G., *219*
Vaughn, T. H., 316 (402), *353*
Veatch, R., *216*
Večeřa, M., *28, 187, 274, 314, 315, 347, 353, 443, 525*
Veibel, S., *527*
Vene, J., *220*
Venkateswarlu, P., *132, 149, 353*
Verdino, A., *150*
Verina, A. D., *275, 353*
Vertalier, S., *443*
Vickers, C., *347*
Vickery, H. B., *420*
Vieböch, F., *422, 443*
Viollier, G., *507*
Virtanen, A. I., *494*
Viswanathan, R., *353*
Viswanathan, T. S., *349*
Vogt, R. R., *442*
Vojtech, F., *274*
Volpi, A., *421*
Volynskii, N., *315, 353*
Vonesch, E. E., *527*
Voráček, J., *344*
Vrbaški, Ž., *525*
Vrchlabsky, M., *272*
Vries, G. de, *274*
Vries, J. X. de, *216*
Vulterin, J., *527*
Vysochina, L. D., *315*

Waber, J. T., *82*
Waddington, G., *563*
Wade, P., *132, 353*
Wagner, A. F., *417*
Wagner, E. C., *149, 220*, 276 (136, 174), *312, 313, 315*, 316 (306), *350, 364, 376*

Author Index

Wagner, G., *421*
Wagner, H., *28, 182, 274, 315, 353, 364*
Wagner, R., *219*
Wajzer, J., *419*
Wake, R. G., *218*
Wake, W. C., *506*
Waldman, H., *525*
Waldmann, H., *28*
Waldrep, P. G., *564*
Waldschmidt, M., *343*
Walisch, W., *507*
Walker, G. T., *132*
Walker, R. D., *147*
Walker, R. W., *274*
Wallace, H. S., *218*
Wallberg-Olausson, B., 175, *184*
Walter, R. N., 276 (243), 302 (244), *315*
Walton, H. F., *150*
Walton, W. W., 383, *409*
Wansink, E. H., *421*
Warburg, O., *494, 507*
Wark, S., *421*
Warren, V. L., *313*
Warren, W. B., *128*
Wasicki, R., *563*
Waters, W. A., *25, 26,* 50 (41, 42), 51 (41, 42), 73 (41, 42), 74 (41, 42), *81,* 88 (81, 82), *100, 130, 148,* 151 (121, 122), 152 (121, 122), 158 (121, 122), 161 (121, 122), 174 (121, 122), *184, 185,* 188 (130, 131), *218, 271,* 276 (155, 156), *312, 349, 365,* 367 (43, 44), *375, 408, 419, 442, 452, 493, 506, 525, 546*
Watson, H., 424 (54), 438 (54), *441*
Wear, G. E. C., *183,* 377, *407*
Weatherby, H. J., *525*
Webb, A. D., *524*
Webb, J. R., *145, 216,* 354, *365*
Weber, A. P., 316 (418), *353*
Webster, J. H., *129*
Weezel, G. J. van, *524*
Weiblen, D. G., *267, 345*
Weil, H., *274*
Weill, C. E., *187*
Weil-Malherbe, H., *421*
Weiner, N., 103 (94), *129,* 217
Weinstein, B., *525*
Weiss, F. T., *523, 527*
Weissberger, A., *26,* 551 (92), 560 (91), *564*

Weisz, H., *342*
Welch, C. M., *366*
Well, I. C., *150*
Weller, O. E., *524*
Wellington, E. F., *421*
Wells, J. C., *344*
Wellwood, G. W., 172 (71), *183*
Welwart, Y., *127, 144*
Wenger, P., *145, 150*
Wenger, P. E., *28*
Wennerstrand, B., *129*
Wenske, H. H., *148*
Werner, A., 316 (419), *353, 507*
Werner, O., 84 (63), *99, 101*
Wernimont, G., *150*
Wesenbeek, W., *408*
Wesseley, F., *524*
West, P. W., *129, 366, 418*
West, T. S., *23, 28, 47, 98, 127, 144, 215, 265, 309, 343, 416, 421, 440, 494, 521*
Weyer, F. G., *149*
Weygand, C., 158 (210), *187,* 274, 316 (419), *353, 507*
Weymouth, F. J., *127*
Whiffen, D. H., *23*
Whisman, M. L., *310*
White, D. C., *315*
White, E. P., *443*
White, E. V., *443*
White, J. C., *364*
White, J. W., Jr., *547,* 551 (96), *564*
White, L. M., 219, 220, 316 (420, 421), *353, 547*
White, T., *443*
White, T. T., *274*
Whitehead, T. P., *218*
Whitehouse, M. W., *311*
Whiteley, A. H., *527*
Wiberley, J. S., *101, 314, 564*
Wichmann, H. I., *344*
Wickbold, R., *353*
Wiegand, C., *564*
Wieland, T., *421*
Wiele, M. B., *149*
Wiese, H. F., *421*
Wiesenberger, E., *49, 453, 527*
Wilcox, C. W., *350*
Wilkenfeldt, J. W., 14 (187), 15 (187), 19 (187), *28*
Wilkinson, H. A., *218*

Wilkinson, R. H., *150*
Willard, H. H., *353*
Willard, M. L., *562*
Willenberg, H., *272*
Williams, A. F., *82*
Williams, H. A., *353*
Williams, H. G., *313*
Williams, H. R., *128*
Williams, M., *28, 132, 215*
Williams, P. A., *269*
Williams, R. J., 516 (259), *527*
Williamson, H. G., *351*
Willits, C. O., 3 (179-184), 4 (205, 206, 207), *26, 28, 29,* 30 (93), *49,* 58 (63, 64), 59 (63, 64), 60 (63), 62 (63, 70, 71), 63 (63), 64 (63), 65 (63), 66 (63), 67 (63), 68 (63, 64), 69 (64), 70 (64), 73 (63, 64), 74 (63), 75 (63, 64), *81, 82,* 88 (121, 126, 127), 91 (121, 126, 127), 92 (125), 93 (125), *100, 101,* 103 (149, 225, 226), 104 (150), 105 (150), 110 (150), *130, 132,* 134 (224, 225), *149,* 151 (211, 212), 154 (211, 213), 160 (189, 190), 161 (190, 212), 164 (212), 171 (190), 172 (190), *186, 187,* 188 (144, 145, 146, 213-221), 189 (218), 190 (144, 214, 216, 218, 221), 191 (144, 214, 216, 218, 221), 192 (144, 194, 214, 216, 218), 193 (194), 195 (194), 196 (194), 199 (194), 200 (194), 205 (221), 207 (221), 211 (194), *218, 220,* 221 (383, 384), 225 (351), 228 (351), 230 (351, 385), 235 (382-385), 236 (382-385), 238 (351), 240 (351), 253 (383, 384, 385), 255 (351), 257 (351, 382), 258 (351), *272, 274, 275,* 276 (169, 246), 277 (169), 278 (169, 211), 280 (211), 281 (211), 282, 287 (212), 295 (247), 297 (212), 298 (212), 302 (212, 247), *313, 314, 315,* 317 (380, 381), 318 (381), 319 (381), 322 (381), 324 (381), *352, 353,* 356 (91), 360 (90), *366,* 368 (58), 369 (58), 370 (59), 371 (59), *376,* 377 (101), 378 (102), 390 (102), 393 (90, 102), 401 (90), *409, 421,* 447 (45, 46), *453, 507,* 512 (176), 514, *525, 527*
Wilson, A., 540 (83a), 543 (83a), *547*
Wilson, C. L., *24, 29, 545, 547*
Wilson, H. A. B., *441*
Wilson, H. N., *315,* 535 (85), *547*

Wilzbach, K. E., *275*
Winans, W. R., *182*
Winefordner, J. D., *349*
Wingo, W. J., *132, 220*
Winkler, L. W., *150*
Winter, O. B., *353*
Winteringham, F. P. W., *132*
Wintersteiner, O., 367 (65), 368 (65), 371 (64), *376*
Wiper, A., *346*
Wirth, W., *271*
Wise, E. N., *132*
Wise, R. W., *187*
Wiseman, H. G., 316 (398), *352*
Wiseman, W. A., 383 (91), 389 (91), *409*
Wiser, W. C., *344*
Witekowa, S., *29*
Witten, B., *366*
Woiwod, A. J., *494*
Wolfgang, J. K., *187*
Wolfrom, M. L., *441, 452, 453*
Wolkowitz, H., *417*
Wollish, E. G., *131, 132, 351, 420*
Wolstadt, R., *149, 376*
Wong, R., *266*
Wood, J. H., *375*
Wood, L. K., *346*
Woodland, W. C., *184*
Woollard, L. D., *348*
Wormley, T. G., 1, *29*
Woy, R., *366*
Wreath, A. R., *364*
Wright, C. D., *522*
Wright, G. F., *266, 442, 443, 527*
Wright, R., *547*
Wurzschmitt, B., *187, 275, 315*
Wust, H., *187, 220*
Wyatt, G. H., 83 (128), *101, 131*
Wyld, G. E. A., *418*
Wynne, E. A., *345*

Xuong, N. D., *522*

Yagi, H., 316 (358), *351*
Yagi, Y., *315*
Yakubik, M. G., *421*
Yamaji, I., *315*
Yamamoto, S., *269*
Yamamura, S. S., *310, 421, 527*
Yamane, T., *545*

Yarden, A., *345*
Yasuda, S. K., *350, 353*
Yeh, C. S., 222, 275
Yoe, J. H., *526*
Yokoo, M., *527*
Yoshimura, C., *376*
Youden, W. J., 56 (72), *82*
Yudasina, A. G., *315*
Yudis, M. D., *523*
Yuska, H., 103 (190), *131*
Yutzy, H. C., *347*

Zabrodina, A. S., *150, 267, 275*
Zabrodina, K. S., *524*
Zacherl, M. K., *29, 273, 353, 443*
Zahn, V., 158 (56), *183*
Zahner, R. J., *218, 266*
Zaki, R., *145*
Zakrzewski, Z., *220*
Zalta, J. P., *218*
Zamanov, R. K., *348*
Zarembo, J. E., *271, 312*

Zaugg, H. E., *507, 527*
Zdybek, G., *315*
Zeigler, C. A., *522*
Zeile, K., *421, 494*
Zeisel, S., *443*
Zelle, M. R., *127*
Zenchelsky, S. T., *563*
Zerewitinoff, Th., 514 (265-269), *527*
Zimin, A. V., *275, 353*
Zimmerman, P. W., *353*
Zimmermann, W., 45 (94), *49,* 155 (219), *187, 275,* 276 (255, 257), *315, 409*
Zinneke, F., 205 (255), *220, 315*
Zöhler, A., *310*
Zombory, L., *150*
Zscheile, F. P., *547,* 551 (96), *564*
Zuazaga, G., 103 (128), *130,* 190 (117), 191 (117), *218*
Zuckerman, J. L., *130*
Zuehlke, C. W., *150*
Zyka, J., *351, 521, 527*

SUBJECT INDEX

A

Abderhalden pistol drier(s), 88-91
  use of, for drying samples and for determination of moisture, 91
Absorption of carbon dioxide, from ampuls, vials, 485-488
  in carbon determination, 468, 469
  in manometric apparatus, 468, 469
Absorption tube(s), 238, 239, 258
  filling of, for carbon-hydrogen determination, 239, 240, 246
  Vigreux type, 400, 401
  weighing of, for carbon-hydrogen determination, 253, 254
  wiping of, for carbon-hydrogen determination, 252, 253
Acetals, determination of, 518
Acetic acid, vapor pressure of, table, 503, curve, 503
  use of, in determination of nitrogen, 209, 475, 477
  in determination of unsaturation, 495
  (see also Glacial acetic acid)
Acetic acid—potassium acetate—bromine solution, 423
  use of, in alkimide determination, 509
  in alkoxyl determination, 431, 432
Acetic anhydride, 512
  use of, in determination of hydroxyl groups, 513
Acetone, determination of, 93
  use of, in molecular weight determination, 536
Acetyl (acyl) apparatus, 445, 446
Acetyl group(s), determination of, 444-453
  additional information related to, table of, 450, 451
  N-acetyl, 444-453
  O-acetyl, 444-453
  types of compounds which interfere, 444, 451
  various methods, 450, 451
Acetylene linkage (triple bond), 516, 520
Acid(s) (acid groups) determination of, 410, 421

Acid amides, determination of, 516, 519, 520
Acidimetric (direct neutralization) methods for determination of carboxyl groups, 414
Active hydrogen, determination of, 514-516, 518, 519
  additional information related to, table of, 517-520
  general methods, 518
  using aluminum nitride, 519
  deuterium, 519
  diazomethane, 519
  Grignard reagent, 514, 519
  lithium aluminum hydride, 519
  tritium, 519
Acyl apparatus, 445, 446
Acyl groups, determination of (see also Acetyl, Formyl), 444-453
Additional information for the various chapters, 21, 22, 42-45, 79, 93, 115, 137, 171, 211, 255-259, 304, 332-334, 540-543
Additional Information on References Related to Various Chapters, Tables, 21, 22, 46, 79, 96, 97, 124-126, 138-143, 178, 181, 212-215, 260-264, 305-308, 335-342, 362-364, 373, 374, 404-406, 413-416, 438-440, 450, 451, 490-492, 504, 505, 517-520, 543, 544, 561, 562
Ainsworth microchemical balance, 35-39
Air conditioning, 12
Air filter, 300
Alcoholic silver nitrate, 434
  use of, in alkimide determination, 509
  in alkoxyl determination, 434-436
Alcohol(s), determination of (see Hydroxyl groups)
Alcohol of crystallization, elimination of, 93
Aldehyde(s), determination of, 516, 518
Aliphatic chain(s), effect of, on Dumas determination, 152
Alizarin Red S (see Sodium alizarin sulfonate)

Subject Index

Alkali and alkaline earth metal(s), effect of, on determination of carbon-hydrogen, 222, 223, 255
Alkaline hydrazine, use of, in carbon determination, 456, 457, 467
Alkaline permanganate, use of, in manometric determination, 475, 479-482
Alkimide apparatus, 433, 435, 509, 510
Alkimide group(s), (N-Methyl, N-Ethyl) determination of, 508-512, 517
　additional information related to, table of, 517-520
Alkoxyl apparatus,
　modified Clark, 425-430
　Steyermark, 433, 435
　various other types, 440
Alkoxyl group(s), determination of, 422-443
　additional information related to, table of, 438-440
　false results in absence of, 439
　gravimetric method, 433-443
　iodometric method, 422-432, 438-443
　related groups and higher homologs, 439
　various methods and conditions, 438, 439
Alkyl group(s), determination of, attached to,
　carbon, 516, 518
　nitrogen, 508-512, 517, 518
　oxygen, 422-443
　sulfur, 438
Alkyl halide(s), formation of, in determination of,
　alkimide groups, 508-512
　alkoxyl groups, 422-443
Allowable error(s), 171, 209, 255, 286, 323, 359, 369, 373, 412, 432, 437, 450, 473, 534, 539
Aluminum oxide, use of, in determination of carbon-hydrogen, 222, 224, 230-235
Amide(s), determination of, 519, 520
　determination of nitrogen in, 151-187, 188-220
Amine(s) (secondary and tertiary), determination of, 516, 519, 520
　determination of nitrogen in,
　　Dumas method, 151-187

　　Kjeldahl method, 188-220
　　Van Slyke method, 454, 474-485
Amino acid(s), determination of, 415, 491
Aminoid nitrogen, determination of, 151-187, 188-220, 491
Ammonium bromide, use of, in simultaneous halogen determination, 326
Ammonium iodide, use of,
　in alkimide determination, 509
　in simultaneous halogen determination, 326
Ammonium phosphomolybdate, 354
Amperometric methods, 364
Amperometric titrations, 124, 339, 342, 364, 406
Ampul testing by manometric apparatus, 485-488
Amyl alcohol, use of, in micronitrometers, 175
Analysis report(s) (*see* Laboratory report sheets)
Anhydrone (Dehydrite, magnesium perchlorate), use of, as water absorbent, 223, 228, 229, 246, 393
Anschütz thermometers, 552
　correction for exposed stem, 554
Antimony, determination of, 140
Antimony, effect of, on carbon-hydrogen determination, 222
Apparatus (*see* table at end of each chapter, "Additional Information Related to Chapter (*1 through 22*)", 21, 22, 46, 79, 96, 97, 124-126, 138-140, 178-181, 212-215, 260-264, 305-308, 335-342, 362-364, 373, 374, 404-406, 413-416, 438-440, 450, 451, 490-492, 504, 505, 517-520, 543, 544, 561, 562
Apparatus for the various chapters, additional types, 138, 181, 212, 259, 260, 261, 305, 306, 335, 340, 405, 416, 440, 450, 492, 505, 520, 544
Apparent dissociation constants, 416
Arsenic, determination of, 367-376
　additional information related to, table of, 373, 374
　Carius gravimetric method, 370-376
　Carius volumetric method, 367-369, 373-376

Subject Index

Arsenic, determination by,
  arsine test, 374
  Gutzeit test, 374
  Marsh test, 374
  trichloride, distillation of, 374
Arsine, use of, as a means of determination of arsenic, 374
Asbestos, 223, platinized, 224-235
  use of, in combustion tubes, 224-235
Ascarite, use of, as absorbent, 223, 228, 238-240, 246, 393
Ashing technique, method of determining metals, 133-150
Assembling,
  carbon-hydrogen train, 245-250
  catalytic combustion apparatus for halogen and sulfur, 298-301
  Dumas nitrogen train, 164, 165
  oxygen determination apparatus, 394, 401
Assembling and cleaning the balance (*see* Balance)
Assembling Pregl catalytic combustion train, 298-301
Assembly,
  carbon-hydrogen determination, 245-250
  catalytic combustion, halogen and sulfur, 298-301
  crucible filter, 371, 372
  Dumas nitrogen determination, 164, 165
  filtration, 319
Atomic and molecular weights and their multiples (some of them), 572, 573
Aurate(s), determination of gold in, 133-150
Audiohm thermistor attachment for Kofler micro melting point apparatus, 550, 551
Automatic combustion apparatus, 117, 259
Automatic time switch, 278, 280
Automatic titration methods, 124, 332, 333, 339, 406
Azo compounds, analysis of, for nitrogen, 189, 209
  determination of, 517
Azotometer (*see* Micronitrometer)

B

Baffle, Kofler, 550
Balance
  additional information related to subject, 46
  analytical, 30, 46
  assay, 30, 46
  electronic, 46
  micro-, 30, 45, 46
  microchemical, 30, 46
    assembling, 41
    capacities, table of, 30
    cleaning, 41
    descriptive information, table of, 38, 39
    determination of,
      precision, 55-57
      sensitivity, 55
      zero reading and deflection, 50-55
    illumination of, 19
    mounting, 41
    parts of, 31-40
    performance of, 30, 79
    room for, 13
    table for, 13-20, 22
    weighing on, 50-82
  semimicrochemical, 30, 46
  ultramicro-, 45
Barcroft-Warburg apparatus, 492, 505
Barium, determination of, 133-150, 361
  simultaneous, with phosphorus, 361
Barium chloride,
  standardization of, 0.01N, 110
  use of, in determination of sulfur, 278
Barium hydroxide, use of, to prevent foaming of potassium hydroxide, 175
Barium sulfate,
  treatment of, 287-290
  weighing of, 287-290
Barometer, correction of, reading for temperature, 168
Bases, determination of, 519
Beakers, 96
Beam of balance, 32
Becker microchemical balance, 36-39
Benzoyl group(s), determination of, 451
Beryllium, determination of, 140
Beta-ray absorption method for hydrogen, 264
Bismuth, determination of, 140
Black paper, use of, to protect silver halides, 320
Blank analysis (blank tests), 5, 472, 483
Blast lamp, 281
Blood and urine analysis, 488, 489

Subject Index

Boat(s) (*see also* Combustion boats)
  Hayman, 58, 59
  platinum, 58-60
  porcelain, 59
Boiling point, determination of, 552-554, 561
  additional information related to, table of, 561, 562
  apparatus for, 552, 553, 561
Boiling point apparatus, 552, 553, 561
Boiling point capillaries (*see* Capillaries)
Boiling point rise method, of molecular weight determination (*see* Ebullioscopic method)
Bomb(s) (*see also* Carius combustion tubes), 138, 264, 337, 341, 363
Bomb methods (*see also* individual determinations)
  Carius, 276-291, 316-323, 359-362
  other, 337, 341, 363, 373
  Parr, 333, 334, 337, 363
Books relating to subject, 21, 46, 79, 96, 124, 138, 212, 260, 305, 335, 340, 362, 373, 404, 413, 438, 450, 490, 504, 517, 543, 561
Boric acid,
  use of, in Kjeldahl nitrogen determination, 190, 197
Borneol, use of, in determination of molecular weight, 529, 535
Bornylamine, use of, in determination of molecular weight, 529, 535
Boron, determination of, 140
  compounds,
    determination of alkoxyl on, 438
    determination of carbon-hydrogen on, 262
Bottle(s),
  dropping, for mercury, 464
  Mariotte, 240-242, 246, 247
  tare (tare flasks), 60
  weighing, 60-64
    glass, 60-63
    metal, 63, 64
Bromination, use of, for determination of unsaturation, 495, 505
Bromine,
  determination of, 316-326, 332-339, 342-353 (*see also numerous* items regarding determination, such as apparatus, modifications, conditions, types of compounds, etc.)
    additional information on, 332-339
    table of, 335-339
  Carius combustion, 316-323, 326, 336
  catalytic combustion, 323-326, 337
  fusion, 333, 334, 337
  oxidation with sulfuric acid, silver, and potassium dichromates, 337
  Parr bomb fusion, 333, 334, 337
  presence of other halogens, 326, 336
  Schöniger flask combustion, 337
  simultaneous,
    bromine and chlorine, 326, 336
    with carbon and hydrogen, 336
    with sulfur, 336
  titration methods, 338, 339, 413
  wet combustion, 337
  use of, in determinations of,
    alkimide, 508
    alkoxyl, 423
    oxygen, 397, 398, 403
    sulfur, 291, 295
Bromocresol green-methyl red indicator, 103
Brush, camel's hair, use of, 74
Bubble counter-U-tube, 228, 229
  filling of, 228
  use of, for determination of,
    carbon-hydrogen, 228, 245-250
    halogens, 324
    oxygen, 393
    sulfur, 297
    unsaturation, 497
Bubbles adhering to mercury surface, prevention of, in micronitrometers, 162, 181
Bunge microchemical balance, 32-40, 43
Burette(s), 104, 125
  automatic, 104, 116
  calibration of, 125
  Linderstrøm-Lang and Holter, 123
  Rehberg, 122
Burner(s), used for determination of,
  alkoxyl, 430
  oxygen, 383
  [*see also* Furnace(s)]
By-pass tube and stopcocks, 391

C

C-methyl group(s), determination of, 516, 518
Cadmium, determination of, 140

Cadmium sulfate, use of, in alkoxyl determination, 434
Cahn microbalance, 45
Calcium, determination of, 133, 140
Calcium carbide, use of, in determination of moisture, 97
Calculation of,
  factors, 567, 568
  normality of solutions, 109, 111, 112
  percentages from empirical formula, and vice versa, 568-570
  results, 92, 109, 111, 136, 137, 168-171, 178, 208, 254, 260, 286, 290, 303, 322, 323, 359, 369, 373, 396, 404, 412, 413, 432, 437, 449, 450, 472, 473, 483-485, 501-504, 511-514, 533, 534, 539, 542, 543, 565-566
Calculations, 565-602
Calculators, 260
Calibration, of micronitrometers (microazotometers) (*see also* Micronitrometer) 162
  of pipettes, 116, 125, 126
  of weights, 57, 79
Calomel, use of, in micronitrometer, 152, 162, 175
Camphene, use of, in molecular weight determination, 529, 535
Camphenilone, use of, in molecular weight determination, 529, 535
Camphoquinone, use of, in molecular weight determination, 529, 535
Camphor, use of, in determination of molecular weight, 529, 535
Capillaries, filling of, 76
  for boiling point determination, 552
  for molecular weight determination, 531, 532
  for weighing, 76
Capillary burette(s), 122
Capillary pipette(s), 117-120, 558, 559
Capillary tube(s) for hydroxyl determination, 513
Caprylic alcohol, use of, in manometric apparatus to prevent foaming, 475
Capsules, for weighing of liquids, 78
Carbon,
  isotopic, determination of, 264, 491
  manometric determination of, 454-474, 490-494
  additional information related to, table of, 490-492
  Wyex compact black, use of, in oxygen determination, 379
Carbon and hydrogen, determination of, 221-275 (*see also* various *numerous* items, such as, tube fillings, pressure regulation, temperatures, modifications, types of compounds, pieces of apparatus, etc.)
  additional information on, 255-264
  table of, 260-264
  alkali and alkaline earth salts, 222
  arsenic, antimony, mercury, etc., compounds, 222, 262
  automatic apparatus for, 236, 259
  boron-compounds, 262
  chromatographic procedures, 264
  empty tube method, 259, 263
  fluoro-compounds, 222, 262
  mercury compounds, 222, 262
  metallic compounds, 222, 223, 262
  phosphorus compounds, 223, 262
  selenium compounds, 222
  silicon compounds, 262
  simultaneous, with halogens, nitrogen, and sulfur, 264
  sulfur compounds, 222
  various procedures (titrimetric, colorimetric, etc.), 264
Carbon dioxide,
  absorption of (*see* Absorption of carbon dioxide)
  elimination of, for acid-base titrations, 111, 412
  gasometers, 172
  generators, 155, 180
    cylinders, 174, 175, 180
  purification of, for use in Dumas determination, 155, 174
  use of, alkimide determination, 510
    alkoxyl determination, 425
    for conditioning hydriodic acid, 424
    in Dumas method, 154, 155
Carbon dioxide cylinder, 174, 180
Carbon disulfide, use of, in micronitrometer, 175
Carbon-hydrogen combustion, 221-275
  first combustion, 251
  second combustion, 252
  sweep out, 252

Subject Index

Carbon monoxide scrubber tube, 393
Carbon tetrachloride, use of, as standard for specific gravity determination, 556, 557
Carbonyl group(s), determination of, 516, 520
Carboxyl group(s), determination of, 410-421
    additional information related to, table of, 413-416
    groups interfering with, 410, 411
Carius combustion furnace, 278
Carius combustion tube(s), 278, 280, 281
    opening, 285, 286
    sealing, 283-285
Carius method,
    for arsenic, 367-376
    for halogens, 316-323, 336
    for phosphorus, 359-362
    for sulfur, 276-290, 306
Carrier gas for oxygen determination, purification of, 379, 405
Case and releasing mechanisms of balance, 34
Catalysts, for hydrogenation, palladium, 496, platinum, 496
    for Kjeldahl nitrogen determination, 188, 214
Catalytic combustion method(s),
    for carbon-hydrogen, 221-275
    for halogen, 323-326, 336, 337, 340, 341
    for sulfur, 295-304, 306, 307
Cation exchange methods,
    determination of acyl groups by, 451
    determination of phosphorus by, 364
Centrifuge (microcentrifuge), use for filling capillaries, 77
Centrifuge accessories, 96
Centrifuge tubes, use of, for microcrystallization, 83-86, 96
    for volumetric titrations, 113
    with conical bottom, plain, 83, 84
    with conical bottom, stoppered, 83, 85
    with conical bottom, stoppered, grad., 83, 86
    with cylindrical bottom, plain, 83, 86
Cerimetry, 124, 374
Chamois skin, 42, 244
    use of, for wiping various objects to be weighed, 74, 244, 252, 253

Charges (see Electrostatic charges)
Charging (weighing) tube(s), 70
Chloric acid, use of, for digestion purposes, 307, 373
Chlorine, determination of, 316-326, 332-339, 342-353 (see also numerous items regarding determination, such as apparatus, modifications, conditions, types of compounds, etc.)
    additional information on, 332-339
    table of, 335-339
    Carius combustion, 316-323, 326, 336
    catalytic combustion, 323-326, 337
    fusion, 333, 334, 337
    gravimetric, 316-326, 338
    oxidation with sulfuric acid, silver, and potassium dichromates, 337
    Schöniger flask combustion, 337
    simultaneous, with
        bromine, 326, 336
        carbon, hydrogen and sulfur, 336
        sulfur, 336
    titrimetric, 338, 339, 413
    wet combustion, 337
Chromatographic methods, 140, 214, 264, 339, 342, 364, 374, 406, 414, 439, 491
Chromatographic procedures, 264
Chromic acid combustion, 307, 337, 455, 456
Chromium, determination of, 141
Cinnamyl chloride, use of, in determination of moisture, 97
Clamp,
    for holding ball and socket joints, 162, 201, 202
    for holding top on steam generator, 196
Clamping device for micro-Parr bomb assembly, 333, 334
Cleaning of apparatus,
    alkoxyl (ethoxyl, methoxyl), 437
    Kjeldahl,
        one-piece model, 206, 207
        Pregl (Parnas-Wagner), 208
    oxygen, 396
    Van Slyke, 472
Cleaning deposited carbon from reaction tube in oxygen apparatus, 396
Cobalt, determination of, 141

Cold stage for determination of freezing point, 550
Coleman nitrogen analyzer, 177
Collaboration, 2
Collaborative studies, 2, 178, 188, 260, 305, 335, 340, 362, 404, 438, 450, 543
Colorimetric determination(s), 264
Colorimetric equipment, 125
Colorimetric methods, 139, 214, 264, 308, 338, 339, 341, 342, 363, 374, 406, 414, 415, 451
Colorimetric procedures, 264, 414, 415
Combustion apparatus, 155-158, 235-238, 259, 298, 382-391
Combustion boats, 58-61
  platinum,
    size A, 58-61
    size B, 59-61
    size C, 59-61
  porcelain 59
Combustion fluid, for manometric carbon determination, 455, 456
Combustion furnace(s),
  Carius, 278, 279
  for arsenic, 368
  for carbon-hydrogen, 235-237
  for halogens, 317, 324
  for nitrogen, 155-158, 181
  for phosphorus, 360
  for oxygen, 382-391, 399
  for sulfur, 278, 297
Combustion method(s),
  Carius, 276-290, 316-353, 359-366, 367-376
  for arsenic, 367-376
  for carbon-hydrogen, 221-275, 259
  for halogens, 316-353
  for metals, 133-150
  for nitrogen, 151-187
  for oxygen, 377-409
  for phosphorus, 359-366
  for sulfur, 276-315
Combustion temperature(s),
  carbon-hydrogen determination, 236, 247, 259
  nitrogen determination, 155
  oxygen determination, 378
  Schöniger combustion, 291
  sulfur and halogens determinations, 276-315, 316-366

Combustion train,
  for carbon-hydrogen determination, 245-248, 259
  for halogens determination, 324
  for nitrogen determination, 154, 164, 165
  for oxygen determination, 380, 381, 394, 401
  for sulfur determination, 298-301
Combustion tube(s),
  for arsenic determination, 368
  for carbon-hydrogen determination, 230
    conditioning of, 245-248, 250
    filling of, 230-235
    pressure regulation of, 248, 249
  for Carius method, 278, 280, 281
  for Dumas determination, 158, 171, 173
    filling of, 158
      permanent filling, 158
      temporary filling, 159, 163
  for halogens determination, 317, 324
  for manometric apparatus, 463
  for oxygen determination, 391, 392
    conditioning of, 394, 401
    filling of, 392
  for phosphorus determination, 360
  for sulfur determination, 297, 304
  with inner spiral, 297, 304
Combustion unit, 157, 158
Complexometric method (EDTA), 124, 139, 364, 416
Compressed air supply, 225, 226
Conditioning carbon-hydrogen combustion tube, 250
Conductometric methods (see Conductometric titrations)
Conductometric titrations, 124, 342, 406, 415
Connecting combustion tube with extraction chamber (manometric apparatus), 465, 466
Connecting tube and cup, 462, 463
Constant load principle for balance construction, 40
Constant temperature chamber, 399, 400
Contact star(s), platinum, 298, 299
Copper, determination of, 133, 141
  use of, in Kjeldahl nitrogen determination, 188
    in Dumas nitrogen determination, 153

Subject Index 646

Copper oxide, 224
  use of, in determination of carbon-hydrogen, 230, 235
    nitrogen, 153, 175
    conditioning of, 153
    oxygen, 392
Copper oxide-lead chromate mixture, 224, 230-235
Copper oxide oxidation tube, 392
Copper salt(s), analysis of, for carbon-hydrogen, 223
Copper turning(s), 153
  use of, in determination of nitrogen, 153, 158
    oxygen, 377, 397
Correction for zero reading of balance, 53-55
Correction table for micronitrometer, 169
Corrections, 5, 168
Coulometric method, use for, determination of arsenic, 374
  for determination of unsaturation, 505
Counterweight(s) (*see also* Tare flasks or bottles), 65, 73
Critical determination(s), 7
Crucible, Neubauer (porcelain filter crucible and cover), 371
Crucible filter assembly, 371, 372
Cryoscopic method, for the determination of molecular weight, 528-535, 544
Crystallization, 96
  dry, Hooker, 83
Cup, aluminum, for holding capsules, 78
Cuvette, 105, 106
  use of, for determination of sulfur, 282
    for standardization of barium chloride, 110
Cyclohexanol, use of, in molecular weight determination, 529, 535
p-Cymene, use of, in heating mortar, 258

D

Decomposition of alpha amino groups, 478
Deflection, of balance (*see also* Zero reading), 50-56
Dehydrite (Anhydrone, magnesium perchlorate), 223
  use of, for absorption of water, 223, 378
    for carbon-hydrogen determination, 223, 228, 238-240

Density, determination of (*see* Specific gravity)
Depression of melting or freezing point method (cryoscopic), for determination of molecular weight, 528-535, 544
  additional information related to, table of, 543, 544
Desiccator(s), 97
  metal block and, 68-70
  with metal insert, 68-70
Dewar flask (thermos flask, bottle), 155
Dialkyl sulfide group(s), determination of, 516, 520
Dichromate, use of,
  in determination of carbon-hydrogen, 222, 255
  in determination of halogens, 337
  in determination of nitrogen, 152
  in manometric determination of carbon, 455, 456
Digestion flask(s), Kjeldahl, 192, 211
  use of, in determination of phosphorus, 356
Digestion rack, 193
  use of, in determination of nitrogen, 193
    phosphorus, 356
Direct neutralization, 410, 415
Direct titration, use of, in determination of,
  bromine, 413
  carboxyl groups, 413-421
  chlorine, 413
  hydrogen, ionic, 413-421
  neutralization equivalent, 413-421
  sulfur, 413
Disodium biphenyl method for halogens, 338, 341
Dissociation constants, determination of, 416
Distillation, for fractionation purposes, 96
  in determination of, acyl groups, 445-453
    alkimide groups, 510, 511
    carboxyl groups, 416
    fluorine, 328-332
    nitrogen, 196
Distillation apparatus, for fluorine determination, 328-331
  for Kjeldahl determination, one-piece model, 196
    Pregl (Parnas-Wagner), 199

Distilled water reservoir, 465
Disulfide group(s), determination of, 516, 520
Dithiocarbamate group(s), determination of, 516, 520
Double bond(s), determination of (*see* Unsaturation)
Draft(s), elimination of, in balance room, 13
Drier(s), 88-91
Dropping bottle, for mercury, used for manometric procedures (*see also* Bottle), 464
Dry crystallization, Hooker (*see* Crystallization), 83
Dry ice, use of, as source of carbon dioxide, in Dumas method, 154, 180
Drying chamber (*see* Abderhalden pistol driers)
Drying pistol (*see* Abderhalden pistol driers)
Drying tubes (*see* Absorption tubes)
Dumas determination of nitrogen, 151-187 (*see also* Nitrogen)
 additional information on, 171-181
 table of, 178-181
 Gustin automatic modification, 177
 on fluoro-compounds, 164, 179
 simultaneous, with other elements, 179
Dumas method (*see also* Nitrogen), 151 (*see also* individual subjects, such as, filling of tubes, temperatures, variations, etc.)
 first combustion, 165, 166
 second combustion, 165, 167
 sweep out, 164, 165

E

Ebullioscopic method, for the determination of molecular weight (boiling point rise), 528, 543
EDTA titrations, 124, 139, 308, 339, 342, 364
EDTA titration methods, see EDTA titrations
Effect of heat on balances (*see also* Heat), 13
Effusion method, for molecular weight determination, 544

Ejection of air from manometric apparatus, 466
Ejection of unabsorbed gases in manometric apparatus, 470
Electric clock motor(s), use of, 155
Electric furnace(s), (*see* Combustion furnaces)
Electric heater, Hankes type, 430
Electrodeposition methods, 140
Electrolytic method, use of, for determination of,
 arsenic, 374
 carboxyl groups, 415
 hydrogen, ionic, 415
Electronic balance(s), 46 (*see also* Balances)
Electrostatic (static) charge(s), 74
 effect of, 74
 elimination of, 74, 253, 259
Elimination of bubbles sticking to mercury surface in micronitrometer, 152, 175, 181
Empirical formula(s), calculation of, 568-570
 use of, for calculation of percentages, 568-570
Empty tube technique(s), 259, 306, 307, 337
α-Epoxy group(s), determination of, 519
Errors, 404
Esters, determination of, 516, 518
Ester saponification, determination of, (*see* Saponification number)
Ethanol, neutral, 411
 use of, for determination of neutralization equivalent and carboxyl groups, 411, 412
 for standardization of alkalies, 111
Ethanolamine-sodium method, for determination of halogens, 341
Ethoxyl group(s) (*see* Alkoxyl)
Ethyl alcohol (*see* Ethanol)
Ethyl-benzene, 557
 use of, as standard for specific gravity determination, 556, 557
Ethyl bromide, use of, in molecular weight determination, 536
Ethyl ether, use of, in molecular weight determination, 536

Ethyl formate, use of, in molecular weight determination, 536
Ethyl group(s), determination of,
N-
O- } (see Alkimide and Alkoxyl)
S-
Ethylene, determination of (see Unsaturation)
Ethylimide group(s) (see Alkimide)
Evaporation, 93, 97
Evaporator, 93
Exaltone, use of, in molecular weight determination, 529, 535
Exchange resins, use of, in determination of nitrogen, 179
Explosive substance(s), combustion of, 264
Extraction, 97
Extraction of carbon dioxide and reading of P_1 on the manometric apparatus, 470, 471

F

Factor(s), 137, 254, 286, 290, 322, 359, 369, 373, 396, 404, 437, 511
  method of calculating, 567, 568
  table of, 474, 484, 571
    α-amino nitrogen, 484
    carbon, 474
    gravimetric, 571
    volumetric, 571
False results in absence of alkoxyl groups, 439
Fatty acid(s), determination of, 410-421
Felt (see Flannel)
Fiberglas, use of, in filling bubble counter-U-tube, 225, 228
Filter crucible (see Crucible)
Filter paper carriers for flask combustion procedures, 294
Filter stick, platinum, 302, 303
  porcelain, 287
Filter tube(s), 317
  cleaning of, 317-319, 322, 356
  drying of, 318, 322, 358
  weighing of, 319, 322, 358, 359
  wiping of, 319, 322, 358
Filtration, 97, 358
Filtration assembly, 319
  for arsenic determination, 371, 372
  for halogens determination, 319
  for phosphorus determination, 356

Filtration procedure(s), 97
  for arsenic, 372
  for barium, 361
  for halogens, 321, 322, 325
  for phosphorus, 358-361
  for sulfur, 287-289
  using centrifuge tubes, 83, 96
Fisher-Davidson Gravitometer, 554-557
Flame photometric methods, 139, 363
Flannel, or felt, 244
  use of, for wiping absorption tubes, 252, 253
Flask(s), Kjeldahl digestion, 192, 211
  tare, 65, 244
  volumetric, 83, 113, 114, 126
Flask combustion,
  electrical ignition unit for, 293
  use of, in determination of
    arsenic, 373
    halogens, 328, 337, 341
    metals, 139
    phosphorus, 363
    sulfur, 291-295, 306
Fluoride, titration of, 112, 113
Fluorine,
  determination of, 326-334, 340-353 (see also numerous items regarding determination, such as apparatus, modifications, conditions, types of compounds, etc.)
  additional information on, 333-335, table of, 340-342
  distillation apparatus for, 328-331
  in absence of arsenic, mercury, or phosphorus, 331
  in presence of arsenic, mercury, or phosphorus, 332, 340
  table of, 340-342
  effect of, on determination of,
    carbon-hydrogen, 222, 230-235, 251
    carboxyl, 416
    halogens, 316, 336
    nitrogen, 164, 179, 214
    oxygen, 377, 405
    phosphorus, 354, 362
    sulfur, 305
Fluoro-compounds, analysis of,
  for acyl (acetyl, formyl) groups, 444
  for alkoxyl (ethoxyl, methoxyl) groups, 423

for bromine, chlorine, iodine, 316-326, 332-339, 342-353
for carbon-hydrogen, 222, 230-235, 262
for carboxyl, 416
for nitrogen, 164, 179, 214
for oxygen, 377, 405
for phosphorus, 359-362
for sulfur, 276-315
Fluosilicic acid,
  distillation of, 328-332, 342
  titration of, 332
Folic acid type of compound, determination of nitrogen in, 188, 189
Forceps, 67, 68, with
  conical tapered holders, 68
  platinum tips, 67
Fork, 71
  use of, in weighing of absorption tubes (carbon-hydrogen), 244
Formaldehyde, conversion of methanol to, as a test for methanol, 93
Formic acid, use of, in iodometric determinations, 398, 403, 424
  in Kjeldahl nitrogen determination, 191
Formula(s), general, 565
  gasometric, 565
  gravimetric, 565
  molecular weight, 566
    hydrogen number, 566
    Neumayer (isothermal distillation), 543
    neutralization equivalent, 566
    Rast, 566
    Signer (isothermal distillation), 566
  volumetric, 565
Formyl group(s), determination of,
  additional information related to, table of, 450, 451
  N-formyl, 444-453
  O-formyl, 444-453
  types of compounds which interfere, 444, 451
  various methods, 450, 451
Fractionation, 97
Freezing point, determination of, 550, 561
  (see also Melting point)
  apparatus for, 549-551
Friedrich bottle, 391
  (see also Pressure regulator, Friedrich type)

Furnace(s) (see Combustion furnaces)
Fusion, use of, in determination of nitrogen, 179
Fusion method(s), for determination of,
  halogens, 333, 337, 341
  arsenic, 373
  oxygen, 406
  phosphorus, 363
  sulfur, 307

G

Gamma (see Microgram)
Gas analysis, 491, 516, 520
Gas chromatographic procedures, 264, 439, 451, 491
Gas solubility studies, 489, 491
Gasometer, 172
Gasometric determination of nitrogen (see Nitrogen, Dumas determination of)
General, miscellaneous information, various chapters (for example, for chlorine, see table at end of Chapter 11, i.e., Table 23), 21, 126, 138, 178, 212, 259, 260, 305, 335, 336, 340, 362, 373, 404, 416, 438, 439, 450, 491, 504
General formulas, 565, 566
  gasometric, 565
  gravimetric, 565
  manometric, 565
  volumetric, 565
Generator(s),
  carbon dioxide, 171, 174, 180
  steam, 196-199
Glacial acetic acid, use of,
  in manometric determination, 475
  in molecular weight determination, 536
  in unsaturation determination, 495
Glass bead(s) for burette drainage tube, 104
  for fluorine determination, 332
  for tare purposes, 71
  for tare bottles (counterpoise weights), 71, 244
Glass cement (see Kroenig glass)
Glass cutting [see Grinding (cutting) wheel, use of]
Glass plunger(s) (plunger rods), 513
Glass rod, use of,
  for acyl determination, 447
  for hydroxyl determination, 513
  for molecular weight determination, 531

Subject Index

Glycols and polyhydric alcohols, determination of alkoxyl groups on, 439
Gold, determination of, 133, 141
Gold chloride solution, use of, in alkimide determination, 509
Gold salt(s), analysis of, for carbon-hydrogen, 223
Graded glass tube(s) and plunger combination (melting capillaries, Rast tubes), 513
Gravimetric Carius method,
  for halogens (see Carius method, for halogens; Bromine, Chlorine, Iodine)
  for sulfur (see Carius method, for sulfur)
  for phosphorus (see Carius method, for phosphorus)
Gravimetric factor(s) (see Factors)
Gravimetric oxygen procedure (see Oxygen)
Gravimetric procedures,
  for alkoxyl (ethoxyl, methoxyl) groups, 433-437, 439
  for arsenic, 370-376
  for barium, 361, 362
  for halogens (bromine, chlorine, fluorine, iodine), 316-326, 338, 341
  for metals, 140
  for oxygen, 378-396, 404-409
  for phosphorus, 354-366
  for sulfur, 287-290, 295-303
Gravitometer (see Fisher-Davidson Gravitometer)
Grease (Sisco #300), use of, 152
  in manometric apparatus, 458
  in micronitrometer, 152, 162
Grignard reagent, use of, in determination of active hydrogen (see Active hydrogen)
Grinding (cutting) wheel, 282
  use of, for opening combustion tubes (Carius), 285, 286
Ground joints, use of, on Dumas nitrogen apparatus, 158-163, 181
Grounding, use of, for eliminating electrostatic charges, 74, 253, 259
Group determination(s) (see under specific name)

Groups capable of reacting with Grignard reagent, determination of (see Active hydrogen)
Guard tube, 240, 241, 246, 247
Gutzeit test, use of, for determination of arsenic, 374

H

Halogen(s) (see Bromine, Chlorine, Iodine, Fluorine) (see also numerous items regarding determination, such as apparatus, modifications, conditions, types of compounds, etc.)
Halogenation, use of, in determination of unsaturation, 505
Hanger(s) (see Pans, of balance)
Hankes heater (see Electric heater)
Hayman combination, 58-63
Heat, sensitivity of balance to, 13
Heating, 12
Heating mortar, 238, 258
  use of, for carbon-hydrogen determinations, 245-250, 258
Heavy-walled impregnated connectors, 243, 244
Heavy-walled rubber tubing, 243, 464
Helium, use of, in determination of oxygen, 379, 397
Hempel pipette, 476
  filling of, 476
Heraeus combustion apparatus, 237
Heterometric titrations, 124, 139
High-frequency discharge, use of, for elimination of electrostatic charge, 74, 259
High-frequency methods (see High-frequency titrations)
High-frequency titrations, 124, 339, 342, 415
Historical data, 1, 22
Homogeneity of samples, 87
Humidity, relative, ideal, for elimination of electrostatic charges, 12, 74
Hydrazine, use of,
  in carbon determination, 456, 457
  in halogen determination, 323
Hydrazine sulfate, 323
Hydrazines, analysis of, for nitrogen, 151, 189, 209
  determination of, 520

Hydrazones, analysis of, for nitrogen, 151, 189, 209
Hydriodic acid, 424
  treatment of, 424, 439
  use of, for alkimide determination, 508
  for alkoxyl determination, 422-443
Hydrochloric acid, standard, 0.01N, 112, 192
  standardization of, 112
  treatment of concentrated acid for use in arsenic determination, 367
Hydrochloric acid, use of, in Kjeldahl nitrogen determination, 191
Hydrochloric acid of crystallization, 93
Hydrogen, active, determination of (see Active hydrogen)
  determination of,
    gravimetric (see Carbon and hydrogen)
  beta-ray absorption, 264
  interference of, in determination of oxygen, 399, 405
  ionic, determination of (see Neutralization equivalent)
Hydrogen cylinder, 496
Hydrogen number, determination of, (see Unsaturation)
Hydrogen peroxide (Superoxol, Perhydrol), 296
Hydrogenation, 495-507
  use of, in determination of nitrogen, 179
  in determination of sulfur, 307
Hydrogenation apparatus, 497-499
Hydrogenation methods, 179, 307, 406, 495-507
Hydroxamic acids, determination of, 517
Hydroxyl group(s), determination of, 512-514, 519
  additional information related to, table of, 517-520
  micromethod, 512-514
  semimicromethod, 514
Hydroxylamine number, determination of, 518
Hydroxyquinoline, (8-), use of, for determination of magnesium, 490
Hygroscopic sample(s) weighed in boat(s), 74, 75
Hygroscopic substance(s),
  drying of, 62, 91
  weighing of, 74-76, 62

I

Illuminated titration stand assembly, 104-106
  use of, for determination of sulfur, 282
  for standardization of barium chloride, 110
Illumination of balances, 19
Indicator(s), 103, 104, 126, 213, 308, 339
  bromocresol green-methyl red, 103, 191
  methyl red, 103
  methyl red-methylene blue, 103, 191
  phenolphthalein, 104
  sodium alizarin sulfonate, 103
  starch, 103
  tetrahydroxyquinone (THQ), 104
  thymolphthalein, 103
Indirect method(s) based on manometric combustion of organic precipitates, 489, 490
  determination of,
    magnesium, 490
    phosphorus, 489, 490
    sulfur, 490
Inertia block-isolator type table, 13-20
Insulation, weatherproofing, 13
  windows, 14
Interferences of oxygen determination, 377
  fluorine, 377, 405
  hydrogen, 399, 405
  metals, 377
  phosphorus, 377
  sulfur, 379, 405
Introduction (laboratory set-up, general information, test compounds, etc.), 1
  additional information related to subject, 21, 22
Iodine, determination of, 316-325, 332-339, 342-353 (see also *numerous* items regarding determination, such as apparatus, modifications, conditions, types of compounds, etc.)
  additional information on, 332, 339
  table of, 335-339
  by Carius combustion, 316-323, 325, 336
  by catalytic combustion, 323-325, 337
  by fusion, 333, 334, 337
  by Schöniger combustion, 337
  by titration methods, 338, 339

radioactive, 339
simultaneous, with other halogens, 336
use of, in determination of unsaturation, 495, 505
Iodine pentoxide, use of, in determination of oxygen, 396, 397, 400, 401, 405, 406
Iodine solution, standard, 0.01N, 109
standardization of, 109
Iodine valve, 505
Iodoform reaction, use of, for detection of acetone, 93
Iodometric determination(s) (*see* under individual determination)
Ion exchange methods, 339, 342
Ionic halogen, determination of, 413
Ionic hydrogen (*see* Carboxyl group; Neutralization equivalent)
Iron, analysis of, salts, for carbon-hydrogen, 223
determination of, 133, 141
wire, knurled, 71
Iron powder, use of, in Kjeldahl nitrogen determination, 191
Isocyanate group(s), determination of, 516, 520
Isopropylidene groups, determination of, 516, 518
Isothermal distillation procedure for determination of molecular weight, 535-547
additional information on, 540-544
table of, 543, 544
Neumayer apparatus for, 540, 541
Signer apparatus for, 536, 537
Isothiocyanate group(s), determination of, 516, 520
Isotopic carbon, determination of, 264, 491
Isoxazoles, analysis of, for nitrogen, 189, 209

K

Karl Fischer reagent, use of,
for determination of unsaturation, 505
for determination of water, 94, 97
Ketones, determination of, 516, 518
Kipp generator, 171, 180
Kjeldahl determination of nitrogen, 188-220 (*see also* Nitrogen and *numerous* items regarding the determination, such as apparatus, conditions, types of compounds to be analyzed, modifications, etc.)
additional information on, 211-215
table of, 212-215
digestion conditions, 214
elimination of titration, 214
fluoro-compounds, 188, 214
manometric method, 488
N-C compounds, 188, 205-209
N-N compounds, 189, 209, 214
N-O compounds, 189, 209, 210, 214
nitrates, 210
omission of distillation, 214
other methods, 215
results of collaborative study on, 188-190
Kjeldahl-type digestion, 139, 188-220, 357, 358, 362, 373
Knife edge(s), of balance, 32, 33, 38, 39
Knurled iron wire, 71, 244
Kofler micro hot stage, 549-551
use of, for determination of melting point, 548-552
for determination of molecular weight, 544
Kroenig glass cement, 153, 225

L

Laboratory plan, 10, 12
Laboratory report sheet(s), (microchemical reports), 10, 11
Laboratory set-up, 10-21
additional information related to subject, 21, 22
Lactic acid, use of, in manometric determinations, 458, 470
Lambda (*see* Microliter)
Lead, determination of, 141
Lead chromate, 224
use of, in determination of carbon-hydrogen, 230-235
Lead peroxide (lead dioxide), 224
use of, in determination of carbon-hydrogen, 230-235
omission of, 261, 262
substitutes for, 261, 262
Lead shot, 71
Leveling bulb attachment, manometric apparatus, 469
Lighting fixture(s), grounding against, to dissipate electrostatic charge, 74, 253

653   Subject Index

Liquid(s), weighing of, 76-78
  non-volatile, 76
  volatile, 76-78
Literature cited (*see* References)
Lithium, determination of, 142
Logarithm(s), nitrogen reduction, table of (*see* Nitrogen reduction table)
  table of, 584-601

M

Magnesia mixture, use of, in determination of arsenic, 370, 372
Magnesium, determination of, 133, 142
  by ashing, 133
  by manometric procedure, 490
Magnesium aluminate, use of, in carbon-hydrogen combustion tube for fluoro compounds, 234, 235
Magnesium nitride, use of, in determination
  of nitrogen, 179
  of oxygen, 377
Magnesium oxide, use of, in determination of carbon-hydrogen, 222, 224, 230-235, 251
Magnesium perchlorate (Anhydrone, Dehydrite), use of, in carbon-hydrogen determination, 223, 228, 229, 239, 240
  in oxygen determination, 378
Manganese, determination of, 142
Manifold, for use in Kjeldahl digestion, 195
  for vacuum driers, 91
Manifold support, 195
Manometric apparatus, 458-465, 480-482, 485-489, 492
  ampul testing attachment, 485-487
  Barcroft, 492
  Barcroft-Warburg, 492
  Van Slyke, 458-465, 480-482
  vial testing attachment, 488, 489
  Warburg, 492
Manometric determination(s), 454-494
  additional information related to, table of, 490-492
  ampul and vial testing, 485-488
  blood and urine, 488
  for carbon, 264, 454-474, 490-494
  for carboxyl, 415
  for magnesium, 140, 490

  for nitrogen, 488, 491
  alpha-amino, 474-485, 490-494
  for oxygen, 377, 406, 491
  for phosphorus, 364, 489
  for sulfur, 307, 490
  gas solubility, 489
Marble, use of, in nitrogen determination, 180
Mariotte bottle, 240, 242, 243
  use of, in determination of,
    carbon-hydrogen, 240, 242, 245-252
    halogens, 324
    oxygen, 380, 381, 393
    sulfur, 298
Marsh test, use of, for determination of arsenic, 374
Measurement of alkaline hydrazine solution into extraction chamber, 467
Measuring the nitrogen in manometric apparatus, 483
Melting point, determination of, 548-552, 561
Melting point apparatus,
  micro (Kofler), 549-551
  Thomas-Hoover, 529-531
Melting point bath, 529, 552, 553
Melting point capillaries for use in molecular weight determination, 531, 532
Mercapto group(s) (mercaptans) determination of, 516, 520
Mercuric oxide, use of, in Kjeldahl nitrogen determination, 190
Mercury, determination of, 142
Mercury, effect of, on analysis of other elements, 222, 262
  use of, in Dumas determination, 152
    in Kjeldahl nitrogen determination, 188
    in manometric apparatus, 458, 500
Mercury bottle with capillary, small, 464
Mercury dropping bottle, 466
Mercury valve for use with dry ice generator, 172
Metal(s), determination of, 133-150
  additional information, 137-143
  table of, 138-143
Metal cooling block, 68-70
Metallic compounds, effect of, on carbon-hydrogen determination, 222, 223, 262

Subject Index

Methane, formation of,
  in active hydrogen determination, 514
  in Dumas determination, 152
Methanol of crystallization, determination of, 93
Methoxyl apparatus (*see* Alkoxyl apparatus)
Methoxyl groups (*see* Alkoxyl groups)
Methyl acetate, use of, in molecular weight determination, 536
Methyl group(s),
  determination of (*see* under individual group)
  effect of, on determination
    of carbon-hydrogen, 222
    of nitrogen, 151
Methyl group(s) attached to carbon, determination of, 516, 518
Methyl red, 103
Methyl red-methylene blue indicator, 103
Methylene chloride, use of, in molecular weight determination, 536
Methylimide (*see* Alkimide groups)
Mettler microbalance, 45
  ultramicro balance, 45
Mettler microchemical balance, 37-40
Microazotometer (*see* Micronitrometer)
Microbalance(s), 30, 45
Microbubble(s), 164, 165
Microburette(s), (*see* Burettes)
Microburner, 430
Microcentrifuge tube(s), 83-86
  with conical bottom, plain, 83, 84
  with conical bottom, stoppered, 83, 85
  with conical bottom, stoppered, grad., 83, 86
  with cylindrical bottom, plain, 83, 86
Microchemical balance (*see also* Balance), 30, 46
Microchemical laboratory, plan for (*see also* Plan for laboratory), 10
Microcrystallization (*see* Crystallization)
Microdiffusion methods, 339, 416
Microdistillation, 96, 416
Microevaporator, 94
Microgram (gamma), 2
Microgram procedures (*see* Ultramicro-, submicro methods)
Micro hot stage (*see* Kofler micro hot stage)
Microliter, 2

Micromuffle, 134, 137, 138
Micronitrometer (microazometer), 161, 180, 181, 492
  calibration of, 162
  use of two, 181
Micropipettes, calibration of, 116, 125, 126
  (*see also* Pipettes)
Microscope, use of, for melting points, 548-552
Microspatula(s), 66, 67
  (*see also* Spatulas)
Microtitration techniques, 102-132
  additional information, 115-126
  table of additional information, 124-126
Microtitrators, 125
Micro weighing pipettes, density-type (pycnometers), 558, 559
Modification of carbon-hydrogen method, for salts of alkali and alkaline earth metals, 255
Modification of Kjeldahl method for N—N, N=N, NO, NO_2 linkages, 209, 211, 214
Modification(s), of apparatus (*see* Additional Information for and Additional Information on References Related to each chapter)
Modification of various methods (*see* table on additional information on references related to the various chapters), 179, 213, 214, 262, 263, 491
Modified Abderhalden drying apparatus, 88-91
  parts of,
    desiccator bulb, 90
    drying chamber, 89
Modified Clark alkoxyl apparatus, 426-430
Moisture, determination of, 91-93
  additional information on, 93-95
  table of, 96, 97
  use of (for determination of)
    calcium carbide, 97
    cinnamyl chloride, 97
    Karl Fischer, 92, 94, 95, 97
    naphthyl phosphorus oxychloride, 97
    neutron scattering, 97
    succinyl chloride, 97
Molecular refraction, determination of, 562
Molecular weight, determination of,
  additional information on, 540-544
  table of 543, 544

by boiling point rise (ebullioscopic), 543
by cryoscopic (Rast) method, 528-535, 544-547
by depression of melting or freezing point method (cryoscopic), 528-535, 544-547
by effusion method, 544
by isothermal (Signer) method, 535-539, 544
by Neumayer method, 540-543, 544
by neutralization equivalent, 410-421
by other determinations, 528, 543, 544
by vapor pressure comparison method, 544
by vaporimetric method, 544
Molybdate reagent, preparation, 355
use of, in phosphorus determination, 358, 361
Mortar (see Heating mortar)
Mounting the balance (see Balance)
Muffle (see also Micromuffle), 134
use of, for conditioning copper oxide, 153
for determination of arsenic, 371

N

Naphthyl phosphorus oxychloride, use of, in determination of moisture, 97
Needle, valve,
glass, 180
metal, 160
use of in,
acyl determination, 447
Dumas method, 160, 180
N-ethyl group(s), determination of (see Alkimide groups)
N-methyl group(s), determination of, (see Alkimide groups)
Neubauer crucible (see Crucible)
Neumayer method for determination of molecular weight, 540-543
Neutral ethanol (see Ethanol)
Neutralization equivalent, determination of, 410-421
additional information related to, table of, 413-416
as a means of determination of molecular weights, 410
Neutron absorption methods, 339
Nickel, determination of, 142

Ninhydrin reaction, 491
use of, for determination of amino acids, 491
Nitrates, analysis of, for nitrogen, 152, 179, 188, 210
determination of, 516, 517
Nitrate(s), modification of Dumas method for analysis of, 152, 179
modification of Kjeldahl method for analysis of, 188, 210
Nitric acid, use of, in determination of,
alkimide groups, 509
alkoxyl groups, 434, 436
arsenic, 367
halogens, 317
metals, 133
phosphorus, 355
sulfur, 277
Nitric-sulfuric acid mixture, 355
Nitriles,
determination of, 517
determination of nitrogen in, 151, 188, 191
Nitro compound(s), determination of nitrogen in, 152, 179, 188, 209, 214
determination of, 516, 517
Nitrogen, determination of (see also Dumas, Kjeldahl, as well as numerous individual items—modifications, tube fillings, temperatures, apparatus involved, types of compounds to be analysed, etc.)
additional information on,
Dumas method, 171-181
table of, 178-181
Kjeldahl method, 211-215
table of, 212-215
α-amino (manometric),
table of, 490-492
alpha-amino, 474-485, 490 494
aminoid, 188-220
chromatographic method, 214, 491
colorimetric method, 179, 214
Dumas, 151-187
exchange resins method, 179
flame photometry method, 179
fluoro-compounds, 164, 179, 214
fusion method, 179
hydrogenation method, 179
iodometric method, 179, 214

Kjeldahl, 188-220
  magnesium nitride method, 179
  manometric method, 179, 474-485, 490-494
  Kjeldahl, 488
  N-C compounds, 152
  N-N compounds, 152, 188, 214
  N-O compounds, 152, 188, 214
  spectrophotometric method, 214
  various methods, 215, 490, 491
  micronitrometer, 161
    calibration, 162
    correction tables, 169
  reduction table, 574-583
  use of, for conditioning of hydriodic acid, 424
    for oxygen determination, 379, 397
    for unsaturation determination, 496
Nitrogen reduction table, 574-583
m-Nitro-methyl benzoate, preparation of, as means of detecting methanol, 93
p-Nitro phenyl hydrazone, use of, for detection of acetone, 93
Nitrometer (see Micronitrometer)
Nitrometry, use of, for alkoxyl determination, 439
Nitroso group(s), determination of, 516, 517
N-methyl group(s), determination of (see Alkimide groups)
N—N, N=N, NO, and NO_2 compounds, analysis of for nitrogen, 151, 152, 189, 209
Nomenclature, 2, 21
Nonaqueous titrations, 124, 125, 339, 416, 544
Nuclear magnetic resonance methods, 342

O

Oertling microchemical balance, 44, 45
Optical methods, 338, 339, 363, 560, 561
Optical rotation, determination of, 560, 561
Orange brown glass filter plate, 105, 106
  use of, for determination of sulfur, 282
    for standardization of barium chloride, 110
Ordinary laboratory oven(s), 91, 289, 320
Organic acid(s) (see Carboxyl group, determination of)
Oscillometric methods, 342

Osmotic pressure, use of, for molecular weight determination, 544
Oven, drying devices, 91, 97
Oxidation method for carboxyl groups, 415
Oxidation tube, 392, 400
Oxidation tube fillings for determination of oxygen,
  copper oxide, 392, 406
  hopcalite (manganese dioxide), 406
  iodine pentoxide, 400, 405, 406
  mercuric oxide, 406
Oximes, analysis of, for nitrogen, 189, 209
  determination of, 517
Oxygen, determination of, 377-409 (see also numerous items regarding determination, such as apparatus, types of compounds, modifications, interferences, etc.)
  additional information related to, table of, 404-406
  comparison of methods, 377, 404
  gravimetric, 378-396, 404-409
  hydrogenation methods, 406
  isotopic, 406
  iodometric, 396-409
  manometric, 377, 406, 491
  volumetric, 396-409
Oxygen, determination of, in presence of,
  fluorine, 377, 405
  hydrogen, 399, 400, 405
  metals, 377, 406
  phosphorus, 377
  sulfur, 379, 392, 405
Oxygen cylinder, use of, in determination of
  carbon-hydrogen, 225
  halogens, 324, 328
  sulfur, 297
Oxygen flask combustion (see Flask combustion)
Oxyhydrogen flame methods, 306, 307, 337, 341

P

Palladium, determination of, 133-150
  use of (oxide) as catalyst for hydrogenation (determination of unsaturation), 496
Palladium (or palladium on asbestos), use of, for eliminating hydrogen interference in determination of oxygen, 398-405

Palladium tube, 399, 400
Pan(s), of balance (hangers), 34
Parr bomb assembly, use of, for determination of fluorine, 333, 334
Paraffin-lined bottle, for storing molybdate reagent, 355
Percentage, calculation of, 565, 568, 569
Perchloric acid, 328
  ashing with, 139, 307
  distillation of fluosilicic acid with, 332
Perchloric acid combustion methods, 139, 307, 362
Perchloric acid, use of, in determination of fluorine, 328, 332
Perhydrol (Superoxol) (*see* Hydrogen peroxide)
Peroxide group(s), determination of, 516, 519
pH meter, 106
  use of, for titration of fluoride, 112
Phenol, 423
  use of, in alkimide determination, 509
  in alkoxyl determination, 423, 433
Phenol(s), effect of, on determination of neutralization equivalent and carboxyl, 410
Phenolphthalein, 104
Phosphoric acid of crystallization, 93
Phosphoric acid, use of, in manometric determination, 458
Phosphorus,
  determination of, 354-366
    additional information related to, table of, 362-364
    as quinoline phosphomolybdate (8-quinolinol phosphomolybdate), 363
    as strychnine phosphomolybdate, 363
    fluorine-free compounds, 354-359
    fluoro-containing compounds, 359-362
    simultaneously,
      with barium, 361, 362
      with sulfur, 306, 362
  effect of, on
    carbon-hydrogen determination, 223, 255, 262
    oxygen determination, 377
    sulfur determination, 276-315
Photoelectric filter photometer, 106
  use of in determination of fluorine, 328, 331
  use of in standardization of fluoride, 112
Photometric titration, 139, 374
Physical chemical technique for determination of oxygen, 406
Physical constant(s) (properties), determination of, 548-564
  additional information related to, table of, 561, 562
  boiling point, 552-554, 561
  density, 554-559, 561
  melting point, 548-552, 561
  molecular refraction, 562
  optical rotation, 560, 561
  refractive index, 560, 562
  specific gravity, 554-559, 561
  viscosity, 562
Pig (piggie) (*see also* Weighing bottle), 60-64
  Hayman, 63
  aluminum, 63
  glass, 60
  metal, 63, 64
Pinene dibromide, use of, in molecular weight determination, 529, 535
Pipe, grounding against, to dissipate electrostatic charges, 74, 259
Pipettes, 83, 96, 125
  calibration of, 116, 125, 126
  density (pynometer or weighing), 116, 558, 559
  Folin, 116
  micro-, 116
  microliter, 116-119
  with cylindrical tips, 113, 115
  self-filling, self-adjusting, 116
  precision, 116, 125
  stop-cock, 463, 464
  washout, 120
Pitchblende (radioactive material), use of, to eliminate electrostatic charges, 74, 259
Plan for laboratory, 10, 12
Platinized asbestos, 224, 496
Platinum, asbestos, 224
  boat (*see also* Boat, platinum), 58-61
  catalyst, 496
  contact star, 298, 299
  cylinder, 134, 251
  determination of, 133, 142
  filter stick, 302, 303

forceps (tips), 67
hook (wire), 73
sleeve, 134, 251
tetrahedra, 431, 466
wire, 73
Platinum, effect of, on Kjeldahl nitrogen determination, 205
Platinum salt(s), analysis of, for carbon-hydrogen, 223
Plunger(s), glass, 513, 531
Pointer, of balance, 32
Pointer scale, of balance, 32
Polarimeter, 560
 tube, 560
Polarograph, use of, in determination of,
 carboxyl, 415
 copper, 140
 halogens, 339
 neutralization equivalent, 415
 phosphorus, 364
Polarographic methods (polarography), 140, 339, 364, 415
Porcelain filter crucible and cover (Neubauer), 371
Porcelain filter stick, 287
Porcelain sulfur crucible, 287
Potassium, determination of, 133, 142
Potassium acetate-glacial acetic acid solution, use of, in determination of oxygen, 398, 403
Potassium biiodate (biniodate), 102
 use of, for standardization of bases, 102
 of thiosulfate, 102, 108
Potassium chlorate, use of, in analysis of samples, 76, 77, 151, 152
Potassium dichromate, 222
 use of, in determination of carbon-hydrogen, 222, 224, 251, 255
 in determination of nitrogen (Dumas), 152
 in manometric carbon determination, 456
Potassium hydrosulfide methods for determination of carboxyl groups, 415
Potassium hydroxide,
 use of, in carbon monoxide scrubber tube (oxygen determination), 397
 in micronitrometer (Dumas determination), 152
Potassium iodate, use of, in determination of

acetyl, 445, 449
carbon (manometric), 455, 456
formyl, 445, 449
Potassium iodide, use of, in iodometric determinations,
 acetyl, 445, 447
 alkimide, 509
 alkoxyl, 424
 formyl, 445, 447
 oxygen, 398
 solution, 30%, for cleaning filter tubes, 317
Potassium permanganate, alkaline, use of in manometric determination, 475
Potassium sulfate, 190
 standard solution, 0.01N, 110, 278
Potentiometric methods, 124, 140, 339, 364, 374, 415, 544
Potentiometric titrations, 124, 140, 406, 544
Precision (reproducible sensitivity) (see also Balance)
 determination of, of balance, 55-57
Precision stopcock, 171, 172
 (see also stopcock)
Pregl catalytic combustion method, 295-304
 for halogens, 323-326
 for sulfur, 295-304
  titration procedure, in absence of nitrogen, halogens, and phosphorus, 303
Preheater type of microcombustion furnace, 390, 496
Preheater, 257
 use of, in determination of carbon-hydrogen, 257
Preliminary ejection of air from manometric apparatus, 466
Preparation of samples for analysis, 83-101
 additional information on, 93-95
 table of, 96, 97
Pressure regulation, carbon-hydrogen train, 248, 249
 oxygen apparatus, 394, 402
 Pregl catalytic combustion, 298
Pressure regulator, 225-228, 255, 256
 use of, in carbon-hydrogen determination, 225-261

in Dumas nitrogen determination, 155, 172, 180
in halogen determination, 324
in oxygen determination, 391, 394, 402
in sulfur determination, 298
Pressure regulator, Friedrich type, 391, 496
use of, in oxygen determination, 394
in unsaturation determination, 496
Pressure tube(s) (see Carius combustion tubes)
Prevention of foaming of potassium hydroxide, 152, 175, 181
Prevention of sticking of bubbles to mercury surface, 162, 175, 181
Primary amino nitrogen, in alpha-amino aliphatic acid, determination of (see Manometric determinations)
Primary standards, 126
Problems arising from weighing of absorption tubes, 261
Projection system, for balance, 32, 33, 39
Propionic anhydride, 433
use of, in alkimide determination, 509
in alkoxyl determination, 433
Protein(s), effect of, on manometric procedures, 475
Purification, 97
Purification tube(s),
use of, in oxygen determination, 392
in unsaturation determination, 496
Purifications of carrier gas for oxygen determination, 379, 405
Pycnometer, 558, 559
Pyrazine(s), determination of nitrogen in, 188
Pyrex glass rod (or beads), use of, in, acyl determination, 447
fluorine determination, 332
Pyridine, use of, for hydroxyl determination, 512
Pyrimidine(s), determination of nitrogen in, 151, 188

Q

Quartz, capillaries, 395, 402
chips, 379
combustion tubes, 230, 391, 392
wool, 379
Quinoline phosphomolybdate (8-quinolinol phosphomolybdate),
phosphorus determined as, 363

R

Rack, digestion, Kjeldahl, 193, 194
wire, 70, 244
Radioactive carbon, determination of, 264, 491
Radioactive iodine, 339
Radioactive material(s), use of, to eliminate electrostatic charges, 74, 259
Radioactive methods, 140
Radium tube(s), spent, use of, to eliminate electrostatic charges, 74, 259
Rast molecular weight method, 528-535
tubes, 531
Reading of pressures in manometric apparatus,
p_2, 471
p_0, 483
p_1, 470, 471, 483
Recommendations, 2
Reduced copper purification tube, 392
Reducing valve, 225, 297, 391
References, 21-29, 46-49, 79-82, 96-101, 124-132, 138-150, 178-187, 212-220, 260-275, 305-315, 335-353, 362-366, 373-376, 404-409, 413-421, 438-443, 450-453, 490-494, 505-507, 517-527, 543-547, 561-564, 602
Refractive index, determination of, 560, 562
Reproducible sensitivity, of balance (see Balance, determination of precision)
Rest point, of balance, 50
Reticle, of balance, 32
Return of purified nitrogen to manometric apparatus, 480-482
Review(s), 22, 138, 212, 260, 305, 335, 340, 373, 404, 413, 438, 450, 490, 517, 543
Rider, of balance, 34
Rider carrier, of balance, 34
Rigid, combination type balance table, 19, 20
Rubber connector(s), 243, 246, 247
impregnation of, 243
lubrication of, 244, 245
marking of, 246, 248
Rubber stopper(s),
one hole, 172, 298, 300
solid, 172, 356, 357
Rubber tip(s), 464

Subject Index   660

Rubber tubing,
  heavy-walled, 243, 464
  thin-walled, 243

S

Safety valve, 391
Salicylic acid, use of, in Kjeldahl nitrogen determination, 191
S-alkyl groups (*see also* Alkyl groups), determination of, 438
Salt(s) of alkali and alkaline earth metals, determination of carbon-hydrogen in, 222, 250-255
Salts, use of, as source of carbon dioxide, 180
Sample,
  introducing of, into Dumas tube, 181
  preparation of, for analysis, 83-101
  size of, practical, 2, 31
    theoretical minimum, 31
  weighing of, 73
Sample containers, 88
Saponification equivalent, determination of (*see* saponification number)
Saponification number, determination of, 516, 518
Sargent combustion apparatus, 236
Sargent manometric apparatus, 459
Scale, of balance, various, 32
Scale indicating reaction period required for complete decomposition of alpha-amino acids, 478
Schöniger combustion (*see also* Flask combustion), 291-295, 331, 332
Scrubber solutions, various, for alkoxyl determination, 439
Scrubbing towers (bottles), use of, 379, 496
Selenium, determination of, 143,
  use of, in Kjeldahl determination, 188, 214
Semicarbazide(s), determination of, 520
Semicarbazide, determination of nitrogen in, 152, 189
Semimicrochemical balances, 30, 45
Sensitivity of microchemical balances, 30, 38, 39, 55-57
Separation, 97
Shot, lead, use of, for tare purposes, 71, 244

Signer (isothermal distillation) apparatus, 536, 537
Signer method for determination of molecular weight, 535-539
  additional information related to, table of, 543, 544
Silica, determination of, 143
Silicon, determination of, 143
  determination of alkoxyl on compounds containing, 438
Silver, determination of, 133, 143
  use of, for determination of alkoxyl, 439
    for determination of sulfur, 307
Silver, use of, in determination of carbon and hydrogen, 223, 230-235
Silver condenser tube(s), 204
Silver halide precipitate(s), dissolving of, 322
  drying of, 322
  filtering of, 321, 322
  treatment of (silver iodide) with nitric acid, 436
  weighing of, 322
Silver iodide, treatment of, in alkoxyl determination, 423, 433-437
Silver nitrate, alcoholic (*see* Alcoholic silver nitrate)
  use of, in determination of halogen, 317, 324
Silver perchlorate, use of, in determination of fluorine, 328, 332
Silver permanganate, use of, in carbon-hydrogen determination, 259
Silver salt(s), analysis of, for carbon-hydrogen, 223
Silver sulfate,
  use of, in determination of acetyl, 444, 447
  in determination of formyl, 444, 447
Silver wire,
  use of, in carbon-hydrogen determination, 230-235
Simultaneous determinations (*see* tables on additional information on references related to the various chapters), 139, 179, 264, 306, 336, 340, 362, 373, 405, 438
Simultaneous determination of alkoxyl and alkimide, 438, 509

Simultaneous determination of barium and phosphorus, 361, 362
Simultaneous determination of chlorine and bromine, 326, 336
Simultaneous determination of ethoxyl and methoxyl, 438
Siphon receiver and inner container, 289
Small mercury bottle with capillary, 464
Soda-lime, use of, as absorbent, 457, 461
Sodium, determination of, 133, 143
Sodium acetate solution, 424
  use of, in alkoxyl determination, 424
    in oxygen determination, 398
Sodium-alcohol (and other solvents) method for determination of halogens, 338
Sodium alizarin sulfonate, 103
Sodium fluoride, 112
  standard solution, 0.01N, 112
  standardization of, 112
Sodium hydroxide, carbonate free, 111, 456, 457
  on asbestos (see Ascarite) 223
  five % solution, use of, in pressure regulators, 223, 226
  standard solution, 0.01N, 111
  standardization of, 111
Sodium hydroxide-thiosulfate solution, 192
  use of, in Kjeldahl method, 192
Sodium nitrite, 475
  use of, in manometric determination of alpha-amino nitrogen, 478, 479
Sodium nitroprusside test for acetone, 93
Sodium peroxide, use of, in determination of fluorine, 327
Sodium thiosulfate, crystalline, use of, in Kjeldahl nitrogen determination, 191
Sodium thiosulfate, standard solution, 0.01N, 108
  standardization, 108
  use of, in alkimide determination, 508, 509
    in acyl (acetyl, formyl) determination, 445
    in alkoxyl determination, 424, 434
    in arsenic determination, 368
    in oxygen determination, 398
Solid(s), drying of, 91-93, 97
  weighing of, 73-76
Soltys active hydrogen apparatus, 515

Solvent(s),
  for molecular weight determination,
    by isothermal distillation, 536, 542
    by Rast (cryoscopic), 529, 535
  of crystallization, 92
    determination of, 92
Solvents, various, for alkoxyl determination, 439
Spatula(s) (microspatulas), 66, 67
Specific gravity, determination of, 554-559, 561
  additional information related to, table of, 561, 562
  using gravitometer, 554-558
  using micro weighing pipettes, density type (pycnometers), 558, 559
Spectrophotometric equipment, 125
Spectrophotometric methods, 139, 214, 308, 338, 339, 341, 342, 363, 414, 415, 439, 451, 505, 544
Stand, for manometric apparatus, 460
  illuminated titration (see Illuminated titration stand assembly)
Standard solution(s), 102-115, 126
  additional information, 115-126
  table of, 124-126
  barium chloride, 0.01N, 110
  hydrochloric acid, 0.01N, 112, 126
  iodine, 0.01N, 109
  potassium biiodate (biniodate), 102
  potassium sulfate, 0.01N, 110, 126
  sodium fluoride, 0.01N, 112, 126
  sodium hydroxide, 0.01N, 111
  sodium thiosulfate, 0.01N, 108, 126
  thorium nitrate, 0.01N, 112
Standard(s), use of, in molecular weight determination (see also Test substances, compounds, samples), 536
Standard deviation of balance, 56
Standardization, 2-5
Standardization program(s), 2
Standards (see Test compounds, samples, substances)
Starch indicator, 103
  use of, in determination of,
    acyl groups, 445
    arsenic, 368
    oxygen, 398
Static charge(s) (see Electrostatic charges),
  effect of, 74
  elimination of, 12, 74, 253, 259

Subject Index

Statistical studies, 260
Steam generator, 196-199, 201, 328
Steam trap, 199-202
Steaming apparatus, 123
Stirring devices, 125
Stirrup(s) (suspensions), of balance, 33, 35
Stopcock, three-way, for use in Dumas method, 171
Stopcock cylinder, 461, 464
Stopcock grease, 152, 488
Stopcock grease for manometric apparatus, 488
Stopcock pipette, 464
Storm window, heated, for balance room, 13
Strontium, determination of, 133, 143
Strychnine phosphomolybdate, phosphorus determined as, 363
Sublimation,
 during carbon-hydrogen determination, 251
 use of, for purification, 84, 96
Sublimation apparatus, vacuum, 84, 96
Submicro methods (see Ultramicro-, submicro methods)
Submicrogram methods (see Ultramicro-, submicro methods)
Succinyl chloride, use of, in determination of moisture, 97
Sulfate, titration of, 110, 126
Sulfhydryl group(s), determination of (see Mercapto, Mercaptans)
Sulfonamide(s), Dumas modification for analyzing, 152
Sulfur, detection of elementary, 549
 determination of, 276-315 (see also numerous items regarding determination, such as apparatus, types of compounds, modifications, interferences, etc.)
 additional information on, 304-308
 table of, 305-308
 bomb fusion, 307
 Carius method, 276-290, 306
 catalytic combustion, 295-304, 306, 307
 direct titration, 413
 fusion in general, 307
 gravimetric, 287-290, 308
 hydrogenation, 307
 manometric method, 307
 oxy-hydrogen flame, 306, 307
 perchloric acid digestion, 307
 Schöniger combustion, 291-295, 306
 silver gauze, 307
 simultaneous, carbon, hydrogen, halogen, sulfur, 264, 306
 effect of, on carbon-hydrogen determination, 221, 222, 255
Sulfur compound, interference of, in determination of oxygen, 379, 405
Sulfur compounds, group determinations on, 516, 520
Sulfuric acid, 191
 standardization of, 20% excess of SO_3, 456
Superoxol (see Hydrogen peroxide)
Suspension(s), of balance (see Stirrups)
System for filling capillaries, 79

T

Table(s),
 additional information related to chapters, 21, 22, 46, 79, 96, 97, 124-126, 138-143, 178-181, 212-215, 260-264, 305-308, 335-342, 362-364, 373, 374, 404-406, 413-416, 438-440, 450, 451, 490-492, 504, 505, 517-520, 543, 544, 561, 562
 atomic and molecular weights and their multiples (some of them), 572, 573
 balance data (information), 38, 39
 calibration of micronitrometers (correction), 169
 carbon factors, 474
 Carius combustion tube data, 281
 combustion boats, 59
 empirical formula calculation series, 570
 gravimetric factors, 571
 historical developments, 1
 logarithms, 584-601
 nitrogen, alpha-amino, calculation factors, 484
 nitrogen reduction, 574-583
 precision of balance, 56
 precision of balance–sample weight relationship, 31
 sample size, various determinations, 2, 31
 sensitivity of balances, 30, 38, 39, 55

solvents for Rast method, 535
vapor pressure of acetic acid, 503
test substances (test compounds, test samples), 8, 9
volumetric factors, 571
weighing bottles, 61
Table for balance, 13-20, 22
    inertia block-isolator type, 14-20
    rigid type, 19, 20
Tare flasks (bottles), 65, 244
Temperature correction for Dumas method, 168
Test compounds (samples, substances), 6, 8, 9, 22, 260
    additional information related to subject, 21, 22
Tests, blank,
    (see Blank tests)
Tetrahydroxyquinone indicator (THQ), 104, 278
Thallium, determination of, 143
Thermal decomposition (combustion or reaction) tube, 377, 391, 392
Thermal decomposition tube fillings, 392, 405
Thermal decomposition tube temperatures, 378-381, 394, 405
Thermistors, use of,
    for determination of melting point, 550, 551
    for determination of molecular weight, 540-544
Thermometer(s), Auschütz, 552
    correction for exposed stem, 554
Thermos flask (bottle), use of, for generation of carbon dioxide, Dumas determination, 155
Thiosemicarbazides, determination of, 520
Thiuramdisulfide, determination of, 516, 520
Thomas combustion furnace, 157
Thomas manometric apparatus, 460
Thorium, determination of, 143
Thorium nitrate,
    standard solution, 0.01N, 112
    standardization of, 112, 113
Thymolphthalein, 103
Tin, determination of, 143
Tin foil, use of, in alkoxyl determination, 434

Titration stand, illuminated (see Illuminated titration stand assembly)
Titration vessel, for Karl Fischer determination, 95
Titrations, 102-132
    amperometric, 124, 339, 342
    automatic, 124, 332, 333, 339
    cerimetry, 124
    conductometric, 124, 342, 415
    EDTA, complexometric, 124, 139, 308, 339, 342, 416
    high-frequency, 124, 339, 342, 416
    heterometric, 124, 139
    nonaqueous, 124, 339, 416
    potentiometric, 124, 140, 339, 415
    general applications, 126, 416
    oscillometric, 342
    ultramicro-, submicro-, 414
Titrimetric procedures, 102-132, 264, 451
p-Toluene sulfonic acid, use of, in acyl determination, 445, 448
Trace determinations, 405
Transfer of nitrogen and nitric oxide to Hempel pipette, 479, 480
Transformer, use of, constant voltage, 11
    variable auto- 136, 197
Treatment of barium sulfate precipitate, 288-290
Tube(s) (see under individual names)

U

Ultramicro Kjeldahl method(s), 214
Ultramicro-, submicro methods, 179, 214, 259, 263, 306, 336, 340, 362, 373, 414, 438, 490, 491, 504, 517, 543
Ultraviolet light, use of, to eliminate electrostatic charges, 74, 259
Unsaturation (double bonds), determination of, 495-507
    additional information related to,
    table of, 504, 505
Uranium, determination of, 143
Urea, decomposition of, manometric determination, 475, 478

V

Vacuum desiccator(s), use of, 357, 358
Vacuum oven(s), 91
Vacuum system, use of, for acyl determination, 447

Subject Index

for drying of samples, 88-92
for molecular weight determination, 536
Vanadium pentoxide, 224
  use of, in carbon-hydrogen determination, 222, 224, 255
  use of, in Dumas nitrogen determination, 152
Vapor pressure, determination of, 561
Vapor pressure, use of liquids with high, for isothermal distillation method of molecular weight determination, 536
Vapor pressure of acetic acid, table, curve, 503
Vapor pressure comparison method for molecular weight determination, 544
Vaporimetric method(s), for determination of molecular weight, 544
Vial testing with manometric apparatus, 485-489
Vials, sample containers, 88
Vibration(s), dampening out of, in balance table construction, 13-21
Vibration isolation, efficiency of, calculation, 15
Vinyl ether, determination of, 516, 519
Viscose liquid(s), determination of specific gravity of, 557
Viscosity, determination of, 562
Volatile compounds, determination of unsaturation on, 505
  weighing of, in general, 76-78
Voltage regulation, 10
Volume correction for Dumas method, 168
Volumetric factor(s), table of, 571
Volumetric flasks (see Flasks, volumetric)
Volumetric procedures,
  for acyl (acetyl, formyl) groups, 444-453
  for alkoxyl (ethoxyl, methoxyl) groups, 422-432, 439
  for arsenic, 367-369, 373-376
  for carboxyl groups, 415, 416
  for fluorine, 326-332, 341, 342
  for halogens, 413
  for metals, 139
  for oxygen, 396-409
  for phosphorus, 363
  for sulfur, 276-287, 291-295, 303, 304, 307, 308, 413
Vorisek test, use of, as means of indirect detection of methanol, 93

W

Warburg apparatus, 492, 505
Wash bottle(s), 288
Washing extraction chamber, 472
Water, heavy, use of, for active hydrogen determination, 514, 515, 519
Water (moisture), determination of, 91-93, 97
  distilled, for neutralization equivalent and carboxyl determinations, 411
Water absorption tube (see Absorption tubes)
Water suction system, 465
Weatherproofing, 13
Weighing bottle, 60, 61
  various types, 60-64
Weighing bottle, outside cap, 60, 61
Weighing bottle, pig-type, metal, 59, 63, 64
  size A, 59, 63
  size B, 59, 64
  size C, 59, 64
Weighing bottle, pig-type, with outside cap, 61, 62
Weighing bottle with two caps, 62, 63
Weighing cup, 60
Weighing on microchemical balance, 50-82
  absorption tubes, 253
  additional information, 79
  charging tube, 70
  filter tubes, 318, 319, 322, 358, 359
  pig, 74
  sample, 73
    hygroscopic, 74, 75
    liquid, 76-78
    non-hygroscopic, 73, 75
    solid, 73-76
  performance, 79
  problems arising, 79, 261
  weighing bottles (pigs), 74
  weighing (charging) tubes, 70, 75
Weighing of samples, 73-79
Weighing techniques, 50, 79
Weighing (charging) tubes, 70
Weighing vessels, 60-64, 70, 76, 78, 79
Weight(s), calibration of, 57, 79
Wet combustion(s), for arsenic, 367-376
  for carbon, 263, 454-474, 491
  for halogens, 316-323, 336-338, 341
  for nitrogen (Kjeldahl), 188-220

for phosphorus, 354-366
for sulfur, 276-291, 307
Window, heated storm, for balance room, 13, 14

X

X-ray fluorescence methods, 139, 308, 339, 363
X-ray spectrophotometric methods, 139, 308
X-ray techniques, 308

Z

Z-tube, 172
Zero point of balance, 50-55
Zero reading of balance, 50-55
Zinc, determination of, 143
  use of, in determination of nitrogen, 191
Zinc dust, use of, in Kjeldahl nitrogen determination, 191
Zirconium, determination of, 143